Harnessing Biotechnology Tools for Product Development

Patrick Omoregie Isibor • Paul Akinniyi Akinduti
Editors

Harnessing Biotechnology Tools for Product Development

Selected Papers from the International Biotechnology Conference Exhibition and Workshop (IBCEW 2024)

Springer

Editors
Patrick Omoregie Isibor
Department of Biological Sciences
Covenant University
Ota, Nigeria

Paul Akinniyi Akinduti
Department of Medical Laboratory Science
Babcock University
Ilishan-Remo, Ogun State, Nigeria

ISBN 978-3-031-86001-0   ISBN 978-3-031-86002-7 (eBook)
https://doi.org/10.1007/978-3-031-86002-7

This Springer imprint is published by the registered company Springer Nature Switzerland AG
The registered company address is: Gewerbestrasse 11, 6330 Cham, Switzerland

Contents

About the Editors

Patrick Omoregie Isibor, Ph.D., is a specialist in pollution studies and ecotoxicology and a lecturer/researcher in the Department of Biological Sciences at Covenant University. He received his Ph.D. and M.Sc. in Pollution Studies and Ecotoxicology (Environmental Quality Management) from the University of Benin and a B.Sc. in Zoology from Ambrose Alli University. Dr. Isibor's research interests include ecotoxicology, hydrobiology, bioaccumulation, biosequestration, biodiversity conservation, and aquatic ecology. He is a member of the Association for Environmental Impact Assessment of Nigeria (AEIAN), the International Association of Risk and Compliance Professionals (IARCP), and the African Society for Toxicological Sciences (ASTS). Dr. Isibor is currently an editor for the *African Journal of Health, Safety, and Environment* and a reviewer for several reputable international journals. He is also an editor of the books titled Biotechnological Approaches to Sustainable Development Goals (Springer, 2023), Nanochitosan-Based Enhancement of Fisheries and Aquaculture: Aligning with Sustainable Development Goal 14—Life Below Water (Springer, 2024), Environmental Nanotoxicology: Combatting the Minute Contaminants (Springer, 2024), and Arctic Marine Ecotoxicology: Climate Change, Pollutants, and Their Far-Reaching Effects (Springer, 2024)

Paul Akinniyi Akinduti, Ph.D., is a lecturer and researcher at the Department of Medical Laboratory Science, Babcock University, Ilishan, Ogun State. He received his Ph.D. in Medical Microbiology and Bacteriology from Olabisi Onabanjo University. He is a Fellow of the Medical Laboratory Science Council of Nigeria and a recipient of the World Academy of Sciences and Bill and Melinda Gates Foundation Travel Awards. Dr. Akinduti's areas of research interest include antimicrobial agents, antimicrobial resistance, molecular biology, bacterial pathogenesis, antimicrobial susceptibility testing, bacterial antibiotic resistance, bacteriology, antibacterial activity, bacterial drug resistance, diagnostic microbiology, molecular bacteriology, molecular microbiology, foodborne diseases, molecular typing, microbial genetics, bacterial toxins, bacterial conjugation, and pathogenesis.

Chapter 1
Antimalarial Drug Resistance Markers in *Plasmodium falciparum* Among Pregnant Women in Nigeria: A Systematic Review

Abimbola David Akinyosoye and Olayemi Oluseun Akinnola

1 Introduction

The global prevalence of malaria, a tropical disease (Duguma et al., 2022) caused by infected female anopheles mosquito, *Plasmodium falciparum* (pf) (Zekar & Sharman, 2023), was estimated to be 207 million in 2023 (Inklaar et al., 2022), with Nigeria accounting for 27% of the cases (Anjorin et al., 2023). Over 600,000 deaths were also recorded in the same year, with Nigeria contributing 24% to the death burden (Anjorin et al., 2023). Vulnerable groups or high-risk groups such as immunosuppressed, immunocompromised individuals, exposed individuals, children below the age of 5, and pregnant women are often more susceptible to the devastating effects of this disease (Mirzohreh et al., 2022; Bamikole & Bamikole, 2024). Pregnant women are at risk of this infection due to the temporarily compromised immune system as a result of the pregnancy condition, a phenomenon known as pregnancy-associated malaria (PAM) (Cardona-Arias & Carmona-Fonseca, 2024). The presence of the asexual parasite stage in the blood or placenta of a pregnant woman defines pregnancy-associated malaria (PAM) (Chua et al., 2021). It is common in regions with malaria vector and parasite pressure; examples are tropical regions such as Africa. Of over 11.8 million women exposed to malaria in 2023, 400,000 thousand were from Nigeria, with 11% of maternal deaths associated with malaria recorded (Bello & Ayede, 2019; Anjorin et al., 2023).

A. D. Akinyosoye (✉) · O. O. Akinnola
Department of Biological Sciences, Covenant University, Ota, Ogun State, Nigeria

Covenant Applied Informatics and Communication Africa Centre of Excellence (CApIC-ACE), Ota, Ogun State, Nigeria
e-mail: abimbola.akinyosoyepgs@stu.cu.edu.ng

P. O. Isibor, P. A. Akinduti (eds.), *Harnessing Biotechnology Tools for Product Development*, https://doi.org/10.1007/978-3-031-86002-7_1

Plasmodium falciparum is the species most commonly linked with pregnancy-associated malaria (PAM); it poses significant risks to both the mother and the foetus, especially in primigravidae (Lufele et al., 2017; Manirakiza et al., 2017; Berhe et al., 2023). Other risk factors aside from gravidity, according to various studies, include the maternal age of the mother (Gontie et al., 2020). According to Bakken & Iversen (2021), babies born during pregnancy associated with malaria are often underweight in cases where there is no spontaneous abortion, preterm births, and stillbirths. Other complications include congenital malaria (Cardona-Arias & Carmona-Fonseca, 2024), hepatic dysfunction for both mother and foetus (Romero et al., 2021), severe anaemia (Rouamba et al., 2021), and foetal heart rate abnormalities (Chua et al., 2021). Symptoms of severe complications in pregnancy-associated malaria (PAM) include altered consciousness, convulsions, shock, pulmonary oedema, irregular bleeding, jaundice, and haemoglobinuria (Seal et al., 2012; Yagnik et al., 2019). Parasite detection through microscopy remains essential for pregnant women in low-resource settings; however, due to limited expertise and infrastructure, rapid diagnostic tests (RDTs) are primarily used in most endemic areas (Fitri et al., 2022). Nonetheless, identifying parasites with RDTs is challenging because of the transmission of parasites lacking the *PfHRP2* and *PfHRP3* genes (Boush et al., 2020).

The significant challenge in eradicating this tropical disease is the reliance on drug therapy due to the delays in effective vaccine discovery in combating the *P. falciparum* malaria parasite. Drug-resistant *P. falciparum* strains emerge and spread due to an over-reliance on medications, mitigating eradication efforts. For pregnant women with uncomplicated malaria, the World Health Organisation (WHO) recommends two doses of intermittent preventive treatment with sulphadoxine-pyrimethamine (IPTp-SP) (Figueroa-Romero et al., 2022), which has been shown to improve maternal outcomes. However, resistance to SP has hindered the effectiveness of this programme. Single nucleotide polymorphisms (SNPs) in the dihydropteroate synthase (*Pfdhps*) and dihydrofolate reductase (*Pfdhfr*) genes further complicate the use of this medication (Adegbola et al., 2023). In 2001, the WHO recommended the use of combination therapies, including an artemisinin derivative, as a proactive measure against emerging resistant strains (Kokori et al., 2024). However, the efficacy of artemisinin-based treatments is threatened by the recent emergence of artemisinin resistance (*K13*) in various African and Asian countries.

Surveillance of antimalarial drug resistance is essential for the early detection of resistant strains and for controlling their spread through timely adjustments in treatment policies. Surveillance can be done using in vitro, in vivo, and markers studies (Wang et al., 2020). In vivo tests involving host, drug, and antimalarial molecule factors, though the gold standard, are challenging to implement, cumbersome, very expensive, and time consuming (Maji, 2018). In vitro assays can be utilised to study the inhibition of parasite growth when exposed to various concentrations of anti-malarial drugs (Charman et al., 2020). However, it is important to note that these assays do not consider host immunity and drug pharmacokinetics. Thus, they may not accurately reflect the in vivo scenario where parasites can survive in therapeutic drug concentrations (Radohery et al., 2022). Moreover, conducting these tests can be costly and requires live parasites, trained personnel, and laboratory facilities for in vitro culture (Calderaro et al., 2024). Alternatively, analysing drug resistance markers associated with resistance in previous studies can provide an overview of the prevalence of drug resistance during a specific period. Given the size of the population, conducting studies on malaria resistance in pregnant women in Nigeria using in vitro or in vivo methods would be highly costly and time consuming. Therefore, molecular marker analysis presents a more feasible option. A systematic review was undertaken to gather insight into the increasing antimalarial resistance markers among pregnant women in Nigeria and establish a baseline for future research for possible targets for drug development and control of transmission of malaria parasites among pregnant women.

2 Methodology

The search was conducted across electronic databases, including Scopus, Google Scholar, and PubMed, from January 2014 to April 2024. Key search terms such as "antimalarial", "pregnant women", "resistance markers", "malaria", and "Nigeria"were employed. Additional relevant articles were identified through the references of the retrieved articles. Samples about malaria infection in pregnant women in Nigeria were included if they met the following criteria: (a) analysis of one or more antimalarial resistance markers and (b) exclusion of original publications, short reports, and review articles. Full-text articles were thoroughly reviewed to ensure adherence to the selection criteria. Out of the initial 49 papers identified, 10 were reviews and studies conducted outside Nigeria, while 27 were in vitro and in vivo investigations. After careful screening, 12 additional articles that met the inclusion criteria were incorporated. The articles were categorised based on parameters such as year of collection, year of publishing, sample size and characteristics, analysis of molecular markers, and identification of polymorphisms. The search process adhered to the PRISMA guidelines (Frost et al., 2022), as illustrated in Fig. 1.1.

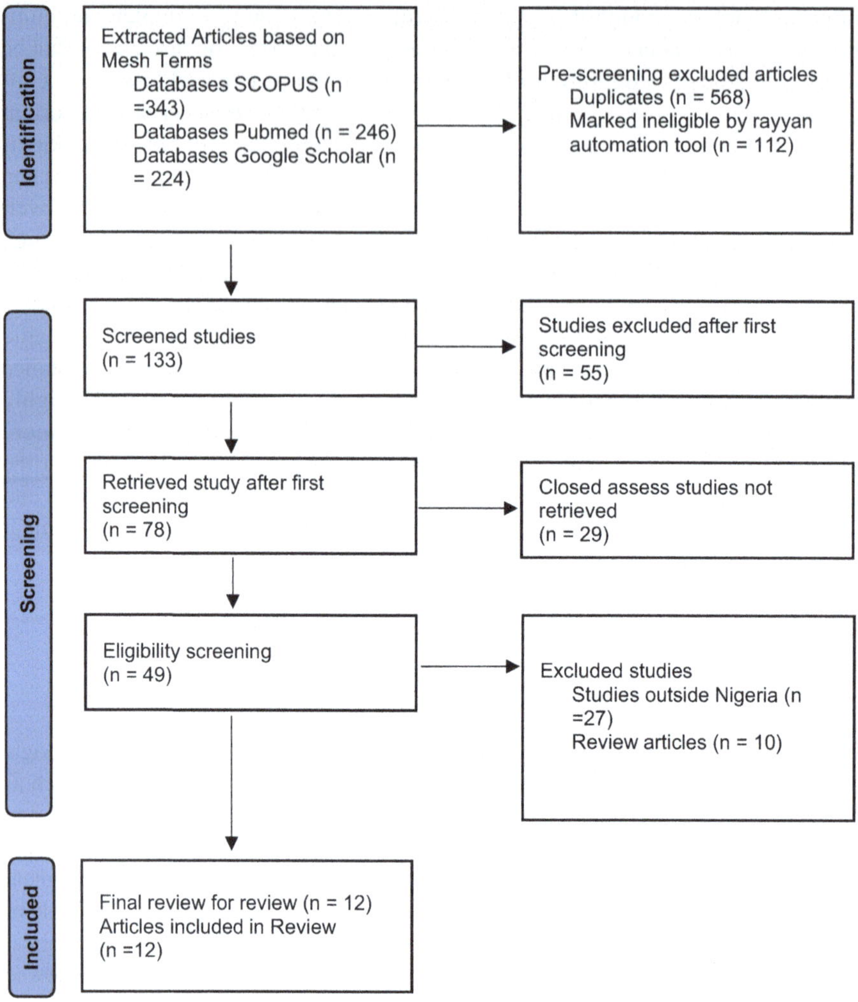

Fig. 1.1 PRISMA checklist for literature search and screening

3 Results

The overall articles as shown in Table 1.1. included in this review were samples collected from 2014 to 2024 (Iwalokun et al., 2015; Agomo et al., 2016; Oguike et al., 2016; Monjol & Useh, 2017; Esu et al., 2018; Lawal et al., 2018; Ojurongbe et al., 2018; Fagbemi et al., 2020; Quan et al., 2020; Abubakar et al., 2022; Azeez & Akinbo, 2022; Oboh et al., 2022; Ibekpobaoku et al., 2024). Each study included Nigerian women who had contracted *Plasmodium falciparum* malaria in Nigeria. Sixty-seven per cent of the studies reviewed were conducted in southwest Nigeria, with more studies conducted in Lagos (Four), while 17% were conducted in

Table 1.1 Summary of findings

Authors and year of publication	Location	Period of collection	Mode of collection	Sample size	Identified markers
Agomo et al. (2016)	Lagos	2014	Dried blood spot	54	*Pfcrt, pfmdr1, pfdhfr*
Quan et al. (2020)	Lagos	October 2010 to September 2011 and January 2013 to February 2014	Dried blood spot	242	*pfdhps, pfdhfr*
Ibekpobaoku et al. (2024)	Lagos	November 2019 to March 2021	Pricked finger for blood smears and dried blood spot	121	*pfdhps, pfdhfr, pfmdr1, pfcrt*
Monjol and Useh (2017)	Calabar	2016	Venous blood	369	*Pfcrt,*
Esu et al. (2018)	Calabar	October 2013 to November 2014	dried blood spot	459	*pfdhps, pfdhfr*
Oguike et al. (2016)	Ibadan, Borno, Enugu	May 2003 and October 2004, 2010 and 2012, 2010 and 2012	Filter paper blood spots and placental impressions	100, 142, 225	*Pfdhps, pfdhfr*
Abubakar et al. (2022)	Sokoto	February,2020 to May, 2020	Dried blood spots	200	*Pfmdr1, Pfcrt*
Azeez and Akinbo (2022)	Oyo	2021	venous blood	316	*Pfmdr1, Pfcrt*
Lawal et al. (2018)	Niger	December 2016	Pricked finger and venous blood	250	*Pfmdr1*
Ojurongbe et al. (2018)	Osun	June and December 2015	Pricked finger	200	*pfdhfr*
Iwalokun et al. (2015)	Lagos	April and July 2011	finger-pricked and dried blood spots	107	*pfdhps, pfdhfr*
Fagbemi et al. (2020)	Ogun	June 2018 and August 2019	Dried blood spots	406	*Pfdhfr, Pfdhps*

southeast and Northeast Nigeria as well. The investigations focused on identifying genetic markers associated with antimalarial drug resistance, including the *K13*-propeller gene, *pfcrt, pfdhps, pfdhfr,* and *pfmdr1*. Among these genes, *pfdhfr* and *pfdhps* were the most frequently analysed, as they are linked to resistance against sulphadoxine-pyrimethamine (SP). Eight articles (69.2%) specifically examined these genes (Iwalokun et al., 2015; Agomo et al., 2016; Oguike et al., 2016; Esu et al., 2018; Ojurongbe et al., 2018; Fagbemi et al., 2020; Quan et al., 2020; Ibekpobaoku et al., 2024). Several studies (Agomo et al., 2016; Monjol & Useh, 2017; Azeez & Akinbo, 2022; Ibekpobaoku et al., 2024) examined *Pfcrt* resistance in pregnant women. *Pfmdr1* was examined in four publications (33.3%). This gene is linked to resistance to multiple drugs, including mefloquine (MQ), lumefantrine

(LMF), AMQ, and potentially artemisinin. Notably, none of the studies included in this review investigated the *K13*-propeller gene associated with artemisinin resistance. Table 1.1 provides an overview of the studies included in the review.

3.1 **Pfcrt** ***Gene***

Four studies of the 12 studies were analysed for the *pfcrt* gene, as shown in Table 1.2. The studies where this gene was analysed were conducted in the south of Nigeria, mainly Lagos (50%), Oyo (25%), and Calabar states (25%). Two of the four studies examined the haplotypes (CVMNK, CVIET) associated with specific mutations at amino acid positions 72–76, which are linked to amodiaquine (AMQ) resistance. Four studies documented mutant genes, including the K76T mutation, while one study recorded the K78T mutant gene. Table 1.2 summarises the main findings of these articles.

3.2 **pfdhps** ***and*** **pfdhfr** ***Genes***

Both genes are associated with susceptibility to the sulphadoxine-pyrimethamine (SP) combination, so they are often studied together. Table 1.3 highlights the number of reviewed studies that analysed *pfdhps* and *pfdhfr* genes. Fifty per cent of the reviewed studies examining the presence of these antimalarial drug resistance genes in pregnant women were conducted in Lagos. Two studies focused solely on *pfdhfr*, while only one examined the *pfdhps* gene. The most common point mutations on the *dhfr* gene were found at positions 51, 59, and 108, while on the *dhps* gene, mutations were observed at positions 436, 437, 540, 581, and 613.

Table 1.2 Summary of studies involving *pfcrt* genes

Authors and year of publication	Location	Period of collection	Codon region of mutation	Mutant strain detected	Rate of occurrence (%)	Haplotypes found	Mutant 76, 78T (%)	
Agomo et al. (2016)	Lagos	2014	72–76	*Pfcrt* 76	75.9	CVMNK, CVIET	13/54, 41/54	24 76
Ibekpobaoku et al. (2024)	Lagos	November 2019 to March 2021	72–76	*Pfcrt* 76	14.3	CVMNK CVIET	–	–
Monjol and Useh (2017)	Calabar	2016	76	*PfCRT* K76 *PfCRT* 76T	72 28	–	–	–
Azeez and Akinbo (2022)	Oyo	2021	78	*Pfcrt* 78T	91.5	–	–	–

Table 1.3 Summary of studies involving *pfdhfr* and *pfdhps* genes

Authors and year of publication	Location	Period of collection	Codon region of mutation	Mutant strain detected	Rate of occurrence (%)	Rate of occurrence					
						pfdfhr mutations	pfdhps mutations	Wild type mutations	*Pfdhfr* mutations (%)	*pfdhps* mutations (%)	Wild type mutations (%)
Agomo et al. (2016)	Lagos	2014	*Pfdhfr* 16,50,51,59, 108,140,164	*Pfdhfr* 108	25	ACIRNVI, ACICNSVI	–	ACNCSVI	66.7 6.7	–	26.7
Quan et al. (2020)	Lagos	October 2010 to September 2011 and January 2013 to February 2014	*Pfdhfr* (N51I, C59R and S108N) *Pfdhps* (A437G, A613S, S436A, A581G, I431V, K540E)	–	–	CIRNI, CICNI CNRNI	SGKAA VAGKGS	CNCSI	93.8 0.4 1.2	44.3 19.8	3.3
Ibekpobaoku et al. (2024)	Lagos	November 2019 to March 2021	*Pfdhfr* (codons 51, 59, 108, 164) *Pfdhps* (codons 436, 437, 540, 581, 613)	–	–	NCNI IRNL	SAKAS	–	57.1 28.6	42.9	–
Esu et al. (2018)	Calabar	October 2013 to November 2014	*Pfdhfr* (codons 51, 59, 108, 164) *Pfdhps* (436, 437, 581, and 613	*Pfdhfr* 108 N and *Pfdhps* 437G	–	–	–	–	–	–	–

(continued)

Table 1.3 (continued)

Authors and year of publication	Location	Period of collection	Codon region of mutation	Mutant strain detected	Rate of occurrence (%)	Rate of occurrence					
						pfdfhr mutations	pfdhps mutations	Wild type mutations	*Pfdhfr* mutations (%)	*pfdhps* mutations (%)	Wild type mutations (%)
Oguike et al. (2016)	Ibadan, Borno, Enugu	May 2003 and October 2004, 2010 and 2012, 2010 and 2012	*pfdhfr* 6, 50, 51, 59, 108, 140 and 164 *pfdhps* 431, 436, 437, 540, 581 and 613	–	–	ISGKAA	ACIRNVI	–	42	86	–
						IAGKAA,	ACICNVI		5	1	
						IAGKASIAAKAA	ACNCSVI		4	2	
						IAAKGA	ACNRNVI		8.2	3	
						IFAKAS	ACNCNVI		2	2	
						ISAKAA			2		
						ISGEAA			1		
						ISGKAS			1		
						IYAKAS			1		
						IAAKGS			1		
						ICAKAA			1		
						IAGKGA			1		
						IAGKGS			1		
						ISGKGA			1		
						ISGKGS			1		
						VAGKAA			6		
						VAGKAS			2.3		
						VAGKGS			27		
						VAGKGA			7		
						VSGKAA			1		
						VSGKGA			1		
						VSGKGS			2		
Ojurongbe et al. (2018)	Osun	June and December 2015	*pfdhfr* (N51I, C59R, S108N and I1464L)	*Pfdhfr* (N51I, C59R, and S108N)	63%	–	–	–	–	–	–

Iwalokun et al. (2015)	Lagos	April and July 2011	*pfdhfr* (N51I, C59R and S108N) *pfdhps* (A437G and K540E)	–	–	–	–	–	–	–	–
Fagbemi et al. (2020)	Ogun	June 2018 and August 2019	*Pfdhfr* N51I, C59R, C50R, I164L and S108N *pfdhps* A437G, A581G, A436A, and A613S	*Pfdhfr* N51I, C59R, and S108N	98%	CICNICIRNI	SGKAA SGKGA AAKGAAGKAS AGKGA AGKGS	CNCSI SAKAA	2 98	2 44 2 28 14	0

3.3 pfmdr1 *Gene*

Among the 12 studies, the *pfmdr1* gene was examined in four, as shown in Table 1.3. The studies where this gene was analysed were conducted in the south of Nigeria, mainly Lagos (50%), Oyo (25%), and the Northern region as well in Niger state (25%). Five-point mutations at codon locations 186, 184, 1034, 1042, and 1246 were investigated. The N86Y mutation was present in 75% of the studies, while the 184F and 1042D mutations were found in 50% and 25%, respectively (Table 1.4).

3.4 K13-*propeller Gene*

Despite its recent association with artemisinin resistance, this gene has not been examined in samples taken from pregnant women in Nigeria.

4 Discussion

The literature search yielded 12 molecular studies conducted in Nigeria, with none analysing markers in more than three states within a region. These investigations comprised samples from 54 to 459 pregnant women (with a mean of 220). No study primarily focused on one to three states across various regions of Nigeria. None of the studies also completed a study in a whole geo-political region of Nigeria, even though it was stated otherwise in the topics, and none of the studies completed a whole national survey in the country. Given the population of Nigeria (Africa's most populous nation, with an average of ten million pregnancies annually as of 2019 (Lambonmung et al., 2023)), as well as the endemicity and regional variations in vector and parasite spread patterns, the findings from the majority of these studies should be more broadly applicable to the country as a whole, while also taking local context into account.

In the southwest region of Nigeria, research conducted by Agomo et al. (2016), Fagbemi et al. (2020), and Ojurongbe et al. (2018) revealed significant molecular findings. The incidence rates of pfdfr and *pfdhps* were reported as 98%, 63%, and 25%, respectively. This high frequency of double mutants in *pfdfr* and *pfdhps*, indicating resistance to sulphadoxine, underscores the non-effectiveness of the antimalarial drug's active component, administered to pregnant patients for treatment and prevention. However, the World Health Organization recommends against using medication if resistance prevalence exceeds 10%, suggesting the need for medication substitution in one location while maintaining its use in another. This highlights the importance of medical providers not relying solely on regional outcomes when deciding whether to administer antimalarial medication to pregnant women. Instead, representative sampling should be employed for nationwide surveillance of markers

Table 1.4 Summary of studies involving the *pfmdr1* gene

Authors and year of publication	Location	Year of collecting	Codon region of mutation	Mutant strain detected	Rate of occurrence (%)	Haplotypes found			Occurrence rate		
						Single mutant	Double mutant	Wild mutant	Single mutant (%)	Double mutant (%)	Wild mutant (%)
Agomo et al. (2016)	Lagos	2014	*Pfmdr1* 86, 184, 1034, 1042, 1246	*Pfmdr* 186Y, 184F	25 28.1	YYSND NFSND	YFSND	NYSND	17.9 21.4.	7.1	53.6
Ibekpobaoku et al. (2024)	Lagos	November 2019 to March, 2021	*Pfmdr1* (86, 184, 1042, 1246)	*Pfmdr1* 86Y, 184 F and 1042D	85.7, 71.4 14.3	YYND NFND YYDD	YFND	–	All 14.3	57.1	–
Azeez and Akinbo (2022)	Oyo	2021	*Pfmdr1* 86Y	*Pfmdr1* 86Y	97.6	–	–	–	–	–	–
Lawal et al. (2018)	Niger	December 2016	*Pfmdr1* 86–184, 1034, 1042	None	–	–	–	–	–	–	–

and resistance patterns. The lack of a cross-country and cross-region representative study is a significant disadvantage; thus, therapy policies are often employed based on geographical results from the studies in that region. Supporting studies among pregnant women simultaneously in the country's Southeast and North Central regions may likely yield different results. However, the primary challenge often lies in the high cost of surveillance studies, particularly those involving molecular research, which may need more reagents and equipment locally. This hampers the reproducibility and validation of research findings. Nonetheless, WHO experts recommend conducting resistance surveillance in endemic countries every 2 years (Ippolito et al., 2021).

The *P. falciparum* chloroquine resistance transporter gene (*pfcrt*) K76T mutation has been consistently linked to chloroquine (CQ) resistance (Mulenga et al., 2021). According to Agomo et al. (2016) and Ibekpobaoku et al. (2024), *P. falciparum* carrying the K76T mutation in this study were commonly associated with two major haplotypes, CVMNK and CVIET, which are indicative of amodiaquine (AMQ) resistance. These haplotypes were identified in Lagos, southwest Nigeria, suggesting potential dissemination of drug resistance due to frequent population migration across borders (Ajibaye et al., 2024; Oboh et al., 2018). The figures in the occurrence of resistance to CQ show there might not be a total withdrawal from the drug administration to pregnant women in the area; it would be prudent to monitor this haplotype's occurrence continuously.

Parasite resistance to sulphadoxine-pyrimethamine (SP) medications is associated with mutations in dihydropteroate synthase (*dhps*) and dihydrofolate reductase (*dhfr*) genes (Quan et al., 2020). According to the previous studies, pyrimethamine resistance is correlated with mutations at *pfdfr* 108, N51I, and C59R (Agomo et al., 2016; Esu et al., 2018; Fagbemi et al., 2020; Ojurongbe et al., 2018), and *Pfdhps* 437G mutation (Esu et al., 2018). Haplotypes in these genetic marker mutations range from wild to Quint mutants. The rate of occurrence of these haplotypes identified and collated from the studies was relatively high. Mvumbi et al. (2015) suggested discontinuing sulphadoxine-pyrimethamine (SP) as a malaria treatment in malaria-endemic areas where high resistance rates have been reported. However, it is difficult to determine whether SP resistance is increasing or decreasing due to variations in study methods, locations, and timeframes. Hence, future studies must be conducted with all these factors, such as methodology, locations, and timeframes of sample collection, in mind. Additionally, from the reviewed articles, compared to other regions of Nigeria, the southwest region appears to have higher levels of SP resistance, which was further substantiated by Quan et al. (2020) due to pressure on the drug as prophylaxis for pregnant women in the region. None of the studies evaluating SP use for intermittent preventative therapy (IPTp) in pregnant women assessed the impact of using SP in high-resistance areas.

The *P. falciparum* multidrug resistance 1 (*pfmdr1*) gene, particularly the N86Y allele (Azeez & Akinbo, 2022), the N86Y allele, and *pfmdr* 184F (Agomo et al., 2016), and *Pfmdr1* 86Y, 184 F, and 1042D (Ibekpobaoku et al., 2024), are associated with resistance to chloroquine (CQ), mefloquine, quinine, and artemisinin in *P. falciparum*. Mutations at the *pfmdr1* N86Y and Y184F sites were commonly observed, and in Ibekpobaoku et al. (2024), mutations in Y184F, 1042D, and N86Y occurred simultaneously. Notably, the native samples contained the N1042D mutation, which research suggests makes parasites more susceptible to pyronaridine (Bai et al., 2018).

Monitoring the emergence of artemisinin-resistant strains in Nigeria is crucial when employing artemisinin-based combination therapy (ACT), despite the absence of *K13* as a molecular marker of drug resistance in the reviewed studies among pregnant women. This is further stated by the marker identification in a study by Ajogbasile et al. (2022). Notably, this marks the first instance of a resistant molecular marker being identified before the resistant strains had disseminated to all malaria-endemic areas. Despite the lack of validation for *K13* markers and the absence of polymorphisms found in Southeast Asia, new mutations of *K13* have been reported in the surrounding countries of Ghana (Dieng et al., 2023). Further research and surveillance are warranted to assess the impact of these novel mutations on susceptibility to artemisinin.

5 Conclusion

This review identified only 12 studies, most of which covered only specific regions of Nigeria. Nigeria, being a large country, exhibits diverse responses to malaria medication, especially among vulnerable groups, including pregnant women whose compromised immune systems make them more susceptible. This review revealed a significant gap in regional and sub-regional levels of drug resistance, especially among vulnerable groups in Nigeria. As a result, it is recommended that targeted policy and educational interventions be implemented to raise awareness of these consequences among pregnant women and develop effective countermeasures. Effective monitoring of antimalarial drug resistance in pregnant women requires regional and nationwide surveillance efforts; thus, this current review provides an opportunity to conduct a nationwide survey to address the limitations identified in the review. This is very important for preventing emerging artemisinin resistance in some parts of Asia and Africa.

Acknowledgements The authors acknowledge the management of Covenant Applied Informatics and Communication Africa Centre of Excellence (CApIC-ACE), Covenant University, for the financial support in the publication of this study.

References

Abubakar, H., Yabo, A. B., Bandiya, H. M., & Shuaibu, A. B. (2022). Molecular evaluation of Plasmodium falciparum drug resistant genes among pregnant women attending ANC at Maryam Abacha women and children hospital (MAWCH) Sokoto, Nigeria. *Journal of Current Biomedical Research, 2*(5, September-October), 552–567. https://doi.org/10.54117/JCBR.V2I5.10

Adegbola, A. J., Ijarotimi, O. A., Ubom, A. E., Adesoji, B. A., Babalola, O. E., Hocke, E. F., Hansson, H., Mousa, A., Bolaji, O. O., Alifrangis, M., & Roper, C. (2023). A snapshot of the prevalence of dihydropteroate synthase-431V mutation and other sulfadoxine-pyrimethamine resistance markers in Plasmodium falciparum isolates in Nigeria. *Malaria Journal, 22*(1). https://doi.org/10.1186/S12936-023-04487-5

Agomo, C. O., Oyibo, W. A., Sutherland, C., Hallet, R., & Oguike, M. (2016). Assessment of markers of antimalarial drug resistance in Plasmodium falciparum isolates from pregnant women in Lagos. *Nigeria. PLOS ONE, 11*(1), e0146908. https://doi.org/10.1371/JOURNAL.PONE.0146908

Ajibaye, O., Olukosi, Y. A., Oriero, E. C., Oboh, M. A., Iwalokun, B., Nwankwo, I. C., Nnam, C. F., Adaramoye, O. V., Chukwemeka, S., Okanazu, J., Gabriel, E., Balogun, E. O., & Amambua-Ngwa, A. (2024). Detection of novel Plasmodium falciparum coronin gene mutations in a recrudescent ACT-treated patient in South-Western Nigeria. *Frontiers in Cellular and Infection Microbiology, 14*, 1366563. https://doi.org/10.3389/FCIMB.2024.1366563/BIBTEX

Ajogbasile, F. V., Oluniyi, P. E., Kayode, A. T., Akano, K. O., Adegboyega, B. B., Philip, C., Ogbulafor, N., Okafor, H. U., Oguche, S., Wammanda, R. D., Mokuolu, O. A., Folarin, O. A., & Happi, C. T. (2022). Molecular profiling of the artemisinin resistance Kelch 13 gene in Plasmodium falciparum from Nigeria. *PLoS One, 17*(2). https://doi.org/10.1371/JOURNAL.PONE.0264548

Anjorin, E. T., Olulaja, O. N., Osoba, M. E., Oyadiran, O. T., Ogunsanya, A. O., Akinade, O. N., & Inuojo, J. M. (2023). Overtreatment of malaria in the Nigerian healthcare setting: Prescription practice, rationale and consequences. *The Pan African Medical Journal, 45*(111), 45. https://doi.org/10.11604/PAMJ.2023.45.111.31780

Azeez, M. M., & Akinbo, F. O. (2022). Characterisation of markers of chloroquine resistance in plasmodium falciparum among pregnant women in Oyo state: Any hope of chloroquine resistance reversal? *Biomedical And Biotechnology Research Journal, 6*(2), 216–223. https://doi.org/10.4103/BBRJ.BBRJ_3_22

Bai, Y., Zhang, J., Geng, J., Xu, S., Deng, S., Zeng, W., Wang, Z., Ngassa Mbenda, H. G., Zhang, J., Li, N., Wu, Y., Li, C., Liu, H., Ruan, Y., Cao, Y., Yang, Z., & Cui, L. (2018). Longitudinal surveillance of drug resistance in plasmodium falciparum isolates from the China-Myanmar border reveals persistent circulation of multidrug resistant parasites. *International Journal for Parasitology: Drugs and Drug Resistance, 8*(2), 320. https://doi.org/10.1016/J.IJPDDR.2018.05.003

Bakken, L., & Iversen, P. O. (2021). The impact of malaria during pregnancy on low birth weight in East-Africa: a topical review. *Malaria Journal, 20*(1), 1–9. https://doi.org/10.1186/S12936-021-03883-Z/TABLES/2

Bamikole, T. H., & Bamikole, T. H. (2024). Confronting malaria – Addressing a critical health crisis among vulnerable groups in Nigeria. *Neglected Tropical Diseases - Unsolved Debts for the One Health Approach.* https://doi.org/10.5772/INTECHOPEN.1002999

Bello, F. A., & Ayede, A. I. (2019). Prevalence of malaria Parasitaemia and the use of malaria prevention measures in pregnant women in Ibadan, Nigeria. *Annals of Ibadan Postgraduate Medicine, 17*(2), 124. /pmc/articles/PMC7358809/.

Berhe, A. D., Doritchamou, J. Y. A., & Duffy, P. E. (2023). Malaria in pregnancy: Adverse pregnancy outcomes and the future of prevention. *Frontiers in Tropical Diseases, 4*, 1229735. https://doi.org/10.3389/FITD.2023.1229735/BIBTEX

Boush, M. A., Djibrine, M. A., Mussa, A., Talib, M., Maki, A., Mohammed, A., Beshir, K. B., Mohamed, Z., & Hajissa, K. (2020). Plasmodium falciparum isolate with histidine-rich protein 2 gene deletion from Nyala City, Western Sudan. *Scientific Reports, 10*(1), 12822. https://doi.org/10.1038/S41598-020-69756-8

Calderaro, A., Piccolo, G., & Chezzi, C. (2024). The laboratory diagnosis of malaria: A focus on the diagnostic assays in non-endemic areas. *International Journal of Molecular Sciences, 25*(2). https://doi.org/10.3390/IJMS25020695

Cardona-Arias, J. A., & Carmona-Fonseca, J. (2024). Prospective study of malaria in pregnancy, placental and congenital malaria in Northwest Colombia. *Malaria Journal, 23*(1), 1–8. https://doi.org/10.1186/S12936-024-04948-5/FIGURES/2

Charman, S. A., Andreu, A., Barker, H., Blundell, S., Campbell, A., Campbell, M., Chen, G., Chiu, F. C. K., Crighton, E., Katneni, K., Morizzi, J., Patil, R., Pham, T., Ryan, E., Saunders, J., Shackleford, D. M., White, K. L., et al. (2020). An in vitro toolbox to accelerate anti-malarial drug discovery and development. *Malaria Journal, 19*(1), 1–27. https://doi.org/10.1186/S12936-019-3075-5/TABLES/10

Chua, C. L. L., Khoo, S. K. M., Ong, J. L. E., Ramireddi, G. K., Yeo, T. W., & Teo, A. (2021). Malaria in pregnancy: From placental infection to its abnormal development and damage. *Frontiers in Microbiology, 12*, 777343. https://doi.org/10.3389/FMICB.2021.777343/BIBTEX

Dieng, C. C., Morrison, V., Donu, D., Cui, L., Amoah, L., Afrane, Y., & Lo, E. (2023). Distribution of Plasmodium falciparum K13 gene polymorphisms across transmission settings in Ghana. *BMC Infectious Diseases, 23*(1), 1–10. https://doi.org/10.1186/S12879-023-08812-W/TABLES/2

Duguma, T., Nuri, A., & Melaku, Y. (2022). Prevalence of malaria and associated risk factors among the Community of Mizan-Aman Town and its Catchment Area in Southwest Ethiopia. *Journal of Parasitology Research, 2022*. https://doi.org/10.1155/2022/3503317

Esu, E., Tacoli, C., Gai, P., Berens-Riha, N., Pritsch, M., Loescher, T., & Meremikwu, M. (2018). Prevalence of the Pfdhfr and Pfdhps mutations among asymptomatic pregnant women in Southeast Nigeria. *Parasitology Research, 117*(3), 801–807. https://doi.org/10.1007/S00436-018-5754-5/METRICS

Fagbemi, K. A., Adebusuyi, S. A., Nderu, D., Adedokun, S. A., Pallerla, S. R., Amoo, A. O. J., Thomas, B. N., Velavan, T. P., & Ojurongbe, O. (2020). Analysis of sulphadoxine–pyrimethamine resistance-associated mutations in plasmodium falciparum isolates obtained from asymptomatic pregnant women in Ogun State, Southwest Nigeria. *Infection, Genetics and Evolution, 85*, 104503. https://doi.org/10.1016/J.MEEGID.2020.104503

Figueroa-Romero, A., Pons-Duran, C., & Gonzalez, R. (2022). Drugs for intermittent preventive treatment of malaria in pregnancy: Current knowledge and way forward. *Tropical Medicine and Infectious Disease, 7*(8). https://doi.org/10.3390/TROPICALMED7080152

Fitri, L. E., Widaningrum, T., Endharti, A. T., Prabowo, M. H., Winaris, N., & Nugraha, R. Y. B. (2022). Malaria diagnostic update: From conventional to advanced method. *Journal of Clinical Laboratory Analysis, 36*(4). https://doi.org/10.1002/JCLA.24314

Frost, A. D., Hróbjartsson, A., & Nejstgaard, C. H. (2022). Adherence to the PRISMA-P 2015 reporting guideline was inadequate in systematic review protocols. *Journal of Clinical Epidemiology, 150*, 179–187. https://doi.org/10.1016/J.JCLINEPI.2022.07.002

Gontie, G. B., Wolde, H. F., & Baraki, A. G. (2020). Prevalence and associated factors of malaria among pregnant women in Sherkole district, Benishangul Gumuz regional state, West Ethiopia. *BMC Infectious Diseases, 20*(1). https://doi.org/10.1186/S12879-020-05289-9

Ibekpobaoku, A. N., Oboh, M. A., Faal, F., Adeniji, E., Ajibaye, O., Idowu, E. T., & Amambua-Ngwa, A. (2024). Sub-microscopic Plasmodium falciparum infections and multiple drug resistant single nucleotide polymorphic alleles in pregnant women from southwestern Nigeria. *BMC Research Notes, 17*(1), 1–9. https://doi.org/10.1186/S13104-024-06763-2

Inklaar, M. R., Barillas-Mury, C., & Jore, M. M. (2022). Deceiving and escaping complement - the evasive journey of the malaria parasite. *Trends in Parasitology, 38*(11), 962. https://doi.org/10.1016/J.PT.2022.08.013

Ippolito, M. M., Moser, K. A., Kabuya, J.-B. B., Cunningham, C., & Juliano, J. J. (2021). Antimalarial drug resistance and implications for the WHO global technical strategy. *Current Epidemiology Reports, 8*(2), 46. https://doi.org/10.1007/S40471-021-00266-5

Iwalokun, B. A., Iwalokun, S. O., Adebodun, V., & Balogun, M. (2015). Carriage of mutant dihydrofolate reductase and dihydropteroate synthase genes among plasmodium falciparum isolates recovered from pregnant women with asymptomatic infection in Lagos, Nigeria. *Medical Principles and Practice, 24*(5), 436–443. https://doi.org/10.1159/000430987

Kokori, E., Olatunji, G., Akinboade, A., Akinoso, A., Egbunu, E., Aremu, S. A., Okafor, C. E., Oluwole, O., & Aderinto, N. (2024). Triple artemisinin-based combination therapy (TACT): Advancing malaria control and eradication efforts. *Malaria Journal, 23*(1), 1–7. https://doi.org/10.1186/S12936-024-04844-Y/FIGURES/1

Lambonmung, A., Acheampong, C. A., & Langkulsen, U. (2023). The effects of pregnancy: A systematic review of adolescent pregnancy in Ghana, Liberia, and Nigeria. *International Journal of Environmental Research and Public Health, 20*(1), 605. https://doi.org/10.3390/IJERPH20010605

Lawal, B., Shittu, O. K., Abubakar, A., & Kabiru, A. Y. (2018). Human genetic markers and structural prediction of Plasmodium falciparum multidrug Resistance Gene (pfmdr1) for ligand binding in pregnant women attending general hospital Minna. *Journal of Environmental and Public Health, 2018*, 3984316. https://doi.org/10.1155/2018/3984316

Lufele, E., Umbers, A., Ordi, J., Ome-Kaius, M., Wangnapi, R., Unger, H., Tarongka, N., Siba, P., Mueller, I., Robinson, L., & Rogerson, S. (2017). Risk factors and pregnancy outcomes associated with placental malaria in a prospective cohort of Papua New Guinean women. *Malaria Journal, 16*(1), 427. https://doi.org/10.1186/S12936-017-2077-4/TABLES/4

Maji, A. K. (2018). Drug susceptibility testing methods of antimalarial agents. *Tropical Parasitology, 8*(2), 70. https://doi.org/10.4103/2229-5070.248695

Manirakiza, A., Serdouma, E., Ngbalé, R. N., Moussa, S., Gondjé, S., Degana, R. M., Bata, G. G. B., Moyen, J. M., Delmont, J., Grésenguet, G., & Sepou, A. (2017). A brief review on features of falciparum malaria during pregnancy. *Journal of Public Health in Africa, 8*(2), 190–201. https://doi.org/10.4081/JPHIA.2017.668

Mirzohreh, S. T., Safarpour, H., Pagheh, A. S., Bangoura, B., Barac, A., & Ahmadpour, E. (2022). Malaria prevalence in HIV-positive children, pregnant women, and adults: A systematic review and meta-analysis. *Parasites & Vectors, 15*(1). https://doi.org/10.1186/S13071-022-05432-2

Monjol, B. E., & Useh, M. F. (2017). Detection of plasmodium falciparum chloroquine resistance transporter (PfCRT) mutant gene amongst malaria-infected pregnant women in Calabar, Nigeria. *Annals of Parasitology, 63*(4), 323–330. https://doi.org/10.17420/AP6304.119

Mulenga, M. C., Sitali, L., Ciubotariu, I. I., Hawela, M. B., Hamainza, B., Chipeta, J., & Mharakurwa, S. (2021). Decreased prevalence of the Plasmodium falciparum Pfcrt K76T and Pfmdr1 and N86Y mutations post-chloroquine treatment withdrawal in Katete District, Eastern Zambia. *Malaria Journal, 20*(1), 329. https://doi.org/10.1186/S12936-021-03859-Z

Mvumbi, D. M., Kayembe, J. M., Situakibanza, H., Bobanga, T. L., Nsibu, C. N., Mvumbi, G. L., Melin, P., De Mol, P., & Hayette, M. P. (2015). Falciparum malaria molecular drug resistance in the Democratic Republic of Congo: A systematic review. *Malaria Journal, 14*(1), 1–6. https://doi.org/10.1186/S12936-015-0892-Z/TABLES/4

Oboh, M. A., Ndiaye, D., Antony, H. A., Badiane, A. S., Singh, U. S., Ali, N. A., Bharti, P. K., & Das, A. (2018). Status of artemisinin resistance in malaria parasite Plasmodium falciparum from molecular analyses of the Kelch13 gene in Southwestern Nigeria. *BioMed Research International, 2018*. https://doi.org/10.1155/2018/2305062

Oboh, M. A., Faal, F., Adeniji, O. E., Correa, S., Amawu, A. U., Ogban, E., Heinz, E., Hughes, G., Meremikwu, M. M., & Amambua-Ngwa, A. (2022). Multiple plasmodium falciparum drug resistance polymorphisms identified in a pregnant woman with severe malaria and a concomitant spontaneous abortion in Cross River, Nigeria, West Africa. *Malaria Journal, 21*(1), 1–5. https://doi.org/10.1186/S12936-022-04176-9/FIGURES/1

Oguike, M. C., Falade, C. O., Shu, E., Enato, I. G., Watila, I., Baba, E. S., Bruce, J., Webster, J., Hamade, P., Meek, S., Chandramohan, D., Sutherland, C. J., Warhurst, D., & Roper, C. (2016). Molecular determinants of sulfadoxine-pyrimethamine resistance in plasmodium falciparum in Nigeria and the regional emergence of dhps 431V. *International Journal for Parasitology: Drugs and Drug Resistance, 6*(3), 220–229. https://doi.org/10.1016/J.IJPDDR.2016.08.004

Ojurongbe, O., Nguetse, C. N., Fayemiwo, S. A., Falade, C. O., Ojurongbe, T. A., Thomas, B. N., Meyer, C. G., & Velavan, T. P. (2018). High prevalence of dihydrofolate reductase gene mutations in plasmodium falciparum parasites among pregnant women in Nigeria after reported use of sulfadoxine-pyrimethamine. *Pathogens and Global Health, 112*(2), 86–92. https://doi.org/10.1080/20477724.2017.1422615

Quan, H., Igbasi, U., Oyibo, W., Omilabu, S., Chen, S. B., Shen, H. M., Okolie, C., Chen, J. H., & Zhou, X. N. (2020). High multiple mutations of Plasmodium falciparum-resistant genotypes to sulphadoxine-pyrimethamine in Lagos, Nigeria. *Infectious Diseases of Poverty, 9*(1), 1–11. https://doi.org/10.1186/S40249-020-00712-4/TABLES/8

Radohery, G. F. R., Walz, A., Gumpp, C., Cherkaoui-Rbati, M. H., Gobeau, N., Gower, J., Davenport, M. P., Rottmann, M., McCarthy, J. S., Möhrle, J. J., Rebelo, M., Demarta-Gatsi, C., & Khoury, D. S. (2022). Parasite viability as a measure of in vivo drug activity in preclinical and early clinical antimalarial drug assessment. *Antimicrobial Agents and Chemotherapy, 66*(7), e0011422. https://doi.org/10.1128/AAC.00114-22

Romero, M., Leiba, E., Carrión-Nessi, F. S., Freitas-De Nobrega, D. C., Kaid-Bay, S., Gamardo, Á. F., Chavero, M., Figuera, L., Camejo-Ávila, N. A., Marcano, M. V., Lopez-Perez, M., & Forero-Peña, D. A. (2021). Malaria in pregnancy complications in Southern Venezuela. *Malaria Journal, 20*(1), 1–8. https://doi.org/10.1186/S12936-021-03728-9/FIGURES/2

Rouamba, T., Samadoulougou, S., Ouédraogo, M., Hien, H., Tinto, H., & Kirakoya-Samadoulougou, F. (2021). Asymptomatic malaria and anaemia among pregnant women during high and low malaria transmission seasons in Burkina Faso: Household-based cross-sectional surveys in Burkina Faso, 2013 and 2017. *Malaria Journal, 20*(1), 1–13. https://doi.org/10.1186/S12936-021-03703-4/TABLES/5

Seal, S. L., Mukhopadhay, S., & Ganguly, R. P. (2012). Malaria in pregnancy. *Obstetric Medicine, 5*(1), 2. https://doi.org/10.1258/OM.2011.110063

Wang, S., Xu, S., Geng, J., Si, Y., Zhao, H., Li, X., Yang, Q., Zeng, W., Xiang, Z., Chen, X., Zhang, Y., Li, C., Kyaw, M. P., Cui, L., & Yang, Z. (2020). Molecular surveillance and in vitro drug sensitivity study of plasmodium falciparum isolates from the China–Myanmar Border. *The American Journal of Tropical Medicine and Hygiene, 103*(3), 1100. https://doi.org/10.4269/AJTMH.20-0235

Yagnik, K., Farooqi, B., Mandernach, M. W., Cannella, A. P., & Kalyatanda, G. (2019). Severe case of plasmodium falciparum malaria in a pregnant woman from Nigeria. *Case Reports in Infectious Diseases, 2019*, 1–3. https://doi.org/10.1155/2019/2630825

Zekar, L., & Sharman, T. (2023). *Plasmodium falciparum Malaria. StatPearls*. https://www.ncbi.nlm.nih.gov/books/NBK555962/

Chapter 2
Assessment of Antibiotics Susceptibility Pattern of Gram-negative Bacteria Associated with Deteriorated Carrot and Cabbage from Markets in Ota, Ogun State, Nigeria

Margaret Ikhiwili Oniha, Abimbola Omolara Hassan, Stephen Oluwanifise Oyejide, Motunrayo Oluwabunmi Aina, Olugbenga Samson Taiwo, John Shade Olorunshola, and Eze Frank Ahuekwe

1 Introduction

Vegetables are among the basic and nourishing foods for humans that help maintain good health and prevent diseases (Ülger et al., 2018). Vegetables are priced for their supply of nutrients that contain bioactive compounds including vitamins, dietary fibre, minerals and phytochemicals (Slavin & Lloyd, 2012), which have been implicated in lowering the risk of chronic illnesses, such as obesity, diabetes, some malignancies and cardiovascular diseases (Remizeb et al., 2018).

Carrot (*Daucus carota*, family Apiaceae) is a vegetable commonly identified in Nigeria, Africa, Europe and southwest Asia (Iorizzo et al., 2020). Carrot is an orange, purple, black, red, white or yellow root vegetable with edible taproot, stems and leaves. It is endowed with nutrients, dietary fibre and health-promoting bioactive substances such as carotenoids, flavonoids, polyacetylenes, vitamins and minerals that serve as immune system boosters, anticancer agents and antioxidants. Furthermore, studies have shown that carrots are effective against diabetes, high cholesterol, cardiovascular diseases, hypertension and skin ageing (Varshney & Mishra, 2022).

M. I. Oniha (✉) · A. O. Hassan · M. O. Aina · O. S. Taiwo · E. F. Ahuekwe
Department of Biological Sciences, Covenant University, Ota, Ogun State, Nigeria
e-mail: margaret.oniha@covenantuniversity.edu.ng

S. O. Oyejide
Department of Cell Biology and Genetics, University of Lagos, Akoka, Lagos State, Nigeria

J. S. Olorunshola
Department of Biological Sciences, Caleb University, Imota, Lagos State, Nigeria

P. O. Isibor, P. A. Akinduti (eds.), *Harnessing Biotechnology Tools for Product Development*, https://doi.org/10.1007/978-3-031-86002-7_2

Cabbage (*Brassica oleracea*, family Brassicaceae) is a ready-to-eat vegetable available in several parts of the world. Soliman (2019) reported that cabbages contain potassium and fibre that support cardiovascular health and facilitate digestion. Cabbages contain abundant vitamins C and K, which are essential for a robust immune system, and contain a substance identified as sinigrin that facilitates weight loss in humans (Dias, 2019).

Several studies have reported high microbial loads associated with vegetables, including pathogens, parasites, bacteria, moulds and viruses (Luna-Guevara et al., 2019; Rahman et al., 2021). Post-harvest contamination can occur through the reuse of soil that has previously germinated fruits and produce (Giordano et al., 2021). Other sources of contamination include animal sewage, irrigation aqua, fungicides, insecticides, grime, insects, untamed and domestic animals, operating, unsterilized garnering equipment, transport vessels, rinse aqua, ice, conveyance vehicles and operating equipment (Horn & Bhunia, 2018).

Vegetables sold in open markets typically have greater microbial loads that constitute a risk to human haleness (Ajiboye & Emmanuel, 2021; Rahman et al., 2021). A prevalent group of food-borne bacterial pathogens that have been identified in fresh vegetables are the coliform bacteria such as *Escherichia coli, Salmonella* spp., *Listeria monocytogenes* and *Klebsiella* spp. (Degaga et al., 2022). These pathogens could be antimicrobial-resistant, infect persons when the vegetables are digested, leading to an extended hospital stay with a substandard prognosis, enhanced frequency of bloodstream infections and escalated mortality. The effects of antimicrobial-resistant pathogens are global, and there is evidence of an increasing trend in the prevalence of antibiotic resistance in foodborne bacteria especially in vegetables (World Health Organization, 2014; David & Oluyege, 2015; Golly et al., 2016). Given the widespread consumption of these vegetables (carrots and cabbages) in most semi-urban and urban communities in Nigeria, surveillance studies are needed to determine the safety of vegetables consumed in local markets. Hence, this study aimed to identify Gram-negative bacterial pathogens from fresh carrots and cabbages sold in Ota markets in Ogun State as well as to deduce the antibiotic susceptibility patterns of the isolated bacteria.

2 Materials and Methods

2.1 Sample Collection

A total of thirty-six (36) healthy carrot and cabbage samples were randomly obtained from three vending markets (Ota, Atan and Iju markets) in the Ado-Odo, Ota LGA of Ogun State in Nigeria and labelled accordingly. Carrot and cabbage samples were immediately transferred in Ziploc bags to the microbiology laboratory for evaluation. The samples were surface sterilized with sterile distilled water before being pulverized using a sterile mortar and pestle.

2.2 Bacterial Isolation and Characterization

All samples were isolated using serial dilution and the pour-plate technique to obtain pure isolates. One gram of each of the 36 carrot and cabbage samples was homogenized in 9 ml of sterilized distilled water in separate test tubes. Standard tenfold serial dilutions were prepared and 1 ml of the required dilution of sample suspensions was inoculated into labelled solidified MacConkey agar medium. Every plate was immediately incubated at 37 °C for 24 h thereafter the bacterial colonies were counted as well as recorded. Each bacterial colony was identified using standard microbiological methods that include morphological features, Gram staining reaction and conventional biochemical characteristics. Pure bacterial isolates were inoculated onto freshly prepared nutrient agar medium slants in McCartney bottles, cultivated at 37 °C for 24 h, in addition to being stored in the refrigerator at 4 °C for further assays.

2.3 Antibiotic Susceptibility Evaluation

Antibiotic susceptibility assays were conducted using the Kirby-Bauer disc diffusion method following the Clinical Laboratory Standard Institute (CLSI) guidelines. All antibiotic discs used included cefuroxime (CXM), ceftriaxone (CRO), cefotaxime (CTX), ofloxacin (OFX), gentamicin (GM), nalidixic acid (NA), augmentin (AUG), imipenem (IPM), nitrofurantoin (NF), ampiclox (ACX), levofloxacin (LBC) and cefixime (ZEM). To carry out this test, sterile swabs were suspended in normal saline containing the bacteria isolates that matched McFarland's standard (0.5) and then smoothly streaked onto plate surfaces containing Mueller-Hinton agar medium. Sterile forceps were used to transfer the antibiotic discs for the placement of Gram-negative isolates on each surface of the plates for incubation at 37 °C for 24–48 hours. After incubation, the plates were examined for inhibition. Each of the zones of inhibition was evaluated and elucidated according to Clinical Laboratory Standard Institute guidelines (Clinical and Laboratory Standards Institute, 2021).

3 Results

The total number of vegetables included in this study is 36 (Carrots, $n = 24$; Cabbage, n-12) as shown in Table 2.1.

Carrot and cabbage samples from the three markets in Ado-Odo, Ota LGA were investigated for microbial standards using the aerobic plate count. All the aggregate aerobic bacteria existing in carrots ranged from 1.0 to 17.8×10^6 cfu/g and from 1.8 to 7.1×10^6 cfu/g for cabbage (Table 2.2).

Table 2.1 Total quantity of vegetables amassed

Sample type	Number of samples collected			Total
	Sango market	Oja Ota	Iju market	
Carrot	8	8	8	24
Cabbage	4	4	4	12
Total				36

Table 2.2 Aggregate aerobic bacteria accounted for in the vegetable samples analysed (cfu/g)

Sample type	cfu/g
Carrot	$1.0–17.8 \times 10^6$
Cabbage	$1.8–7.1 \times 10^6$

Nine bacterial strains isolated from the combined plate count of cultures obtained from the produce samples were determined to belong to the genera of *Salmonella*, *Klebsiella*, *Escherichia*, *Pseudomonas*, *Proteus* and *Shigella* based on morphological characteristics such as size, shape, margin, pigment, opacity, form and biochemical characteristics (Tables 2.2 and 2.3). Results presented in Fig. 2.1 showed that *Escherichia coli* (34%) recorded the most prevalent bacterium isolated, closely accompanied by *Pseudomonas* spp. (22%), while *Proteus* spp., *Salmonella* spp., *Shigella* spp. and *Klebsiella* spp. were the least prevalent (11%) (Table 2.4).

The findings of the antibiotic sensitivity evaluations on the bacterial isolates situated in Table 2.5 portray that all the bacteria isolates were sensitive to the antibiotics; ofloxacin and levofloxacin (100%) and resistant to augmentin, cefotaxime and imipenem (100%) with varying degrees of sensitivity or resistance to other antibiotics.

4 Discussion

With an ever-growing global population, food losses are a principal concern worldwide with approximately one-third of food for human consumption recorded as either lost or wasted (Ishangulyyev et al., 2019). Carrots and cabbages are food vegetables that are prone to microbial attack owing to their nutritive components and moisture content present in them. The enumeration and identification of microorganisms in a food material are induced through microbial analysis which provides an attestation of the sample's quality and degree of possible risk to susceptible consumers. In this study, different pathogenic bacterial species isolated from vegetable

Table 2.3 Morphological characteristics of bacteria isolates from the vegetable samples

Isolate codes	Size	Shape	Margin	Elevation	Pigment	Opacity	Form
CAO1	Punctiform	Rod	Entire	Raised	Colourless	Translucent	Circular
CAA2	Large	Short rod	Entire	Raised	Cream	Translucent	Circular
CAA3	Small	Short rod	Entire	Raised	Cream	Translucent	Circular
CAO4	Small	Short rod	Entire	Flat	Green	Opaque	Circular
CAA5	Small	Rod	Entire	Convex	Greyish white	Opaque	Circular
CAI6	Punctiform	Rod	Entire	Raised	Cream	Opaque	Circular
CAO7	Small	Rod	Entire	Convex	Colourless	Translucent	Circular
CBA8	Small	Rod	Entire	Convex	Pink	Opaque	Circular
CBI9	Small	Short rod	Entire	Convex	Cream	Opaque	Circular

Key: *CAO* Carrot Ota, *CAA* Carrot Atan, *CAI* Carrot Iju, *CBA* Cabbage Atan, *CBI* Cabbage Iju

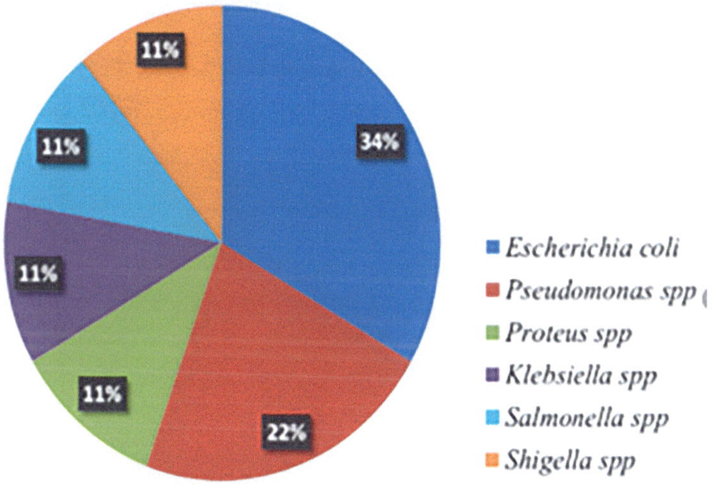

Fig. 2.1 Percentage of occurrence of Gram-negative bacteria isolated from carrots and cabbages

samples were recorded of which *E. coli* and *Pseudomonas* spp. were recorded as the most abundant.

Aerobic plate count (APC) is a procedure used to estimate the aggregate aerobic microorganisms present in a food specimen (Owolabi & Ichoku, 2014). Although it provides a general idea of microbiological quality, it cannot distinguish between harmless natural microflora, spoilage organisms and potentially harmful pathogens. The existence of aerobic organisms in eatables indicates favourable conditions for microbial growth. In this study, all carrot and cabbage samples analysed, regardless of vendor or market source, had average contamination proportions exceeding 1×10^5 cfu/g. The New South Wales (NSW) Food Authority (2009) set a standard limit of $<1 \times 10^5$ cfu/g for bacterial counts for vegetables and ready-to-eat foods (Owolabi & Ichoku, 2014). Therefore, our findings imply that the vegetables

Table 2.4 Biochemical characteristics of the bacteria cultures

Isolate code	Gram Staining reaction	Oxidase	Catalase	Urease	H_2S	Indole	Motility	MR	VP	Citrate	Glucose	Lactose	Galactose	Sucrose	Fructose	Organism identified
CAO1	–	–	+	–	+	–	+	+	–	–	+	–	+	–	–	*Salmonella* spp.
CAA2	–	–	+	+	–	–	–	–	+	+	+	+	+	+	+	*Klebsiella* spp.
CAA3	–	–	+	–	–	+	+	+	–	–	+	+	+	+	+	*Escherichia coli*
CAO4	–	+	+	+	–	–	+	–	–	+	+	–	–	–	+	*Pseudomonas* spp.
CAA5	–	–	+	+	+	+	+	+	–	+	+	–	–	–	+	*Proteus* spp.
CAI6	–	–	+	–	–	+	+	+	–	–	+	+	+	+	+	*Escherichia coli*
CAO7	–	–	+	–	–	–	–	+	–	–	+	–	+	–	–	*Shigella* spp.
CBA8	–	+	+	+	–	–	+	–	–	+	+	–	–	–	+	*Pseudomonas* spp.
CBI9	–	–	+	–	–	+	+	+	–	–	+	+	+	+	+	*Escherichia coli*

Key: *CAO* Carrot Ota, *CAA* Carrot Atan, *CAI* Carrot Iju, *CBA* Cabbage Atan, *CBI* Cabbage Iju, *MR* methyl red, *VP* Voges Proskaeur, *(+)* positive, *(–)* negative

Table 2.5 Antibiotics susceptibility test for Gram-negative bacteria isolates

	Gram-negative isolates ($n = 6$)	
Antibiotics	Sensitive (%)	Resistant (%)
Ampiclox (ACX)	16.7	83.3
Augmentin (AUG)	0	100
Cefotaxime (CTX)	0	100
Ceftriaxone (CRO)	33.3	66.7
Cefuroxime (CXM)	16.7	83.3
Gentamicin (GN)	50	50
Imipenem (IMP)	0	100
Levofloxacin (LBC)	100	0
Nalidixic acid (NA)	16.7	83.3
Nitrofurantoin (NF)	67	33
Ofloxacin (OFX)	100	0
Cefixime (ZEM)	66.7	33.3

analysed in this study are not suitable for human manducation without additional measures taken by food handlers and consumers. Previous reports in Nigeria have also documented microbial contamination of vegetables, which is consistent with our results. These detections underline the significance of implementing proper hygiene implementations and quality benchmark measures through vegetable production and handling to ensure consumer safety (Eni et al., 2010; Abdullahi & Abdulkareem, 2010; Aparecida de Oliveira et al., 2011; Ali et al., 2011; Owolabi & Ichoku, 2014).

Six genera of Gram-negative bacteria were cultured from the samples that included *Escherichia coli Pseudomonas* spp., *Proteus* spp., *Salmonella* spp., *Shigella* spp. and *Klebsiella* spp. *Escherichia coli* had the highest bacteria count compared to the other Gram-negative bacilli isolated in this research which corroborates the detections of Obajuluwa et al. (2021). The presence of food spoilage organisms or opportunistic pathogens that contaminate food can indirectly pose a risk to public health. These pathogenic bacteria can develop antibiotic resistance which poses a public health challenge to humans. Bacterial contamination in vegetable samples may be the result of the poor hygiene habits of market vendors. Many bacteria species isolated are members of the Enterobacteriaceae family, a cluster of Gram-negative intestinal bacteria that are extremely pathogenic to humans and animals. The isolation of these enteric bacteria in this study signifies the public health importance of poor-quality vegetables being circulated in the study regions. The source from which the veggies were obtained may have yielded immense bacterial loads in a few samples. Furthermore, the use of wastewater for irrigation or untreated manure for fertilizer, as well as the unclean environment where vegetables are grown and marketed, may be responsible for the high bacterial load obtained in the vegetable samples (Alegbeleye et al., 2018). Generally, these species of the Enterobacteriaceae family are indicators of faecal contamination which has been implicated in numerous human infections. Identified *Escherichia coli* and *Proteus* are notorious for their ability to infect the human lower respiratory tract and also engender chronic diseases specifically when antimicrobial-resistant strains are associated. *Pseudomonas* spp. is a common Gram-negative soil bacterium that can engender bacteraemia, pneumonia and other nosocomial infections. It can also result in septic shock and is the cause of cystic fibrosis in people. The carrot samples had a higher bacterial load than the cabbage because as a root vegetable with ridges on its skin, it creates room for trapping bacteria.

The isolated Gram-negative bacilli were sensitive to ofloxacin and levofloxacin but were resistant to augmentin, cefotaxime and imipenem which corroborates the results. The primary cause of resistance of Gram-negative bacteria's resistance to a variety of drugs, such as beta-lactams, quinolones and other antibiotics, is the structure of the outer membrane (Miller, 2016). It is fundamental to note that resistant isolates contain resistance determinants encoding multidrug outpouring siphons from the paramount originator superfamily as well as resistance genes such as beta-lactamase genes, quinolone resistance causations, the regulatory system regulating antibiotic efflux and the supervisory system moderating antibiotic outflow (opposition-nodulation-division multidrug outflow siphons). This correlates with the reported findings of Igbeneghu and Abdu (2014) and Richter et al. (2019).

Microbiologically guarded vegetables are crucial to augment the haleness benefits associated with their manducation. Washing vegetables with aqua complemented with differing concentrations of acetic or citric acid is effective for decontamination before consumption (Owolabi & Ichoku, 2014).

5 Conclusion

The results of the current research on ready-to-eat vegetables gathered from three vends in Ado-Odo Ota LGA revealed that *Escherichia coli* and *Pseudomonas* spp. associated with these produces are potential carriers of antimicrobial resistance genes, and as such posing a substantial risk for causing serious health challenges to humans when these vegetables are consumed. To minimize the risk associated with the spread of human illness through these ready-to-eat vegetables, the hazard points should be controlled from the field to harvesting, transportation, distribution in retail markets, processing at home or food service facilities to storage. Also, the government and relevant agencies should provide regular training on the need for appropriate hygiene practices by handlers, sellers and consumers of ready-to-eat vegetables while explicating the implications of consuming contaminated vegetables. Thus, further studies through the application of biotechnology tools could foster improved standards and policies by policymakers.

References

Abdullahi, I. O., & Abdulkareem, S. (2010). Bacteriological quality of some ready to eat vegetables as retailed and consumed in Sabon-Gari, Zaria, Nigeria. *Bayero Journal of Pure and Applied Sciences., 3*(1), 173–175.

Ajiboye, A. E., & Emmanuel, T. (2021). Assessment of bacterial contamination in ready-to-eat fruits and vegetables sold at Oja-Oba market, Ilorin. *African Journal of Biomedical Research, 24*(2), 203–209.

Alegbeleye, O. O., Singleton, I., & Sant'Ana, A. S. (2018). Sources and contamination routes of microbial pathogens to fresh produce during field cultivation: A review. *Food Microbiology, 73*, 177–208.

Ali, M., Khan, M. R., & Saha, M. L. (2011). Antibiotic-resistant patterns of bacterial isolates from ready-to-eat (RTE) street-vended fresh vegetables and fruits in Dhaka city. *Bangladesh Journal of Science Research., 24*(2), 127–134.

Aparecida de Oliveira, M., Maciel de Souza, V., Bergamini, A. M. M., & De Martinis, E. C. P. (2011). Microbiological quality of ready-to-eat minimally processed vegetables consumed in Brazil. *Food Control, 22*, 1400–1403.

Clinical and Laboratory Standards Institute. (2021). *Performance standards for antimicrobial susceptibility testing approved standard M100-S23*. Clinical and Laboratory Standards Institute.

David, O. M., & Oluyege, A. O. (2015). Antibiotic resistance and plasmid carriage among Salmonella typhi isolated from food and hands of food handlers in Nigeria university. *International Journal of Current Microbiology and Applied Science., 4*(3), 906–914.

Degaga, B., Sebsibe, I., Belete, T., & Asmamaw, A. (2022). (2022) microbial quality and safety of raw vegetables of fiche town, Oromia, Ethiopia. *Journal of Environmental and Public Health., 5*, 523–528.

Dias, J. S. (2019). Nutritional quality and effect on disease prevention of vegetables. *Food and Nutrition Sciences., 10*(4), 369–402.

Eni, A. O., Oluwawemitan, I. A., & Oranusi, U. S. (2010). Microbial quality of fruits and vegetables sold in Sango Ota, Nigeria. *African Journal of Food Sciences., 4*(5), 291–296.

Giordano, M., Petropoulos, S. A., & Rouphael, Y. (2021). The fate of nitrogen from soil to plants: Influence of agricultural practices in modern agriculture. *Agriculture, 11*(10), 944–951.

Golly, M. K., Salifu, P. S., & Mills-Robertson, F. C. (2016). Resistance of bacteria isolates from cabbage (Brassica oleracea), carrots (Daucus carota), and lettuce (Lactuca sativa) in the Kumasi Metropolis of Ghana. *International Journal of Nutrition and Food Sciences., 5*(4), 297–303. https://doi.org/10.11648/j.ijnfs.20160504.20

Horn, N., & Bhunia, A. K. (2018). Food-associated stress primes foodborne pathogens for the gastrointestinal phase of infection. *Frontiers in Microbiology, 9*(2), 19–62.

Igbeneghu, O. A., & Abdu, A. B. (2014). Multiple antibiotic-resistant bacteria on fluted pumpkin leaves, an herb of therapeutic value. *Journal of Health, Population and Nutrition., 32*(2), 176–182.

Iorizzo, M., Curaba, J., Pottorff, M., Ferruzzi, G., Simon, C., & Pablo, F. (2020). Carrot anthocyanins genetics and genomics: Status and perspectives to improve its application for the food colorant industry. *Genes, 11*(8), 906–918.

Ishangulyyev, R., Kim, S., & Lee, S. H. (2019). Understanding food loss and waste-why are we losing and wasting food? *Food, 8*, 1–23.

Luna-Guevara, J. J., Arenas-Hernandez, M. M. P., Martínez de la Peña, C., Silva, J. L., & Luna-Guevara, M. L. (2019). The role of Pathogenic E. coli in fresh vegetables: Behavior, contamination factors, and preventive measures. *International Journal of Microbiology., 2019*, 1–10.

Miller, S. I. (2016). Antibiotic resistance and regulation of the Gram-negative bacterial outer membrane layer by host innate immune molecules. *MBio, 15*, 41–61.

Obajuluwa, A. F., Nwadili, C. C., & Adeshina, G. O. (2021). Isolation and susceptibility profile of Bacteria isolated from mobile phones of pharmacy students of Ahmadu Bello University, Zaria, Nigeria. *Nigerian Journal of Microbiology., 35*(2), 5822–5828.

Owolabi, J. B., & Ichoku, C. K. (2014). Evaluation of antibiotic resistance pattern of gram-positive Bacilli isolated from ready-to-eat vegetables sold in Ota Metropolis, Nigeria. *Covenant Journal of Physical and Life Sciences., 2*(2), 138–149.

Rahman, M., Alam, M.-U., Luies, S. K., Kamal, A., Ferdous, S., Lin, A., Sharior, F., Khan, R., Rahman, Z., Parvez, S. M., Amin, N., Hasan, R., Tadesse, B. T., Taneja, N., Islam, M. A., & Ercumen, A. (2021). Contamination of fresh produce with antibiotic-resistant Bacteria and associated risks to human health: A scoping review. *International Journal of Environmental Research and Public Health., 19*(1), 360–361.

Remizeb, F., Septembre-Malaterreb, A., & Pouchereta, P. (2018). Fruits and vegetables, as a source of nutritional compounds and phytochemicals: Changes in bioactive compounds during lactic fermentation. *Food Research International, 104*, 86–99.

Richter, L., Du Plessis, E. M., Duvenage, S., & Korsten, L. (2019). Occurrence, identification, and antimicrobial resistance profiles of extended-spectrum and AmpC β-lactamase-producing Enterobacteriaceae from fresh vegetables retailed in Gauteng Province, South Africa. *Foodborne Pathogens and Disease., 16*(6), 421–427.

Slavin, J. L., & Lloyd, B. (2012). Health benefits of fruits and vegetables. *Advances in Nutrition., 3*(4), 506–516.

Soliman, G. A. (2019). Dietary fiber, atherosclerosis, and cardiovascular disease. *Nutrients, 11*(5), 11–55.

Ülger, T. G., Songur, A. N., Çırak, O., & Pınar Çakıroğlu, F. (2018). *Role of vegetables in human nutrition and disease prevention* (pp. 7–32). IntechOpen.

Varshney, K., & Mishra, K. (2022). An analysis of health benefits of carrot. *International Journal of Innovative Research in Engineering & Management., 9*(1), 211–214.

World Health Organization (WHO). (2014). *Antimicrobial resistance, global report on surveillance* (pp. 1–6). WHO Press.

Chapter 3
Advancements in *Bacillus anthracis* Diagnostic and Treatment Techniques

Obinna Chukwuemeka Nwinyi, Alaba Tolulope Ajayi, and Abigael Tolulope Owolabi

1 Introduction

B. anthracis is a rod-shaped, aerobic, non-motile Gram-positive bacterium that causes anthrax. The bacteria thrive well when cultured on blood or nutrient agar and incubated at a temperature of 37 °C. *Bacillus anthracis* develops in two distinct forms, namely the active vegetative and inactive spore forms, which are biologically inclined (Chateau et al., 2020). However, the vegetative form is mostly involved in the tissue of infected hosts, which is accountable for the disease pathology. The spore form is observed when the host yields to the infection, which converts the vegetative form into an inert resistant spore (Zhu et al., 2018). The relative tolerance of *B. anthracis* spores to extreme environmental conditions (for example, rain, heat, radiation, and some disinfectants) has been investigated to be used as a potential biowarfare agent (Doganay et al., 2023). While uncommon in humans, anthrax ultimately affects herbivores such as horses, cattle, sheep, goats, deer, and bison. Carnivores can also contract anthrax if they consume infected animal tissues. The principal route of anthrax infection in wildlife and livestock is found in grazed vegetated regions where previous anthrax outbreaks were once recorded (Romero-Alvarez et al., 2020). Anthrax outbreaks remain prevalent in many desert grasslands, steppes, and tundras around the world, and have recently been discovered as a major cause of wildlife mortality in tropical forests (Mukarati et al., 2020). Anthrax infection in humans occurs via three main routes: cutaneous, respiratory, and gastrointestinal (Fig. 3.1). Cutaneous anthrax is found to cause prominent symptoms observed in humans, accounting for more than 92% of cases. It is commonly transmitted by spores entering through skin abrasions or wounds of individuals having direct

O. C. Nwinyi (✉) · A. T. Ajayi · A. T. Owolabi
Department of Biological Sciences, College of Sciences and Technology, Covenant University, Ota, Nigeria
e-mail: obinna.nwinyi@covenantuniversity.edu.ng

P. O. Isibor, P. A. Akinduti (eds.), *Harnessing Biotechnology Tools for Product Development*, https://doi.org/10.1007/978-3-031-86002-7_3

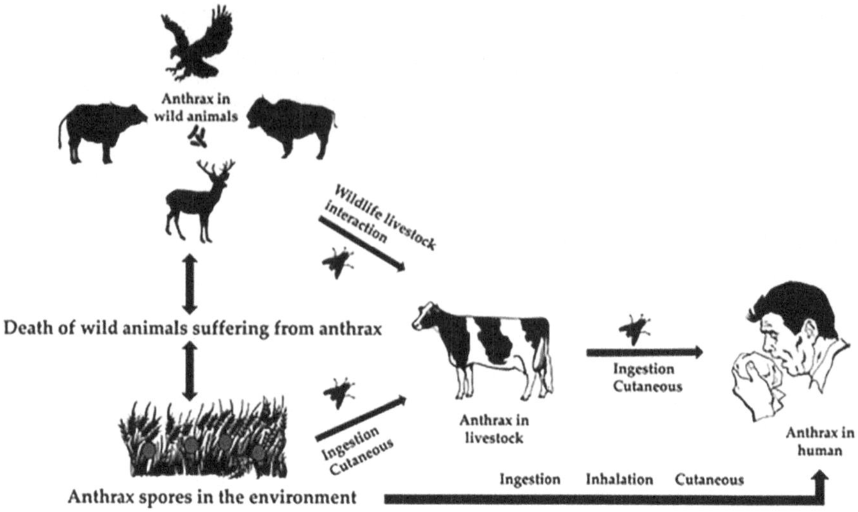

Fig. 3.1 The life cycle of *Bacillus anthracis* at the animal–environment–human interfaces. Source: Subedi et al. (2024)

contact (handling, touching) with contaminated animal items such as meat, wool, skins, and leather, and more rarely through insect bites (Kutmanova et al., 2020).

2 Biology of *Bacillus anthra*cis

*Bacillus anthra*cis is a Gram-positive, rod-shaped bacterium with sharp edges measuring about 3–5 μm in length and 1–2 μm in width. Smear preparation of blood from patient samples shows blue clustered in pairs, chains, or single cells when observed under the microscope (Figs. 3.2 and 3.3) (Chateau et al., 2020; Jiranantasak et al., 2022). *B. anthracis* endospores appear oval located in the centre and have the same diameter as the vegetative form. These endospores show up mostly when the bacteria are situated outside the host (Bakhteeva & Timofeev, 2022). The resistance of *B. anthracis* to antibiotics is associated with several attributes. One is the cell envelope, which is known to have a complex composition and the presence of a peptidoglycan layer of the bacteria. *B. anthracis* is found to grow on most culture media, but thrives best on blood agar free of antibiotics (Norris et al., 2020). A more distinct and pronounced growth can be observed in an aerobic environment, with temperatures ranging from 15 °C to 40 °C.

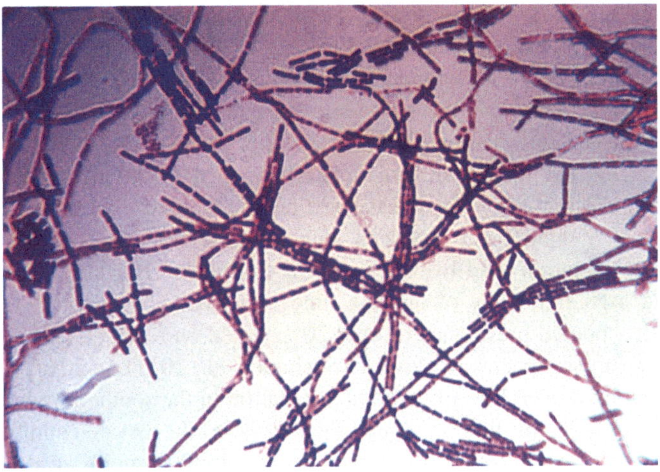

Fig. 3.2 A photomicrograph of *Bacillus anthracis* bacteria using Gram's stain technique. Source: CDC (2024)

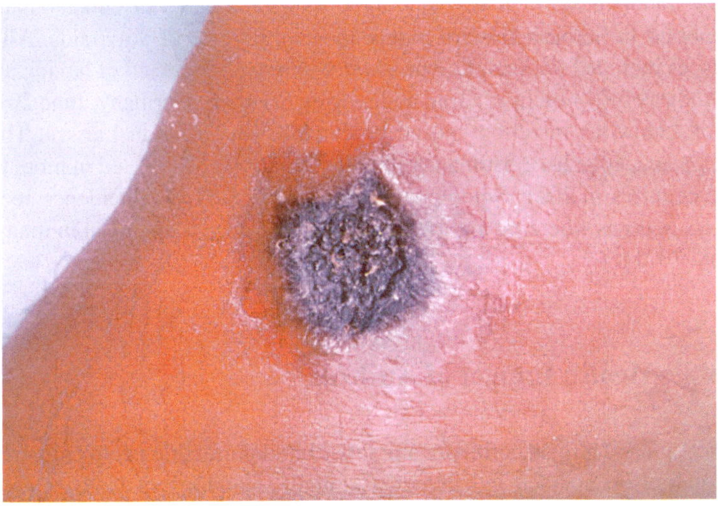

Fig. 3.3 Anthrax lesion on the skin of the forearm caused by the bacterium *Bacillus anthracis*. Source: CDC Public health Image Libarary https://phil.cdc.gov/details.aspx?pid=2226

3 Incidence of *Bacillus anthracis*

Anthrax outbreaks are linked to ecological and socioeconomic conditions that promote the disease's spread. Precipitation, temperature, soil qualities (calcium, pH, moisture, mineral composition), vegetation, host density, and elevation are some examples of well-studied ecological conditions Furthermore, increased rainfall may promote vegetation growth, providing more food for herbivores, which serve as hosts for anthrax. This can lead to higher animal densities and potentially more frequent outbreaks. Warm temperatures favour the persistence and growth of *Bacillus anthracis* spores in the soil while stressing animals and weakening their immune systems. Additionally, calcium-rich soils enhance the survival of anthrax spores, as calcium stabilizes their structures, extending their viability. Socioeconomic determinants, on the other hand, have not been adequately investigated, despite indications of a relationship between some human activities (cultivation, land removal, and irrigation) and livestock anthrax outbreaks (Blackburn et al., 2017). However, a high frequency of anthrax illness in livestock has been reported in Africa, with a pooled prevalence of over 20%, based on research on the spread of anthrax in herbivore species worldwide (Sushma et al., 2021). In 2017, Kyrgyzstan Republic recorded over 35% of the total population in the southern region infected with human anthrax, this was accounted for because of the elevated number of natural disasters due to mudslides, earthquakes, and floods (Luong et al., 2023). In Zimbabwe, year 2008, anthrax joined the severe cholera outbreak in a chaotic country where real numbers of illness victims were impossible to ascertain. World Health Organization reported over 200 human cases with 8 (eight) confirmed deaths, which are likely understated. In June 2011, three suspected cases of cutaneous anthrax were reported in Serbia's Eastern independence province of Vojvodina. All of the occurrences included the slaughter of a heifer that was suspected of having anthrax. A fatal anthrax infection in a heroin addict from Southern Germany, June 2012, and a case of cutaneous infection caused by *Bacillus anthracis* in Larissa, Thessaly, Central Greece, July 2012. Human anthrax outbreaks were detected in more than 20 Chinese counties in 2021. Some occurred in historically low-incidence locations, and others in areas where no anthrax infections had been recognized in many years (Yu et al., 2021).

4 Public Health Impact of Anthrax

Anthrax primarily targets herbivore animals. Human infection is typically transmitted through close contact with infected animals. Handling of leathers processed from animal hair poses a significant risk for anthrax (Croicu, 2019). The infection may also spread through contact with diseased meat, which is rampant among abattoir workers (Hamutyinei Dhliwayo et al., 2022). Cases of cutaneous anthrax have been reported to be prevalent among people working directly with animals. Anthrax may be employed in bioterrorism often using aerosols. Specific countries in the world still face a significant risk of anthrax outbreaks, which periodically cause fatalities (Zhang et al., 2022). Estimates suggest that each cow infected with anthrax in Africa could cause up to ten human cases (Muturi et al., 2018). In the early 1900s, there were over 129 cases of cutaneous anthrax every year in the United States of America, but there has been a significant decrease over the years as there are just about two occurrences annually (Hendricks et al., 2022). Anthrax is a rare disease that is common among agricultural workers which accounts for 90–95% of anthrax infections (Doganay et al., 2023). Gastrointestinal anthrax outbreak is associated with tainted meat; however, this is uncommon. Aerosol bioweapons are projected to produce a high proportion of inhaled anthrax due to the motive behind the spread of the spores. In 2021, bioterrorists made use of anthrax spore as a means of weapon using anthrax-contaminated mail, and 11 cases of inhalation anthrax were linked to the mail, however; 6 out of the 11 patients recovered after treatment, having a 45% fatality rate. The various form of anthrax infection has varying mortality, with skin anthrax ranging from 6 to 20% in untreated patients and less than 1% in those receiving antibiotics. Inhalation anthrax is known to have a significant death rate even with adequate treatment (Luong et al., 2023). Previously, the case fatality rate for this kind was estimated to be close to 100%. One study reported a slight shift from a 100 to 97% death rate once patients reached the tensed stage, regardless of therapy. However, more holistic treatment approaches may assist in reducing mortality rates (Hendricks et al., 2022). In some anthrax outbreaks, adults are documented to exhibit asymptomatic attributes while children have a greater fatality rate (Hendricks et al., 2022).

5 Forms of Anthrax in Humans

Human anthrax is traditionally classified as either non-industrial or industrial reflecting on how the occupation of the individual led to the exposure (Doganay et al., 2023). Non-industrial form of anthrax results from non-restricted contact with infected animals or their corpse after death, while the industrial form is associated with occupation, mostly occurring in farmers, butchers, knackers, and veterinarians, all of these people have direct contact with the animal by-products which is processed into wool, hair, hides, bones, or other animal products (Khairullah et al., 2024). Anthrax can also be contracted through a deliberate act such as bioterrorism. Human anthrax can also be classified by the route by which the disease was acquired, distinguishing between cutaneous anthrax and others, Cutaneous anthrax, initially documented by Maret in 1752 and Fournier in 1769, typically occurs among people who have encounters with infected animals, particularly farmers and veterinarians, with contamination occurring through skin sores acquired through a break in the skin. Intestinal (gastrointestinal) anthrax is contracted from the ingestion of contaminated food, primarily from the meal of animals infected with the disease or conceivably from the ingestion of contaminated water (Kutmanova et al., 2020), and inhalation (pulmonary) anthrax from breathing in air with *B. anthracis* spores which are immediately taken up through the lungs to the lymph nodes, where they develop, invade the phagocytes, and infiltrate into the bloodstream and lesions of the skin follow (Kutmanova et al., 2020). Lesions left unattended to without proper treatment may affect the local lymph nodes of the host (Doganay et al., 2023).

6 Transmission of *Bacillus anthracis*

B. anthracis infections can enter the body via the openings of the body. Insects, dogs, wild birds, and most carnivores are various vectors that act as vehicles for *B. anthracis*. Humans are at increased risk when in direct contact with animal faeces (Alam et al., 2022). Anthrax contracted through inhalation is prevalent among wool sorters working in the wool and hair industries (Maison et al., 2023). Anthrax is not limited to only workers working in industries, it was observed that people who have access to animal cemeteries, burial sites, and biothermal vents are carriers of anthrax spores after diagnosis (Finke et al., 2020). In Russia, certain areas were termed "stationary anthrax-affected areas" due to higher populations with epizootic anthrax spores, mostly near farms, meadows, and natural areas (Nikolakakis et al., 2023). The vegetative phase of *B. anthracis* is always active in bloody faeces and organs of anthrax-suspected animals, which aids transmission and the spread of anthrax diseases (Maison et al., 2023). Infection can occur through direct contact with infected corpses, undercooked foods, and exposure to spore-contaminated items. Sverdlovsk's incidence of anthrax outbreak in 1979 was traceable to mishandling of laboratory specimens which led to thousands of individuals being infected with anthrax. The disease is rapidly transmitted through aerosols, which enter the lungs deeply (Chugh, 2019).

7 Risk Factors

Recent research has drawn attention to the role of climate change in the resurgence of anthrax, particularly in regions where permafrost is thawing. This phenomenon has been observed in areas such as Siberia, where dormant anthrax spores, long buried in frozen soil, are being re-exposed as the ground thaws. For example, the 2016 anthrax outbreak in Siberia, Russia, was linked to unusually high temperatures that caused the melting of permafrost and released spores from carcasses of infected reindeer buried decades earlier (Ezhova et al., 2021). This discovery raises concerns that as global temperatures rise, more outbreaks could occur in historically anthrax-endemic regions. Areas that were once considered low-risk may face increased vulnerability, not only in terms of livestock but also in terms of human populations living near these regions. Governments and global health organizations are beginning to factor in the effects of climate change when developing new models for predicting anthrax outbreaks.

A novel and promising area of research involves genetic engineering to create disease-resistant livestock. Scientists have begun experimenting with CRISPR-Cas9 technology to modify the genes of livestock animals, particularly in regions with endemic diseases (Wang et al., 2022). By enhancing the natural immunity of animals through gene editing, researchers hope to reduce infections such as anthrax in livestock and thereby cut down human exposure to contaminated animal products.

One of the most promising developments in modern vaccine technology is the emergence of mRNA vaccines, as showcased during the COVID-19 pandemic (Chavda et al., 2022). These vaccines offer several advantages, including faster production times, lower production costs, and the ability to tailor vaccines to specific variants of a pathogen. While current anthrax vaccines are effective, they often require multiple doses and present logistical challenges, especially in low-resource settings. This has spurred interest in applying mRNA technology to anthrax (Georgopoulos & James, 2024). The potential for mRNA-based vaccines is considerable because they can be created quickly in response to outbreaks and may provide greater immune responses with fewer doses (González-Cueto et al., 2024). This is particularly important in areas where vaccine distribution is a challenge. The production process for mRNA vaccines does not rely on live pathogens, so future vaccines could be developed without the need for extensive biosafety precautions, reducing costs and production time (Gote et al., 2023). Developing such vaccines will require significant investment in research, but the long-term benefits could be substantial. The faster and more flexible production process could revolutionize how we respond to both natural anthrax outbreaks and bioterrorism threats.

One other concern is the risk of anthrax-contaminated animal products, such as wool, hides, or meat, entering global markets (Alam et al., 2022). In some countries, the regulation and inspection of these products may not be rigorous, increasing the risk of cross-border contamination. For example, wool or hides from anthrax-infected animals may be exported to other countries without sufficient biosecurity measures in place, introducing the pathogen to new areas (Alam et al., 2022).

Additionally, as globalization increases the movement of livestock for trade, there is a growing need for international cooperation to monitor and control anthrax outbreaks in animals. The World Organization for Animal Health (OIE) and the World Health Organization (WHO) must collaborate more closely to develop and enforce stronger guidelines for anthrax monitoring and vaccination in animals, particularly in countries where the disease is endemic. In the future, developing global tracking systems for livestock and animal products could be a game-changer. These systems would allow authorities to monitor the movement of potentially contaminated goods or animals in real time, enabling swift quarantine and decontamination efforts.

8 Symptoms Associated with Anthrax Disease

Eighty per cent (80%) of anthrax cases are recognized as being cutaneous. These lesions often start with a tingle that develops on exposed skin. The clinical symptoms advance through three notable stages: popular, vesicular, and eschar stages (that is, the wound healing stage). The eschar stage appears between 2 and 5 days after the hemorrhagic vesicles dry out, resulting in a depressed black scab with redness and swelling vesicles. Anthrax lesions are normally innocuous, although the oedema surrounding them might cause pain (Finke et al., 2020). In severe cases of cutaneous anthrax diseases, severe septicaemia is known to set in. However, early treatment and intervention minimize the risk of death of the host (Blackburn et al., 2021). The pulmonary kind of anthrax is infrequent and typically causes mild to non-specific symptoms including fever, tiredness, mild cough, and chest pain. The early symptoms of the infection may be confused for flu. Symptoms such as hypoxia, dyspnoea, high body temperature, and radiographic evidence of mediastinal enlargement appear within 2–6 days (Bower et al., 2022). Meningitis surfaces with a nearly 100% mortality rate if not treated or delayed (Thompson et al., 2022). Antibiotics provided adequately throughout the prodromal (preceding the more noticeable symptoms) period in good time have resulted in much lower mortality (Eshraghi et al., 2020). The oropharyngeal type of anthrax led to noteworthy outbreaks in low-income countries after the consumption of infected animal flesh. Some of the symptoms observed include fever and sore throat, mouth sores, lymphadenopathy-induced neck inflammation, dysphagia, severe localized swelling, ulceration, septicaemia, and sometimes loss of life (Doganay et al., 2023).

9 Laboratory Diagnosis of *Bacillus anthracis*

Bacillus species are quite hard and usually survive transport to the laboratory either in freshly collected specimens or in a standard transport medium. To effectively diagnose the presence of *Bacillus anthracis*, laboratory tests such as direct staining or culture of samples from malignant pustules, sputum, blood, or patient discharge are required (Golnari et al., 2024). *B. anthracis* differs from *B. cereus* in several ways, including its ability to form capsules, susceptibility to penicillin, non-motility, and ability to lyse bacteriophage (Baldwin, 2020). Only certain laboratories are assigned for the diagnosis of anthrax, this is due to the stringent biological safety criteria that need to be in place. Polymerase chain reaction (PCR) and enzyme-linked immunosorbent assay (ELISA) are the specific rapid screening techniques used (Zasada, 2020). Prior identification can be done in the hospital using direct Gram staining from vesicular fluid and cerebrospinal fluid of the lesion of the skin (Hoorali et al., 2020). These samples were initially identified by culturing them on blood agar to observe the cell morphology. The blood samples of patients who tested positive can be used to isolate the virus more effectively (Mechaly et al., 2018).

Real-Time Polymerase Chain Reaction (RT-PCR) has solidified its position as a gold standard in diagnosing *Bacillus anthracis*, owing to its unparalleled specificity and sensitivity. RT-PCR detects the presence of bacterial DNA by amplifying specific gene sequences, making it highly effective in identifying *B. anthracis*, even in low quantities (Braun et al., 2021). The technique is particularly valuable in both clinical settings and emergency response scenarios due to its rapid turnaround time and reliability.

Biosensor technology has emerged as another powerful and highly sensitive tool for detecting biological markers associated with *Bacillus anthracis*, the causative agent of anthrax (Wang et al., 2021). These devices can identify spores, toxins, and other biological indicators of *B. anthracis* presence in clinical, environmental, or food samples (Mir et al., 2023). Biosensors offer a range of benefits, including rapid detection, portability, and the ability to provide real-time data, making them crucial for both public health monitoring and biodefense. Biosensors work by converting a biological response—such as the presence of specific toxins or spores—into an electrical, optical, or thermal signal, which is then quantified (Mir et al., 2023). This allows for a highly precise detection of *B. anthracis*, even at low concentrations, which is critical in early diagnosis or contamination surveillance.

Next-generation sequencing (NGS) has transformed pathogen detection by providing unprecedented accuracy and depth in detecting and characterizing bacterial species, including *Bacillus anthracis*, the causal agent of anthrax (Gwinn et al., 2019). NGS allows extensive study by sequencing a pathogen's whole genome, which not only confirms its presence but also gives valuable information on strain variation, virulence factors, and drug resistance profiles. This degree of data is critical for successful epidemic response, epidemiological tracking, and treatment recommendations. NGS for *Bacillus anthracis* detection involves fragmenting the pathogen's DNA into smaller sequences and reading them concurrently (parallel sequencing) (Cummings et al., 2009).

10 Treatment of Anthrax

Early diagnosis of anthrax can be difficult and pose a challenge due to its rarity and this often leads to misinterpretation of diagnosis (Aminu et al., 2020). The treatment of anthrax is challenging and mostly takes longer than expected. However, some antibiotics such as penicillin, tetracycline, and erythromycin are implemented for the treatment of anthrax, though it is always prescribed in large doses (Doganay et al., 2023). Additional antibiotics include clindamycin, imipenem, chloramphenicol, rifampin, aminoglycosides, fluoroquinolones, and vancomycin which target the protein synthesis of the bacteria. Several antibiotics have been confirmed to build resistance against *B. anthracis*; some include ceftazidime, cefotaxime, sulfamethoxazole, aztreonam, cefuroxime, and trimethoprim (Stark, 2022). Untreated inhalation type of anthrax is almost usually lethal, even with early antibiotics therapy (Hendricks et al., 2022). Prompt antibiotic administration and supportive care are crucial for the proper management of inhalation anthrax.

11 Preventive Measures Against Anthrax

To prevent the spread of anthrax, livestock, one of the agents that transmit the disease, is subjected to routine vaccines done annually. Livestock are vaccinated with purified or dead bacterial toxins (Sarker et al., 2020). Most vaccines are administered to susceptible livestock in endemic areas (Sahoo et al., 2020). Rapid steps are necessary to mitigate anthrax outbreaks to limit environmental pollution and human exposure, and effectively diagnose and treat sick animals (Doganay et al., 2023). A laid down procedure by the government bodies is duly followed for the report of any smelly, slimy, or dull meat obtained from this livestock, hence appropriate measures are taken immediately. Direct contact with suspected sick animals is restricted and autopsies or surgeries are not allowed to be done on such animals. The disposal of carcasses is done by approved health personnel, which is done with care and caution (Sidwa et al., 2020). The public is advised to patronize only legally certified slaughterhouses. Here, meats are properly checked and certified to be fit for consumption before being sold to the masses. Other preventive measures are also put in place such as proper cooking of animal meats and washing hands with soap and water before and after handling meats (Seid et al., 2020). A reporting channel is made available for the public to report any sickness of deceased livestock (Mwakapeje et al., 2018). Cows with noticeable symptoms should be quarantined promptly and given high doses of penicillin for the next 45 days (Stark, 2022). Additionally, persons are not permitted to transport sick animals outside the area and must promptly clean themselves with soap or disinfectant after coming into contact with sick animals. Dead livestock infected with anthrax should be incinerated or buried at a minimum depth of two metres and properly disinfected (Alam et al., 2022). The depth of less than two metres can cause long-term contamination through water bodies and pose greater risks in the future (Biloivan et al., 2020). Anthrax spores can spread during natural disasters including floods, landslides, and earthquakes, increasing the risk of infection in herbivore animals (Zorigt et al., 2021). Avoiding locations with anthrax spores will help prevent the spread of the disease. To raise public awareness regarding anthrax, it is recommended that a questionnaire be administered, which is used to conduct a knowledge, attitude, and practice survey first. Furthermore, public health organizations must improve through close collaboration with the public by monitoring, inspecting, and packaging animal products, as this will reduce the danger of food contamination.

12 Conclusion

Although anthrax is a forgotten disease of the past, it is now becoming a re-emerging disease in many parts of the world such as the Middle East, Central Asia, and African countries, and even developed Western countries such as the United States. *Bacillus anthracis* continues to pose a severe public health danger due to its possible use as a biowarfare and the continued risk of being an occupational disease. Early diagnosis and quick intervention of antibiotic therapy are essential for successful treatment. Continuous research for new diagnostic approaches and vaccines is vital for minimizing the impact of anthrax diseases.

References

Alam, M., Kamal, M., Rahman, M., Kabir, A., Islam, M., & Hassan, J. (2022). Review of anthrax: A disease of farm animals. *Journal of Advanced Veterinary and Animal Research., 9*(2), 323–324. https://doi.org/10.5455/javar.2022.i599

Aminu, O. R., Lembo, T., Zadoks, R. N., Biek, R., Lewis, S., Kiwelu, I., Mmbaga, B. T., Mshanga, D., Shirima, G., Denwood, M., & Forde, T. L. (2020). Practical and effective diagnosis of animal anthrax in endemic low-resource settings B.R. Beechler (ed.). *PLOS Neglected Tropical Diseases., 14*(9), 1–5. https://doi.org/10.1371/journal.pntd.0008655

Bakhteeva, I., & Timofeev, V. (2022). Some peculiarities of anthrax epidemiology in herbivorous and carnivorous animals. *Life., 12*(6), 870–875. https://doi.org/10.3390/life12060870

Baldwin, V. M. (2020). You Can't B. Cereus—A review of Bacillus cereus strains that cause anthrax-like disease. *Frontiers in Microbiology, 11*, 1731. https://doi.org/10.3389/fmicb.2020.01731

Biloivan, O. V., Stegniy, B. T., Gerilovych, A. P., Arefiev, V. L., Wölfel, R., Schwarz, J., Popp, C., & Grass, G. (2020). Screening of possibly anthrax-contaminated burial sites in eastern and southern Ukraine. *Agricultural Science and Practice., 7*(3), 3–14. https://doi.org/10.15407/agrisp7.03.003

Blackburn, J. K., Kenu, E., Asiedu-Bekoe, F., Sarkodie, B., Kracalik, I. T., Bower, W. A., Stoddard, R. A., & Traxler, R. M. (2021). High case-fatality rate for human anthrax, northern Ghana, 2005–2016. *Emerging Infectious Diseases., 27*(4), 1216–1219. https://doi.org/10.3201/eid2704.204496

Blackburn, J. K., Matakarimov, S., Kozhokeeva, S., Tagaeva, Z., Bell, L. K., Kracalik, I. T., & Zhunushov, A. (2017). Modeling the ecological niche of *Bacillus anthracis* to map anthrax risk in Kyrgyzstan. *The American Journal of Tropical Medicine and Hygiene., 96*(3), 550–556. https://doi.org/10.4269/ajtmh.16-0758

Bower, W. A., Hendricks, K. A., Vieira, A. R., Traxler, R. M., Weiner, Z., Lynfield, R., & Hoffmaster, A. (2022). What is anthrax? *Pathogens., 11*(6), 690–700. https://doi.org/10.3390/pathogens11060690

Braun, P., Duy-Thanh Nguyen, M., Walter, M. C., & Grass, G. (2021). Ultrasensitive detection of bacillus anthracis by real-time PCR targeting a polymorphism in multi-copy 16S rRNA genes and their transcripts. *International Journal of Molecular Sciences, 22*(22), 12224. https://doi.org/10.3390/ijms222212224

CDC. (2024). Public Health Image Library. https://phil.cdc.gov/details.aspx?pid=2226

Chateau, A., Van der Verren, S. E., Remaut, H., & Fioravanti, A. (2020). The bacillus anthracis cell envelope: Composition, physiological role, and clinical relevance. *Microorganisms., 8*(12), 1864–1867. https://doi.org/10.3390/microorganisms8121864

Chavda, V. P., Soni, S., Vora, L. K., Soni, S., Khadela, A., & Ajabiya, J. (2022). MRNA-based vaccines and therapeutics for COVID-19 and future pandemics. *Vaccine, 10*(12). https://doi.org/10.3390/vaccines10122150

Chugh, T. (2019). Bioterrorism: Clinical and public health aspects of anthrax. *Current Medicine Research and Practice., 9*(3), 110–111. https://doi.org/10.1016/j.cmrp.2019.05.004

Croicu, A.-M. (2019). An optimal control model to reduce and eradicate anthrax disease in herbivorous animals. *Bulletin of Mathematical Biology, 81*(1), 235–255. https://doi.org/10.1007/s11538-018-0525-0

Cummings, C. A., Bormann Chung, C. A., Fang, R., Barker, M., Brzoska, P. M., Williamson, P., Beaudry, J. A., Matthews, M., Schupp, J. M., Wagner, D. M., Furtado, M. R., Keim, P., & Budowle, B. (2009). Whole-genome typing of Bacillus anthracis isolates by next-generation sequencing accurately and rapidly identifies strain-specific diagnostic polymorphisms. *Forensic Science International: Genetics Supplement Series, 2*(1), 300–301. https://doi.org/10.1016/j.fsigss.2009.08.097

Doganay, M., Dinc, G., Kutmanova, A., & Baillie, L. (2023). Human anthrax: Update of the diagnosis and treatment. *Diagnostics., 13*(6), 1056–1060. https://doi.org/10.3390/diagnostics13061056

Eshraghi, B., Zarrin, Y., & Fazel, M. (2020). Palpebral anthrax, a rare though important condition in villagers: A case report and literature review. *International Journal of Infectious Diseases., 99*, 260–262. https://doi.org/10.1016/j.ijid.2020.07.083

Ezhova, E., Orlov, D., Suhonen, E., Kaverin, D., Mahura, A., Gennadinik, V., Kukkonen, I., Drozdov, D., Lappalainen, H. K., Melnikov, V., Petäjä, T., Kerminen, M., Zilitinkevich, S., Malkhazova, S. M., Christensen, T. R., & Kulmala, M. (2021). Climatic factors influencing the anthrax outbreak of 2016 in Siberia, Russia. *Ecohealth, 18*(2), 217–228. https://doi.org/10.1007/s10393-021-01549-5

Finke, E.-J., Beyer, W., Loderstädt, U., & Frickmann, H. (2020). Review: The risk of contracting anthrax from spore-contaminated soil—A military medical perspective. *European Journal of Microbiology and Immunology., 10*(2), 29–63. https://doi.org/10.1556/1886.2020.00008

Georgopoulos, A. P., & James, L. M. (2024). Anthrax vaccines in the 21st century. *Vaccine, 12*(2), 159. https://doi.org/10.3390/vaccines12020159

Golnari, M., Bahrami, N., Milanian, Z., Rabbani Khorasgani, M., Asadollahi, M. A., Shafiei, R., & Fatemi, S. S.-A. (2024). Isolation and characterization of novel Bacillus strains with superior probiotic potential: Comparative analysis and safety evaluation. *Scientific Reports, 14*(1), 1457–1460. https://doi.org/10.1038/s41598-024-51823-z

González-Cueto, E., De La Fuente, J., & López-Camacho, C. (2024). Potential of mRNA-based vaccines for the control of tick-borne pathogens in one health perspective. *Frontiers in Immunology, 15*, 1384442. https://doi.org/10.3389/fimmu.2024.1384442

Gote, V., Bolla, P. K., Kommineni, N., Butreddy, A., Nukala, P. K., Palakurthi, S. S., & Khan, W. (2023). A comprehensive review of mRNA vaccines. *International Journal of Molecular Sciences, 24*(3). https://doi.org/10.3390/ijms24032700

Gwinn, M., MacCannell, D., & Armstrong, G. L. (2019). Next generation sequencing of infectious pathogens. *JAMA, 321*(9), 893. https://doi.org/10.1001/jama.2018.21669

Hamutyinei Dhliwayo, T., Chonzi, P., Madembo, C., Juru, T. P., Chadambuka, A., Gombe, N. T., Chikerema, S., Shambira, G., Umeokonkwo, C. D., & Tshimanga, M. (2022). Anthrax outbreak investigation in Tengwe, Mashonaland West Province, Zimbabwe, 2022 D. Bhattacharya (ed.). *PLoS One, 17*(12), 1–10. https://doi.org/10.1371/journal.pone.0278537

Hendricks, K., Person, M. K., Bradley, J. S., Mongkolrattanothai, T., Hupert, N., Eichacker, P., Friedlander, A. M., & Bower, W. A. (2022). Clinical features of patients hospitalized for all routes of anthrax, 1880-2018: A systematic review. *Clinical Infectious Diseases: An Official Publication of the Infectious Diseases Society of America., 75*(Suppl 3), S341–S353. https://doi.org/10.1093/cid/ciac534

Hoorali, F., Khosravi, H., & Moradi, B. (2020). Automatic bacillus anthracis bacteria detection and segmentation in microscopic images using UNet++. *Journal of Microbiological Methods., 177*, 106056. https://doi.org/10.1016/j.mimet.2020.106056

Jiranantasak, T., Benn, J. S., Metrailer, M. C., Sawyer, S. J., Burns, M. Q., Bluhm, A. P., Blackburn, J. K., & Norris, M. H. (2022). Characterization of bacillus anthracis replication and persistence on environmental substrates associated with wildlife anthrax outbreaks W.D. Walter (ed.). *PLoS One, 17*(9), 1–6. https://doi.org/10.1371/journal.pone.0274645

Khairullah, A. R., Kurniawan, S. C., Effendi, M. H., Widodo, A., Hasib, A., Silaen, O. S. M., Moses, I. B., Yanestria, S. M., Gelolodo, M. A., Kurniawati, D. A., Ramandinianto, S. C., Afnani, D. A., Riwu, K. H. P., & Ugbo, E. N. (2024). Anthrax disease burden: Impact on animal and human health. *International Journal of One Health., 10*(1), 45–55. https://doi.org/10.14202/IJOH.2024.45-55

Kutmanova, A., Doganay, M., & Zholdoshev, S. (2020). Human anthrax in Kyrgyz Republic: Epidemiology and clinical features. *Journal of Infection and Public Health., 13*(8), 1161–1165. https://doi.org/10.1016/j.jiph.2020.02.043

Luong, T., Be, T. B., Hoang, M. D., Hoang, T. T. H., Pham, Q. T., Tran, T. M. H., Ho, H. D., Long, P. T., & Blackburn, J. K. (2023). Incidence and spatial distribution of human and livestock anthrax in Cao Bang Province, Vietnam (2004–2020). *Vector-Borne and Zoonotic Diseases., 23*(5), 306–309. https://doi.org/10.1089/vbz.2022.0072

Maison, R. M., Priore, M. R., Brown, V. R., Bodenchuk, M. J., Borlee, B. R., Bowen, R. A., & Bosco-Lauth, A. M. (2023). Feral swine as indirect indicators of environmental anthrax contamination and potential mechanical vectors of infectious spores. *Pathogens., 12*(4), 622–630. https://doi.org/10.3390/pathogens12040622

Mechaly, A., Vitner, E., Levy, H., Weiss, S., Bar-David, E., Gur, D., Koren, M., Cohen, H., Cohen, O., Mamroud, E., & Fisher, M. (2018). Simultaneous Immunodetection of anthrax, plague, and tularemia from blood cultures by use of multiplexed suspension arrays G.A. Land (ed.). *Journal of Clinical Microbiology., 56*(4), 1–17. https://doi.org/10.1128/JCM.01479-17

Mir, T. U. G., Wani, A. K., Akhtar, N., Katoch, V., Shukla, S., Kadam, U. S., & Hong, J. C. (2023). Advancing biological investigations using portable sensors for detection of sensitive samples. *Heliyon, 9*(12), e22679. https://doi.org/10.1016/j.heliyon.2023.e22679

Mukarati, N. L., Matope, G., De Garine-Wichatitsky, M., Ndhlovu, D. N., Caron, A., & Pfukenyi, D. M. (2020). The pattern of anthrax at the wildlife-livestock-human interface in Zimbabwe A.M. Samy (ed.). *PLOS Neglected Tropical Diseases., 14*(10), 1–6. https://doi.org/10.1371/journal.pntd.0008800

Muturi, M., Gachohi, J., Mwatondo, A., Lekolool, I., Gakuya, F., Bett, A., Osoro, E., Bitek, A., Thumbi, S. M., Munyua, P., Oyas, H., Njagi, O. N., Bett, B., & Njenga, M. K. (2018). Recurrent anthrax outbreaks in humans, livestock, and wildlife in the same locality, Kenya, 2014-2017. *The American Journal of Tropical Medicine and Hygiene., 99*(4), 833–839. https://doi.org/10.4269/ajtmh.18-0224

Mwakapeje, E. R., Høgset, S., Fyumagwa, R., Nonga, H. E., Mdegela, R. H., & Skjerve, E. (2018). Anthrax outbreaks in the humans—livestock and wildlife interface areas of northern Tanzania: A retrospective record review 2006–2016. *BMC Public Health, 18*(1), 106–110. https://doi.org/10.1186/s12889-017-5007-z

Nikolakakis, I., Michaleas, S. N., Panayiotakopoulos, G., Papaioannou, T. G., & Karamanou, M. (2023). The history of anthrax Weaponization in the Soviet Union. *Cureus., 15*, e36800. https://doi.org/10.7759/cureus.36800

Norris, M. H., Zincke, D., Leiser, O. P., Kreuzer, H., Hadfied, T. L., & Blackburn, J. K. (2020). Laboratory strains of Bacillus anthracis lose their ability to rapidly grow and sporulate compared to wildlife outbreak strains N.S. Duesbery (ed.). *PLoS One, 15*(1), 1–12. https://doi.org/10.1371/journal.pone.0228270

Romero-Alvarez, D., Peterson, A. T., Salzer, J. S., Pittiglio, C., Shadomy, S., Traxler, R., Vieira, A. R., Bower, W. A., Walke, H., & Campbell, L. P. (2020). Potential distributions of Bacillus anthracis and Bacillus cereus biovar anthracis causing anthrax in Africa R. Yang (ed.). *PLOS Neglected Tropical Diseases., 14*(3), 1–15. https://doi.org/10.1371/journal.pntd.0008131

Sahoo, K. C., Negi, S., Barla, D., Badaik, G., Sahoo, S., Bal, M., Padhi, A. K., Pati, S., & Bhattacharya, D. (2020). The landscape of anthrax prevention and control: Stakeholders' perceptive in Odisha, India. *International Journal of Environmental Research and Public Health., 17*(9), 3094–3100. https://doi.org/10.3390/ijerph17093094

Sarker, M. S. A., El Zowalaty, M. E., Shahid, M. A. H., Sarker, M. A., Rahman, M. B., Järhult, J. D., & Nazir, K. H. M. N. H. (2020). Maximization of livestock anthrax vaccination coverage in Bangladesh: An alternative approach. *Vaccine, 8*(3), 435–439. https://doi.org/10.3390/vaccines8030435

Seid, K., Shiferaw, A. M., Yesuf, N. N., Derso, T., & Sisay, M. (2020). Livestock owners' anthrax prevention practices and its associated factors in Sekota Zuria district, Northeast Ethiopia. *BMC Veterinary Research., 16*(1), 39–42. https://doi.org/10.1186/s12917-020-2267-0

Sidwa, T., Salzer, J. S., Traxler, R., Swaney, E., Sims, M. L., Bradshaw, P., O'Sullivan, B. J., Parker, K., Waldrup, K. A., Bower, W. A., & Hendricks, K. (2020). Control and prevention of anthrax, Texas, USA, 2019. *Emerging Infectious Diseases., 26*(12), 2815–2824. https://doi.org/10.3201/eid2612.200470

Stark, J. F. (2022). 'A remedy for this dread disease': Achille Sclavo, anthrax and serum therapy in early twentieth-century Britain. *The British Journal for the History of Science., 55*(2), 207–226. https://doi.org/10.1017/S0007087422000012

Subedi, D., Pantha, S., Jyoti, S., Gautam, B., Kaphle, K., Yadav, R. K., Ghimire, S., & Dhakal, S. (2024). Anthrax in humans, animals, and the environment and the one health strategies for anthrax control. *Pathogens, 13*(9), 773. https://doi.org/10.3390/pathogens13090773

Sushma, B., Shedole, S., Suresh, K. P., Leena, G., Patil, S. S., & Srikantha, G. (2021). An estimate of global anthrax prevalence in livestock: A meta-analysis. *Veterinary World., 14*(5), 1263. https://doi.org/10.14202/vetworld.2021.1263-1271

Thompson, J. M., Cook, R., Person, M. K., Negrón, M. E., Traxler, R. M., Bower, W. A., & Hendricks, K. (2022). Risk factors for death or meningitis in adults hospitalized for cutaneous anthrax, 1950–2018: A systematic review. *Clinical Infectious Diseases., 75*(Supplement_3), S459–S467. https://doi.org/10.1093/cid/ciac533

Wang, D., Cui, M., Li, M., & Zhang, X. (2021). Biosensors for the detection of bacillus anthracis. *Accounts of Chemical Research, 54*(24), 4451–4461. https://doi.org/10.1021/acs.accounts.1c00407

Wang, S., Qu, Z., Huang, Q., Zhang, J., Lin, S., Yang, Y., Meng, F., Li, J., & Zhang, K. (2022). Application of gene editing technology in resistance breeding of livestock. *Life, 12*(7), 1070. https://doi.org/10.3390/life12071070

Yu, X., Zhang, L., Fang, M., Liang, H., Zhang, Q., Yin, X., Kou, Z., Yin, W., Li, Z., Zhang, E., Ding, S., & Shandong Center for Disease Control and Prevention, Jinan, Shandong, China, Binzhou Center for Disease Control and Prevention, Binzhou, Shandong, China, Shandong University, Jinan, Shandong, China, & Chinese Center for Disease Control and Prevention, Beijing, China. (2021). Two confirmed patients of anthrax—Binzhou City, Shandong Province, China, august 2021. *China CDC Weekly., 3*(38), 808–810. https://doi.org/10.46234/ccdcw2021.185

Zasada, A. A. (2020). Detection and identification of bacillus anthracis: From conventional to molecular microbiology methods. *Microorganisms., 8*(1), 125–130. https://doi.org/10.3390/microorganisms8010125

Zhang, H., Zhang, E., Guo, M., He, J., Li, W., & Wei, J. (2022). Epidemiological characteristics of human anthrax—China, 2018–2021. *China CDC Weekly., 4*(35), 783–787. https://doi.org/10.46234/ccdcw2022.165

Zhu, D., Sorg, J. A., & Sun, X. (2018). Clostridioides difficile biology: Sporulation, germination, and corresponding therapies for C. difficile infection. *Frontiers in Cellular and Infection Microbiology, 8*, 29. https://doi.org/10.3389/fcimb.2018.00029

Zorigt, T., Ito, S., Isoda, N., Furuta, Y., Shawa, M., Norov, N., Lkham, B., Enkhtuya, J., & Higashi, H. (2021). Risk factors and spatio-temporal patterns of livestock anthrax in Khuvsgul Province, Mongolia W.C. Turner (ed.). *PLoS One, 16*(11), 1–10. https://doi.org/10.1371/journal.pone.0260299

Chapter 4
Biofilm Formation and Antibiotic Profile of *Staphylococcus aureus* Isolated from Food and Fomites in Selected Cafeterias

Adeola Elizabeth Benson, Sharon Orok Duke-Inyang, Tochi Rosemarie Maxwell-Iyama, and Solomon Uche Oranusi

1 Introduction

Foodborne diseases caused by the consumption of food harbouring pathogens are of major public health and food safety concern (Elbehiry et al., 2023). Staphylococcal food poisoning is an intoxication caused by ingesting food samples that have been improperly handled or preserved and contain significant quantities of Staphylococcal enterotoxins (Niveditha et al., 2012). *Staphylococcus aureus* food intoxication can result from the handlers preparing the food as well as from the cooking environment, from improper cooking and handling of the food and cooking large amounts of food at a given time can also contribute to this contamination. *Staphylococcus aureus* is classified as a generalist pathogen due to its wide host range, they can quickly adapt to a variety of hosts, mostly through the acquisition of mobile genetic elements (MGEs) (Matuszewska et al., 2020). *Staphylococcus aureus*'ability to form biofilms and adhere to fomites significantly compromise food safety by diminishing the quality and availability of consumables, thereby elevating health risks for potential consumers. The protection ability of different *Staphylococcus aureus* strains by the formation of biofilms increases the ability of the organisms to cause cross-contamination in food and food surfaces, increasing *Staphylococcus aureus* food poisoning cases (Vázquez-Sánchez et al., 2012). Biofilms of *Staphylococcus aureus* contribute to its antibiotic-resistant patterns. The antibiotic resistance of some *Staphylococcus aureus* strains to different antibiotics such as methicillin (MRSA) is a threat to the health of consumers as the effects of the enterotoxins cannot be controlled, leading to diverse clinical complications (Dweba et al., 2018).

A. E. Benson (✉) · S. O. Duke-Inyang · T. R. Maxwell-Iyama · S. U. Oranusi
Department of Biological Sciences, Microbiology Unit Covenant University, Ota, Ogun State, Nigeria
e-mail: elizabeth.onibokun@covenantuniversity.edu.ng

P. O. Isibor, P. A. Akinduti (eds.), *Harnessing Biotechnology Tools for Product Development*, https://doi.org/10.1007/978-3-031-86002-7_4

Staphylococcal food poisoning (SFP) is a common cause of bacterial food poisoning, and it is often associated with contamination by asymptomatic food handlers. Humans are commonly infected with *Staphylococcus aureus*, either symptomatically or asymptomatically. Most reports of staphylococcal food poisoning have been linked to food contamination during preparation owing to insufficient refrigeration, cooking, heating or poor personal hygiene. SFP manifests through an acute onset of symptoms, including severe nausea, profuse vomiting, and intense abdominal cramping, which may or may not be accompanied by diarrhoea. Symptomatology typically emerges within a rapid timeframe of 2–8 h post-exposure. The condition poses heightened severity in infants, the elderly and immunocompromised persons, with a significant risk of progression to hospitalization if prompt medical intervention is not administered (Wu et al., 2018). Food handlers have been connected to several foodborne diseases, with *Staphylococcus aureus* being one of the most prevalent infections spread through food contaminated by healthy and unhealthy food handlers (Verkaik et al., 2011). Regardless of the efficacy of antibiotics, resistant *Staphylococcus aureus* strains, particularly methicillin-resistant *Staphylococcus aureus* (MRSA), have become a major public health problem (Craft et al., 2019). Worldwide MRSA infection rate is continuously rising, and clinical anti-infective treatment for MRSA has continued to be problematic due to the growth of harmful organisms and the indiscriminate use of antibiotics (Guo et al., 2020). Therefore, the aim of this study was to determine the prevalence and antibiotic profile of biofilm-forming *Staphylococcus aureus* in foods and food surfaces in two cafeterias of a selected Nigerian University, with a view to proffering solution to preservation measures that will prevent biofilm formation and enterotoxin production by food contaminants, thus forestall foodborne diseases.

2 Materials and Methods

2.1 Sample Collection

Sample collection occurred from two cafeterias that are resident in the chosen institution and form the source of all the meals consumed by the students. Five different ready-to-eat food samples (smoothie, meat, fish and hot dog meat and bread), one for each food item, were collected from the two cafeterias and coded as Cafeterias 1 and 2 using sterile universal bottles. Swabs of different surfaces such as chopping boards, show glasses, chaffing dishes, serving spoons and hands of food handlers, two from each cafeteria, were collected and transported immediately to the Microbiology Laboratory of Covenant University. Samples were labelled according to the food and location from which they were collected.

2.2 *Isolation and Identification of* Staphylococcus aureus

Serial dilution of 10^{-1} to 10^{-5} was carried out using sterile distilled water. A 1 mL volume of dilutions 10^{-1}, 10^{-3} and 10^{-5} were plated into pre-sterilized and cooled down mannitol salt agar (MSA) plates using the spread plated method, while the swabs were inoculated directly into the MSA plates using the same method. Both were incubated at 37 °C for 24 h. Two series of sub-culturing were carried out to obtain pure cultures on nutrient agar plates.

Staphylococcus aureus was confirmed using morphological characterization/ Gram staining techniques. This was followed by a series of biochemical tests: catalase, coagulase, mannitol sugar fermentation and haemolysis test.

2.3 *Detection of Biofilm Formation*

A loop full of *Staphylococcus aureus* was inoculated into nutrient broth tubes supplemented with 2% sucrose and glucose, and incubated at 37 °C overnight. A 30 μL volume of broth culture was dispensed into 96-well microtitre plate and incubated at 37 °C for 48 h. Each well was rinsed three times under running water, 50 μL of crystal violet was then added to each well and rinsed with water after 5 min. A 100 μL volume of 100% ethanol was then added to each well, allowed to dry for 30 mins and colour development was observed (peptone water was used as a negative control and biofilm-positive *Pseudomonas aeruginosa* was used as a positive control) (Oche et al., 2020).

2.4 *Antibiotic Sensitivity Testing*

Antibiotic sensitivity testing was carried out on the *Staphylococcus aureus* isolates on Mueller Hinton Agar plates using the disc diffusion method. The actively growing bacterial cell was homogenized in normal saline and standardized using 0.5 McFarland standard. Gram-positive antibiotic discs containing the following antibiotics ceftazidime (30 μg), cloxacilin (10 μg), gentamicin (10 μg), cefuroxine (30 μg), ofloxacin (5 μg), augmentin (10 μg), erythromycin (15 μg), ceftriaxone (30 μg) and oxacillin (1 μg) were placed on petri dishes containing Mueller Hinton Agar with isolates of the *Staphylococcus aureus* that were inoculated unto the plates using sterile swab sticks. The plates were incubated for 24 h and observed for zones of inhibition. The zones of inhibition were measured using a millimetre rule and interpreted as stated in the Clinical Laboratory Standard Institute guidelines (CLSI, 2020).

3 Results

A total of 42 isolates were obtained from the food and surface samples collected. Twenty-seven (27) were isolated from Cafeteria 1, while fifteen (15) were isolated from Cafeteria 2. Gram's staining revealed that only twenty (20) were Gram-positive cocci in clusters, characteristic of *Staphylococcus aureus* (Tables 4.1 and 4.2). All isolates were positive for catalase and coagulase test. However, only 19 were able to ferment mannitol sugar (Table 4.3). Eight isolates exhibited β-haemolysis, twelve (12) exhibited γ-haemolysis, and none exhibited α-haemolysis (Table 4.4). Biofilm formation was observed in eighteen (90%) of the isolates using the microtitre plate method. Of the eighteen, four (22.2%) were strongly positive, ten (55.5%) were positive, while the remaining four (22.2%) were weakly positive (Table 4.5). All isolates (100%) from both cafeterias were resistant to ceftazidime, cloxacillin and augmentin. Similarly, all the isolates were resistant to two or more classes of antibiotics. One isolate from hotdog bread in Cafeteria 1 was resistant to all the antibiotics tested. *Staphylococcus aureus* from a smoothie in Cafeteria 1 also showed resistance to all the antibiotics except ofloxacin. Isolates showed greater sensitivity to ofloxacin (95%), gentamicin (90%) and erythromycin (65%). Ninety-five per cent

Table 4.1 Morphology of suspected *Staphylococcus aureus* from food surfaces and handlers selected cafeterias

S/N	Codes	Source	Morphological identification	Gram reaction
1.	C2Ai	Hand of food handler in Cafeteria 1	Yellow, mucoid, lobate, translucent, slightly raised, circular, with entire margins	Gram-positive cocci in clusters
2.	C2Bii	Chopping board in Cafeteria 2	White, granular, slightly raised, lobate, opaque, with entire margins	Gram-positive cocci in clusters
3.	C1Eii	Chaffing dish in Cafeteria 1	White, granular, translucent, raised, circular with entire margins	Gram-positive cocci in clusters
4.	C2Dii (a)	Show glass in Cafeteria 2	Yellow, mucoid, translucent, slightly raised, circular, entire margins	Gram-positive cocci in clusters
5.	C2Dii(b)	Show glass in Cafeteria 2	White, granular, opaque, raised, entire margins	Gram-positive cocci in clusters
6.	C2Bi	Chopping board in Cafeteria 2	White, translucent, granular, slightly raised, circular, entire margins	Gram-positive cocci in clusters
7.	C2Aii (a)	Hand of food handler in Cafeteria 2	White, circular, translucent, granular, slightly raised, entire margins	Gram-positive cocci in clusters
8.	C2Ci	Cooking utensil in cafeteria 2	White, translucent, circular, granular, slightly raised, entire margins	Gram-positive cocci in clusters
9.	C1Ai (b)	Hand of food handler in Cafeteria 1	White, translucent, circular, slightly raised, granular with entire margins	Gram-positive cocci in clusters
10.	C1Ai (a)	Hand of food handler in Cafeteria 1	White, translucent, circular, slightly raised, granular with entire margins	Gram-positive cocci in clusters

Table 4.2 Morphology of suspected *Staphylococcus aureus* from food samples in selected cafeterias

S/N	Codes	Source	Morphological identification	Gram reaction
11.	C1A-1	Beef in Cafeteria 1 (10^1)	Yellow, circular, undulate, slightly raised, translucent with entire margins	Gram-positive cocci in clusters
12.	C2A-1	Beef in Cafeteria 2 (10^1)	Yellow, circular, lobate, slightly raised, opaque with entire margins	Gram-positive cocci in clusters
13.	C1E-1	Hot dog bread in Cafeteria 1 (10^1)	White, granular, circular, translucent, slightly raised with entire margins	Gram-positive cocci in clusters
14.	C1E-5	Hot dog bread in Cafeteria 1 (10^5)	Yellow, mucoid, circular, translucent, raised with entire margins	Gram-positive cocci in clusters
15.	C1D-1	Hot dog meat in Cafeteria 1 (10^1)	White, granular, slightly raised, undulate, circular with entire margins	Gram-positive cocci in clusters
16.	C1C-1	Smoothie in Cafeteria 1 (10^1)	White, opaque, slightly raised, circular entire colonies	Gram-positive cocci in clusters
17.	C2B-3	Fish in Cafeteria 2 (10^3)	Yellow, opaque, granular, translucent, raised, circular with entire colonies	Gram-positive cocci in clusters
18.	C2D-1 (ii)	Hot dog meat in Cafeteria 2 (10^1)	Yellow, circular, slightly raised, translucent with entire colonies	Gram-positive cocci in clusters
19.	C1B-1 (ii)	Fish in Cafeteria 1 (10^1)	White, scanty growth, slightly raised, granular, circular, translucent with entire colonies	Gram-positive cocci in clusters
20.	C2B-1 (i)	Fish in Cafeteria 1 (10^1)	Yellow, raised, mucoid, circular, opaque with entire colonies	Gram-positive cocci in clusters

of *Staphylococcus aureus* reported were sensitive to ofloxacin, while 100% were resistant to ceftazidime, cloxacilin, augmentin, cefuroxime and oxacillin (Tables 4.6 and 4.7).

4 Discussion

In this study, 20 *Staphylococcus aureus* isolates were obtained. Ten were isolated in foods from both study locations (Cafeterias 1 and 2) within the university. *S. aureus* in foods has been documented in several literatures (Yahya Ahmed et al., 2021; Avila-Novoa et al., 2018; Wang et al., 2017). Of the remaining ten, four were isolated from the hands of food handlers. *S. aureus* is a normal flora of the skin (Ajayi et al., 2017), colonizing both the hands and mucosal membranes of asymptomatic carriers (Bencarodino & Vitali, 2019). Therefore, their presence in the hands of handlers is expected. Six *S. aureus* were reportedly isolated from food surfaces such as cooking utensils and chaffing dishes. Their presence on the utensils could have resulted from contamination by *S. aureus* from the food handlers. There have been reports of *S. aureus* among food handlers which are possible food contamination sources (Castro et al., 2015; Leibler et al., 2016). Nigeria being the largest country in sub-Saharan Africa records an alarming increase in cases of foodborne disease.

Table 4.3 Biochemical identification of *Staphylococcus aureus* isolates from food and fomite samples in selected cafeterias

S/N	Isolate code	Catalase	Coagulase	Mannitol sugar fermentation	
				Acid	Gas
1.	C2Ai	+	+	–	–
2.	C2Bii	+	+	+	+
3.	C1Eii	+	+	+	–
4.	C2Dii (a)	+	+	+	–
5.	C2Dii (b)	+	+	+	–
6.	C2Bi	+	+	+	–
7.	C2Aii (a)	+	+	+	–
8.	C2Ci	+	+	+	–
9.	C1Ai (b)	+	+	+	–
10.	C1Ai (a)	+	+	+	–
11.	C1A-1	+	+	+	+
12	C2A-1	+	+	+	–
13.	C1E-1	+	+	+	–
14.	C1E-5	+	+	+	–
15.	C1D-1	+	+	+	–
16.	C1C-1	+	+	+	–
17.	C2B-3	+	+	+	–
18.	C2D-1 (ii)	+	+	+	–
19.	C1B-1 (ii)	+	+	+	–
20.	C2B-1 (i)	+	+	+	–

The environmental conditions of these tropics encourage the spread of *S. aureus* (Odetokun et al., 2023). *Escherichia coli* 0157:H7, *Salmonella*, *Shigella* and *S. aureus* are the major pathogens of public health importance involved in food-borne disease outbreaks (Odo et al., 2021). Meat sources harboured 50% (which is the highest percentage in this study) of the *S. aureus* isolated in this study. Wu et al. (2018) reported that animal flesh is a widely known reserve for *S. aureus*. Several factors may contribute to the contamination of meat and meat products by bacteria, such as the type of meat, processing/cooking method and addition of accessory ingredients such as vegetables (Shawish & Al-Humam, 2016). The presence of *Staphylococcus aureus* in meat is a significant concern due to its prevalence, antimicrobial resistance and enterotoxin production (Osada et al., 2022). The presence of *S. aureus* was also detected in fish and hot dog bread, accounting for 20% of the cases. These are ready-to-eat products, and *S. aureus* has been previously reported in ready-to-eat products (Dittmann et al., 2017). The remaining ten per cent of the *S. aureus* isolated in this study were from smoothies. Smoothies are milk-based food products and milk can also serve as a transmission vehicle for *S. aureus*. Similarly, Dittmann et al. (2017) reported the presence of *S. aureus* in four out of five dairies sampled. In another study on ice cream (which is another milk-based product) six different ice cream brands from Lagos State were analysed for

Table 4.4 Haemolysis patterns of *Staphylococcus aureus* isolates from food and fomite samples in selected cafeterias

S/n	Isolate code	Alpha	Beta	Gamma
1.	C2Ai	–	–	+
2.	C2Bii	–	–	+
3.	C1Eii	–	–	+
4.	C2Dii (a)	–	–	+
5.	C2Dii (b)	–	–	+
6.	C2Bi	–	–	+
7.	C2Aii (a)	–	–	+
8.	C2Ci	–	–	+
9.	C1Ai (b)	–	–	+
10.	C1Ai (a)	–	+	–
11.	C1A-1	–	+	–
12.	C2A-1	–	–	+
13.	C1E-1	–	+	–
14.	C1E-5	–	+	–
15.	C1D-1	–	+	–
16.	C1C-1	–	–	+
17.	C2B-3	–	–	+
18.	C2D-1 (ii)	–	+	–
19.	C1B-1 (ii)	–	+	–
20.	C2B-1 (i)	–	+	–

Escherichia coli, *Klebsiella* spp., *Bacillus* spp., *Enterobacter* and *S. aureus*. Interestingly, *S. aureus* had the highest prevalence (100%) compared to the other isolates (Ojo-Omoniyi & Babatunde, 2018). In most cases, milking is done by hand; therefore, the possibility of contaminating the milk through the hands of the farmers abounds (Mekonnen et al., 2018). Dittmann et al. (2017) noted that milk contamination by *S. aureus* is largely by cross-contamination between handlers and equipment thus indicating the absence of good manufacturing practices. The presence of *S. aureus* in milk and milk-based products is still a major public health concern (Gebremedhin et al., 2022).

The result presented in this study showed a beta and gamma haemolytic pattern. Dittmann et al. (2017) observed a higher beta haemolytic pattern (43.9%) in the *S. aureus* isolated. Haemolysin is an important virulence factor in *S. aureus* isolates (Zhang et al., 2016); it is the most characterized virulence trait in *S. aureus* and plays a significant role in the pathogenesis of staphylococcal diseases (Divyakolu et al., 2019). Therefore, we may infer that the *S. aureus* reported in this study may not be virulent. However, this may need to be further confirmed.

The result on biofilm formation is similar to Chen et al. (2020), where they reported 72.16% of *S. aureus* isolated to be positive for biofilm formation using a similar method employed in this study. A similar study by Avila-Novoa et al. (2018) also reported that 90.4% of *S. aureus* isolates were positive for biofilm formation by employing the same microtitre plate method. *S. aureus* although a normal flora can

Table 4.5 Biofilm formation in *Staphylococcus aureus* isolates from food and fomite samples in selected cafeterias

S/n	Isolate code	Biofilm formation
1.	C2Ai	+++
2.	C2Bii	++
3.	C1Eii	++
4.	C2Dii (a)	++
5.	C2Dii (b)	++
6.	C2Bi	++
7.	C2Aii (a)	+++
8.	C2Ci	+
9.	C1Ai (b)	+
10.	C1Ai (a)	++
11.	C1A-1	++
12.	C2A-1	+++
13.	C1E-1	+++
14.	C1E-5	++
15.	C1D-1	++
16.	C1C-1	++
17.	C2B-3	+
18.	C2D-1 (ii)	–
19.	C1B-1 (ii)	+
20.	C2B-1 (i)	–

KEY: +++ strong positive, ++ positive, + weak positive, – negative

become an opportunistic human pathogen causing serious problems in the food and medical sector when involved in biofilm formation (Sun et al., 2012). Mounting evidence demonstrated by den Reijer et al. (2016) suggests that the haemolytic pattern may contribute to biofilm formation in *S. aureus*. It is suggested that *S. aureus* showing beta haemolytic patterns is most likely to be positive for biofilm formation. However, this is not as presented in this study as only 40% of the isolates showed beta haemolysis, whereas 90% were positive for biofilm formation.

From the antibiotic sensitivity test, the *S. aureus* isolates showed 100% resistance to ceftazidime, cloxacillin, augmentin, cefuroxime and oxacillin. Oranusi et al. (2020) reported a similar antibiotic resistance pattern in *S. aureus* from the chairs, shelves and indoor environments of the main administrative buildings at Covenant University. Of the 18 *S. aureus* species reported, the highest resistance was observed in ceftazidime (88.8%), cloxacillin (77.7%), augmentin (61.1%) and cefuroxime (55.5%), respectively. Ninety-five per cent and 80% resistance was recorded against ceftriaxone and erythromycin, respectively. However, the most significant sensitivity was observed in ofloxacin (95%) and gentamycin (90%). This is also similar to Oranusi et al. (2020), with the isolates showing greater sensitivity to both ofloxacin and gentamycin at 88.8% and 72.2%, respectively. In a recent systematic review studying the prevalence of multidrug-resistant bacteria in food and drinks in Nigeria for 5 years, 131 out of 286 (45.8%) *Staphylococcus* spp. were multidrug resistant (MDR) next only to *Salmonella* spp. which has 84.5%

Table 4.6 Antibiotic patterns of *Staphylococcus aureus* isolates from food and fomite samples in selected cafeterias

S/n	Isolate code	CAZ 30 μg	CXC 10 μg	GEN 10 μg	CRX 30 μg	OFL 5 μg	AUG 10 μg	ERY 15 μg	CTR 30 μg	OXA 1 μg
1.	C2Ai	R	R	S	R	S	R	I	R	R
2.	C2Bii	R	R	S	R	S	R	I	R	R
3.	C1Eii	R	R	S	R	S	R	R	R	R
4.	C2Dii (a)	R	R	R	R	S	R	R	R	R
5.	C2Dii (b)	R	R	R	R	R	R	R	R	R
6.	C2Bi	R	R	S	R	S	R	R	R	R
7.	C2Aii (a)	R	R	S	R	S	R	R	R	R
8.	C2Ci	R	R	S	R	S	R	I	R	R
9.	C1Ai (b)	R	R	S	R	S	R	R	R	R
10.	C1Ai (a)	R	R	S	R	S	R	R	R	R
11.	C1A-1	R	R	S	R	S	R	R	R	R
12.	C2A-1	R	R	S	R	S	R	R	R	R
13.	C1E-1	R	R	S	R	S	R	R	R	R
14.	C1E-5	R	R	S	R	S	R	R	R	R
15.	C1D-1	R	R	S	R	S	R	I	R	R
16.	C1C-1	R	R	S	R	S	R	R	R	R
17.	C2B-3	R	R	S	R	S	R	R	R	R
18.	C2D-1 (ii)	R	R	S	R	S	R	R	R	R
19.	C1B-1 (ii)	R	R	S	R	S	R	R	R	R
20.	C2B-1 (i)	R	R	S	R	S	R	R	S	R

KEY: *S* sensitive, *I* intermediate, *R* resistant, *CAZ* ceftazidime, *CXC* cloxacillin, *GEN* gentamicin, *CRX* cefuroxime, *OFL* ofloxacin, *AUG* augmentin, *ERY* erythromycin, *OXA* oxacillin, *CTR* ceftriaxone

MDR (Iyanuoluwa et al., 2021). Resistance to a wide range of antibiotics by *Staphylococcus aureus* is known. This characteristic makes it difficult to manage cases of *S. aureus* infections (Saba et al., 2017). The isolates in this study showed complete resistance to ofloxacin, a variant of methicillin. Methicillin resistance in *S. aureus* is significant as it is reported to be responsible for 20–80% of all nosocomial infections (Gurung et al., 2020). High antibiotic resistance among *S. aureus* may be due to the indiscriminate use of antibiotics and refusal to follow standard treatment guidelines for *S. aureus* infections (Mendem et al., 2016).

5 Conclusion

This study reported the presence of more *Staphylococcus aureus* in the food items sampled from Cafeteria 2. Although Cafeteria 1 appeared to have more staphylococcus present in the cooking materials, this did not eventually reflect in the food items as the cooking process may have adequately eliminated this threat. The presence of *Staphylococcus* in the hands of more handlers in Cafeteria 2 and in

Table 4.7 Zone of inhibition (mm) of antibiotic-resistant *Staphylococcus aureus* isolates from food and fomite samples in selected cafeterias

S/n	Isolate code	CAZ 30 µg	CXC 10 µg	GEN 10 µg	CRX 30 µg	OFL 5 µg	AUG 10 µg	ERY 15 µg	CTR 30 µg	OXA 1 µg
1.	C2Ai			26		24		21		
2.	C2Bii			31	16	19		15	9	
3.	C1Eii			22		21		11		
4.	C2Dii (a)					23				
5.	C2Dii (b)									
6.	C2Bi			22		28				
7.	C2Aii (a)			22		25				
8.	C2Ci			23		24		14		
9.	C1Ai (b)			19		24		12		
10.	C1Ai (a)			22		26		10		
11.	C1A-1			21		24		8		
12.	C2A-1			19		26				
13.	C1E-1			20		22				
14.	C1E-5			21		22		12		
15.	C1D-1			18		24		19		
16.	C1C-1			22		28		11		
17.	C2B-3			20		22		9		
18.	C2D-1 (ii)			21		23		12		
19.	C1B-1 (ii)			18		23		9		
20.	C2B-1 (i)			23		23			25	

KEY: *CAZ* ceftazidime, *CXC* cloxacillin, *GEN* gentamicin, *CRX* cefuroxime, *OFL* ofloxacin, *AUG* augmentin, *ERY* erythromycin, *OXA* oxacillin, *CTR* ceftriaxone

equipment for displaying ready-to-eat foods showed that the contamination of the food was more in the handling process after cooking. Therefore, this study recommends adherence to proper hygienic practices in handling ready-to-eat foods, particularly in Cafeteria 2, and also to discourage the misuse of antibiotics among the students to curb antibiotic resistance. Standard food storage practices are advised to prevent food contamination by pathogens and forestall foodborne diseases.

References

Ajayi, A. A., Bibi, O., Oniha, N. I., Atolagbe, O. M., Anosike, S. O., & Onibokun, E. A. (2017). Studies on Staphylococcus aureus isolated from pimples. *Pakistan Journal of Biological Sciences., 20*, 350–354.

Avila-Novoa, M. G., Iñíguez-Moreno, M., Solís-Velázquez, O. A., González-Gomez, J. P., Guerrero-Medina, P. J., & Gutiérrez-Lomelí, M. (2018). Biofilm formation *by Staphylococcus aureus* isolated from food contact surfaces in the dairy industry of Jalisco, Mexico. *Journal of Food Quality., 2018*(1), 1746139.

Bencarodino, D., & Vitali, L. A. (2019). *Staphylococcus aureus* carriage among food handlers in Pasta Company: Patterns of virulence and resistance to linezolid. *Food Control, 96*, 351–356.

Castro, A., Santos, C., Meireles, H., Silva, J., & Teixaira, P. (2015). Food handlers as potential sources of dissemination of virulent strains of *Staphylococcus aureus* in the community. *Journal of Infection and Public Health., 9*, 153–160.

Chen, Q., Xie, S., Lou, X., Cheng, S., Liu, X., Zheng, W., & Wang, H. (2020). Biofilm formation and prevalence of adhesion genes among *Staphylococcus aureus* isolates from different food sources. *Microbiologyopen., 9*(1), e00946.

Clinical and Laboratory Standard Institute (CLSI) (2020). *Performance Standards for Anti-Microbial Susceptibility Testing 30*, M100.

Craft, K. M., Nguyen, J. M., Berg, L. J., & Townsend, S. D. (2019). Methicillin-resistant *Staphylococcus aureus* (MRSA): antibiotic-resistance and the biofilm phenotype. *MedChemComm., 10*(8), 1231–1241.

den Reijer, P. M., Haisma, E. M., Lemmens-den Toom, N. A., Willemse, J., Koning, R. A., & Demmers, J. A. (2016). Detection of alpha-toxin and other virulence factors in biofilms of *Staphylococcus aureus* on polystyrene and a human epidermal model. *Public Library of Science ONE, 11*, e0145722.

Dittmann, K. K., Chaul, L. T., Lee, S. H., Corassin, C. H., Fernandes de Oliveira, C. A., Pereira De Martinis, E. C., & Oxaran, V. (2017). *Staphylococcus aureus* in some Brazilian dairy industries: changes of contamination and diversity. *Frontiers in Microbiology, 8*, 2049.

Divyakolu, S., Chikkala, R., Ratnakar, K. S., & Sritharan, V. (2019). Hemolysins of *Staphylococcus aureus*—an update on their biology, role in pathogenesis and as targets for anti-virulence therapy. *Advances in Infectious Diseases, 9*(2), 80–104.

Dweba, C. C., Zishiri, O. T., & El Zowalaty, M. E. (2018). Methicillin-resistant *Staphylococcus aureus*: Livestock-associated, antimicrobial, and heavy metal resistance. *Infection and Drug Resistance, 11*, 2497.

Elbehiry A, Abalkhail A, Marzouk E, Elmanssury A. E., Almuzaini A. M., Alfheeaid H, Alshahrani M. T., Huraysh N, Ibrahem M, Alzaben F, Alanazi F, Alzaben M, Anagreyyah S. A., Bayameen A. M., Draz A, Abu-Okail A. (2023) An Overview of the Public Health Challenges in Diagnosing and Controlling Human Foodborne Pathogens. *Vaccines (Basel). 11*(4), 725. https://doi.org/10.3390/vaccines11040725

Gebremedhin, E. Z., Ararso, A. B., Borana, B. M., Kelbesa, K. A., Tadese, N. D., Marami, L. M., & Sarba, E. J. (2022). Isolation and identification of Staphylococcus aureus from milk and milk products, associated factors for contamination, and their antibiogram in Holeta, Central Ethiopia. *Veterinary Medicine International, 1*, 6544705.

Guo, Y., Song, G., Sun, M., Wang, J., & Wang, Y. (2020). Prevalence and therapies of antibiotic-resistance in *Staphylococcus aureus*. *Frontiers in Cellular and Infection Microbiology, 10*, 107.

Gurung, R. R., Maharjan, P., & Chhetri, G. G. (2020). Antibiotic resistance pattern of *Staphylococcus aureus* with reference to MRSA isolates from pediatric patients. *Future Science OA, 6*(4), FSO464.

Leibler, J. H., Jordan, J. A., Brownstein, K., Lander, L., Price, L. B., & Perry, M. J. (2016). *Staphylococcus aureus* nasal carriage among beef packing workers in Mid-Western United States slaughter house. *Public Library of Science One, 11*(2), e0148789.

Matuszewska, M., Murray, G. G., Harrison, E. M., Holmes, M. A., & Weinert, L. A. (2020). The evolutionary genomics of host specificity in Staphylococcus aureus. *Trends in Microbiology, 28*(6), 465–477.

Mekonnen, S. A., Lam, T. J. G. M., Hoekstra, J., Rutten, V. P., Tessema, T. S., Broens, E. M., & Koop, G. (2018). Characterization of Staphylococcus aureus isolated from milk samples of dairy cows in small holder farms of North-Western Ethiopia. *BMC Veterinary Research, 14*(1), 1–8.

Mendem, S. K., Gangadhara, T. A., Shivannavar, C. T., & Gaddad, S. M. (2016). Antibiotic resistance patterns of Staphylococcus aureus: A multicenter study from India. *Microbial Pathogenesis., 98*, 167–170.

Iyanuoluwa, M., Adeola, O., & Solomon, O. (2021). Prevalence of multi-drug resistant bacteria associated with foods and drinks in Nigeria (2015–2020): A systematic review. *Italian Journal of Food Safety., 10*(4), 9417.

Niveditha, S., Pramodhini, S., Umadevi, S., Kumar, S., & Stephen, S. (2012). The isolation and the biofilm formation of uropathogens in the patients with catheter associated urinary tract infections (UTIs). *Journal of Clinical and Diagnostic Research., 6*(9), 1478.

Oche, D. A., Abdulrahim, U., Oheagbulem, A. S., & Olayinka, B. O. (2020). Isolation of biofilm producing methicillin-resistant Staphylococcus aureus from hospitalized orthopaedic patients in Kano State, Nigeria. *Nigerian Journal of Basic and Applied Sciences, 28*(1), 66–74.

Odetokun, I. A., Adetona, M. A., Ade-Yusuf, R. O., Adewoye, A. O., Ahmed, A. N., Ghali-Mohammed, I., Al-Mustapha, A. I., & Fetsch, A. (2023). *Staphylococcus aureus* contamination of animal-derived foods in Nigeria: a systematic review, 2002—2022. *Food Safety and Risk, 10*(1), 6.

Odo, S. E., Uchechukwu, C. F., & Ezemadu, U. R. (2021). Foodborne Diseases and Intoxication in Nigeria: Prevalence of Escherichia coli 0157: H7, Salmonella, Shigella and *Staphylococcus aureus*. *Journal of Advances in Microbiology., 20*(12), 84–94.

Ojo-Omoniyi, O. A., & Babatunde, O. A. (2018). Microbiological quality assessment of branded retail ice cream products in Lagos Metropolis, Southwest Nigeria. *UNILAG Journal of Medicine, Science and Technology, 6*(1), 57–68.

Oranusi, S. U., Akinkunmi, O., Onibokun, E. A., Olopade, B. K., Oshamika, O. O., & Obafemi, Y. D. (2020). The Microbial Burden of Interior Surfaces and Spaces in Main Administrative Buildings of Covenant University. *Tropical Journal of Natural Product Research., 4*(11), 985–989.

Osada, M., Aung, M. S., Urushibara, N., Kawaguchiya, M., Ohashi, N., Hirose, M., & Kobayashi, N. (2022). Prevalence and Antimicrobial Resistance of Staphylococcus aureus and Coagulase-Negative Staphylococcus/Mammaliicoccus from Retail Ground Meat: Identification of Broad Genetic Diversity in Fosfomycin Resistance Gene fosB. *Pathogens, 11*(4), 469.

Saba, C. K. S., Amenyona, J. K., & Kpordze, S. W. (2017). Prevalence and pattern of antibiotic resistance of *Staphylococcus aureus* isolated from door handles and other points of contact in public hospitals in Ghana. *Antimicrobial Resistance and Infection Contro.l, 6*(1), 1–6.

Shawish, R. R., & Al-Humam, N. A. (2016). Contamination of beef products with staphylococcal classical enterotoxins in Egypt and Saudi Arabia. *GMS Hygiene and Infection Control, 11*, Doc08.

Sun, J. L., Zhang, S. K., Chen, X. X., Chen, J. Y., & Han, B. Z. (2012). Growth properties of *Staphylococcus aureus* in biofilm formed on polystyrene plate. *African Journal of Microbiology Research., 6*(13), 3284–3291.

Vázquez-Sánchez, D., López-Cabo, M., Saá-Ibusquiza, P., & Rodríguez-Herrera, J. J. (2012). Incidence and characterization of Staphylococcus aureus in fishery products marketed in Galicia (Northwest Spain). *International Journal of Food Microbiology., 157*(2), 286–296.

Verkaik, N. J., Benard, M., Boelens, H. A., De Vogel, C. P., Nouwen, J. L., Verbrugh, H. A., Melles, D. C., Van Belkum, A., & Van Wamel, W. J. B. (2011). Immune evasion cluster-positive bacteriophages are highly prevalent among human *Staphylococcus aureus* strains, but they are not essential in the first stages of nasal colonization. *Clinical Microbiology and Infection., 17*(3), 343–348.

Wu, S., Huang, J., Wu, Q., Zhang, J., Zhang, F., Yang, X., Wu, H., Zeng, H., Chen, M., Ding, Y., & Wang, J. (2018). *Staphylococcus aureus* isolated from retail meat and meat products in China: incidence, antibiotic resistance and genetic diversity. *Frontiers in Microbiology., 9*, 2767.

Wang, W., Baloch, Z., Jiang, T., Zhang, C., Peng, Z., Li, F., & Xu, J. (2017). Enterotoxigenicity and antimicrobial resistance of *Staphylococcus aureus* isolated from retail food in China. *Frontiers in Microbiology., 8*, 2256.

Yahya Ahmed, M., Abdalbagi Ali, H., Mohammed Taher Gorish, B., Omer Ali, S., Saif Aldein Abdalrhim, E., Hamza Mergani, M., & Altayb, H. N. (2021). Molecular detection of Staphylococcal enterotoxins and mecA genes products in selected food samples collected from different areas in khartoum state. *International Journal of Microbiology., 2021*(1), 5520573.

Zhang, H., Zheng, Y., Gao, H., Xu, P., Wang, M., Li, A., & Du, H. (2016). Identification and characterization of Staphylococcus aureus strains with an incomplete hemolytic phenotype. *Frontiers in Cellular and Infection Microbiology., 6*, 146.

Chapter 5
Demographic Impact of Diarrheagenic Enterotoxigenic *Escherichia coli* Infection on Children Under 5 Years Old in Sub-Saharan Africa

Paul Akinniyi Akinduti and Oluwapelumi Sola Oyesile

1 Introduction

One of the model organisms that has been thoroughly researched is *Escherichia coli* (Idalia & Bernardo, 2017). One of the first commensal bacteria to colonize the intestines after birth is *Escherichia coli*. *E. coli* may be responsible for debilitating illnesses in immunocompromised individuals and healthy individuals whose natural body barriers have been compromised (Leimbach et al., 2013).

Intestinal pathogenic *E. coli* (InPEC) and Extraintestinal pathogenic *E. coli* (ExPEC) are the two main categories into which pathogenic *E. coli* can be classified (Denamur et al., 2021).

Based on tissue tropism, availability of virulence factors, clinical symptoms, mechanisms of infection, and host–cell interactions, *E. coli* may be designated into a myriad of pathotypes. These include:

(i) Enterotoxigenic *E. coli* (ETEC), one of the causative agents of travelers' diarrhea
(ii) Enterohemorrhagic *E. coli* (EHEC)
(iii) Septicemia-associated *E. coli* (SEPEC)
(iv) Uropathogenic *E. coli* (UPEC)
(v) Diffusely adherent *E. coli* (DAEC)
(vi) Enteropathogenic *E. coli* (EPEC)
(vii) Enteroaggregative *E. coli* (EAEC)

P. A. Akinduti (✉)
Department of Medical Laboratory Science, Babcock University, Ilishan, Ogun State, Nigeria
e-mail: akindutip@babcock.edu.ng

O. S. Oyesile (✉)
Department of Biological Sciences, Covenant University, Ota, Ogun State, Nigeria
e-mail: paul.oluwapelumi.oyesilepgs@stu.cu.edu.ng

P. O. Isibor, P. A. Akinduti (eds.), *Harnessing Biotechnology Tools for Product Development*, https://doi.org/10.1007/978-3-031-86002-7_5

(viii) Adherent-invasive *E. coli* (AIEC)
(ix) Neonatal meningitis *E. coli* (NMEC)
(x) Enteroinvasive *E. coli* (EIEC)

Different evolutionary lineages of *E. coli* can also be distinguished, including clade I, D1, D2, E, and B1, B2. Nondisease-causing *E. coli* that live along the gastrointestinal mucosa is primarily represented by Group A. Commensal strains and several strains associated with the EHEC pathotype are found in Phylogroup B1. Phylogroup E, D2, and D1 represent InPEC. Category B2 houses several ExPEC members.

This bacterial species' genome is very malleable, which allows it to undergo genetic alterations that result in gene gain and loss and the formation of disease-causing strains from mutualistic strains (Baumgart et al., 2021).

Numerous antibiotics that humans have been using since the 1930s are no longer effective against *E. coli* (Tadesse et al., 2012). The elevated incidence of acquisition of genetic information and transfer of genetic information from generation to generation of *E. coli* strains could contribute to this. Antibiotic resistance may have multiple causes, but human activities and growing antibiotic mismanagement are thought to be the main culprits (Chokshi et al., 2019). Antibiotic classes, including sulfonamides, aminoglycosides, and many more, have been mostly resistant to evolution in *E. coli* strains. *E. coli* strains that produce AmpC are prominent in environmental contamination and human intestinal colonization in impecunious territories (Mahmudul Hassan, 2021). According to Iseppi et al. (2020), more and more reports link AmpC- and ESBL-producing *E. coli* to serious illnesses. We are limited to the antibiotics of last resort, such as polymyxins and carbapenems. Furthermore, *E. coli* strains expressing carbapenem-hydrolyzing oxacillinase-48 (Oxa-48) have also been identified in Europe (Nordmann, 2014), where clinical samples from Germany yielded 134 cases of *E. coli* bacteria bearing the OXA-48 mutation OXA-244 (Pfeifer et al., 2012). Additionally, 119 *E. coli* strains isolated from different European nations were shown to harbor this same variant (Bakthavatchalam et al., 2016). Similarly, a class of zinc-requiring metallo-β-lactamases can hydrolyze a wide variety of β-lactams, including all penicillins, known as New Delhi metallo-β-lactamase (NDM-1) and its closely related enzymes (Williamson et al., 2012). Additionally, resistant strains that carry over 20 NDM variants and New Delhi metallo-beta-lactamase 1 (NDM-1) have spread and are linked to illnesses in numerous regions of the world (Farhat & Khan, 2020). Globalization of trade and increased international travel have made resistant germs a global hazard to public health. Similarly, antibiotics are frequently administered to food animals for various purposes, such as mass preventive therapy or growth enhancement (Naveen et al., 2018). The lack of legislation, surveillance, controls, and resources in impecunious nations has led to an increase in the amount of *E. coli* that are resistant to fluoroquinolones and produce extended-spectrum β-lactamases, making the problem considerably worse (Founou et al., 2017). The resistance of very virulent strains must be prevented, and international cooperation is needed to provide substitute options. About 700,000 deaths annually are already attributed to antibiotic-resistant infections, and by 2050, ten million fatalities annually are predicted (Ventola, 2015). The creation of innovative therapeutic strategies will remain crucial. Vaccination is another sustainable preventive strategy that can help restrict the spread of *E. coli* that are resistant to antibiotics.

2 Diarrheagenic *E. coli* Pathotypes

Although diarrhea is a worldwide issue, impecunious nations account for the majority of its morbidity and mortality. Dehydration, weight loss, and loss of appetite can result from diarrhea, which is defined as passing loose stools more than thrice a day (Hebbelstrup Jensen et al., 2014). A sizable portion of diarrheal episodes in African nations is caused by diarrheagenic *E. coli* (Manhique-Coutinho et al., 2022). About 30–40% of acute diarrhea episodes in children under the age of five in underdeveloped nations are caused by diarrheagenic *Escherichia coli* (DEC) (Miliwebsky et al., 2016).

Though they share similar steps in the mechanism of pathogenesis, such as intestinal mucosa adherence and the presence of plasmids that aid bacterial virulence, the majority of these are caused by InPEC (Rojas-Lopez et al., 2018).

Enterotoxigenic *E. coli* (ETEC) is a virulent bacterial strain of *E. coli* characterized by the development of heat-stable (ST) along with heat-labile (LT) enterotoxins, which cause diarrhea.

ETEC contributes significantly to bacterial diarrheal illnesses and is the most prevalent etiological agent of traveler's diarrhea, contributing significantly to intestinal morbidity and diarrheal maladies, mostly in infants from impecunious nations (Anderson et al., 2019). Particularly in endemic regions, children under five are vulnerable to ETEC, which caused an estimated 100 million bouts of diarrhea and 60,000 fatalities in 2015 (Bagamian et al., 2020). Expanded impacts of mild ETEC diarrhea and asymptomatic infections may have lasting ramifications for children's development, significantly aggravating the burden of these infections in addition to high rates of morbidity and death (Anderson et al., 2019). It has been demonstrated that initial ETEC infection protects humans and animals from reinfection with related ETEC strains (Carlin & Svennerholm, 2020).

The majority of ETEC diarrhea cases resolve on their own and do not need antibiotics; nevertheless, treatment is frequently administered empirically based on the illness's severity and potential for consequences. According to O'Ryan et al. (2014) children who suffer from nutritional inadequacies may present with more severe symptoms and necessitate an antibacterial agent. Antibiotic medication may also benefit patients with chronic ETEC-caused diarrhea (Becker et al., 2013). This makes the use of antimicrobials and other effective treatments for diarrhea crucial, despite the fact that the death rate from diarrheal illness has significantly decreased (Liu et al., 2016), the rise in antibiotic resistance over time has changed the way ETEC infections are treated with antibiotics (Steffen et al., 2015). Tetracycline, cotrimoxazole, and ampicillin are among the commonly prescribed drugs for treating ETEC infections; however, resistance to these drugs has increased throughout the past 10 years (DuPont & Steffen, 2017). Another worry is that predicted climatic changes can increase the risk of morbidity and death from enteric illnesses like ETEC (Colston et al., 2022).

There is an absence of efficient defenses against ETEC diarrhea (Zhang & Sack, 2015). Many impecunious nations fail to attain improved sanitation and potable water, which could lower the frequency of diarrheal illnesses. Due to ETEC strains developing antimicrobial resistance at an alarmingly high pace, antibiotics used to treat ETEC infections are becoming less and less effective (Lääveri et al., 2018). To avoid ETEC diarrhea, vaccinations have been seen as a cost-effective countermeasure (Walker & Dull, 2017).

Projections from the Maternal and Child Epidemiology Estimation Group (MCEE), alongside data from the Institute for Health Metrics and Evaluation (IHME), indicate that enterotoxigenic *Escherichia coli* (ETEC) is responsible for an estimated 220 million global cases of diarrheal illness, with approximately 75 million occurrences in children under 5 years of age. These episodes are responsible for 18,700–42,000 mortalities in infants under five (Hosangadi et al., 2019). The World Health Organization has classified multidrug-resistant Enterobacteriaceae, which includes *Escherichia coli*, as a high-priority microorganism that presents a significant risk to the health of humans (Breijyeh et al., 2020).

Recently, the Global Enteric Multicenter Study posited that the most significant pathogens linked to diarrheal infections in infants below the age of five in impecunious countries were ETEC strains that produced a combination of heat-stable enterotoxin (ST) and heat-labile toxin (LT) or exclusively heat-stable enterotoxin (ST) (Kotloff et al., 2017). According to recent research, ETEC strains that produce heat-labile and heat-stable enterotoxins (ST and LT) are strongly linked to diarrhea in children under five. In contrast, strains that only produce LT have a negligible effect (Platts-Mills et al., 2015).

ETEC infections are responsible for the majority of empirical antibiotic use, which may hasten the emergence of antibiotic resistance (Frost et al., 2023). Furthermore, the indiscriminate use of antimicrobials sustains elevated resistance rates of *E. coli* strains, particularly ETEC, to antibiotics (Kantele et al., 2023). The symptoms of ETEC infection may vary from moderate infections that resolve without treatment to infections that have symptoms similar to cholera, with the most common symptom being watery diarrhea that lasts 3–5 days. Children residing in underdeveloped nations and travelers to such nations often get diarrhea from it (Croxen et al., 2013). Childhood stunting is positively correlated with recurrent and frequent nonfatal diarrhea episodes. Stunting in childhood increases the risk of dying from diseases such as pneumonia, measles, and diarrhea. A child's capacity for learning, level of schooling attained, and potential earnings may all be impacted by childhood stunting, which also has an impact on cognitive development (Guerrant et al., 2021).

Sukwa et al. (2024) posited that children living with HIV were roughly thrice more likely than children without HIV to have ETEC diarrhea, and children who got their potable water from a well, a borehole, or a public tap were more than two times as likely to get ETEC diarrhea as children who had piped water into their home. Additionally, compared to children whose mothers had less than a secondary school education, children whose mothers had completed at the very least senior high school had a 50% lower incidence of ETEC infections.

Chisenga et al. (2018) in Zambia posited that among children under 5, ETEC is the third most often found pathogen. This highlights the prevalence of ETEC worldwide in impecunious environments, which is mostly caused by issues like inadequate sanitation and restricted access to potable water.

In a study by Sukwa et al. (2024), they posited that LT toxin was expressed by 50% of the ETEC strains, ST toxin was expressed by 33% of the ETEC strains, and others expressed both toxins. It has been demonstrated that the distribution of toxins varies by region, with some strains being more prevalent than others everywhere. It has also been shown that *Escherichia coli* strains containing certain virulence factors have been obtained from human controls who are in good condition, indicating that the presence of a particular virulence factor does not always indicate its involvement in clinical disease (Cabal et al., 2016).

There is an elevated risk of ETEC infections in HIV-positive children. According to a study by Basile et al. (2021) conducted during the HIV epidemic, there was an elevated incidence of diarrhea in children living with HIV when compared to children living without HIV. Additionally, compared to children who are not living with HIV, children who live with HIV are more likely to experience severe, chronic, and recurrent diarrhea (Bejide et al., 2023). These results further support the idea that immunocompromised children are more susceptible to enteric infections, especially when paired with the existing body of knowledge about the consequences of diarrhea, such as malnutrition. This correlation brought attention to the increased susceptibility of children living with HIV to gastrointestinal infections, highlighting the need for focused medical therapy.

The correlation between poor water sources and elevated incidence of ETEC diarrhea corroborates data about world health, highlighting the significance of water quality in diarrheal illnesses. Iron deficiency anemia, malnutrition, diarrheal mortality, environmental enteropathy, and morbidity are all strongly correlated with poor water sanitation and hygiene (WASH) (Chakravarty et al., 2017). In many impecunious nations, at least 11.4% of deaths are due to water sanitation and hygiene-associated factors (Nyambe & Yamauchi, 2022).

The perks of educating mothers in relation to fewer infections due to ETEC emphasize the vital role that elevated health literacy in mothers plays in illness prevention. According to more extensive studies on public health, this indicates that better maternal education leads to healthier children.

Prior research has demonstrated a positive link between a mother's educational attainment and her child's health markers like wasting and weight (De Buhr & Tannen, 2020). Additionally, the usage of medical services, taking iron supplements during pregnancy, and childhood immunization have all been favorably correlated with maternal education (Vikram & Vanneman, 2020).

In impoverished nations, where there is insufficient infrastructure for proper waste disposal and supply of potable water, ETEC infections are acquired via the intake of tainted water and edibles. The propensity for the generation of virulence factors such as heat-stable (ST) and/or heat-labile (LT) enterotoxins is typical of ETEC (Zhang et al., 2022). One essential step for ETEC to start exerting its toxicity is by colonization of the intestinal epithelium. The colonization factors (CFs), in addition to the two kinds of enterotoxins, constitute a significant source of ETEC pathogenicity (Zhang et al., 2022). Enterotoxins and fimbrial adhesins are produced by ETEC, which are vital mediators of its virulence. Fimbrial adhesins facilitate bacterial colonization by enhancing bacterial adhesion to the host epithelial cell surface (Vidal et al., 2019). Water and electrolyte deposition in the intestinal lumen is a result of the successful production of enterotoxins, which starts after the colonization of the intestinal epithelium by ETEC using colonization factors. There are an estimated hundreds of millions of ETEC infections among newborns and infants in impecunious countries annually (Khalil et al., 2018). Travelers visiting ETEC endemic regions who do not have access to basic sanitary facilities and clean water are also heavily affected. Empirical antibiotic usage is driven by ETEC infections, which may encourage resistance to antimicrobial agents (Khalil et al., 2021).

Most prevalent colonization factors (CFs) have been identified as candidates for ETEC vaccination. In order to supplement current methods, research into additional conserved antigens has been spurred by the discovery of over 30 target antigens that differ from one another antigenically (Vidal et al., 2019). In the hunt for novel ETEC virulence components, two potential new antigens that could help in the creation of new highly functional vaccines are EatA mucinase and EtpA adhesin. These two highly conserved proteins are unique to the ETEC pathovars and are found in varied geographical regions (Kuhlmann et al., 2019).

Therefore, the present study reviews the potential impact of diarrheagenic enterotoxigenic *Escherichia coli* infection demography in children under the age of 5 years in sub-Saharan Africa.

3 Incidence of Diarrheagenic Enterotoxigenic *Escherichia coli* (DEC) in Sub-Saharan Africa (SSA)

Based on estimates from the Institute for Health Metrics and Evaluation (IHME) and the Maternal and Child Epidemiology Estimation Group (MCEE), ETEC has been linked to over 200 million diarrheal cases worldwide, a third of which involved children under five.

These episodes are thought to have given rise to 18,700–42,000 mortalities in infants under five (Hosangadi et al., 2019). About 30–40% of diarrhea episodes and 70,000 diarrhea-related fatalities in children under five are caused by ETEC, one of the main causes of moderate-to-severe diarrhea (Abbasi et al., 2020).

Up to 40% of travelers' diarrhea and about a quarter of diarrheal cases in infants in impecunious countries are caused by ETEC (Ashkenazi et al., 2016). ETEC infections in these impecunious countries have been shown to be a major factor in the intricate relationship between undernourishment and recurrent episodes of diarrheal sickness in young children (Anderson et al., 2019).

The Global Enteric Multicenter Study posited that the most significant pathogens linked to diarrheal infections in infants below the age of five in impecunious countries were ETEC strains that produced a combination of heat-stable enterotoxin (ST) and heat-labile toxin (LT) or just heat-stable enterotoxin (ST) (Kotloff et al., 2017).

Children are often infected numerous times at an early age in places where ETEC is prevalent, and they later acquire immunity. An estimated 220,000,000 cases and over 50,000 fatalities occur worldwide each year (Anderson et al., 2019), with the majority of infections happening in underdeveloped nations among infants below the age of 5 (Khalil et al., 2018).

Traveler's diarrhea, which is estimated to cause 10,000,000 episodes of diarrhea per year among tourists to endemic impoverished countries, has been mostly linked to ETEC infection in industrialized nations (Bourgeois et al., 2016). Little children in the impecunious countries in Africa, Asia, Latin America, and the Middle East bear a higher burden of the disease.

Diarrhea and dehydration: ETEC infection typically leads to watery stool, which can be deleterious and prolonged. In infants and toddlers, this can quickly lead to dehydration, which can be life-threatening if not promptly treated (Levine et al., 2020). Prolonged diarrhea caused by ETEC can lead to malnutrition in children, particularly in areas where food insecurity is already a concern. Malnutrition can further weaken the child's immune system and increase susceptibility to other infections (Gazi et al., 2022). Growth impairment contributes to chronic or recurrent ETEC infections and can interfere with a child's growth and development. Malabsorption of nutrients due to diarrhea and the metabolic demands of fighting infection can hinder normal growth (Koshy et al., 2022). The impact on cognitive development severely prolonged or recurrent bouts of diarrhea in early childhood can affect cognitive development and academic achievement. Although there is overlap, the factors causing inadequate growth may not be the same as those causing delayed neurodevelopment. Lower neurodevelopmental scores in children from impecunious families have been linked to systemic inflammatory cytokines and febrile illness (Jiang et al., 2014). In some toddlers, abnormal facial-recognition reactions have been linked to chronic inflammation (Xie et al., 2019). In impecunious territories, diarrheal disease is one of the vital etiologies of inflammation, research on infants from these countries has shown a direct correlation between diarrheal sickness and delayed neurodevelopment (Pinkerton et al., 2016).

Malnutrition and dehydration associated with ETEC infection can impair brain function and cognitive abilities (Koshy et al., 2022). Children with ETEC infections can transmit the bacteria to other children and family members, contributing to the dispersal of the illness within societies.

3.1 Impact of Clonally Diverse DEC Strains on SSA Children

Improvements in economic development, the availability of clean water, and improved sanitation have led to a decline in the mortality rates from diarrhea for ETEC. This decline has not been matched by a significant decline in the morbidity associated with diarrhea, which still has a detrimental effect on infant health in many impecunious countries.

ETEC infection is initially accompanied by watery diarrhea, which causes fast dehydration and is often accompanied by nausea, vomiting, and cramping in the abdomen (Smith et al., 2014) . After a 10–72 h incubation period, the disease normally lasts 3–5 days and goes away without the need for antibiotic treatment; however, symptoms may linger for 2–3 weeks. ETEC infections are often mild and do not require treatment, and they also lack the clinical differentiation from cholera. In children from underdeveloped nations, the synergy of diarrhea and hunger creates a vicious cycle that worsens symptoms and ultimately results in mortality among several children in SSA.

Somalia has a high mortality rate due to increased diarrheal deaths and an increased amount of children with stunted growth. Nations in sub-Saharan Africa, particularly Niger, Angola, and Chad have shown increased mortalities due to ETEC. Mortality due to ETEC diarrhea and other pathogenic illnesses in infants below age five was seen to be highest in the African region, followed by the Southeast Asian region, and then followed by the Eastern Mediterranean Region (Anderson et al., 2019).

There is increased diarrheal danger for infants below age five in certain central and eastern African nations, particularly a large part of DR Congo and Northern Kenya (Bagamian et al., 2020).

4 Pathogenesis of ETEC

Two steps must occur in the ETEC pathogenesis for an infection to start. In order for ETEC to produce enterotoxin, the initial step involves the colonization of the small intestinal epithelium by adhesins. Second, enterotoxins such as LT and ST are the primary agents discharged into the host's epithelia (Zhang et al., 2022). As is commonly known, heat-labile toxins and heat-stable toxins induce diarrhea in hosts by increasing the synthesis of cyclic nucleotides, which results in the loss of water and electrolytes.

4.1 Increased Expression of Colonization Factors

Various required colonization factors (CFs) on the surface bacteria allow ETEC to adhere to the duodenal epithelium. Pili or fimbriae are polymers of proteins that are extracellularly linked to fimbria; they make up the bulk of CFs. Fimbrial, fibrillar, helical, and afimbrial are the four morphological types of CFs (Zhang et al., 2022).

ETEC encodes three primary groups of colonization factor antigens (CFAs) that elicit human diarrhea: CFA/IV, CFA/I, and CFA/II. CFA/I is the only member in the first family. CS6 alone is expressed by CFA/IV strains, it may also be expressed in conjunction with CS5 or CS4, and coli surface (CS) antigen 3 (CS3) is encoded alone by CFA/II strains or in conjunction with CS2 or CS1. Fibrillae that are flexible and thin, with a diameter of 2–4 nm, are called CS3, and hard fibrillae ~6–7 nm in diameter are called CS5, CS2, CS4, CFA/I, and CS1. CS6 structure and shape are nondescript (Vidal et al., 2019).

In ETEC, more than 25 CFs with a variety of antigens and structural features—including fimbrial and fibrillar structures—have been found. Of these, CS1 to CS6 (coli surface antigen) together with CFA/I (colonization factor antigen) are the seven that are typically more common than the others (Bourgeois et al., 2016). Although most colonization factor receptors are undisclosed, colonization factors are bound to conjugates of glycoprotein on the surface of host cells and also in the small intestine, particularly the mucus fraction.

In the past, there existed complexity and uneven naming practices for CFs. The acronym "CS" (coli surface antigen) is used to identify ETEC CFs, which is then followed by an Arabic number, in an effort to standardize and streamline the nomenclature of CFs; plasmids encode most of the ETEC colonization factors (Croxen et al., 2013). Based on the CF assembly procedure, ETEC CFs are divided into two types: Type IV pili and chaperone usher (CU) pili. Many bacteria that are Gram-negative build pili via the CU route, which is also how most ETEC pili are produced (Hospenthal et al., 2017). The proper functioning of the CU pathway depends on two proteins: a protein found on the outer membrane protein known as "usher," which organizes and gathers complexes that are chaperone-subunits that form the pilus, together with a chaperone protein found on the periplasm that encourages folding of pilins, hinders polymerization of the periplasm, and directs ushering. Numerous biological activities, including host-bacterium adhesion, signal transduction, DNA absorption, biofilm formation, and twitching motility, are facilitated by Type IV pili (T4P) (Debnath et al., 2023).

The most prevalent colonization factors among ETEC isolates from infants below the age of five who had diarrheal infections in impecunious countries were CFA/I and CS1–CS6 (Vidal et al., 2019). Nonfimbrial adhesins have also been linked to the pathogenesis, including TibA, an autotransporter that is glycosylated, Tia and EtpA, an outer membrane protein, and a molecular bridge, respectively, connect cell receptors from the host to ETEC flagella tips (Stevens & Frankel, 2014).

In addition to the canonical virulence factors, several more recently discovered plasmid-encoded proteins appear to be conserved within the ETEC pathovar (Kuhlmann et al., 2019). Among these is the EatA autotransporter protein, a member of the serine protease autotransporters of the Enterobacteriaceae (SPATE) family (Ruiz-Perez et al., 2011). EatA, which shares approximately 74% identity with the SepA autotransporter from *Shigella flexneri* (Kumar et al., 2014), has recently been shown to degrade MUC2, the major gel-forming mucin secreted by goblet cells in both the small and large intestine (Johansson et al., 2013).

4.2 Heat-Labile Enterotoxin

The virulence plasmid's *eltAB* operon codes for the high molecular weight enterotoxin known as LT. Heat-labile enterotoxin is a heterohexameric molecule consisting of pentamer B that binds to the glycoconjugates on epithelia and one A subunit that is catalytic in nature. LT is categorized mainly into two depending on its genetic sequence and antigenic capacity. Type I LT is found in pigs (LT-Ip) along with humans (LT-Ih), whereas type II LT (LT-IIa, b, c) is separated chiefly found in animals (Mirhoseini et al., 2018). The different LT synthesized by ETEC shows a wide variety of features. LT also functions as an adjuvant by increasing vaccine antigenicity, promoting bacterial adhesion and intestinal colonization, and transporting foreign molecules into cells (Duan et al., 2019). LT induces apoptosis in epithelial cells of the intestine, leading to a decrease in the viability of intestinal epithelium (Lu et al., 2017). Recent research posited that LT may decrease intestinal vitamin C uptake, increasing intestinal susceptibility to infection (Subramenium et al., 2019).

4.3 Heat-Stable Enterotoxin

In impecunious nations, the etiologic agent of infantile diarrhea is mostly heat-stable enterotoxin (ST) (Kotloff et al., 2017). Unlike LT, ST is a peptide that is short and nonimmunogenic and is divided into two groups called STa and STb according to its structure and function. In addition to having three disulfide links formed by its six cysteine residues, STa is structurally and sequence comparable to guanylin and uroguanylin, which also have two disulfide linkages. The extracellular ligand binding domain of STa was linked to the protein via receptors of GC-C that are found expressed on the epithelium of the small intestine, particularly on the brush border membrane. GTP's conversion to cyclic GMP (cGMP) involves intracellular catalysis by receptors of GC-C, which is activated by this binding and causes a sharp increase in cGMP (Zhang et al., 2022). Diarrhea is mediated by cGMP accumulation in two ways. The alternative is to indirectly inhibit the Na/H brush border exchanger NHE3, which lowers the reabsorption of the brush border (Chen et al., 2019). Ultimately, water, salt ions, and accumulation in the lumen of the intestine lead to diarrhea. However, recent studies present a different perspective; signaling by a second messenger within the cell has shown that extracellular cGMP could be involved in infection since, in human jejunal organoid monolayers, ST-induced cGMP was primarily transferred into the basolateral region instead of remaining intracellular, potentially complicating the mechanism of ETEC (Foulke-Abel et al., 2020).

4.4 Impact of Public Health Antimicrobial Stewardship in Controlling DEC Infectivity

Most cases of ETEC infections are caused by consuming tainted food or drink. Humans are the only reservoir for ETEC infections (Ferdous et al., 2021). Contaminated water, including potable water and surface water from areas without proper sewage infrastructure and sanitation, is the primary cause of ETEC infection (Kristanti et al., 2022). Lately, however, various dietary products, such as imported herbs and vegetables from endemic areas, have also been linked to outbreaks or rare, sporadic cases in industrialized nations.

Adults need an infectious dosage of ETEC ranging from 10^8 to 10^{10} cells, while lower doses may be necessary to infect sensitive groups such as newborns (Levine et al., 2019). The pathogen that causes travelers' diarrhea (TD) that is most often isolated is called enterotoxigenic *Escherichia coli*, or ETEC. TD is self-limiting, but it can be quite debilitating as well. One of the common quinolone antibiotics prescribed to TD patients is ciprofloxacin. However, the alarmingly high levels of antibiotic resistance have made finding new medications urgently necessary. Ciprofloxacin targets gyrA, a protein found only in prokaryotes and crucial to the survival of bacteria. Reduced affinity for quinolones is explained by enhanced quinolone resistance in vivo, linked to several mutations in the gyrA region that determine quinolone resistance (Mehla & Ramana, 2016).

ETEC diarrhea is self-limiting, but it can be quite debilitating as well. One of the common quinolone antibiotics prescribed is ciprofloxacin. However, the alarmingly high levels of antibiotic resistance call for new drugs. Reduced affinity for quinolones is explained by enhanced quinolone resistance in vivo, linked to several mutations in the gyrA region that determine quinolone resistance (Mehla & Ramana, 2016). As regards the ETEC infection diagnosis, the highest resistance rates were observed for ciprofloxacin, amoxicillin/clavulanate, ampicillin, sulfamethoxazole/trimethoprim, and nalidixic acid. A high proportion of ETEC strains were resistant to nalidixic acid (40%), ceftriaxone (62.5%), ampicillin (50%), and macrolides (87.5%). Remarkably, quinolones and tetracycline were effective against many ETEC infections (Johura et al., 2024).

5 Geno-Surveillance and Prevention of DEC Infectivity in SSA Children

ETEC is one of the most prevalent etiological agents of diarrheal infections in children in Africa and South Asia. The O6, O78, O8, O128, and O153 are the most often implicated serogroups in outbreaks and occasional cases (Feng & Reddy, 2014). The extent, spread, epidemiology, and resistance patterns of bacterial microorganisms that cause diarrhea in Africa remain abysmally comprehended, even though newer studies show the troubling unfolding of antimicrobial resistance to common diarrheal etiological agents, such as ETEC. This is primarily because many cases go unreported, untreated, and undiscovered (Kalule et al., 2023).

Oral rehydration and other interventions have reduced the number of fatal cases of infectious diarrhea over the last several decades. However, ETEC still causes significant acute morbidity and related consequences, such as malnourishment and stunted development (Bagamian et al., 2020). To avoid electrolyte loss and dehydration in treating ETEC infections, oral rehydration is used with supportive care. It has been shown that fluoroquinolones work well for ETEC travelers' diarrhea (Roussel et al., 2017). Still, it should be administered with care due to the potential for adverse effects and the global increase in antibiotic resistance. Antimotility medications may be recommended, but they must be used cautiously as they may extend the period of time that bacteria or their toxins remain in the gut. Potable water, proper sanitation that separates human waste from excrement, and proper hand hygiene before eating or preparing food all have a favorable impact on people's health and their social and economic well-being. (Ejemot-Nwadiaro et al., 2015). Hindrances in getting reliable estimates of ETEC infection processes inhibit estimating the global disease burden (Prüss-Ustün et al., 2019). An estimated one million or more people die from infectious illnesses each year as a result of unsafe WASH practices, with children under the age of 5, bearing a disproportionate share of this burden (Murray et al., 2020).

ETEC infections are not routinely diagnosed, leading to under-reporting of cases. Diarrheal illnesses by ETEC are due to poor access to potable water and due to abysmal health conditions. Ameliorating better health conditions and vaccine production processes can be very efficacious in controlling infections. Vaccination will limit the number of people infected and prevent the spread of ETEC in sub-Saharan Africa.

6 Vaccine Strategies against ETEC

It is currently acknowledged that vaccination is the most effective way to prevent ETEC infection, since public hygiene issues in impecunious countries cannot be promptly addressed (Seo & Zhang, 2020). Following immunization, individuals ought to be protected against prevalent strains, at minimum (Seo & Zhang, 2020).

In order to better inform donors and stakeholders in public health about the overall public health benefits of creating, approving, and deploying an efficacious ETEC vaccine, ETEC has been designated as a priority target for more thorough epidemiology, health, and economic impact studies (Hasso-Agopsowicz et al., 2021).

Vaccination may be a potential method of preventing ETEC infections, given the high death rates and morbidity in the infant population and the lasting effects of diarrheal infections on infant maturation. Several approaches have been developed and used in the last 20 years to generate an ETEC vaccine, either to stop ETEC from attaching to the mucosa of the duodenum conversely to bestow vaccination using LT-targeting IgA antibodies (Rojas-Lopez et al., 2018).

The most advanced of these approaches, ETVAX, is undergoing phase two clinical trials (Qadri et al., 2020). This approach combines hybrid LT/CTB with or without adjuvant with inactivated whole-cell vaccination strains that overexpress CS6, CS5, CS3, and CFA/I antigens (Clements & Norton, 2018). Since ETEC strains vary in their O:H serotypes, distinct fimbrial antigens found in the most common ETEC pathotypes have been employed in this technique to offer broad protection. Excellent safety was shown in an age-descending experiment conducted in Bangladesh using this vaccination. Swedish adults and infants from Bangladesh showed faster and greater mucosal antibody responses upon adding the dmLT adjuvant (Qadri et al., 2020). In theory, a broad-acting ETEC vaccine that expresses CFA/IV, CFA/I, and CFA/II antigens should be able to protect against most ETEC strains found globally (Barry & Levine, 2019). Another vaccine candidate is live attenuated *E. coli* that expresses ETEC fimbriae. The prototype strain *E. coli* strain E1392–75-2A, shown in clinical trials to contain the fimbriae CS3 and CS1 but is devoid of the genes encoding ST and LT (Turner et al., 2011).

Although it has been challenging to determine how protection from ETEC infections in endemic locations relates to anti-CF antibodies, controlled human infection trials provide strong evidence that passive immunization against CF tip adhesin molecules or CFs can inhibit diarrheal ETEC infections (Savarino et al., 2017). CF-like chaperone-usher-pilus operons are encoded by the majority of ETEC strains (Sahl et al., 2017), and 20–50% of isolates do not possess any of the 25 recognized and immunologically unique CFs that have been identified thus far (Vidal et al., 2019). The possibly confusing creation of a vaccine that was only meant to protect against CFs led to the investigation of other antigens, such as toxoids, which, used together, could provide a wide range of protection (Taxt et al., 2016).

According to Lundgren et al. (2016), all adult ETVAX studies have demonstrated that the vaccine is not harmful, with no appreciable variations in side effects among participants of the study. Adults from ETEC endemic areas can safely use ETVAX (Akhtar et al., 2019). Clinical investigations have shown that ETVAX has a positive preventative effect, but it does not confer immunity against ETEC strains that generate heat-stable toxins, so the genuine preventive advantage may be limited. Recent research indicates that ETEC-producing ST causes mild or extreme diarrhea in infants more quickly than heat-labile toxin-producing ETEC (Kotloff et al., 2017). Furthermore, more than two-thirds of the ETEC strains recovered from patients express ST alone or in conjunction with LT, according to a prior study. Therefore, in order to develop antiadhesin immunity and antitoxin immunity against LT and ST, the most common colonization factor antigens should be included in many doses of a vaccine to provide broad protection (Seo & Zhang, 2020).

7 Conclusion

ETEC infections are widely associated with developing countries due to insufficient infrastructure for proper waste disposal and supply of potable water. Other factors influencing ETEC infections are poor detection and diagnostic facilities required for efficient diagnosis and effective therapeutic management. More recently discovered compounds specific to the ETEC pathovar may provide insight into ETEC pathogenicity, even though classical antigens have received the majority of pathogenesis investigations and vaccine development efforts.

While canonical antigens have received much attention in pathogenesis research and vaccine development, other recently discovered molecules specific to the ETEC pathovar may also help us comprehend ETEC virulence.

The routine diagnosis of ETEC infections should be emphasized in clinical laboratories, particularly for high-risk populations.

Rapid Kits for ETEC infections should be deployed to resource-limited locations to aid diagnosis.

Vaccines for the prevention of ETEC infections, including ETVAX, ShigETEC, and CVD 1208S-122, should be included in primary health care services. In adults and children. ETVAX has been shown to be safe and immunogenic. The administered dose produced noticeably stronger IgA responses.

References

Abbasi, E., Mondanizadeh, M., Van Belkum, A., & Ghaznavi-Rad, E. (2020). Multi-drug-resistant diarrheagenic escherichia coli Pathotypes in pediatric patients with gastroenteritis from Central Iran. *Infection and Drug Resistance, 13*, 1387–1396. https://doi.org/10.2147/IDR.S247732

Akhtar, M., Chowdhury, M. I., Bhuiyan, T. R., Kaim, J., Ahmed, T., Rafique, T. A., Khan, A., Rahman, S. I. A., Khanam, F., Begum, Y. A., Sharif, M. Z., Islam, L. N., Carlin, N., Maier, N., Fix, A., Wierzba, T. F., Walker, R. I., Bourgeois, A. L., Svennerholm, A.-M., Qadri, F., & Lundgren, A. (2019). Evaluation of the safety and immunogenicity of the oral inactivated multivalent enterotoxigenic Escherichia coli vaccine ETVAX in Bangladeshi adults in a double-blind, randomized, placebo-controlled phase I trial using electrochemiluminescence and ELISA assays for immunogenicity analyses. *Vaccine, 37*, 5645–5656. https://doi.org/10.1016/j.vaccine.2018.11.040

Anderson, J. D., Bagamian, K. H., Muhib, F., Amaya, M. P., Laytner, L. A., Wierzba, T., & Rheingans, R. (2019). Burden of enterotoxigenic Escherichia coli and shigella nonfatal diarrheal infections in 79 low-income and lower middle-income countries: A modelling analysis. *The Lancet Global Health, 7*, e321–e330. https://doi.org/10.1016/S2214-109X(18)30483-2

Ashkenazi, S., Schwartz, E., & O'Ryan, M. (2016). Travelers' diarrhea in children: What have we learnt? *The Pediatric Infectious Disease Journal, 35*, 698–700. https://doi.org/10.1097/INF.0000000000001145

Bagamian, K. H., Anderson, J. D., Muhib, F., Cumming, O., Laytner, L. A., Wierzba, T. F., & Rheingans, R. (2020). Heterogeneity in enterotoxigenic Escherichia coli and shigella infections in children under 5 years of age from 11 African countries: a subnational approach quantifying risk, mortality, morbidity, and stunting. *The Lancet Global Health, 8*, e101–e112. https://doi.org/10.1016/S2214-109X(19)30456-5

Bakthavatchalam, Y., Anandan, S., & Veeraraghavan, B. (2016). Laboratory detection and clinical implication of oxacillinase-48 like carbapenemase: The hidden threat. *Journal of Global Infectious Diseases, 8*, 41. https://doi.org/10.4103/0974-777X.176149

Barry, E. M., & Levine, M. M. (2019). A tale of two bacterial enteropathogens and one multivalent vaccine. *Cellular Microbiology, 21*, e13067. https://doi.org/10.1111/cmi.13067

Basile, F. W., Fedele, M. C., & Lo Vecchio, A. (2021). Gastrointestinal diseases in children living with HIV. *Microorganisms, 9*, 1572. https://doi.org/10.3390/microorganisms9081572

Baumgart, L. A., Lee, J. E., Salamov, A., Dilworth, D. J., Na, H., Mingay, M., Blow, M. J., Zhang, Y., Yoshinaga, Y., Daum, C. G., & O'Malley, R. C. (2021). Persistence and plasticity in bacterial gene regulation. *Nature Methods, 18*, 1499–1505. https://doi.org/10.1038/s41592-021-01312-2

Becker, S. L., Vogt, J., Knopp, S., Panning, M., Warhurst, D. C., Polman, K., Marti, H., Von Müller, L., Yansouni, C. P., Jacobs, J., Bottieau, E., Sacko, M., Rijal, S., Meyanti, F., Miles, M. A., Boelaert, M., Lutumba, P., Van Lieshout, L., N'Goran, E. K., Chappuis, F., & Utzinger, J. (2013). Persistent digestive disorders in the tropics: Causative infectious pathogens and reference diagnostic tests. *BMC Infectious Diseases, 13*, 37. https://doi.org/10.1186/1471-2334-13-37

Bejide, O. S., Odebode, M. A., Ogunbosi, B. O., Adekanmbi, O., Akande, K. O., Ilori, T., Ogunleye, V. O., Nwachukwu, V. U., Grey-Areben, A., Akande, E. T., & Okeke, I. N. (2023). Diarrheal pathogens in the stools of children living with HIV in Ibadan, Nigeria. *Frontiers in Cellular and Infection Microbiology, 13*, 1108923. https://doi.org/10.3389/fcimb.2023.1108923

Bourgeois, A. L., Wierzba, T. F., & Walker, R. I. (2016). Status of vaccine research and development for enterotoxigenic Escherichia coli. *Vaccine, 34*, 2880–2886. https://doi.org/10.1016/j.vaccine.2016.02.076

Breijyeh, Z., Jubeh, B., & Karaman, R. (2020). Resistance of gram-negative bacteria to current antibacterial agents and approaches to resolve it. *Molecules, 25*, 1340. https://doi.org/10.3390/molecules25061340

Cabal, A., García-Castillo, M., Cantón, R., Gortázar, C., Domínguez, L., & Álvarez, J. (2016). Prevalence of Escherichia coli virulence genes in patients with diarrhea and a subpopulation of healthy volunteers in Madrid, Spain. *Frontiers in Microbiology, 7*, 641. https://doi.org/10.3389/fmicb.2016.00641

Carlin, N., & Svennerholm, A.-M. (2020). Oral vaccines for enterotoxigenic Escherichia coli. In *Mucosal vaccines* (pp. 563–578). Elsevier. https://doi.org/10.1016/B978-0-12-811924-2.00032-8

Chakravarty, I., Bhattacharya, A., & Das, S. (2017). Water, sanitation and hygiene: The unfinished agenda in the World Health Organization South-East Asia Region. *WHO South-East Asia Journal of Public Health, 6*, 22. https://doi.org/10.4103/2224-3151.213787

Chen, T., Lin, R., Avula, L., Sarker, R., Yang, J., Cha, B., Tse, C. M., McNamara, G., Seidler, U., Waldman, S., Snook, A., Bijvelds, M. J. C., De Jonge, H. R., Li, X., & Donowitz, M. (2019). NHERF3 is necessary for *Escherichia coli* heat-stable enterotoxin-induced inhibition of NHE3: Differences in signaling in mouse small intestine and Caco-2 cells. *American Journal of Physiology-Cell Physiology, 317*, C737–C748. https://doi.org/10.1152/ajpcell.00351.2018

Chisenga, C. C., Bosomprah, S., Makabilo Laban, N., Mwila-Kazimbaya, K., Mwaba, J., Simuyandi, M., & Chilengi, R. (2018). Aetiology of diarrhea in children under five in Zambia detected using luminex xTAG gastrointestinal pathogen panel. *Pediatric Infectious Diseases Open Access, 3*(2), 1–6. https://doi.org/10.21767/2573-0282.100064

Chokshi, A., Sifri, Z., Cennimo, D., & Horng, H. (2019). Global contributors to antibiotic resistance. *Journal of Global Infectious Diseases, 11*, 36. https://doi.org/10.4103/jgid.jgid_110_18

Clements, J. D., & Norton, E. B. (2018). The mucosal vaccine adjuvant LT(R192G/L211A) or dmLT. *mSphere, 3*, e00215–e00218. https://doi.org/10.1128/mSphere.00215-18

Colston, J. M., Zaitchik, B. F., Badr, H. S., Burnett, E., Ali, S. A., Rayamajhi, A., Satter, S. M., Eibach, D., Krumkamp, R., May, J., Chilengi, R., Howard, L. M., Sow, S. O., Jahangir Hossain, M., Saha, D., Imran Nisar, M., Zaidi, A. K. M., Kanungo, S., Mandomando, I., Faruque, A. S. G., Kotloff, K. L., Levine, M. M., Breiman, R. F., Omore, R., Page, N., Platts-Mills, J. A., Ashorn, U., Fan, Y., Shrestha, P. S., Ahmed, T., Mduma, E., Yori, P. P., Bhutta, Z., Bessong, P., Olortegui, M. P., Lima, A. A. M., Kang, G., Humphrey, J., Prendergast, A. J., Ntozini, R., Okada, K., Wongboot, W., Gaensbauer, J., Melgar, M. T., Pelkonen, T., Freitas, C. M., & Kosek, M. N. (2022). Associations between eight earth observation-derived climate variables and Enteropathogen infection: An independent participant data meta-analysis of surveillance studies with broad Spectrum nucleic acid diagnostics. *GeoHealth, 6*, e2021GH000452. https://doi.org/10.1029/2021GH000452

Croxen, M. A., Law, R. J., Scholz, R., Keeney, K. M., Wlodarska, M., & Finlay, B. B. (2013). Recent advances in understanding enteric pathogenic Escherichia coli. *Clinical Microbiology Reviews, 26*, 822–880. https://doi.org/10.1128/CMR.00022-13

De Buhr, E., & Tannen, A. (2020). Parental health literacy and health knowledge, behaviours and outcomes in children: A cross-sectional survey. *BMC Public Health, 20*, 1096. https://doi.org/10.1186/s12889-020-08881-5

Debnath, A., Sabui, S., & Chatterjee, N. S. (2023). Structural and functional characterization of colonization factors AIBI-CS6 and AIIBII-CS6 of enterotoxigenic Escherichia coli. *Protein Expression and Purification, 203*, 106201. https://doi.org/10.1016/j.pep.2022.106201

Denamur, E., Clermont, O., Bonacorsi, S., & Gordon, D. (2021). The population genetics of pathogenic Escherichia coli. *Nature Reviews. Microbiology, 19*, 37–54. https://doi.org/10.1038/s41579-020-0416-x

Duan, Q., Xia, P., Nandre, R., Zhang, W., & Zhu, G. (2019). Review of newly identified functions associated with the heat-labile toxin of Enterotoxigenic Escherichia coli. *Frontiers in Cellular and Infection Microbiology, 9*, 292. https://doi.org/10.3389/fcimb.2019.00292

DuPont, H. L., & Steffen, R. (2017). Use of antimicrobial agents for treatment and prevention of travellers' diarrhea in the face of enhanced risk of transient fecal carriage of multi-drug resistant enterobacteriaceae: Setting the stage for consensus recommendations. *Journal of Travel Medicine, 24*, S57–S62. https://doi.org/10.1093/jtm/tax040

Ejemot-Nwadiaro, R. I., Ehiri, J. E., Arikpo, D., Meremikwu, M. M., & Critchley, J. A. (2015). Hand washing promotion for preventing diarrhea. *Cochrane Database of Systematic Reviews, 2015*, CD004265. https://doi.org/10.1002/14651858.CD004265.pub3

Farhat, N., & Khan, A. U. (2020). Evolving trends of New Delhi Metallo-betalactamse (NDM) variants: A threat to antimicrobial resistance. *Infection, Genetics and Evolution, 86*, 104588. https://doi.org/10.1016/j.meegid.2020.104588

Feng, P. C. H., & Reddy, S. P. (2014). Prevalence and diversity of enterotoxigenic Escherichia coli strains in fresh produce. *Journal of Food Protection, 77*, 820–823. https://doi.org/10.4315/0362-028X.JFP-13-412

Ferdous, J., Rashid, R. B., Sultana, R., Saima, S., Jahan Prima, M., Begum, A., & Mackie Jensen, P. K. (2021). Is it human or animal? The origin of pathogenic *E. coli* in the drinking water of a low-income Urban Community in Bangladesh. *Tropical Medicine and Infectious Disease, 6*, 181. https://doi.org/10.3390/tropicalmed6040181

Foulke-Abel, J., Yu, H., Sunuwar, L., Lin, R., Fleckenstein, J. M., Kaper, J. B., & Donowitz, M. (2020). Phosphodiesterase 5 (PDE5) restricts intracellular cGMP accumulation during enterotoxigenic *Escherichia coli* infection. *Gut Microbes, 12*, 1752125. https://doi.org/10.1080/19490976.2020.1752125

Founou, R. C., Founou, L. L., & Essack, S. Y. (2017). Clinical and economic impact of antibiotic resistance in developing countries: A systematic review and meta-analysis. *PLoS One, 12*, e0189621. https://doi.org/10.1371/journal.pone.0189621

Frost, I., Sati, H., Garcia-Vello, P., Hasso-Agopsowicz, M., Lienhardt, C., Gigante, V., & Beyer, P. (2023). The role of bacterial vaccines in the fight against antimicrobial resistance: An analysis of the preclinical and clinical development pipeline. *Lancet Microbe, 4*, e113–e125. https://doi.org/10.1016/S2666-5247(22)00303-2

Gazi, M. A., Alam, M. A., Fahim, S. M., Wahid, B. Z., Khan, S. S., Islam, M. O., Hasan, M. M., Hasan, S. M. T., Das, S., Mahfuz, M., Haque, R., & Ahmed, T. (2022). Infection with Escherichia coli pathotypes is associated with biomarkers of gut enteropathy and nutritional status among malnourished children in Bangladesh. *Frontiers in Cellular and Infection Microbiology, 12*, 901324. https://doi.org/10.3389/fcimb.2022.901324

Guerrant, R. L., Bolick, D. T., & Swann, J. R. (2021). Modeling enteropathy or diarrhea with the top bacterial and protozoal pathogens: Differential determinants of outcomes. *ACS Infectious Diseases, 7*, 1020–1031. https://doi.org/10.1021/acsinfecdis.0c00831

Hasso-Agopsowicz, M., Lopman, B. A., Lanata, C. F., Rogawski McQuade, E. T., Kang, G., Prudden, H. J., Khalil, I., Platts-Mills, J. A., Kotloff, K., Jit, M., Riddle, M. S., Pavlinac, P. B., Luz, P. M., Pitzer, V. E., Breiman, R. F., & Giersing, B. K. (2021). World Health Organization expert working group: Recommendations for assessing morbidity associated with enteric pathogens. *Vaccine, 39*, 7521–7525. https://doi.org/10.1016/j.vaccine.2021.11.033

Hebbelstrup Jensen, B., Olsen, K. E. P., Struve, C., Krogfelt, K. A., & Petersen, A. M. (2014). Epidemiology and clinical manifestations of enteroaggregative Escherichia coli. *Clinical Microbiology Reviews, 27*, 614–630. https://doi.org/10.1128/CMR.00112-13

Hosangadi, D., Smith, P. G., Kaslow, D. C., & Giersing, B. K. (2019). WHO consultation on ETEC and Shigella burden of disease, Geneva, 6–7th April 2017: Meeting report. *Vaccine, 37*, 7381–7390. https://doi.org/10.1016/j.vaccine.2017.10.011

Hospenthal, M. K., Costa, T. R. D., & Waksman, G. (2017). A comprehensive guide to pilus biogenesis in gram-negative bacteria. *Nature Reviews. Microbiology, 15*, 365–379. https://doi.org/10.1038/nrmicro.2017.40

Idalia, V.-M. N., & Bernardo, F. (2017). *Escherichia coli* as a model organism and its application in biotechnology. In A. Samie (Ed.), *Escherichia coli—recent advances on physiology*. Pathogenesis and Biotechnological Applications. https://doi.org/10.5772/67306

Iseppi, R., Di Cerbo, A., Messi, P., & Sabia, C. (2020). Antibiotic resistance and virulence traits in vancomycin-resistant enterococci (VRE) and extended-spectrum β-lactamase/AmpC-producing (ESBL/AmpC) enterobacteriaceae from humans and pets. *Antibiotics, 9*, 152. https://doi.org/10.3390/antibiotics9040152

Jiang, N. M., Tofail, F., Moonah, S. N., Scharf, R. J., Taniuchi, M., Ma, J. Z., Hamadani, J. D., Gurley, E. S., Houpt, E. R., Azziz-Baumgartner, E., Haque, R., & Petri, W. A. (2014). Febrile illness and pro-inflammatory cytokines are associated with lower neurodevelopmental scores in Bangladeshi infants living in poverty. *BMC Pediatrics, 14*, 50. https://doi.org/10.1186/1471-2431-14-50

Johansson, M. E. V., Sjövall, H., & Hansson, G. C. (2013). The gastrointestinal mucus system in health and disease. *Nature Reviews. Gastroenterology & Hepatology, 10*, 352–361. https://doi.org/10.1038/nrgastro.2013.35

Johura, F.-T., Sultana, M., Sadique, A., Monira, S., Sack, D., Sack, R., Alam, M., & Chakraborty, S. (2024). The antimicrobial resistance of Enterotoxigenic Escherichia coli from diarrheal patients and the environment in two geographically distinct rural areas in Bangladesh over the years. *Microorganisms, 12*, 301. https://doi.org/10.3390/microorganisms12020301

Kalule, J. B., Bester, L. A., Banda, D. L., Derra, F. A., Chikuse, F., Tessema, S. K., Africa PGI Foodborne Diseases Genomic Surveillance Focus Group, & Foster-Nyarko, E. (2023). Molecular epidemiology of diarrheagenic *Escherichia coli* in Africa: A systematic review and meta-analysis. *medRxiv*, 2023–2010. https://doi.org/10.1101/2023.10.11.23296874

Kantele, A., Riekkinen, M., Jokiranta, T. S., Pakkanen, S. H., Pietilä, J.-P., Patjas, A., Eriksson, M., Khawaja, T., Klemets, P., Marttinen, K., Siikamäki, H., Lundgren, A., Holmgren, J., Lissmats, A., Carlin, N., & Svennerholm, A.-M. (2023). Safety and immunogenicity of ETVAX®, an oral inactivated vaccine against enterotoxigenic *Escherichia coli* diarrhea: A double-blinded, randomized, placebo-controlled trial amongst Finnish travellers to Benin, West Africa. *Journal of Travel Medicine, 30*, taad045. https://doi.org/10.1093/jtm/taad045

Khalil, I., Walker, R., Porter, C. K., Muhib, F., Chilengi, R., Cravioto, A., Guerrant, R., Svennerholm, A.-M., Qadri, F., Baqar, S., Kosek, M., Kang, G., Lanata, C., Armah, G., Wierzba, T., Hasso-Agopsowicz, M., Giersing, B., & Louis Bourgeois, A. (2021). Enterotoxigenic Escherichia coli (ETEC) vaccines: Priority activities to enable product development, licensure, and global access. *Vaccine, 39*, 4266–4277. https://doi.org/10.1016/j.vaccine.2021.04.018

Khalil, I. A., Troeger, C., Blacker, B. F., Rao, P. C., Brown, A., Atherly, D. E., Brewer, T. G., Engmann, C. M., Houpt, E. R., Kang, G., Kotloff, K. L., Levine, M. M., Luby, S. P., MacLennan, C. A., Pan, W. K., Pavlinac, P. B., Platts-Mills, J. A., Qadri, F., Riddle, M. S., Ryan, E. T., Shoultz, D. A., Steele, A. D., Walson, J. L., Sanders, J. W., Mokdad, A. H., Murray, C. J. L., Hay, S. I., & Reiner, R. C. (2018). Morbidity and mortality due to shigella and enterotoxigenic Escherichia coli diarrhea: The global burden of disease study 1990–2016. *The Lancet Infectious Diseases, 18*, 1229–1240. https://doi.org/10.1016/S1473-3099(18)30475-4

Koshy, B., Srinivasan, M., Gopalakrishnan, S., Mohan, V. R., Scharf, R., Murray-Kolb, L., John, S., Beulah, R., Muliyil, J., & Kang, G. (2022). Are early childhood stunting and catch-up growth associated with school age cognition?—Evidence from an Indian birth cohort. *PLoS One, 17*, e0264010. https://doi.org/10.1371/journal.pone.0264010

Kotloff, K. L., Platts-Mills, J. A., Nasrin, D., Roose, A., Blackwelder, W. C., & Levine, M. M. (2017). Global burden of diarrheal diseases among children in developing countries: Incidence, etiology, and insights from new molecular diagnostic techniques. *Vaccine, 35*, 6783–6789. https://doi.org/10.1016/j.vaccine.2017.07.036

Kristanti, R. A., Hadibarata, T., Syafrudin, M., Yılmaz, M., & Abdullah, S. (2022). Microbiological contaminants in drinking water: Current status and challenges. *Water, Air, and Soil Pollution, 233*, 299. https://doi.org/10.1007/s11270-022-05698-3

Kuhlmann, F. M., Martin, J., Hazen, T. H., Vickers, T. J., Pashos, M., Okhuysen, P. C., Gómez-Duarte, O. G., Cebelinski, E., Boxrud, D., Del Canto, F., Vidal, R., Qadri, F., Mitreva, M., Rasko, D. A., & Fleckenstein, J. M. (2019). Conservation and global distribution of non-canonical antigens in Enterotoxigenic Escherichia coli. *PLoS Neglected Tropical Diseases, 13*, e0007825. https://doi.org/10.1371/journal.pntd.0007825

Kumar, P., Luo, Q., Vickers, T. J., Sheikh, A., Lewis, W. G., & Fleckenstein, J. M. (2014). EatA, an immunogenic protective antigen of enterotoxigenic Escherichia coli, degrades intestinal mucin. *Infection and Immunity, 82*, 500–508. https://doi.org/10.1128/IAI.01078-13

Lääveri, T., Vilkman, K., Pakkanen, S., Kirveskari, J., & Kantele, A. (2018). Despite antibiotic treatment of travellers' diarrhea, pathogens are found in stools from half of travellers at return. *Travel Medicine and Infectious Disease, 23*, 49–55. https://doi.org/10.1016/j.tmaid.2018.04.003

Leimbach, A., Hacker, J., & Dobrindt, U. (2013). *E. coli* as an all-rounder: The thin line between commensalism and pathogenicity. In U. Dobrindt, J. H. Hacker, & C. Svanborg (Eds.), *Between pathogenicity and commensalism, current topics in microbiology and immunology* (pp. 3–32). Springer. https://doi.org/10.1007/82_2012_303

Levine, M. M., Barry, E. M., & Chen, W. H. (2019). A roadmap for enterotoxigenic *Escherichia coli* vaccine development based on volunteer challenge studies. *Human Vaccines & Immunotherapeutics, 15*, 1357–1378. https://doi.org/10.1080/21645515.2019.1578922

Levine, M. M., Nasrin, D., Acácio, S., Bassat, Q., Powell, H., Tennant, S. M., Sow, S. O., Sur, D., Zaidi, A. K. M., Faruque, A. S. G., Hossain, M. J., Alonso, P. L., Breiman, R. F., O'Reilly, C. E., Mintz, E. D., Omore, R., Ochieng, J. B., Oundo, J. O., Tamboura, B., Sanogo, D., Onwuchekwa, U., Manna, B., Ramamurthy, T., Kanungo, S., Ahmed, S., Qureshi, S., Quadri, F., Hossain, A., Das, S. K., Antonio, M., Saha, D., Mandomando, I., Blackwelder, W. C., Farag, T., Wu, Y., Houpt, E. R., Verweiij, J. J., Sommerfelt, H., Nataro, J. P., Robins-Browne, R. M., & Kotloff, K. L. (2020). Diarrheal disease and subsequent risk of death in infants and children residing in low-income and middle-income countries: Analysis of the GEMS case-control study and 12-month GEMS-1A follow-on study. *The Lancet Global Health, 8*, e204–e214. https://doi.org/10.1016/S2214-109X(19)30541-8

Liu, L., Oza, S., Hogan, D., Chu, Y., Perin, J., Zhu, J., Lawn, J. E., Cousens, S., Mathers, C., & Black, R. E. (2016). Global, regional, and national causes of under-5 mortality in 2000–15: An updated systematic analysis with implications for the sustainable development goals. *The Lancet, 388*, 3027–3035. https://doi.org/10.1016/S0140-6736(16)31593-8

Lu, X., Li, C., Li, C., Li, P., Fu, E., Xie, Y., & Jin, F. (2017). Heat-labile enterotoxin-induced PERK-CHOP pathway activation causes intestinal epithelial cell apoptosis. *Frontiers in Cellular and Infection Microbiology, 7*, 244. https://doi.org/10.3389/fcimb.2017.00244

Lundgren, A., Jertborn, M., & Svennerholm, A.-M. (2016). Induction of long term mucosal immunological memory in humans by an oral inactivated multivalent enterotoxigenic Escherichia coli vaccine. *Vaccine, 34*, 3132–3140. https://doi.org/10.1016/j.vaccine.2016.04.055

Mahmudul Hassan, M. (2021). Scenario of antibiotic resistance in developing countries. In M. Mareş, S. Hua Erin Lim, K.-S. Lai, & R.-T. Cristina (Eds.), *Antimicrobial resistance—A one health perspective*. IntechOpen. https://doi.org/10.5772/intechopen.94957

Manhique-Coutinho, L., Chiani, P., Michelacci, V., Taviani, E., Bauhofer, A. F. L., Chissaque, A., Cossa-Moiane, I., Sambo, J., Chilaúle, J., Guimarães, E. L., Salência, J., Cassocera, M., Bero, D. M., Langa, J. P., & De Deus, N. (2022). Molecular characterization of diarrheagenic Escherichia coli isolates from children with diarrhea: A cross-sectional study in four provinces of Mozambique. *International Journal of Infectious Diseases, 121*, 190–194. https://doi.org/10.1016/j.ijid.2022.04.054

Mehla, K., & Ramana, J. (2016). Travelers' diarrhea-associated enterotoxigenic *Escherichia coli* gyrA mutants and quinolone antibiotic affinity: A molecular dynamics simulation and residue interaction network analysis. *OMICS: A Journal of Integrative Biology, 20*, 635–644. https://doi.org/10.1089/omi.2016.0104

Miliwebsky, E., Schelotto, F., Varela, G., Luz, D., Chinen, I., & Piazza, R. M. F. (2016). Human diarrheal infections: diagnosis of diarrheagenic Escherichia coli pathotypes. In A. G. Torres (Ed.), *Escherichia coli in the Americas* (pp. 343–369). Springer International Publishing. https://doi.org/10.1007/978-3-319-45092-6_15

Mirhoseini, A., Amani, J., & Nazarian, S. (2018). Review on pathogenicity mechanism of enterotoxigenic Escherichia coli and vaccines against it. *Microbial Pathogenesis, 117*, 162–169. https://doi.org/10.1016/j.micpath.2018.02.032

Murray, C. J. L., Aravkin, A. Y., Zheng, P., Abbafati, C., Abbas, K. M., Abbasi-Kangevari, M., Zhang, Y., Zhang, Z.-J., Zhao, J. T., Zhao, X.-J. G., Zhao, Y., Zhou, M., Ziapour, A., Zimsen, S. R. M., Brauer, M., Afshin, A., & Lim, S. S. (2020). Global burden of 87 risk factors in 204 countries and territories, 1990–2019: A systematic analysis for the global burden of disease study 2019. *The Lancet, 396*, 1223–1249. https://doi.org/10.1016/S0140-6736(20)30752-2

Naveen, V., Siddiq, A., & Chandana, G. (2018). A study on drug utilization pattern of cephalosporins in general medicine and surgical inpatient department. *International Journal of Current Pharmaceutical Research, 10*, 33. https://doi.org/10.22159/ijcpr.2018v10i3.27225

Nordmann, P. (2014). Carbapenemase-producing enterobacteriaceae: Overview of a major public health challenge. *Medecine et maladies infectieuses, 44*, 51–56. https://doi.org/10.1016/j.medmal.2013.11.007

Nyambe, S., & Yamauchi, T. (2022). Peri-urban water, sanitation and hygiene in Lusaka, Zambia: Photovoice empowering local assessment via ecological theory. *Global Health Promotion, 29*, 66–76. https://doi.org/10.1177/1757975921995713

O'Ryan, G. M., Ashkenazi-Hoffnung, L., O'Ryan-Soriano, M. A., & Ashkenazi, S. (2014). Management of acute infectious diarrhea for children living in resource-limited settings. *Expert Review of Anti-Infective Therapy, 12*, 621–632. https://doi.org/10.1586/14787210.2014.901168

Pfeifer, Y., Schlatterer, K., Engelmann, E., Schiller, R. A., Frangenberg, H. R., Stiewe, D., Holfelder, M., Witte, W., Nordmann, P., & Poirel, L. (2012). Emergence of OXA-48-type carbapenemase-producing enterobacteriaceae in German hospitals. *Antimicrobial Agents and Chemotherapy, 56*, 2125–2128. https://doi.org/10.1128/AAC.05315-11

Pinkerton, R., Oriá, R. B., Lima, A. A. M., Rogawski, E. T., Oriá, M. O. B., Patrick, P. D., Moore, S. R., Wiseman, B. L., Niehaus, M. D., & Guerrant, R. L. (2016). Early childhood diarrhea predicts cognitive delays in later childhood independently of malnutrition. *The American Journal of Tropical Medicine and Hygiene, 95*, 1004–1010. https://doi.org/10.4269/ajtmh.16-0150

Platts-Mills, J. A., Babji, S., Bodhidatta, L., Gratz, J., Haque, R., Havt, A., McCormick, B. J., McGrath, M., Olortegui, M. P., Samie, A., Shakoor, S., Mondal, D., Lima, I. F., Hariraju, D., Rayamajhi, B. B., Qureshi, S., Kabir, F., Yori, P. P., Mufamadi, B., Amour, C., Carreon, J. D., Richard, S. A., Lang, D., Bessong, P., Mduma, E., Ahmed, T., Lima, A. A., Mason, C. J., Zaidi, A. K., Bhutta, Z. A., Kosek, M., Guerrant, R. L., Gottlieb, M., Miller, M., Kang, G., & Houpt, E. R. (2015). Pathogen-specific burdens of community diarrhea in developing countries: A multisite birth cohort study (MAL-ED). *The Lancet Global Health, 3*, e564–e575. https://doi.org/10.1016/S2214-109X(15)00151-5

Prüss-Ustün, A., Wolf, J., Bartram, J., Clasen, T., Cumming, O., Freeman, M. C., Gordon, B., Hunter, P. R., Medlicott, K., & Johnston, R. (2019). Burden of disease from inadequate water, sanitation and hygiene for selected adverse health outcomes: An updated analysis with a focus on low- and middle-income countries. *International Journal of Hygiene and Environmental Health, 222*, 765–777. https://doi.org/10.1016/j.ijheh.2019.05.004

Qadri, F., Akhtar, M., Bhuiyan, T. R., Chowdhury, M. I., Ahmed, T., Rafique, T. A., Khan, A., Rahman, S. I. A., Khanam, F., Lundgren, A., Wiklund, G., Kaim, J., Löfstrand, M., Carlin, N., Bourgeois, A. L., Maier, N., Fix, A., Wierzba, T., Walker, R. I., & Svennerholm, A.-M. (2020). Safety and immunogenicity of the oral, inactivated, enterotoxigenic Escherichia coli vaccine ETVAX in Bangladeshi children and infants: A double-blind, randomised, placebo-controlled phase 1/2 trial. *The Lancet Infectious Diseases, 20*, 208–219. https://doi.org/10.1016/S1473-3099(19)30571-7

Rojas-Lopez, M., Monterio, R., Pizza, M., Desvaux, M., & Rosini, R. (2018). Intestinal pathogenic Escherichia coli: Insights for vaccine development. *Frontiers in Microbiology, 9*, 440. https://doi.org/10.3389/fmicb.2018.00440

Roussel, C., Sivignon, A., De Wiele, T. V., & Blanquet-Diot, S. (2017). Foodborne enterotoxigenic *Escherichia coli*: From gut pathogenesis to new preventive strategies involving probiotics. *Future Microbiology, 12*, 73–93. https://doi.org/10.2217/fmb-2016-0101

Ruiz-Perez, F., Wahid, R., Faherty, C. S., Kolappaswamy, K., Rodriguez, L., Santiago, A., Murphy, E., Cross, A., Sztein, M. B., & Nataro, J. P. (2011). Serine protease autotransporters from *Shigella flexneri* and pathogenic *Escherichia coli* target a broad range of leukocyte glycoproteins. *Proceedings of the National Academy of Sciences, 108*, 12881–12886. https://doi.org/10.1073/pnas.1101006108

Sahl, J. W., Sistrunk, J. R., Baby, N. I., Begum, Y., Luo, Q., Sheikh, A., Qadri, F., Fleckenstein, J. M., & Rasko, D. A. (2017). Insights into enterotoxigenic Escherichia coli diversity in Bangladesh utilizing genomic epidemiology. *Scientific Reports, 7*, 3402. https://doi.org/10.1038/s41598-017-03631-x

Savarino, S. J., McKenzie, R., Tribble, D. R., Porter, C. K., O'Dowd, A., Cantrell, J. A., Sincock, S. A., Poole, S. T., DeNearing, B., Woods, C. M., Kim, H., Grahek, S. L., Brinkley, C., Crabb, J. H., & Bourgeois, A. L. (2017). Prophylactic efficacy of Hyperimmune bovine colostral antiadhesin antibodies against enterotoxigenic Escherichia coli diarrhea: A randomized, double-blind, placebo-controlled, phase 1 trial. *The Journal of Infectious Diseases, 216*, 7–13. https://doi.org/10.1093/infdis/jix144

Seo, H., & Zhang, W. (2020). Development of effective vaccines for enterotoxigenic Escherichia coli. *The Lancet Infectious Diseases, 20*, 150–152. https://doi.org/10.1016/S1473-3099(19)30631-0

Smith, J. L., Fratamico, P. M., & Gunther, N. W. (2014). Shiga toxin-producing Escherichia coli. In *Advances in applied microbiology* (pp. 145–197). Elsevier. https://doi.org/10.1016/B978-0-12-800262-9.00003-2

Steffen, R., Hill, D. R., & DuPont, H. L. (2015). Traveler's diarrhea: A clinical review. *JAMA, 313*, 71. https://doi.org/10.1001/jama.2014.17006

Stevens, M. P., & Frankel, G. M. (2014). The locus of enterocyte effacement and associated virulence factors of Enterohemorrhagic *Escherichia coli*. *Microbiology Spectrum, 2*, 2.4.16. https://doi.org/10.1128/microbiolspec.EHEC-0007-2013

Subramenium, G. A., Sabui, S., Marchant, J. S., Said, H. M., & Subramanian, V. S. (2019). Enterotoxigenic *Escherichia coli* heat labile enterotoxin inhibits intestinal ascorbic acid uptake via a cAMP-dependent NF-κB-mediated pathway. *American Journal of Physiology-Gastrointestinal and Liver Physiology, 316*, G55–G63. https://doi.org/10.1152/ajpgi.00259.2018

Sukwa, N., Bosomprah, S., Somwe, P., Muyoyeta, M., Mwape, K., Chibesa, K., Luchen, C. C., Silwamba, S., Mulenga, B., Munyinda, M., Muzazu, S., Chirwa, M., Chibuye, M., Simuyandi, M., Chilengi, R., & Svennerholm, A.-M. (2024). The incidence and risk factors for enterotoxigenic *E. Coli* diarrheal disease in children under three years old in Lusaka, Zambia. *Microorganisms, 12*, 698. https://doi.org/10.3390/microorganisms12040698

Tadesse, D. A., Zhao, S., Tong, E., Ayers, S., Singh, A., Bartholomew, M. J., & McDermott, P. F. (2012). Antimicrobial drug resistance in *Escherichia coli* from humans and food animals, United States, 1950–2002. *Emerging Infectious Diseases, 18*, 741–749. https://doi.org/10.3201/eid1805.111153

Taxt, A. M., Diaz, Y., Aasland, R., Clements, J. D., Nataro, J. P., Sommerfelt, H., & Puntervoll, P. (2016). Towards rational design of a toxoid vaccine against the heat-stable toxin of Escherichia coli. *Infection and Immunity, 84*, 1239–1249. https://doi.org/10.1128/IAI.01225-15

Turner, A. K., Stephens, J. C., Beavis, J. C., Greenwood, J., Gewert, C., Randall, R., Freeman, D., & Darsley, M. J. (2011). Generation and characterization of a live attenuated enterotoxigenic Escherichia coli combination vaccine expressing six colonization factors and heat-labile toxin subunit B. *Clinical and Vaccine Immunology, 18*, 2128–2135. https://doi.org/10.1128/CVI.05345-11

Ventola, C. L. (2015). The antibiotic resistance crisis: Part 1: Causes and threats. *P & T: A Peer-Reviewed. Journal for Formulary Management, 40*, 277–283.

Vidal, R. M., Muhsen, K., Tennant, S. M., Svennerholm, A.-M., Sow, S. O., Sur, D., Zaidi, A. K. M., Faruque, A. S. G., Saha, D., Adegbola, R., Hossain, M. J., Alonso, P. L., Breiman, R. F., Bassat, Q., Tamboura, B., Sanogo, D., Onwuchekwa, U., Manna, B., Ramamurthy, T., Kanungo, S., Ahmed, S., Qureshi, S., Quadri, F., Hossain, A., Das, S. K., Antonio, M., Mandomando, I., Nhampossa, T., Acácio, S., Omore, R., Ochieng, J. B., Oundo, J. O., Mintz, E. D., O'Reilly, C. E., Berkeley, L. Y., Livio, S., Panchalingam, S., Nasrin, D., Farag, T. H., Wu, Y., Sommerfelt, H., Robins-Browne, R. M., Del Canto, F., Hazen, T. H., Rasko, D. A., Kotloff, K. L., Nataro, J. P., & Levine, M. M. (2019). Colonization factors among enterotoxigenic Escherichia coli isolates from children with moderate-to-severe diarrhea and from matched controls in the global enteric multicenter study (GEMS). *PLoS Neglected Tropical Diseases, 13*, e0007037. https://doi.org/10.1371/journal.pntd.0007037

Vikram, K., & Vanneman, R. (2020). Maternal education and the multidimensionality of child health outcomes in India. *Journal of Biosocial Science, 52*, 57–77. https://doi.org/10.1017/S0021932019000245

Walker, R., & Dull, P. (2017). Combination vaccine strategies to prevent enteric infections. *Vaccine, 35*, 6790–6792. https://doi.org/10.1016/j.vaccine.2017.06.076

Williamson, D. A., Sidjabat, H. E., Freeman, J. T., Roberts, S. A., Silvey, A., Woodhouse, R., Mowat, E., Dyet, K., Paterson, D. L., Blackmore, T., Burns, A., & Heffernan, H. (2012). Identification and molecular characterisation of New Delhi metallo-β-lactamase-1 (NDM-1)- and NDM-6-producing enterobacteriaceae from New Zealand hospitals. *International Journal of Antimicrobial Agents, 39*, 529–533. https://doi.org/10.1016/j.ijantimicag.2012.02.017

Xie, W., Kumar, S., Kakon, S. H., Haque, R., Petri, W. A., & Nelson, C. A. (2019). Chronic inflammation is associated with neural responses to faces in bangladeshi children. *NeuroImage, 202*, 116110. https://doi.org/10.1016/j.neuroimage.2019.116110

Zhang, W., & Sack, D. A. (2015). Current progress in developing subunit vaccines against enterotoxigenic Escherichia coli-associated diarrhea. *Clinical and Vaccine Immunology, 22*, 983–991. https://doi.org/10.1128/CVI.00224-15

Zhang, Y., Tan, P., Zhao, Y., & Ma, X. (2022). Enterotoxigenic Escherichia coli: Intestinal pathogenesis mechanisms and colonization resistance by gut microbiota. *Gut Microbes, 14*, 2055943. https://doi.org/10.1080/19490976.2022.2055943

Chapter 6
Genetic Modulation of Pgh1 in Multidrug Resistance *Plasmodium falciparum* to Antimalarial Drugs Used as Prophylaxis in Pregnancy

Paul Akinniyi Akinduti and Oluwaferanmi Esther Ojo

1 Introduction

Plasmodium species, the principal etiological agents of malaria, a blood-borne infectious disease, constitute the most significant threat among endemic pathogens contributing to the prevalence of this condition. Every year, malaria affects about 25 million pregnant mothers in both high and low-endemic areas (WHO, 2020). Primigravidae and multigravidae are equally vulnerable to malaria infection (Khaja & Sequeira, 2021). Whether a malaria infection is symptomatic or not, it must be promptly diagnosed and treated to avoid low birth weight in neonates and maternal anemia (WHO, 2015). Similarly, in areas where the malaria parasite is common, like sub-Saharan Africa (SSA), prevention and treatment of malaria are crucial parts of prenatal care (WHO, 2015). However, the cost of treating pregnant women for malaria has continued to rise, both individually and nationally (Rogerson et al., 2018).

From the 1940s till the 2000s, basically the most widely used antimalarial in Africa was chloroquine (CQ), which served as *P. falciparum* malaria's first-line

P. A. Akinduti (✉)
Department of Biological Sciences, College of Sciences and Technology, Covenant University, Ota, Ogun State, Nigeria
e-mail: paul.akindutip@babcock.edu.ng

Covenant Applied Informatics and Communication Africa Centre of Excellence (CApIC-ACE), Ota, Ogun State, Nigeria

Department of Medical Laboratory Sciences, UniversityBabcock UniversityIllisshan, Nigeria

O. E. Ojo
Department of Biological Sciences, College of Sciences and Technology, Covenant University, Ota, Ogun State, Nigeria

Covenant Applied Informatics and Communication Africa Centre of Excellence (CApIC-ACE), Ota, Ogun State, Nigeria

P. O. Isibor, P. A. Akinduti (eds.), *Harnessing Biotechnology Tools for Product Development*, https://doi.org/10.1007/978-3-031-86002-7_6

treatment (Zhao et al., 2021). As chloroquine resistance spread in several African nations in the latter part of the 1990s and the beginning of the 2000s, chloroquine was substituted by artemisinin-based combination treatments (ACTs) (Zhao et al., 2021). In most African countries, artesunate-amodiaquine (AS-AQ), artemether-lumefantrine (AL), and dihydroartemisinin-piperaquine (DHA-PPQ) are the first-line treatments for *P. falciparum* (WHO, 2019a). Sulfadoxine-pyrimethamine (SP) is advised by the WHO for intermittent preventative treatment (IPTp and IPTi) of malaria in expectant mothers and younger children.

Drug-resistant parasites are a major challenge in Africa and globally in the fight against malaria (Quakyi et al., 2019). Antimalarial medication resistance is largely caused by *P. falciparum* Pgh1 regulation, and it poses a major risk to expectant mothers and their fetuses. Research has indicated that effective management of resistance requires an in-depth understanding of how evolving antimalarial medication strategies affect resistance (Hemming-Schroeder et al., 2018). For the purpose of creating efficient therapies and enhancing pregnancy outcomes in regions where malaria is prevalent, understanding the intricate connection between Pgh1 and a severe *falciparum* infection during pregnancy is crucial (Wicht et al., 2020b). The use of chemotherapy techniques, such as prophylactic and preventative therapy with antimalarial medications, can be useful treatments for the cure and prevention of malaria, to maintain consistent prevention (Plowe, 2022). Aryl-amino alcohol's modes of action and resistance are not entirely known, though there are various proposed mechanisms of action; researches are still ongoing.

Therefore, investigating changes in *P. falciparum* resistance to antimalarial drug markers in order to direct policy is very crucial. This study reviews the implication of *P. falciparum* Pgh1 gene (P-glycoprotein homologue 1) expression in multidrug resistance to aryl-amino alcohols used as prophylactics in pregnancy.

2 Epidemiology of Malaria in Pregnancy

Malaria transmission is moderate to high in sub-Saharan Africa, where an estimated 11 million out of 38 million (29%) pregnant mothers were exposed to the disease in 2018 (WHO, 2019a, b). An estimated 33.8 million births occurred in 2020 in 33 African nations with moderate to high malaria transmission rates, of which 11.6 million (34%) resulted from malaria exposure (WHO, 2021). Additionally, the predicted number of low birth-weight children delivered in 2020 was a result of pregnant women contracting malaria (WHO, 2022). It was also estimated that 39% of pregnant women in West Africa had malaria infection and projected that 41% of pregnant women in the Central Africa region and 27% of pregnant women in East and Southern Africa contracted malaria (WHO, 2022). According to estimates, *P. falciparum* results in 200,000 stillborn babies and 50,000 maternal deaths annually (WHO, 2022). The second trimester is when malaria prevalence in expectant mothers rises the most. Pregnancy-related malaria is another helpful indicator for community-based malaria surveillance (Brunner et al., 2019) with being young and

Table 6.1 Mechanism of action of aryl-amino alcohol antimalarials

Drug	Drug targets and mechanism	Site of action	References
Quinine	Purine nucleoside phosphorylase: An enzyme that prevents hemozoin from forming on its own. It prevents glycolysis and the formation of proteins and functions against *Plasmodium vivax* and *P. malariae* as a blood schizonticide and gametocytocidal agent	Digestive vacuole	Sanchez et al. (2008b)
Mefloquine	The mode of action is not yet fully entirely understood. According to a few publications, it binds directly to *P. falciparum's* cytoplasmic ribosome (80S-ribosome) to limit the production of parasite proteins, resulting in schizonticidal effects and causes damage to the parasite's membrane	Ribosome	Wong et al. (2017)
Halofantrine	Halofantrine is an orally administered blood schizontocide that is active against both chloroquine-sensitive and chloroquine-resistant plasmodia	Red blood cells	Scholar and Pratt (2000)
Chloroquine	The buildup of heme in the food vacuole is caused by chloroquine's inhibition of hemozoin production. After that, the liberated heme lyses the membrane, which kills the parasite and inhibits DNA and RNA synthesis	Food vacuole	Coban (2020)

primigravida mothers being common risk factors (Martínez-Perez et al., 2018; Berry et al., 2018). Table 6.1 shows the mechanism of action of aryl-amino alcohol antimalarials.

2.1 Declining Efficacy of Aryl-Amino as Prophylaxis for Malaria in Pregnancy

2.1.1 Quinine

For four centuries, people have used quinine, an alkaloid produced from cinchona, to cure malaria. Quinine is the earliest antimalarial medication and a potent blood schizonticidal drug. It has no effect on *P. vivax* or *P. ovale's* liver stages. *P. falciparum*, which is resistant to several medications including chloroquine, that are treated with it, is the cause of acute malaria attacks. When artesunate is unavailable, quinine is typically utilized with another medication, such as doxycycline. Quinine is also prescribed with clindamycin for first-trimester *P. falciparum* infections in pregnant women that are not complex, as well as for infections caused by *P. vivax* that are resistant to chloroquine (WHO, 2015).

Quinine has been an effective antimalarial medication for about 200 years, since its isolation and purification. The development of treatment resistance in malaria parasites has proven to be the largest barrier to control efforts. Quinine resistance was first documented in 1910, and its resistance usually occurs gradually and incompletely. There have been reports of reduced *P. falciparum* sensitivity to quinine in Asia and South America, despite the fact that this seems unlikely in Africa. Over the past three decades, quinine treatment-associated rates of malaria recurrence have not significantly changed, indicating that quinine's effectiveness has remained mostly stable.

The pharmacokinetic characteristics of quinine can cause treatment to be ineffective. Age, pregnancy, immune system, and disease severity all affect how a treatment works. Moreover, higher dosages are required when the patient's health improves since there is an increase in clearance and volume of distribution. It has been suggested that treating resistant *P. falciparum* infections with a larger dose might be beneficial on the third day of therapy. This protocol is hardly commonly used, though the quality of the drug is another significant factor contributing to treatment failure as substandard medications are a global issue. Studies on antimalarial medications found in several African nations have revealed that a significant fraction of them are of poor quality. Issues encompass elevated concentrations of contaminants, reduced than specified concentrations of the active component, and, occasionally, the total nonexistence of the active component. A further factor contributing to treatment failure is noncompliance by patients. Quinine has three daily dosages for 7 days of treatment; nevertheless, quinine's side effects may make compliance difficult. Nonetheless, quinine remains the advised therapy for *P. falciparum* infections and malaria in the first trimester of pregnancy. It is also an essential part of many medication regimens used in combination therapy (Achan et al., 2011).

2.1.2 Mefloquine

Mefloquine was produced by the U.S. Army in the late 1970s. It belongs to the family of antimalarial drugs called aryl-amino alcohol and has blood schizonticidal properties (Gonzalez et al., 2014).

Mefloquine is a derivative of 4-quinolinemethanol and is chemically related to quinine. When treating *P. falciparum* that is resistant to other medications, it works well. Mefloquine is the recommended medication for prophylaxis in malaria-endemic areas, with doxycycline serving as a standard substitute for anyone unable to take mefloquine. Pregnant women of all gestational ages who are visiting regions where malaria is endemic are advised to take mefloquine by the US Centers for Disease Control (CDC) and the World Health Organization (WHO) (CDC, 2019). Since the 1990s, mefloquine has been used orally as a highly efficient blood schizontocide for the treatment of acute *falciparum* malaria, as well as for malaria prophylaxis, under the trade name Lariam. This long-acting antimalarial drug is effective against *P. falciparum* that is resistant to chloroquine and sulfadoxine-pyrimethamine. Changes in the neurological system are among the major negative effects of mefloquine. Mefloquine, an antimalarial medication, is currently advised by the Centers for Disease Control and Prevention (CDC) for pregnant women as a therapy for malaria, as well as a preventative measure against malaria infection in all trimesters. Mefloquine was not previously advised for treating malaria in expectant mothers (CDC, 2019).

Mefloquine has many of the same characteristics that make sulfadoxine-pyrimethamine an ineffective antimalarial for sporadic prophylactic therapy during pregnancy (Gonzalez et al., 2014). It has been determined to be the best candidate for intermittent preventive treatment due to its long serum half-life (median of prophylactic dosages of 12–17 days), low rates of *Plasmodium* resistance in sub-Saharan Africa, ease of administration (one dose), appropriate animal research on the profile of reproductive toxicity, and pregnant women with a well-defined pharmacokinetic profile (Gonzalez et al., 2014). The US Centers for Disease Control and Prevention (CDC) and the French Reference Center on Teratogenic Agents (CRAT) are among the expert agencies that believe that mefloquine is suitable for chemoprophylaxis for pregnant women traveling to high-risk areas, regardless of gestational age.

2.1.3 Halofantrine

The blood schizontocide halofantrine, when taken orally, is effective against plasmodia that are resistant to and sensitive to chloroquine. When treating acute malarial attacks, it is phenanthrene methanol that is used in place of quinine and mefloquine and its use is limited by significant toxicity (Scholar and Pratt, 2000). The Walter Reed Army Institute of Research afterward developed halofantrine, an aryl-amino alcohol antimalarial medication that was initially recognized by the World War II Chemotherapy program as a possible antimalarial agent (Schuster and Canfield, 1989). It was initially sold as Halfan by Smith Kline and French in 1984. Halofantrine's chemical structure is [R,S]-1,3-dichloro-α-[2-{dibutylamino} ethyl]-6-fluoromethyl 1–9-phenanthrene methanol hydrochloride, with a molecular weight of 500.4 g/mol and its molecular formula is C26H30O2F3NO. Quinine, mefloquine, and lumefantrine are examples of additional class II schizonticidal medications that have a chemical structure with halofantrine, a racemate with two enantiomers. The drug was initially administered to treat acute malaria, but it was later found that some patients also had a potentially deadly prolongation of the QT interval, which may result in abrupt cardiac death and torsades de pointes, due to the drug's variable pharmacokinetic profile, especially its strong food effect. The medication was taken off the market as a result. Halofantrine was available in a number of oral administration forms, such as 250 mg tablets, flavored solutions, and as an intravenous formulation (Price et al., 2017).

3 Drug Resistance in Malaria Parasite During Pregnancy

As long as the subject is exposed to enough drugs at the site of action, the World Health Organization (WHO) provides an explanation for the parasite strain's capacity to survive and/or multiply even after consuming and absorbing medicine at doses that are comparable to or more than those that are typically advised but still within the subject's tolerance as antimalarial substance resistance. Resistance to antimalarial medications arises due to the parasites' selection for genetic alterations or gene amplifications that reduce sensitivity (Reyburn, 2010). A change in the composition, amount, or structure of a protein seems to be the origin of the resistance. Genetic modifications such as gene amplification or single nucleotide polymorphisms (SNPs) carry out protein alteration.

Malaria resurfaced in the early 1960s with the emergence of chloroquine resistance, and it persisted for decades in most countries. The aryl-amino alcohols, mefloquine, and halofantrine, as well as the chloroquine analog amodiaquine, were discovered by the U.S. Army through high-throughput screening and drug development activities. Since the late 1980s, their effectiveness has also decreased due to the expansion of parasites that have developed resistance. More recently, combination treatments have been employed in Southeast Asia for another derivative of chloroquine called piperaquine, which was created because of its ability to eliminate chloroquine resistance as a bis 4-aminoquinoline under the Chinese National Malaria Elimination Program (Blasco et al., 2017).

Various targets and modes of action are present in antimalarial medicines. Many medications like quinine, amodiaquine, chloroquine, and mefloquine act on heme in the parasite's feeding vacuole. By doing this, they prevent hemoglobin from polymerizing, which could have detrimental effects on the *plasmodium* parasite. In tropical locations, where there is a high requirement for treatment, the most often used medications are antimalarials. Which is sometimes made worse by noncompliant patients, this exerts enormous pressure on drug selection, forcing *P. falciparum* parasites to develop resistance mechanisms. The main way that parasites become resistant to antimalarial drugs is when they are subjected to high drug pressure, which causes mutational changes and an increase in the parasite medication targets' copy numbers (Anthony & Parija, 2016). Controlling and containing drug-resistant genotypes is becoming more difficult due to the rise of cross-resistant medications and multidrug resistance involving distinct genetic markers.

Antimalarial medication resistance is emerging as the most formidable obstacle to the effectiveness of antimalarial therapy, prompting scientists to constantly investigate solutions to this issue. There will always be resistant parasite strains, necessitating the ongoing production of new compounds. Clinical trials for new medications with unique mechanisms of action are starting (Kumar et al., 2018).

3.1 Pgh1 Expression in Plasmodium falciparum

The most popular antimalarial medication for many years, chloroquine (CQ), is now essentially ineffective due to resistance that has been documented in areas where malaria is prevalent; in fact, chloroquine resistance (CQR) has been implicated as a major cause of the increased worldwide malaria incidence (Greenwood, 2002). Recent research has connected two genes to resistance to the antimalarial medication; chloroquine, *Plasmodium falciparum* multidrug resistance 1 transporter (pfmdr1), which modifies CQR and resistance to quinoline methanols and similar compounds, and *Plasmodium falciparum* chloroquine resistance transporter (pfcrt), which imparts CQR in *P. falciparum* (Lakshmanan et al., 2015).

According to Johnson et al. (2008), pfmdr1, which encodes *P. falciparum* P-glycoprotein homologous protein 1 (Pgh1), was discovered more than 10 years prior to the discovery of pfcrt. Genetic studies have revealed that, in contrast to pfcrt, several multidrug resistances have been associated with mutations in pfmdr1. These mutations have been found to have a greater impact on a range of other antimalarial medications, including mefloquine, halofantrine, and artemisinin used to treat parasites, than they do on CQ susceptibility (Sidhu et al., 2006). Studies conducted on *P. falciparum* have focused on this phenomenon and observed that pfmdr1 expression and CQR were correlated (Foote et al., 1989).

Subsequent investigations, however, were unable to support this finding. Consequently, it has been established beyond dispute that, in drug-pressed laboratory lines and field isolates, resistance to mefloquine is more strongly correlated with pfmdr1 copy number and Pgh1 expression than with CQR (Nelson et al., 2005). Following proof of Pgh1's capacity to pump an anticancer medication, Saliba et al. (2008) reported pgh1 capacity to transport several antimalarial drugs using the oocyte system. Unlike the case of vinblastine, when expression of all Pgh1 variants resulted in a similar reduction in the accumulation of the antimalarials tested between the different forms of the protein.

As the wild-type pfmdr1 allele expression does not significantly affect drug accumulation in the instance of halofantrine, expression of the various pfmdr1 alleles linked to increased antimalarial sensitivity significantly reduced drug accumulation. Up to four amino acid changes, occurring at five distinct polymorphic locations were present in the mutant alleles that set them apart from the wild-type form. It is implied that halofantrine can be transported by variants of the protein linked to increased drug sensitivity, but not by the wild-type version (or, if it does, the transport happens too slowly to make a significant impact on the accumulation of drugs in oocytes) (Wicht et al., 2020a, b). XR-9576 and ONT-093, two mammalian P-glycoprotein inhibitors, were shown to reverse the decrease in halofantrine accumulation seen in oocytes that contain a Pgh1 mutant. Furthermore, it was directly found that oocytes expressing the variation Pgh1 had an increased (XR-9576-sensitive) efflux of halofantrine. The pattern of accumulation for quinine and chloroquine differed significantly from that of halofantrine. Drug accumulation for these two medications was significantly (albeit somewhat) reduced when the wild-type pfmdr1 allele was expressed in oocytes; this reduction was counteracted by XR-9576 and ONT-093. Using a synthetic construct that was unique from Pgh1's wild-type form in just one amino acid at one of the five polymorphism locations, the oocyte quinine accumulation at 400 nm was somewhat reduced when the new type of Pgh1 was expressed, but not at the lower concentration of 100 nm. Conversely, there was no appreciable effect on quinine or chloroquine accumulation when the expression was observed for pfmdr1 alleles from parasite strains with modified drug sensitivity. The research study by Saliba et al. (2008) provides fresh information on the potential impact of Pgh1 on antimalarial drug sensitivity. It demonstrates that halofantrine is transported by Pgh1 mutations linked to enhanced susceptibility to chloroquine and quinine, although wild-type Pgh1 is not susceptible to either drug. During the oocyte tests, the medicines that were carried were pushed from the area in the oocyte cytoplasm, where the protein's ATP binding domains are visible, and into the extracellular solution (Saliba et al., 2008). Thus, it is anticipated that the transferred species will be pumped from the cytoplasm into the digestive vacuole via Pgh1 on the parasite's digestive vacuole membrane, increasing the drug's intravacuolar concentration. If the target mechanisms of halofantrine, quinine, and chloroquine are inside the digestive vacuole, therefore, Pgh1-induced increases in intravacuolar concentration ought to render the parasite more susceptible to these drugs (Saliba et al., 2008).

3.2 Pgh1 Protein and its Role in Drug Resistance

Within the feeding vacuole of *P. falciparum*, a PfMDR1 transporter is produced in response to the pfmdr1 gene. The organism possesses a P-glycoprotein homologue-1 (Pgh1) transporter that is resistant to several drugs. This transporter is structurally similar to the human multidrug resistance gene (Adamu et al., 2020). The degree of quinoline antimalarial drug sensitivity of *P. falciparum* strains has been associated with mutations in the Pfcrt and Pfmdr1 genes including mefloquine, lumefantrine, amodiaquine, and chloroquine (Agomo et al., 2016).

P-glycoprotein homologue 1 (Pgh1), also known as the *P. falciparum* multidrug resistance 1 (pfmdr1) gene, has been connected to parasite susceptibility, which is situated on chromosome 5 (Veiga et al., 2016). Single-nucleotide polymorphisms (SNPs) for pfmdr1 have been identified; the top five are N86Y, Y184F, S1034C, N1042D, and D1246Y (Wicht et al., 2020a, b). Previous research has found that for these five SNPs, the most common mutation in Asian and African parasites, codon 86 (N86Y), substitutes tyrosine for asparagine and increases the parasite's resistance to CQ and AQ while also making it more susceptible to a variety of first-line antimalarial medications, such as LMF, MFQ, and dihydroartemisinin (DHA) (Huang et al., 2016; Veiga et al., 2016). Fascinatingly, a few years after ACT was implemented, earlier research also revealed a significant decrease in the 86Y point mutation (Maiga et al., 2021). Thus, monitoring shifts in the frequency of pfmdr1 SNPs may be able to detect resistance to ACT before it becomes apparent.

Reduced inflow of certain antimalarial medications and their intracellular accumulation are linked to mutations in pfmdr1 (Adamu et al., 2020). Single nucleotide polymorphisms (SNPs), which are particular changes in the pfmdr1 gene, are linked to the resistance to aminoquinolines (Sekihara et al., 2018). Changes in the parasite's susceptibility to antimalarial medications have been associated with many codons in pfmdr1; in sub-Saharan Africa, for example, the codons N86Y, Y184F, and D1246Y are particularly linked to modifications in the parasite's response to lumefantrine (LUM) and amodiaquine (AQ) (Adamu et al., 2020). Although the pfmdr1 86Y allele was found to be strongly linked to resistance to both amodiaquine (AQ) and chloroquine (CQ) (Folarin et al., 2011), the presence of 1246Y alleles was discovered to provide resistance to quinine (QN) and increase the parasites' susceptibility to artemisinin (ART), halofantrine (HF), and mefloquine (MQ). The identification of the gene encoding the transporter for chloroquine resistance (Pfcrt) revolutionized the scientific understanding of malaria defense. Other polymorphic amino acids can be found between lysine and threonine; however, the most reliable genetic marker among them all is a mutation in the Pfcrt codon K76T gene, which may be crucial for coding the resistance composition. Nevertheless, changes resulting in chloroquine resistance might happen anywhere along codons 72–76. A mutation in the codon N86Y chromosomal causes the multidrug-resistant *P. falciparum* strain Pfmdr-1 to be linked to resistance to medications to quinine, halofantrine, chloroquine, and mefloquine. Some studies have found that a single nucleotide polymorphism (SNP) at codon 86 (N86Y) that results in a base shift from asparagine to tyrosine is connected to chloroquine resistance in both vitro and in vivo settings (Olasehinde et al., 2019).

3.3 Arylaminoalcohols and 4-Aminoquinolines Susceptibility in **P. falciparum***: A Modulator of Pfmdr1*

According to Veiga et al. (2016), two transporters located in the digestive vacuole, *P. falciparum* chloroquine resistance transporter (pfCRT) and pfmdr1, which generates the protein P-glycoprotein homologue-1 (Pgh-1), are largely responsible for the development of drug resistance. PfMDR1 is the gene that codes for the *P. falciparum* multidrug resistance 1, a transporter homolog for P-glycoprotein, pfmdr1, is another element that contributes to resistance to heme-targeting antimalarials (Valderramos et al., 2010). Similar to PfCRT, PfMDR1 is found on the membrane of the digesting vacuole (DV), although transport is assumed to be directed inward toward the DV.

Six predicted helices are followed by a hydrophilic nucleotide-binding pocket in each of the two membrane-spanning homologous domains of PfMDR1. Topologically, PfMDR1 is similar to a normal P-glycoprotein-type ABC transporter (Sanchez et al., 2008a). The cytosolic side of the DV appears to have this binding domain, which may be where it initially interacts with antimalarial drugs. An investigation into the function of the transporter proteins pfcrt and pfmdr1 in connection to quinine's in vitro action resulted from the observation that parasites are more susceptible to mefloquine, halofantrine, lumefantrine, quinine, and artemisinin derivatives when their pfmdr1 copy number is reduced. Two potential transporters that localize to the digestive vacuole membrane are encoded by the Pfcrt and Pfmdr1 genes, which weigh 45 and 162 kDa, respectively. Pfcrt has ten transmembrane domains and belongs to the "drug metabolite transporter family," while Pfmdr1, on the other hand, is made up of two homologous halves, each of which has six putative transmembrane domains in addition to a conserved nucleotide-binding domain (ATB binding domain). Polymorphisms, mostly at codons N1042D, S1034C, or

D1246Y, were linked to increased resistance to quinine in Pfmdr1 transfection experiments (Okombo et al., 2010). Field isolates have also supported this association (Andriantsoanirina et al., 2010; Nsobya et al., 2010), although other studies found no correlation (Price et al., 2006). According to Chaijaroenkul et al. (2010), a higher Pfmdr1 number decreases sensitivity to QN and other aryl-amino quinolines such as lumefantrine and mefloquine, Pfmdr1 hence influences quinine action.

4 Impacts of Drug Resistance on Malaria Treatment Outcomes in Pregnancy

Malaria is a major cause of morbidity and death in Nigeria and other endemic countries, especially among pregnant women (USAID, 2016). As per the definition provided by WHO (WHO/CDS/CSR/DRS/2001.4), antimalarial drug resistance refers to the capacity of a parasite strain to endure and proliferate even after being administered and absorbed with a drug at a dosage that is within the subject's tolerance range and above usual recommendations. This means that the medication is taken according to the prescribed dosage and schedule. Despite quinine being used for almost 400 years, there is no evidence of global resistance, despite a number of studies indicating the formation of quinine resistance in vivo. An analysis of clinical trials examining quinine's antimalarial effect, for example, revealed that, over a 30-year span, from the 1970s to early 2000, the drug's effectiveness remained constant in the majority of endemic areas. In contrast, mefloquine, pyrimethamine, and chloroquine's effectiveness significantly declined during the same time frame (Myint et al., 2004). Africa has demonstrated resistance, with reports of a reduced cure rate of 80–60% (Achan et al., 2009). Nevertheless, further examination of these data suggests that in vivo resistance in Africa might not be as great as previously stated.

Malaria infection in pregnancy is a serious obstetric issue that puts both the unborn infant and expectant mother at risk. Clinical study results showed that 25% of women who had placental malaria also got malaria during childbirth (WHO, 2015). Malaria has equally terrible consequences on infants who get it congenitally. According to estimates, maternal malaria is the cause of 1 in 5 low birth weight (≤2500 g) births. Around 100,000 infants die as a result of this every year, which inevitably leads to poor infant development and survival (Bassey et al., 2015; WHO, 2015). Malaria is a public health emergency; however, in recent years, the parasites have become resistant to the majority of widely used preventive and therapeutic antimalarial drugs (WHO, 2015). Pregnancy-related physiological characteristics and malarial pathology work in concert to increase the possibility of adverse consequences for the mother and the fetus (USAID, 2016).

Compared to multigravidae, primigravidae had a much greater frequency of malaria. This outcome is in line with other comparable studies conducted in tropical regions where malaria is common, which found that primigravidae had a greater incidence of *P. falciparum* malaria than multigravidae (USAID, 2016). While primigravidae are still susceptible to malaria due to incomplete immunity development, multigravidae's decreased malaria prevalence may be linked to their development of pre-immunity with higher parity (Uneke et al., 2008). Pregnant women in their second trimester had the greatest rate of malaria compared to those in their first and third trimesters. Similar findings were reported by authors in different sub-Saharan African locations, with women reporting the maximum parasite density and frequency of malaria in their second trimester (USAID, 2016).

There have been reports of malaria resistance in Nigeria in the past to antimalarial drugs based on artemisinin and chloroquine (Olasehinde et al., 2014). The WHO Guidelines, which strongly advise quinine in conjunction with clindamycin for 7 days, are in conflict with the use of quinine alone (WHO, 2015). Quinine usage in pregnancy has disadvantages even if it is thought to be safe in the first trimester when taken at the recommended dosage (WHO, 2015; Saito et al., 2018). Pregnant women had higher quinine clearance than nonpregnant women (Tarning et al., 2013). In comparison to ACT (Saito et al., 2017) or quinine plus clindamycin regimens (Saito et al., 2018), five randomized studies were meta-analyzed, and the results showed that a seven-day course of quinine monotherapy was associated with a higher risk of gametocyte carrier emergence risk and a delayed rate of malaria parasite clearance. Quinine's gametocidal action against *falciparum* malaria has been shown to be less effective than ideal at stopping the disease's spread in endemic locations (Achan et al., 2011). Since parenteral artesunate was found to be more effective in preventing deaths from severe malaria than quinine in a randomized comparative trial (Dondorp et al., 2010), quinine is no longer advised as the treatment of choice for severe malaria. This is because resistance markers linked to the quinoline family of drugs, including quinine (Pfmdr N86Y and Pfmdr Y184F), were found. Early studies showed that quinine and quinine structural analogs including lumefantrine, mefloquine, halofantrine, and chloroquine exhibited in vitro cross-resistance. This finding raised the possibility that these medications shared certain resistance mechanisms (Uhlemann & Krishna, 2005).

4.1 Consequences of Antimalaria Resistance Used as Prophylaxis

Pregnant women in high-transmission regions are more vulnerable to malaria infection because of altered hormones and weak immune systems, enhancement of mosquito attraction, and placental sequestration of the parasite (Rogerson et al., 2007). The degree of a woman's pre-acquired immunity to malaria and other factors, including the number of previous pregnancies in which she was exposed to the virus and her elevated risk as a primigravida, determine how severe pregnancy-related malaria's effects will be. According to Desai et al. (2007), acquired immunity is reliant on the degree of transmission of malaria in the area where one lives.

Fetal anemia, congenital malaria, and a higher risk of adverse pregnancy outcomes, such as miscarriage, stillbirth, and early delivery, are among the impacts on the fetus and baby. Furthermore, low birth weight and intrauterine growth retardation can result from placental infection; these outcomes can then have a negative impact on the development of the child's cognitive and behavioral abilities (WHO, 2022).

4.1.1 Miscarriage

By definition, as stated by the World Health Organization, a miscarriage occurs when a fetus dies in the first two trimesters of pregnancy, or before 28 weeks of gestation. The gestational age of viability, which is often defined as between 24 and 28 weeks, has been widely accepted to be the age at which pregnancy losses are classified as miscarriages. Preventing miscarriages is important for women and their future pregnancies since they are associated with a higher risk of preterm birth (PTB), fetal development limitation, and other obstetric difficulties in subsequent pregnancies (Lancet, 2021). One of the malaria in pregnancy (MiP) consequences that has received less research is miscarriage, in part because endemic locations have lower rates of early pregnancy clinic attendance (Huynh et al., 2015). In the first trimester, a single case of either *vivax* or *falciparum* malaria, whether symptomatic or not, may increase the risk of miscarriage, according to research done on the Thai–Myanmar border (Moore et al., 2016). Miscarriage was more common approximately in 16% of pregnancies among women with *falciparum* malaria infection, as compared with 11% in those who were infected with *vivax* and 9% in those who were not. A cohort in Mali had a loss rate of 12.1% in 43 out of 358 pregnancies; 65.1% of these miscarriages happened in the first trimester of pregnancy (Gaoussou et al., 2022). Although the prevalence of first-trimester miscarriages in a Mali region where malaria is endemic was demonstrated, the study did not assess malaria infection. High levels of parasitemia (aOR 1.49 (1.25–1.78) for every tenfold rise in parasitemia) and infected women who have severe malaria (adjusted odds ratio (aOR) 3.63, 95% confidence interval (CI) 1.15–11.46) are at an increased risk of miscarriage (McGready et al., 2012).

4.1.2 Stillbirth

Prenatal malaria is a key cause of 98% of the estimated 2.6 million stillbirths that take place each year in countries with low and medium incomes (Lawn et al., 2016). 20% of stillbirths in sub-Saharan Africa are thought to be caused by *P. falciparum* malaria during pregnancy (Moore et al., 2017a, b). Stillbirths are pregnancy losses that happen beyond 28 weeks of pregnancy and are classified as happening beyond the gestational age of viability. Numerous complications that arise from placental infection, such as reduced placental perfusion, anemia in the mother, and premature delivery, might result in malaria-induced stillbirth (Yatich et al., 2010). An increased incidence of stillbirth was predicted by malaria infection in observational research conducted in Mali, with an adjusted hazard ratio of 3.87 (Mahamar et al., 2021) *P. falciparum* found and treated during pregnancy was likewise linked to stillbirth, but less so than untreated infection, according to a meta-analysis including nineteen nations (Moore et al., 2017a, b). The risks of stillbirth were higher when *P. vivax* malaria was discovered after birth, but not at the time of diagnosis and treatment. The time of infection is important as well; in Uganda, malaria was found to be linked to a two-fold increased risk of stillbirth during the first 2 weeks following delivery (De Beaudrap et al., 2013). Among pregnant women in Sudan who reported having malaria in the first or second trimester, the chance of stillbirth was considerably higher. Prenatal data from expectant mothers along the Thai-Burmese border, however, indicate that a first-trimester malaria episode does not always increase the risk of a subsequent stillbirth (McGready et al., 2012). As a result, it is unclear how exactly first-trimester malaria causes stillbirth and if this varies between parasite strains or communities. Insecticide-treated nets (ITNs) and intermittent preventative therapy (IPTp) have the potential to reduce stillbirths by 22%, and preventative interventions help to reduce stillbirths and the difficulties that go along with them (Ishaque et al., 2011). Prenatal care visits are particularly crucial; in Malawi, stillbirth was linked to less than five visits in regions with high malaria prevalence (Kalanda et al., 2006). Even with the extensive use of IPTp-SP, among primigravidae, malaria was common and raised the chance of stillbirth around four times in a longitudinal cohort of pregnant women in Mali. These helpful techniques do not, therefore, fully address the MiP issue (Mahamar et al., 2021).

4.1.3 Perinatal Death

The term "perinatal death" includes stillbirth and early neonatal death, and it describes a baby's death that takes place between the 28th week of gestation and 7 days after birth (Getiye & Fantahun, 2017). In regions where malaria is common, perinatal death rates are frequently higher, with an estimated prenatal mortality rate of 34.7 per 1000 babies (Breman et al., 2004). Since many prenatal fatalities in malaria-endemic countries go undetected, it is thought that true perinatal mortality rates are significantly greater than those that are recorded (van Geertruyden et al., 2004). A study conducted in the Democratic Republic of the Congo (previously known as Zaire) discovered that low birth weight (LBW; risk ratio = 12.4), as well as perinatal death, were doubled by maternal malaria (Nyirjesy et al., 2023). *Falciparum* and *vivax* MiP raise mortality risk by 2.55 and 1.98 times, respectively, on the Thai-Myanmar border (Moore et al., 2017a, b). Placental malaria infection is known to raise the risk of LBW, indicating a connection between perinatal mortality and malaria (Mahamar et al., 2021). Moreover, decreased uteroplacental blood flow has been demonstrated to be predictive of perinatal mortality in research including pregnant women in Kenya (Dorman et al., 2002). Third-trimester malaria infection has been associated with abnormal uterine artery flow velocity waveforms, indicating that the pathophysiology of maternal erythrocytes infected in the placenta may be connected to altered uteroplacental blood flow. When *falciparum* infections are successfully prevented, primigravidae have a 27% lower risk of perinatal death (Desai et al., 2007). Chloroquine prophylaxis was demonstrated to protect against perinatal mortality (risk ratio = 0.38), even after it was no longer effective as an antimalarial (Nyirjesy et al., 2023). While more complete perinatal mortality statistics are required to precisely assess the efficacy of prevention efforts, it is evident that novel strategies are required to lessen the incidence of pregnancy-related malaria sequelae.

5 Current Strategies and Considerations for Monitoring Gene Mutation in Antimalaria Used as Prophylaxis

In order to effectively treat malaria and restrict its transmission, therapeutic activity and drug resistance processes can be better understood by delving further into the drug resistance genes and their molecular markers. The K76T mutation in *P. falciparum* has been shown to be responsible for the chloroquine resistance transporter gene (pfcrt) and is intimately linked to CQ resistance. *P. falciparum* that has this mutation typically also carries the mutations M74I, N75E, and A220S (Sa and Twu, 2010). An increased tendency for mutations at pfcrt gene positions 74–76 was also seen in the set of malaria specimens from both China and other countries. Furthermore, Yingjiang County in Yunnan Province and the Kachin area of Myanmar were reported to have mutations at loci Q271E, N326S, I356T, and R371I. These results may be explained by the regular population mobility in border regions, which can disperse drug resistance to parasites (Wang et al., 2020).

Y184F and N86Y mutations in pfmdr1 were often found, and in the majority of cases, these mutations happened at the same time, according to the data gathered. Although the F1226Y mutation was only found in a small number of samples, studies have shown that it improves the resistance of parasites to lumefantrine and quinine (Wang et al., 2020; Duan et al., 2022). Significantly, research has shown that parasites carrying the N1042D mutation are more susceptible to pyronaridine. Nevertheless, the imported African samples included the D1246Y mutation, suggesting that the evolution of malaria resistance varies by area (Zhao et al., 2021). Parasite resistance to SP medications is linked to mutations in dihydropteroate synthase (dhps) and dihydrofolate reductase (dhfr) (Triglia et al., 1997). According to the study, pyrimethamine resistance was linked to mutations in the pfdhfr gene's N51I, C59R, S108N, and I164L (Happi et al., 2005). These pfdhfr and pfdhp locations that we gathered have a significant mutation frequency. Studies conducted in Southeast Asia (Myanmar, Thailand, and Cambodia) have found comparable trends to this phenomenon (Huang et al., 2012). *P. vivax*, an indigenous species in Southeast Asia, is thought to be resistant to SP. In contrast to native samples, Lower mutation frequency at I164L and greater mutation frequency (>80%) in N51I, C59R, and S108N of pfdhfr are seen in imported *falciparum* malaria samples from Africa. It could suggest that high levels of pyrimidine resistance have evolved in native parasites. Therefore, it is imperative that the equivalent resistance to imported malaria in Africa be constantly monitored. It has been shown that the kelch13 (K13) propeller domain of *P. falciparum* is associated with artemisinin resistance (Miotto et al., 2013). It has been established that artemisinin resistance is closely associated with the pfK13 gene mutations at the F446I, N458Y, M476I, Y493H, R539T, I543T, P553L, R561H, and C580Y sites (Feng et al., 2019). It was discovered that the pfK13 gene's F446I mutation predominates in the border regions of Yunnan Province, China, and Myanmar. Additionally, imported African samples had higher frequencies of pfK13 mutations at the A578S and P574L loci. Despite the fact that there haven't been any native cases of malaria in China since 2017, due to human importation from malaria-endemic countries in Southeast Asia and Africa, as well as cross-border malaria transmission by *Anopheles* mosquitoes in the China-Myanmar border area, it is challenging to manage malaria and monitor treatment resistance (Feng et al., 2019).

5.1 Conclusion

The resistance mechanisms against substances that interfere with parasite digestive vacuole activities, such as the synthesis of hemozoin and the breakdown of hemoglobin, are mostly associated with changes in the transmembrane proteins PfCRT and PfMDR1 that misdirect the anti-malaria activities from the site of action. *P. falciparum* Pgh1 modulates multi-drug resistance to quinine, mefloquine, and halofantrine and underscores the complexity of drug resistance mechanisms in malaria parasites. The most cost-effective and feasible antimalaria approach for pregnant women includes the use of IPTp-SP and ITNs, and also access to prompt diagnosis and treatment, which can be made widely available, affordable, and supported by government programs in many malaria-endemic countries, including Nigeria. Further studies on the regulation and function of Pgh1 could provide valuable insights into developing drug targets for the elimination of drug-resistant malaria strains. Understanding the interplay between parasite Pgh1 modulation and drug resistance is essential for advancing malaria treatment and control efforts globally. The type of haplotype that predominates in field isolates is influenced by a combination of fitness costs.

References

Achan, J., Tibenderana, J. K., Kyabayinze, D., Mangen, F. W., Kamya, M. R., Dorsey, G., D'Alessandro, U., Rosenthal, P. J., & Talisuna, A. O. (2009). Effectiveness of quinine versus artemether-lumefantrine for treating uncomplicated *falciparum* malaria in Ugandan children: Randomised trial. *BMJ, 339*(7715), 283–289.

Achan, J., Talisuna, A. O., Erhart, A., Yeka, A., Tibenderana, J. K., Baliraine, F. N., Rosenthal, P. J., & D'Alessandro, U. (2011). Quinine, an old anti-malarial drug in a modern world: Role in the treatment of malaria. *Malaria Journal., 10*(144), 1–12.

Adamu, A., Jada, M. S., Haruna, H. M. S., Yakubu, B. O., Ibrahim, M. A., Balogun, E. O., Sakura, T., Inaoka, D. K., Kita, K., Hirayama, K., Culleton, R., & Shuaibu, M. N. (2020). *Plasmodium falciparum* multidrug resistance gene-1 polymorphisms in Northern Nigeria: Implications for the continued use of artemether-lumefantrine in the region. *Malaria Journal., 19*(1), 439–459.

Agomo, C. O., Oyibo, W. A., Sutherland, C., Hallet, R., & Oguike, M. (2016). Assessment of Markers of Antimalarial Drug Resistance in *Plasmodium falciparum* Isolates from Pregnant Women in Lagos, Nigeria. *PLoS One., 11*(1), 1–10.

Andriantsoanirina, V., Ménard, D., Rabearimanana, S., Hubert, V., Bouchier, C., Tichit, M., Bras, J. L., & Durand, R. (2010). Association of microsatellite variations of *Plasmodium falciparum* Na+/H+ exchanger (Pfnhe-1) gene with reduced in vitro susceptibility to quinine: Lack of confirmation in clinical isolates from Africa. *American Journal of Tropical Medicine and Hygiene., 82*(5), 782–787.

Anthony, H. A., & Parija, S. C. (2016). Antimalarial drug resistance: An overview. *Tropical Parasitology., 6*(1), 30–41.

Bassey, G., Nyengidiki, T. K., & John, C. T. (2015). Prevalence of placenta *Plasmodium* parasitemia and pregnancy outcome in asymptomatic patients at delivery in a university teaching hospital in Nigeria. *Nigeria Journal of Clinical Practice., 18*(1), 27–32.

Berry, I., Walker, P., Tagbor, H., Bojang, K., Coulibaly, S. O., Kayentao, K., Williams, J., Oduro, A., Milligan, P., Chandramohan, D., Greenwood, B., & Cairns, M. (2018). Seasonal dynamics of malaria in pregnancy in West Africa: Evidence for carriage of infections acquired before pregnancy until first contact with antenatal care. *American Journal of Tropical Medicine and Hygiene., 98*(2), 534–542.

Breman, J. G., Alilio, M. S., & Mills, A. (2004). Conquering the intolerable burden of malaria: what's new, what's needed: a summary. *The American Journal of Tropical Medicine and Hygiene, 71*(2 Suppl), 1–15.

Blasco, B., Leroy, D., & Fidock, D. A. (2017). Antimalarial drug resistance: Linking *Plasmodium falciparum* parasite biology to the clinic. *Nature Medicine., 23*(8), 917–928.

Brunner, N. C., Chacky, F., Mandike, R., Mohamed, A., Runge, M., Thawer, S. G., Ross, A., Vounatsou, P., Lengeler, C., Molteni, F., & Hetzel, M. W. (2019). The potential of pregnant women as a sentinel population for malaria surveillance. *Malaria Journal., 18*(1), 370–388.

CDC. (2019). Guidelines for treatment of malaria in the United States. Retrieved Jun 15, 2020, from https://www.cdc.gov/malaria/diagnosis_treatment/index.html

Chaijaroenkul, W., Wisedpanichkij, R., & Na-Bangchang, K. (2010). Monitoring of in vitro susceptibilities and molecular markers of resistance of *plasmodium falciparum* isolates from Thai-Myanmar border to chloroquine, quinine, mefloquine and artesunate. *Acta Tropica., 113*(2), 190–194.

Coban, C. (2020). The host targeting effect of chloroquine in malaria. *Current Opinions in Immunology., 66*(1), 98–107.

De Beaudrap, P., Turyakira, E., White, L. J., Nabasumba, C., Tumwebaze, B., Muehlenbachs, A., Guérin, P. J., Boum, Y., McGready, R., & Piola, P. (2013). Impact of malaria during pregnancy, on pregnancy outcomes in a Ugandan prospective cohort with intensive malaria screening and prompt treatment. *Malaria Journal., 12*(1), 139–139.

Desai, M., ter Kuile, F. O., Nosten, F., McGready, R., Asamoa, K., Brabin, B., & Newman, R. D. (2007). Epidemiology and burden of malaria in pregnancy. *Lancet Infectious Diseases., 7*(2), 93–104.

Dondorp, A. M., Fanello, C. I., Hendriksen, I. C. E., Gomes, E., Seni, A., Chhaganlal, K. D., Bojang, K., Olaosebikan, R., Anunobi, N., Maitland, K., Kivaya, E., Agbenyega, T., Nguah, S. B., Evans, J., Gesase, S., Kahabuka, C., Mtove, G., Nadjm, B., Deen, J., Mwanga-Amumpaire, J., Nansumba, M., Karema, C., Umulisa, N., Uwimana, A., Mokuolu, O. A., Adedoyin, O. T., Johnson, W. B. R., Tshefu, A. K., Onyamboko, M. A., Sakulthaew, T., Pan Ngum, W., Silamut, K., Stepniewska, K., Woodrow, C. J., Bethell, D., Wills, B., Oneko, M., Peto, T. E., von Seidlein, L., Day, N. P. J., & White, N. J. (2010). Artesunate versus quinine in the treatment of severe *falciparum* malaria in African children (AQUAMAT): An open-label, randomised trial. *Lancet, 376*(9753), 1647–1657.

Dorman, E. K., Shulman, C. E., Kingdom, J., Bulmer, J. N., Mwendwa, J., Peshu, N., & Marsh, K. (2002). Impaired uteroplacental blood flow in pregnancies complicated by *falciparum* malaria. *Ultrasound in Obstetrics and Gynecology., 19*(2), 165–170.

Duan, M., Bai, Y., Deng, S., Ruan, Y., Zeng, W., Li, X., Wang, X., Zhao, W., Zhao, H., Sun, K., Zhu, W., Wu, Y., Miao, J., Kyaw, M. P., Yang, Z., & Cui, L. (2022). Different in vitro drug susceptibility profile of *Plasmodium falciparum* isolates from two adjacent areas of Northeast Myanmar and molecular markers for drug resistance. *Tropical Medicine and Infectious Disease., 7*(12), 442–445.

Folarin, O. A., Bustamante, C., Gbotosho, G. O., Sowunmi, A., Zalis, M. G., Oduola, A. M. J., & Happi, C. T. (2011). In vitro amodiaquine resistance and its association with mutations in pfcrt and pfmdr1 genes of *Plasmodium falciparum* isolates from Nigeria. *Acta Tropica., 120*(3), 224–230.

Foote, S. J., Thompson, J. K., Cowman, A. F., & Kemp, D. J. (1989). Amplification of the multidrug resistance gene in some chloroquine-resistant isolates of *P. falciparum. Cell, 57*(6), 921–930.

Gaoussou, S., Attaher, O., Swihart, B., Traore, M., Diarra, S., Soumbounou, I. H., Ndiaye, O., Issiaka, D., Morrison, R., Mahamar, A., Duffy, P. E., Dicko, A., & Fried, M. (2022). Pregnancy outcomes in a malaria-exposed Malian cohort of women of child-bearing age. *Frontiers in Medicine (Lausanne)., 9*(1), 106–1538.

Getiye, Y., & Fantahun, M. (2017). Factors associated with perinatal mortality among public health deliveries in Addis Ababa, Ethiopia, an unmatched case control study. *BMC Pregnancy and Childbirth, 17*(1), 245.

Gonzalez, R., Hellgren, U., Greenwood, B., & Menéndez, C. (2014). Mefloquine safety and tolerability in pregnancy: A systematic literature review. *Malaria Journal., 13*(1), 75–75.

Greenwood, B. (2002). The molecular epidemiology of malaria. *Tropical Medicine and International Health., 7*(12), 1012–1021.

Happi, C., Gbotosho, G., Folarin, O., Akinboye, D., Yusuf, B., Ebong, O., Sowunmi, A., Kyle, D., Milhous, W., Wirth, D., & Oduola, A. (2005). Polymorphisms in Plasmodium falciparum dhfr and dhps genes and age related in vivo sulfadoxine–pyrimethamine resistance in malaria-infected patients from Nigeria. *Acta Tropica., 95*(3), 183–193.

Hemming-Schroeder, E., Umukoro, E., Lo, E., Fung, B., Tomás-Domingo, P., Zhou, G., Zhong, D., Dixit, A., Atieli, H., Githeko, A., Vardo-Zalik, A., & Yan, G. (2018). Impacts of antimalarial drugs on *Plasmodium falciparum* drug resistance markers, Western Kenya, 2003–2015. *American Journal of Tropical Medicine and Hygiene., 98*(3), 692–699.

Huang, B., Wang, Q., Deng, C., Wang, J., Yang, T., Huang, S., Su, X., Liu, Y., Pan, L., Li, G., Li, D., Zhang, H., Bacar, A., Abdallah, K. S., Attoumane, R., Mliva, A. M. S. A., Zheng, S., Xu, Q., Lu, F., Guan, Y., & Song, J. (2016). Prevalence of crt and mdr-1 mutations in *Plasmodium falciparum* isolates from Grande Comore Island after withdrawal of chloroquine. *Malaria Journal., 15*(1), 414–414.

Huynh, B.-T., Cottrell, G., Cot, M., & Briand, V. (2015). Burden of malaria in early pregnancy: A neglected problem? *Clinical Infectious Diseases., 60*(4), 598–604.

Ishaque, S., Yakoob, M. Y., Imdad, A., Goldenberg, R. L., Eisele, T. P., & Bhutta, Z. A. (2011). Effectiveness of interventions to screen and manage infections during pregnancy on reducing stillbirths: A review. *BMC Public Health, 11*(Suppl 3), S3–S3.

Johnson, D. J., Owen, A., Plant, N., Bray, P. G., & Ward, S. A. (2008). Drug-regulated expression of *Plasmodium falciparum* P-glycoprotein homologue 1: A putative role for nuclear receptors. *Antimicrobial Agents and Chemotherapy., 52*(4), 1438–1445.

Kalanda, B. F., Verhoeff, F. H., Chimsuku, L., Harper, G., & Brabin, B. J. (2006). Adverse birth outcomes in a malarious area. *Epidemiology and Infection., 134*(3), 659–666.

Khaja, K. A. J., & Sequeira, R. P. (2021). Drug treatment and prevention of malaria in pregnancy: A critical review of the guidelines. *Malaria Journal., 20*(1), 1–62.

Kumar, S., Bhardwaj, T. R., Prasad, D., & Singh, R. K. (2018). Drug targets for resistant malaria: Historic to future perspectives. *Biomedicine and Pharmacotherapy., 104*(1), 8–27.

Lakshmanan, V., Bray, P. G., Verdier-Pinard, D., Johnson, D. J., Horrocks, P., Muhle, R. A., Alakpa, G. E., Hughes, R. H., Ward, S. A., Krogstad, D. J., Sidhu, A. B., & Fidock, D. A. (2015). A critical role for PfCRT K76T in *Plasmodium falciparum* verapamil-reversible chloroquine resistance. *EMBO Journal., 24*(13), 2294–2305.

Lawn, J. E., Blencowe, H., Waiswa, P., Amouzou, A., Mathers, C., Hogan, D., Flenady, V., Frøen, J. F., Qureshi, Z. U., Calderwood, C., Shiekh, S., Jassir, F. B., You, D., McClure, E. M., Mathai, M., & Cousens, S. (2016). Stillbirths: Rates, risk factors, and acceleration towards 2030. *Lancet, 387*(10018), 587–603.

Mahamar, A., Andemel, N., Swihart, B., Sidibe, Y., Gaoussou, S., Barry, A., Traore, M., Attaher, O., Dembele, A. B., Diarra, B. S., Keita, S., Dicko, A., Duffy, P. E., & Fried, M. (2021). Malaria infection is common and associated with perinatal mortality and preterm delivery despite widespread use of chemoprevention in Mali: An observational study 2010 to 2014. *Clinical Infectious Diseases., 73*(8), 1355–1361.

Maiga, H., Grivoyannis, A., Sagara, I., Traore, K., Traore, O. B., Tolo, Y., Traore, A., Bamadio, A., Traore, Z. I., Sanogo, K., Doumbo, O. K., Plowe, C. V., & Djimde, A. A. (2021). Selection of pfcrt K76 and pfmdr1 N86 coding alleles after uncomplicated malaria treatment by artemether-lumefantrine in Mali. *International Journal of Molecular Sciences., 22*(11), 6057.

Martínez-Perez, G., Lansana, D. P., Omeonga, S., Gupta, H., Breeze-Barry, B., González, R., Bardají, A., Sarukhan, A., Goteh, J. D. K., Tody, E., Cisteró, P., Benda, B., Kercula, J. D., Kibungu, F. D., García-Sípido, A. M., Bassat, Q., Tarr-Attia, C. K., & Mayor, A. (2018). Prevalence of *Plasmodium falciparum* infection among pregnant women at first antenatal visit in post-Ebola Monrovia, Liberia. *Malaria Journal., 17*(1), 357.

McGready, R., Lee, S. J., Wiladphaingern, J., Ashley, E. A., Rijken, M. J., Boel, M., Simpson, J. A., Paw, M. K., Pimanpanarak, M., Mu, O., Singhasivanon, P., White, N. J., & Nosten, F. H. (2012). Adverse effects of *falciparum* and *vivax* malaria and the safety of antimalarial treatment in early pregnancy: A population-based study. *Lancet Infectious Diseases., 12*(5), 388–396.

Miotto, O., Almagro-Garcia, J., Manske, M., MacInnis, B., Campino, S., Rockett, K. A., Amaratunga, C., Lim, P., Suon, S., Sreng, S., Anderson, J. M., Duong, S., Nguon, C., Chuor, C. M., Saunders, D., Se, Y., Lon, C., Fukuda, M. M., Amenga-Etego, L., & Kwiatkowski, D. P. (2013). Multiple populations of artemisinin-resistant Plasmodium falciparum in Cambodia. *Nature Genetics., 45*(6), 648–655.

Moore, K. A., Simpson, J. A., Paw, M. K., Pimanpanarak, M., Wiladphaingern, J., Rijken, M. J., Jittamala, P., White, N. J., Fowkes, F. J. I., Nosten, F., & McGready, R. (2016). Safety of artemisinins in first trimester of prospectively followed pregnancies: An observational study. *Lancet Infectious Diseases., 16*(5), 576–583.

Moore, K. A., Fowkes, F. J. I., Wiladphaingern, J., Wai, N. S., Paw, M. K., Pimanpanarak, M., Carrara, V. I., Raksuansak, J., Simpson, J. A., White, N. J., Nosten, F., & McGready, R. (2017a). Mediation of the effect of malaria in pregnancy on stillbirth and neonatal death in an area of low transmission: Observational data analysis. *BMC Medicine., 15*(1), 98.

Moore, K. A., Simpson, J. A., Scoullar, M. J. L., McGready, R., & Fowkes, F. J. I. (2017b). Quantification of the association between malaria in pregnancy and stillbirth: A systematic review and meta-analysis. *Lancet Global Health., 5*(11), e1101–e1112.

Myint, H. Y., Tipmanee, P., Nosten, F., Day, N. P., Pukrittayakamee, S., Looareesuwan, S., & White, N. J. (2004). A systematic overview of published antimalarial drug trials. *Transactions of the Royal Society of Tropical Medicine and Hygiene., 98*(2), 73–81.

Nelson, A. L., Purfield, A., McDaniel, P., Uthaimongkol, N., Buathong, N., Sriwichai, S., Miller, R. S., Wongsrichanalai, C., & Meshnick, S. R. (2005). pfmdr1 genotyping and in vivo mefloquine resistance on the Thai-Myanmar border. *America Journal of Tropical Medicine and Hygiene., 72*(5), 586–592.

Nsobya, S. L., Kiggundu, M., Nanyunja, S., Joloba, M., Greenhouse, B., & Rosenthal, P. J. (2010). In vitro sensitivities of *Plasmodium falciparum* to different antimalarial drugs in Uganda. *Antimicrobial Agents and Chemotherapy., 54*(3), 1200–1206.

Nyirjesy, P., Kavasya, T., Axelrod, P., & Fischer, P. R. (2023). Malaria during pregnancy: Neonatal morbidity and mortality and the efficacy of chloroquine chemoprophylaxis. *Clinical Infectious Diseases., 16*(1), 127–132.

Okombo, J., Kiara, S. M., Rono, J., Mwai, L., Pole, L., Ohuma, E., Borrmann, S., Ochola, L. I., & Nzila, A. (2010). In vitro activities of quinine and other antimalarials and pfnhe polymorphisms in *Plasmodium* isolates from Kenya. *Antimicrobial Agents and Chemotherapy., 54*(8), 3302–3307.

Olasehinde, G. I., Ojurongbe, O., Adeyeba, A. O., Fagade, O. E., Valecha, N., Ayanda, I. O., Ajayi, A. A., & Egwari, L. O. (2014). In vitro studies on the sensitivity pattern of *Plasmodium falciparum* to anti-malarial drugs and local herbal extracts. *Malaria Journal., 13*(1), 63.

Olasehinde, G. I., Diji-Geske, R. I., Fadina, I., Arogundade, D., Darby, P., Adeleke, A., Dokunmu, T. M., Adebayo, A. H., & Oyelade, J. (2019). Epidemiology of *Plasmodium falciparum* infection and drug resistance markers in Ota Area, Southwestern Nigeria. *Infection and Drug Resistance., 12*(1), 1941–1949.

Plowe, C. V. (2022). Malaria chemoprevention and drug resistance: A review of the literature and policy implications. *Malaria Journal., 21*(1), 20–22.

Price, R. N., Uhlemann, A.-C., van Vugt, M., Brockman, A., Hutagalung, R., Nair, S., Nash, D., Singhasivanon, P., Anderson, T. J. C., Krishna, S., White, N. J., & Nosten, F. (2006). Molecular and pharmacological determinants of the therapeutic response to artemether-lumefantrine in multidrug-resistant *Plasmodium falciparum* malaria. *Clinical Infectious Diseases., 42*(11), 1570–1577.

Price, R. N., Simpson, J. A., & McCarthy, J. S. (2017). Kucers' the use of antibiotics. In *Halofantrine* (7th ed., p. 6). ImprintCRC Press.

Quakyi, I., Tornyigah, B., Houze, P., Kusi, K. A., Coleman, N., Escriou, G., Laar, A., Cot, M., Fobil, J., Asare, G. Q., Deloron, P., Anang, A. K., Cottrell, G., Ofori, M. F., & Ndam, N. T. (2019). High uptake of Intermittent Preventive Treatment of malaria in pregnancy is associated with improved birth weight among pregnant women in Ghana. *Scientific Reports., 9*(1).

Reyburn, H. (2010). New WHO guidelines for the treatment of malaria. *BMJ., 340*(1), c2637.

Rogerson, S. J., Hviid, L., Duffy, P. E., Leke, R. F., & Taylor, D. W. (2007). Malaria in pregnancy: Pathogenesis and immunity. *Lancet Infectious Diseases., 7*(2), 105–117.

Rogerson, S. J., Desai, M., Mayor, A., Sicuri, E., Taylor, S. M., & van Eijk, A. M. (2018). Burden, pathology, and costs of malaria in pregnancy: New developments for an old problem. *Lancet Infectious Diseases., 18*(4), e107–e118.

Sa, J. M., & Twu, O. (2010). Protecting the malaria drug arsenal: halting the rise and spread of amodiaquine resistance by monitoring the PfCRT SVMNT type. *Malaria Journal., 9*(1).

Saito, M., Gilder, M. E., Nosten, F., McGready, R., & Guérin, P. J. (2017). Systematic literature review and meta-analysis of the efficacy of artemisinin-based and quinine-based treatments for uncomplicated *falciparum* malaria in pregnancy: Methodological challenges. *Malaria Journal., 16*(1), 488–488.

Saito, M., Gilder, M. E., McGready, R., & Nosten, F. (2018). Antimalarial drugs for treating and preventing malaria in pregnant and lactating women. *Expert Opinion on Drug Safety., 17*(11), 1129–1144.

Saliba, K. J., Lehane, A. M., & Kirk, K. (2008). A polymorphic drug pump in the malaria parasite. *Molecular Microbiology., 70*(4), 775–779.

Sanchez, C. P., Rotmann, A., Stein, W. D., & Lanzer, M. (2008a). Polymorphisms within PfMDR1 alter the substrate specificity for anti-malarial drugs in *Plasmodium falciparum. Molecular Microbiology., 70*(4), 786–798.

Sanchez, C. P., Stein, W. D., & Lanzer, M. (2008b). Dissecting the components of quinine accumulation in *Plasmodium falciparum. Molecular Microbiology., 67*(5), 1081–1093.

Scholar & Pratt. (2000). Thirteen chemotherapy of malaria. Oxford Academic. 3 (13), 375–418.

Schuster, B. G., & Canfield, C. J. (1989). Halofantrine in the treatment of multi-drug resistant malaria. *Parasitology Today., 1*, 3–13.

Sekihara, M., Tachibana, S.-I., Yamauchi, M., Yatsushiro, S., Tiwara, S., Fukuda, N., Ikeda, M., Mori, T., Hirai, M., Hombhanje, F., & Mita, T. (2018). Lack of significant recovery of chloroquine sensitivity in *Plasmodium falciparum* parasites following discontinuance of chloroquine use in Papua New Guinea. *Malaria Journal., 17*(1), 434–434.

Sidhu, A. B., Uhlemann, A. C., Valderramos, S. G., Valderramos, J. C., Krishna, S., & Fidock, D. A. (2006). Decreasing pfmdr1 copy number in *Plasmodium falciparum* malaria heightens susceptibility to mefloquine, lumefantrine, halofantrine, quinine, and artemisinin. *Journal of Infectious Diseases., 194*(4), 528–535.

Tarning, J., Kloprogge, F., Dhorda, M., Jullien, V., Nosten, F., White, N. J., et al. (2013). Pharmacokinetic properties of artemether, dihydroartemisinin, lumefantrine, and quinine in pregnant women with uncomplicated *Plasmodium falciparum* malaria in Uganda. *Antimicrobial Agents and Chemotherapy., 57*(10), 5096–5103.

The Lancet. (2021). Miscarriage: Worldwide reform of care is needed. *Lancet, 397*(10285), 1597.

Triglia, T., Menting, J. G. T., Wilson, C., & Cowman, A. F. (1997). Mutations in dihydropteroate synthase are responsible for sulfone and sulfonamide resistance in Plasmodium falciparum. *Proceedings of the National Academy of Sciences., 94*(25), 13944–13949.

Uhlemann, A. C., & Krishna, S. (2005). Antimalarial multi-drug resistance in Asia: Mechanisms and assessment. *Current Topics in Microbiology and Immunology., 295*(1), 39–53.

Uneke, C. J., Iyare, F. E., Sunday-Adeoye, H., & Ajayi, J. A. (2008). Evaluation of maternal malaria at childbirth using rapid diagnostic test and its relationship with birthweight and fetal hemoglobin levels in Nigeria. *Internet Journal of Gynecology and Obstetrics., 10*(1), 2–7.

USAID. (2016). President's malaria initiative Nigeria. Malaria operational plan FY. Retrieved Feb 13, 2017, from https://www.pmi.gov/docs/default-source/default-document-library/malaria-operational-plans/fy16/fy-2016-nigeria-malaria-operational-plan.pdf?sfvrsn=13

Valderramos, S.G., Valderramos, J-C., Musset, L., Purcell, L. A., Mercereau-Puijalon, O., Legrand, E., & Fidock, D. A. (2010). Identification of a mutant PfCRT-mediated chloroquine tolerance phenotype in *Plasmodium falciparum*. PLOS Pathogens. 6 (5), e1000887.

Veiga, M. I., Dhingra, S. K., Henrich, P. P., Straimer, J., Gnädig, N., Uhlemann, A.-C., Martin, R. E., Lehane, A. M., & Fidock, D. A. (2016). Globally prevalent PfMDR1 mutations modulate *Plasmodium falciparum* susceptibility to artemisinin-based combination therapies. *Nature Communications, 7*(2041–1723), 11553–11553.

Van Geertruyden, J., Thomas, F., Erhart, A., & D'alessandro, U. (2004). THE CONTRIBUTION OF MALARIA IN PREGNANCY TO PERINATAL MORTALITY. *American Journal of Tropical Medicine and Hygiene., 71*(2), 35–40.

Wang, C., Pan, R., Wan, X., Tan, Y., Xu, L., Ho, C. S., & Ho, R. C. (2020). Immediate Psychological Responses and Associated Factors during the Initial Stage of the 2019 Coronavirus Disease (COVID-19) Epidemic among the General Population in China. *International Journal of Environmental Research and Public Health., 17*(5), 1729.

WHO. (2015). *Guidelines for the treatment of malaria* (3rd ed.). World Health Organization. Retrieved June 15, 2020, from https://www.who.int/malaria/publications/atoz/9789241549127/en/

WHO. (2019a). *Malaria: Intermittent preventive treatment in pregnancy (IPTp). Last update: 21 October 2019.* World Health Organization. Retrieved June 8, 2020, from https://www.who.int/malaria/areas/preventive_therapies/pregnancy/en/

WHO. (2019b). *World malaria report 2019.* World Health Organization. Retrieved June 8, 2020, from https://www.who.int/publications-detail/world-malaria-report-2019

WHO. (2020). *World malaria report 2020.* World Health Organization. Retrieved May, 29 2021, from https://www.who.int/publications/i/item/9789240015791

WHO. (2021). *World malaria report, 2021.* World Health Organization.

Wicht, K. J., Mok, S., & Fidock, D. A. (2020a). Molecular mechanisms of drug resistance in *Plasmodium falciparum* malaria. *Annual Review of Microbiology., 74*(1), 431–454.

Wicht, P. J., Mohandas, N., & Ghislaine, J. D. (2020b). Antimalarial drug resistance linked to P-glycoprotein homologue (Pgh1) modulation in *Plasmodium falciparum. Malaria Journal., 19*(1), 223–224.

Wong, W., Bai, X. C., Sleebs, B. E., Triglia, T., Brown, A., Thompson, J. K., Jackson, K. E., Hanssen, E., Marapana, D. S., Fernandez, I. S., Ralph, S. A., Cowman, A. F., Scheres, S. H. W., & Baum, J. (2017). Mefloquine targets the *Plasmodium falciparum* 80S ribosome to inhibit protein synthesis. *Nature Microbiology., 13*(2), 1–9.
Yatich, N. J., Funkhouser, E., Ehiri, J. E., Agbenyega, T., Stiles, J. K., Rayner, J. C., Turpin, A., Ellis, W. O., Jiang, Y., Williams, J. H., Afriyie-Gwayu, E., Phillips, T., & Jolly, P. E. (2010). Malaria, intestinal helminths and other risk factors for stillbirth in Ghana. *Infectious Diseases in Obstetrics and Gynecology., 2010*(1064–7449), 350763.
Zhao, D., Zhang, H., Ji, P., Li, S., Yang, C., Liu, Y., Qian, D., Deng, Y., Wang, H., Lu, D., Zhou, R., & Zhao, Y. (2021). Surveillance of antimalarial drug-resistance genes in imported *Plasmodium falciparum* isolates from Nigeria in Henan, China, 2012–2019. *Frontiers in Cellular and Infection Microbiology., 11*(1), 644576.

Chapter 7
Next-Generation Biotechnology Tool in Mitigation of Resistance to Carbamates Insecticides in *Anopheles gambiae*

Wisdom Deborah Cleanclay and Azeez Blessing Opeyemi

1 Introduction

Infectious tropical diseases are spreading rapidly worldwide, and arthropod vectors such as mosquitoes, which cause malaria, play a major role in contributing to this global burden (Franklinos et al., 2019; Rohr et al., 2019). Diseases caused by these vectors are re-emerging in areas where they were once eliminated and emerging in new areas. These diseases' frequency and geographic range are rising globally (Franklinos et al., 2019).

Malaria, an infectious disease, stands as one of the most widespread vector-borne illnesses, with its origins extending back to the earliest periods of human civilization. (Talapko et al., 2019). It is caused by the protozoon *Plasmodium* which has several subspecies. This protozoon has a lifecycle that involves a mosquito vector and a vertebrate host. It was Sir Ronald Ross in the nineteenth century who discovered that mosquitoes are responsible for transmitting malaria; he was also able to identify the particular species of mosquito responsible for transmitting this disease (Schneider, 2021).

The fight against malaria began in 1955 by eliminating mosquitoes using dichlorodiphenyltrichloroethane (DDT) (Talapko et al., 2019). According to Talapko et al. (2019), preventing insect bites is the best way of preventing malaria. However, this preventive approach is being threatened by the increasing resistance of mosquitoes to insecticides (Shaw & Catteruccia, 2019).

Insecticide-treated nets (ITNs) and indoor residual spraying (IRS) are the two strategies for controlling mosquito vectors recommended by the WHO (Fikadu &

W. D. Cleanclay · A. B. Opeyemi (✉)
Department of Biochemistry, Covenant University, Ota, Ogun State, Nigeria

Covenant Applied Informatics and Communication, Africa Centre of Excellence (CApIC-ACE), Covenant University, Ota, Ogun State, Nigeria
e-mail: blessing.azeezpgs@stu.cu.edu.ng

P. O. Isibor, P. A. Akinduti (eds.), *Harnessing Biotechnology Tools for Product Development*, https://doi.org/10.1007/978-3-031-86002-7_7

Ashenafi, 2023). For IRS, five insecticides have been proven safe and efficient for use by the WHO: pyrethroids, organochlorines, carbamates, organophosphates, and neonicotinoids (Fikadu & Ashenafi, 2023). The most common way to prevent the disease is by combatting the vector, and this is one of the main aims of public health.

2 *Anopheles* as the Primary Malaria Vector

Of all the anthropophilic mosquitoes there are, those of the genus *Anopheles* are the deadliest. The infective female *Anopheles* mosquitoes carry the parasite *Plasmodium* which is responsible for causing malaria and has a record of killing about 400,000 people annually (Ahmed et al., 2020). The high preference of female *Anopheles* mosquitoes for humans is because they require a blood meal for reproduction to take place, the protein in human blood serves as a nutrient for egg development in these mosquitoes (Ahmed et al., 2020). There are a minimum of 25 known species complexes of *Anopheles* species worldwide, which include over 135 sibling species, with the majority located in Asia. These sibling species consist of closely related, phenotypically identical species (Tananchai et al., 2019). The species within these complexes play various roles in the transmission of malaria, acting as primary vectors, secondary (incidental) vectors, or nonvectors (Tananchai et al., 2019). Depending on their epidemiological involvement in malaria transmission, members of a complex can be major primary vectors, secondary (incidental) concerns, or nonvectors.

Vector management is the primary strategy for combating malaria; this is because protecting the host from exposure to the vector prevents transmission of the parasite (Suh et al., 2023). The two main interventions are indoor residual spraying (IRS) and long-lasting insecticidal nets (LLIN), which have succeeded in averting approximately 78% of malaria-related deaths (Opondo et al., 2019). These two interventions help prevent transmission by taking advantage of malaria vectors' resting and biting behaviours.

3 Mosquito Control Strategies

Long-lasting insecticidal nets (LLINs) are mosquito nets produced from materials that have had insecticide added to them or bonded around their fibres (Hiruy et al., 2023). It protects human hosts from mosquito bites through chemical and physical barriers (Chanda et al., 2020; Villalta et al., 2021). The average shelf life of LLINs is 3 years (Musa et al., 2020; Hiruy et al., 2023).

Indoor residual spraying (IRS) refers to the application of insecticides to the interior walls of houses. It works by reducing the life span of endophagic and endophilic female *Anopheles* mosquitoes that dwell indoors (Chanda et al., 2020). The use of IRS in vector control started during the Global Malaria Eradication Campaign which implemented the organochloride, dichlorodiphenyltrichloroethane (DDT) (Tangena et al., 2020). Unlike insecticide-treated nets (ITNs), which typically use pyrethroids, the WHO has approved 16 pesticide formulations from the five main insecticide classes for IRS (Tangena et al., 2020).

Pyrethroids are natural compounds derived from the flowers of the *Tanacetum cinerariifolium* plant (Hołyńska-Iwan & Szewczyk-Golec, 2020). Pyrethroids are naturally more potent but unstable compounds due to their photodegradability which synthesize derivatives that are resistant to radiation (Hołyńska-Iwan & Szewczyk-Golec, 2020). Pyrethroid insecticides are neurotoxins that target insects' receptor sites of voltage-gated sodium channels. It causes excitatory paralysis in insects by binding to their sodium channels (Ahamad & Kumar, 2023). In addition, they alter their nerve cells' membrane potential, causing hyperexcitability. These changes cause "knockdown," an impairing but not fatal effect, in insects (Ahamad & Kumar, 2023). Examples of pyrethroids that are inexpensive and readily available for use include deltamethrin, permethrin, and alpha-cypermethrin, and they are the main insecticides that are employed for use in LLINs nets (Brake et al., 2022).

Carbamates inhibit the enzyme acetylcholinesterase, by preventing the hydrolysis of acetylcholine (Moreira et al., 2022). This inhibition stimulates the postsynaptic neuron membrane and kills the insect. They were identified to have insecticidal activity in 1931 but were primarily manufactured for vector control in the late 1950s (Karunaratne & Surendran, 2022). Carbamates, derived from carbamic acid, have a wide range of subtypes such as aldicarb, carbofuran, carbaryl, and propoxur (Zulfa et al., 2022).

Organochlorines are a class of chlorinated hydrocarbon chemicals that were first discovered during the 1940s. For example, dichlorodiphenyltrichloroethane (DDT) is the most widely used chlorinated insecticide (Gan et al., 2021). Other organochlorines include chlordane, chlorobenzilate, lindane, and dieldrin (Gan et al., 2021). The mode of action involves the modification of ion exchange by disrupting sodium and potassium channels in nerve axons in the peripheral and central nervous systems, leading to diminished action potentials (Leemans et al., 2019).

Organophosphates are phosphoric acid esters that are direct or indirect lipophilic inhibitors of acetylcholinesterase (*AChE*) (Ganie et al., 2022). They had wide application as pesticides in 1950 following the development of resistance to organochlorides and other pesticides (Ahamad et al., 2023; Ahamad et al., 2023). Organophosphates are irreversible inhibitors of acetylcholinesterase-1 (*AChE1*), which breaks down acetylcholine (Ach). Nerve signals are transmitted by ACh, which is produced from presynaptic neurons and binds to ACh receptors on the postsynaptic membrane (Ganie et al., 2022). By hydrolysing ACh into inactive metabolites, acetate anion, and choline, *AChE* discontinues the signal transmission. Organophosphates obstruct this mechanism, causing prolonged nerve signalling and an ACh build-up in the synaptic cleft (Ganie et al., 2022).

Neonicotinoids are a family of insecticides that are important in resistance management, their mechanism of action is unique in comparison to insecticides that are currently in use (Mouhamadou et al., 2019). By specifically targeting the nicotinic acetylcholine receptor (nAChR) in invertebrates, they cause paralysis and death by interfering with excitatory cholinergic neurotransmission. Examples include nitenpyram, acetamiprid, thiacloprid, dinotefuran, etc. Neonicotinoids have been approved for use in indoor residual spraying (IRS) in vector control (Chukwuekezie et al., 2020).

4 Insecticide Resistance

Resistance to the four insecticides (carbamates, organochlorines, organophosphates, and pyrethroids) recommended by the World Health Organization Pesticide Evaluation Scheme (WHOPES) in *Anopheles* mosquitoes was first documented in Africa in the 1950s (Chukwuekezie et al., 2020; Suh et al., 2023). A fifth class, called neonicotinoids, has been recorded to be effective against mosquitoes and is now being used for vector control in many countries (Chukwuekezie et al., 2020). These insecticides proved efficient in controlling malaria vectors between the years 2000 and 2015; however, their efforts were rendered ineffective due to the emerging resistance of mosquitoes to them. Except for the WHO European region, resistance to these insecticides has been documented in every WHO region (Chukwuekezie et al., 2020), which is why the WHO encourages the monitoring of malaria mosquitoes to insecticides which would serve as a guide for malaria control programs (Namountougou et al., 2019).

Insecticide resistance occurs when selected mosquitoes exposed to insecticides can thrive and procreate in that environment (Pwalia et al., 2019). Mosquitoes use four mechanisms to bypass the effect of the insecticides, which include: target site resistance, metabolic resistance, cuticular resistance, and behavioural resistance (Suh et al., 2023). The most common mechanisms among all four are target site resistance and metabolic resistance.

Target site resistance involves an alteration in the insecticide binding site caused by a mutation in the genes that encode the target protein, reducing the sensitivity of the target site to the insecticides, preventing catalysis, and rendering the insecticides ineffective (Pwalia et al., 2019; Chanda et al., 2020). In malaria vectors, there are two prevalent forms of resistance mechanisms related to target sites: alterations in the voltage-gated sodium channel (Vgsc), namely L1014F or L1014S, which result in pyrethroid resistance, and a mutation in the acetylcholinesterase gene (*AChE1*), G119S, which leads to resistance against carbamates and organophosphates (Pwalia et al., 2019). Insecticides like pyrethroids and DDT target sodium channels, causing the repetitive discharge of signals by the nervous system, leading to a depolarization of the nerve membrane and eventually the death of the insect (Liu, 2015). The mutation occurs in the sodium channel receptor and confers knockdown resistance. Another binding site is the protein acetylcholinesterase, a neurotransmitter that is targeted by carbamates and organophosphate insecticides (Liu, 2015). The mutation occurs in the acetylcholinesterase enzyme and confers *AChE1* resistance.

Metabolic resistance involves the breakdown of insecticides by specific enzymes before getting to the binding site, these enzymes are usually overexpressed and aid in detoxifying the insecticides into nontoxic compounds (Khan et al., 2020). The major families of enzymes responsible for carrying out this action are the cytochrome P-450 monooxygenases, esterases, and glutathione s-transferases (Chanda et al., 2020). The overexpression of enzymes in this family has been reported in some mosquito species such as *Anopheles*, *Culex*, and *Aedes* (Adolfi et al., 2019; Amelia-Yap et al., 2019; Xu et al., 2019). Several mechanisms trigger the upregulation of these enzymes such as the copy number variation or detoxification gene regulation by cis- and trans-regulatory elements (Huang et al., 2023).

Cuticular resistance involves the reduction in the rate at which insecticides penetrate through the skin of the mosquitoes (Vontas et al., 2020). In cuticular resistance, the insecticides penetrate the skin at a very slow rate; this is seen particularly in mosquitoes that are resistant to the insecticide, whereas the susceptible absorbs it faster. An example of cuticular resistance was seen in the resistance of *Anopheles gambiae* to pyrethroid and DDT facilitated by the overexpression of the cytochrome P450 enzymes, cuticular protein, low complexity glycine-rich 3 (CPLCG3), and cytochrome P450 reductases (CPRs) (Balabanidou et al., 2019). These enzymes help to deposit cuticular hydrocarbons into the epicuticle which helps provide a waterproof barrier thereby strengthening the cuticle (Balabanidou et al., 2019).

In a study conducted by Sparks in 1989, he clearly defined behavioural resistance as "an evolved behaviour that decreases an insect's exposure to toxic substances or enables it to survive in environments that would typically be toxic and lethal" (Siddiqui et al., 2023). These insects do this by changing their behaviour and avoiding regions that have been sprayed with insecticides, to survive (Siddiqui et al., 2023). As insecticides are used more frequently, mosquitoes might find it impossible to avoid contact. In such situations, if mosquitoes are not immediately killed by the insecticide (due to sublethal doses or the presence of physiologically resistant genotypes, for instance), they may develop quantitative behavioural resistance (Carrasco et al., 2019).

5 Carbamate Insecticides

Carbamates are esters of carbamic acid that have a simple structure and are known to be safer than organophosphates, they have less persistence, low bioaccumulation, and degrade easily in the environment and hence are widely used in agriculture, as well as in industries and homes (Mishra et al., 2021). In modern agriculture, they are commonly used as herbicides, fungicides, nematicides, insecticides, etc. because of their broad range of biological activity (Mishra et al., 2021). Carbamates are characterized by the R–O–CONH–R functional group. Examples of carbamate insecticides include propoxur, bendiocarb, aldicarb, etc.

5.1 Mechanism of Action of Carbamate Insecticide

Carbamate insecticides work by inhibiting the enzyme acetylcholinesterase (*AChE*) (EC 3.1.1.7) competitively, this reaction is reversible (Adedeji et al., 2020). Acetylcholinesterase is an enzyme that plays a role in terminating the action of acetylcholine, a neurotransmitter in cholinergic synapses used by nerve cells and muscles for communication (Matošević & Bosak, 2020). It plays this role by catalysing the hydrolysis of acetylcholine (Ach) into acetyl and choline, which is mostly found in muscles and nerves. The inhibition of the *AChE* enzyme by carbamate causes the accumulation of ACh and hence continuous transmission of signals between the nerve cells (Mishra et al., 2021). This could lead to overstimulation, continuous muscle contraction, excessive neuronal firing, and eventually mosquito death.

5.2 Resistance to Carbamate Insecticides in Anopheles Gambiae

Due to the growing resistance of mosquitoes to pyrethroid insecticides which has been for public health purposes such as in IRS and singly in LLINs, there has been an interest in using alternative insecticides such as carbamates to improve vector control (Elanga-Ndille et al., 2019). However, studies in recent years have shown the increasing resistance of *Anopheles gambiae* to carbamate insecticides, which is a threat to malaria control efforts.

Target-site resistance and metabolic resistance are the primary resistance mechanisms to carbamates that have been identified to date (Keïta et al., 2020). A mutation in a single nucleotide characterizes the target site resistance, while overexpression of detoxifying genes like the cytochrome P450s characterizes the metabolic resistance. The gene that is mutated in carbamate resistance is the acetylcholinesterase-1 (*AChE1*) gene, while the detoxification genes involved are the cytochrome P450 genes (Keïta et al., 2020).

The *AChE1* gene encodes the instructions for the acetylcholinesterase enzyme, which breaks down acetylcholine, a neurotransmitter, in the synaptic cleft (Matošević & Bosak, 2020). In resistant mosquitoes, a mutation in the *AChE1* gene causes a switch of glycine (GGC) with serine (AGC) at the 199th position, altering the structure and function of the *AChE1* enzyme (Keïta et al., 2020). Due to this change in structure, the binding affinity of carbamate insecticides is reduced, thereby decreasing its ability to inhibit the *AChE1* enzyme.

Reports of carbamate resistance to carbamates among *Anopheles gambiae* mosquitoes have been documented in various regions in sub-Saharan Africa. In Cameroon, researchers were able to establish a link between resistance to carbamate insecticides and a mutation in G119S, in the acetylcholinesterase (*AChE1*) enzyme, within a population of *Anopheles gambiae* from a single site located in a forested region of central Cameroon (Elanga-Ndille et al., 2019). Another study examined how wild populations of *Anopheles gambiae s.l.* across Cameroon responded to carbamate and organophosphate insecticides. The study also examined the prevalence and distribution of the G119S mutation across the country (Binyang et al., 2022).

In Nigeria, the first study of resistance to carbamate in *Anopheles gambiae* was first documented by Oduola et al. (2012). However, no relationship was seen between carbamate resistance and the G119S mutation, but an association was seen with the metabolic enzymes. It was not until 2020, that a study established an association between carbamate insecticides and the G119S mutation in Lagos (Fagbohun et al., 2020). Subsequently, many studies have been carried out to show a link between carbamate resistance and the G119S mutation (Ikpo et al., 2021; Ekerette et al., 2023). A study in Mali reveals that wild populations of *Anopheles arabiensis*, *Anopheles coluzzii*, and *Anopheles gambiae* (s.s.) resisted bendiocarb. Additionally, the research examines the frequency of the G119S mutation alleles in these wild populations of *Anopheles gambiae (s.l.)* (Keïta et al., 2020).

A study in Guinea found that the G119S Ace-1 mutation strongly predicts survival in bioassays testing resistance to bendiocarb and provides resistance to carbamate insecticides. Additionally, mosquitoes that survived bendiocarb exposure showed significantly higher expression of a cytochrome P450 enzyme, CYP6P3, suggesting that this enzyme may play a role in the reduced susceptibility to carbamates observed in Faranah (Stica et al., 2019).

A study in Togo reported that *Anopheles gambiae, Anopheles coluzzii, and the mixed type* were discovered to have mutated insensitive acetylcholinesterase. The prevalence of this mutation in the *Anopheles* population indicates a decreased susceptibility to both organophosphate and carbamate insecticides, which are also commonly used by farmers. The frequencies of the Ace-1 allele in *Anopheles coluzzii* showed a decrease in Kovié between 2013 and 2016 compared to Baguida between 2012 and 2013. Previous reports of the allele were made at Kovié and Nangbéto, Lomé, and other locations. Regrettably, in Kovié, the duplication of the copy number of the ace-1 allele was first observed and linked to carbamate resistance, implying the strengthening and dissemination of this resistance. This poses a severe threat to resistance, which is why the situation is concerning (Amoudji et al., 2019).

6 Challenges, Recommendations, Future Perspectives, and Conclusion

Reports of gene duplication in mosquitoes have been documented and this is characterized by the presence of both resistant and susceptible copies of the *AChE1* gene. Their resistance to carbamate insecticides is increased by this genetic arrangement, which enables them to withstand insecticide treatments that would otherwise be effective. Normally, mutations might lower an organism's fitness in insecticide-free environments because the resistant gene may have a negative impact on the mosquito's overall health or reproductive success. However, the availability of resistant and susceptible copies due to gene duplication reduces these fitness costs. This means that resistant mosquitos can grow and survive even in the absence of insecticide.

There is an urgent need for additional studies to map the distribution of key insecticide resistance alleles in the *AChE1* genes and gene duplications associated with metabolic and target-site resistance across different regions. These studies should focus on *Anopheles gambiae* complex species and aim to identify regional variations in resistance mechanisms to inform malaria control strategies better This information is critical for determining the scope of the resistance problem and designing targeted solutions. Insecticides or integrated pest management strategies that can overcome resistance must be developed to tackle the problem of gene duplication. This could entail producing genetic strategies to lower the prevalence of resistance, mixing chemical and nonchemical treatments, or rotating different types of insecticides.

7 Conclusion

The increasing resistance to carbamates and other insecticides emphasizes the importance of continued study and adaptation of control measures. To address these issues, a complete approach is required, which involves the development of novel insecticides from novel chemical classes that act on previously untargeted molecular pathways, such as those affecting neuropeptides or juvenile hormone regulation. These new compounds would circumvent existing resistance mechanisms by targeting different physiological processes in mosquitoes.

Tackling insecticide resistance is crucial to maintaining the efficacy of mosquito control measures and advancing efforts against malaria transmission. Addressing this challenge can sustain the effectiveness of current interventions. One promising avenue is the application of gene-editing technologies, such as CRISPR-Cas9, to engineer mosquitoes with genetic modifications that reduce their ability to develop resistance. This innovative approach could lead to the creation of vector populations less prone to resistance mechanisms, ultimately enhancing the durability of control strategies and contributing to the long-term eradication of malaria.

References

Adedeji, E. O., Ogunlana, O. O., Fatumo, S., Beder, T., Ajamma, Y., Koenig, R., & Adebiyi, E. (2020). Anopheles metabolic proteins in malaria transmission, prevention and control: A review. *Parasites and Vectors, 13*(1), 1–30. https://doi.org/10.1186/s13071-020-04342-5

Adolfi, A., Poulton, B., Anthousi, A., Macilwee, S., Ranson, H., & Lycett, G. J. (2019). Functional genetic validation of key genes conferring insecticide resistance in the major African malaria vector, Anopheles gambiae. *Proceedings of the National Academy of Sciences of the United States of America, 116*(51), 25764–25772. https://doi.org/10.1073/pnas.1914633116

Ahamad, A., Ahamad, J., & Naim, M. J. (2023). Current perspective on pesticides: Their classification, behaviour, potential use and toxic effects. *Journal of Angiotherapy, 7*(1). https://doi.org/10.25163/angiotherapy.719360

Ahamad, A., & Kumar, J. (2023). Pyrethroid pesticides: An overview on classification, toxicological assessment and monitoring. *Journal of Hazardous Materials Advances, 10*, 100284. https://doi.org/10.1016/j.hazadv.2023.100284

Ahmed, R., Dey, D., & Tyagi, V. (2020). Mosquito vector borne disease-malaria and public awareness in Sonitpur and Udalguri districts of Assam, India. *International Journal of Mosquito Research, 7*(5), 26–31. https://doi.org/10.22271/23487941.2020.v7.i5a.475

Amelia-Yap, Z. H., Sofian-Azirun, M., Chen, C. D., Wayan Suana, I., Lau, K. W., Elia-Amira, N. M. R., Haziqah-Rashid, A., Tan, T. K., Lim, Y. A. L., & Low, V. L. (2019). Pyrethroids use: Threats on metabolic-mediated resistance mechanisms in the primary dengue vector Aedes aegypti (Diptera: Culicidae). *Journal of Medical Entomology, 56*(3), 811–816. https://doi.org/10.1093/JME/TJZ007

Amoudji, A. D., Ahadji-Dabla, K. M., Hien, A. S., Apétogbo, Y. G., Yaméogo, B., Soma, D. D., Bamogo, R., Atcha-Oubou, R. T., Dabiré, R. K., & Ketoh, G. K. (2019). Insecticide resistance profiles of Anopheles gambiae s.l. in Togo and genetic mechanisms involved, during 3-year survey: Is there any need for resistance management? *Malaria Journal, 18*(1), 1–12. https://doi.org/10.1186/s12936-019-2813-z

Balabanidou, V., Kefi, M., Aivaliotis, M., Koidou, V., Girotti, J. R., Mijailovsky, S. J., Patricia Juárez, M., Papadogiorgaki, E., Chalepakis, G., Kampouraki, A., Nikolaou, C., Ranson, H., & Vontas, J. (2019). Mosquitoes cloak their legs to resist insecticides. *Proceedings of the Royal Society B: Biological Sciences, 286*(1907), 20191091. https://doi.org/10.1098/rspb.2019.1091

Binyang, A. J., Elanga-Ndille, E., Tene-Fossog, B., Ndo, C., Nouage, L., Assatse, T., Fotso-Toguem, Y., Tabue, R., Zeukeng, F., Nguiffo, D. N., Etang, J., Njiokou, F., & Wondji, C. S. (2022). Distribution of acetylcholinesterase (Ace-1 R) target-site G119S mutation and resistance to carbamates and organophosphates in Anopheles gambiae sensu lato populations from Cameroon. *Parasites and Vectors, 15*(1), 1–13. https://doi.org/10.1186/s13071-022-05174-1

Brake, S., Gomez-Maldonado, D., Hummel, M., Zohdy, S., & Peresin, M. S. (2022). Understanding the current state-of-the-art of long-lasting insecticide nets and potential for sustainable alternatives. *Current Research in Parasitology and Vector-Borne Diseases, 2*, 100101. https://doi.org/10.1016/j.crpvbd.2022.100101

Carrasco, D., Lefèvre, T., Moiroux, N., Pennetier, C., Chandre, F., & Cohuet, A. (2019). Behavioural adaptations of mosquito vectors to insecticide control. *Current Opinion in Insect Science, 34*, 48–54. https://doi.org/10.1016/j.cois.2019.03.005

Chanda, J., Saili, K., Phiri, F., Stevenson, J. C., Mwenda, M., Chishimba, S., Mulube, C., Mambwe, B., Lungu, C., Earle, D., Bennett, A., Eisele, T. P., Kamuliwo, M., Steketee, R. W., Keating, J., Miller, J. M., & Sikaala, C. H. (2020). Pyrethroid and carbamate resistance in anopheles funestus Giles along Lake Kariba in southern Zambia. *American Journal of Tropical Medicine and Hygiene, 103*(2), 90–97. https://doi.org/10.4269/ajtmh.19-0664

Chukwuekezie, O., Nwosu, E., Nwangwu, U., Dogunro, F., Onwude, C., Agashi, N., Ezihe, E., Anioke, C., Anokwu, S., Eloy, E., Attah, P., Orizu, F., Ewo, S., Okoronkwo, A., Joseph, A., Ikeakor, I., Haruna, S., & Gnanguenon, V. (2020). Resistance status of Anopheles gambiae (s.l.) to four commonly used insecticides for malaria vector control in south-East Nigeria. *Parasites and Vectors, 13*(1), 1–10. https://doi.org/10.1186/s13071-020-04027-z

Ekerette, I. B., Dimkpa, S. O., Ebere, N., & Nyenke, C. U. (2023). Acetylcholinesterase and knockdown resistant mutations in anopheles Gambiae S. L (Culicidae: Diptera). *In Three Vegetations of the Niger Delta Region, 1151*(1), 1–15.

Elanga-Ndille, E., Nouage, L., Ndo, C., Binyang, A., Assatse, T., Nguiffo-Nguete, D., Djonabaye, D., Irwing, H., Tene-Fossog, B., & Wondji, C. S. (2019). The g119s acetylcholinesterase (Ace-1) target site mutation confers carbamate resistance in the major malaria vector anopheles gambiae from Cameroon: A challenge for the coming IRS implementation. *Genes, 10*(10), 1–14. https://doi.org/10.3390/genes10100790

Fagbohun, I. K., Idowu, E. T., Otubanjo, O. A., & Awolola, T. S. (2020). First report of AChE1 (G119S) mutation and multiple resistance mechanisms in Anopheles gambiae s.s. in Nigeria. *Scientific Reports, 10*(1), 1–9. https://doi.org/10.1038/s41598-020-64412-7

Fikadu, M., & Ashenafi, E. (2023). Malaria: An overview. *Infection and Drug Resistance, 16*, 3339–3347. https://doi.org/10.2147/IDR.S405668

Franklinos, L. H. V., Jones, K. E., Redding, D. W., & Abubakar, I. (2019). The effect of global change on mosquito-borne disease. *The Lancet Infectious Diseases, 19*(9), e302–e312. https://doi.org/10.1016/S1473-3099(19)30161-6

Gan, S. J., Leong, Y. Q., bin Barhanuddin, M. F. H., Wong, S. T., Wong, S. F., Mak, J. W., & Ahmad, R. B. (2021). Dengue fever and insecticide resistance in Aedes mosquitoes in Southeast Asia: A review. *Parasites and Vectors, 14*(1), 1–19. https://doi.org/10.1186/s13071-021-04785-4

Ganie, S. Y., Javaid, D., Hajam, Y. A., & Reshi, M. S. (2022). Mechanisms and treatment strategies of organophosphate pesticide induced neurotoxicity in humans: A critical appraisal. *Toxicology, 472*, 153181. https://doi.org/10.1016/j.tox.2022.153181

Hiruy, H. N., Irish, S. R., Abdelmenan, S., Wuletaw, Y., Zewde, A., Woyessa, A., Haile, M., Chibsa, S., Lorenz, L., Worku, A., Yukich, J., Berhane, Y., & Keating, J. (2023). Durability of long-lasting insecticidal nets (LLINs) in Ethiopia. *Malaria Journal, 22*(1), 1–13. https://doi.org/10.1186/s12936-023-04540-3

Hołyńska-Iwan, I., & Szewczyk-Golec, K. (2020). Pyrethroids: How they affect human and animal health? *Medicina (Lithuania), 56*(11), 1–5. https://doi.org/10.3390/medicina56110582

Huang, X., Kaufman, P. E., Athrey, G. N., Fredregill, C., Alvarez, C., Shetty, V., & Slotman, M. A. (2023). Potential key genes involved in metabolic resistance to malathion in the southern house mosquito, Culex quinquefasciatus, and functional validation of CYP325BC1 and CYP9M12 as candidate genes using RNA interference. *BMC Genomics, 24*(1), 1–15. https://doi.org/10.1186/s12864-023-09241-4

Ikpo, A. U., Nwaorgu, O. C., Irikannu, K. C., Aniefuna, C. O., Anumba, J. U., & Ezemuoka, L. C. (2021). Insecticide resistance and target site mutations (G119S ace-1 and L1014F kdr) in Anopheles gambiae s.l. in Enugu state, southeastern Nigeria. *Journal of Entomology and Zoology Studies, 9*(4), 111–117. https://doi.org/10.22271/j.ento.2021.v9.i4b.8788

Karunaratne, S. H. P. P., & Surendran, S. N. (2022). Mosquito control: A review on the past, present and future strategies. *Journal of the National Science Foundation of Sri Lanka, 50*(Special Issue), 277–292. https://doi.org/10.4038/jnsfsr.v50i0.11244

Keïta, M., Kané, F., Thiero, O., Traoré, B., Zeukeng, F., Sodio, A. B., Traoré, S. F., Djouaka, R., Doumbia, S., & Sogoba, N. (2020). Acetylcholinesterase (ace-1R) target site mutation G119S and resistance to carbamates in Anopheles gambiae (sensu lato) populations from Mali. *Parasites and Vectors, 13*(1), 1–9. https://doi.org/10.1186/s13071-020-04150-x

Khan, S., Uddin, M. N., Rizwan, M., Khan, W., Farooq, M., Shah, A. S., Subhan, F., Aziz, F., Rahman, K. U., Khan, A., Ali, S., & Muhammad. (2020). Mechanism of insecticide resistance in insects/pests. *Polish Journal of Environmental Studies, 29*(3), 2023–2030. https://doi.org/10.15244/pjoes/108513

Leemans, M., Couderq, S., Demeneix, B., & Fini, J. B. (2019). Pesticides with potential thyroid hormone-disrupting effects: A review of recent data. *Frontiers in Endocrinology, 10*. https://doi.org/10.3389/fendo.2019.00743

Liu, N. (2015). Insecticide resistance in mosquitoes: Impact, mechanisms, and research directions. *Annual Review of Entomology, 60*, 537–559. https://doi.org/10.1146/annurev-ento-010814-020828

Matošević, A., & Bosak, A. (2020). Carbamate group as structural motif in drugs: A review of carbamate derivatives used as therapeutic agents. *Arhiv za Higijenu Rada i Toksikologiju, 71*(4), 285–299. https://doi.org/10.2478/aiht-2020-71-3466

Mishra, S., Pang, S., Zhang, W., Lin, Z., Bhatt, P., & Chen, S. (2021). Insights into the microbial degradation and biochemical mechanisms of carbamates. *Chemosphere, 279*, 130500. https://doi.org/10.1016/j.chemosphere.2021.130500

Moreira, S., Silva, R., Carrageta, D. F., Alves, M. G., Seco-Rovira, V., Oliveira, P. F., & de Lourdes Pereira, M. (2022). Carbamate pesticides: Shedding light on their impact on the male reproductive system. *International Journal of Molecular Sciences, 23*(15), 1–13. https://doi.org/10.3390/ijms23158206

Mouhamadou, C. S., De Souza, S. S., Fodjo, B. K., Zoh, M. G., Bli, N. K., & Koudou, B. G. (2019). Evidence of insecticide resistance selection in wild anopheles coluzzii mosquitoes due to agricultural pesticide use. *Infectious Diseases of Poverty, 8*(1), 1–8. https://doi.org/10.1186/s40249-019-0572-2

Musa, J. J., Moore, S. J., Moore, J., Mbuba, E., Mbeyela, E., Kobe, D., Swai, J. K., & Odufuwa, O. G. (2020). Long-lasting insecticidal nets retain bio-efficacy after 5 years of storage: Implications for malaria control programmes. *Malaria Journal, 19*(1), 1–12. https://doi.org/10.1186/s12936-020-03183-y

Namountougou, M., Soma, D. D., Kientega, M., Balboné, M., Kaboré, D. P. A., Drabo, S. F., Coulibaly, A. Y., Fournet, F., Baldet, T., Diabaté, A., Dabiré, R. K., & Gnankiné, O. (2019). Insecticide resistance mechanisms in Anopheles gambiae complex populations from Burkina Faso, West Africa. *Acta Tropica, 197*(April), 105054. https://doi.org/10.1016/j.actatropica.2019.105054

Oduola, A. O., Idowu, E. T., Oyebola, M. K., Adeogun, A. A., Olojede, J. B., Otubanjo, O. A., & Awolola, T. S. (2012). Evidence of carbamate resistance in urban populations of Anopheles gambiae s.s. Mosquitoes resistant to DDT and deltamethrin insecticides in Lagos. *International Pest Control, 54*(4), 206.

Opondo, K. O., Jawara, M., Cham, S., Jatta, E., Jarju, L., Camara, M., Sanneh, F., Gaye, P. M., Jadama, L., Ceesay, S., Njie, E., Assogba, B. S., Kandeh, B., & D'Alessandro, U. (2019). Status of insecticide resistance in Anopheles gambiae (s.l.) of The Gambia. *Parasites and Vectors, 12*(1), 1–8. https://doi.org/10.1186/s13071-019-3538-0

Pwalia, R., Joannides, J., Iddrisu, A., Addae, C., Acquah-Baidoo, D., Obuobi, D., Amlalo, G., Akporh, S., Gbagba, S., Dadzie, S. K., Athinya, D. K., Hadi, M. P., Jamet, H. P., & Chabi, J. (2019). High insecticide resistance intensity of Anopheles gambiae (s.l.) and low efficacy of pyrethroid LLINs in Accra, Ghana. *Parasites and Vectors, 12*(1), 1–9. https://doi.org/10.1186/s13071-019-3556-y

Rohr, J. R., Barrett, C. B., Civitello, D. J., Craft, M. E., Delius, B., DeLeo, G. A., Hudson, P. J., Jouanard, N., Nguyen, K. H., Ostfeld, R. S., Remais, J. V., Riveau, G., Sokolow, S. H., & Tilman, D. (2019). Emerging human infectious diseases and the links to global food production. *Nature Sustainability, 2*(6), 445–456. https://doi.org/10.1038/s41893-019-0293-3

Schneider, K. A. (2021). Charles Darwin meets Ronald Ross: A population-genetic framework for the evolutionary dynamics of malaria. *Infectious Diseases and Our Planet*, 149–191. https://doi.org/10.1007/978-3-030-50826-5_6

Shaw, W. R., & Catteruccia, F. (2019). Vector biology meets disease control: Using basic research to fight vector-borne diseases. *Nature Microbiology, 4*(1), 20–34. https://doi.org/10.1038/s41564-018-0214-7

Siddiqui, J. A., Fan, R., Naz, H., Bamisile, B. S., Hafeez, M., Ghani, M. I., Wei, Y., Xu, Y., & Chen, X. (2023). Insights into insecticide-resistance mechanisms in invasive species: Challenges and control strategies. *Frontiers in Physiology, 13*, 1–18. https://doi.org/10.3389/fphys.2022.1112278

Stica, C., Jeffries, C. L., Irish, S. R., Barry, Y., Camara, D., Yansane, I., Kristan, M., Walker, T., & Messenger, L. A. (2019). Characterizing the molecular and metabolic mechanisms of insecticide resistance in Anopheles gambiae in Faranah, Guinea. *Malaria Journal, 18*(1), 1–15. https://doi.org/10.1186/s12936-019-2875-y

Suh, P. F., Elanga-Ndille, E., Tchouakui, M., Sandeu, M. M., Tagne, D., Wondji, C., & Ndo, C. (2023). Impact of insecticide resistance on malaria vector competence: A literature review. *Malaria Journal, 22*(1), 1–11. https://doi.org/10.1186/s12936-023-04444-2

Talapko, J., Škrlec, I., Alebić, T., Jukić, M., & Včev, A. (2019). Malaria: The past and the present. *Microorganisms, 7*(6), 179. https://doi.org/10.3390/MICROORGANISMS7060179

Tangena, J. A. A., Hendriks, C. M. J., Devine, M., Tammaro, M., Trett, A. E., Williams, I., Depina, A. J., Sisay, A., Herizo, R., Kafy, H. T., Chizema, E., Were, A., Rozier, J., Coleman, M., & Moyes, C. L. (2020). Indoor residual spraying for malaria control in sub-Saharan Africa 1997 to 2017: An adjusted retrospective analysis. *Malaria Journal, 19*(1), 1–15. https://doi.org/10.1186/s12936-020-03216-6

Tananchai, C., Manguin, S., Bangs, M. J., & Chareonviriyaphap, T. (2019). Malaria Vectors and Species Complexes in Thailand: Implications for Vector Control. *Trends in Parasitology, 35*(7), 544–558. https://doi.org/10.1016/j.pt.2019.04.013

Villalta, E. L., Soto Bravo, A. M., Vizcaino, L., Dzuris, N., Delgado, M., Green, M., Smith, S. C., Lenhart, A., & Macedo de Oliveira, A. (2021). Evaluation of the durability and use of long-lasting insecticidal nets in Nicaragua. *Malaria Journal, 20*(1), 1–13. https://doi.org/10.1186/s12936-021-03604-6

Vontas, J., Katsavou, E., & Mavridis, K. (2020). Cytochrome P450-based metabolic insecticide resistance in anopheles and Aedes mosquito vectors: Muddying the waters. *Pesticide Biochemistry and Physiology, 170*(April), 104666. https://doi.org/10.1016/j.pestbp.2020.104666

Xu, N., Sun, X. H., Liu, Z. H., Xu, Y., Sun, Y., Zhou, D., Shen, B., & Zhu, C. L. (2019). Identification and classification of differentially expressed genes in pyrethroid-resistant Culex pipiens pallens. *Molecular Genetics and Genomics, 294*(4), 861–873. https://doi.org/10.1007/S00438-018-1521-7/METRICS

Zulfa, R., Lo, W. C., Cheng, P. C., Martini, M., & Chuang, T. W. (2022). Updating the insecticide resistance status of Aedes aegypti and Aedes albopictus in Asia: A systematic review and meta-analysis. *Tropical Medicine and Infectious Disease, 7*(10). https://doi.org/10.3390/tropicalmed7100306

Chapter 8
Phylogenetic Analyses of Selected Underutilized Members of the Fabaceae Family Based on DNA Barcodes

Olawole Odun Obembe, Tobi Tejumade Obadire, Oluwakemi Adetutu Bello, and Oluwadurotimi Samuel Aworunse

1 Introduction

The family Fabaceae encompasses a wide array of cultivated plants, including food crops, animal feed, ornamental plants, medicinal plants, timber, and wood products, holding significant economic value (Graham & Vance, 2003). Distinguishing features of the Fabaceae family, also known as Leguminosae, include typical legume fruits and the ability to fix nitrogen symbiotically (Moss et al., 2022). These fruits, commonly called legumes or pods, split upon drying, releasing enclosed seeds (Britannica, 2023). The family Fabaceae is categorized into grain legumes (pulses) and pasture/forage legumes. Grain legumes are renowned for their dietary protein and micronutrient contributions, offering a superior source of protein compared to globally consumed staples like maize, and are rich in essential micronutrients (De Jager et al., 2019).

In sub-Saharan Africa (SSA), grain legumes are classified into major and minor species based on utility and economic significance. Major legumes are well established, with good domestication, cultivation, and utilization practices, whereas minor species, often termed orphan or neglected legumes, remain underutilized due to limited research, breeding, and awareness (Paliwal et al., 2021). These species include winged bean (*Psophocarpus tetragonolobus* L.), pigeon pea (*Cajanus cajan* L.), lablab (*Lablab purpureus* L.), lima bean (*Phaseolus lunatus* L.), jack and sword bean (*Canavalia* sp.), mung bean (*Vigna mungo* L.), Bambara groundnut (*Vigna subterranea* L.), kersting's groundnut (*Macrotyloma geocarpum* Harms), African yam bean (AYB) (*Sphenostylis stenocarpa* Harms), and rice bean (*Vigna angularis*

O. O. Obembe · T. T. Obadire (✉) · O. A. Bello · O. S. Aworunse
UNESCO Chair on Plant Biotechnology, Plant Science Research Cluster Department of Biological Sciences, Covenant University, Ota, Ogun State, Nigeria
e-mail: tobi.obadirepgs@stu.cu.edu.ng

P. O. Isibor, P. A. Akinduti (eds.), *Harnessing Biotechnology Tools for Product Development*, https://doi.org/10.1007/978-3-031-86002-7_8

L.), among others. These species hold untapped potential and are accessible via gene banks across African nations, serving various purposes from food to genetic improvement (Popoola et al., 2020).

Phylogenetics explores the evolutionary history and relationships among individuals or groups of organisms, including populations, species, or higher taxonomic units (Haber & Joel, 2022). These relationships are inferred using phylogenetic methods based on observed heritable traits, such as DNA sequences, protein (amino acid) sequences, or morphological characteristics. A way to visualize the phylogenetic relationship between or among species is by constructing a phylogenetic tree based on their identity values. Genetic variability within crop germplasm is crucial in breeding efforts and trait improvement (Oludare et al., 2024). DNA barcoding, employing short DNA sequences or barcodes, facilitates taxonomic characterization and phylogenetic analysis of organisms. Recommended by the Consortium for the Barcode of Life (CBOL) Plant Working Group, the chloroplast genes rbcL and matK serve as core barcode loci, though the ideal locus for DNA barcoding remains debated, with varying effectiveness across taxa (Okoth et al., 2016; Hashim et al., 2021; Ho et al., 2021).

This study sought to elucidate the phylogenetic relationships of selected legumes using rbcL and matK genes as DNA barcodes, evaluating their efficiency in classifying the Fabaceae family.

2 Materials and Method

A list of conserved underutilized legumes containing 16 species was obtained from the Genetic Resource Center at the International Institute of Tropical Agriculture (IITA). They include African Yam bean (*Sphenostylis stenocarpa*), Bambara groundnut (*Vigna subterranean*), Cowpea (*Vigna unguiculata*), Green gram/Mung beans (*Vigna radiata*), Groundnut (Arachis hypogaea), Jack bean (*Canavalia ensiformis*), Kersting's groundnut (*Macrotyloma geocarpum*), Lablab bean (*Lablab purpureus*), Lima bean (*Phaseolus lunatus*), Velvet bean (*Mucuna pruriens*), Pigeon pea (*Cajanus cajan*), Horse gram (*Macrotyloma uniflorum*), Sword bean (*Canavalia gladiata*), Wild vigna (*Vigna heterophylla*), common bean/kidney bean (*Phaseolus vulgaris*), and Winged bean (*Psophocarpus tetragonolobus*).

This study employed amino acid and nucleotide sequences of the 16 plant species and the complete chloroplast genome of 11 species. The rbcL and MatK gene sequences of 15 species were retrieved from the NCBI (GenBank) database (2000–2024). Kersting's groundnut rbcL and matK genes were sequenced.

The seeds of Kersting's groundnut (*Macrotyloma geocarpum*) also known as *Kerstingiella geocarpa* were collected from the Genetic Resource Center at International Institute of Tropical Agriculture and cultivated in the screen house. Fresh leaves were harvested and used for DNA extraction after 21 planting days.

2.1 Primer Designing

The nucleotide sequences of rbcL and matK from 15 underutilized legumes were obtained in the FASTA format from (NCBI National Centre for Biotechnology Information) GenBank database (https://www.ncbi.nlm.mih.gov/genbank/). Consensus sequences were generated for each gene and utilized for primer design. The primers were designed using NCBI's web-based Primer-BLAST (Basic Local Alignment Search Tool). These primers, as shown in Table 8.1, were utilized to amplify the rbcL and matK genes of *Macrotyloma geocarpum*.

2.2 Amplification and Sequencing

Genomic DNA was extracted from young fresh leaves of *Macrotyloma geocarpum* using the Quick-DNA Plant/Seed Kits (Zymo Research, Catalogue No. D6020), following the manufacturer's guidelines. The quality and quantity of the extracted DNA were assessed with a Thermo Scientific™ NanoDrop™ One Microvolume UV-Vis Spectrophotometer. Target regions for two genes were amplified via polymerase chain reaction (PCR) using OneTaq® Quick-Load® 2× Master Mix (NEB, Catalogue No. M0486), under conditions outlined in Table 8.2. PCR was conducted on an Eppendorf Mastercycler nexus gradient 230, beginning with an initial denaturation at 95 °C for 5 min, followed by 35 cycles of 95 °C for 30 s, 53 °C for 1 min, and 68 °C for 1.5 min, with a final extension at 68 °C for 10 min. Amplicons producing a single discrete band of the expected size on 1% agarose gel electrophoresis were purified using the enzymatic ExoSAP method. Purified PCR amplicons were sequenced using Sanger sequencing with primers and the Nimagen Brilliant Dye™ Terminator Cycle Sequencing Kit V3.1 (BRD3–100/1000), following the manufacturer's instructions: https://www.nimagen.com/products/Sequencing/Capillary-Electrophoresis/BrilliantDye-Terminator-Cycle-Sequencing-Kit/. The labeled products were cleaned with the ZR-96 DNA Sequencing Clean-up Kit (Catalogue No. D4053): http://www.zymoresearch.com/downloads/dl/file/id/52/d4052i.pdf before being injected into the Applied Biosystems ABI 3500XL Genetic Analyzer equipped with a 50 cm array and POP7: https://www.thermofisher.com/order/catalog/product/4406016, for sequence data collection.

Table 8.1 Primer sequences

Gene	Primer sequence 5′ → 3′	
rbcL	GGGACTTATGTCACCACAAACA	F
	CCTTCCATACTTCACAAGCAGC	R
matK	GGACTAGCTTATGGTCATGAGTCC	F
	TGCAACATTTGACTTCGTACTACTG	R

Table 8.2 Condition for polymerase chain reaction

Component	Volumes for a 12.5 μL reaction
Template DNA	2 μL
10 μM forward primer	0.25 μL
10 μM reverse primer	0.25 μL
One Taq quick load 2× master mix with standard buffer	6.25 μL
Nuclease free water	4.25 μL

2.3 *Sequence Analysis*

The sequenced genes, provided in the ABI prism format, were processed using BioEdit software version 7.2.5 (Hall, 1999). This software was used to trim and remove background noise based on the strength and clarity of the signal peaks in the chromatogram. Forward and reverse sequences of the *rbcL* and *matK* genes were aligned using Codon Code Aligner Software (http://www.codoncode.com/) and combined into a consensus sequence (contig). These contigs were then used as queries in a BLASTn search at https://blast.ncbi.nlm.nih.gov/Blast.cgi to find homologous sequences in the NCBI nucleotide database. The appropriate reading frame was identified using Geneious Prime 2024.0.2, and the nucleotide sequences of both genes were translated into amino acid sequences using the web-based translation tool ExPaSy: https://web.expasy.org/translate/.

2.4 *Phylogenetic Analysis*

Sequences of varying lengths were retrieved and trimmed at both ends to standardize their length using BioEdit software. This software was also employed to perform Multiple Sequence Alignment (MSA) for each species' amino acid sequences, generating gene consensus sequences. The consensus sequences of the rbcL gene and their corresponding matK genes from each species were merged and aligned using the MUSCLE algorithm within the Molecular Evolutionary Genetics Analysis (MEGA) 11 software (Tamura et al., 2021). These multiple sequence alignment results were then utilized for phylogenetic analysis. MEGA 11 was used to determine the phylogenetic relationships among species by constructing evolutionary trees. Additionally, the complete chloroplast genomes of 11 species were aligned using the MUSCLE algorithm in MEGA 11, and a phylogenetic tree was constructed using the neighbor-joining algorithm. MEGA 11 is widely used in molecular biology, genetic research, biotechnology, and bioinformatics.

To validate the efficacy of the single DNA barcode for classifying the selected members of the Fabaceae family, the consensus sequences of each gene per species were also employed for the taxonomic classification of the legumes. A bootstrap analysis with 1000 replicates was conducted to evaluate clade support using the maximum parsimony, neighbor-joining, and maximum likelihood algorithms.

Bootstrap values were categorized as high (85%), moderate (70–85%), weak (50–69%), or very weak (<50%) based on the criteria set by Wardani et al. (2022). The DNA barcode-based classification was compared with the taxonomic classification derived from the chloroplast genome.

3 Results

In this study, five sequence entries were retrieved for each species spanning the years 2000–2024 for the rbcL gene, except for *C. ensiformis, M. uniflorum, P. lunatus, P. tetragonolobus, V. heterophylla, V. radiata*, and *V. subterranean*, which had less than five entries. Similarly, only *A. hypogaea* and *M. pruriens* had five sequence entries for the matK gene across the years, while other species had less than five entries, as depicted in Table 8.3.

The obtained nucleotide sequence of the rbcL gene from *M. geocarpum* was analyzed using BLASTn, revealing the closest match to *M. uniflorum* with 99.92% identity (accession number LT576832.1), followed by *S. stenocarpa* with 98.44% identity (LT576860.1), and *L. purpureus* with 98.44% identity (LT576828.1). These matches exhibited zero E-value and 100% query coverage. Subsequent analysis with BLASTp using amino acid sequences revealed the highest identity with *M.*

Table 8.3 List of plant species and number of retrieved sequences used in this study

Botanical name	Genus	Common name	MatK	rbcL
			Number of entries	Number of entries
Sphenostylis stenocarpa	*Sphenostylis*	African yam bean	1	5
Arachis hypogaea	*Arachis*	Groundnut	5	5
Cajanus cajan	*Cajanus*	Pigeon pea	4	5
Canavalia ensiformis	*Canavalia*	Jack bean	3	2
Canavalia gladiata		Sword bean	3	5
Lablab purpureus	*Lablab*	Hyacinth bean	4	5
Mucuna pruriens	*Mucuna*	Velvet bean	5	5
Macrotyloma uniflorum	*Macrotyloma*	Horse gram	3	3
Phaseolus lunatus	*Phaseolus*	Lima bean	2	4
Phaseolus vulgaris		Common/kidney beans	4	5
Psophocarpus tetragonolobus	*Psophocarpus*	Winged bean	2	1
Vigna heterophylla	Vigna	Wild Vigna	4	4
Vigna radiata		Mung bean	3	4
Vigna unguiculata		Cowpea	3	5
Vigna subterranean		Bambara groundnut	2	–

uniflorum (100%, accession number SBO07513.1), followed by *L. purpureus* (99.39%, QRI60657.1), and *C. gladiata* (99.09%, KAK7299388.1).

Likewise, the analysis of the matK gene's nucleotide sequence using BLASTn indicated the highest identity with *M. uniflorum* (99.65%, EU717410.1), followed by *S. stenocarpa* (96.58%, AY582977.1) and *L. purpureus* (95.76%, MN966634.1). For amino acid sequences of matK, the highest identity was again with *M. uniflorum* (99.26%), followed by *L. purpureus* (89.96%, YP_010046450.1), with an E-value of 5e-165.

The matK sequence of *M. geocarpum* comprised 849 bases with a nucleotide composition of adenine 32%, cytosine 15%, guanine 12%, and thymine 42%. This sequence has been deposited in the NCBI gene bank with the assigned accession number PP782344. The rbcL sequence, also deposited into the gene bank and assigned accession number PP797597, consisted of 1281 bases with a nucleotide composition of adenine 27%, cytosine 18%, guanine 24%, and thymine 31%.

Phylogenetic trees provide valuable insights into genetic classification, biodiversity, and evolutionary relationships. Eghit phylogenetic trees were constructed using the neighbor-joining (NJ), maximum parsimony (MP), and maximum likelihood (ML) algorithms based on aligned sequences of matK, rbcL, and complete chloroplast genomes.

A double DNA barcode (matK and rbcL genes) was used to classify the species studied. The matK gene sequence of each species was merged with the respective rbcL gene sequence (matK + rbcL) to construct a tree using a neighbor-joining algorithm. It was observed that the sequences were classified randomly, and most species of the same genus were not placed in a clade, as shown in Fig. 8.1. To clarify the classification of the species, a complete chloroplast genome was used to construct a phylogenetic tree using the neighbor-joining algorithm, as shown in Fig. 8.2. It was observed that species of the same genus were grouped together. Moreover, the genome sequence of *V. unguiculata, C. cajan,* and *P. tetragonolobus* was placed in the same clade on the tree. Furthermore, rbcL and matK genes were used independently to classify the species and confirm the individual gene's effectiveness in classifying the members of the Fabaceae family. All phylogenetic analyses of the matK gene showed that the species in the *Canavalia* and *Macrotyloma* genera and those in the *Vigna* genus clustered together based on their genera, as shown in the trees. Moreover, Fig. 8.3 depicts separate clades in the neighbor-joining phylogenetic tree derived from matK gene sequences. The matK sequence of *V. unguiculata* was found in the same group with *C. cajan* and *S. stenocarpa* sequences; nevertheless, *C. cajan* shares a high level of similarity with *V. unguiculata*, as demonstrated by a high bootstrap value in both trees created by neighbor-joining and maximum parsimony in Fig. 8.4. This shows that *C. cajan*, an underutilized legume, is closely related to *V. unguiculata*. However, the tree constructed using the maximum likelihood algorithm placed *V. unguiculata* in the same group as *Macrotyloma* species, although it showed a weak bootstrap value. It was further observed that the matK gene sequences of *Phaseolus vulgaris,* another major legume, were found in a clade with *Lablab purpureus* in all the trees, suggesting the relatedness of the two species (Fig. 8.5).

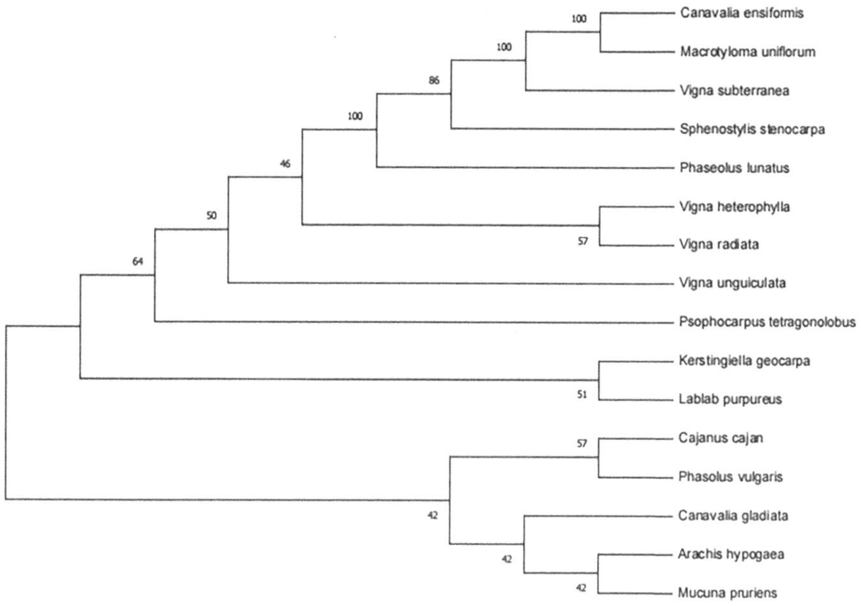

Fig. 8.1 Neighbor-joining tree of Fabaceae species based on matK + rbcL sequences

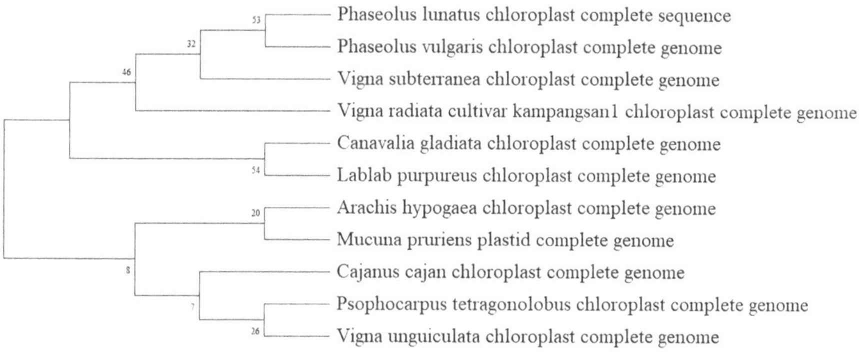

Fig. 8.2 Neighbor-joining tree of Fabaceae species based on the chloroplast genome

In the phylogenetic trees constructed using rbcL amino acid sequences (Figs. 8.6, 8.7, and 8.8), it was observed that the species are randomly grouped, and no species from the same genus are placed together or in the same clade. The rbcL gene sequence of *V. unguiculata* clustered with varying underutilized legumes in the three analyses but close to *S. stenocarpa* in maximum parsimony and maximum likelihood trees but with weak bootstrap values. *Phaseolus vulgaris* was found in the same clade as *V. unguiculata* and *S. stenocarpa* in the MP and ML trees but with *C. cajan* in the NJ tree.

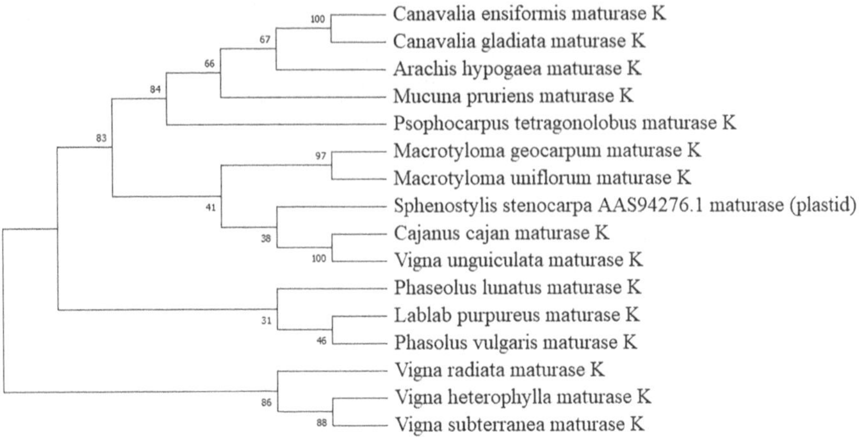

Fig. 8.3 Neighbor-Joining tree of Fabaceae species based on the matK dataset

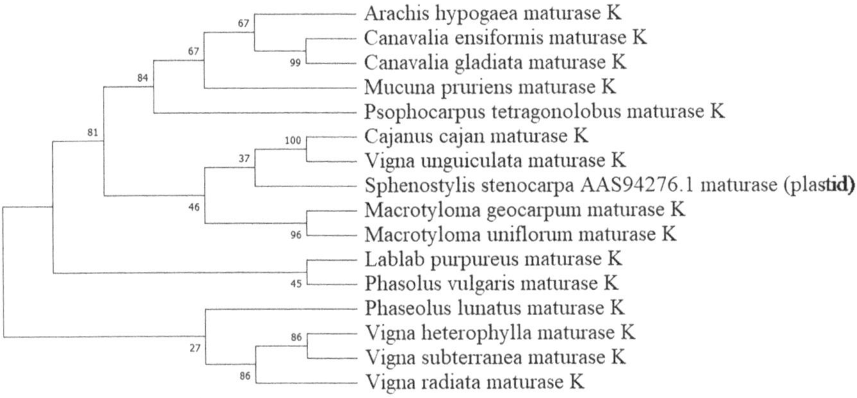

Fig. 8.4 Maximum parsimony tree of Fabaceae species based on the matK dataset

4 Discussion

The chloroplast is an organelle within plant cells responsible for conducting photosynthesis in plants and algae. Its genome exhibits distinctive features, including uniparental inheritance, a conserved sequence structure in coding regions, multiple variable sites, and a typical quadripartite genome arrangement. In vascular plants, the chloroplast genome is highly conserved and is a crucial tool in phylogenetic, evolutionary, and population genetics studies by utilizing coding and noncoding regions (Hong et al., 2020).

The rbcL gene is recognized for its conserved sequences across various plant species and evolves significantly slower than other coding regions. In contrast, the matK gene undergoes rapid evolution. It possesses a sequence similar to the animal

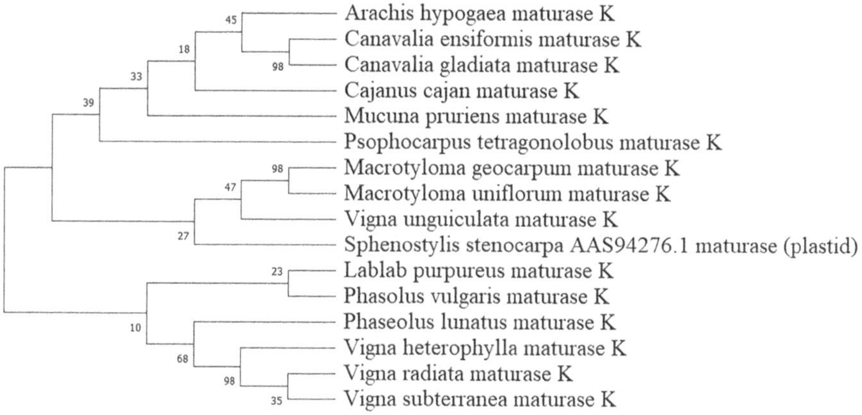

Fig. 8.5 Maximum likelihood tree of Fabaceae species based on the matK dataset

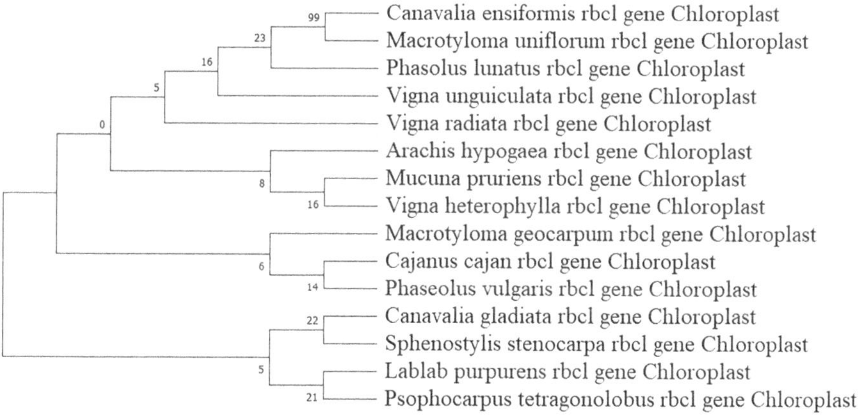

Fig. 8.6 Neighbor-joining tree of Fabaceae species based on the rbcL dataset

cytochrome oxidase subunit I (COI), which is frequently used as a primary barcode region for animal identification (Singh et al., 2019; Ho et al., 2021).

In this study, multiple sequence alignment (MSA) of matK gene sequences revealed high substitution rates within species, indicative of its rapid evolutionary rate. This observation is consistent with previous research reporting elevated substitution rates in the matK gene (Duffy et al., 2009; Labar et al., 2021; Ho et al., 2021; Su'udi, 2023). Su'udi (2023) emphasizes the universality of the matK gene, attributing its effectiveness as a DNA barcode to its high nucleotide substitution rates. Conversely, MSA of rbcL gene sequences demonstrated high conservation within species, a characteristic supported by studies across various plant families (Singh et al., 2019; Pere et al., 2023; Bello et al., 2023; Bare et al., 2024). Notably, matK exhibits nucleotide sequence substitutions at three times the rate and amino acid

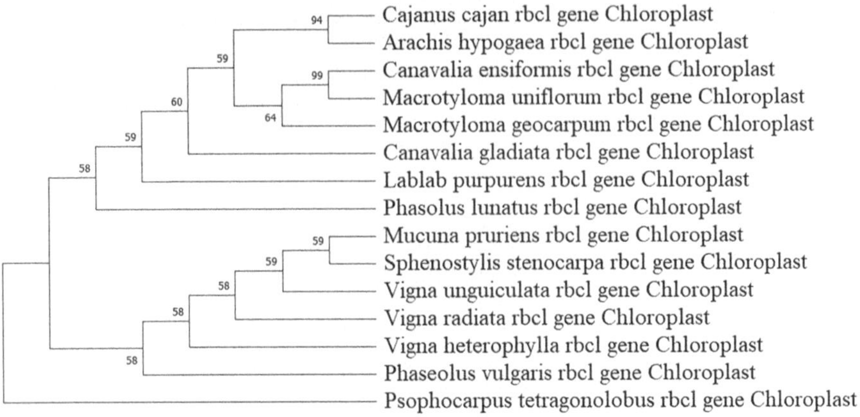

Fig. 8.7 Maximum parsimony tree of Fabaceae species based on the rbcL dataset

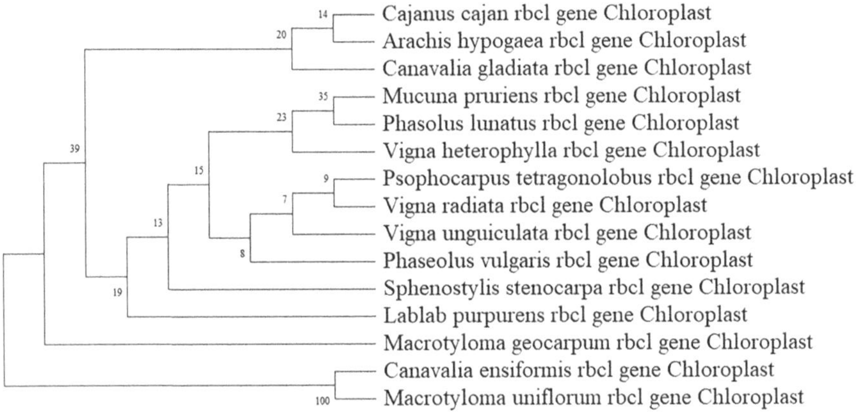

Fig. 8.8 Maximum likelihood tree of Fabaceae species based on the rbcL dataset

sequence substitutions at six times the rate of rbcL, further highlighting its evolutionary significance (Su'udi, 2023).

Plant DNA barcodes, which are suitable for classifications above the species level, facilitate the analysis of germplasm genetic diversity by revealing genetic relationships both within and between species of the same genus (Wang et al., 2022). In this study, classification based on combined double barcodes resulted in a random grouping of members within the Fabaceae family, indicating that the combination of these two genes is unsuitable for taxonomic or evolutionary classification. While the use of multilocus barcodes (double barcodes) has been shown to enhance species classification in numerous studies (Ho et al., 2021; Trang et al., 2015; Ismail et al., 2020), in this case, the combination of the two barcode genes was less effective in classifying species at the genus level compared to using single barcodes alone. This outcome is consistent with previous findings in various trees and plants (Raveenadar et al., 2018; Ho et al., 2021).

Classification using the matK gene alone extended to both genus and species levels, effectively grouping species within the same genus and identifying the closest underutilized legumes to cowpea. This demonstrates the matK gene's effectiveness in classifying members of the Fabaceae family at both genus and species levels. Ho et al. (2021) stated that the resolving power of a barcode depends on its ability to distinguish and identify species based on interspecific sequence differences. Wardani et al. (2022) reported that matK gene sequences provided the highest interspecific and intraspecific distances and the best accuracy (82.05%) for identifying Fabaceae species. The efficacy of the matK gene in classifying plants at the species and genus levels was further supported by Worthy et al. (2022), who facilitated detailed exploration of interspecies and intraspecies genetics. Amandita et al. (2019) also found that the matK gene achieved higher overall species identification success (78.3%) compared to the rbcL gene (71.4%) in their best-close match study. Similarly, Alasmari (2020) noted the advantage of matK over rbcL in plant classification.

In this study, rbcL gene analyses yielded inconsistent results, as cowpea was grouped with various underutilized legumes in the phylogenetic trees, suggesting that the rbcL gene did not provide a clear and well-defined classification. Notably, although there is no definitive report on the closest underutilized species to cowpea, the analysis conducted in this study using complete chloroplast genomes of 11 species produced results similar to those based on the matK gene.

4.1 Limitations of the Study

Most of the legumes have less than five entries across the proposed years. More so, only 11 complete chloroplast genomes of the 16 studied plants are available on the database due to numerous underutilized legume genomes yet to be sequenced. As a result, it limited the number of taxa available for investigation.

4.2 Conclusion

The study investigated the efficacy of rbcL and matK genes in classifying 16 members of the Fabaceae family and identifying the underutilized legume closest to cowpea. The combination of the two barcode genes was not better in classifying than a single one, and the results suggest that the classification based on the matK gene is superior to that of the rbcL gene, which was corroborated by the results obtained from the complete chloroplast genome. The results of this study clearly showed that pigeon peas (*Cajanus cajan*) are closely related to cowpeas (*Vigna unguiculata*).Conflicts of InterestNo conflicts of interest.

References

Alasmari, A. (2020). DNA-barcoding of some medicinal plant species in Saudi Arabia using rbcL and matK genes. *Phyton-International Journal of Experimental Botany., 89*(4), 1059–1081. https://doi.org/10.32604/phyton.2020.010952

Amandita, F. Y., Rembold, K., Vornam, B., Rahayu, S., Siregar, I. Z., Kreft, H., & Finkeldey, R. (2019). DNA barcoding of flowering plants in Sumatra, Indonesia. *Ecology and Evolution., 9*(4), 1858–1868. https://doi.org/10.1002/ece3.4875

Bare, N. B., Jadhav, P. S., & Ponnuchamy, M. (2024). DNA barcoding for species identification and phylogenetic investigation employing five genetic markers of *Withania coagulans*. *Journal of Applied Biology and Biotechnology., 12*(1), 69–76. https://doi.org/10.7324/JABB.2023.145330

Bello, R. O., Ojokoh, A. O., Babalola, D., Shonde, T. E., & Towobola, F. M. (2023). DNA barcoding: Efficiency of rbcl gene for sequence characterization of Cyperus Esculentus in Nigeria. *Research Square*, 1–13. https://doi.org/10.21203/rs.3.rs-2386798/v1

Britannica, T. Editors of Encyclopaedia (2023, January 10). Fabaceae. Encyclopedia Britannica. https://www.britannica.com/plant/Fabaceae

De Jager, I., Borgonjen-van den Berg, K. J., Giller, K. E., & Brouwer, I. D. (2019). Current and potential role of grain legumes on protein and micronutrient adequacy of the diet of rural Ghanaian infants and young children: Using linear programming. *Nutritional Journal, 18*(12), 1–16. https://doi.org/10.1186/s12937-019-0435-5

Duffy, A. M., Kelchner, S. A., & Wolf, P. G. (2009). Conservation of selection on matK following an ancient loss of its flanking intron. *Gene, 438*(1–2), 17–25.

Graham, H. P., & Vance, C. P. (2003). Legumes: Importance and constraints to greater use. *Journal of Plant Physiology, 131*(3), 872–877. https://doi.org/10.1104/pp.017004

Haber, M., & Joel, V. (2022). Phylogenetic inference. In N. Z. Edward & U. Nodelman (Eds.), *The Stanford encyclopedia of philosophy (fall 2022 edition)*. https://plato.stanford.edu/archives/fall2022/entries/phylogenetic-inference/

Hall, T. A. (1999). BioEdit, a user-friendly biological sequence alignment editor and analysis program for windows 95/98/NT. *Nucleic Acids Symposium Series, 41*, 95–98.

Hashim, A. M., Alatawi, A., Altaf, F. M., Qari, S. H., Elhady, M. E., Osman, G. H., & Abouseadaa, H. H. (2021). Phylogenetic relationships and DNA barcoding of nine endangered medicinal plant species endemic to Saint Katherine protectorate. *Saudi Journal of Biological Science., 28*(3), 1919–1930. https://doi.org/10.1016/j.sjbs.2020.12.043

Ho, V. T., Tran, T. K. P., Vu, T. T. T., et al. (2021). Comparison of matK and rbcL DNA barcodes for genetic classification of jewel orchid accessions in Vietnam. *Journal of Genetic Engineering and Biotechnology, 19*, 93. https://doi.org/10.1186/s43141-021-00188-1

Hong, Z., Wu, Z., Zhao, K., Yang, Z., Zhang, N., Guo, J., Tembrock, L. R., & Xu, D. (2020). Comparative analyses of five complete chloroplast genomes from the genus *Pterocarpus* (Fabaceae). *International Journal of Molecular Sciences., 21*(11), 3758. https://doi.org/10.3390/ijms21113758

Ismail, M., Ahmad, A., Nadeem, M., Javed, M. A., Khan, S. H., Khawaish, I., Sthanadar, A. A., Qari, S. H., Alghnem, S. M., Khan, K. A., Khan, M. F., & Qamer, S. (2020). Development of DNA barcodes for selected *Acacia* species by using rbcL and matK DNA markers. *Saudi Journal of Biological Science, 27*(12), 3735–3742. https://doi.org/10.1016/j.sjbs.2020.08.020

Labar, R., Kar, P., Biswas, P., Sen, A., & Bhattacharya, M. (2021). Evolution of matK gene among the elite tea clones (Camellia sinensis) revealed by nucleotide substitution within the consensus region. *Journal of Applied Biology and Biotechnology, 9*(1), 32–40. https://doi.org/10.7324/JABB.2021.9105

Moss, W. M., Nichols, P. G. H., Foster, K. J., Ryan, M. H., Erskine, W., & Guzzomi, A. L. (2022). A century of subclover: Lessons for sustainable intensification from a historical review of innovations in subterranean clover seed production. *Advances in Agronomy, 171*, 305–339. https://doi.org/10.1016/bs.agron.2021.08.003

Okoth, P., Muoma, J., Emmanuel, M., Omayio, D. O., Angienda, P. O., & Clabe, W. (2016). The potential of DNA barcode-based delineation using seven putative candidate loci of the plastid region in inferring molecular diversity of cowpea at sub-species level. *American Journal of Molecular Biology, 6*, 138–158.

Oludare, A. J., Adebare, I. J., Ibrahim, I. A., Modinat, A. A., Odoh, I., Bola, A. A., & Kioko, J. I. (2024). Phenotypic evaluation of traits and barcode identification using plastid-specific ribulose bisphosphate carboxylase (Rbcl) analysis on lablab beans [*Lablab purpureus* (L.) sweet]. *Agricultural Science Digest, 44*(1), 81–87. https://doi.org/10.18805/ag.AF-729

Paliwal, R., Adegboyega, T., Abberton, M., Faloye, B., & Oyatomi, O. (2021). Potential of genomics for the improvement of underutilized legumes in sub-Saharan Africa. *Legume Science, 3*(3), e69. https://doi.org/10.1002/leg3.69

Pere, K., Mburu, K., Muge, E. K., Wagacha, J. M., & Nyaboga, E. N. (2023). Molecular discrimination and phylogenetic relationships of physalis species based on ITS2 and rbcL DNA barcode sequence. *Crops, 3*(4), 302–319. https://doi.org/10.3390/crops3040027

Popoola, J., Ojuederie, O., Omonhinmin, C., & Adegbite, A. (2020). *Neglected and underutilized legume crops: Improvement and future prospects*. IntechOpen. https://doi.org/10.5772/intechopen.87069

Raveenadar, S., Lee, G. A., Lee, J. R., Lee, K. J., Lee, S. Y., Cho, G. T., Ma, K. H., & Chun, J. W. (2018). DNA barcodes for the assessment of phylogenetic relationships based on CpDNA and NrDNA regions in *Vigna* species. *Plant Breeding and Biotechnology., 6*(3), 285–292. https://doi.org/10.9787/PBB.2018.6.3.285

Singh, A., Garg, D., Negi, G., & Mohanty, A. (2019). In silico analysis of sequence variation in rbcL gene to assess phylogenetic relations in Setaria species. International conference on innovative research in agriculture, food science, forestry, horticulture, aquaculture, animal sciences, biodiversity, ecological sciences and climate change (AFHABEC-2016). https://www.researchgate.net/publication/336989611. pp. 16–18.

Su'udi, M. (2023). The matK mini-barcode as a potential molecular identification tool for medicinal orchids. *Life Science and Biotechnology, 1*(1), 21–26.

Tamura, K., Stecher, G., & Kumar, S. (2021). MEGA 11: Molecular evolutionary genetics analysis version 11. *Molecular Biology and Evolution, 38*, 3022–3027. https://doi.org/10.1093/molbev/msab120

Trang, N. T. P., Duc, N. M., Sinh, N. V., & Triest, L. (2015). Application of DNA barcoding markers to the identification of *Hopea* species. *Genetic and Molecular Research., 14*(3), 9181–9190. https://doi.org/10.4238/2015

Wang, J., Yan, Z., Zhong, P., Shen, Z., Yang, G., & Ma, L. (2022). Screening of universal DNA barcodes for identifying grass species of Gramineae. *Frontiers of Plant Science, 13*, 998863. https://doi.org/10.3389/fpls.2022.998863

Wardani, I. G. A. K., Amandita, F. Y., Moura, C. C. D. M., Gailing, O., & Siregar, I. Z. (2022). Molecular taxonomy via DNA barcodes for species identification in selected genera of Fabaceae. *Journal of Natural Resources and Environmental Management., 12*(1), 112–122. https://doi.org/10.29244/jpsl.12.1.112-122

Worthy, S. J., Bucalo, K., & Perry, E. (2022). The ability of rbcL and matK DNA barcodes to discriminate between montane forest orchids. *Plant Systematics and Evolution, 308*, 19. https://doi.org/10.1007/s00606-022-01809-z

Chapter 9
Probiotic Properties and Safety Profiling of *Lactobacillus* Species Obtained from Fermented Beverages in Ota, Nigeria

Olubukola Oziegbe, Azeta Iziegbe Joy, and Yemisi Dorcas Obafemi

1 Introduction

Fermented beverages have long been a fundamental component of the human diet, renowned for their distinctive flavors and associated health-promoting properties. These beverages are produced through the natural fermentation of food substrates by microorganisms, including bacteria and yeasts (Dimidi et al., 2019). In Nigeria, traditional fermented beverages such as palm wine and *burukutu* (fermented sorghum/millet beer) are popular. Palm wine is a natural alcoholic beverage obtained from the sap of various palm trees, while *burukutu* is a traditional alcoholic drink made from fermented sorghum or millet (Nwaiwu et al., 2020). Fermented beverages are a staple in many cultures, including Nigeria, where they are part of the daily diet and traditional practices. These beverages harbor a diverse range of microorganisms, some of which possess probiotic properties that can confer health benefits to the host (Mgbodile & Nwagu, 2023). Probiotics are live microorganisms that, when administered in adequate amounts, confer health benefits to the host. They are primarily known for their role in maintaining gut health by enhancing gut flora, inhibiting pathogenic bacteria, and modulating the immune system (Plaza-Diaz et al., 2019). Among the various genera of probiotics, *Lactobacillus* species are extensively studied and widely utilized due to their robust probiotic properties and historical use in fermented foods (Al-Yami et al., 2022). *Lactobacillus* species are commonly found in fermented foods and are known for their health-promoting effects. These effects include improvement of intestinal health,

O. Oziegbe · A. I. Joy (✉) · Y. D. Obafemi
Department of Biological Sciences, Covenant University, Ota, Ogun State, Nigeria
e-mail: joy.azetapgs@stu.cu.edu.ng

P. O. Isibor, P. A. Akinduti (eds.), *Harnessing Biotechnology Tools for Product Development*, https://doi.org/10.1007/978-3-031-86002-7_9

enhancement of the immune response, and protection against gastrointestinal infections (Dempsey & Corr, 2022). The ability of *Lactobacillus* species to survive gastrointestinal transit, adhere to intestinal epithelial cells, and inhibit pathogenic microorganisms are key attributes that contribute to their probiotic potential. The consumption of probiotics, live microorganisms that confer health benefits when administered in adequate amounts, has gained significant attention in recent years due to their potential to promote gut health and overall well-being (Han et al., 2021). Among the diverse array of probiotic microorganisms, *Lactobacillus* species have been extensively studied for their potential health-promoting effects. These bacteria are commonly found in fermented foods and beverages, where they play a key role in the fermentation process and contribute to the characteristic flavor and texture of the final product. The growing interest in probiotics, especially *Lactobacillus* species from fermented beverages, is increasingly gaining attention due to their significant role in promoting health and enhancing food quality. The genus *Lactobacillus* is especially notable for its contributions to gut health, immune function, and overall well-being, making it a focal point for probiotic research. This research aims to comprehensively analyze the probiotic potential and safety profile of *Lactobacillus* species isolated from locally fermented beverages in Ota, Nigeria. By investigating their morphological, biochemical, and safety characteristics, this study seeks to bridge the gap between traditional food practices and modern scientific understanding, potentially leading to the development of new probiotic products. Locally fermented beverages are an integral part of the Nigerian culinary culture, and they have been found to harbor a diverse range of microorganisms, some of which may have probiotic potential (Mgbodile & Nwagu, 2023). Fermented beverages, such as kefir, kombucha, and traditional fermented milk products, have been traditionally associated with health benefits, and the presence of *Lactobacillus* species in these beverages has been linked to their probiotic potential (Dimidi et al., 2019). In recent years, there has been a growing interest in understanding the specific probiotic properties of *Lactobacillus* species isolated from fermented beverages, as well as their safety profile for human consumption (Huligere et al., 2023). Several studies have highlighted the potential probiotic properties of *Lactobacillus* species isolated from fermented beverages, including their ability to survive passage through the gastrointestinal tract, adhere to intestinal epithelial cells, modulate the gut microbiota, and exert antimicrobial and immunomodulatory effects (Garcia-Gonzalez et al., 2021). However, the probiotic potential of these strains must be rigorously evaluated to determine their suitability for human consumption (Binda et al., 2020). Furthermore, the safety assessment of *Lactobacillus* species obtained from fermented beverages is crucial to ensure the absence of pathogenic or harmful traits that could pose a risk to human health (Žuntar et al., 2020). While *Lactobacillus* species are generally recognized as safe (GRAS) and have a long history of safe use in food fermentation, it is essential to conduct thorough safety evaluations, including the assessment of antibiotic resistance and hemolysis evaluation (Colautti et al., 2022).

2 Materials and Methods

2.1 *Ethical Approval*

The ethical approval was obtained from the Covenant Health Research Ethics Committee (CHREC), with the approval number CU/HREC/AJI/394/24.

2.2 *Sampling*

Six locally fermented drink samples (palm wine, $n = 3$; and burukutu, $n = 3$) were purchased from local markets in Ota, Ogun State, Nigeria (Fig. 9.1). Upon procurement, the samples were transported under cold chain conditions to the Microbiology Laboratory in the Department of Biological Sciences, Covenant University. All samples were then stored at 4 °C until further analysis.

2.3 *Proximate Analysis of the Fermented Drink Samples*

The proximate analysis of the fermented drink samples was carried out to assess their nutritional composition. The following analysis was done on the fermented drink samples.

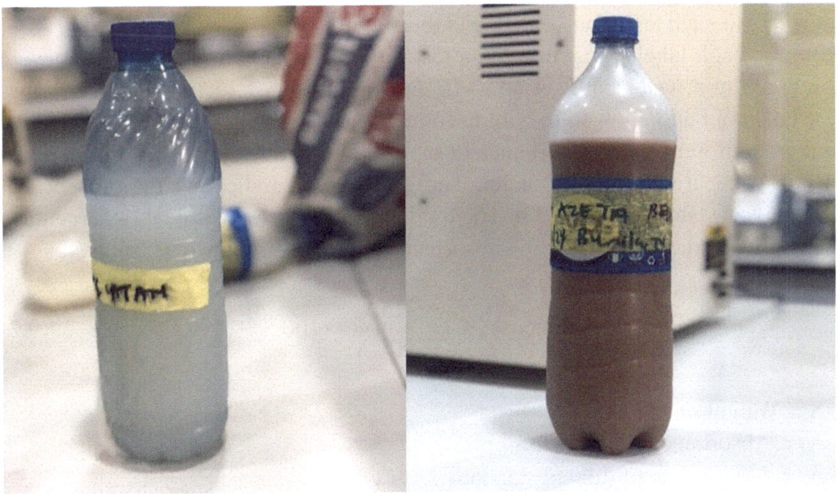

A- Palmwine
(a traditional fermented drink)

B-*Burukutu* (fermented sorghum/millet beer)

Fig. 9.1 Fermented drink samples (palm wine and *burukutu*). (**a**) Palm wine (a traditional fermented drink). (**b**) *Burukutu* (fermented sorghum/millet beer)

2.3.1 Total Ash Content

The ash content was determined using the furnace incineration gravimetric method as previously described by AOAC (2023). A previously weighed porcelain crucible was filled with 5.0 g of the treated sample. After that, the sample was burned to ashes at 550 °C in a muffle furnace. The sample was weighed after it had been fully ashed and cooled in a desiccator. The weight difference between the initial and final weights was used to calculate the weight of the ash that was produced. This weight of ash was then calculated as a percentage of the weight of the sample analyzed as follows:

$$\text{Ash}(\%) = \frac{W_2 - W_1}{\text{Weight of sample}} \times \frac{100}{1} \tag{9.1}$$

where

W_1 = Weight (g) of empty crucible

W_2 = Weight of crucible + Ash

2.3.2 Carbohydrate Content

The carbohydrate content was determined using the method of James. 45 mL of each sample extract were diluted to 450 mL with distilled water. 1 mL of each diluted filtrate was pipetted into different test tubes. Additionally, a test tube containing 1 mL of water was pipetted as a blank, and a test tube containing 1 mL of glucose was pipetted as a standard. 5 mL of newly made 0.10% Anthrone reagent was added to each test tube, sealed, and thoroughly mixed by gently shaking. After setting up each labeled test tube on a test tube rack, the tubes and the rack were submerged in a water bath set at 30 °C for a duration of 12 min. The test tubes were taken out and allowed to cool to room temperature after incubation. Using a blank as a reference, the absorbance of the samples and standard was measured at 630 nm using a spectrophotometer. The green color, indicating the presence of glucose, was stable for about 2 h. The total available carbohydrate as a percentage of glucose is calculated using the absorbance readings as shown as follows:

$$\text{Glucose}(\%) = \frac{25A_1}{X \times A_2} \times 100 \tag{9.2}$$

X = Weight of sample (g)

A_1 = Absorbance of diluted sample

A_2 = Absorbance of diluted standard

2.3.3 Protein Content

This protein content was determined using the Kjeldahl method as described by AOAC (2023). The total nitrogen was determined and multiplied with a factor of 6.25 to obtain protein content. A sample (0.5 g) was mixed with 10 mL of concentrated H_2SO_4 in a digestion. The mixture was heated under a fume cupboard with a tablet of selenium catalyst added until a clear solution known as the digest was achieved. For the analysis, the digest was diluted to 100 mL in a volumetric flask. To neutralize the digest, 10 mL of the digest was combined with an equivalent volume of 45% NaOH solution in a Kjeldahl distillation unit. After the liquid was neutralized, 10 mL of 40% boric acid were added, along with three drops of a mixed indicator (methyl red and bromocresol green). A total of 50 mL of distillate was collected. The collected distillate was titrated from green to a deep red endpoint using 0.02 N EDTA. Additionally, a reagent blank was titrated, distilled, and digested. To calculate the protein content, the total nitrogen content from the titration was multiplied by a factor of 6.25. The following formula was used to determine the nitrogen content and subsequently the protein content:

$$1\,\mathrm{L\,of\,1\,N\,H_2SO_4} = 14\,\mathrm{mg}$$

$$\mathrm{Protein}(\%) = \mathrm{N_2}(\%) \times 6.25 \tag{9.3}$$

$$\mathrm{N2}(\%) = \frac{100}{\mathrm{W}} \times \frac{\mathrm{N} \times 14}{1000} \times \frac{\mathrm{Vt}}{\mathrm{Va}} \times \mathrm{T.B}$$

where

W = Weight of sample (0.5 g)
N = Normality of titrant (0.002 N H_2SO_4)
Vt = Total digest volume (100 mL)
Va = Volume of digest analyzed (10 mL)
T = Sample titer value
B = Blank titer value

2.3.4 Crude Fat Content

The crude fat content was determined by solvent extraction gravimetric method as described by AOAC. A thimble was filled with 5 g of the sample, which had been wrapped in Whatman filter paper, a porous material. The thimble containing the sample was placed in a Soxhlet reflux flask, which was mounted into a weighted extraction flask containing 200 mL of petroleum ether. The upper part of the reflux flask was connected to a water condenser. In order to fill the reflux flask and transfer its oil extract to the boiling flask, the solvent (petroleum ether) was heated, boiled, evaporated, and condensed. For 4 h, this process was permitted to continue repeatedly. The oil extract remained in the flask after the defatted sample was taken out and the solvent was extracted after 4 h. The flask containing the oil extract was dried in the oven at 60 °C for 30 min to remove any residual solvent. It was then cooled in a desiccator and weighed. The weight of the oil (fat) extract was determined by the difference and calculated as a percentage of the weight of the sample analyzed as follows:

$$\text{Fat}\left(\%\right) = \frac{W_2 - W_1}{\text{Weight of sample}} \times \frac{100}{1} \tag{9.4}$$

where
W_1 = Weight (g) of empty extraction flask
W_2 = Weight of flask + oil (fat) extract

2.4 *Enumeration of* Lactobacillus *Species*

Isolation of *Lactobacillus* species from the fermented samples was carried out using the pour plate technique described by Talib et al. 1 mL of each of the fermented drink samples was serially diluted into 9 mL of sterile distilled water. 1 mL of each diluent factor 10^4 and 10^6 was inoculated on de Mann Rogosa and Sharpe (MRS) Agar and incubated anaerobically at 37 °C for 48 h in triplicates. After incubation, distinct colonies are enumerated and subcultured severally into sterile MRS agar. The isolated *lactobacillus* colonies were phenotypically identified and pure *lactobacillus* cultures were kept in MRS slant and stored at 4 °C for further analysis.

2.5 *Phenotypical Identification of* Lactobacillus *Species*

The isolated colonies were evaluated for morphological characteristics, Gram staining, and biochemical tests using the method described by Abid et al. (2022).

2.6 *Analytical Profile Index (API)*

A pure colony of bacteria isolated from 20 h of culture was thoroughly emulsified in sterile 5 mL 0.85% NaCl to make a homogenous suspension. Each ampoule of API was filled with 0.5 mL broth culture suspension, while ampoules containing L-adenine, L-lysin, L-ornithine, hydrogen sulfide, and urea were overlaid with mineral oil to create anaerobiosis. The inoculated strip was incubated at 37 °C for 18–24 h. After the incubation period, each ampoule color change was read, while other reagents were added to L-tryptophan, Indole, Voges-Proskauer, and nitrate tests to obtain the final reaction. The color change seen was interpreted accordingly and scored to form a numerical profile on the result sheet. The values corresponding to positive reactions within each group were put together to form a seven-digit profile number for each of the 20 tests in API 20E tests. The oxidase reaction constitutes the 21st test with its own specific value after the test. The Analytical Profile Index was formed into a nine-digit profile with the addition of the motility index value. This nine-digit value provides the identification code for the bacteria isolate, which was used to extrapolate the bacteria identification in the Analytical Profile Index list.

2.7 *Probiotic Activity*

The identified *Lactobacillus* was screened for probiotic activity including pH, bile, and phenol tolerance.

2.7.1 Acid Tolerance

The acid tolerance of the selected *Lactobacillus* isolates was tested to assess their ability to survive in acidic conditions. The isolates were cultured in MRS broth with varying pH levels at 2.0, 4.0, 5.5, and 7.0. After incubation, the survival rate was determined at 630 nm absorbance for 0 and 24 h. A 100 μL of each *lactobacillus* culture at different pH was inoculated on sterile MRS agar and the plate was incubated for 24 h. The colonies were counted and the survival rate was calculated using Eq. (9.5):

$$\text{pH tolerance}(\%) = \frac{\text{Number of viable cells in cfu / mL}}{\text{Number of initial viable cells}} \times 100 \qquad (9.5)$$

2.7.2 Bile Salt Tolerance

The bile salt tolerance of the selected *Lactobacillus* isolates was evaluated for different concentrations of bile salts. The *Lactobacillus* isolates were exposed to MRS broth containing different concentrations of bile salts (0.4%, 0.6%, 0.8% and 1.0%). After incubation, the absorbance at 630 nm was measured at 0 and 24 h to determine the survival rate of the isolates. A 100 μL of each *lactobacillus* culture at different bile salt concentrations was inoculated on sterile MRS agar and the plate was incubated for 24 h. The colonies were counted, and the survival rate was calculated using Eq. (9.6):

$$\text{Bile tolerance}(\%) = \frac{\text{Number of viable cells in cfu / mL}}{\text{Number of initial viable cells}} \times 100 \qquad (9.6)$$

2.7.3 Phenol Tolerance

The phenol tolerance of the selected *Lactobacillus* isolates was evaluated for different concentrations of phenol. The *Lactobacillus* isolates were exposed to MRS broth containing different concentrations of phenol (0.3%, 0.6%, 0.9%, and 1.0%). After incubation, the absorbance at 630 nm was measured at 0 and 24 h to determine the survival rate of the isolates. A 100 μL of each *lactobacillus* culture at different phenol concentrations was inoculated on sterile MRS agar and the plate was incubated for 24 h. The colonies were counted and the survival rate was calculated using Eq. (9.7):

$$\text{Bile tolerance}\left(\%\right)=\frac{\text{Number of viable cells in cfu / mL}}{\text{Number of initial viable cells}}\times 100 \tag{9.7}$$

2.8 *Safety Evaluation*

2.8.1 Antibiotic Susceptibility Testing

Antibiotic susceptibility of the selected *Lactobacillus* with probiotic properties was evaluated with the agar diffusion technique using various classes of antibiotics that include ceftazidime (30 μg), ciprofloxacin (5 μg), cefuroxime (30 μg), nitrofurantoin (300 μg), gentamicin (10 μg), augmentin (30 μg), cefixime (5 μg), and ofloxacin (5 μg). Each *Lactobacillus* strain was diluted to 0.5 McFarland standard, inoculated onto MRS agar plates, and allowed to dry, before placing the antibiotics-impregnated disks. The plates were incubated at 37 °C for 48 h under anaerobic conditions and the observed zones of inhibition were estimated and interpreted according to the CLSI, 2020 guidelines.

2.8.2 Hemolysis Test

Hemolysis tests were performed on 5% blood agar plates to detect potential hemolytic activity. The presence of clear zones around colonies indicated hemolysis. These tests were conducted to determine the safety profiles of the isolated *Lactobacillus* strains.

3 Results

3.1 *Proximate Analysis of the Samples*

The nutritional component in the fermented drink samples is shown in Table 9.1. The sample BB exhibited the highest ash content (29.45%), while sample BA had the lowest (25.71%). Sample BB has a higher carbohydrate content at 93.95%,

Table 9.1 Proximate analysis of the fermented drink samples

	Nutritional content in samples (%)			
Sample code	Ash content	Carbohydrate	Protein	Lipid
PA	29.23	46.05	0.10	0.40
PB	26.45	37.50	0.10	0.50
PO	29.27	57.40	0.10	0.20
BB	29.45	93.95	0.10	0.75
BA	25.71	78.70	0.20	0.45
BO	25.83	86.20	0.10	1.05

while sample PB contains the least carbohydrates at 37.50%. Protein content in the samples is notably low across all samples, ranging between 0.10% and 0.20%. The highest protein content is observed in sample BA at 0.20%. Lipid content ranged from 0.2% to 1.05%. Sample BO had the highest lipid content (1.05%), indicating it had the highest fat content among the samples. Sample PO had the lowest lipid content at 0.20%.

3.2 *Total Bacterial Count of* Lactobacillus *Isolates from Fermented Drinks*

Table 9.2 presents the total bacterial count of *Lactobacillus* species from fermented drinks. Sample BA exhibited a high *Lactobacillus* count (9.5×10^7 cfu/mL), suggesting favorable conditions for *Lactobacillus* growth and having the highest *Lactobacillus* population among all samples in the fermented drink. In comparison, palm wine (PA) had a lower count (6.7×10^6 cfu/mL), and PB presented the lowest counts for palm wine with (3.7×10^5 cfu/mL) indicating less optimal fermentation. Samples BO and PO showed lower *Lactobacillus* counts with 1.7×10^7 cfu/mL and 0.7×10^9 cfu/mL, respectively.

3.3 *Morphological and Gram Stain Reaction of Bacterial Isolate from Fermented Drink Samples*

Based on the morphological characteristics and the results of the biochemical tests, all isolates were presumptively identified as *Lactobacillus* species. The consistent Gram-positive rod morphology and negative catalase reaction support this identification. Additionally, the positive methyl red test for most isolates aligns with typical *Lactobacillus* metabolic pathways (Table 9.3).

Table 9.2 Total bacterial count of *Lactobacillus* isolates from fermented drinks

Fermented drink	Sample code	Total *Lactobacillus* count (cfu/mL)
Burukutu	BA	9.5×10^7
Palm wine	PA	6.7×10^6
Burukutu	BB	6.5×10^5
Palm wine	PB	3.7×10^5
Burukutu	BO	1.7×10^7
Palm wine	PO	0.7×10^9

BB, BA, BO: Burukutu samples
PA, PO, PB: Palm wine samples

3.4 Identification of Lactobacillus *Species Using API*

Table 9.4 provides the physiological identification of *lactobacillus* organisms isolated from fermented drink samples using an API (Analytical Profile Index) system. Each row corresponds to an isolate and displays the results of various biochemical tests used to identify the bacterial species. *Lactobacillus plantarum* was identified in isolates BB05 and PB03 based on the positive results for numerous carbohydrate fermentations and decarboxylase activities, *Lactobacillus acidophilus* was identified in isolates B006 and PA03, characterized by positive results for certain fermentations and negative results for decarboxylase activities, *Lactobacillus fermentum* was identified in isolates BA01 and BB06, showing similar biochemical profiles, and *Lactobacillus* spp. was identified in isolate BA04, displaying a distinct biochemical pattern compared to the other isolates.

3.5 ARA: Fermentation of Arabinose (Pentose Sugar) Probiotic Activity

The probiotic activity of the *lactobacillus* isolates from the fermented drink samples is presented in Figs. 9.2, 9.3, and 9.4 illustrating the survival rates of different *lactobacillus* isolates at varying pH levels (2, 4, 5.5, and 7). All the screened nine isolates showed good tolerance to pH with survival rates ranging between 89.84 and 198.53%. Isolate PB03 showed higher survival rates at pH 2 and a notable increase in survival rates at pH 4. At pH 5.5, isolate BO02 showed a higher survival rate. The highest survival rates are generally observed at pH 7, which is neutral. All isolates show significantly high survival rates at pH 7, with PA03 having the highest survival rate, which suggests the optimal growth conditions at this pH. Figure 9.3 presents the survival rates of various *lactobacillus* isolates when exposed to different concentrations of bile (0.4%, 0.6%, 0.8%, and 1%). All the screened nine isolates showed good tolerance to bile salt with survival rates ranging between 67.83 and 187.01%. Isolate BO02 showed a higher survival rate at bile 0.4%, while isolate PB02 showed a higher survival rate at bile 0.6%. Isolate BB02 and PB02 showed a

Table 9.3 Morphological and Gram staining of bacterial isolates from fermented drink samples

Isolate code	Fermented drink	Morphology	Gram reaction	Catalase	Sulfide	Indole	Motility	Citrate	Urease	Methyl red	VP	Glucose	Lactose	Galactose	Sucrose	Presumptive identification
PA03	Palm wine	Creamy, raised, smooth, circular	Gram-positive rods	–	–	–	–	–	–	+	–	+	+	+	+	*Lactobacillus species*
PB02	Palm wine	Creamy, raised, wavy, circular	Gram-positive rods	–	–	–	–	–	–	+	–	+	+	+	+	*Lactobacillus species*
BB06	Burukutu	Whitish, flat, entire, circular	Gram-positive rods	–	–	–	–	–	–	–	–	+	+	+	+	*Lactobacillus species*
BO06	Burukutu	Whitish, flat, entire, circular	Gram-positive rods	–	–	–	–	–	–	+	–	+	+	+	+	*Lactobacillus species*
PB03	Palm wine	Whitish, flat, smooth, circular	Gram-positive rods	–	–	–	–	–	–	+	–	+	+	+	+	*Lactobacillus species*
BB05	Burukutu	Whitish, flat, wavy, irregular	Gram-positive rods	–	–	–	–	–	–	–	–	+	+	+	+	*Lactobacillus species*
BO01	Burukutu	Creamy, convex, wavy, circular	Gram-positive rods	–	–	–	–	–	–	–	–	+	+	+	+	*Lactobacillus species*
BO02	Burukutu	Creamy, flat, wavy, irregular	Gram-positive rods	–	–	–	–	–	–	+	–	+	+	+	+	*Lactobacillus species*
BA03	Burukutu	White, flat, wavy, irregular	Gram-positive rods	–	–	–	–	–	–	+	–	+	+	+	+	*Lactobacillus species*

Key: + positive, – negative

Table 9.4 *Lactobacillus* spp. identification using API

S/N	Label	ONPG	ADH	LDC	ODC	CIT	H_2S	URE	TDA	IND	VP	GEL	GLU	MAN	INO	SOR	RHA	SAC	MEL	AMY	ARA	OX	Isolate identification
1	BB05	+	+	+	+	–	–	+	+	–	–	+	+	+	+	+	–	+	–	+	+	–	*Lactobacillus plantarum*
2	B006	+	+	–	–	+	–	–	–	–	–	–	+	+	–	–	+	+	+	–	+	–	*Lactobacillus acidophilus*
3	BA04	–	+	+	–	–	–	–	–	–	+	+	+	+	+	–	+	–	–	+	+	–	*Lactobacillus* spp.
4	PB03	+	+	+	+	–	–	+	+	–	–	+	+	+	+	+	–	+	–	+	+	–	*Lactobacillus plantarum*
5	BA01	+	+	+	+	+	–	+	–	–	–	–	+	+	–	–	+	+	+	–	+	–	*Lactobacillus fermentum*
6	BB06	+	+	+	+	+	–	+	–	–	–	–	+	+	–	–	+	+	+	–	+	–	*Lactobacillus fermentum*
7	PA03	+	+	–	–	+	–	–	–	–	–	–	+	+	–	–	+	+	+	–	+	–	*Lactobacillus acidophilus*

Table 9.4 (continued)

Notes:

1. **ONPG**: test for β-galactosidase enzyme by hydrolysis of the substrate o-nitrophenyl-b-D-galactopyranoside
2. **ADH**: decarboxylation of the amino acid arginine by arginine dihydrolase
3. **LDC**: decarboxylation of the amino acid lysine by lysine decarboxylase
4. **ODC**: decarboxylation of the amino acid ornithine by ornithine decarboxylase
5. **CIT**: utilization of citrate as only carbon source
6. **H2S**: production of hydrogen sulfide
7. **URE**: test for the enzyme urease
8. **TDA (Tryptophan deaminase)**: detection of the enzyme tryptophan deaminase: Reagent—ferric chloride
9. **IND**: indole test—production of indole from tryptophan by the enzyme tryptophanase. Reagent—Indole is detected by the addition of Kovac's reagent
10. **VP**: the Voges-Proskauer test for the detection of acetoin (acetyl methylcarbinol) produced by fermentation of glucose by bacteria utilizing the butylene glycol pathway
11. **GEL**: test for the production of the enzyme gelatinase which liquefies gelatin
12. **GLU**: fermentation of glucose (hexose sugar)
13. **MAN**: fermentation of mannose (hexose sugar)
14. **INO**: fermentation of inositol (cyclic polyalcohol)
15. **SOR**: fermentation of sorbitol (alcohol sugar)
16. **RHA**: fermentation of rhamnose (methyl pentose sugar)
17. **SAC**: fermentation of sucrose (disaccharide)
18. **MEL**: fermentation of melibiose (disaccharide)
19. **AMY**: fermentation of amygdalin (glycoside)

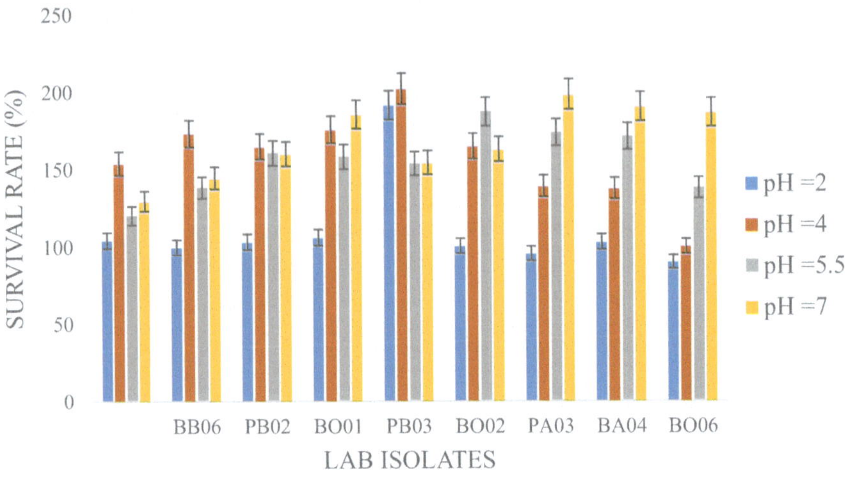

Fig. 9.2 pH tolerance

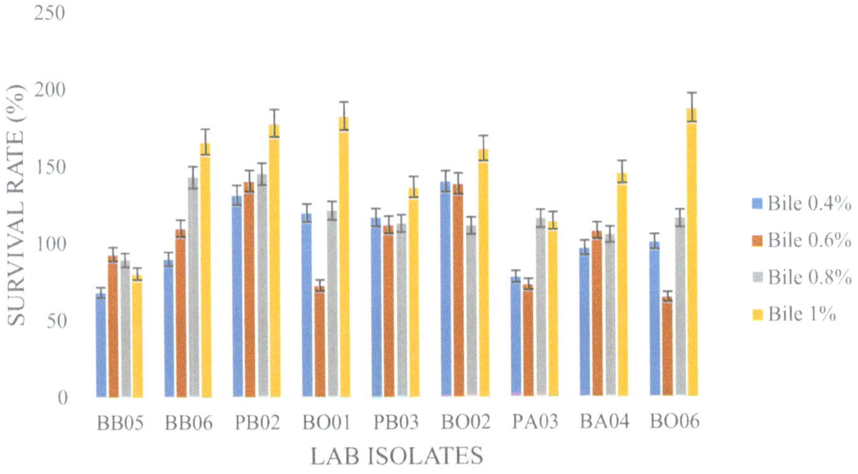

Fig. 9.3 Bile tolerance

higher survival rate at bile 0.8%, with 142.44% and 144.47%, respectively. Isolate BO06 showed the highest survival rate at bile 1%. Figure 9.4 presents the survival rates of various *lactobacillus* isolates when exposed to different concentrations of phenol (0.3%, 0.6%, 0.9%, and 1%). All the screened nine isolates showed good tolerance to phenol with survival rates ranging between 203 and 245.64%. Isolate PA03 showed a higher survival rate at phenol 0.3% and 0.6%. Also, isolate BO06 showed a high survival rate at phenol 0.9%, while isolate BB06 showed a high survival rate at phenol 1%.

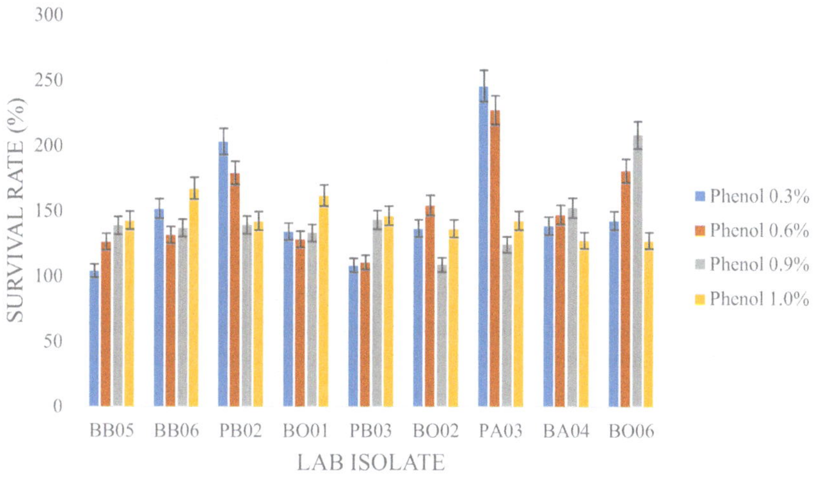

Fig. 9.4 Phenol tolerance

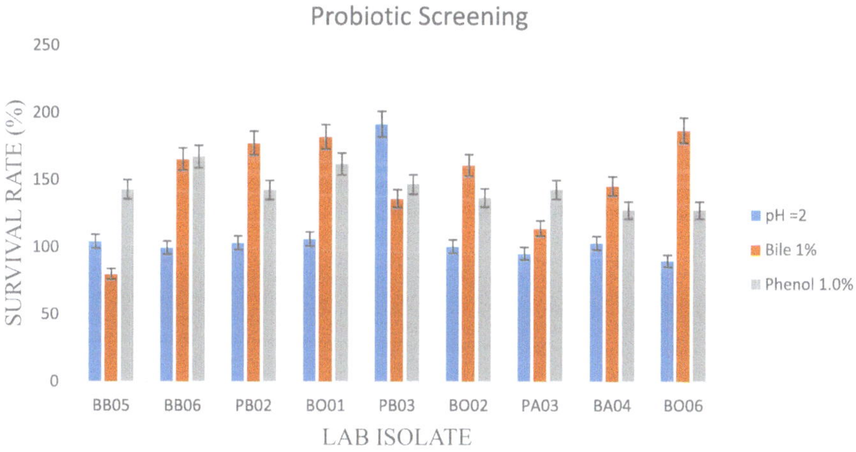

Fig. 9.5 Probiotic activity of the LAB isolates from fermented drinks

In Fig. 9.5, the survival rates of various *lactobacillus* isolate under different extreme conditions: low pH (pH = 2), high bile concentration (1%), and high phenol concentration (1%) are presented. At pH 2, isolate PB03 showed a high survival rate, which is the most acidic environment. At bile 1%, isolate BO06 has a high survival rate, which indicates the isolates' ability to tolerate bile salts in the gastro-intestinal tract. At phenol 1%, isolate BB06 showed a high survival rate.

3.6 Safety Evaluation

3.6.1 Hemolysis of *Lactobacillus* Isolates

The hemolytic result for the *lactobacillus* species is presented in Table 9.5. The result showed that 77.8% of the *lactobacillus* isolates exhibited gamma hemolysis, while 22.2% exhibited alpha hemolysis. Isolates BO02 and BB05 exhibited α-hemolysis (partial hemolysis), while other isolates (BA04, PB03, PB02, BB06, PA03, BO01, BO06) all exhibited ′Υ-hemolysis (no hemolysis), hence safe for probiotic use. However, no isolates exhibited β-hemolysis, which is a positive finding as β-hemolysis is associated with complete lysis of red blood cells and is a characteristic of many pathogenic bacteria. The absence of β-hemolytic activity among the isolates indicates that none of the *Lactobacillus* isolates are likely to be pathogenic.

3.6.2 Antibiotic Susceptibility Test of the *Lactobacillus* Isolates

Table 9.6 presents the antibiotic susceptibility of *Lactobacillus* isolates from fermented drink samples. The isolates BB06, PB02, and PB03 show the highest sensitivity, particularly to gentamicin and imipenem, while isolates BO2, PB02, PB03, and PA03 exhibit resistance to a broad range of antibiotics including cefuroxime, cefotaxime, and ceftriaxone, indicating a significant pattern of resistance among these isolates.

Table 9.5 Hemolysis test carried out on the *Lactobacillus* isolates from fermented drink samples

Isolate code	α-Hemolysis	β-Hemolysis	′Υ-Hemolysis
BA04	–	–	+
PB03	–	–	+
PB02	–	–	+
BO02	+	–	–
BB06	–	–	+
PA03	–	–	+
BO01	–	–	+
BB05	+	–	–
BO06	–	–	+

(+) *Positive* presence of hemolysis, (–) *Negative* no hemolysis

Table 9.6 Antibiotic susceptibility test carried out on *Lactobacillus* isolates from fermented drink samples

Isolate code	AZN (15)	OFX (5)	CXM (30)	GN (10)	CTX (25)	CRO (45)	ZEM (5)	LBC (5)	AUG (30)	CIP (5)	IMP (10)	ERY (15)
BO06	S	I	R	S	R	S	I	S	I	I	I	I
BB05	I	I	S	S	S	S	S	S	I	S	R	S
BB06	S	S	S	S	S	I	R	S	I	S	I	I
PA03	S	I	R	S	R	R	I	S	I	S	R	R
BO2	S	S	R	S	R	R	R	S	R	S	R	R
PB02	R	S	R	S	R	R	S	S	S	R	R	I
PB03	R	S	S	S	R	R	S	S	R	I	S	R
BA04	S	S	R	R	R	I	S	S	S	S	R	I

Note: The zone of inhibitions was interpreted as R (Resistance) = 0– 14 mm; I (Intermediate) = 15–19 mm; S (Sensitive) = Above 20 mm

AUG amoxicillin clavulanate, *CTX* cefotaxime, *CRO* ceftriaxone sulbactam, *ZEM* cefixime, *LBC* levofloxacin, *CIP* ciprofloxacin, *ERY* erythromycin, *GN* gentamicin, *AZN* azithromycin, *OFX* ofloxacin, *CXM* cefuroxime, *IMP* imipenem

4 Discussion

The rich diversity of fermented foods and beverages in developing countries, particularly in Africa, presents a valuable resource for the discovery and isolation of novel probiotics. This diversity is not only geographical but is deeply rooted in the ecological and cultural variation across the continent (Mokoena et al., 2016). In Africa, fermented products like palm wine and millet-based beverages such as burukutu are integral to the diet, with each region contributing unique fermentation practices that are influenced by local ecological factors such as soil composition, climate, and indigenous plant species (Driscoll & Damm, 2023). For example, palm wine fermentation may vary significantly based on the type of palm tree species, the climatic conditions of the region (rainforest versus savanna), and the microbial community present in the environment (Djeni et al., 2020). Similarly, the fermentation of millet or sorghum-based drinks like burukutu is influenced by the local farming practices, soil properties, and water quality, all of which contribute to the diversity of indigenous microorganisms involved in the fermentation process (Motlhanka et al., 2018; Bayoï & Etoa, 2021). These factors create distinct microbial environments, leading to the potential for novel probiotic strains that are adapted to specific ecological niches. Most probiotic foods contain lactic acid bacteria belonging to the genera *Lactobacillus*, which have the capacity to stimulate and regulate various aspects of the immune response (Obafemi et al., 2022). The proximate analysis of the fermented drink samples revealed significant variations in their nutritional composition, particularly in terms of ash, carbohydrate, protein, and lipid content. Sample BB exhibited the highest ash content (29.45%) and carbohydrate content (93.95%), while sample PB had the lowest carbohydrate content (37.50%). The high carbohydrate content in *burukutu* (fermented sorghum/millet beer) samples aligns with the fact that this beverage is produced through the fermentation of sorghum or millet, which are carbohydrate-rich grains. These findings are in agreement with Nwaiwu et al. (2020), who reported that fermented beverages such as *burukutu* (fermented sorghum/millet beer) typically have high carbohydrate levels due to the breakdown of starches during fermentation. Conversely, palm wine samples showed moderate carbohydrate content, likely due to their sugar-rich base derived from palm sap, as also observed in similar studies by Sarkar et al. (2023). The total bacterial count varied significantly across the samples, with *burukutu* (BA) showing the highest *Lactobacillus* count of 9.5×10^7 cfu/mL, while palm wine (PB) had the lowest count of 3.7×10^5 cfu/mL. This variation suggests that sample BA (*burukutu:* fermented sorghum/millet beer) provided more favorable conditions for *Lactobacillus* growth, possibly due to differences in the fermentation process or initial microbial load in the raw materials. This observation is consistent with other studies on fermented beverages, such as that of Obafemi et al. (2022), which reported that locally fermented beverages vary greatly in microbial content depending on factors like storage conditions and the substrate used. The lower bacterial count in PB may indicate less optimal fermentation or potential microbial suppression during the storage or handling processes. The acid tolerance of the *Lactobacillus* isolates was evaluated by measuring survival rates at different pH levels. Isolate PB03 exhibited the highest survival rate at pH 2 (191.52%), indicating excellent acid tolerance and the potential to survive in the acidic environment of the stomach. All isolates showed significantly higher survival rates at neutral pH (pH 7), with PA03 having the highest survival rate at pH 7 (198.53%).

Compared to previous studies, these findings are in line with Madana and Sathiavelu (2024), who reported similar high survival rates of *Lactococcus lactis* strains in low pH environments, with survival rates ranging between 72% and 91%. The strong acid tolerance observed in this study suggests that these isolates can survive gastric transit and potentially confer probiotic benefits. Probiotic bacterial strains were tested for tolerance to bile salts and phenol. The isolates survived bile salt concentrations ranging from 0.4% to 1%, with survival rates between 109 and 187%, similar to Hoque, where most *Lactobacillus* strains survived up to 0.3% bile salt. For phenol tolerance, the isolates survived concentrations ranging from 0.3% to 1%, with survival rates between 100 and 186%, also aligning with Hoque's findings, where most strains survived up to 0.3% phenol. The evaluation was to determine if the isolate could survive and pass through the gastrointestinal tract to colonize and confer associated health benefits to the host. High absorbance values after 24 h in low pH (pH values <4.0) conditions indicate good acid tolerance, suggesting that the isolates can survive the acidic gastric environment (Amenu & Bacha, 2023), and this aligns with a previous study by Madana and Sathiavelu (2024), who reported that *Lactococcus lactis strain MKL8 isolated from the herb Murraya koenigii observed high absorbance values after 24 h in low pH conditions (pH < 4.0), indicating good acid tolerance and the ability to survive the acidic gastric environment. The Lactococcus lactis strain MKL8 exhibited a survival rate of 72–91% in low pH conditions, demonstrating a robust acid tolerance*. The screening for probiotic activities such as acid tolerance, bile salt resistance, and phenol tolerance indicates that these isolates could survive and proliferate under conditions simulating the gastrointestinal environment (Yasmin et al., 2020). In previous research, various lactic acid bacteria (LAB) have been evaluated for their probiotic characteristics, including their ability to withstand acidic conditions and bile salts. For instance, a study on *Lactobacillus* strains isolated from Koozeh cheese demonstrated notable acid and bile tolerance, which is essential for their survival in the gastrointestinal tract (Hassanzadazar et al., 2012). The ability to tolerate low pH and bile salts is critical for probiotic efficacy, as it ensures that the probiotic *lactobacillus* can survive in the intestine (Khushboo et al., 2023). Also, the isolates' resilience to phenol, a common metabolite in the gut, further supports their potential as effective probiotics (Bernatek et al., 2022).

The hemolysis test is crucial for assessing the safety of probiotic strains. None of the isolates exhibited β-hemolysis, which is a positive finding as β-hemolysis is associated with a complete lysis of red blood cells and is a characteristic of many pathogenic bacteria (Zhang et al., 2016). The absence of β-hemolytic activity among the isolates indicates that none of the *Lactobacillus* isolates are likely to be pathogenic. This suggests their suitability for use as probiotics (Gebre et al., 2023). The absence of β-hemolysis confirms their nonpathogenic nature, in line with FAO guidelines. This finding is consistent with prior research by Khushboo et al. (2023), Das et al. (2022a, b), Wei et al. (2022), and Wang et al. (2018), who also reported nonhemolytic *Lactobacillus* species in various foods. The antibiotic susceptibility profiles suggest that the isolates are suitable for probiotic use. BB06 (*L. fermentum*) and PB03 (*L. plantarum*) exhibited the highest sensitivity (Halder et al., 2017). *L. plantarum* BB05 was susceptible to most antibiotics but resistant to imipenem and showed intermediate resistance to azithromycin, ofloxacin, and augmentin. These results align with findings from Anisimova et al. (2022). Ultimately, the *Lactobacillus* isolates from fermented drink samples exhibit promising probiotic properties, including the ability to withstand acidic and bile conditions, essential for survival in the gastrointestinal tract (Somashekaraiah et al., 2019). The results of this study contribute to the understanding of the safety and functionality of *Lactobacillus* strains in fermented foods and their potential therapeutic applications. Probiotic *Lactobacillus* species isolated from locally fermented beverages in Ota, Nigeria, hold significant potential for enhancing both the nutritional value and health benefits of these traditional drinks.

5 Conclusion

The *Lactobacillus* species (*L. acidophilus, L. plantarum*, and *L. fermentum*) isolated from fermented beverages in Ota, Nigeria, exhibit promising probiotic properties and safety profiles, making them suitable for use in functional foods and dietary supplements. These findings support the traditional consumption of these beverages and suggest the potential for developing probiotic-rich products that can contribute to gut health and overall well-being. Further studies are recommended to explore the specific strains and their health benefits.

Acknowledgments The authors appreciate the International Biotechnology Conference, Exhibition, and Workshop—Hybrid Conference (IBCEW).

Ethics Statement This study was approved by the Covenant Health Research Ethics Committee (CHREC), Covenant University, Ogun State, Nigeria, with reference number CU/HREC/AJI/394/24.

Authors' Information Oziegbe, Olubukola: Department of Biological Sciences, Covenant University, Ota, Nigeria.

Azeta, I. Joy: Department of Biological Sciences, Covenant University, Ota, Nigeria.

Obafemi, D. Yemisi: Department of Biological Sciences, Covenant University, Ota, Nigeria.

Consent for Publication All authors read and approved the final manuscript for publication in the International Biotechnology Conference, Exhibition, and Workshop—Hybrid Conference.

Availability of Data and Materials The author confirmed that all the relevant data and materials are available within the manuscript.

Competing Interests The authors have declared none.

References

Abid, S., Farid, A., Abid, R., Rehman, M. U., Alsanie, W. F., Alhomrani, M., Alamri, A. S., Asdaq, S. M. B., Hefft, D. I., Saqib, S., & Muzammal, M. (2022). Identification, biochemical characterization, and safety attributes of locally isolated *Lactobacillus fermentum* from Bubalus bubalis (buffalo) milk as a probiotic. *Microorganisms, 10*(5), 954.

Al-Yami, A. M., Al-Mousa, A. T., Al-Otaibi, S. A., & Khalifa, A. Y. (2022). *Lactobacillus* species as probiotics: Isolation sources and health benefits. *Journal of Pure and Applied Microbiology, 16*, 2270–2291.

Amenu, D., & Bacha, K. (2023). Probiotic potential and safety analysis of lactic acid bacteria isolated from Ethiopian traditional fermented foods and beverages. *Annals of Microbiology, 73*(1), 37.

Anisimova, E., Gorokhova, I., Karimullina, G., & Yarullina, D. (2022). Alarming antibiotic resistance of lactobacilli isolated from probiotic preparations and dietary supplements. *Antibiotics, 11*(11), 15–57.

Bayoï, J. R., & Etoa, F. X. (2021). Physicochemical changes occurring during long-time fermentation of the indigenous alcoholic sorghum-based beverages brewed in northern Cameroon. *Beverages, 7*(2), 39.

Bernatek, M., Żukiewicz-Sobczak, W., Lachowicz-Wiśniewska, S., & Piątek, J. (2022). Factors determining effective probiotic activity: Evaluation of survival and antibacterial activity of selected probiotic products using an "in vitro" study. *Nutrients, 14*(16), 3323.

Binda, S., Hill, C., Johansen, E., Obis, D., Pot, B., Sanders, M. E., Tremblay, A., & Ouwehand, A. C. (2020). Criteria to qualify microorganisms as "probiotic" in foods and dietary supplements. *Frontiers in Microbiology, 11*, 1662.

Colautti, A., Arnoldi, M., Comi, G., & Iacumin, L. (2022). Antibiotic resistance and virulence factors in *Lactobacilli*: Something to carefully consider. *Food Microbiology, 103*, 103934.

Das, T. K., Pradhan, S., Chakrabarti, S., Mondal, K. C., & Ghosh, K. (2022a). Current status of probiotic and related health benefits. *Applied Food Research., 2*(2), 100–185.

Das, S., Vishakha, K., Banerjee, S., Bera, T., Mondal, S., & Ganguli, A. (2022b). A novel probiotic strain of Lactobacillus fermentum TIU19 isolated from Haria beer showing both in vitro antibacterial and antibiofilm properties upon two multi resistant uro-pathogen strains. *Current Research in Microbial Sciences, 3*, 100150. https://doi.org/10.1016/j.crmicr.2022.100150

Dempsey, E., & Corr, S. C. (2022). *Lactobacillus* spp. for gastrointestinal health: Current and future perspectives. *Frontiers in Immunology, 13*, 840245.

Dimidi, E., Cox, S. R., Rossi, M., & Whelan, K. (2019). Fermented foods: Definitions and characteristics, impact on the gut microbiota and effects on gastrointestinal health and disease. *Nutrients, 11*(8), 1806.

Djeni, T. N., Kouame, K. H., Ake, F. D., Amoikon, L. S., Dje, M. K., & Jeyaram, K. (2020). Microbial diversity and metabolite profiles of palm wine produced from three different palm tree species in Côte d'Ivoire. *Scientific Reports, 10*(1), 1715.

Driscoll, J., & Damm, J. C. (2023). A first draught: Pitfalls and potentials in the archaeological chemistry of beer. *Chemistry in the Service of Archaeology*, 11–63.

Garcia-Gonzalez, N., Battista, N., Prete, R., & Corsetti, A. (2021). Health-promoting role of *Lactiplantibacillus plantarum* isolated from fermented foods. *Microorganisms, 9*(2), 349.

Gebre, T. S., Emire, S. A., Chelliah, R., Aloo, S. O., & Oh, D. H. (2023). Isolation, functional activity, and safety of probiotics from Ethiopian traditional cereal-based fermented beverage,"Borde". *LWT, 184*, 115076.

Halder, D., Mandal, M., Chatterjee, S. S., Pal, N. K., & Mandal, S. (2017). Indigenous probiotic Lactobacillus isolates presenting antibiotic like activity against human pathogenic bacteria. *Biomedicine, 5*(2), 31.

Han, S., Lu, Y., Xie, J., Fei, Y., Zheng, G., Wang, Z., Liu, J., Lv, L., Ling, Z., Berglund, B., & Yao, M. (2021). Probiotic gastrointestinal transit and colonization after oral administration: A long journey. *Frontiers in Cellular and Infection Microbiology, 11*, 609722.

Hassanzadazar, H., Ehsani, A., Mardani, K. & Hesari, J. (2012). Investigation of antibacterial, acid and bile tolerance properties of lactobacilli isolated from Koozeh cheese. In Veterinary research forum (3, 3, 181). Faculty of Veterinary Medicine, Urmia University, .

Huligere, S. S., Chandana Kumari, V. B., Alqadi, T., Kumar, S., Cull, C. A., Amachawadi, R. G., & Ramu, R. (2023). Isolation and characterization of lactic acid bacteria with potential probiotic activity and further investigation of their activity by α-amylase and α-glucosidase inhibitions of fermented batters. *Frontiers in Microbiology, 13*, 1042263.

Khushboo, Karnwal, A., & Malik, T. (2023). Characterization and selection of probiotic lactic acid bacteria from different dietary sources for development of functional foods. *Frontiers in Microbiology, 14*, 1170725.

Madana, S. T., & Sathiavelu, M. (2024). Probiotic evaluation, adherence capability and safety assessment of Lactococcus lactis strain isolated from an important herb "Murraya koenigii". *Scientific Reports, 14*.

Mgbodile, F. C., & Nwagu, T. N. T. (2023). Probiotic therapy, African fermented foods and food-derived bioactive peptides in the management of SARS-CoV-2 cases and other viral infections. *Biotechnology Reports, 38*, e00795.

Mokoena, M. P., Mutanda, T., & Olaniran, A. O. (2016). Perspectives on the probiotic potential of lactic acid bacteria from African traditional fermented foods and beverages. *Food & Nutrition Research, 60*(1), 29630.

Motlhanka, K., Zhou, N., & Lebani, K. (2018). Microbial and chemical diversity of traditional non-cereal based alcoholic beverages of sub-Saharan Africa. *Beverages, 4*(2), 36.

Nwaiwu, O., Aduba, C. C., Igbokwe, V. C., Sam, C. E., & Ukwuru, M. U. (2020). Traditional and artisanal beverages in Nigeria: Microbial diversity and safety issues. *Beverages, 6*(3), 53.

Obafemi, Y. D., Oranusi, S. U., Ajanaku, K. O., Akinduti, P. A., Leech, J., & Cotter, P. D. (2022). African fermented foods: Overview, emerging benefits, and novel approaches to microbiome profiling. *npj Science of Food, 6*(1), 15.

Plaza-Diaz, J., Ruiz-Ojeda, F. J., Gil-Campos, M., & Gil, A. (2019). Mechanisms of action of probiotics. *Advances in Nutrition, 10*, S49–S66.

Sarkar, T., Mukherjee, M., Roy, S., & Chakraborty, R. (2023). Palm sap sugar an unconventional source of sugar exploration for bioactive compounds and its role on functional food development. *Heliyon, 9*(4), e14788.

Somashekaraiah, R., Shruthi, B., Deepthi, B. V., & Sreenivasa, M. Y. (2019). Probiotic properties of lactic acid bacteria isolated from neera: A naturally fermenting coconut palm nectar. *Frontiers in Microbiology, 10*, 448547.

Wang, J., Yang, K., Liu, M., Zhang, J., Wei, X., & Fan, M. (2018). Screening for potential probiotic from spontaneously fermented non-dairy foods based on in vitro probiotic and safety properties. *Annales de Microbiologie, 68*, 803–813. https://doi.org/10.1007/s13213-018-1386-3

Wei, C., Luo, K., Wang, M., Li, Y., Pan, M., Xie, Y., et al. (2022). Evaluation of potential probiotic properties of a strain of lactobacillus plantarum for shrimp farming: From beneficial functions to safety assessment. *Frontiers in Microbiology, 13*, 854131. https://doi.org/10.3389/fmicb.2022.854131

Yasmin, I., Saeed, M., Khan, W. A., Khaliq, A., Chughtai, M. F. J., Iqbal, R., Tehseen, S., Naz, S., Liaqat, A., Mehmood, T., & Ahsan, S. (2020). In vitro probiotic potential and safety evaluation (hemolytic, cytotoxic activity) of *Bifidobacterium* strains isolated from raw camel milk. *Microorganisms, 8*(3), 354.

Zhang, F., Jiang, M., Wan, C., Chen, X., Chen, X., Tao, X., Shah, N. P., & Wei, H. (2016). Screening probiotic strains for safety: Evaluation of virulence and antimicrobial susceptibility of *Enterococci* from healthy Chinese infants. *Journal of Dairy Science, 99*(6), 4282–4290.

Žuntar, I., Petric, Z., Kovacevíc, D. B., & Putnik, P. (2020). Safety of probiotics: Functional fruit beverages and nutraceuticals. *Food, 9*(7), 947.

Chapter 10
Hospital-Based Retrospective Study of *Plasmodium falciparum* and *Salmonella enterica* Co-Infection on Gender Demography and Leukocytes

Samuel Adeniyi Oyegbade, Emmanuel Ojochegbe Mameh, Chioma Jane Agunsoye, Adenike Olugbemi Adewoye, Felix Akinyemi Akintaju, Olusegun Popoola, and Paul Akinniyi Akinduti

1 Introduction

Malaria is a serious worldwide health concern that is highly endemic in Asia and Africa (Kayiba et al., 2024), affecting more than 120 million people globally and largest proportion of 120 million in Nigeria (Patel et al., 2024). Global malaria eradication remains unachieved, with transmission persistently sustained across numerous endemic regions (Dokunmu et al., 2019). Asymptomatic infections are frequent in hyperendemic locations, and clinical illness manifestations resulting

S. A. Oyegbade · E. O. Mameh
Department of Biological Sciences, Covenant University, Ota, Ogun State, Nigeria

Covenant Applied Informatics and Communication Africa Centre of Excellence (CApIC-ACE), Ota, Ogun State, Nigeria

C. J. Agunsoye
Covenant University Medical Centre, Ota, Ogun State, Nigeria

A. O. Adewoye
Medical Microbiology/Parasitology Department, State Hospital, Ota, Ogun State, Nigeria

F. A. Akintaju
Department of Chemical Pathology and Immunology, Olabisi Onabanjo University, Ago-Iwoye, Ogun State, Nigeria

O. Popoola
Golden Gate University School of Law, San Francisco, CA, USA

P. A. Akinduti (✉)
Department of Biological Sciences, Covenant University, Ota, Ogun State, Nigeria
e-mail: akindutip@babcock.edu.ng

Department of Medical Laboratory Science, Babcock University, Illishan, Ogun State, Nigeria

P. O. Isibor, P. A. Akinduti (eds.), *Harnessing Biotechnology Tools for Product Development*, https://doi.org/10.1007/978-3-031-86002-7_10

from malaria parasite infection range widely, from mild to severe (Bousema et al., 2014). A significant factor hindering malaria elimination strategies is the continuous emergence of Plasmodium falciparum (*P. falciparum*) parasites resistant to all known existing malaria drugs and currently available insecticides in the malaria vector (Olasehinde et al., 2019; Adedeji et al., 2020). The most prevalent and deadly malaria parasite, Plasmodium falciparum, is known to cause chronic asymptomatic infection, which cause systemic or organ morbidity in infected persons (Tadesse et al., 2018; Adedeji et al., 2020). The co-infection of typhoid infection called "typhoid" is caused by *Salmonella typhi* (*Salmonella enterica* serotype Typhi) (Slater et al., 2019). The consumption of contaminated water or food has been described as major etiological source of typhoid transmission and its fecal-oral spread is commonly observed among children (Ashour & Othman, 2020; Silva Viana et al., 2024). The co-infection of *P. falciparum* malaria and typhoid fever are among the major enteric two leading infections related to of poverty with severe health and socioeconomic impacts (Mohammed & Sadiq, 2020; Okolo et al., 2023). In several cases, the two infections present similar clinical symptoms including high fever, rigor, weakness, headache, malaise, fatigue, loss of appetite, and anemia, causing misdiagnosis and poor treatment (Wilairatana et al., 2021). Despite the prevalence of typhoid, most clinical laboratory utilizes the Widal reaction for the sero-detection of typhoid fever in resource-limited settings and the result usually gives rise to low specificity and sensitivity leading to false diagnosis, over-prescription of antibiotics, and increased morbidity (Ashour & Othman, 2020; Diwaker et al., 2024). Due to emphasis on malaria, co-infection with typhoid is often neglected and this causes significant complications that could lead to death in some severe cases. A retrospective four-year study of the *prevalence of P. falciparum malaria and typhoid infection* among the population strata instigate the evaluation of its implication on gender demography. Female gender categories are speculated to be more affected with the *P. falciparum* and typhoid co-infection resulting to maternal anemia, fever, miscarriage, stillbirth, and even death (Pam et al., 2015). The increased bacterial superinfection in malaria affects the immune system primarily impaired leukocyte and macrophage activity due to phagocytosis and subsequent malaria pigmentation (Lyke et al., 2003). Considering malaria and typhoid fever endemicity in southwest Nigeria and its prevalence among the populace, evaluating these diseases over the last four years in clinical settings will aid in designing precise and appropriate diagnosis, practical intervention strategies, and community epidemiological surveillance (Diwaker et al., 2024; Okoror et al., 2024). This study aimed to evaluate the clinical implications of *Plasmodium falciparum* and typhoid fever co-infection on gender demography and leukocytes pattern in hospital-based retrospective study.

2 Materials and Method

Ethical Approval: The study received approval from the Covenant Health Research Ethics Committee (CHREC /432/2024) to obtained data from the patients' records and the identity of each patient remains anonymous with confidentiality given the intricacy and sensitivity of the study.

Study Location: The study retrieved the 4-year laboratory data from the Covenant Health University Health Centre, Ota Nigeria. This health facility is located within the Covenant University Ota Ogun State Nigeria which is a semi-urban region of the State bordering the Benin Republic and Lagos State with average patients' attendance population of 355 per month including adult and children. It has 104 bed facilities providing primary healthcare services and ten specialist areas for medical interventions. Based on the health care services provided, malaria diagnosis is a crucial investigation for new and admitted patients from Ota and other adjoining cities and villages known for malaria endemicity mostly in the rainy season of the year (Olukanni et al., 2014).

Study Population: This is hospital-based retrospective study conducted to estimate the prevalence, distribution of *P. falciparum* malaria infection, and co-infection with typhoid fever. The study populations were in- and outpatients of the hospital for routine and special medical health care services from January 2020 to December 2023. The retrieved patients' data included age, gender, and hematological parameters, *P. falciparum* and typhoid fever results. All the selected patients were registered in the laboratory logbook of the health center and were malaria-suspected individuals who consented to provide blood samples for blood rapid diagnostic test, blood film microscopy, and sero-detection by Widal reaction.

Inclusion and Exclusion Criteria: All patients who provided full records of their age, sex, year, and date of diagnosis were included in the study. The names and addresses of the patients were recorded as anonymous for ethical reasons. Patients with incomplete information and diagnosis other than malaria and typhoid were excluded.

Data Collection Tools: Prior to the retrieval of the patients' data from the laboratory logbook, the data collectors were trained on the use of data extraction method, recording on data extraction Excel sheet, variable of interest, recording of hematological parameters, plasmodium species of interest, and objective of the study. The demographic data specifically the age and sex were retrieved from lab logbook in respect to the patient's diagnosis, *P. falciparum* RDT data, blood film identified *P. falciparum* species, and Widal test results according to respective month and year in prepared and well-outlined Microsoft Excel sheets. The variables of interest were transferred from the laboratory logbook to the prepared data extraction Excel sheet. The quality of the extracted study variables was further assessed for consistency of monthly and yearly recorded patients' sex, age, diagnosis, *P. falciparum* detected, typhoid diagnosis, and comment from the results. To ensure adequate and precise recording, the study team supervised and validated the entry of the data.

Data Processing and Analysis: Prior to analysis, retrieved data were coded, checked for completeness, and re-assessed for proper entry on Microsoft Excel sheet. Age and gender demography were analyzed with descriptive statistics and assessed for significant differences with paired t-test. The association of *P. falciparum* infection and typhoid co-infection with WBC parameters, age, and PCV in respective year was analyzed with Pearson's test taking $p < 0.05$ as significant and presented with Sankey diagram. The yearly *P. falciparum* and typhoid co-infection variables were analyzed for significance using ANOVA, considering $p < 0.05$.

3 Results

The obtained data recovered from metadata of the patients with *Plasmodium falciparum* infections in the four-year period of 2020–2023 showed a disproportional rate. For retrospective years examined for *P. falciparum* malaria infection, from the 917 cases of malaria recorded, 48.4% were diagnosed to be *P. falciparum* malaria-positive male patients (444/917) compared to 46.0% female-positive patients (422/917) and the remaining 5.6%, both male and female were tested negative in the Year 2020 ($p = 0.52$; Fig. 10.1a). Lower rates of 46.2% (1061/2297) *P. falciparum* positive cases among males compared to 48.0% positive female patients 48.0 (1102/2297) and negative cases of 5.8% in both genders were recorded in Year 2021 ($p = 0.5$; Fig. 10.1b). Significant and higher rates of 47.9% (1067/2228) *P. falciparum* malaria-positive men were observed in year 2022 compared to 36.5% (814/2228) female patients tested positive, while 15.6% (75/2228) of both genders tested negative ($p = 0.023$; Fig. 10.1c). Significantly higher rates of 51.1% (1194/2335) male patients tested positive for *P. falciparum* than 42.0% (979/2335) female patients tested positive for Plasmodium falciparum malaria infection, while 6.9% (71/2335) were negative ($p = 0.02$; Fig. 10.1d). Over the years, males were more infected compared to women.

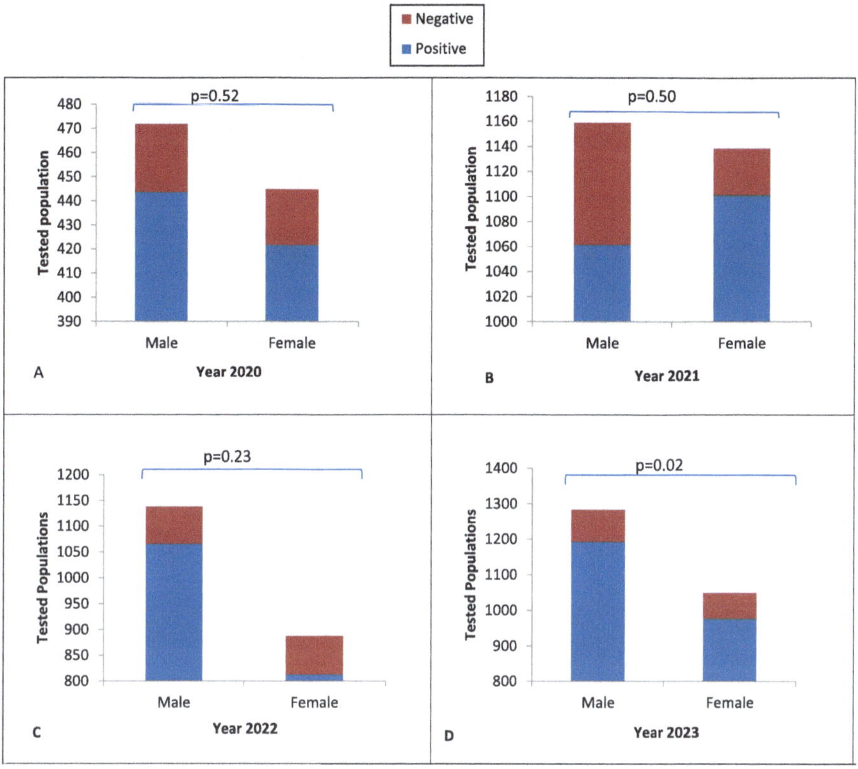

Fig. 10.1 *P. falciparum* infection proportion based according to gender distribution**Co-Infection Rates of *Plasmodium Falciparum* and *Salmonella Enterica* over the Years 2020–2023**: The comparative estimation of co-infection rates of *P. falciparum* and typhoid was examined within the period of 2020–2023 among the tested patients. Among the recorded patient's population, 55% and 45% male (55%) and female (45%) were recorded among the 2020 positive population of 866 (Fig. 10.2a) and male (48%) and female (52%) among the 2021 year tested population ($n = 2163$) (Fig. 10.2b) were observed to be *P. falciparum* and typhoid co-infected. The male (45%) and female (55%) in 2022-year population ($n = 1881$) (Fig. 10.2c) and male (44%) and female (46%) patients in 2023 population ($n = 2173$) were co-infected with *P. falciparum* and typhoid fever (Fig. 10.2d). Relatively high rates of *P. falciparum* and typhoid co-infection were observed in the studied period

4 Discussion

The co-infection of malaria and typhoid fever is a major public health concern in tropical regions (Mutua et al., 2015). Both infections were reported to be enhanced through declined population demography caused by low socioeconomic condition that have given rise to poverty, malnutrition, poor environmental sanitations, poor personal hygiene, inadequate health facilities, poor social service, and low levels of education (Ia & Mo, 2020). The prevalence of malaria and co-infection with typhoid in the last four years was evaluated to provide guideline and trend of

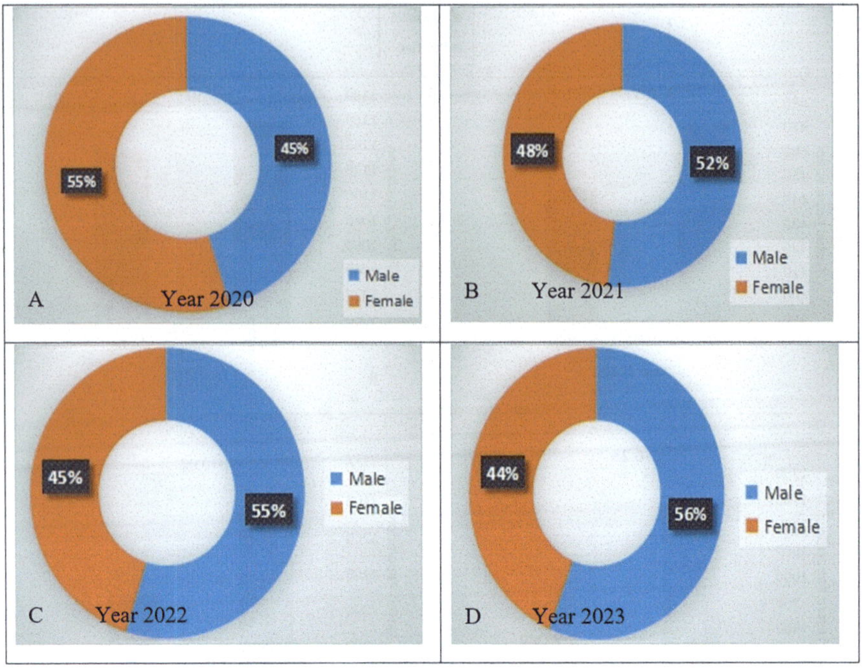

Fig. 10.2 Co-infection rates *Plasmodium falciparum* and *Salmonella enterica* over the studied years**Changes in Leukocyte Rates with Yearly Co-infection and Overall Infection Prevalence Rates:** Sankey diagram (Fig. 10.3a) showed the yearly distribution of *P. falciparum* and typhoid co-infection among male subjects with increasing rates of WBC, neutrophil, and lymphocytes while there is decreasing rates of PCV and monocytes as from year 2020 to 2023. There is no change in co-infection rates based on the median age. Among the female category, WBC, PCV, decrease while there is no change in neutrophil, lymphocyte, and monocyte counts in *P. falciparum* and typhoid co-infection over the studied years (Fig. 10.2b). Similarly, the average age of the female subjects with identified co-infection decreases over the year. The co-infection pattern further reveals the trend of cellular immune responses based on the leukocyte level. The observed significant differences in yearly prevalence rates of *P. falciparum* are disproportional over the studied years (Fig. 10.4a). Considering the co-infection of *P. falciparum* and typhoid, a significant increase in prevalence rates as from 2020 till 2023 was recorded (Fig. 10.4b)

transmission and appropriate prevention. The implication of gender demography facilitating the infectivity of *Plasmodium falciparum* infection indicated a four-year disproportional rate which is evident by the higher rates of co-infection among male compared to female. This finding was similarly reported by Awosolu et al. (2021) presenting malaria infection pattern among the gender with higher malaria prevalence and mean parasite density among males than their female subjects. Related reports from other studies in malaria-endemic areas such as Ethiopia and Chile emphasized more male infection rates (Gebretsadik et al., 2018; Escobar et al., 2020). Considering the previous reports on the sex-specific difference in malaria burden, males are more frequently bitten by malaria-carrying mosquitoes due to behavioral differences such as outdoor activities, not sleeping under a net,

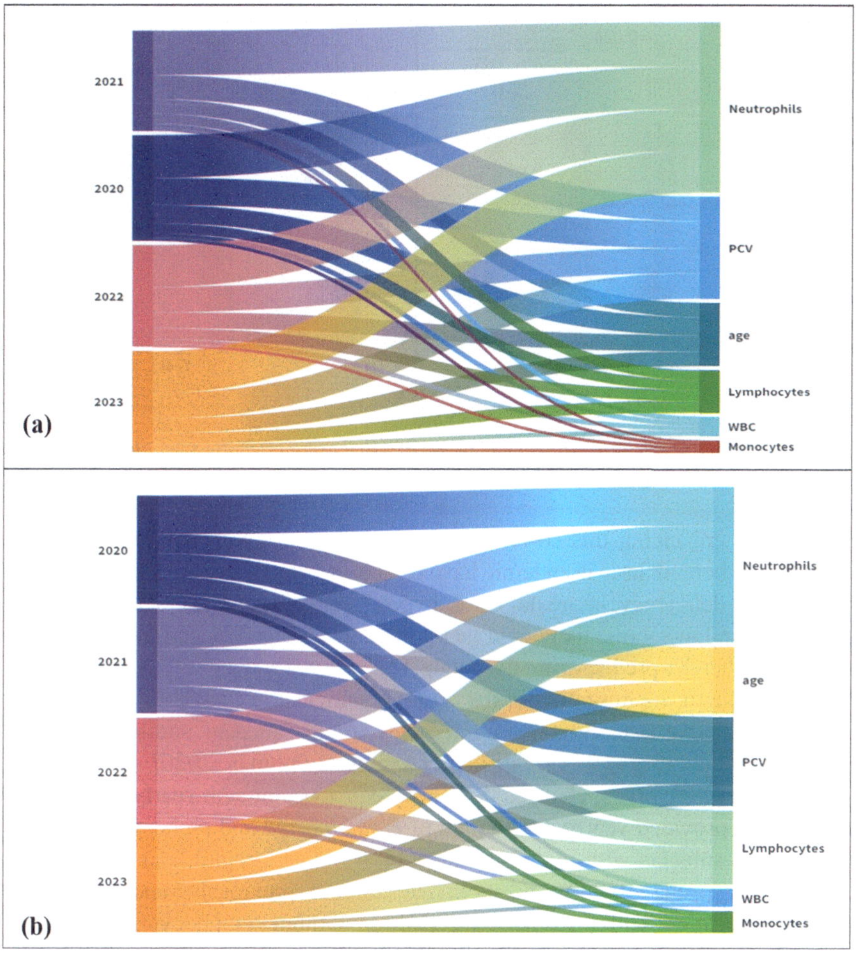

Fig. 10.3 Yearly distribution of *Plasmodium falciparum* and *Salmonella enterica* co-infection among (**a**) male and (**b**) female group

showing lackadaisical attitude toward malaria prevention (Kahar et al., 2023; Ayodele, 2024).

The comparative rates of *Plasmodium falciparum* and *Salmonella enterica* co-infection rates within the period of 2020 to 2023 among the population suggested that both male and female are usually affected and present relatively high rates of *P. falciparum* and typhoid co-infection. Similar reports of malaria and typhoid fever co-infection was presented in studies from Nigeria (Pradhan, 2011; Omoya, 2017; Nas et al., 2017), Zambia (Mwansa et al., 2017), Burkina Faso (Maltha et al., 2014), and Cameroon (Netongo et al., 2022; Sohanang Nodem et al., 2023). Despite the paucity of this co-infection among gender in Nigeria, increasing morbidity and occasional death were being recorded in previous studies (Onyido et al., 2014). The

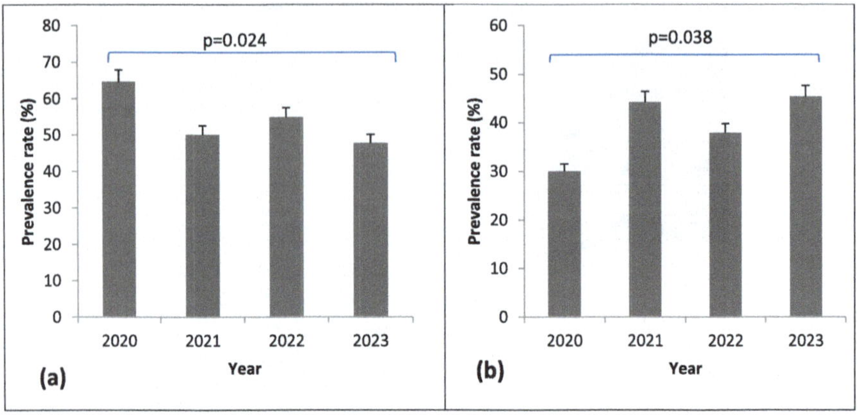

Fig. 10.4 Overall yearly prevalence rates of (**a**) *P. falciparum* infection rates (**b**) *Plasmodium falciparum* and *Salmonella enterica* Co-infection rates

mechanisms enhancing this co-infection were not well confirmed but the studied population where malaria is endemic have possibility of malaria infection leading to complement deficiency, anorexia, rigor, hypoglycemia, metabolic acidosis, hyperlactatemia, and metabolic that could facilitate reduce immunity and iron deposition in liver cell thereby predisposing the patient to *Salmonella typhi* infection (Mawson, 2013). In spite of the limitation of the Widal serological test which is prone to low predictive positive detection, both infections usually present similar symptoms making definitive diagnosis very difficult (Ashour & Othman, 2020).

The observation on the changes in leukocyte proportion with yearly co-infection rates further provided yearly distribution of *P. falciparum* and typhoid co-infection among male subjects with increasing rates of WBC, neutrophil, and lymphocytes but decreasing rates of PCV and monocytes. It was previously reported that the polymorphonuclear leukocyte response with or without lymphocytes and macrophages to *Salmonella typhi* infection induces cytotoxic phagocytosis which is usually bacterial-serotype specific, playing a significant role in removal of the bacteria cell (Salerno-Goncalves et al., 2019; Lamichhane et al., 2024). In cases of malaria infection, macrophages functionally stimulate major histocompatibility for antigen presentation for cellular immune clearance (Kumar et al., 2020). Co-infection is usually accompanied by elevated hemolysis, resulting in low PCV and acute anemia (Ndako et al., 2020). Previously reported poor medical services and diagnosis due to annual decline in socioeconomic condition (Kinuthia et al., 2012; Odikamnoro et al., 2018) may aggravate the decreasing PCV in severe malaria and typhoid co-infection.

Recording no changes in co-infection rates based on the median age suggest that all ages were exposed to the infection and age disparities have not contributed to the infectivity rates among the population. Co-infection of malaria and typhoid fever may continue to increase among the populations in endemic zone such as Ota, Nigeria, and its surrounding settlements. The age and gender demography may be affected by these infections, causing severe health and socioeconomic impacts.

5 Conclusion

The prevalence of malaria and typhoid co-infection is expected to increase over the years among the populace due to malaria endemicity in this location. Effective grassroot public awareness on the preventive and control measures of malaria and typhoid and Empirical treatment should be guided by comprehensive laboratory diagnostics, with emphasis on confirmatory typhoid culture and antimicrobial susceptibility testing, to ensure optimal therapeutic outcomes. These co-infection data are critical for the development of targeted preventive guidelines aimed at mitigating mortality and morbidity associated with these pathogens.

Acknowledgments The authors appreciate the management of the Covenant Medical Health Centre for assisting in data collation and storage and Covenant Applied Informatics and Communication Africa Centre of Excellence (CApIC-ACE) for their support.

Conflict of Interest The authors declare that there is no conflict of interest.

Author Contributions Each author has made contribution as follows: Oyegbade SA, Agunsoye Chioma J, Adewoye Adenike, and Akinduti P.A provide conception of the study.

Oyegbade SA, Mameh EO, and Agunsoye C supervised curation of the data from register logbook.

Oyegbade SA, Agunsoye C, Adewoye AO, Akintaju F, Popoola O, and Akinduti P.A interpreted the data. All authors read and approved the final manuscript.

Funding Covenant University Centre for Research and Innovation Development (CUCRID) provided funding support for the study and its publication. The mentioned support institution had no role in study design, data collection and analysis, interpretation of data, decision to publish, or preparation of the manuscript.

References

Adedeji, E. O., Ogunlana, O. O., Fatumo, S., Beder, T., Ajamma, Y., Koenig, R., & Adebiyi, E. (2020). Anopheles metabolic proteins in malaria transmission, prevention and control: A review. *Parasites & Vectors, 13*(1). https://doi.org/10.1186/s13071-020-04342-5

Ashour, D. S., & Othman, A. A. (2020). Parasite–bacteria interrelationship. *Parasitology Research, 119*(10), 3145–3164. https://doi.org/10.1007/s00436-020-06804-2

Awosolu, O. B., Yahaya, Z. S., & Farah Haziqah, M. T. (2021). Prevalence, parasite density and determinants of falciparum malaria among febrile children in some Peri-urban communities in southwestern Nigeria: A cross-sectional study. *Infection and Drug Resistance, 14*, 3219–3232. https://doi.org/10.2147/idr.s312519

Ayodele, I. V. (2024). Geospatial mapping of malaria in the lowland forest ecological zone of Ondo State, Nigeria. *Journal of Applied Geographical Studies, 4*(1), 1–16.

Bousema, T., Okell, L., Felger, I., & Drakeley, C. (2014). Asymptomatic malaria infections: Detectability, transmissibility and public health relevance. Nature reviews. *Microbiology, 12*(12), 833–840. https://doi.org/10.1038/nrmicro3364

Diwaker, A., Tiwari, A., Jain, S., Rupali, K. A., Ram, J., Singh, S., & Kishore, D. (2024). Enteric fever and the diagnostic tools: Defining the accuracy. *Frontiers in Bacteriology, 3*, 1332180.

Dokunmu, T. M., Adjekukor, C. U., Yakubu, O. F., Bello, A. O., Adekoya, J. O., Akinola, O., Amoo, E. O., & Adebayo, A. H. (2019). Asymptomatic malaria infections and *P. falciparum* mdr1 mutations in an endemic area of Nigeria. *Malaria Journal, 18*(1). https://doi.org/10.1186/s12936-019-2833-8

Escobar, D. F., Lucchi, N. W., Abdallah, R., Valenzuela, M. T., Udhayakumar, V., Jercic, M. I., & Chenet, S. M. (2020). Molecular and epidemiological characterization of imported malaria cases in Chile. *Malaria Journal, 19*(1). https://doi.org/10.1186/s12936-020-03353-y

Gebretsadik, D., Feleke, D. G., & Fiseha, M. (2018). Eight-year trend analysis of malaria prevalence in Kombolcha, South Wollo, north-central Ethiopia: A retrospective study. *Parasites and Vectors, 11*(1). https://doi.org/10.1186/s13071-018-2654-6

Ia, S.-O., & Mo, A. (2020). Prevalence of malaria and typhoid coinfection in relation to haematological profile of university students in Akure, Nigeria. *Journal of Infectious Diseases and Epidemiology, 6*(5). https://doi.org/10.23937/2474-3658/1510166

Kahar, F., Setiadi, Y., Widiyanto, S. D., Ardiyansyah, D., & Qomariyah, N. (2023). Malaria: Transmission, diagnosis, treatment and prevention in Indonesia. In *Malaria-transmission, diagnosis and treatment*. IntechOpen.

Kayiba, N. K., Nitahara, Y., Tshibangu-Kabamba, E., Mbuyi, D. K., Kabongo-Tshibaka, A., Kalala, N. T., et al. (2024). Malaria infection among adults residing in a highly endemic region from the Democratic Republic of the Congo. *Malaria Journal, 23*(1), 82.

Kinuthia, G. K., Gicheru, M. M., Ngure, P. K., & Kabiru, E. W. (2012). Lifestyles and practices that enhance malaria and typhoid fever in Njoro District, Kenya. *Journal of Community Health, 37*, 224–233.

Kumar, R., Loughland, J. R., Ng, S. S., Boyle, M. J., & Engwerda, C. R. (2020). The regulation of CD4+ T cells during malaria. *Immunological Reviews, 293*(1), 70–87.

Lamichhane, B., Mawad, A. M., Saleh, M., Kelley, W. G., Harrington, P. J., Lovestad, C. W., et al. (2024). Salmonellosis: An overview of epidemiology, pathogenesis, and innovative approaches to mitigate the antimicrobial resistant infections. *Antibiotics, 13*(1), 76.

Lyke, K. E., Diallo, D. A., Dicko, A., Kone, A., Coulibaly, D., Guindo, A., et al. (2003). Association of intraleukocytic *Plasmodium falciparum* malaria pigment with disease severity, clinical manifestations and prognosis in severe malaria. *American Journal of Tropical Medicine and Hygiene, 69*(3), 253–259.

Maltha, J., Guiraud, I., Kaboré, B., Lompo, P., Ley, B., Bottieau, E., et al. (2014). Frequency of severe malaria and invasive bacterial infections among children admitted to a rural hospital in Burkina Faso. *PLoS One, 9*(2), e89103.

Mawson, A. R. (2013). The pathogenesis of malaria: A new perspective. *Pathogens and Global Health, 107*(3), 122–129.

Mohammed, H. I. M. I., & Sadiq, H. A. (2020). Malaria and typhoid fever: Prevalence, co-infection and socio-demographic determinants among pregnant women attending antenatal care at a primary healthcare facility in Central Nigeria. *International Journal of Pathogen Research, 5*(4), 17–24.

Mutua, J. M., Wang, F. B., & Vaidya, N. K. (2015). Modeling malaria and typhoid fever co-infection dynamics. *Mathematical Biosciences, 264*, 128–144.

Mwansa, F., Gama, A., Kapaya, F., Yard, E., Kumar, R., Chongwe, G., & Sinyange, N. (2017). Typhoid fever outbreak investigation in a malaria endemic community, Solwezi, North-Western province, Zambia, 2017. *Heal Press Zambia Bull, 2*(1), 20–28.

Nas, F. S., Ali, M., & Yahaya, A. (2017). Malaria and typhoid fever co-infection among febrile patients in Kumbotso Local Government Area Kano, Nigeria. *Bayero Journal of Pure and Applied Sciences, 10*(2), 122–125.

Ndako, J. A., Dojumo, V. T., Akinwumi, J. A., Fajobi, V. O., Owolabi, A. O., & Olatinsu, O. (2020). Changes in some haematological parameters in typhoid fever patients attending Landmark University Medical Center, Omuaran-Nigeria. *Heliyon, 6*(5).

Netongo, P. M., Eric, M. B., Chedjou, J. P., Kamdem, S. D., Achonduh-Atijegbe, O., & Mbacham, W. F. (2022). Malaria and typhoid fever co-infection amongst febrile patients in Yaoundé,

Cameroon: Implication in the genetic diversity of Plasmodium falciparum. *Biochemistry and Molecular Biology, 7*(2), 47–53.

Odikamnoro, O. O., Ikeh, I. M., Okoh, F. N., Ebiriekwe, S. C., Nnadozie, I. A., Nkwuda, J. O., & Asobie, G. C. (2018). Incidence of malaria/typhoid co-infection among adult population in Unwana community, Afikpo north local government area, Ebonyi state, Southeastern Nigeria. *African Journal of Infectious Diseases, 12*(1), 33–38.

Okolo, M. L., Adeshina, K., Omatola, C., Mudi, I., & Ugbane, E. (2023). Prevalence of malaria and typhoid fever co-infection among pregnant women attending antenatal clinic in Anyigba, Kogi State, Nigeria. *Microbes and Infectious Diseases, 4*(2), 671–680.

Okoror, L., Bankefa, E. O., Ukhureigbe, O. M., Ajayi, E. O., Osanyilusi, S., & Ogeneh, B. (2024). Misdiagnosis of dengue fever as malaria and typhoid fever and their co-infection in rural areas of Southwest Nigeria. In *Qeios*.

Olasehinde, G., Diji-geske, R., Fadina, I., Arogundade, D., Darby, P., Adeleke, A., Dokunmu, T., Adebayo, A. H., & Oyelade, J. (2019). Epidemiology of *Plasmodium falciparum* infection and drug resistance markers in Ota Area, Southwestern Nigeria. *Infection and Drug Resistance, 12*, 1941–1949. https://doi.org/10.2147/idr.s190386

Olukanni, D. O., Azuh, D. E., Toogun, T. O., & Okorie, U. E. (2014). Medical waste management practices among selected health-care facilities in Nigeria: A case study.

Omoya, F. O. (2017). Co-infection of malaria and typhoid fever among pregnant women attending primary health care centre, Ojo Local Government, Lagos, Nigeria. *Health Science Journal, 11*(2), 1.

Onyido, A. E., Ifeadi, C. P., Umeanaeto, P. U., Irikannu, K. C., Aribodor, D. N., Ezeanya, L. C., et al. (2014). Co-infection of malaria and typhoid fever in Ekwulumili community Anambra State, Southeastern Nigeria. *New York Science Journal, 7*(7), 18–27.

Pam, V. A., Landan, S., Pam, D. D., Gullek, J. F., Okoro, J., Ogbu, K. I., et al. (2015). The prevalence of malaria and typhoid co-infection in pregnant women attending antenatal in Wuse general hospital Abuja, Nigeria.

Patel, P., Bagada, A., & Vadia, N. (2024). Epidemiology and current trends in malaria. *Rising Contagious Diseases: Basics, Management, and Treatments*, 261–282.

Pradhan, P. (2011). Coinfection of typhoid and malaria. *Journal of Medical Laboratory and Diagnosis, 2*(3), 22–26.

Salerno-Goncalves, R., Kayastha, D., Fasano, A., Levine, M. M., & Sztein, M. B. (2019). Crosstalk between leukocytes triggers differential immune responses against Salmonella enterica serovars Typhi and Paratyphi. *PLoS Neglected Tropical Diseases, 13*(8), e0007650.

Silva Viana, I. P., Paulo Vieira, C., Lima Santos Rosario, I., Brizack Monteiro, N., Sousa Vieira, I. R., Conte-Junior, C. A., & Pereira Costa, M. (2024). Typhoid fever and non-typhoidal salmonella outbreaks: A portrait of regional socioeconomic inequalities in Brazil. *Current Microbiology, 81*(2), 57.

Slater, H. C., Ross, A., Felger, I., Hofmann, N. E., Robinson, L., et al. (2019). The temporal dynamics and infectiousness of subpatent *Plasmodium falciparum* infections in relation to parasite density. *Nature Communications, 10*(1), 1–16. https://doi.org/10.1038/s41467-019-09441-1

Sohanang Nodem, F. S., Ymele, D., Fadimatou, M., & Fodouop, S. P. C. (2023). Malaria and typhoid fever coinfection among febrile patients in Ngaoundéré (Adamawa, Cameroon): A cross-sectional study. *Journal of Parasitology Research, 2023*.

Tadesse, F. G., Slater, H. C., Chali, W., Teelen, K., Lanke, K., et al. (2018). The relative contribution of symptomatic and asymptomatic Plasmodium vivax and *Plasmodium falciparum* infections to the infectious reservoir in a low-endemic setting in Ethiopia. *Clinical Infectious Diseases, 66*(12), 1883–1891. https://doi.org/10.1093/cid/cix1123

Wilairatana, P., Mala, W., Klangbud, W. K., Kotepui, K. U., Rattaprasert, P., & Kotepui, M. (2021). Prevalence, probability, and outcomes of typhoidal/non-typhoidal Salmonella and malaria co-infection among febrile patients: A systematic review and meta-analysis. *Scientific Reports, 11*(1), 21889.

Chapter 11
Insecticide Resistance Markers in *Anopheles gambiae* Complex: Trends and Prospects

Margaret Ikhiwili Oniha and Motunrayo Oluwabunmi Aina

1 Introduction

Malaria is an infectious disease spread by mosquitoes that obtain blood meal from humans and other vertebrates. Over 90% of malaria cases and mortalities worldwide occur in the African continent, making malaria a persistent public health challenge especially in tropical and subtropical areas (Wagah et al., 2021). Among the *Plasmodium* species responsible for malaria, *Plasmodium falciparum* is most significantly associated with human infection, particularly in tropical regions (Acford-Plamer et al., 2023). In sub-Saharan Africa, *An. gambiae* and *An. funestus* predominantly infect the host, initiating their life cycle in the liver before progressing to the red blood cells (RBCs), where they induce symptomatic manifestations. The persistent high temperatures, elevated humidity, and increased rainfall in tropical and subtropical regions create optimal conditions for the heightened prevalence of the disease. Stagnant water also creates a suitable breeding site for mosquitoes (Acford-Plamer et al., 2023). Malaria remains a significant source of morbidity and mortality globally, with an estimated 229 million cases and 409,000 deaths in 2019 (WHO, 2022). Vulnerable populations, such as pregnant women and children under five years old, are disproportionately affected by malaria. Several vector control measures have been adopted, including indoor residual spraying (IRS) and long-lasting insecticide-treated nets. Nonetheless, these efforts have been undermined by the emergence and proliferation of insecticide resistance, driven by specific genetic

M. I. Oniha
Department of Biological Sciences, College of Science and Technology, Covenant University, Ota, Ogun State, Nigeria

M. O. Aina (✉)
Covenant Applied Informatics and Communication Africa Centre of Excellence (CApIC-ACE), Ota, Ogun State, Nigeria
e-mail: motunrayo.ainapgs@stu.cu.edu.ng

P. O. Isibor, P. A. Akinduti (eds.), *Harnessing Biotechnology Tools for Product Development*, https://doi.org/10.1007/978-3-031-86002-7_11

markers. *Plasmodium* has a complicated life cycle that includes several phases in the human host and the mosquito vector, making it difficult to control and eradicate. Malaria exerts profound socioeconomic impacts, severely compromising healthcare systems, diminishing productivity, and impeding economic growth in endemic regions. Malaria disproportionately affects the most vulnerable groups, which impedes the achievement of sustainable development goals and prolongs poverty cycles.

1.1 Pathogenicity and Mode of Transmission

Plasmodium species are protozoan parasites that cause malaria with *P. falciparum* and *P. vivax* being the most common in humans primarily transmitted by *Anopheles* mosquito. The life cycle of a mosquito undergoes four stages include egg, larva, pupa, and adult. The larvae of female mosquitoes develop from eggs they deposit in or near water and reside there while feeding on microorganisms (Baer et al. 2007). Larvae undergo multiple molts until reaching the pupal stage when they become adults and stop feeding. After emerging from the pupal case, the adult mosquitoes go to the air to feed and procreate (Baer et al. 2007). Depending on the surroundings, the life cycle can extend from four days to a month (Baer et al. 2007). Microcirculatory anomalies and parasite-induced changes to red blood cells (RBCs) interact intricately in the pathogenesis of malaria. The parasites undergo sexual development into gametocytes, which infect the mosquito's host during blood meals, and erythrocytic schizogony. When erythrocytic-stage schizonts mature, the rupture of erythrocytes is associated to the first signs and symptoms of malaria. It is presumed that the release of parasite component sets off an immunological response in the host. The fever, chills, sweats, weakness, and other systemic symptoms of malaria are mostly engendered by the cytokines, reactive oxygen intermediates, and other cellular products generated during the immunological response. These substances are important in pathogenesis. Malaria symptoms can range from asymptomatic to severe (complex), and digestive problems being the most common. Severe malaria is frequently lethal and can cause damage to multiple organs, including brain malaria. The pathophysiology of malaria is significantly influenced by innate and adaptive immunological processes, which both cause and prevent illness and mortality (Milner et al., 2018). The molecular elements of malaria pathogenesis include molecules produced from parasites such as PfEMP1, which mediate interactions with human receptors on human erythrocyte surfaces such as complement receptor 1 (CR1), ABO blood group antigens, CD31, and heparan sulfate-related glycosaminoglycans (Milner et al., 2018). Humans, mosquitoes, and the parasite interact to cause malaria infection. When female mosquitoes lack the protein and plant sucrose needed to develop their eggs, they change during the egg-laying season. As a result, they feed on human blood and can spread infection by successful biting that involves injecting infectious protozoans (*Plasmodium* species) into the bloodstream. The malaria parasite is injected into the human host's bloodstream via the bite of an infected female *Anopheles* mosquito, which is the main method of malaria transmission (Savi, 2023). Through a blood meal, the mosquito transfers the parasite from an earlier infected individual to a new host, resulting in this transmission. Congenital transmission, blood transfusions, and organ transplantation are further transmission mechanisms (Savi, 2023). To mitigate the risk of malaria transmission, it is essential to implement measures to prevent mosquito bites and adopt appropriate protective strategies.

1.2 Epidemiology

There were 233 million cases of malaria and 580,000 fatalities in 2022; making it a severe public health burden in Africa (WHO, 2022). Endemicity of malaria in Africa accounts for significantly 94% of cases and 95% of fatalities globally. Children under the age of 5 years account for approximately 80% of all malaria mortalities in the WHO African Region in 2022, and they also account for the bulk of cases and fatalities (WHO, 2022). Several *Plasmodium* parasite species induce malaria in Africa, where its epidemiology is complex and regionally variable. According to Nkumama et al. (2017), *Plasmodium falciparum* is the most prevalent and lethal species, whereas *P. vivax, P. ovale,* and *P. malariae* are less common and geographically confined. The condition is preventable and curable, and the parasites are spread through the bites of female *Anopheles* mosquito vector (Nkumama et al., 2017). The WHO areas of South-East Asia, Eastern Mediterranean, Western Pacific, and the Americas reported a considerable number of cases and fatalities from malaria in 2022, with nearly half of the world's population at risk (WHO, 2022). Nonetheless, with 94% and 95% of malaria cases and fatalities, respectively, in the African Region that suffered the most tremendous malaria burden (WHO, 2022). The effectiveness of malaria vectors, level of poverty, human mobility, resistance to insecticides, and antimalarial medications are some factors that affect malaria epidemiology in Africa (Mbama et al., 2022).

In Nigeria, every one in four persons suffer from malaria and this has accounted for 27% of the world's malaria cases (WHO, 2021). Additionally, malaria is responsible for 32% of related mortalities worldwide, with Nigeria accounting for around one in three cases (WHO, 2022). Nigeria's malaria prevalence varies significantly from area to area, with the northern portions of the country having the highest rates. The North West has a prevalence of 34%, whereas both the South West and South East have a cumulative prevalence of 16%. In comparison to urban areas, the prevalence is similarly higher in rural areas (31% vs. 13%) (WHO, 2021). *Plasmodium falciparum* is the most prevalent malaria parasite species, accounting for most cases. *Anopheles gambiae* is the primary vector and *Anopheles* funestus is the secondary vector (WHO, 2021). Between 2010 and 2018, an uneven decline was observed in malaria prevalence in many states, with the southern and central regions experiencing more significant declines than the northern ones (Oyibo et al., 2021). These regional disparities are influenced by socioeconomic position and preventative care availability (WHO, 2021). With malaria being responsible for 60% of outpatient visits and fatality rates of 30%, 11% and 25% from the pediatrics, maternal and infants' categories, respectively. In Nigeria, malaria continues to be a significant source of morbidity and mortality (WHO, 2021). The most vulnerable categories are pregnant women and children under five because of their weak and underdeveloped immune systems (WHO, 2021).

2 Malaria Prevention and Control

One of the most effective methods of malaria prevention is the use of insecticide-treated bed nets (ITNs), which act as a physical barrier against mosquito bites and also kill or repel mosquitoes that come into contact with the insecticide (Lengeler, 2004). Studies have shown that using ITNs can reduce malaria incidence by up to 50% in areas with high coverage (Eisele et al., 2010). To ensure the effectiveness of ITNs, it is crucial to maintain high levels of coverage and to replace worn-out or damaged nets. Indoor residual spraying (IRS) is another essential tool in the fight against malaria, involving the application of insecticides to the walls and other surfaces inside homes (WHO, 2015). IRS can reduce malaria transmission by killing mosquitoes that rest on treated surfaces, and it is particularly effective in areas with high levels of insecticide resistance (Pluess et al., 2010). However, the success of IRS depends on achieving high coverage rates and ensuring that the insecticides used are effective against local mosquito populations (WHO, 2015). Prompt and accurate diagnosis and treatment of malaria cases is also essential for preventing the spread of the disease and reducing the burden of illness and death (WHO, 2021). Rapid diagnostic tests (RDTs) and microscopy are commonly used to diagnose malaria, and effective treatment with artemisinin-based combination therapies (ACTs) can help to clear the infection and prevent the development of severe complications (WHO, 2021). To maximize treatment efficacy, it is imperative to ensure patients adhere to the full course of medication while closely monitoring for the emergence of drug resistance (WHO, 2021).

2.1 Malaria Vector Control

Global efforts to reduce and eradicate malaria must include vector control, which focuses on actions taken against mosquitoes conveying the parasite to prevent them from further spread (Gari & Lindtjørn, 2018). Different approaches have been employed to eliminate malaria, including larval control, source reduction, chemical larviciding, and biological control (Lobo et al., 2017). In targeting mosquitoes that bite and rest indoors, there are two main vector control strategies that encapsulate indoor residual spraying (IRS) and long-lasting insecticidal nets (LLINs) (Lobo et al., 2017). To effectively combat malaria vectors, several approaches employ different insecticides. Indoor residual spraying (IRS) and long-lasting insecticidal nets (LLINs) include pyrethroids, carbamates, organochlorines, and organophosphates (CDC, 2020). These insecticides have the efficacy to reduce mosquitoes and their capacity to spread malaria. In low-income communities, the strategies employed, including insecticide-treated eaves ribbons, also employ insecticides such as transfluthrin to combat malaria vectors (Kaindoa et al., 2021). *Anopheles* mosquitoes can be killed and repelled by transfluthrin, a vapor-phase pyrethroid that offers protection by limiting indoor bites and eliminating a sizable percentage of insects (Kaindoa et al., 2021). This technology offers a supplementary and focused strategy to reduce mosquitoes, especially in locations where standard approaches (LLINs and IRS) may present limitations. Generally, diverse factors including efficacy, resistance management, safety, and cost-effectiveness influence the choice of insecticides for controlling malaria vectors. This emphasizes the significance of ongoing research and evaluation to maximize the use of insecticides in the effective preventing malaria transmission (Kaindoa et al., 2021). These techniques, however, are less successful in controlling mosquitoes that bite, rest outside, or feed on animals, which can result in residual transmission (Gari & Lindtjørn, 2018). Additional vector management methods that target malaria vectors at various phases of their life cycles are required to address residual transmission (Lobo et al., 2017). Biological control, habitat alteration, manipulation, larviciding, and other techniques are some larval source management (LSM) approaches that involve modifying or eliminating mosquito larvae's breeding grounds (Gari & Lindtjørn, 2018). Individuals can prevent mosquito bites by using personal protective methods such as topical repellents, clothing sprayed with insecticide, and spatial/airborne repellents (Gari & Lindtjørn, 2018).

3 Insecticides Resistance in *Anopheles gambiae* Complex

Insecticide resistance associated with *Anopheles* is a major concern in the combat of malaria because it jeopardizes the efficacy of vector control methods such as indoor residual spraying (IRS) and long-lasting insecticidal nets (LLINs) (Chukwuekezie et al., 2020). Studies have published that populations of *A. gambiae complex* have become resistant to diverse insecticides including pyrethroids, carbamates, and organochlorides, across locations such as Gambia, Nigeria, Côte d'Ivoire, Kenya, and South-Western Chad (Chukwuekezie et al., 2020). In Côte d'Ivoire, *An. gambiae* complex has been found to exhibit a broad range of pyrethroids, organochlorides and carbamates' resistance and these resistant populations exhibit high allelic frequencies of the *kdr* L1014F mutation (Camara et al., 2018).

The efficiency of indoor spraying and bed nets treated with insecticides, two essential methods for preventing malaria, is diminished by this resistance. The high levels of resistance presented by *An. gambiae* mosquitoes validates the urgent need for creative approaches to manage resistance and maintain the effectiveness of currently used vector control techniques (Ononamadu et al., 2020). Cross-resistance, the usage of agricultural insecticides, the use of pyrethroids in long-lasting insecticidal nets, and indoor residual spraying are posited as the main causes of resistance. It is imperative to address insecticide resistance in *A. gambiae* complex because of the widespread resistance pyrethroids as well as the establishment of resistance to organophosphates (Ononamadu et al., 2020).

A. arabiensis, the primary malaria vector in Gambia, has been shown in a research to be resistant to DDT, permethrin, and deltamethrin, especially in the coastal areas (Opondo et al., 2019). Additionally, resistance mutations were found, including the *kdr* L1014F mutation, which may have contributed to the observed resistance (Opondo et al., 2019). The voltage-gated sodium channel (Vgsc) gene in *An. funestus* has five non-synonymous mutations (F763L, I768L/M, L788F, and G793C) that may contribute to insecticide resistance (Acford-Palmer et al., 2023). *An. funestus* was shown to be vulnerable to bendiocarb in South-Western Chad while *An. gambiae s.l.* was found to be resistant to it (Opondo et al., 2019). The efficiency of vector control measures in the region may be diminished by this resistance. In *Anopheles stephensi* from Ethiopia, two resistance markers were identified using a multiplex amplicon sequencing assay comprising the N1575Y mutation in the vgsc gene and the A296S mutation in the rdl gene (Acford-Palmer et al., 2023).

3.1 Mechanisms of Action of Insecticides

3.1.1 Pyrethroids

This is a class of insecticides that work by interfering with voltage-gated sodium channels in insects' neurological systems. As a result of this interaction, sodium channels remain open for an extended period of time, causing depolarization, nerve excitation, and final elimination of the insect.

Changes in nerve membrane permeability to sodium and potassium ions that induce the excitation of nerves generated by pyrethroids can be study by electrophysiological approaches such as voltage clamp and intracellular microelectrode methods (Soderlund et al., 2012). By demonstrating the significance of sodium and potassium ion permeability in the mechanism of action of pyrethroids, these techniques have unveiled the ionic foundation of their effects on the nerves.

3.1.2 Carbamates

This class of insecticides functions by inhibiting the enzyme acetylcholinesterase (AChE), a critical component in nerve signal transmission, thereby disrupting the neurological function of *Anopheles* mosquitoes (Chanda et al., 2020). At nerve synapses, AChE is responsible for degrading the neurotransmitter acetylcholine, enabling appropriate nerve signal transmission. Carbamates prevent the breakdown of Acetylcholine by blocking AChE, which engenders a buildup of the neurotransmitter at nerve synapses (Swale et al., 2018). This buildup results in constant nerve stimulation, which paralyzes and eventually kills the insect.

Carbamates are less harmful to non-target organisms, such as honeybees, than other insecticides (Oduola et al., 2012).

3.1.3 Organochlorines

For many years, the control of mosquitoes has been achieved through the use of organochlorine insecticides, such as dichlorodiphenyltrichloroethane (DDT). Organochlorines work on *Anopheles* mosquitoes by obstructing nerve impulse transmission, which causes nervous system disruption (Jayaraj et al., 2016). Organochlorines function by building up within the neuronal membrane and making it more permeable to sodium ions. As a result of this buildup, the neuronal membrane depolarizes, allowing sodium ions to enter and sodium channels to remain open for an extended period of time. This extended opening causes paralysis and eventual fatality of the mosquito by disrupting the usual flow of nerve impulses. Due to their detrimental impacts on non-target species, environmental persistence, and bioaccumulation, organochlorines including DDT have been applied with controversy (Wan-Norafikah et al., 2020). Although many developed countries have outlawed the use of organochlorines, several underdeveloped nations continue to employ them as a mosquito repellent. Concerns over the long-term effects of organochlorines on human health and the environment have been raised regarding their use, and research is being conducted on insecticide substitutes that have less detrimental effects on the environment and human health (Wan-Norafikah et al., 2020).

3.1.4 Organophosphates

Acetylcholinesterase (AChE) is an enzyme that organophosphate insecticides, such as pirimiphos-methyl, block to target the nervous system of *Anopheles* mosquitoes. At nerve synapses, AChE is in charge of degrading the neurotransmitter acetylcholine, enabling appropriate nerve signal transmission (Wanjala & Kweka, 2018). Acetylcholine is not broken down by organophosphates, which inhibits AChE and causes the neurotransmitter to build up at nerve synapses. The result of this buildup is an overstimulation of the neuromuscular system, which paralyzes and kills the insect eventually (Medjigbodo et al., 2022). Organophosphates are well known for killing mosquitoes fast and for being extremely toxic to insects (Kitungulu et al., 2022). They have been extensively utilized in mosquito control initiatives, such as bed nets treated with insecticides and indoor residual spraying. Nonetheless, the application of organophosphates has been linked to *Anopheles* mosquito resistance, potentially diminishing its efficacy in managing mosquito populations.

3.2 *Insecticide Resistance Markers in* Anopheles gambiae *Complex Mosquitoes*

The mosquito species known as *Anopheles gambiae* complex are largely found in sub-Saharan Africa, where it is a major vector of the human–malaria parasite transmission chain. The "*gambiae* complex" refers to multiple morphologically indistinguishable mosquito species, including important malaria vectors that include *Anopheles gambiae sensu stricto*, *Anopheles coluzzii,* and *Anopheles arabiensis* exhibiting varying vectorial efficiency and behaviors which include athropophily and zoophily (Tennessen et al., 2020). *Anopheles gambiae* complex mosquitoes are well suited to the tropical and subtropical climates of sub-Saharan Africa because they usually breed in clear, sunny water bodies such as tiny puddles, ponds, and makeshift rain pools. Their affinity for feeding on humans, their capacity to nest near to human settlement, and their inclination to rest indoors make them extremely effective malaria vectors (Akpan et al., 2018).

Anopheles gambiae complex mosquitoes possess unique genetic variation known as insecticide resistance markers, which are associated with the mosquito's capacity to withstand insecticide exposure. These markers show the insecticide to which a specific mosquito population may have resistant to. Insecticide resistance markers in mosquito populations are genetic polymorphisms that signal the presence of insecticide resistance (Mwagira-Maina et al., 2021). *A. gambiae* develops insecticide resistance markers through a complicated process that has multiple contributing factors. The widespread application of insecticides to combat malaria is a significant contributing factor as it has resulted in the selection of resistant mosquito populations (Ondeto 2012). The existence of stable chromosomal inversion polymorphisms such as the 2La inversion on the left arm of chromosome 2, is another mechanism causing insecticide resistance in *A. gambiae* (Mwagira-Maina et al., 2021). This inversion is attributed to mosquito behaviors and resistance to desiccation. Its existence in populations of *An. gambiae* is associated with ecological plasticity, which enables the species to adapt to varying environmental conditions. Moreover, a primary contributor to insecticide resistance in *A. gambiae* is metabolic resistance mechanisms, such as the overexpression of genes encoding detoxification enzymes (Ibrahim et al., 2023). *Anopheles gambiae* complex mosquitoes have evolved a number of defense mechanisms against insecticides, making efforts to control malaria difficult. Among the most important indicators of insecticide resistance connected to *A. gambiae* complex are:

3.2.1 Knockdown Resistance (KDR) Mutations

The voltage-gated sodium channel gene (Vgsc), which is the target of pyrethroid insecticide, undergoes amino acid substitutions due to the *kdr* mutations, which lessens the channel's susceptibility to insecticide consequently compromising the efficacy of pyrethroid insecticides. The L1014F and L1014S *kdr* mutations are the most well known in *An. gambiae* complex (Wolie et al., 2021). Knockdown Resistance (*kdr*) mutations are biochemical markers that are polymorphic in single nucleotides and polymorphic in restriction fragment length. They identify substitutions in the amino acid sequences that result in refractoriness to particular insecticides. The *kdr* confers resistance to pyrethroids and DDT insecticides due to a mutation at locus 1014 of the voltage-gated sodium channel gene (Vgsc) where a serine or phenylalanine amino acid replaces a leucine amino acid, resulting in either an L1014F (Vgsc-1014F) or L1014S (Vgsc-1014S) mutation (Mwagira-Maina et al., 2021).

3.2.2 The Acetylcholinesterase Enzymes (*AChE*)

The serine hydrolase enzyme acetylcholinesterase (*AChE*) is essential to the neurological systems of insects, especially mosquitoes. At the synaptic cleft, it is in charge of breaking down the neurotransmitter acetylcholine into acetate and choline, which stops nerve signals. When this mechanism is interfered with, acetylcholine builds up resulting in overstimulation and ultimately blockage of neurotransmission. Two *AChE* genes, *AChE*1 and *AChE*2, are found in mosquitoes and are encoded by the ace-1 and ace-2 genes, respectively. In mosquitoes, catalytic acetylcholinesterase activity and *AChE*-induced insecticide resistance are mediated by *AChE*1, which is encoded by the ace-1 gene. Glycine to serine substitution (G119S) in the *ace-1* gene is a naturally occurring mutation that gives mosquito tolerance to organophosphates and carbamates (Mwagira-Maina et al., 2021). *An. gambiae* acetylcholinesterase (*AChE*) enzymes are essential for combating insecticide resistance in malaria. Since it hydrolyzes acetylcholine at cholinergic synapses, *AChE* has been the focus of research into insecticide development. *AChE*-1 and *AChE*-2 proteins are encoded by the two *AChE* genes, *ace-1* and *ace-2*, in *An. gambiae*, with *AChE*-1 being principally responsible for the nervous system (Han et al., 2017). The dynamics of malaria transmission are impacted by the establishment of resistant strains to *AChE*-targeting insecticides, which presents difficulties for mosquito control. The effectiveness of insecticides employed in vector control is impacted by the G119S mutation in the *ace-1* gene, which provides resistance to carbamates and organophosphates (Keïta et al., 2020). This mutation emphasizes the necessity of managing insecticide resistance in *An. gambiae* populations by providing a robust diagnosis of resistance (Keïta et al., 2020).

3.2.3 The *Rdl* Gene

The Resistance to dieldrin gene, or *Rdl* gene, is a genetic locus present in a variety of insect species, including *An. gambiae* mosquitoes. In the past, dieldrin, an organochlorine insecticide, was frequently employed in programs to reduce malaria vectors and agricultural pests (Grau-Bové et al., 2020). *Rdl* gene is an important genetic locus connected to resistance to cyclodiene insecticide, including dieldrin

(Grau-Bové et al., 2020). These insecticides target a gamma aminobutyric acid (GABA) gated chloride channel subunit, which is encoded by this gene. Nicotinic acetylcholine receptors, GABA receptors, and serotonin receptors are all members of the *Cys*-loop ligand-gated ion channel (*CysLGIC*) superfamily, which the *Rdl* gene belongs to. In insects, these channels are required for controlling synaptic transmission and neural excitability (Grau-Bové et al., 2020). *An. gambiae Rdl* locus has independent mutations that have been found to provide resistance to dieldrin insecticides (Liu et al., 2020). These alterations are genetically associated with insecticide resistance in these malaria vector species, such as the substitution of alanine for glycine in the *Rdl* locus.

3.3 Metabolic Detoxification

In the context of malaria control, the emergence of metabolic resistance in *Anopheles gambiae* poses a growing challenge, as it has the potential to compromise the efficacy of insecticide-based vector control strategies. The main cause of this type of resistance is the overexpression of enzymes that include glutathione-S-transferases and cytochrome P450-dependent monooxygenases that detoxify or sequester insecticides (Stica et al., 2019). These enzymes possess the ability to degrade or make ineffective a number of insecticide types, including as carbamates, organophosphates, and pyrethroids (Mavridis et al., 2019). The absence of molecular markers makes it challenging to identify and monitor metabolic resistance, which is one of the major management challenges (Tchouakui et al., 2022). Nonetheless, current research has created quick multiplex gene expression assays to track An. *gambiae* (*s.l.*) levels of key detoxification genes and metabolic resistance (Mavridis et al., 2019). These assays are quick, easy to use, reliable, and provide information for vector control programs that is both operationally useful and grounded in data. Certain genes including the glutathione-S-transferase *GSTE2* and *CYP6P3*, *CYP6M2*, *CYP9K1*, *CYP6P4* and *CYP6Z1* have been associated with metabolic resistance among the *A. gambiae* (Mavridis et al., 2019). It has been demonstrated that the degree of resistance to certain insecticides is correlated with the expression levels of these genes, which can be overexpressed to impart resistance (Stica et al., 2019). The primary categories of detoxifying enzymes implicated in metabolic resistance consist of:

3.3.1 Cytochrome P450s

A broad class of enzymes known as cytochrome P450 monooxygenases is involved in the metabolism of a variety of substances, including insecticides. Certain cytochrome P450 enzymes have the ability to oxidize insecticides in mosquitoes, which causes the insecticides to become inactive and eventually become expelled from the mosquito's body (Mavridis et al., 2019).

3.3.2 Esterases

Esterases are enzymes that accelerate the hydrolysis of insecticide molecules ester linkages, changing them into less harmful substances that mosquitoes can expel. The capacity of mosquitoes to detoxify insecticide may be improved by increased esterase expression or activity (Yusuff 2021).

3.3.3 Glutathione S-transferases (GSTs)

These are the enzymes that catalyze the conversion of insecticide molecules to glutathione molecules, allowing the mosquito to efficiently expel the former from its body. With regards to the detoxification of insecticides, GSTs are very important since they can easily conjugate electrophilic groups with glutathione (Tchouakui et al., 2022). Certain genes in the mosquito's genome encode these detoxifying enzymes, and exposure to insecticides can control the process of enzyme expression. In mosquito populations that are frequently exposed to insecticides due to agricultural practices or public health initiatives, natural selection tends to prefer individuals with genetic variants that confer higher levels of detoxifying enzyme expression or activity. Consequently, populations of mosquitoes that are more metabolically resistant to insecticide may eventually evolve (Yusuff et al., 2021).

4 Current Trends with Insecticide Resistance in *Anopheles* Mosquitoes

A notable rise in pyrethroids resistance is a hallmark of the present trends in insecticide resistance in *Anopheles* mosquitoes in mainland Tanzania. In 2020, more than 80% of *Anopheles gambiae* mosquitoes have developed pyrethroids resistance due to the mosquito species' global proliferation and high levels of resistance (Tungun et al., 2023). The extensive use of pyrethroids in insecticide-based malaria control strategies, such as indoor residual spraying (IRS) and long-lasting insecticidal nets (LLINs), is responsible for the resistance (Tungun et al., 2023). The susceptibility of *Anopheles gambiae* mosquitoes to insecticides, such as permethrin, deltamethrin, lambdacyhalothrin, bendiocarb, and pirimiphos-methyl, has been evaluated using the World Health Organization's (WHO) standard protocols (Tungun et al., 2023). A finding showed a significant inverse relationship between the survey year and the susceptibility status to pyrethroids, indicating a progressive increase in resistance over time (Tungun et al., 2023). In certain situations, using synergists such as piperonyl butoxide (PBO) has been shown to be entirely or substantially restore the efficiency of pyrethroids. This suggests that cytochrome P 450 detoxifying enzymes play a key role in the observed pyrethroids resistance. This emphasizes the necessity of creating and implementing LLINs and next-generation insecticides that can successfully combat resistance mechanisms. The present trends in insecticide resistance in *Anopheles* mosquitoes are characterized by a sharp rise in pyrethroids resistance, underscoring the critical need for efficient methods and tactics for controlling malaria vectors in order to maintain the progress made in the fight against malaria.

4.1 Geographic Variation and Emerging Patterns

Studies conducted in various locations show that resistance patterns varied widely. For example, RNA-Seq investigation in Benin has demonstrated that populations of resistant *Anopheles gambiae* from different locations overexpress salivary gland and cuticular proteins, which contributes to resistance against a variety of insecticides such as pyrethroids and organophosphates (Saizonou et al., 2024). Studies on transcriptomics conducted in Benin have revealed that mosquitoes from various locations overexpress genes linked to insecticide resistance. Notably, there is a correlation between resistance to pyrethroids and organophosphates and increased expression of cuticular proteins and cytochrome P450s, including *CYP6M2*. These results suggest that selection pressure from widespread insecticide usage has led to the development of cross-resistance in mosquitoes in Benin (Saizonou et al., 2024). Transcriptome study of Benin's *Anopheles gambiae* showed upregulation of salivary and cuticular proteins associated with organophosphates and cross-resistant pyrethroids.

In Côte d'Ivoire, agricultural practices have been accountable for insecticide resistance. High frequency of the *Ace1-G280S* mutation, which confers resistance to carbamates and organophosphates, are observed in mosquitoes in vegetable fields. These insecticides are heavily used in agriculture, which has contaminated mosquito breeding grounds and favored the development of resistant genotypes. Furthermore, it has been shown that P450 gene overexpression contributes to metabolic resistance in Côte d'Ivoire's agricultural practices and the insecticide resistance profiles of *Anopheles gambiae sensu lato* (Kouadio et al., 2023).

Extensive research conducted in Kenya's coastal regions reveals notable resistance to many pesticides, such as DDT, permethrin, and deltamethrin. L1014S and L1014F *kdr* mutations are a significant contributing reason to this resistance. These mosquitoes become less vulnerable to pyrethroids as a result of these mutations that change their sodium channels. *Anopheles gambiae* (s.l.) insecticide resistance status in coastal Kenya (Munywoki et al., 2021).

4.2 Prospect of Insecticide Resistance Markers in Anopheles gambiae *Complex*

4.2.1 Gene Drive

Gene drive technology holds significant potential to mitigate Africa's persistent malaria challenge; however, it remains in the early stages of development and testing. The African Union High Level Panel on Emerging Technologies has designated gene drives as a priority technology for the eradication of malaria. By altering the mosquitoes that carry the malaria parasite, gene drives, a sort of genetic modification, can be utilized to reduce and eventually eradicate the disease. The World Health Organization's (WHO) ongoing anti-malaria efforts in Africa stand to benefit greatly from the technology. The development of gene drive modified mosquitoes (GDMMs) may be able to overcome challenges that include drug and insecticide resistances that have impeded efforts to eradicate malaria. They propagate through mosquito populations by skewing the inheritance of genetic modifications. However, as GDMMs are a novel and sophisticated technology, thorough stakeholder consultation with African communities is essential to the evaluation and effective application of this technology. The frameworks for its regulations are still being created. In cooperation with research institutions, African nations including Burkina Faso, Mali, and Uganda are already engaged in field trials and gene drive research. However, governments will need to commit to developing the necessary infrastructure and capability for adoption. Experts emphasize that in addition to currently available interventions such as insecticide-treated nets and enhanced case.

4.2.2 Enhanced Monitoring and Surveillance

Incorporating insecticide resistance markers into monitoring programs enables more rapid and precise identification of resistance patterns within *Anopheles gambiae* populations than the current techniques. Specific genetic alterations linked to insecticide resistance such as mutations in the *kdr* gene (knockdown resistance) and amplification of *CYP6P3* (a cytochrome P450 gene involved in metabolic resistance) can be detected using molecular methods such as PCR and sequencing, which offer comprehensive insights into the emergence and evolution of resistance (Munywoki et al., 2021). Targeted therapies can also be developed by knowing the genetic basis of insecticide resistance. For instance, locations with high insecticide resistance levels can adopt integrated pest management techniques or alternative pesticides. This can ensure that the therapies utilized are still effective against the local mosquito population, which can enhance the efficacy of malaria control programs (Munywoki et al., 2021).

4.2.3 Strategies for Insecticide Resistance Management

Scientists can simulate the dynamics of resistance development under various conditions such as varying insecticide concentrations, environmental factors (temperature and humidity), population size, and genetic diversity within *Anopheles gambiae* populations. This can help guide the development of insecticide resistance management techniques that include alternating insecticide applications, combining treatments, or creating havens where non-resistant mosquitoes can live. These tactics can prolong the effectiveness of insecticide and impede the emergence of resistance (Yin et al., 2021).

4.2.4 Integration with Genomic Data

Combining insecticide resistance marker data with more extensive genomic data from *An. gambiae* populations can offer a thorough understanding of the population dynamics and gene flow. This comprehensive understanding helps reveal potential impediments to gene flow that could be used in control methods and assist forecast how resistance would spread (Munywoki et al., 2021).

5 Conclusion and Recommendations

In order to combat malaria effectively, a comprehensive, multisectoral strategy that combines community involvement, sustainable interventions, and health system development is required. Furthermore, an effective control of malaria requires epidemiological surveillance. Genetic alterations associated with insecticide resistance, such as mutations in the *kdr* gene, conferring knockdown resistance to pyrethroids, and the amplification of cytochrome P450 genes including *CYP6P3*, which contribute to metabolic resistance, can be identified through molecular methods comprising PCR and sequencing. These techniques provide detailed insights into the mechanisms underlying the development and evolution of resistance (Munywoki et al., 2021; Riveron et al., 2018).

Therefore, the integration of advanced biotechnological interventions such as CRISPR-based gene editing to restore insecticide susceptibility in mosquito populations, or the development of novel biopesticides targeting resistant vectors, holds great potential to enhance the efficacy of insecticides. These innovative approaches can play a pivotal role in achieving sustainable malaria control and eradication efforts.

General comment: The Authors have put in a great work in delineating the mechanical modes of action of insecticidal functionalities and the mitigating pathways of mutation in mosquitoes. The report is however yet to close up on this evidence-based presentation with a compelling idea that may solve the challenge addressed. The report holds the potentials to seal some knowledge gaps beyond the current standard of the recommendations and conclusion that read somewhat like a casual advice. Input is thus needed to make this section more scientific. For examples, the pace was set from the beginning to explore the shortcomings in the trends and project into the prospects. A good effort was made to attain this set goal in Sect. 4.2.1 Gene Drive, and Sect. 4.2.2 enhanced Monitoring and Surveillance. A cue should be gleaned from these sections to make a better closure to the report.

References

Acford-Palmer, H., Phelan, J. E., Tadesse, F. G., Kristan, M., Collins, E., Spadar, A., & Campino, S. (2023). Identification of two insecticide resistance markers in Ethiopian Anopheles stephensi mosquitoes using a multiplex amplicon sequencing assay. *Scientific Reports, 13*(1). https://doi.org/10.1038/s41598-023-32336-7

Akpan, G. E., Adepoju, K. A., Oladosu, O. R., & Adelabu, S. A. (2018). Dominant malaria vector species in Nigeria: Modelling potential distribution of Anopheles gambiae sensu lato and its siblings with MaxEnt. *PLoS One, 13*(10), e0204233. https://doi.org/10.1371/journal.pone.0204233

Baer, K., Klotz, C., Kappe, S. H. I., Schnieder, T., & Frevert, U. (2007). Release of hepatic Plasmodium yoelii merozoites into the pulmonary microvasculature. *PLoS Pathogens, 3*(11), e171. https://doi.org/10.1371/journal.ppat.0030171

Camara, S., Koffi, A. A., Ahoua Alou, L. P., Koffi, K. K., Jean, P. K., Kabran, A. K., Mathieu, F. K., Rapheal, N. G., & Cedric, P. (2018). Mapping insecticide resistance in Anopheles gambiae (s.l.) from Côte d'Ivoire. *Parasites and Vectors, 11*, 19. https://doi.org/10.1186/s13071-017-2546-1

CDC (2020). *Centers for Disease Control and Prevention in Côte d'Ivoire*. Retrieved 10 May 2024, from https://www.cdc.gov/globalhealth/countries/cote-d-ivoire/.

Chukwuekezie, O., Nwosu, E., Nwangwu, U., et al. (2020). Resistance status of Anopheles gambiae (s.l.) to four commonly used insecticides for malaria vector control in south-East Nigeria. *Parasites and Vectors, 13*, 152. https://doi.org/10.1186/s13071-020-04027-z

Eisele, T. P., Larsen, D., & Steketee, R. W. (2010). Protective efficacy of interventions for preventing malaria mortality in children in plasmodium falciparum endemic areas. *International Journal of Epidemiology, 39*(suppl_1), i88–i101. https://doi.org/10.1093/ije/dyq026

Gari, T., & Lindtjørn, B. (2018). Reshaping the vector control strategy for malaria elimination in Ethiopia in the context of current evidence and new tools: Opportunities and challenges. *Malaria Journal, 17*(1). https://doi.org/10.1186/s12936-018-2607-8

Grau-Bové, X., Tomlinson, S., O'Reilly, A. O., Harding, N. J., Miles, A., Kwiatkowski, D., et al. (2020). Evolution of the insecticide target Rdl in African anopheles is driven by interspecific and interkaryotypic introgression. *Molecular Biology and Evolution, 37*(10), 2900–2917. https://doi.org/10.1093/molbev/msaa128

Han, L., Hudgens, M. G., Emch, M. E., Juliano, J. J., Keeler, C., Martinson, F., Kamthunzi, P., Tegha, G., Lievens, M., & Hoffman, I. F. (2017). RTS,S/AS01 malaria vaccine efficacy is not modified by seasonal precipitation: Results from a phase 3 randomized controlled trial in Malawi. *Scientific Reports, 7*(1), 7200. https://doi.org/10.1038/s41598-017-07533-w. PMID: 28775306; PMCID: PMC5543056.

Ibrahim, S. S., Muhammad, A., Hearn, J., Weedall, G. D., Nagi, S. C., Mukhtar, M. M., et al. (2023). Molecular drivers of insecticide resistance in the Sahelo-Sudanian populations of a major malaria vector Anopheles coluzzii. *BMC Biology, 21*(1). https://doi.org/10.1186/s12915-023-01610-5

Jayaraj, R., Megha, P., & Sreedev, P. (2016). Organochlorine pesticides, their toxic effects on living organisms and their fate in the environment. *Interdisciplinary Toxicology, 9*(3-4), 90. https://doi.org/10.1515/intox-2016-0012

Kaindoa, E. W., Mmbando, A. S., Shirima, R., Hape, E. E., & Okumu, F. O. (2021). Insecticide-treated eave ribbons for malaria vector control in low-income communities. *Malaria Journal, 20*(1). https://doi.org/10.1186/s12936-021-03945-2

Keïta, M., Kané, F., Thiero, O., Traoré, B., Zeukeng, F., Sodio, A. B., et al. (2020). Acetylcholinesterase (ace-1R) target site mutation G119S and resistance to carbamates in Anopheles gambiae (sensu lato) populations from Mali. *Parasites & Vectors, 13*(1). https://doi.org/10.1186/s13071-020-04150-x

Kitungulu, N., Guyah, B., Webale, M., Shaviya, N., Machani, M., Mulama, D., & Ndenga, B. (2022). Resistance of Anopheles gambiae sensu lato to Pirimiphos-methyl Insecticide in Kakamega County, Highlands of Western Kenya. *African Health Sciences, 22*(1), 589–597. https://doi.org/10.4314/ahs.v22i1.68

Kouadio, F. P. A., Wipf, N. C., Nygble, A. S., Fodjo, B. K., Sadia, C. G., Vontas, J., et al. (2023). Relationship between insecticide resistance profiles in Anopheles gambiae sensu lato and agricultural practices in Côte d'Ivoire. *Parasites & Vectors, 16*(1). https://doi.org/10.1186/s13071-023-05876-0

Lengeler, C. (2004). Insecticide-treated nets for malaria control: Real gains. *Bulletin of the World Health Organization, 82*(2), 84. PMID: 15042228; PMCID: PMC2585896.

Liu, N., Feng, X., & Qiu, X. (2020). RDL mutations in Guangxi Anopheles sinensis populations along the China–Vietnam border: Distribution frequency and evolutionary origin of A296S resistance allele. *Malaria Journal, 19*(1). https://doi.org/10.1186/s12936-020-3098-y

Lobo, N. F., Achee, N. L., Greico, J., & Collins, F. H. (2017). Modern vector control. *Cold Spring Harbor Perspectives in Medicine, 8*(1), a025643. https://doi.org/10.1101/cshperspect.a025643

Mavridis, K., Wipf, N., Medves, S., Erquiaga, I., Müller, P., & Vontas, J. (2019). Rapid multiplex gene expression assays for monitoring metabolic resistance in the major malaria vector Anopheles gambiae. *Parasites & Vectors, 12*(1), 9. https://doi.org/10.1186/s13071-018-3253-2. PMID: 30612581; PMCID: PMC6322220.

Medjigbodo, A. A., Djihinto, O. Y., Salavi, E. B. J., Sonounameto, E. G., Abbey, E., Djossou, L., & Djogbénou, L. S. (2022). Organophosphate insecticide exposure impacts reproductive success in insensitive acetylcholinesterase Anopheles gambiae mosquitoes. *Frontiers in Tropical Diseases, 3*. https://doi.org/10.3389/fitd.2022.903654

Munywoki, D. N., Kokwaro, E. D., Mwangangi, J. M., Muturi, E. J., & Mbogo, C. M. (2021). Insecticide resistance status in Anopheles gambiae (s.l.) in coastal Kenya. *Parasites & Vectors, 14*(1). https://doi.org/10.1186/s13071-021-04706-5

Milner, D. A. (2018). Malaria pathogenesis. *Cold Spring Harbor Perspectives in Medicine, 8*(1), a025569.

Mwagira-Maina, S., Runo, S., Wachira, L., Kitur, S., Nyasende, S., Kemei, B., & Kamau, L. (2021). Genetic markers associated with insecticide resistance and resting behaviour in Anopheles gambiae mosquitoes in selected sites in Kenya. *Malaria Journal, 20*(1). https://doi.org/10.1186/s12936-021-03997-4

Nkumama, I. N., O'Meara, W. P., & Osier, F. H. A. (2017). Changes in malaria epidemiology in Africa and new challenges for elimination. *Trends in Parasitology, 33*(2), 128–140. https://doi.org/10.1016/j.pt.2016.11.006. Epub 2016 Dec 6. PMID: 27939610; PMCID: PMC6995363.

Nkumama, I. N., Odera, D., Musasia, F., Mwai, K., Nyamako, L., Murungi, L., & Osier, F. H. A. (2022). Breadth of fc-mediated effector function delineates grades of clinical immunity following human malaria challenge. *bioRxiv (Cold Spring Harbor Laboratory)*. https://doi.org/10.1101/2022.10.11.511755

Oduola, A. O., Idowu, E. T., Oyebola, M. K., Adeogun, A. O., Olojede, J. B., Otubanjo, O. A., & Awolola, T. S. (2012). Evidence of carbamate resistance in urban populations of Anopheles gambiae s.s. mosquitoes resistant to DDT and deltamethrin insecticides in Lagos, South-Western Nigeria. *Parasites & Vectors, 5*(1). https://doi.org/10.1186/1756-3305-5-116

Ondeto, B. M. (2012). *Main malaria vector distribution and current status of insecticide resistance in Kenya*. Doctoral Dissertation, University of Nairobi.

Ononamadu, C. J., Datit, J. T., & Imam, A. A. (2020). Insecticide resistance profile of Anopheles gambiae mosquitoes: A study of a residential and industrial breeding sites in Kano Metropolis, Nigeria. *Environmental Health Insights, 14*, 117863021989727. https://doi.org/10.1177/1178630219897272

Opondo, K. O., Jawara, M., Cham, S., Jatta, E., Jarju, L., Camara, M., et al. (2019). Status of insecticide resistance in Anopheles gambiae (s.l.) of The Gambia. *Parasites & Vectors, 12*(1). https://doi.org/10.1186/s13071-019-3538-0

Oyibo, W., Ntadom, G., Uhomoibhi, P., Oresanya, O., Ogbulafor, N., Ajumobi, O., & Conway, D. (2021). Geographical and temporal variation in reduction of malaria infection among children under 5 years of age throughout Nigeria. *BMJ Global Health, 6*(2), e004250. https://doi.org/10.1136/bmjgh-2020-004250

Pluess, B., Tanser, F. C., Lengeler, C., & Sharp, B. L. (2010). Indoor residual spraying for preventing malaria. *Cochrane Database of Systematic Reviews, 2010*(4), CD006657. https://doi.org/10.1002/14651858.CD006657.pub2. PMID: 20393950; PMCID: PMC6532743.

Riveron, J. M., Tchouakui, M., Mugenzi, L., Menze, B. D., Chiang, M. C., & Wondji, C. S. (2018). Insecticide resistance in malaria vectors: An update at a global scale. In *Towards malaria elimination-a leap forward*. IntechOpen. https://doi.org/10.5772/intechopen.78375

Saizonou, H., Impoinvil, L. M., Derilus, D., Omoke, D., Okeyo, S., Dada, N., et al. (2024). Transcriptomic analysis of Anopheles gambiae from Benin reveals overexpression of salivary and cuticular proteins associated with cross-resistance to pyrethroids and organophosphates. *BMC Genomics, 25*(1). https://doi.org/10.1186/s12864-024-10261-x

Savi, M. K. (2023). An overview of malaria transmission mechanisms, control, and modeling. *Medical Science, 11*(1), 3. https://doi.org/10.3390/medsci11010003

Soderlund, D. M. (2012). Molecular mechanisms of pyrethroid insecticide neurotoxicity: Recent advances (2012). *Archives of Toxicology, 86*, 165–181. https://doi.org/10.1007/s00204-011-0726-x

Stica, C., Jeffries, C. L., Irish, S. R., Barry, Y., Camara, D., Yansane, I., Kristan, M., Walker, T., & Messenger, L. A. (2019). Characterizing the molecular and metabolic mechanisms of insecticide resistance in Anopheles gambiae in Faranah, Guinea. *Malaria Journal, 18*(1). https://doi.org/10.1186/s12936-019-2875-y

Swale, D. R., Li, Z., Kraft, J. Z., Healy, K., Liu, M., David, C. M., & Foil, L. D. (2018). Development of an autodissemination strategy for the deployment of novel control agents targeting the common malaria mosquito, Anopheles quadrimaculatus say (Diptera: Culicidae). *PLoS Neglected Tropical Diseases, 12*(4), e0006259. https://doi.org/10.1371/journal.pntd.0006259

Tchouakui, M., Assatse, T., Mugenzi, L. M. J., Menze, B. D., Nguiffo-Nguete, D., Tchapga, W., & Wondji, C. S. (2022). Comparative study of the effect of solvents on the efficacy of neonicotinoid insecticides against malaria vector populations across Africa. *Infectious Diseases of Poverty, 11*(1). https://doi.org/10.1186/s40249-022-00962-4

Tennessen, J. A., Ingham, V. A., Toé, K. H., Guelbéogo, W. M., Sagnon, N., Kuzma, R., et al. (2020). A population genomic unveiling of a new cryptic mosquito taxon within the malaria-transmitting Anopheles gambiae complex. *Molecular Ecology, 30*(3), 775–790. https://doi.org/10.1111/mec.15756

Wagah, M. G., Korlević, P., Clarkson, C., Miles, A., Lawniczak, M. K. N., Makunin, A., & Fontaine, M. C. (2021). Genetic variation at the Cyp6m2 putative insecticide resistance locus in Anopheles gambiae and Anopheles coluzzii. *Malaria Journal, 20*(1). https://doi.org/10.1186/s12936-021-03757-4

Wan-Norafikah, O., Khairul-Hazim, K., Nurul-Najiyah, M., Siti-Nurdiyana, M., Nur-Syaahidah, K., & Atiqah-Nazirah, R. (2020). Dispersal of Aedes aegypti L. and Aedes albopictus Skuse (Diptera: Culicidae) in a university campus in Selangor, Malaysia.

Wanjala, C. L., & Kweka, E. J. (2018). Malaria vectors insecticides resistance in different agroecosystems in Western Kenya. *Frontiers in Public Health, 6*. https://doi.org/10.3389/fpubh.2018.00055

Wolie, R. Z., Koffi, A. A., Alou, L. P. A., Sternberg, E. D., N'Nan-Alla, O., Dahounto, A., et al. (2021). Evaluation of the interaction between insecticide resistance-associated genes and malaria transmission in Anopheles gambiae sensu lato in central Côte d'Ivoire. *Parasites & Vectors, 14*(1). https://doi.org/10.1186/s13071-021-05079-5

World Health Organization. (2015). *World malaria report 2015*. World Health Organization.

World Health Organization. (2021). *World malaria report 2021*. World Health Organization.

World Health Organization (2022) World malaria report 2023. Retrieved 6 April 2024, from: https://www.who.int/publication/i/item/9789240086173.

Yin, J., Yamba, F., Zheng, C., Zhou, S., Smith, S. J., Wang, L., et al. (2021). Molecular detection of insecticide resistance mutations in anopheles gambiae from Sierra Leone using multiplex SNaPshot and sequencing. *Frontiers in Cellular and Infection Microbiology, 11*. https://doi.org/10.3389/fcimb.2021.666469

Yusuff, M. (2021). Real-time disease mapping using AI for public health interventions.

Chapter 12
Reports on *Plasmodium falciparum* Histidine-rich Protein 2/3 Gene Deletion and Utilized Methodological Approaches in Nigeria: Systematic Review

Akinnola Olayemi, Kwarpo Zeendi, Oluwaseyi Temitope, and Omonigbehin Comfort

1 Introduction

Malaria-related mortality in the World Health Organization (WHO) African Region fell from eight hundred and eight thousand in the year 2000 to Five Hundred and forty-eight thousand in 2017. Then, it rose to six hundred and four thousand in 2020 due to COVID-19 (WHO, 2023). The female Anopheles mosquito is the vector for transmitting the *Plasmodium* parasite to human hosts. The primary malaria vectors in Nigeria include *Anopheles gambiae*, *Anopheles arabiensis*, *Anopheles machete*, and *Anopheles funestus* (Awosolu et al., 2021). The mosquito is unscathed by the parasite, and its widespread distribution ensures the transmission of the sickness. *Plasmodium falciparum* is the deadliest infective parasite in the world, causing about 300 million clinical episodes and hundreds of thousands of fatalities annually (Brazier et al., 2017). The two most common human Plasmodium parasites are *Plasmodium vivax* and *Plasmodium falciparum* (Manku et al., 2020). *Plasmodium falciparum* is most prevalent in sub-Saharan Africa, and *Plasmodium vivax* is commonplace in other regions like Asia, the Middle East, South America, and some parts of Africa (Rougeron et al., 2022). Inaccurate diagnosis is a problem for efficient patient care and treatment and is critical for controlling and eliminating malaria. Malaria symptoms are nonspecific; a clinical diagnosis cannot be made in areas lacking laboratory capacity for malaria detection (Alemu et al., 2013). Healthcare professionals have received training in managing malaria, and primary

A. Olayemi · K. Zeendi (✉) · O. Temitope · O. Comfort
Department of Biological Sciences, College of Science and Technology, Covenant University, Ota, Ogun State, Nigeria

Covenant Applied Informatics and Communication Africa Centre of Excellence (CApIC-ACE), Ota, Ogun State, Nigeria
e-mail: zeendi.kwarpopgs@stu.cu.edu.ng

P. O. Isibor, P. A. Akinduti (eds.), *Harnessing Biotechnology Tools for Product Development*, https://doi.org/10.1007/978-3-031-86002-7_12

healthcare facilities continue to be the initial point of contact for the community (Ayandipo et al., 2023). Microscopy is still the gold standard for diagnosing malaria (Wardhani et al., 2020). Still, owing to the difficulty in obtaining malaria microscopy, particularly in rural regions such as Nigeria, malaria diagnosis is mainly carried out via rapid diagnostic test (RDT) kits. Every year, 345 million RDTs are sold, and the most common type is *Plasmodium falciparum* histidine-rich protein 2 (HRP2)-based RDTs, which are very sensitive to the level of this antigen in comparison with lactate dehydrogenase (pLDH) RDT (McCaffery et al., 2021). Polymerase chain reaction (molecular methods) can also be utilized to amplify the parasite's deoxyribonucleic acid, making it a sensitive assay (0.004 parasites/μL) (Tambo et al., 2018), but its limitations are attributed to the need for technical skills in clinical settings, high cost, and low turnaround time. It is usually recommended for research (Bharti et al., 2009; Oyegoke et al., 2022). Rapid diagnostic tools, typically utilized in places without power and with less skilled labor, are quick and easy methods for diagnosing malaria. Malaria infection risk depends on transmission areas, prevention measures, and control measures like mosquito nets and prophylaxis. Symptoms occur based on the host's immune system, with efficient responses eliminating parasites, while poor responses allow parasites to progress and spread through vectors (Olasehinde et al., 2010). Since malaria symptoms are typically difficult to distinguish from those of febrile illnesses like viral hepatitis, dengue fever, and leptospirosis and more making malaria infection indistinguishable (Gadia et al., 2017; Gimenez et al., 2021). *Plasmodium falciparum* strains have acquired gene deletions responsible for proteins (pfhrp2 and pfhrp3), which enable them to evade detection by HRP-2-based RDTs, endangering the test's utility in malaria diagnosis. The emergence of parasites with histidine-rich protein 2 (HRP2) gene deletion influences proper malaria diagnosis, leading to failure efforts to curb malaria and prevent an accurate assessment of malaria prevalence (Gamboa et al., 2010). Based on reports by Thomson et al. (2019), Sepúlveda et al. (2018), and Dorado et al. (2016), these particular parasites have since migrated to various regions of Africa and Asia. Berhane et al. (2018) reported that South America and Sub-Saharan Africa have the largest concentrations of PfHRP2-deleted *Plasmodium falciparum* strains. The prevalence of mixed-strain infections in high-transmission areas might be the underlying cause of the frequency of deletions occurring in the *P. f* population. However, if there was sufficient HRP2 in the blood, a positive HRP2 RDT result would be obtained regardless of the patient's genotypes, enabling appropriate case therapy, as advised by Beshir et al. (2022)."False-negative" results have been attributed to RDT results produced as a result of parasites deleting the pfhrp2/3 gene, which may result in undiagnosed and untreated cases leading to delay in antimalarial treatment, possibly putting the patient's life in danger while rendering them a source of continued malaria transmission. In regions where malaria transmission is very high, molecular epidemiology studies are valuable tools for examining the genetic variety of the *Plasmodium falciparum* population (Metoh et al., 2020). Malaria transmission and epidemiology include antigenic variation, such as genetic diversity of Merozoite Surface Proteins (1 and 2), parasite population structure, parasite relatedness, transmission patterns, and resistance to antimalarial drugs such as *pfK13, pfmdr,* and *pfcrt* as emphasized by Dokunmu et al. (2019). The development of genotyping techniques, analytical instruments, and knowledge has led to the creation of operational research initiatives that highlight the practical applications of genetic epidemiology. The investigation of this gene deletion involves a cascade of methods to confirm the presence or absence of the histidine-rich protein 2 and 3 gene deletion in *Plasmodium falciparum*. These studies have been carried out in different regions: Central Vietnam by Thang et al. (2022), the Mount Cameroon region by Njewa et al. (2024), India, and Ethiopia by Golassa et al. (2020). Most studies have utilized RDT to detect *Plasmodium falciparum*, which can be confirmed by microscopy followed by diverse PCR methods. This review aims to systematically identify histidine-rich protein 2/3 gene deletion investigations in *Plasmodium falciparum* carried out and published in Nigeria from 2010 to 2024.

2 Materials and Methods

Abstracts and journal or article titles were utilized for the study. Databases such as MEDLINE, PUBMED, and Google Scholar were used to search the literature for keywords such as "gene deletion," "*Plasmodium falciparum* HRP 2/3 gene deletion," "PfHRP 2/pfhrp3," "PfHRP3/pfhrp3," "malaria Rapid Diagnostic Kits" and "HRP2/3 deletion in Nigeria." The query utilized by Bosco et al. (2020)) was adopted. No ethical approval was obtained for the review.

2.1 Review Objectives

This review assessed the methodological approaches and consistency among articles identified between January 2010 and April 2024 for *P. falciparum* HRP 2/3 gene deletion in Nigeria.

2.1.1 Review Period and Inclusion Criteria for Articles Utilized

The inclusion criterion used for the review was as follows:

(i) Original publication
(ii) Primary data obtained from studies related to HRP2/3 gene deletion surveillance
(iii) Conducted in Nigeria
(iv) Published during the period selected for the review, 2010 to June 2024.

The literature search was extended to include references for more evidence to make the review more robust. The articles were screened and confirmed based on

Table 12.1 Guidelines for confirmation and reporting of plasmodium falciparum HRP2/HRP3 gene deletion

	Criteria	Recommendations by WHO
1	Study site	Cross-sectional, multisite study
2	Target population	Patients presenting to medical facilities with feverish illness who have been diagnosed with *P. falciparum* malaria based on pf-pLDH RDT or microscopy but whose HRP2 RDT result is negative
3	Study site	Preselected public health facilities representing the spectrum of transmission and geographical diversity across the country
4	Study design	A cross-sectional study will be conducted in ten at-risk health institutions to identify pfhrp2/3 gene deletions causing false-negative HRP2 RDT results, with 37 patients enlisted for each facility
5	Sample size	To determine if the prevalence of pfhrp2 deletion is less than or greater than 5%, a sample size of 370 per screening domain (37 per health facility) is advised. Upon enrolment of the 370 *Pf* patients, molecular confirmation of *pfhrp*2 deletions in suspected false-negative cases is expected to follow
6	Methodological approaches	*Plasmodium falciparum* detection using HRP2-based RDTs, microscopy confirmation, and molecular techniques (using primers for targeted genes and additional validated genes to indicate malaria infection)

WHO protocols for establishing and reporting HRP2/3 gene deletion (WHO protocol, 2018; Bosco et al., 2020) (Table 12.1).

3 Results

The preliminary search of the online literature yielded 737 studies from Google Scholar (731) and PubMed (6), as shown in the prism chart (Fig. 12.1). Six Hundred and fifty-three were automatically screened for ineligible studies, and 5 removed as duplicates. In total, of 18 articles were retrieved, 44.4% ($n = 8$) met the criteria and were retained for the systematic review 55.6% ($n = 10$) that did not meet the requirements for the review were systematic reviews, prevalence studies, or assessment studies carried out in other regions or preprint studies and unrelated studies.

4 Discussion

A cross-sectional study design has been recommended for confirming and reporting HRP2/3 gene deletion in *Plasmodium falciparum* among symptomatic patients with higher parasite density, ensuring the quality confirmation of gene deletion in the sample. Among the eight studies, four adhered to the recommendation of both study design and study site, which is the healthcare facility patients' approach to care (Adeka et al., 2024; Krueger et al., 2023; Nassar et al., 2021; Oreh et al., 2022).

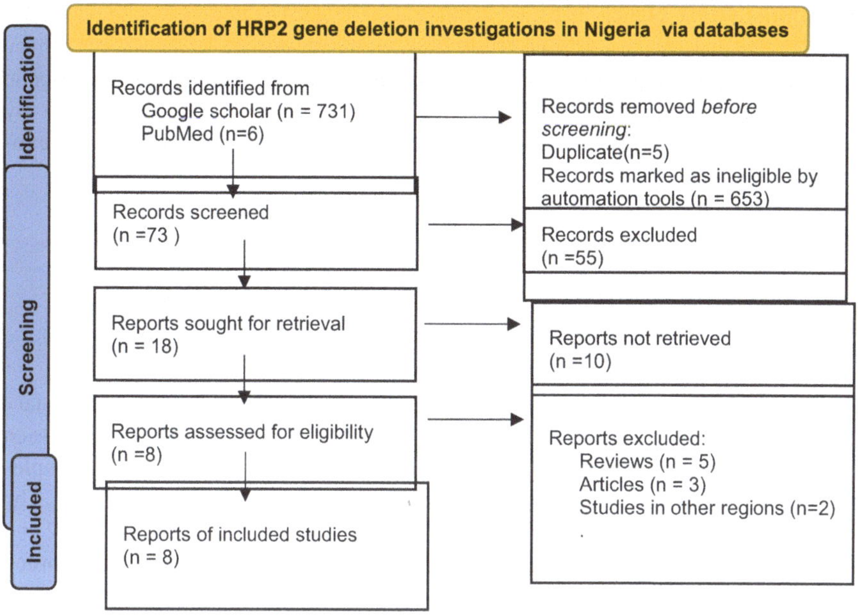

Fig. 12.1 The Prisma Flowchart showing the selection of eligible studies for the review adopted from (Page et al., 2021)

Other studies utilized a prospective cohort study design (Funwei et al., 2019), cohort study design (Ikegbunam et al., 2024; Prosser et al., 2021), or retrospective study design (Krueger et al., 2023). The convenience-based approach was utilized for sampling in one of the studies (Shehu et al., 2021), which often could contain sampling biases and cannot be treated as genuinely random samples (Mayor et al., 2023). Adeka et al. (2024) adhered to the WHO recommendation of spreading the study site by visiting five different locations.

4.1 *Sample Size and Participants for Investigation*

The sample size recommended by the WHO is 370 or 318 for assessing <5%> prevalence. Febrile children (511) were investigated by Funwei et al. (2019) and genotyped (66 samples) for double deletion for malaria-positive initially (35 samples). Another investigation was carried out among 210 individuals, and 30 Nigerian samples were genotyped (Prosser et al., 2021). In the study by Nassar et al. (2021), 50 samples were genotyped symptomatic patients in health facilities. In Kano, Shehu et al. (2021), healthcare facilities were used as sites for sample collection, and 141 symptomatic subjects and 70 samples were genotyped for HRP2 gene deletion only. Another investigation by Oreh et al. (2022), among 370 symptomatic individuals, 175 were positive by microscopy, and 75 were negative by RDT and were utilized for gene deletion investigations. Five communities were visited by Adeka et al. (2024), these areas were identified as high-risk due to their proximity to water bodies and history of bed net usage; 40 samples were genotyped for gene deletion. A double deletion investigation of gene deletion was carried out by Ikegbunam et al. (2022) in 207 symptomatic individuals; 165 samples were genotyped for the deletion.

4.2 HRP2/HRP3 Gene Deletion Prevalence Reported in the Studies

Reports of the prevalence of pfhrp2/3 gene deletions in Nigeria between January 2010 and April 2024 have been published. The prevalence recorded ranges between 0% and 78% with varying sample sizes. The prevalence of 17% among samples genotyped HRP2/3 gene deletion in Abuja was reported by Funwei et al. 2019). The study conducted in Kano by Shehu et al. (2021), showed a prevalence of HRP2 and three gene deletions of 15.7% and 2%, respectively. A 10% prevalence was recorded by Prosser et al. (2021) in a multiregional study among samples imported from Nigeria (30) out of its total population (210). The highest frequency was reported, with a prevalence of 78%, and it was carried out in Osoghbo. According to Oreh et al. (2022), among the 75 false-negative RDT samples, the prevalence of HRP2 and three gene deletions was 7.06% (6 samples). A multisite study by Krueger et al. (2023) involving samples from different locations globally, including Nigeria (no province in Nigeria was indicated), 13.3% of the 30 *Plasmodium falciparum* isolates had HRP2 gene deletion. A multiregional study for West Africa with samples from Nigeria and three other African countries reported 1.9% double gene deletion in 466 collected samples. In 2024, a study published by (Adeka et al., 2024). It was reported that five villages in Makurdi had a 0% prevalence of the HRP2 gene.

The WHO protocol recommends reporting HRP2 and HRP3 gene deletions due to cross-reactivity observed between both homologous genes, affecting the accuracy of reporting HRP2-based RDT kits. Underestimation of gene deletion could occur due to the cross-reactivity of the HRP3 protein in HRP2-based RDTs. Confirmation of RDT-negative samples to malaria infection must be carried out via microscopy followed by PCR methods, which involve assessing the presence and absence of amplicon sizes that can be visualized. Samples subjected to HRP2 and HRP3, gene deletion investigations, should have discordant results between microscopy and RDT, which were further amplified via PCR methods and visualized gene deletions by gel electrophoresis. The pfhrp2/3 gene's exons 1 and 2 are amplified in samples that have been PCR-confirmed to be *P. falciparum* to determine whether the gene is present or absent. The quality of *P. falciparum* parasite DNA suspected for gene deletion can be ascertained via PCR amplification of MSP1 and MSP2 single-copy genes (Agaba et al., 2019). Reporting deletions in flanking genes of the subtelomeric region of pfhrp2/3 is unnecessary. However, studies have shown that whole-genome sequencing has revealed that deletions also exist in flanking genes. Most studies have confirmed microscopy as the gold standard, except (Krueger et al., 2023). Microscopy was used for preliminary confirmation of *Plasmodium falciparum* among mRDT-negative samples (Shehu et al., 2021), followed by PCR methods. Microsatellite analysis and genetic diversity phylogenetic analysis methods were employed to identify haplotypes and polymorphisms, which could enable comparisons between the haplotypes and their distributions across the countries sampled, including Nigeria. Other molecular markers, such as MSP1, MSP2 and glurp, were amplified alongside the 1-EXON-1 and EXON-2 of HRP2 and HRP3.

The absence of the gene after PCR indicates a gene deletion. This was assessed with a negative control for the hrp2/3 genes and positive results for other amplified molecular markers. In a prospective cohort study by Funwei et al. (2019), samples were confirmed via microscopy after negative results were obtained for HRP2-based RDT. 18S RNA was used for molecular confirmation of malaria, and single-copy genes for pf infection (MSP-1 and MSP-2) and nested PCR for Exon 2 of the pf HRP2 and pfHRP3 genes were amplified to assess gene deletion and polymorphisms. Krueger et al. (2023) utilized samples diagnosed with malaria via a pfHRP2-based RDT with a microscopic confirmatory test. Further PCR methods, such as 4-plex quantitative PCR, that target 4 *Plasmodium falciparum* genes were utilized for the study (cytochrome b (pfcytb), ß tubulin (pfbtub), pfhrp2, and pfhrp3), as described by Kreidenweiss et al. (2019). The HRP2/3 gene deletion was validated using DNA obtained from the *P. falciparum* laboratory strains 3D7, Dd2, and HB3, as well as from a non-template control (nuclease-free water). Nassar et al. (2021) and Adeka et al. (2024) carried out *P. falciparum* detection first via the RDT kit (*Pf/ Pv* antigen and pfHRP2 antigen kit), followed by confirmation of malaria infection via microscopy. Both studies investigated single-gene deletion of HRP2 via classic PCR methods, with Nassar et al. (2021) targeting the pfHRP2-2-exon-1 and exon-2 genes and Adeka et al. (2024) lacking information about the target region used for the investigation. Confirmation of malaria parasite (Shehu et al., 2021) PCR methods utilizing multiple single-copy genes (*Pf Cox3, Pf HRP2, Pf HRP3* and *Pf Beta tubulin*). Methodological approaches were not assessed to review Ikegbunam et al. (2022) (Fig. 12.2).

4.2.1 Future Perspectives of HRP2 Gene Deletion Investigations in Nigeria

The World Health Organization (WHO) has been encouraging nations to examine the possibility of HRP2gene deletions that can result in erroneous test findings because of the rapid development of the significance of HRP2-based detection in the management of malaria parasite populations.

Priority is being given to the following regions (Baker et al., 2010):

I. Those with documented discrepancies between HRP2-based RDTs and microscopy

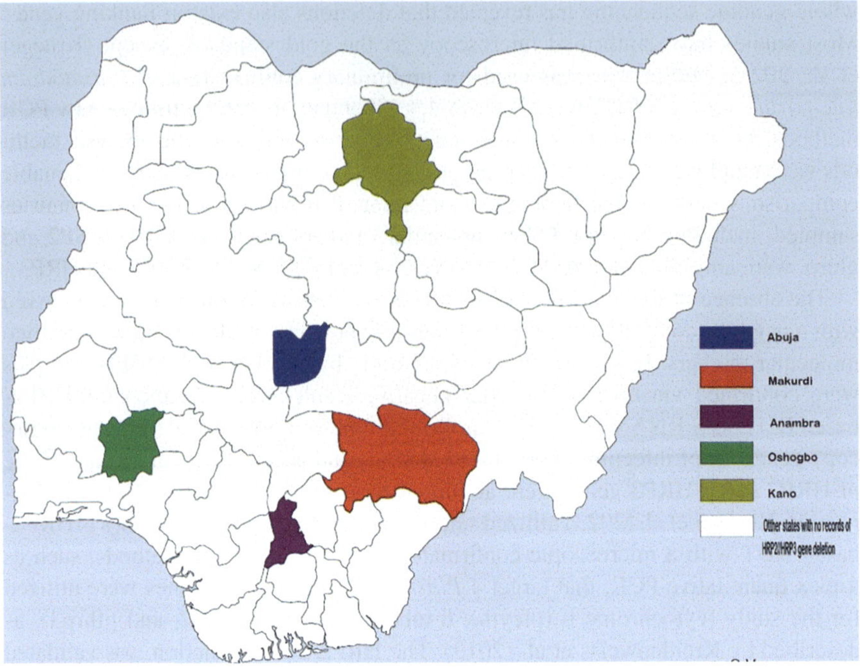

Fig. 12.2 Map showing areas where HRP2/3 gene deletion were reported and areas with no reports

II. Those with infrequent or non-representative reports of HRP2/3 deletions nationwide
III. Those near a region where HRP2/3 deletions have been frequently discovered

Based on WHO protocols for investigating PfHRP2, a minimum of 370 or 318 symptomatic individuals estimated a prevalence of 3.2% (Viana et al., 2017). However, studies have used fewer numbers, like the Eritrean study with 51 and the Zambian study with 28 participants.

4.2.2 Standardization of Methods

The WHO recommend common denominators for the estimation of gene deletion commonly using prevalence recorded by microscopy, which can lead to errors associated with false positives from poor microscopy skills (Agaba et al., 2019). The impact of HRP2 gene deletions could lead to misdiagnosis, delayed treatment, and possible complications, as well as severe malaria cases (Koita et al., 2012).

4.2.3 Ensuring Routine Product Testing Program

The WHO have provided guides to selecting and procuring malaria RDTs to ensure quality and reliable products are provided and distributed in endemic regions like Nigeria, Togo, and Senegal (Koita et al., 2012). This will serve as a guide for procuring and distributing malaria rapid diagnostic tests to ensure standards for malaria diagnosis (Cunningham et al., 2019).

4.2.4 Molecular Surveillance

Exploring malaria genetic epidemiology can be utilized to show how hereditary variables interact with environmental factors to cause disease in communities and have implications for malaria eradication (Dalmat et al., 2019). Traditional epidemiological methods support a significant reduction in malaria incidence in many countries (González-Sanz et al., 2023). Still, high-resolution information on malaria epidemiology in local and regional areas is necessary to target interventions effectively. Only laboratory settings have been used to explore parasite genomes using genetic epidemiology methodologies. Conventional indicators such as case incidence and prevalence and information from patient travel histories have long been utilized to comprehend transmission patterns and intensity. Despite the importance of these data, genetic information from natural populations of parasites can enhance knowledge of malaria transmission and direct control measures (Neafsey et al., 2021; Wesolowski et al., 2018).

Genomic surveillance is required to estimate the prevalence of HRP2&3 gene deletion (Cheng et al., 2014) and examine the effectiveness of HRP2-based rapid diagnosis tools. If the HRP2 deletion frequency exceeds 5%, the WHO advises using alternate diagnostics. At that point, more false-negative tests resulting from HRP2 deletion will occur than from other diagnostics' decreased sensitivity (Vera-Arias et al., 2022). Variation in the occurrence of HRP2 deletion has been observed to vary significantly across regions within countries (Parr et al., 2017; Solano et al., 2015). Therefore, it may be necessary to adjust the selection of diagnostic methods at a more localized level. The practicality of implementing such strategies relies on the structure of the national disease control initiative, hindrances to disease transmission within the country, and additional factors. In areas with few deletions, HRP2-based RDTs are a valuable diagnostic resource (Vera-Arias et al., 2022) (Table 12.2).

5 Conclusion

The study design, nature of participants (asymptomatic and symptomatic), sample size ranging from 30-511 participants, detection of single or double gene deletion (HRP2 OR HRP3 OR BOTH), and methodological approach utilized the MSP1,

Table 12.2 Investigations of HRP2/3 gene deletion carried out between 2010 and 2024 in Nigeria

			Methodology utilized by each study						
S/N	Location of study	Author	Study design	Sample size	Gene reported (HRP2 or HRP3 or BOTH)	Flanking genes Reported	Method of Research	Prevalence	Publication
1.	Nigeria (Abuja)	Funwei et al. (2019)	P	511(S)	Both	No	BM1&2 and HRP2/3	17% (of 66 genotyped)	2019
2.	Regional Study	Prosser et al. (2021)	CH	30	Both		MSP1&2 and HRP2/3 genes	13.3% (of 30 genotyped)	2021
3.	Nigeria (Kano)	Shehu et al. (2021)	CNS	70 (S)	Both.	No	Pf COX gene, PF Hhrp2/3, Pfβ tubulin gene	Pf HRP 2 (15.7%) PfHRP3 (2%) (50 genotyped)	2021
4.	Nigeria (Oshogbo)	Nassar et al. (2021)	CS	323(S)	HRP2	No	Only PCR for HRP2 gene	78% (50 genotyped)	2021
5.	Nigeria (Abuja)	Oreh et al. (2022)	CS	370(S)	Both	No	18 sRNA, qPCR, BM1&2and HRP2&3	HRP2 5.3% (4/75) and HRP3 1.33% (1/75). Double deletion 1.33% (1/75) (75 samples genotyped)	2022
6.	Multi Regional (Onitsha, Nnewi, Awka)	Krueger et al. (2023)	CS & R	466 (A&S)	Both	No	Cytochrome B, β tublin, Pf HRP2, HRP3 genes	1.9%	2023
7.	Nigeria (Makurdi)	Adeka et al. (2024)	CS	374(AS)	HRP2	No	18 sRNA, MSP1 and 2 and HRP2 gene	0%	2024
8.	Nigeria (Nnewi)	Ikegbunam et al. (2024)	CH	207S	Both	No	Not available for review	0.6% for both genes (165 samples genotyped)	2024

S Symptomatic, *A* asymptomatic, *P* prospective cohort, *CS* cross-sectional, *CH* cohort, *R* retro-prospective, *CNS* convenience sampling, *BOTH* HRP2&3 were reported, *No* flanking genes were not genotyped, *BM1&2* MSP1 and MSP2 were amplified

MSP2, quantitative PCR and other genes were included. Most studies adhered to cross-sectional study designs (Nassar et al., 2021; Oreh et al., 2022; Adeka et al., 2024), participants (symptomatic), and sites (healthcare facilities). The level of transmission of malaria was indicated in some studies (Adeka et al., 2024). However, one of the investigations was conducted in an environment where risk factors for malaria transmission were assessed. The WHO's recommended spreading of investigation site distribution was not followed by all studies representative of the malaria parasite population in a given country in all epidemiological circumstances. The study found disparities across the studies that provided scant information about the prevalence and condition of pfhrp2/3 gene deletion in Nigeria. To detect gene deletions in those areas, more research is required to determine which parts of Nigeria are at greater risk of contracting malaria. The World Health Organization advises starting surveys and monitoring programs to enable early identification and containment of parasites with pfhrp2/3 gene deletions.

Acknowledgements I acknowledge the opportunity the Covenant University Centre of Research Innovation and Development gave me to present the review during its 11th International Biotechnology Conference Exhibition & Workshop 2024 (IBCE&W) tagged "Harnessing Biotechnology Tools for Product Development."

Funding This research work is sponsored by the Covenant Applied Informatics and Communication African Centre of Excellence (CApIC-ACE) of Covenant University, Ota, Ogun State, Nigeria.

References

Adeka, P., Imandeh, G. N., Ikpa, T. F., & Okafor, I. D. (2024). Prevalence of asymptomatic falciparum malaria, risk factors, and absence of PfHRP2 gene deletion in Makurdi, Nigeria. *South Asian Journal of Research in Microbiology, 18*(1), 1–7. https://doi.org/10.9734/sajrm/2024/v18i1338

Agaba, B. B., Yeka, A., Nsobya, S., Arinaitwe, E., Nankabirwa, J., Opigo, J., Mbaka, P., Lim, C. S., Kalyango, J. N., Karamagi, C., & Kamya, M. R. (2019). Systematic review of the status of pfhrp2 and pfhrp3 gene deletion, approaches and methods used for its estimation and reporting in Plasmodium falciparum populations in Africa: Review of published studies 2010–2019. *Malaria Journal, 18*(1). https://doi.org/10.1186/S12936-019-2987-4

Alemu, A., Shiferaw, Y., Addis, Z., Mathewos, B., & Birhan, W. (2013). Effect of malaria on HIV/AIDS transmission and progression. *Parasites & Vectors, 6*(1), 18. https://doi.org/10.1186/1756-3305-6-18

Awosolu, O. B., Yahaya, Z. S., & Farah Haziqah, M. T. (2021). Prevalence, parasite density and determinants of falciparum malaria among febrile children in some peri-urban communities in Southwestern Nigeria: A cross-sectional study. *Infection and Drug Resistance, 14*, 3219–3232. https://doi.org/10.2147/IDR.S312519

Ayandipo, E. O., Babatunde, D., Afolayan, O., Kalejaye, O., & Obembe, T. (2023). Assessing the knowledge and practices of primary healthcare workers on malaria diagnosis and related challenges in view of COVID-19 outbreak in a Nigerian Southwestern metropolis. *PLOS Global Public Health, 3*(1), e0000625. https://doi.org/10.1371/journal.pgph.0000625

Baker, J., Ho, M.-F., Pelecanos, A., Gatton, M., Chen, N., Abdullah, S., Albertini, A., Ariey, F., Barnwell, J., Bell, D., Cunningham, J., Djalle, D., Echeverry, D. F., Gamboa, D., Hii, J., Kyaw, M. P., Luchavez, J., Membi, C., Menard, D., et al. (2010). Global sequence variation in the histidine-rich proteins 2 and 3 of Plasmodium falciparum: Implications for the performance of malaria rapid diagnostic tests. *Malaria Journal, 9*(1), 129. https://doi.org/10.1186/1475-2875-9-129

Berhane, A., Anderson, K., Mihreteab, S., Gresty, K., Rogier, E., Mohamed, S., Hagos, F., Embaye, G., Chinorumba, A., Zehaie, A., Dowd, S., Waters, N. C., Gatton, M. L., Udhayakumar, V., Cheng, Q., & Cunningham, J. (2018). Major threat to malaria control programs by Plasmodium falciparum lacking histidine-rich protein 2. *Eritrea. Emerging Infectious Diseases, 24*(3), 462–470. https://doi.org/10.3201/EID2403.171723

Beshir, K.B., Parr, J.B., Cunningham, J., Cheng, Q. & Rogier, E. (2022) Screening Strategies and Laboratory Assays to Support Plasmodium falciparum Histidine-Rich Protein Deletion Surveillance: Where We are and What is Needed. Malaria Journal. 21(1), 201.

Bharti, A. R., Letendre, S. L., Patra, K. P., Vinetz, J. M., & Smith, D. M. (2009). Malaria diagnosis by a polymerase chain reaction-based assay using a pooling strategy. *The American Journal of Tropical Medicine and Hygiene, 81*(5), 754–757. https://doi.org/10.4269/ajtmh.2009.09-0274

Bosco, A. B., Anderson, K., Gresty, K., Prosser, C., Smith, D., Nankabirwa, J. I., Nsobya, S., Yeka, A., Opigo, J., Gonahasa, S., Namubiru, R., Arinaitwe, E., Mbaka, P., Kissa, J., Won, S., Lee, B., Lim, C. S., Karamagi, C., Cunningham, J., et al. (2020). Molecular surveillance reveals the presence of pfhrp2 and pfhrp3 gene deletions in Plasmodium falciparum parasite populations in Uganda, 2017–2019. *Malaria Journal, 19*, 1–14. https://doi.org/10.1186/s12936-020-03362-x

Brazier, A. J., Avril, M., Bernabeu, M., Benjamin, M., & Smith, J. D. (2017). Pathogenicity Determinants of the Human Malaria Parasite Plasmodium falciparum Have Ancient Origins. *MSphere, 2*(1). https://doi.org/10.1128/msphere.00348-16

Cheng, Q., Gatton, M.L., Barnwell, J., Chiodini, P., McCarthy, J., Bell, D. & Cunningham, J. (2014) Plasmodium falciparum Parasites Lacking Histidine-Rich Protein 2 and 3: A Review and Recommendations for Accurate Reporting. Malaria Journal. 13(1), 283.

Cunningham, J., Jones, S., Gatton, M. L., Barnwell, J. W., Cheng, Q., Chiodini, P. L., Glenn, J., Incardona, S., Kosack, C., Luchavez, J., Menard, D., Nhem, S., Oyibo, W., Rees-Channer, R. R., Gonzalez, I., & Bell, D. (2019). A review of the WHO malaria rapid diagnostic test product testing programme (2008-2018): Performance, procurement and policy. *Malaria Journal, 18*(1). https://doi.org/10.1186/S12936-019-3028-Z

Dalmat, R., Naughton, B., Kwan-Gett, T. S., Slyker, J., & Stuckey, E. M. (2019). Use cases for genetic epidemiology in malaria elimination. *Malaria Journal, 18*(1), 163. https://doi.org/10.1186/s12936-019-2784-0

Dokunmu, T. M., Adjekukor, C. U., Yakubu, O. F., Bello, A. O., Adekoya, J. O., Akinola, O., Amoo, E. O., & Adebayo, A. H. (2019). Asymptomatic malaria infections and Pfmdr1 mutations in an endemic area of Nigeria. *Malaria Journal, 18*(1), 218. https://doi.org/10.1186/s12936-019-2833-8

Dorado, E. J., Okoth, S. A., Montenegro, L. M., Diaz, G., Barnwell, J. W., Udhayakumar, V., & Solano, C. M. (2016). Genetic Characterisation of Plasmodium falciparum Isolates with Deletion of the pfhrp2 and/or pfhrp3 Genes in Colombia: The Amazon Region, a Challenge for Malaria Diagnosis and Control. *PLoS ONE, 11*(9). https://doi.org/10.1371/JOURNAL.PONE.0163137

Funwei, R., Nderu, D., Nguetse, C., & Tropica, B. T.-A. (2019). *Molecular surveillance ofpfhrp2 and pfhrp3 genes deletion in Plasmodium falciparum isolates and the implications for rapid diagnostic tests in Nigeria*. Elsevier. Retrieved October 13, 2024, from https://www.sciencedirect.com/science/article/pii/S0001706X1831177X?casa_token=kFY4RXmSyvgAAAAA:HRHChf_ei696WuWiug4fDWOfL-FtutiUXkdOMU2fsFh6jvlLm9IWHDcXGU8KEr16xtn16dSRAUM

Funwei, R., Nderu, D., Nguetse, C.N., Thomas, B.N., Falade, C.O., Velavan, T.P. & Ojurongbe, O. (2019) Molecular Surveillance of Pfhrp2 and Pfhrp3 Genes Deletion in Plasmodium falciparum Isolates and the Implications for Rapid Diagnostic Tests in Nigeria. Acta Tropica. 196, 121–125.

Gadia, C. L. B., Manirakiza, A., Tekpa, G., Konamna, X., Vickos, U., & Nakoune, E. (2017). Identification of pathogens for differential diagnosis of fever with jaundice in the Central African Republic: A retrospective assessment, 2008-2010. *BMC Infectious Diseases, 17*(1), 1–5. https://doi.org/10.1186/S12879-017-2840-8/TABLES/3

Gamboa, D., Ho, M.-F., Bendezu, J., Torres, K., Chiodini, P. L., Barnwell, J. W., Incardona, S., Perkins, M., Bell, D., McCarthy, J., & Cheng, Q. (2010). A large proportion of P. falciparum isolates in the Amazon region of Peru lack pfhrp2 and pfhrp3: implications for malaria rapid diagnostic tests. *PLoS One, 5*(1), e8091. https://doi.org/10.1371/journal.pone.0008091

Gimenez, A. M., Marques, R. F., Regiart, M., & Bargieri, D. Y. (2021). Diagnostic Methods for Non-Falciparum Malaria. *Frontiers in Cellular and Infection Microbiology, 11*. https://doi.org/10.3389/FCIMB.2021.681063/FULL

Golassa, L., Messele, A., Amambua-Ngwa, A., & Swedberg, G. (2020). *Extensive deletion and sequence variation of Plasmodium falciparum histidine rich protein 2/3 (pfhrp2/3) genes in Ethiopia: Implication for RDT-based malaria diagnosis and control*. https://doi.org/10.21203/rs.3.rs-22606/v1

González-Sanz, M., Berzosa, P., & Norman, F. F. (2023). Updates on malaria epidemiology and prevention strategies. *Current Infectious Disease Reports, 25*(7), 1. https://doi.org/10.1007/S11908-023-00805-9

Ikegbunam, M. N., Anagu, L. O., Duru, C., Nworu, C. S., Esimone, C. O., & Ojurongbe, O. (2022). Genetic diversity and allelic frequency of antigenic markers in Plasmodium falciparum isolates from Nnewi district in Nigeria. *The Journal of Infection in Developing Countries, 16*(03), 557–563. https://jidc.org/index.php/journal/article/view/14815

Ikegbunam, M., Maurer, M., Abone, H., Ezeagwuna, D., Sandri, T. L., Esimone, C., Ojurongbe, O., Woldearegai, T. G., Kreidenweiss, A., Held, J., & Fendel, R. (2024). Evaluating malaria rapid diagnostic tests and microscopy for detecting plasmodium infection and status of Plasmodium falciparum histidine-rich protein 2/3 gene deletions in Southeastern Nigeria. *The American Journal of Tropical Medicine and Hygiene, 110*(5), 902–909. https://doi.org/10.4269/AJTMH.23-0690

Koita, O. A., Doumbo, O. K., Ouattara, A., Tall, L. K., Konaré, A., Diakité, M., Diallo, M., Sagara, I., Masinde, G. L., Doumbo, S. N., Dolo, A., Tounkara, A., Traoré, I., & Krogstad, D. J. (2012). False-negative rapid diagnostic tests for malaria and deletion of the histidine-rich repeat region of the hrp2 gene. *The American Journal of Tropical Medicine and Hygiene, 86*(2), 194–198. https://doi.org/10.4269/ajtmh.2012.10-0665

Kreidenweiss, A., Trauner, F., Rodi, M., Koehne, E., Held, J., Wyndorps, L., Manouana, G. P., McCall, M., Adegnika, A. A., Lalremruata, A., Kremsner, P. G., Fendel, R., & Sandri, T. L. (2019). Monitoring the threatened utility of malaria rapid diagnostic tests by novel high-throughput detection of Plasmodium falciparum hrp2 and hrp3 deletions: A cross-sectional, diagnostic accuracy study. *EBioMedicine, 50*, 14. https://doi.org/10.1016/J.EBIOM.2019.10.048

Krueger, T., Ikegbunam, M., Lissom, A., Sandri, T. L., Ntabi, J. D. M., Djontu, J. C., Baina, M. T., Lontchi, R. A. L., Maloum, M., Ella, G. Z., Agonhossou, R., Akoton, R., Djogbenou, L., Borrmann, S., Held, J., Ntoumi, F., Adegnika, A. A., Kremsner, P. G., & Kreidenweiss, A. (2023). Low prevalence of Plasmodium falciparum histidine-rich protein 2 and 3 gene deletions—A multiregional study in central and West Africa. *Pathogens (Basel, Switzerland), 12*(3), 455. https://doi.org/10.3390/pathogens12030455

Manku, R. R., Sharma, A., & Panchbhai, A. (2020). *Malaria detection and classification*. http://arxiv.org/abs/2011.14329

Mayor, A., Ishengoma, D. S., Proctor, J. L., & Verity, R. (2023). Sampling for malaria molecular surveillance. *Trends in Parasitology, 39*(11), 954–968. https://doi.org/10.1016/j.pt.2023.08.007

McCaffery, J. N., Nace, D., Herman, C., Singh, B., Sompwe, E. M., Nkoli, P. M., Ngoyi, D. M., Kahunu, G. M., Halsey, E. S., & Rogier, E. (2021). Plasmodium falciparum pfhrp2 and pfhrp3 gene deletions among patients in the DRC enrolled from 2017 to 2018. *Scientific Reports (Nature Publisher Group), 11*(1). https://doi.org/10.1038/s41598-021-02452-3

Metoh, T. N., Chen, J.-H., Fon-Gah, P., Zhou, X., Moyou-Somo, R., & Zhou, X.-N. (2020). Genetic diversity of Plasmodium falciparum and genetic profile in children affected by uncomplicated malaria in Cameroon. *Malaria Journal, 19*(1), 115. https://doi.org/10.1186/s12936-020-03161-4

Nassar, A. S., Bakarey, A. S., Abdulazeez, A. A., & Fayemi, O. O. (2021). Diagnostic performance of Hrp2 based diagnostic test kits and frequency of Pfhrp2 gene deletion in Plasmodium falciparum isolates of Osogbo, Southwestern Nigeria. *Journal of Advances in Medicine and Medical Research, 20–25*. https://doi.org/10.9734/jammr/2021/v33i730870

Neafsey, D. E., Taylor, A. R., & MacInnis, B. L. (2021). Advances and opportunities in malaria population genomics. *Nature Reviews Genetics, 22*(8), 502–517. https://doi.org/10.1038/s41576-021-00349-5

Njewa, B., Eyong, E. E. J., & Ebai, C. B. (2024). Malaria parasitaemia and its impact on biological parameters among children <16 years old attending the Nkwen District Hospital. *Cameroon. MalariaWorld Journal, 15*. https://doi.org/10.5281/zenodo.10731943

Olasehinde, G. I., Ajay, A. A., Taiwo, S. O., Adekeye, B. T., & Adeyeba, O. A. (2010). Prevalence and management of Falciparum malaria among infants and children in Ota, Ogun state, Southwestern Nigeria. *African Journal of Clinical and Experimental Microbiology, 11*(3). https://doi.org/10.4314/ajcem.v11i3.57773

Oreh, C., Esimone, C., Nworu, C., & Beshir, K. (2022). Prevalence of HRP2/3 Gene Deletion in Plasmodium falciparum Parasites in Abuja, Nigeria. *International Journal of Infectious Diseases, 116*, S125–S126. https://doi.org/10.1016/J.IJID.2021.12.297

Oyegoke, O., Maharaj, L., Akoniyon, O., et al. (2022). *Malaria diagnostic methods with the elimination goal in view*. Springer. Retrieved February 14, 2023, from doi: 10.1007/s00436-022-07512-9.

Page, M. J., McKenzie, J. E., Bossuyt, P. M., Boutron, I., Hoffmann, T. C., Mulrow, C. D., Shamseer, L., Tetzlaff, J. M., Akl, E. A., Brennan, S. E., Chou, R., Glanville, J., Grimshaw, J. M., Hróbjartsson, A., Lalu, M. M., Li, T., Loder, E. W., Mayo-Wilson, E., McDonald, S., et al. (2021). The PRISMA 2020 statement: An updated guideline for reporting systematic reviews. *BMJ, 372*. https://doi.org/10.1136/BMJ.N71

Parr, J. B., Verity, R., Doctor, S. M., Janko, M., Carey-Ewend, K., Turman, B. J., Keeler, C., Slater, H. C., Whitesell, A. N., Mwandagalirwa, K., Ghani, A. C., Likwela, J. L., Tshefu, A. K., Emch, M., Juliano, J. J., & Meshnick, S. R. (2017). Pfhrp2-deleted Plasmodium falciparum parasites in the Democratic Republic of the Congo: A national cross-sectional survey. *The Journal of Infectious Diseases, 216*(1), 36–44. https://doi.org/10.1093/infdis/jiw538

Prosser, C., Gresty, K., Ellis, J., Meyer, W., Anderson, K., Lee, R., & Cheng, Q. (2021). Plasmodium falciparum histidine-rich protein 2 and 3 gene deletions in strains from Nigeria, Sudan, and South Sudan. *Emerging Infectious Diseases, 27*(2), 471–479. https://doi.org/10.3201/EID2702.191410

Rougeron, V., Boundenga, L., Arnathau, C., Durand, P., Renaud, F., & Prugnolle, F. (2022). A population genetic perspective on the origin, spread and adaptation of the human malaria agents Plasmodium falciparum and Plasmodium vivax. *FEMS Microbiology Reviews, 46*(1), fuab047. https://doi.org/10.1093/femsre/fuab047

Sepúlveda, N., Phelan, J., Diez-Benavente, E., Campino, S., Clark, T. G., Hopkins, H., Sutherland, C., Drakeley, C. J., & Beshir, K. B. (2018). Global analysis of Plasmodium falciparum histidine-rich protein-2 (pfhrp2) and pfhrp3 gene deletions using whole-genome sequencing data and meta-analysis. *Infection, Genetics and Evolution: Journal of Molecular Epidemiology and Evolutionary Genetics in Infectious Diseases, 62*, 211–219. https://doi.org/10.1016/J.MEEGID.2018.04.039

Shehu, D., Aminu, F., Danlami, S., & Mashi, J. (2021). Investigation of HRPF2 gene deletion in Plasmodium falciparum in Northwestern Nigeria. *Acta Biologica Marisiensis, 4*, 10–20. https://doi.org/10.2478/abmj-2021-0002

Solano, C. M., Okoth, S. A., Abdallah, J. F., Pava, Z., Dorado, E., Incardona, S., Huber, C. S., de Oliveira, A. M., Bell, D., Udhayakumar, V., & Barnwell, J. W. (2015). Deletion of Plasmodium falciparum histidine-rich protein 2 (pfhrp2) and histidine-rich protein 3 (pfhrp3) genes in Colombian parasites. *PLoS ONE, 10*(7), e0131576. https://doi.org/10.1371/journal.pone.0131576

Tambo, M., Mwinga, M., & Mumbengegwi, D. R. (2018). Loop-mediated isothermal amplification (LAMP) and Polymerase Chain Reaction (PCR) as quality assurance tools for Rapid Diagnostic Test (RDT) malaria diagnosis in Northern Namibia. *PLoS ONE, 13*(12), e0206848. https://doi.org/10.1371/journal.pone.0206848

Thang, N. D., Rovira-Vallbona, E., Binh, N. T. H., Dung, D. V., Ngoc, N. T. H., Long, T. K., Duong, T. T., Martin, N. J., & Edgel, K. A. (2022). Surveillance of pfhrp2 and pfhrp3 gene deletions among symptomatic Plasmodium falciparum malaria patients in Central Vietnam. *Malaria Journal, 21*(1), 371. https://doi.org/10.1186/s12936-022-04399-w

Thomson, R., Beshir, K. B., Cunningham, J., Baiden, F., Bharmal, J., Bruxvoort, K. J., Maiteki-Sebuguzi, C., Owusu-Agyei, S., Staedke, S. G., & Hopkins, H. (2019). pfhrp2 and pfhrp3 gene deletions that affect malaria rapid diagnostic tests for Plasmodium falciparum: Analysis of archived blood samples from 3 African countries. *The Journal of Infectious Diseases, 220*(9), 1444–1452. https://doi.org/10.1093/infdis/jiz335

Vera-Arias, C. A., Holzschuh, A., Oduma, C. O., Badu, K., Abdul-Hakim, M., Yukich, J., Hetzel, M. W., Fakih, B. S., Ali, A., Ferreira, M. U., Ladeia-Andrade, S., Sáenz, F. E., Afrane, Y., Zemene, E., Yewhalaw, D., Kazura, J. W., Yan, G., & Koepfli, C. (2022). High-throughput Plasmodium falciparum hrp2 and hrp3 gene deletion typing by digital PCR to monitor malaria rapid diagnostic test efficacy. *ELife, 11*, e72083. https://doi.org/10.7554/eLife.72083

Viana, G. M. R., Okoth, S. A., Silva-Flannery, L., Barbosa, D. R. L., de Oliveira, A. M., Goldman, I. F., Morton, L. C., Huber, C., Anez, A., Machado, R. L. D., Camargo, L. M. A., Valle, S. C. N., do Póvoa, M. M., Udhayakumar, V., & Barnwell, J. W. (2017). Histidine-rich protein 2 (pfhrp2) and pfhrp3 gene deletions in Plasmodium falciparum isolates from select sites in Brazil and Bolivia. *PLoS ONE, 12*(3), e0171150. https://doi.org/10.1371/journal.pone.0171150

Wardhani, P., Butarbutar, T. V., Adiatmaja, C. O., Betaubun, A. M., Hamidah, N., & Aryati. (2020). Performance comparison of two malaria rapid diagnostic test with real time polymerase chain reaction and gold standard of microscopy detection method. *Infectious Disease Reports, 12*(Suppl. 1), 8731. https://doi.org/10.4081/idr.2020.8731

Wesolowski, A., Taylor, A. R., Chang, H.-H., Verity, R., Tessema, S., Bailey, J. A., Alex Perkins, T., Neafsey, D. E., Greenhouse, B., & Buckee, C. O. (2018). Mapping malaria by combining parasite genomic and epidemiologic data. *BMC Medicine, 16*(1), 190. https://doi.org/10.1186/s12916-018-1181-9

World Malaria Report (2023). Retrieved January 13, 2024, from https://www.who.int/publications-detail-redirect/9789240086173

Chapter 13
Antimalarial Drug Resistance: Newer Approaches to Mitigate the Process

Solomon Uche Oranusi and Queen Elizabeth Sule

1 Introduction

Malaria is a potentially fatal illness that infects humans by pathogens carried via female *Anopheles* mosquitoes. Approximately 94% of all malaria cases and deaths occur in Sub-Saharan Africa (SSA) (Kolawole et al., 2022). In recent years, Sub-Saharan Africa experienced a rapid propagation of antimalarial resistance (Oladipo et al., 2022).

About 241 million new incidents of malaria disease and 627,000 malaria-related fatalities occurred in eighty-five countries in 2020, representing a 69,000 rise from the year before. African Children under 5 years constitute about two-thirds of all malaria-related deaths (Ashley & Poespoprodjo, 2020). In most of Sub-Saharan Africa, *Plasmodium falciparum* is the most prevalent of the five human malaria parasites, namely *P. vivax, P. malariae, P. ovale, P. knowlesi, and P. vivax,* which are predominant in sub-Saharan Africa (Adesegun et al., 2020). The emergence and spread of drug-resistant parasites have presented significant challenges to attempts to manage and eradicate malaria, even though antimalarial medications have been the mainstay of these efforts (Ibrahim et al., 2020). In recent years, Sub-Saharan Africa has seen a rise in the need for more innovative strategies to slow the spread of antimalarial medication resistance (Alegana et al., 2020). Unfortunately, due to

S. U. Oranusi
Department of Biological Sciences, College of Science and Technology, Covenant University, Ota, Ogun State, Nigeria

Q. E. Sule (✉)
Department of Biological Sciences, College of Science and Technology, Covenant University, Ota, Ogun State, Nigeria

Covenant Applied Informatics and Communication Africa Centre of Excellence (CApIC-ACE), Covenant University, Ota, Ogun State, Nigeria
e-mail: queen.sulepgs@stu.cu.edu.ng

P. O. Isibor, P. A. Akinduti (eds.), *Harnessing Biotechnology Tools for Product Development*, https://doi.org/10.1007/978-3-031-86002-7_13

the widespread use of antimalarial drugs, particularly artemisinin. Drug-resistant strains of *Plasmodium falciparum* have evolved and developed more resistant to the contemporary antimalarial medications globally (Saba et al., 2022).

Globally, malaria remains a significant public health burden, with over 229 million cases reported annually. The majority of these cases—approximately 95%—are concentrated in 29 countries within sub-Saharan Africa. Notably, Nigeria (27%), the Democratic Republic of the Congo (12%), Uganda (5%), Mozambique (4%), and Niger (3%) collectively account for 51% of the global malaria incidence (Badmos et al., 2021). sub-Saharan Africa continues to experience the highest prevalence, largely due to favorable conditions for mosquito proliferation and limited access to effective malaria control measures (Amadi et al., 2019).

More than half of the global population remains at risk of malaria infection, with vulnerable populations, such as children under five and pregnant women, disproportionately affected by severe outcomes of the disease (Talapko et al., 2019). The high morbidity and mortality rates within these groups underscore the urgent need for targeted interventions and sustained control efforts across malaria-endemic regions (Reddy et al., 2022). Conquest, colonization, the development of agriculture, and the concentration of people in urban areas all contributed to the spread of illness (Miedaner & Garbelotto, 2024).

2 Antimalarial Drug Resistance

Resistance to antimalarial drugs is a significant challenge in combating malaria, as it can lead to failure to treat and the potential spread of the disease (Amelo & Makonnen, 2021). Drug resistance in malaria therapy is a growing concern in the global malaria prevention and control utilized for prevention and treatment, including chloroquine (CQ), sulfadoxine-pyrimethamine (SP), mefloquine (MQ), and artemisinins (Shibeshi et al., 2020). The emergence of resistant parasites poses significant challenges to controlling malaria, as it reduces the effectiveness of available treatments and threatens progress made towards eliminating the disease as abuse is pervasive pervasive (Dagen et al., 2021) Resistance mechanisms vary among different antimalarial drugs. For example, CQ resistance results from changes in the structure of the *Plasmodium falciparum* parasite's hemoglobin receptor protein, PfCRT, which alters its ability to sequester heme pigment within red blood cells. SP resistance arises due to alterations in dihydrofolate reductase (Pfdhfr) and dihydropteroate synthase (Pfdhps) genes (Mbengue et al., 2015). Resistance mefloquine involves alterations in the *P. falciparum* gene pfcrt and the transporter protein PfMDR1. Artemisinin resistance is more complex and multifactorial, with potential contributions from genetic factors such as altered expression or function of K13 propeller domain proteins, increased efflux pump activity, and reduced sensitivity to oxidative stress (Paloque et al., 2021).

The proliferation of drug resistance in malaria is driven by several interconnected factors, including insufficient adherence to prescribed treatment regimens, subtherapeutic drug dosages, and the widespread use of counterfeit medications containing inadequate concentrations of active compounds (Wassmer & Grau, 2017). These elements create conditions that allow plasmodium parasites to adapt and survive despite drug intervention (Greene et al., 2024).

To mitigate this growing threat, researchers are focusing on developing novel antimalarial agents that target distinct biochemical pathways, thereby reducing the likelihood of resistance (Aguiar et al., 2019). Furthermore, implementing combination therapies where multiple antimalarial drugs are administered concurrently has emerged as a strategic approach to delay resistance development (de Haan et al., 2022). This method reduces the probability that parasites will evolve mechanisms to evade drug efficacy completely, as they must simultaneously resist multiple agents, which is considerably more challenging for the parasite's adaptive capacity (Uwimana et al., 2020). These combined efforts represent a multifaceted approach to preserving the long-term effectiveness of antimalarial treatment (Van Der Pluijm et al., 2020).

Nowadays, in most malaria-affected areas, *P. falciparum* is highly chloroquine-resistant (Imwong et al., 2020). Additionally, resistance to SP is rampant and escalating. Mefloquine resistance is limited to regions where it has been extensively utilized, such as Vietnam, Cambodia, and Thailand, and it has emerged after six years of methodical application. Only a few regions of Indonesia, Papua, New Guinea, and the surrounding places have significant chloroquine resistance; the distribution and prevalence of *Plasmodium vivax* resistance have received less research attention. *Plasmodium vivax* exhibits a broader range of SP resistance (Shibeshi et al., 2021).

2.1 *Resistance to Antimalarial Drugs: Genetic Basis*

Mutation of genes confers resistance to antimalarial drugs (while maintaining parasite survival), which occurs independently of the medicine administered (Siddiqui et al., 2021). It is thought that several genes contribute to *P. falciparum's* resistance to chloroquine; changes in the PfCRT gene are usually related to the onset of resistance. PfMDR1 mutation, a different transporter gene, can affect the degree of resistance seen in lab settings when PfCRT mutations are present (Mbengue et al., 2015). It is unclear, therefore, exactly how PfMDR1 mutations affect how the body reacts to chloroquine therapy (Wicht et al., 2020). On the other hand, the per-parasite likelihood of single-point mutations evolving de novo in the genes that code for dihydrofolate reductase(dhfr) confer resistance to pyrimethamine and cytochrome b (cytB) for atovaquone resistance (Sà et al., 2020) (Table 13.1).

Table 13.1 Antimalarial drug resistance genes

Antimalarial drugs	Resistance genes	Mechanism of resistance	Year of resistance emergence
Chloroquine (CQ)	*Pfcrt, pfmdr1*	Point mutation, Efflux pumping	Since 1950s
Amodiaquine (AQ)	*Pfcrt, pfmdr1*	Point mutation, Efflux pumping	Since 1990s
Mefloquine (MQ)	*Pfmdr1, pfdhfr, pfdhps*	Point mutation	Since 1990s
Proguanil (Cycloguanil)	*Pfdhfr*	Point mutation	Since 1990s
Sulfadoxine pyrimethamine (SP)	Pfdhfr+ pfdhps, pfmrpS	Point mutation	Since 1970s
Atovaquone	*Pfcytb*	Point mutation	2008
Lumefantrine	*pfmdr1*	Efflux pumping	2006
Artemisinin	*K13, pfcrt*	Point mutation	2008

Source: Tchekounou et al. (2022)

2.2 *Drivers of Antimalarial Drug Resistance*

Understanding the parameters that influence the development and spread of resistance is essential for successfully solving the problem associated with resistance to antimalarial medication (Rasmussen et al., 2021). The first occurs when a parasite becomes less sensitive to treatment; the second occurs when parasites with one or more mutations that offer a certain level of immunity against a drug's action survive, are selected for, and subsequently spread (WHO, 2022). Two types of resistance-causing factors have been advanced: factors relating to treatment and background factors (Agrawal et al., 2022). The environment and intrinsic factors related to the parasite, host, and medications administered are examples of background drivers. The factors that determine the frequency, dosage, and duration of a parasite population's interactions with a particular medication are known as treatment-related drivers (Masserey et al., 2022).

2.2.1 Background-Related Factor

Underlying causes of resistance include the extent to which a mutation influences drug susceptibility and its ability to spread, which the genetic background of the parasite can influence (Stokes et al., 2021). Even without treatment stress, resistant parasites may be better able to compete with other parasites due to specific genetic backgrounds or extra mutations. Several other variables dependent on context influence the establishment and distribution of resistance (Nguyen et al., 2023). The capability of local mosquitoes to disseminate the resistant strains, human migration from areas with resistance, and climate change, which modifies the humidity, rainfall, and temperature and may change the geographical location and seasonality of distribution are some of these factors (Anoopkumar & Aneesh, 2021).

2.2.2 Treatment-Related Factors

Resistance in parasites results from drug stress, which also encourages their spread. Various antimalarial medications are administered to the malaria parasites to cure patients by eliminating the parasites (Siddiqui et al., 2021). On the other hand, persistent or recurring exposure to medications can lead to the selection of parasites less susceptible to treatment in successive doses parasite population's sensitivity to medication depends on several factors, such as the availability of high-quality diagnostic and therapeutic services and additional antimalarial medication use, such as chemotherapy prevention (Egwu et al., 2021). The resistance to treatment can persist and spread when blood levels of the drug are subtherapeutic. Several things can expose parasites to subtherapeutic drug concentrations, such as using substandard or counterfeit medications; using non-pharmaceutical forms of Artemisia, like Artemisia tea (WHO, 2022); not adhering to treatment plans; or using antimalarial medications at inadequate dosage monotherapy.

3 Resistance Mechanisms

3.1 Multi-Drug Resistance, the pfmdr1 Gene and P-glycoprotein

When cells that have developed resistance to a single treatment also resist various medications structurally unrelated, this phenomenon is called multidrug resistance (MDR) (Catalano et al., 2022). This resistance is mediated in many mammalian cell lines by P-glycoprotein (Pgp) or the multidrug-resistance (MDR) transporter. After the Pgp gene was cloned, researchers discovered it resembled transport proteins linked to bacterial membranes (Si et al., 2023). Pgps normally have two homologous domains, one with a hydrophilic nucleotide-binding fold after an aquaphobic membrane-associated section with six transmembrane helices. A linkage region connects the two domains. Whether the two portions of Pgp work together to provide a conduit for the expulsion of medicines is unknown (Shrestha et al., 2021). It has been demonstrated that mdr gene mutation, mRNA expression, and gene amplification raise resistance levels. The protein's high phosphorylation levels and N-glycosylation may be related to how it operates (Catalano et al., 2022). Evidence suggests that developing a complex chloroquine-heme within the acidic vacuole of the parasite is the reason for chloroquine's antimalarial adverse effects (Sanchez et al., 2022). In resistant *P. falciparum* strains, the level of hemoglobin breakdown and the heme molecule's characteristics remain unchanged compared to sensitive strains. The key factor influencing chloroquine's efficacy is its concentration within the parasite's acidic vacuole (Mohamed et al., 2020). Studies have shown reduced chloroquine accumulation in resistant parasites, likely due to a lower rate of drug uptake, possibly caused by an elevated vacuolar pH. Therapy with bafilomycin A1 has been noticed to enhance chloroquine resistance (Onyesom et al., 2022). Chloroquine resistance, akin to multidrug resistance in mammalian cells, is characterized by decreased drug accumulation and reversible resistance in the presence of verapamil. Other antimalarials like quinine and mefloquine can also be associated with hemin similarly. The intracellular buildup of hydrophobic medications such as arylamino alcohols and endoperoxide antimalarials is crucial for their efficacy (Shafik et al., 2022).

3.2 Structure of the **pfmdr1** *Gene*

The MDR proteins, the cystic fibrosis transmembrane conductance regulator protein, and the sulfonylurea receptor are members of the ABC (ATP-binding cassette) family of proteins that have a conserved structure of a six-transmembrane segment coupled to a nucleotide-binding fold that is present throughout nature (Khunweeraphong & Kuchler, 2021). Sequences from *P. falciparum* matching counterparts of the mdr transporters that induce multidrug resistance in mammalian cell lines were found (Khunweeraphong & Kuchler, 2021). Pfmdr1 was isolated and then sequenced; it encodes for the plasmodial homolog of the MDR gene of mammals in *P. falciparum.* Known as Pgh1 (P-glycoprotein homolog 1), it is a distinctive member of the ABC transporter superfamily, a polypeptide of approximately 162 kD with a maintained structure that includes two domains consisting of six predicted transmembrane segments coupled to a nucleotide-binding fold joined by a linker region (Adamu et al., 2020). After that, Pgh1 was found to be contained in the parasite's vacuole during the nonsexual phase of the parasite, where it controls the absorption of intracellular drugs. It is a calcium-dependent protein kinase that phosphorylates abundantly on serine and threonine residues and binds ATP (Adamu et al., 2020).

Gene polymorphisms can also result in drug resistance in the case of the pfmdr1 gene through two different mechanisms: mutation and gene amplification. Pfmdr1 was discovered to be amplified by several distinct events, indicating that it originated in multiple distinct locations (Al-Mekhlafi et al., 2022). A resistant genotype is probably more effectively spread across an entire population by point mutation inheritance than gene amplification. Alterations in the pfmdr1 gene were discovered by sequencing multiple lab strains from various regions. Five locations (codons 86, 184, 1034, 1042, and 1246) showed polymorphism. This study found a significant relationship between the chloroquine-resistant phenotype and pfmdr1 gene mutation (Khunweeraphong & Kuchler, 2021).

3.3 *PfCRT as the Major Factor of Amodiaquine and Chloroquine Resistance*

The DV accumulates less chloroquine in chloroquine-resistant parasites because of an energy-dependent drug efflux mechanism (Stanley & Rajkumari, 2023)—mutation in the pfcrt gene, which codes for resistance to chloroquine transporter. The DV membrane is home to the 49-kDa drug/metabolite transporter protein PfCRT, which facilitates chloroquine's efflux out of the DV and out from its heme target (Wicht et al., 2020)—Cryo-EM imaging at a resolution of 3.2 Å. As PfCRT undergoes conformational changes during its transport cycle, this transporter, which is made up of 424 amino acids and ten transmembrane helix domains, forms a negatively charged central cavity that is thought to accommodate charged positively drugs and molecules contained in the digestive vacuole for transportation to the cytosol (Coppée et al., 2020).

3.4 *Chloroquine Resistance*

Trans-stimulation outflow investigations support the transporter or carrier paradigm for PfCRT-mediated drug efflux. This demonstrates that chlorine outside the cell drives the outflow of preloaded chloroquine inside the cell. Specific mutations that characterize a particular PfCRT isoform impact this process's kinetics, efficacy, and method (Willems et al., 2023). Chloroquine-resistant alterations in pfcrt-modified duplicates are adequate to bestow phenotypes to chloroquine on various gene-based backgrounds, as demonstrated by genomic conversion via pfcrt allelomorphic exchange. A crucial amino acid modification found in alleles resistant to chloroquine is the lysine-to-threonine transition (K76T), frequently coupled with many mutations specific to a particular location (Tang et al., 2020). According to the cryo-EM structure of the PfCRT for the 7G8 isoform, all five of its alterations align with the core drug-binding cavity. At least four of these mutations are required to confer chloroquine resistance, suggesting that the extra amino acid modifications have a dependent role (Wicht et al., 2020). In contrast to the African/Asian CVIET haplotypes, the South American SVMNT PfCRT haplotypes are equally connected to amodiaquine resistance, even though the 7G8, GB4, and Dd2 PfCRT isoforms are known to be resistant to chloroquine. This implies that 7G8 backgrounds are the main ones where cross-resistance between amodiaquine and chloroquine is seen (Wicht et al., 2020).

3.5 Resistance to Antifolate

Antifolate drugs interfere with the synthesis of DNA and RNA in the parasite by inhibiting dihydrofolate reductase and dihydropteroate synthetase, the two crucial enzymes for folate metabolism, e.g., sulfadoxine-pyrimethamine resistance to antifols (cycloguanil and pyrimethamine) in *P. falciparum* and *P. vivax* is caused by successive accumulation of changes in dhfr (Fernández-Villa et al., 2019). Every mutation causes a progressive decrease in susceptibility. Sequential changes in the gene dhps, which codes for the target enzyme dihydropteroate synthase, also lead to resistance to the sulfonamides and sulfones, frequently given in combination with antifols. Mutations of one Point in the cytochrome b-coding gene cytB lead to atovaquone resistance (Cisse et al., 2009). Only a specific combination of proguanil (chloroguanide) and atovaquone is used. In this combination, proguanil appears to be the critical player operating the membrane of the mitochondria rather than the proguanil metabolite cycloguanil, which inhibits dhfr (Ramani et al.,2022). It is unknown whether and how resistance to proguanil's mitochondrial activity arises. While PfATPase6 has been identified as the target for artemisinins, initial research has not yet linked polymorphisms in the gene encoding this enzyme to decreased susceptibility to artemisinins (Ramani et al., 2022). Point mutations at seven locations in the DHPS gene confer resistance to pyrimethamine, and five sites in the DHFR gene confer resistance to sulfadoxine. *Plasmodium* susceptibility to SP is based on this genetic component (Roux et al., 2021). Many point mutations, for instance, have been found for Plasmodium falciparum dihydropteroate synthase (pfdhps). Resistance to sulfadoxine is conferred by the combinations of serine amino acid (Ser) → The amino acid Alanine (Ala) at codon 436, Alanine (Ala) → Glycine (Gly) at codon 437, lysine (Lys) → Glutamic acid (Glu) at codon 540, Alanine (Ala) → Glycine (Gly) at codon 581, and serine (Ser) → Phenylalanine (Phe) at codon 436—coupled with either Alanine (Ala) → Threonine (Thr) or Alanine (Ala) → Serine (Ser) at codon 613 confer resistance to sulfadoxine (Roux et al., 2021). *Plasmodium falciparum* dihydrofolate reductase (pfdhfr) has been modified by a single alteration at position 108 (S108N) from Serine (Ser) to Asparagine (Asn). There are several mutations found in Plasmodium species that can lead to pyrimethamine resistance, including Asparagine (Asn) → Isoleucine (Ile) at codon 51 and Cysteine (Cys) → Arginine (Arg) at codon 59 (C59R), a Serine (Ser) → Threonine (Thr) mutation at position 108 (S108N), and an Alanine (Ala) → Valine (Val) alteration at location 16 (Roux et al., 2021).

3.6 Artemisinin Resistance Mechanism

Mutations of K13 have been associated with the altered cycle of cell dynamics in K13 C580Y medical strains, leading to more extended periods of growth from the ring-to-trophozoite stage, potentially affecting Fe (II)PPIX-mediated drug activity

(Chenet et al., 2016). These alterations may destabilize the protein, deregulate PfPK4 phosphorylation, activate stress response pathways, and modulate growth through eIF2α phosphorylation (Xie et al., 2020). K13 mutations may upregulate the unfolded protein response (UPR) to remove damaged proteins and reduce ubiquitinated protein levels as a component of an increased cellular stress reaction (Gnädig et al., 2020).

There has been evidence of artemisinin and proteasome inhibitor interaction, suggesting a potential strategy to combat resistance to artemisinin (Sharma & Ali 2022), The synergy between K13-mutant and wild-type parasites suggests a possible way to overcome artemisinin resistance, revealing that K13 mutations do not confer resistance by regulating proteasomal function (Ndwiga et al., 2021). On the contrary, K13 mutations may protect against the proteostatic effects of artemisinin by decreasing drug binding to and polyubiquitination of PI3K (Dhorda et al., 2021). This could result in increased quantities of PI3P-positive vesicles that are important for functions of cells, such as the UPR and protein folding (Owoloye et al., 2021).

In vitro studies with edited K13-mutant parasites showed increased resistance to DHA-induced inhibition of the action proteasome and maintenance used for protein conversion (Raman et al., 2019). The tolerance of early rings to greater concentrations of artemisinin compared to subsequent trophozoite stages underscores the crucial part of hemoglobin destruction and degradation in artemisinin activation and cell death (Small-Saunders et al., 2024). Disruption of the process by blocking falciparum 2a or introducing the cysteine protease inhibitor E64 further shows that reduced Fe(II)PPIX-mediated drug action can decrease parasite sensitivity to artemisinin (Woodrow & White, 2017). Additionally, artemisinin-resistant K13 mutant parasites exhibited reduced fitness and developmental progression when inhibited from natural amino acid sources, indicating the effects of K13 alterations on the breaking down of hemoglobin and amino acid release (Gnädig et al., 2020).

Several studies conducted in 18 sub-Saharan Africa (SSA) countries have found many K13-propeller mutations at extremely low frequencies. These countries include the Democratic Republic of Madagascar, Ethiopia, Gabon, Congo, Ghana, Kenya, Malawi, Mali, Rwanda, Zambia, Togo, Uganda, Gambia, and Senegal (Balikagala et al., 2021). K13 alterations have been found in 14 isolates from Guyana (1.6%; cystine (Cys) to the amino acid (Tyr) (C580Y) mutation), three isolates from Papua New Guinea (1.3%; cystine (Cys) to tyrosine (Tyr) (C580Y) mutation), and two isolates from each of Rwanda (Proline (Pro) to Leucine (Leu) (P574L) and Alanine (Ala) to the amino acid (Val) (A675V) mutations). These findings are revealed by recent African studies (Miotto et al., 2020). The most prevalent C580Y mutation is in significant concentrations in Vietnam, Thailand, and Western Cambodia (Zhu & Zhou, 2022). Though phenylalanine (Phe) to isoleucine (Ile) F446I and, more recently, glycine (Gly) to serine (Ser) Gh533S predominate in southern China, near the China–Myanmar border, Yunnan, with 56% and 44% prevalence rates, respectively, glutamic acid (Glu) to glutamine (Gln) E252Q is present in Myanmar and western Thai border (Shibeshi et al., 2020) (Table 13.2).

Table 13.2 PfK13 mutation associated with artemisinin resistance in Africa and Asia

Mutation	Region/country reported	Classification	Reference
R561H	Rwanda, Uganda, Tanzania	Validated	Conrad et al. (2023)
A675V	Uganda	Validated	Emiru et al. (2023)
C469Y/F	Rwanda, Southern Pakistan	Validated	Emiru et al. (2023)
P411l	Rwanda	Validated	Emiru et al. (2023)
R622I	Eritrea, Ethiopia, Sudan	Validated	De la fuente et al. (2023)
C580Y	Thailand, Rwanda, China Tanzania	Validated	Owoleye et al. (2021)
P441L	Cambodia, Laos	Associated	Miotto et al. (2020)
G449A	Myanmar	Associated	Owoloye (2021)
P527H	Myanmar	Associated	WHO (2022)
G538V	Thailand	Associated	WHO (2022)
N458Y	Cambodia, Thailand, Myanmar, China	Associated	WHO (2022)

Source: Assefa et al. (2024)

4 Newer Approaches and Future Prospects to Mitigate Antimalarial Drug Resistance in Africa

4.1 WHO Strategy

World Health Organization (WHO) has finalized a strategy to address the threat of drug resistance to antimalarial drugs in Africa, established in November 2022 during World Antimicrobial Awareness Week. This regional strategy outlines a four-pillar approach for countries, regional bodies, and global partners to alleviate the dangers and answer to the development and distribution of resistance; the WHO strategies include:

1. Strengthening surveillance of antimalarial drug efficacy and resistance,
2. Optimization and better regulation of the use of diagnostics and therapeutics to limit drug pressure through pre-emptive measures,
3. Reaction to resistance by limiting the spread of antimalarial drug-resistant parasites, and
4. Stimulating research and innovation to leverage existing tools better and develop new tools against resistance.

4.2 Developing Analogs of Existing Drugs

The amino alcohols (mefloquine, lumefantrine, and halofantrine), naphthoquinones (atovaquone), and sesquiterpene trioxanes (artemisinin derivatives) are among the limited chemical compounds that have been recognized. They have been used as a novel class of antimalarial drugs during the past 20 years (Tibon et al., 2020). Numerous monetary, social, and scientific obstacles exist in identifying and leading to the optimal utilization of a novel chemical series as an effective antimalarial agent. Developing equivalents of currently available antimalarial medications is a significant and straightforward way to plan new chemotherapeutical interventions, which can address the concerns above more significantly. However, these hurdles include toxicity, cost, and ethical considerations (Aguiar et al., 2019). The most remarkable instances of this category are the growth of trioxanes and 4-aminoquinoline derivatives (Kolawole et al., 2022). Clinical trials are being conducted to treat uncomplicated malaria with chloroquine (CQ) equivalents created by connecting 4,7-dichloroquinoline with ferroquine, monoalkynes, 1,2,4-trioxane, and a methanogenic CQ analog, which has an artemisinin-like nucleus (Tang et al., 2020).

4.3 Compounds Analysis from Natural Products

For thousands of years, natural items have been employed as traditional medicine, adding to the toolkit of contemporary medicine (Fontinha et al., 2020). Only natural resources have been used to identify many significant and effective antimalarial medications, including artemisinin (and its semisynthetic derivatives), atovaquone, tetracycline, erythromycin, and clindamycin (a derivative of the natural chemical lincomycin) (Ekiert et al., 2021). To produce antimalarial drugs from natural sources, several challenges must be addressed, such as moderate activity, toxic effects, and the identification of physical and chemical and biological properties (Habibi et al., 2022). Additionally, isolating a single drug candidate from crude oil typically results in losing antimalarial potential (Fontinha et al., 2020). Nonetheless, ongoing global research is being conducted to create antimalarial drugs from natural ingredients, some of which are undergoing clinical testing. Argemone Mexicana's emergence as a strong antimalarial is the most notable recent example (Tse et al., 2019).

4.4 Drug Repurposing

Scientists are adopting an exciting drug repurposing strategy to fight malaria, cancer, and diabetes. This approach involves finding new uses for existing medications (Yadav et al., 2021). It is faster and cheaper than developing entirely new drugs because these existing drugs have already been proven safe, and much is already known about how they work in the body. Significant advancements in the treatment of malaria have been made possible via drug repurposing (Rudrapal et al., 2023). It has been demonstrated that several medications, once prescribed for distinct illnesses, also work well against the malaria parasite. These include pharmaceuticals that decrease cholesterol, such as Atorvastatin, antifungal treatments like Itraconazole, and HIV therapies, such as Lopinavir-Tipranavir. These medications provide new avenues for the fight against the illness because they can target the parasite at various stages of its life cycle (Fontinha et al., 2020).

Several antibiotics, including clindamycin, tetracycline, azithromycin, doxycycline, and fosmidomycin, are also used as combination malaria therapy in medical settings. When coupled with pyrimethamine, sulfadoxine—one of the most significant antibacterial agents—has the most incredible collaborative antimalarial effects. Several nations have included this combination in antimalarial plans (Elaagip et al., 2022).

4.5 *Identification of New Drug Targets and Synthesis of Their Selective Inhibitors*

A significant obstacle in the setting of malaria resistance to multiple drugs is the identification of novel therapeutic targets for antimalarial drug discovery (Ali et al., 2021). The emergence of drug resistance to a specific antimalarial also diminishes the effectiveness of related medications, a phenomenon known as "cross-resistance," decreasing the alternatives available for case management to a more significant degree. Therefore, finding a few new crucial targets and creating fresh pharmacophores to block them (Routt et al., 2019). The identification of a novel therapeutic target was made possible by complete genome sequencing, which is being investigated to categorize a safe, crucial target, and validate it and to create several innovative and harmless treatments against these discovered targets (Fayyazi et al., 2020). These include apicoplast, type II fatty acid synthesis (FASII), and its various pathways, enzymes of the folate pathway, the shikimate pathway, and many more from different essential pathways of the malaria parasite (Batista et al., 2020). Several critical pathways and specific targets are being investigated to determine and validate a harmless, vital target, as well as to create several innovative and harmless (Oyola et al., 2016). Therapies against these discovered targets include type II fatty acid synthesis (FASII), apicoplast and its various pathways, the shikimate pathway, enzymes of the folate pathway, and many more from diverse essential malaria parasite pathways (Batista et al., 2020).

Virtual examination of chemical databases from many sources, or high volumes of in vitro screening, is an emerging technique for leading optimization and validation (Afolabi et al., 2022). This strategy will help decrease or postpone the likelihood of drug resistance in both newly created other possibilities and currently available antimalarial drugs, in addition to being advantageous in developing new antimalarial medications or combinations due to their distinct action mechanism (Batista et al., 2020).

4.6 Identification of Resistance-Reversal Agents

Drug overuse, mutations in target genes found, and transporter gene alterations underlie recent increased emergence of antimalarial drug resistance (Schäfer et al., 2023). Drug discovery is a laborious process primarily constrained by the time required for lead optimization and clinical trials and their high expense and stringent regulatory requirements (Batista et al., 2020). To reactivate the antimalarials into the mainstream, it is crucial to maximize the presently available ones using appropriate methodologies. However, toxic effects and an absence of action mechanisms make the proof of identity of resistance reversal drugs a problematic endeavor as well. For most conventional medications, such as CQ (verapamil, chlorpromazine, promethazine, chlorpheniramine, and citalopram), mefloquine, quinine, and quinidine, numerous resistance-reversal agents have been found (NP30) (Batista et al., 2020).

Drug resistance and decreasing effectiveness are both significantly influenced by antibiotic-metabolizing enzymes. Traditional medications such as mefloquine, quinine, and quinidine may have lower bioavailability due to increased metabolism, which could reduce their efficacy and increase the risk of drug resistance (Batista et al., 2020). The condition can be reversed by blocking certain drug-metabolizing enzymes with inhibitors, and the medicine may continue to work. However, toxicity should be considered as it could be a big worry for these combos. This resistance-reversal strategy was successfully used in the study (Batista et al., 2020). The mixture of clarithromycin with mefloquine/quinine/quinidine counter to multidrug-resistant malaria is helpful against MDR malaria owed to the resistance-reversal activity of ketoconazole for mefloquine.

4.7 Combination Therapy

The WHO prohibited single therapy for malaria to lower the likelihood of resistance to the drug since most antimalarial medications are starting to show signs of resistance to the drug (Ehrlich et al., 2021).

The idea behind the combination is that the chance of resistance to both drugs at the same cell division is their individual probabilities products when two drugs are used with diverse modes of activity and, consequently, different mechanisms of resistance (Rasmussen et al., 2021). Combining more than one blood schizontocidal medication with different mechanisms of action at the same time is known as combination therapy with antimalarial medications. Combinations can increase treatment effectiveness and postpone the emergence of individual components' resistance due to several biochemical targets (Bassat et al., 2022). Global recommendations suggest employing artemisinin-based combinations (ACTs) as a primary choice therapy for mild falciparum malaria (Orege et al., 2021). These compounds work quickly and effectively against the malaria parasite at various stages of its life cycle. Because the half-lives of the artemisinin derivatives are shorter, it is advised that other ACT partner medications have a longer half-life. Artemisinin analogs are completely protected since they are excreted quickly, whereas companion medications are eliminated more slowly (Duffey et al., 2021). Further complicating the situation and significantly hindering the discovery of novel treatments is resistance or prolonged clearance of parasites against commonly used antimalarial combinations, such as ACTs (Adoke et al., 2021).

4.8 Omics-Based Strategy

Recently, scientists have turned to a powerful "omics" approach to understand malaria better. This approach combines various techniques like genomics, proteomics, and metabolomics, providing a comprehensive picture of the parasite's biology (Singh et al., 2021). Advancements in automation and data analysis, omics allow researchers to identify the parasite's weak spots and the proteins crucial for its survival, paving the way for potential drug targets (Zhu & Zhou, 2022). Another significant leap forward is the use of cell-based tests to screen potential drugs against various phases of the parasite's life cycle (sexual and asexual blood stages) (Al-Shehri et al., 2019). Scientists can even create drug-resistant versions of the parasite in the lab and then analyze its entire genome to pinpoint potential targets (Zhou et al., 2021). This method has successfully identified promising targets like phenylalanine tRNA synthetase (PheRS) and translation elongation factor 2(eEF2). However, this approach can be time-consuming for uncharacterized mutant genes, highlighting the importance of readily available functional genomics data. By mapping the essential pathways required for the parasite's survival (*P. falciparum*) and analyzing the gene activity of both the host and the parasite during infection, researchers have discovered promising avenues for drug development. Additionally, deciphering the malaria parasite's epigenome, which controls gene expression, has further aided in identifying these essential pathways (Zhou et al., 2021). Studying the host's genes in the infection process is also responsible for identifying candidates of new genes potentially connected to complicated malaria. A particularly successful omics technique is the IVIEWGA (in vitro Evolution and Whole-Genome Analysis) method. This method has been instrumental in discovering numerous promising targets for novel antimalarial drugs (Cowell & Winzeler, 2019).

4.9 Immunotherapy

Immunotherapy is emerging as a powerful weapon against various diseases, including malaria (Furtado et al., 2021). A recent advancement in immunotherapy is checkpoint blockade. This approach utilizes laboratory-made molecules called monoclonal antibodies (mAbs) to disrupt interactions between the immune system's "off switches" and their targets (Naran et al., 2018). These "off switches" usually dampen the immune response, so blocking them allows the immune system to mount a more potent attack against the malaria parasite. Critical targets for these mAbs include CTLA4, PD-L1, and others (Naran et al., 2018). Preclinical studies have yielded encouraging results. One research revealed that a modified protein called PD-L2 fused with an antibody fragment could significantly reduce malaria infection and even protect against reinfections (Naran et al., 2018). Combining specific mAbs targeting PD-L1 and LAG3 has also shown promise, leading to increased parasite clearance, enhanced immune cell function, and higher antibody levels (Naran et al., 2018). Similarly, studies suggest that blocking another protein called OX40 signaling strengthens the immune response, leading to better parasite clearance in nonlethal infections (Othman et al., 2018). Nevertheless, it is crucial to understand that preclinical studies have also shown that targeting certain immune system checkpoints may increase T-cell activity and potentially lead to a severe complication of malaria called cerebral malaria (Othman et al., 2018).

5 Conclusion

Anti-malarial drug resistance in sub-Saharan Africa remains a critical concern, impacting treatment efficacy and public health. Studies emphasize the need for enhanced molecular surveillance to combat resistance effectively. Efforts to strengthen surveillance, regulate drug use, limit resistance spread, and promote research and innovation are crucial. While ACTs are currently effective, vigilance is essential to prevent widespread resistance; the WHO's new strategy aims to address these challenges through tailored interventions, emphasizing the importance of strong health systems and investments in primary healthcare. Overall, combating anti-malarial drug resistance in sub-Saharan Africa requires a multifaceted approach involving surveillance, regulation, research, and innovation to safeguard treatment efficacy and public health.

The reliance on artemisinin-based combination therapies (ACTs) in Africa makes the threat of treatment failure particularly concerning, as alternative drugs are limited. Vigorous measures are necessary to protect the efficacy of existing treatments. The consequences of antimalarial drug resistance underscore the urgent need for robust surveillance, optimized drug use, containment strategies, and research and innovation to combat resistance effectively and safeguard public health in sub-Saharan Africa.

References

Adamu, A., Jada, M. S., Haruna, H. M. S., Yakubu, B. O., Ibrahim, M. A., Balogun, E. O., et al. (2020). *Plasmodium falciparum* multidrug resistance gene-1 polymorphisms in Northern Nigeria: Implications for the continued use of artemether-lumefantrine in the region. *Malaria Journal, 19*(1). https://doi.org/10.1186/s12936-020-03506-z

Adesegun, O. A., Adeyemi, O. O., Ehioghae, O., Rabor, D. F., Binuyo, T. O., Alafin, B. A., et al. (2020). Current trends in the epidemiology and management of enteric fever in Africa:

A literature review. *Asian Pacific Journal of Tropical Medicine, 13*(5), 204. https://doi.org/10.4103/1995-7645.283515

Adoke, Y., Zoleko-Manego, R., Ouoba, S., et al. (2021). A randomized, double-blind, phase 2b study to investigate the efficacy, safety, tolerability and pharmacokinetics of a single-dose regimen of ferroquine with artefenomel in adults and children with uncomplicated *Plasmodium falciparum* malaria. *Malaria Journal, 20*(1), 222. https://doi.org/10.1186/s12936-021-03749-4

Afolabi, O. J., Oluwafemi, O. R., & Oniya, M. O. (2022). Pfmdr 1 and Kelch 13 genes distribution among children that are 5 years and below in Akure, Nigeria. *Journal of Parasitic Diseases*. https://doi.org/10.21203/rs.3.rs-1906304/v1

Agrawal, P. K., Agrawal, C., & Blunden, G. (2022). Artemisia extracts and artemisinin-based antimalarials for COVID-19 management: Could these be effective antivirals for COVID-19 treatment? *Molecules, 27*, 3828. https://doi.org/10.3390/molecules27123828

Aguiar, A. C. C., De Sousa, L. R. F., Da Silva Garcia, C. R., Oliva, G., & Güido, R. V. C. (2019). New molecular targets and strategies for antimalarial discovery. *Current Medicinal Chemistry., 26*(23), 4380–4402.

Alegana, V. A., Okiro, E. A., & Snow, R. W. (2020). Routine data for malaria morbidity estimation in Africa: Challenges and prospects. *BMC Medicine, 18*(1). https://doi.org/10.1186/s12916-020-01593-y

Ali, F., Wali, H., Jan, S., Zia, A., Aslam, M., Ahmad, I., Afridi, S. G., Shams, S., & Khan, A. (2021). Analysing the essential proteins set of *Plasmodium falciparum* PF3D7 for novel drug targets identification against malaria. *Malaria Journal, 20*(1). https://doi.org/10.1186/s12936-021-03865

Al-Mekhlafi, H. M., Madkhali, A. M., Abdulhaq, A. A., Atroosh, W. M., Ghzwani, A. H., Zain, K. A., et al. (2022). Polymorphism analysis of pfmdr1 gene in Plasmodium falciparum isolates 11 years post-adoption of artemisinin-based combination therapy in Saudi Arabia. *Scientific Reports, 12*(1). https://doi.org/10.1038/s41598-021-04450-x

Al-Shehri, H., Power, B. J., Archer, J., Cousins, A., Atuhaire, A., Adriko, M., et al. (2019). Non-invasive surveillance of *Plasmodium* infection by real-time PCR analysis of ethanol preserved faeces from Ugandan school children with intestinal schistosomiasis. *Malaria Journal, 18*(1). https://doi.org/10.1186/s12936-019-2748-4

Amadi, A. T., Ezeonu, I. M., & Akoma, O. N. (2019). Antimalarial drug resistance: An existential burden for the developing world. *Microbiology Research Journal International, 27*, 1–16. https://doi.org/10.9734/mrji/2019/v27i430104

Amelo, W., & Makonnen, E. (2021). Efforts made to eliminate drug-resistant malaria and its challenges. *BioMedical Research International, 20*(1), 553–564.

Anoopkumar, A. N., & Aneesh, E. M. (2021). A critical assessment of mosquito control and the influence of climate change on mosquito-borne disease epidemics. *Environment Development and Sustainability, 24*(6), 8900–8929. https://doi.org/10.1007/s10668-021-01792-4

Ashley, E. A., & Poespoprodjo, J. R. (2020). Treatment and prevention of malaria in children. *Child & Adolescent Health, 4*(10), 775–789.

Assefa, A., Fola, A. A., & Tasew, G. (2024). The emergence of *Plasmodium falciparum* strains with artemisinin partial resistance in East Africa and the Horn of Africa: Is there a need to panic? *Malaria Journal, 23*(1). https://doi.org/10.1186/s12936-024-04848-8

Badmos, A. O., Alaran, A. J., Adebisi, Y. A., Bouaddi, O., Onibon, Z., Dada, A., et al. (2021). What sub-Saharan African countries can learn from malaria elimination in China. Tropical. *Medicine and Health, 49*(1). https://doi.org/10.1186/s41182-021-00379-z

Balikagala, B., Fukuda, N., Ikeda, M., Katuro, O. T., Tachibana, S. I., Yamauchi, M., et al. (2021). Evidence of artemisinin-resistant malaria in Africa. *New England Journal of Medicine, 385*(13), 1163–1171. https://doi.org/10.1056/nejmoa2101746

Bassat, Q., Maïga-Ascofaré, O., May, J., Clain, J., Mombo-Ngoma, G., Groger, M., & Ramharter, M. (2022). Challenges in the clinical development pathway for triple and multiple drug combinations in the treatment of uncomplicated *Falciparum malaria. Malaria Journal, 21*(1). https://doi.org/10.1186/s12936-022-04079-9

Batista, F. A., Gyau, B., Vilacha, J. F., Bosch, S. S., Lunev, S., Wrenger, C., & Groves, M. R. (2020). New directions in antimalarial target validation. *Expert Opinion on Drug Discovery, 15*(2), 189–202. https://doi.org/10.1080/17460441.2020.1691996

Catalano, A., Iacopetta, D., Ceramella, J., Scumaci, D., Giuzio, F., Saturnino, C., et al. (2022). Multidrug resistance (MDR): A widespread phenomenon in pharmacological therapies. *Molecules/Molecules Online/Molecules Annual, 27*(3), 616. https://doi.org/10.3390/molecules27030616

Chenet, S. M., Akinyi Okoth, S., Huber, C. S., Chandrabose, J., Lucchi, N. W., Talundzic, E., Krishnalall, K., Ceron, N., Musset, L., Macedo de Oliveira, A., Venkatesan, M., Rahman, R., Barnwell, J. W., & Udhayakumar, V. (2016). Independent emergence of the *Plasmodium falciparum* Kelch propeller domain mutant allele C580Y in Guyana. *The Journal of Infectious Diseases, 213*(9), 1472–1475. https://doi.org/10.1093/infdis/jiv752

Cisse, B., Cairns, M., Faye, E., NDiaye, O., Faye, B., Cames, C., et al. (2009). Randomized trial of piperaquine with sulfadoxine-pyrimethamine or dihydroartemisinin for malaria intermittent preventive treatment in children. *PLoS ONE, 4*(9), e7164. https://doi.org/10.1371/journal.pone.000716

Conrad, M. D., Asua, V., Garg, S., Giesbrecht, D., Niaré, K., Smith, S., et al. (2023). Evolution of partial resistance to artemisinins in malaria parasites in Uganda. *New England Journal of Medicine, 389*(8), 722–732. https://doi.org/10.1056/nejmoa2211803

Coppée, R., Sabbagh, A., & Clain, J. (2020). Structural and evolutionary analyses of the *Plasmodium falciparum* chloroquine resistance transporter. *Scientific Reports, 10*(1). https://doi.org/10.1038/s41598-020-61181-1

Cowell, A. N., & Winzeler, E. A. (2019). Advances in omics-based methods to identify novel targets for malaria and other parasitic protozoan infections. *Genome Medicine, 11*(1). https://doi.org/10.1186/s13073-019-0673-3

Dagen, M. (2020). History of malaria and its treatment. In *Elsevier eBooks* (pp. 1–48). https://doi.org/10.1016/b978-0-08-101210-9.00001-9

de Haan, F., Boon, W. P. C., Amaratunga, C., & Dondorp, A. M. (2022). Expert perspectives on the introduction of Triple Artemisinin-based Combination Therapies (TACTs) in Southeast Asia: A Delphi study. *BMC Public Health, 22*(1). https://doi.org/10.1186/s12889-022

Dhorda, M., Amaratunga, C., & Dondorp, A. M. (2021). Artemisinin and multidrug-resistant *Plasmodium falciparum*—A threat for malaria control and elimination. *Current Opinion in Infectious Diseases, 34*(5), 432–439.

Duffey, M., Blasco, B., Burrows, J. N., Wells, T. N., Fidock, D. A., & Leroy, D. (2021). Assessing risks of *Plasmodium falciparum* resistance to select next-generation antimalarials. *Trends in Parasitology, 37*(8), 709–721. https://doi.org/10.1016/j.pt.2021.04.006

Egwu, C. O., Pério, P., Augereau, J., Tsamesidis, I., Benoit-Vical, F., & Reybier, K. (2021). Resistance to artemisinin in falciparum malaria parasites: A redox-mediated phenomenon. *Free Radical Biology and Medicine, 179*, 317–327. https://doi.org/10.1016/j.freeradbiomed.2021.08.016

Ehrlich, H. Y., Bei, A. K., Weinberger, D. M., Warren, J. L., & Parikh, S. (2021). Mapping partner drug resistance to guide antimalarial combination therapy policies in sub-Saharan Africa. *Proceedings of the National Academy of Sciences, 118*(29). https://doi.org/10.1073/pnas.2100685118

Ekiert, H., Świątkowska, J., Klin, P., Rzepiela, A., & Szopa, A. (2021). Artemisia annua—Importance in traditional medicine and current state of knowledge on the chemistry, biological activity and possible applications. *Planta Medica, 87*(08), 584–599. https://doi.org/10.1055/a-1345-9528

Elaagip, A., Absalon, S., & Florentin, A. (2022). Apicoplast dynamics during *Plasmodium* cell cycle. *Frontiers in Cellular and Infection Microbiology, 12*. https://doi.org/10.3389/fcimb.2022.864819

Emiru, T., Getachew, D., Murphy, M., Sedda, L., Ejigu, L. A., Bulto, M. G., et al. (2023). Evidence for a role of Anopheles stephensi in the spread of drug- and diagnosis-resistant malaria in Africa. *Nature Medicine, 29*(12), 3203–3211. https://doi.org/10.1038/s41591-023-02641-9

Fayyazi, N., Esmaeili, S., Taheri, S., Ribeiro, F. F., Scotti, M. T., Scotti, L., et al. (2020). Pharmacophore modeling, synthesis, scaffold hopping, and biological β-hematin inhibition interaction studies for anti-malaria compounds. *Current Topics in Medicinal Chemistry, 19*(30), 2743–2765. https://doi.org/10.2174/1568026619666191116160326

Fernández-Villa, D., Aguilar, M. R., & Rojo, L. (2019). Folic acid antagonists: antimicrobial and immunomodulating mechanisms and applications. *International Journal of Molecular Sciences, 20*(20), 4996. https://doi.org/10.3390/ijms20204996

Fontinha, D., Moules, I., & Prudêncio, M. (2020). Repurposing drugs to fight hepatic malaria parasites. *Molecules/Molecules Online/Molecules Annual, 25*(15), 3409. https://doi.org/10.3390/molecules25153409

Furtado, R., Chorro, L., Zimmerman, N., Guillen, E., Spaulding, E., Chin, S. S., Daily, J. P., & Lauvau, G. (2021). Blockade of LAG-3 in PD-L1-Deficient mice enhances clearance of blood-stage malaria independent of humoral responses. *Frontiers in Immunology, 11*. https://doi.org/10.3389/fimmu.2020.576743

Gnädig, N. F., Stokes, B. H., Edwards, R. L., Kalantarov, G. F., Heimsch, K. C., Kuderjavy, M., et al. (2020). Insights into the intracellular localization, protein associations, and artemisinin resistance properties of *Plasmodium falciparum* K13. *PLoS Pathogens, 16*(4), e1008482. https://doi.org/10.1371/journal.ppat.1008482

Greene, H. C., Makovi, K., Abdul-Mumin, R., Bansal, A., & Frimpong, J. A. (2024). Challenges in the distribution of antimicrobial medications in community dispensaries in Accra, Ghana. *PLoS ONE, 19*(5), e0281699. https://doi.org/10.1371/journal.pone.0281699

Habibi, P., Shi, Y., Grossi-De-Sa, M. F., & Khan, I. (2022). Plants as sources of natural and recombinant antimalaria agents. *Molecular Biotechnology, 64*(11), 1177–1197. https://doi.org/10.1007/s12033-022-00499-9

Ibrahim, V. H., Rabo, M. D., & Ahmed, E. H. (2020). Effect of malaria incidence and malaria control on health outcome and human capital development in Nigeria: An econometric analysis. *Deleted Journal, 10*(6). https://doi.org/10.30845/ijhss.v10n6a10

Imwong, M., Dhorda, M., Tun, K. M., Thu, A. M., Phyo, A. P., Proux, S., et al. (2020). Molecular epidemiology of resistance to antimalarial drugs in the Greater Mekong subregion: An observational study. *The Lancet Infectious Diseases, 20*(12), 1470–1480. https://doi.org/10.1016/s1473-3099(20)30228-0

Khunweeraphong, N., & Kuchler, K. (2021). Multidrug resistance in mammals and fungi—From MDR to PDR: A Rocky road from atomic structures to transport mechanisms. *International Journal of Molecular Sciences, 22*(9), 4806. https://doi.org/10.3390/ijms22094806

Kolawole, E., Ayeni, E., Abolade, S., Ugwu, S., Awoyinka, T., Ofeh, A., & Okolo, B. (2022). Malaria endemicity in Sub-Saharan Africa: Past and present issues in public health. *Microbes and Infectious Diseases /Microbes and Infectious Diseases*. https://doi.org/10.21608/mid.2022.150194.1346

La Fuente, I. M., Benito, M. J. S., Ousley, J., De Bartolomé Gisbert, F., García, L., González, V., et al. (2023). Screening for K13-propeller mutations associated with artemisinin resistance in *Plasmodium falciparum* in Yambio County (Western Equatoria State, South Sudan). *American Journal of Tropical Medicine and Hygiene, 109*(5), 1072–1076. https://doi.org/10.4269/ajtmh.23-0382

Masserey, T., Lee, T., Golumbeanu, M., Shattock, A. J., Kelly, S. L., Hastings, I. M., & Penny, M. A. (2022). Biological, epidemiological, and treatment factors influence the establishment and spread of drug-resistant *Plasmodium falciparum. eLife, 11*. https://doi.org/10.7554/elife.77634

Mbengue, A., Bhattacharjee, S., Pandharkar, T., Liu, H., Estiu, G., Stahelin, R. V., et al. (2015). A molecular mechanism of artemisinin resistance in *Plasmodium falciparum* malaria. *Nature, 520*(7549), 683–687. https://doi.org/10.1038/nature14412

Miedaner, T., & Garbelotto, M. M. (2024). Human-mediated migration of plants, their pathogens, and parasites. *Journal of Plant Pathology*. https://doi.org/10.1007/s42161-024-01589-0

Miotto, O., Sekihara, M., Tachibana, S., Yamauchi, M., Pearson, R. D., Amato, R., et al. (2020). Emergence of artemisinin-resistant *Plasmodium falciparum* with kelch13 C580Y mutations on the island of New Guinea. *PLoS Pathogens, 16*(12), e1009133. https://doi.org/10.1371/journal.ppat.1009133

Mohamed, A. O., Hussien, M., Mohamed, A., Suliman, A., Elkando, N. S., Abdelbagi, H., et al. (2020). Assessment of Plasmodium falciparum drug resistance molecular markers from the Blue Nile State, Southeast Sudan. *Malaria Journal, 19*(1). https://doi.org/10.1186/s12936-020-03165-0

Naran, K., Nundalall, T., Chetty, S., & Barth, S. (2018). Principles of immunotherapy: Implications for treatment strategies in cancer and infectious diseases. *Frontiers in Microbiology, 9*, 3158. https://doi.org/10.3389/fmicb.2018.03158

Ndwiga, L., Kimenyi, K. M., Wamae, K., Osoti, V., Akinyi, M., et al. (2021). A review of the frequencies of *Plasmodium falciparum* Kelch 13 artemisinin resistance mutations in Africa. *International Journal for Parasitology, Drugs and Drug Resistance., 16*, 155–161. https://doi.org/10.1016/j.ijpddr.2021.06.001

Nguyen, T. D., Tran, T. N., Parker, D. M., White, N. J., & Boni, M. F. (2023). Antimalarial mass drug administration in large populations and the evolution of drug resistance. *PLoS Global Public Health, 3*(7), 1–25.

Oladipo, H. J., Tajudeen, Y. A., Oladunjoye, I. O., Yusuff, S. I., Yusuf, R. O., Oluwaseyi, E. M., et al. (2022). Increasing challenges of malaria control in sub-Saharan Africa: Priorities for public health research and policymakers. *Annals of Medicine and Surgery, 81*. https://doi.org/10.1016/j.amsu.2022.104366

Onyesom, I., Elu, C., Acha, J., Ojugbeli, E., Ochonogor, E., & Udumebraye, L. (2022). Evaluation of antiplasmodial activity and inhibitions of lipid peroxidation and parasite haem polymerization activity by medicinal phytochemicals of Phyllanthus amarus. *Nigerian Journal of Parasitology, 43*(2), 293–302. https://doi.org/10.4314/njpar.v43i2.12

Orege, J. I., Adeyemi, S. B., Tiamiyu, B. B., Akinyemi, T. O., Ibrahim, Y. A., & Orege, O. B. (2021). Artemisia and Artemisia-based products for COVID-19 management: Current state and future perspective. *Advances in Traditional Medicine, 23*(1), 85–96. https://doi.org/10.1007/s13596-021-00576-5

Othman, A. S., Franke-Fayard, B. M., Imai, T., Van Der Gracht, E. T. I., Redeker, A., Salman, A. M., et al. (2018). OX40 stimulation enhances protective immune responses induced after vaccination with attenuated malaria parasites. *Frontiers in Cellular and Infection Microbiology, 8*, 247. https://doi.org/10.3389/fcimb.2018.00247

Oyola, S. O., Ariani, C. V., Hamilton, W. L., Kekre, M., Amenga-Etego, L. N., Ghansah, A., Rutledge, G. G., Redmond, S., Manske, M., Jyothi, D., Jacob, C. G., Otto, T. D., Rockett, K., Newbold, C. I., Berriman, M., & Kwiatkowski, D. P. (2016). Whole genome sequencing of *Plasmodium falciparum* from dried blood spots using selective whole genome amplification. *Malaria Journal, 15*(1). https://doi.org/10.1186/s12936-016-1641-7

Owoloye, A., Olufemi, M., Idowu, E. T., & Oyebola, K. M. (2021). Prevalence of potential mediators of artemisinin resistance in African isolates of Plasmodium falciparum. *Malaria Journal, 20*(1). https://doi.org/10.1186/s12936-021-03987-6

Paloque, L., Coppée, R., Stokes, B. H., Gnädig, N. F., Niaré, K., Augereau, J., et al. (2021). Mutation in the *Plasmodium falciparum* BTB/POZ domain of K13 protein confers artemisinin resistance. *Antimicrobial Agents and Chemotherapy, 66*(1). https://doi.org/10.1128/aac.01320-21

Raman, J., Kagoro, F. M., Mabuza, A., Malatje, G., Reid, A., Frean, J., & Barnes, K. I. (2019). Absence of kelch13 artemisinin resistance markers but strong selection for lumefantrine-tolerance molecular markers following 18 years of artemisinin-based combination therapy use in Mpumalanga Province, South Africa (2001–2018). *Malaria Journal, 18*(1), 280. https://doi.org/10.1186/s12936-019-2911-y

Ramani, A., Sangwan, P., Tewari, S., Duhan, J., Mittal, S., & Kumar, V. (2022). Comparative evaluation of complete and partial pulpotomy in mature permanent teeth with symptomatic irreversible pulpitis: A randomized clinical trial. *International Endodontic Journal, 55*(5), 430–440. https://doi.org/10.1111/iej.13714

Rasmussen, C., Alonso, P., & Ringwald, P. (2021). Current and emerging strategies to combat antimalarial resistance. *Expert Review of Anti-infective Therapy, 20*(3), 353–372. https://doi.org/10.1080/14787210.2021.1962291

Reddy, V., Weiss, D. J., Rozier, J., Ter Kuile, F. O., & Dellicour, S. (2022). Global estimates of the number of pregnancies at risk of malaria from 2007 to 2020: A demographic study. *The Lancet Global Health, 11*(1), e40–e47. https://doi.org/10.1016/s2214-109x(22)00431-4

Rout, S., & Mahapatra, R. K. (2019). In silico study of M18 aspartyl amino peptidase (M18AAP) of Plasmodium vivax as an antimalarial drug target. *Bioorganic & Medicinal Chemistry, 27*(12), 2553–2571. https://doi.org/10.1016/j.bmc.2019.03.039

Roux, A. T., Maharaj, L., Oyegoke, O., Akoniyon, O. P., Adeleke, M. A., Maharaj, R., & Okpeku, M. (2021). Chloroquine and sulfadoxine–pyrimethamine resistance in Sub-Saharan Africa—A review. *Frontiers in Genetics, 12*, 668574. https://doi.org/10.3389/fgene.2021.668574

Rudrapal, M., Chetia, D., & Bhattacharya, S. (2023). Development of phytomedicines as novel antimalarial lead molecules: Progress towards successful antimalarial drug discovery. In *IntechOpen eBooks*. https://doi.org/10.5772/intechopen.108729

Sà, J. M., Cannon, M. V., Caleon, R. L., Wellems, T. E., & Serre, D. (2020). Single-cell transcription analysis of *Plasmodium vivax* blood-stage parasites identifies stage- and species-specific expression profiles. *PLoS Biology, 18*(5), e3000711. https://doi.org/10.1371/journal.pbio.3000711

Saba, N., Balwan, W. K., & Mushtaq, F. (2022). Burden of malaria—A journey revisited. *Scholars Journal of Applied Medical Sciences, 10*(6), 934–939. https://doi.org/10.36347/sjams.2022.v10i06.013

Sanchez, C. P., Manson, E. D. T., Cubel, S. M., Mandel, L., Weidt, S. K., Barrett, M. P., & Lanzer, M. (2022). The Knock-Down of the chloroquine resistance transporter PfCRT is linked to oligopeptide handling in *Plasmodium falciparum. Microbiology Spectrum, 10*(4). https://doi.org/10.1128/spectrum.01101-22

Schäfer, T. M., De Carvalho, L. P., Inoue, J., Kreidenweiss, A., & Held, J. (2023). The problem of antimalarial resistance and its implications for drug discovery. *Expert Opinion on Drug Discovery, 19*(2), 209–224. https://doi.org/10.1080/17460441.2023.2284820

Shafik, S. H., Richards, S. N., Corry, B., & Martin, R. E. (2022). Mechanistic basis for multidrug resistance and collateral drug sensitivity conferred to the malaria parasite by polymorphisms in PfMDR1 and PfCRT. *PLoS Biology, 20*(5), e3001616. https://doi.org/10.1371/journal.pbio.3001616

Sharma, S., & Ali, M. E. (2022). How do the mutations in the PfK13 protein promote anti-malarial drug resistance? *Journal of Biomolecular Structure and Dynamics/Journal of Biomolecular Structure & Dynamics, 41*(15), 7329–7338. https://doi.org/10.1080/07391102.2022.2120539

Shibeshi, W., Alemkere, G., Mulu, A., & Engidawork, E. (2021). Efficacy and safety of artemisinin-based combination therapies for the treatment of uncomplicated malaria in pediatrics: a systematic review and metaanalysis. *BMC Infectious Diseases, 21*(1). https://doi.org/10.1186/s12879-021-06018-6

Shibeshi, M. A., Kifle, Z. D., & Atnafie, S. A. (2020). Antimalarial Drug Resistance and Novel Targets for Antimalarial Drug Discovery Infection and Drug Resistance, Volume 13, 4047–4060. https://doi.org/10.2147/idr.s279433

Shrestha, B., Shah, Z., Morgan, A. P., Saingam, P., Chaisatit, C., Chaorattanakawee, S., et al. (2021). Distribution and temporal dynamics of *Plasmodium falciparum* chloroquine resistance transporter mutations associated with piperaquine resistance in Northern Cambodia. *The Journal of Infectious Diseases, 224*(6), 1077–1085. https://doi.org/10.1093/infdis/jiab055

Si, K., He, X., Chen, L., Zhang, A., Guo, C., & Li, M. (2023). The structure of *Plasmodium falciparum* multidrug resistance protein 1 reveals an N-terminal regulatory domain. *Proceedings of the National Academy of Sciences, 120*(32). https://doi.org/10.1073/pnas.2219905120

Siddiqui, F. A., Liang, X., & Cui, L. (2021). *Plasmodium falciparum* resistance to ACTs: Emergence, mechanisms, and outlook. *International Journal for Parasitology, Drugs and Drug Resistance, 16*, 102–118. https://doi.org/10.1016/j.ijpddr.2021.05.007

Singh, A., Kalamuddin, M., Maqbool, M., Mohmmed, A., Malhotra, P., & Hoda, N. (2021). Quinoline carboxamide core moiety-based compounds inhibit *P. falciparumfalcipain-2*: Design, synthesis and antimalarial efficacy studies. *Bioorganic Chemistry, 108*, 104514. https://doi.org/10.1016/j.bioorg.2020.104514

Small-Saunders, J. L., Sinha, A., Bloxham, T. S., Hagenah, L. M., Sun, G., Preiser, P. R., et al. (2024). tRNA modification reprogramming contributes to artemisinin resistance in *Plasmodium falciparum. Nature Microbiology, 9*(6), 1483–1498. https://doi.org/10.1038/s41564-024-01664-3

Stanley, P., & Rajkumari, N. (2023). Burden of drug resistance in vivax malaria in India—A brief update. *International Journal of Tropical Disease & Health, 44*(6), 17–35. https://doi.org/10.9734/ijtdh/2023/v44i61411

Stokes, B. H., Dhingra, S. K., Rubiano, K., Mok, S., Straimer, J., Gnädig, N. F., et al. (2021). *Plasmodium falciparum* K13 mutations in Africa and Asia impact artemisinin resistance and parasite fitness. *eLife, 10*, e66277. https://doi.org/10.7554/elife.66277

Talapko, J., Škrlec, I., Alebić, T., Jukić, M., & Včev, A. (2019). Malaria: The past and the present. *Microorganisms, 7*(6), 179. https://doi.org/10.3390/microorganisms7060179.13212-x

Tang, Y., Ye, Q., Huang, H., & Zheng, W. (2020). An overview of available antimalarials: Discovery, mode of action and drug resistance. *Current Molecular Medicine, 20*(8), 583–592. https://doi.org/10.2174/1566524020666200207123253

Tchekounou, C., Zida, A., Zongo, C., Soulama, I., Sawadogo, P. M., Guiguemde, K. T., Sangaré, I., Guiguemde, R. T., & Traore, Y. (2022). Antimalarial drugs resistance genes of Plasmodium falciparum: a review. PubMed, 68(2), 215–225. https://doi.org/10.17420/ap6802.427

Tibon, N. S., Ng, C. H., & Cheong, S. L. (2020). Current progress in antimalarial pharmacotherapy and multi-target drug discovery. *European Journal of Medicinal Chemistry, 188*, 111983. https://doi.org/10.1016/j.ejmech.2019.111983

Tse, E. G., Korsik, M., & Todd, M. H. (2019). The past, present and future of anti-malarial medicines. *Malaria Journal, 18*(1), 93. https://doi.org/10.1186/s12936-019-2724-z

Uwimana, A., Legrand, E., Stokes, B. H., Ndikumana, J. M., Warsame, M., Umulisa, N., et al. (2020). Emergence and clonal expansion of in vitro artemisinin-resistant *Plasmodium falciparum* kelch13 R561H mutant parasites in Rwanda. *Nature Medicine, 26*(10), 1602–1608. https://doi.org/10.1038/s41591-020-1005-2

Van Der Pluijm, R. W., Tripura, R., Hoglund, R. M., Phyo, A. P., Lek, D., Islam, A. U., et al. (2020). Triple artemisinin-based combination therapies versus artemisinin-based combination therapies for uncomplicated *Plasmodium falciparum* malaria: A multicentre, open-label, randomised clinical trial. *The Lancet, 395*(10233), 1345–1360. https://doi.org/10.1016/s0140-6736(20)30552-3

Wassmer, S. C., & Grau, G. E. R. (2017). Severe malaria: What's new on the pathogenesis front? *International Journal for Parasitology, 47*(2–3), 145–152. https://doi.org/10.1016/j.ijpara.2016.08.002

Wicht, K. J., Mok, S., & Fidock, D. A. (2020). Molecular mechanisms of drug resistance in *Plasmodium falciparum* malaria. *Annual Review of Microbiology, 74*(1), 431–454. https://doi.org/10.1146/annurev-micro-020518-115546

Willems, A., Kalaw, A., Ecer, A., Kotwal, A., Roepe, L. D., & Roepe, P. D. (2023). Structures of Plasmodium falciparum Chloroquine Resistance Transporter (PfCRT) isoforms and their interactions with chloroquine. *Biochemistry, 62*(5), 1093–1110. https://doi.org/10.1021/acs.biochem.2c00669

World Health Organization. (2022). World malaria report 2022. World Health Organization.

Woodrow, C. J., & White, N. J. (2017). The clinical impact of artemisinin resistance in Southeast Asia and the potential for future spread. *FEMS Microbiology Reviews, 41*(1), 34–48. https://doi.org/10.1093/femsre/fuw037

Xie, S. C., Ralph, S. A., & Tilley, L. (2020). K13, the cytostome, and artemisinin resistance. *Trends in Parasitology, 36*(6), 533–544. https://doi.org/10.1016/j.pt.2020.03.006

Yadav, K., Shivahare, R., Shaham, S. H., Joshi, P., Sharma, A., & Tripathi, R. (2021). Repurposing of existing therapeutics to combat drug-resistant malaria. *Biomedicine & Pharmacotherapy, 136*, 111275. https://doi.org/10.1016/j.biopha.2021.111275

Zhou, M., Varol, A., & Efferth, T. (2021). Multi-omics approaches to improve malaria therapy. *Pharmacological Research, 167*, 105570. https://doi.org/10.1016/j.phrs.2021.105570

Zhu, P., & Zhou, B. (2022). The antagonizing role of heme in the antimalarial function of artemisinin: Elevating intracellular free heme negatively impacts artemisinin activity in *Plasmodium falciparum*. *Molecules/Molecules Online/Molecules Annual, 27*(6), 1755. https://doi.org/10.3390/molecules27061755

Chapter 14
Impact of Climate Change on the Invasiveness of Mosquitoes in Nigeria

Olubukola Oziegbe, Deborah Oluwadamilola Esho, and Christiana Chiamaka Okeke

1 Introduction

The spread of mosquito-borne diseases is on the rise owing to the invasive nature of mosquitoes. They spread by invading new habitats causing disease outbreaks (Drakou et al., 2020). Due to incessant increase in morbidity and mortality by mosquito invasion, World Health Organization referred to mosquitoes as the most dangerous arthropods (WHO, 2022a, b, c, d). Climate change, globalization, and human migration have resulted in the proliferation of these extremely invasive mosquitoes (Onen et al., 2023). The transmission of mosquito-borne diseases relies on several climatic factors such as temperature, humidity, and precipitation alongside other human or anthropogenic activities such as urbanization, migration, and poor housing conditions and the use of fossil fuels which increase greenhouse emissions which tend to worsen climate change shortly (Fagbohun et al., 2020; Masson-Delmotte et al., 2021). Also, the rate of unpredictable rainfall patterns is expected to rise; therefore, dry areas may start to experience excessive rainfall which leads to increased humidity that provides the suitable environment for mosquito breeding. This will lead to proliferation of vectors in new or unexpected areas where they were previously not found, hence its invasiveness (Nsereko et al., 2020). As climate change continues, a shift in the geographical distribution of mosquitoes is expected which will increase transmission of diseases and necessitate new ways to combat it (Ryan et al., 2020). Climate changes and global warming are predicted to negatively affect human, animal, and environmental ecosystems (El-Sayed & Kamel, 2020).

Mosquitoes are expected to emerge and invade new habitats in several places, especially in Africa, propagating greater disease outbreaks (El-Sayed & Kamel,

O. Oziegbe · D. O. Esho (✉) · C. C. Okeke
Department of Biological Sciences, College of Science and Technology, Covenant University, Ota, Ogun State, Nigeria
e-mail: deborah.eshopgs@stu.cu.edu.ng

P. O. Isibor, P. A. Akinduti (eds.), *Harnessing Biotechnology Tools for Product Development*, https://doi.org/10.1007/978-3-031-86002-7_14

2020). Since the effect of climate change differs in each region, it is imperative to study the impact of climate change on mosquitoes in Nigeria, especially *Anopheles* species, the major malaria vector (Okon et al., 2021). It is also essential to evaluate projected climate trends and climate change's impact on the current vector control strategies available in Nigeria such as the Insecticide Treated Nets and Indoor Residual Spraying.

Climate change affects all parts of human life including agriculture and spread of diseases. Mosquitoes are essential vectors of various deadly diseases; it is therefore imperative to study whether climate change can increase the spread of mosquitoes and consequently the increase in mosquito-borne disease transmission in Nigeria.

1.1 Mosquito-Borne Diseases

They are transmitted through the biting of infected female mosquitoes. West Nile virus, dengue fever, chikungunya virus, yellow fever, Rift valley fever, Zika virus, lymphatic filariasis, and malaria are the most common diseases mosquitoes transmit. Other arthropods, such as ticks and flies, can spread diseases like tick-borne encephalitis (Sanfelice, 2022). *Aedes*, *Culex*, and *Anopheles* mosquitoes are the most significant vectors of mosquito-borne diseases and they are extensively distributed throughout Africa, Europe, Asia, and South America (Onen et al., 2023). Worldwide, illnesses spread by mosquitoes affect about 700 million people and cause more than a million deaths yearly (Piovezan-Borges et al., 2022).

1.1.1 West Nile Virus

A positive (+) strand enveloped RNA flavivirus belonging to the family *Flaviviridae* is the cause of the mosquito-borne disease known as West Nile (McDonald et al., 2021). Although humans, dogs, horses, and other animals can also contract West Nile Virus, it is mainly liable for bird infections, particularly in crows and bluejays (Ahlers & Goodman, 2018). The notable carriers of the West Nile Virus are mosquitoes of the genus *Culex*, particularly species that affect a diversity of avian and mammalian species, such as *Culex quinquefasciatus*, *Culex pipiens*, *Culex nigripalpus,* and *Culex stigmatosoma*, among others (Mencattelli et al., 2022). A 2020 study detected immunoglobulin M for West Nile Virus in 5% of the patients tested in Kaduna, Nigeria (Ma'aji et al., 2020).

1.1.2 Dengue Fever

The Dengue virus, a positive (+) stranded RNA virus, from the Flavivirus genus and family *Flaviviridae* is responsible for causing dengue fever (Roy & Bhattacharjee, 2021). Virus transmission is primarily via being bitten by an infected female *Aedes aegypti* or *Aedes albopictus* mosquitoes (Onen et al., 2023). Recording roughly 10,000 deaths and 100 million infections yearly in up to 130 nations, dengue is among the foremost causes of mosquito-related disease and deaths in the tropics and subtropics. In 2021 alone, about 390 million confirmed cases were recorded across the globe (transmitted by *Aedes* species). Symptoms usually include fever, headache, muscle pain, if prolonged and untreated, dengue can lead to shock, hemorrhage, and death. Preventive steps, such as clearing up mosquito breeding sites, are crucial to minimize the spread of the virus (Messina et al., 2019; Onen et al., 2023; World Health Organization, 2022a, b, c, d). The presence of dengue virus has been stated by several researchers in several regions of the nation, including northern Nigeria. There have been reports of a 20.5% prevalence of antibodies of the dengue virus (anti-DENV IgG) in the Guinea Savannah of Nigeria (Omatola et al., 2021). Similarly, a high frequency of dengue IgM has been observed in people with febrile fever in Kano, in northwest Nigeria (Abdulaziz et al., 2020).

1.1.3 Yellow Fever

Yellow fever is brought about by a flavivirus usually found in tropical Africa, and also South and North America. The disease can cause human outbreaks and also widespread epizootics in non-human primates (Mboussou et al., 2019). *Aedes albopictus* and *Aedes aegypti* are the primary carriers of the yellow fever virus in Africa (Kamgang et al., 2019). Both sylvatic and urban cycles are involved in the transmission of Yellow Fever. Most documented cases in South America are related to sylvatic cycles. In comparison, reports of the intermediate cycle which entails the spread of yellow fever from humans to monkeys or from humans to humans via mosquito bites have only been documented in Africa (Onen et al., 2023). Yellow fever virus (YFV) affects about 200,000 people worldwide; 30,000 people die from the virus each year, 90% of which are in Africa (CDC, 2021). In 2018, about 70,000 deaths in Africa alone were attributed to yellow fever (WHO, 2021a, b). Nigeria has experienced periodic outbreaks of YFV infections for more than 60 years, predominantly in the southern region of the nation where the virus was first discovered (WHO, 2021a, b). In recent times, over half of the local governments across the country have reported experiencing outbreaks of YFV diseases, with Benue, Enugu, Delta, Bauchi, and Ebonyi states still experiencing an ongoing infection (CDC, 2021). In Nigeria, there were 1312 recorded cases of yellow fever in the first eight months of 2021 alone (WHO, 2021a, b).

1.1.4 Chikungunya

The Chikungunya virus is an alphavirus largely disseminated via infected *Aedes* (subgenus Stegomyia) mosquito species, *Aedes aegypti* and *Aedes furcifer*, among others. The Chikungunya virus is common in Southeast Asia, the tropics and subtropics of Africa, where there are transmission cycles between *Aedes* species and non-human primates (Onen et al., 2023). In Nigeria, *A. aegypti* and *A. albopictus* are the most important vectors of Chikungunya. The disease is usually characterized by joint pains, rashes, headaches, and fever (Mac et al., 2022). Mac et al. (2022) performed a cross-sectional study to determine the seroprevalence of Chikungunya in Nigeria to predict the hidden burden of the disease. The location of this study traverses 3 teaching hospitals in Nigeria in three states, namely Nasarawa, Abia, and Kaduna. Their results indicated that Nasarawa had the highest seroprevalence of 64.9%, and Kaduna came in second with a seroprevalence of 67%. A lower value of 48% was recorded in Abia state (Mac et al., 2022).

1.1.5 Rift Valley Fever

Rift Valley fever virus, belonging to the Phlebovirus family *Bunyaviridae*, affects both humans and domestic animals (Onen et al., 2023). Mosquitoes and one-on-one contact with infected animals are the two ways the virus is transmitted (Ahmed et al., 2020). According to Ndiaye et al. (2016), mosquito species like *Aedes vexans*, *Culex poicilipes*, and *Culex quinquefasciatus* are the key vectors that spread Rift Valley Fever. Certain *Mansonia* and *Anopheles* species are also known to transmit Rift Valley fever (Ndiaye et al., 2016). Although Rift Valley is more predominant in sub-Saharan Africa, recent reports have indicated its presence in Saudi Arabia and the Arabian Peninsula (Onen et al., 2023). Nigeria has been found to harbor the yellow fever, dengue, and Rift Valley fever viruses (Fagbami & Onoja, 2018). Rift Valley fever is an occupational hazard for butchers, abattoir workers, and people in close contact with animals. Oragwa et al. (2022) conducted a serological investigation to detect the virus Northern Nigeria's life stock keepers, butchers, and abattoir workers (Borno, Benue, and Sokoto state, respectively). According to their result, 19.9% of the total samples were positive for Rift Valley Fever antibodies. By state, Sokoto had the highest prevalence (21.7%), followed by Borno (19.8%) and Benue (17.4%). By occupation, butchers had the highest seropositivity, livestock keepers came next while abattoir workers had the least (Oragwa et al., 2022).

1.1.6 Zika Virus

A flavivirus is also responsible for the Zika virus disease (Abushoufa et al., 2021). Zika virus outbreaks were documented in Micronesia and Brazil in 2007 (Gatherer & Kohl, 2016). Between 2015 and 2017, the United States of America also reported roughly 6000 active cases of the disease (Lee et al., 2018). Similar to other members of the flavivirus, Zika is transmitted by the *Aedes* genus, specifically species such as *A. africanus, A. aegypti, A. albopictus,* and *A. hensilli* (Plourde & Bloch, 2016). Zika virus disease can also be contracted sexually, from mother to child, via infected blood transfusions or organ transplants (Brent et al., 2016). In Nigeria, Zika is often misdiagnosed as malaria which could lead to severe health risks in affected people.

Zika has also been associated with birth defects and microcephaly in neonates (Mac et al., 2023).

1.1.7 Lymphatic Filariasis (LF)

Infection with the microfilaria of *Wuchereria bancrofti*, *Brugia malayi*, or *B. timori* results in lymphatic filariasis (Lourens & Ferrell, 2019). Based on geographical distribution, numerous mosquito species are vectors of this disease, most commonly: *Culex*, *Anopheles*, *Aedes*, *Mansonia*, and *Coquillettidia* species (Onen et al., 2023). LF affects over 120 million individuals globally (Turner et al., 2017). Lymphatic filariasis threatens approximately a billion people in Asia, Africa, and some regions of South America and the Caribbean which is endemic in 73 countries (Turner et al., 2017). With approximately 134 million being susceptible to contracting LF, Nigeria is among the countries endemic for this disease. In 2011, a mass drug administration campaign was carried out in Abuja, one of the states where lymphatic filariasis is endemic. The campaign involved the distribution of Ivermectin and Albendazole (Amanyi-Enegela et al., 2022).

1.1.8 Malaria

More than 25% of malaria cases worldwide and 19% of malaria-related mortalities occur in Nigeria (WHO, 2021a, b). When an infected female *Anopheles* mosquito sucks blood, *Plasmodium* penetrates the host via the saliva of the insect and causes malaria (Okwa, 2019). Human malaria parasites include *Plasmodium falciparum*, *Plasmodium malariae*, *Plasmodium ovale*, *Plasmodium knowlesi*, and *Plasmodium vivax* (Sato, 2021). In tropical and temperate regions, *P. falciparum* and *P. vivax* are the most prevalent species (WHO, 2023). According to the documented cases of malaria worldwide, *Plasmodium falciparum* is responsible for about 93.5%; *Plasmodium vivax* accounts for 3%; and the remaining three *Plasmodium* parasites (*P. malariae*, *P. ovale*, and *P. knowlesi*) account for 3.5% of cases (WHO, 2020). *Plasmodium falciparum* is the cause of 99.7% of malaria cases in Africa. However, *Plasmodium vivax* and *Plasmodium falciparum* account for roughly 46% and 53% of cases, in Southeast Asia, respectively (WHO, 2020). According to WHO estimates, Nigeria recorded 68 million cases of malaria in 2021, which accounted for 27% of the burden worldwide and 28% in the WHO African Region. Additionally, 194,000 deaths from severe malaria were recorded in 2021 alone, 80% of these deaths occurring in children under 5 (WHO, 2022a, b, c, d). Also, 31% of global malaria mortality and 40% of mortality in the WHO African Region are attributed to Nigeria making malaria a serious public health crisis in the country (WHO, 2022a, b, c, d). The threat of malaria transmission is present throughout the whole country all year. Regardless, the northern areas of the nation have the highest rates of malaria incidence (WHO, 2022a, b, c, d).

Figure 14.1 shows the prevalence of yellow fever, dengue, and malaria in Nigeria for the year 2023 (NCDC, 2023a, b; WHO, 2023). Malaria was confirmed in all 36 states of the country, as well as the Federal Capital Territory while yellow fever was confirmed in 12 states according to the Nigeria Center for Disease Control. Sokoto was the only state that recorded an outbreak of dengue fever in 2023. Table 14.1 shows the climatic data and malaria occurrence for all the states in Nigeria for the year 2021. Data for this table was extracted from WHO Report on Malaria in Nigeria 2022 and Weather and Climate (n.d.) database.

Mosquito-borne diseases are grossly ignored and considered to be neglected diseases. This is a huge problem as about 80% of the African population is exposed to them. Replete data on malaria are available in literature; however, the other mosquito-borne diseases are under-researched even though they can also be life-threatening. In Africa, especially Nigeria, the majority of these mosquito-borne febrile diseases are often misdiagnosed or assumed to be malaria, due to similarities in their symptoms and inadequate diagnostic tools and trained laboratory personnel. Many healthcare providers continue to prescribe antimalarial medications, despite the patient not recovering. In this case, it is sometimes referred to as fever with no known cause or still malaria. In light of these recurring practices, other mosquito-borne diseases can be grossly ignored till there is an outbreak leading to an epidemic and consequently high morbidity (Sow et al., 2016; Ngouanet et al., 2022).

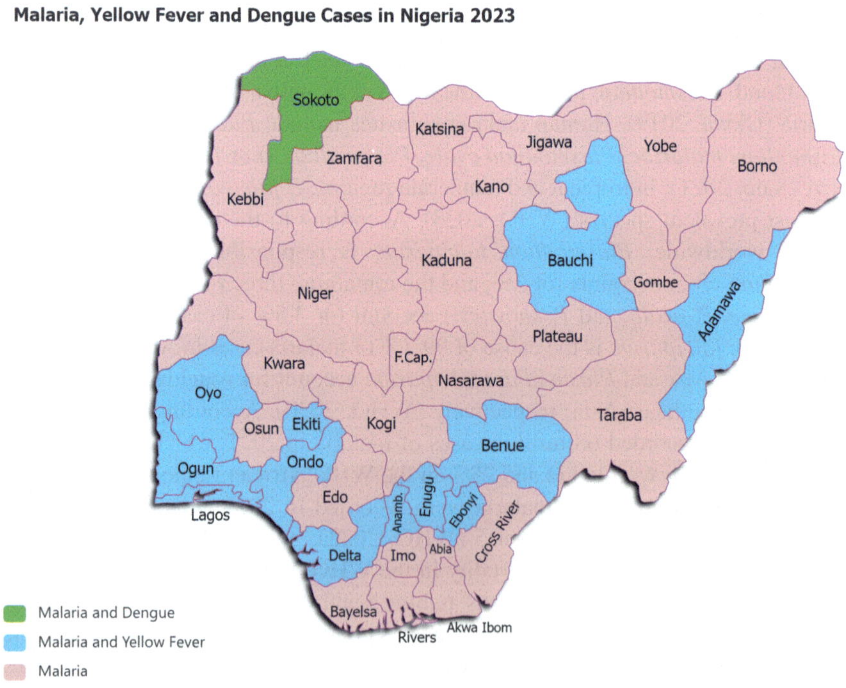

Fig. 14.1 Malaria, dengue, and yellow fever confirmed cases in 2023

Table 14.1 Climatic data and malaria occurrence for the geographical zones in Nigeria for year 2021

Zones	States	Population	Estimated malaria cases	Rainfall (MM)	Temperature (Celsius °C)
North Central	BENUE	6,627,000	1,704,000	1314.5	29.38
	KOGI	5,163,000	1,558,000	1221.3	29.88
	KWARA	3,685,000	1,150,000	1094.9	29.54
	NASARAWA	2,912,000	789,000	1217.4	29.39
	NIGER	6,185,000	2,174,000	1101.7	31.39
	PLATEAU	4,782,000	1,370,000	1042.7	26.60
	ABUJA	5,338,000	1,438,000	1 215.0	29.77
North East	ADAMAWA	4,881,000	1,339,000	1074.9	30.81
	BAUCHI	7,683,000	3,039,000	889.3	28.97
	BORNO	6,887,000	1,987,000	660.0	32.53
	GOMBE	3,793,000	1,363,000	851.0	30.54
	TARABA	3,493,000	1,047,000	1446.6	30.32
	YOBE	3,878,000	1,119,000	547.2	31.29
North West	JIGAWA	6,696,000	2,559,000	589.4	31.23
	KADUNA	9,525,000	3,636,000	1175.0	28.46
	KANO	15,300,000	6,088,000	813.5	30.75
	KATSINA	9,039,000	3,157,000	705.6	30.89
	KEBBI	5,148,000	2,074,000	845.9	31.26
	SOKOTO	5,769,000	1,778,000	586.4	31.67
	ZAMFARA	5,259,000	1,669,000	813.6	30.22
South East	ABIA	4,243,000	1,271,000	2070.9	28.68
	ANAMBRA	6,322,000	1,262,000	1667.3	28.99
	EBONYI	3,294,000	1,046,000	1719.9	29.34
	ENUGU	5,091,000	1,177,000	1548.2	28.27
	IMO	6,300,000	1,734,000	2095.0	28.88
South South	AKWA IBOM	6,443,0000	1,935,000	2432.1	28.47
	BAYELSA	2,617,000	673,000	2701.6	28.64
	CROSS RIVER	4,442,000	1,283,000	2284.6	28.50
	DELTA	6,596,000	1,598,000	2088.8	28.64
	EDO	4,822,000	1,191,000	1577.2	28.78
	RIVERS	8,584,000	1,954,000	2414.2	26.55
South West	EKITI	3,792,000	1,230,000	1333.5	27.90
	LAGOS	14,618,000	2,556,000	1535.8	28.67
	OGUN	6,105,000	1,709,000	1332.3	29.34
	ONDO	5,392,000	1,639,000	1572.4	28.42
	OSUN	5,481,000	1,789,000	1361.8	28.35
	OYO	9,215,000	2,869,000	1088.0	28.82

Source: WHO (2022a, b, c, d)

An important unifying factor for majority of the medically relevant mosquito-borne diseases is that they are caused by either flavivirus (yellow fever, dengue fever, Zika, west Nile virus), or togavirus (chikungunya). A unique feature of these virus families is their genome. They possess an RNA genome that enables them to adapt both their host and the surrounding environmental conditions. It is also interesting to note that many mosquito-borne diseases associated with epidemics belong to the RNA virus group. From 2004 to 2006, antigenic drift was observed to facilitate transmission of the chikungunya virus, during the outbreak that occurred in the Indian Island Ocean. The point mutation caused a replacement of Alanine with Valine in the E1 Structural protein of the virus (Schuffenecker et al., 2006; Agarwal et al., 2017; Abudurexiti et al., 2019; Ngouanet et al., 2022).

1.2 Prevalent Mosquito Species in Nigeria

1.2.1 Anopheles

There are currently 35 known *Anopheles* species in Nigeria (Okwa, 2019). In Nigeria, the *Anopheles gambiae* complex, which includes *A. gambiae* sensu stricto (ss), *A. coluzzii*, and *A. arabiensis*, is the main malaria vector. The *A. funestus* complex includes *A. funestus* sensu stricto (ss) is also a significant vector. In Nigeria, *A. hancocki*, *A. coustani*, *A. pharoensis*, and *A. longipalpis* are secondary vectors. Similar to *A. gambiae* sensu stricto, *A. funestus* sensu stricto is broadly distributed, extremely endophilic, anthropophilic, and anthropophagic. Genetically and physiologically, both complexes are polymorphic. There is also a possibility that *A. funestus* ss and *A. gambiae* ss can live in the same habitat despite *A. funestus* ss having fewer options than *A. gambiae* ss (Okwa, 2019).

Each sibling species in both complexes has a different behavior, and this reflects differences in their roles in the spread of malaria (CDC, 2004). A significant obstacle to malaria control programs is the presence of species complexes with morphologically similar sibling or isomorphic forms, which necessitate the identification of vectors through molecular methods. Failure to identify the sibling species will waste limited resources and time in controlling non-malaria vectors. There used to be no trustworthy data available in many parts of Nigeria regarding the presence or absence of vectors; however, diverse malaria vectors with varying bionomics and vector competencies in Nigeria have been established. Effective control measures necessitate knowledge of vector proportion, behavior, and abundance (Okwa, 2019). Figures 14.2 and 14.3 show the distribution of *Anopheles gambiae* sensu lato (sl) and *Anopheles gambiae* sensu stricto (ss) under current climatic conditions across the six geopolitical zones in Nigeria. South West has the highest prevalence while the North East and North Central have the lowest species prevalence. Data for this map was extracted from Akpan et al. (2018).

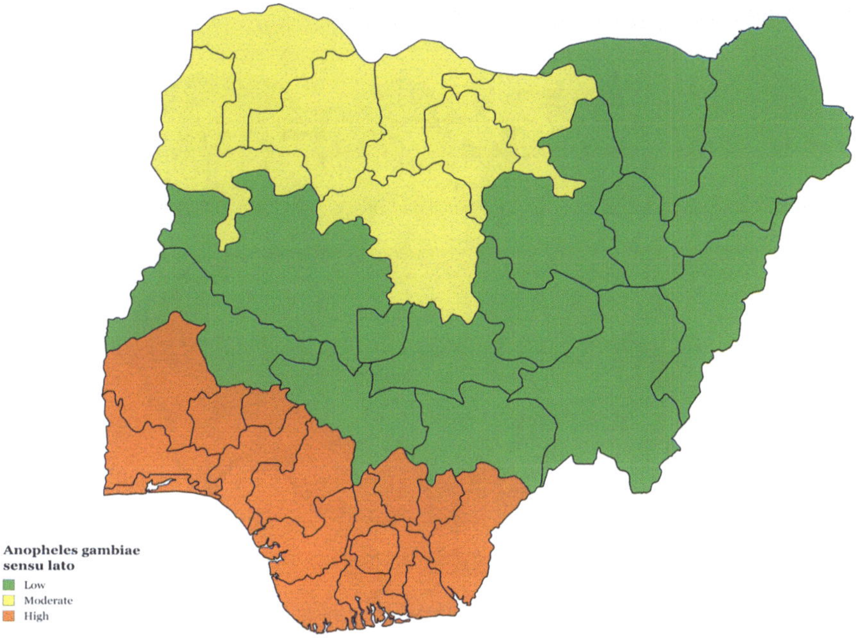

Fig. 14.2 Prevalence of *Anopheles gambiae* sensu lato under current climate in Nigeria

Anopheles stephensi

In recent times, the Asian malaria vector *Anopheles stephensi* has spread throughout Africa and become an invasive species (Carter et al., 2018; Ahn et al., 2023). It was first discovered in a Djibouti seaport in 2012; it was later found in neighboring Ethiopia in 2018, close to seaports in Sudan in 2019, and in Somalia also in 2019 (Ahn et al., 2023; Carter et al., 2021). It has been hypothesized that the emergence of the species was most likely aided by maritime commerce given its unique biological characteristics and identification in seaports (Ahn et al., 2023). An additional obstacle to controlling and eradicating malaria on the African continent is the invasive *A. stephensi* that is successfully thriving in diverse African nations (Sinka et al., 2020). The capacity of this vector to adjust and flourish in urban environments may jeopardize attempts to eradicate and control malaria. The World Health Organization (2019) urged African countries to swiftly enhance surveillance efforts to combat the spread of this invasive vector (Carter et al., 2021; WHO, 2023). The distribution of this vector species is influenced by climate-related factors, which may impact malaria incidence and lead to outbreaks (Fouque & Reeder, 2019). The ideal habitats for *A. stephensi* are areas with warm temperatures and high humidity. Temperature variations affect *A. stephensi* populations directly and are essential to its habitat and reproduction (Whittaker et al., 2023). The best locations for *A. stephensi* spawning and hatching are unnatural habitats created by human activity

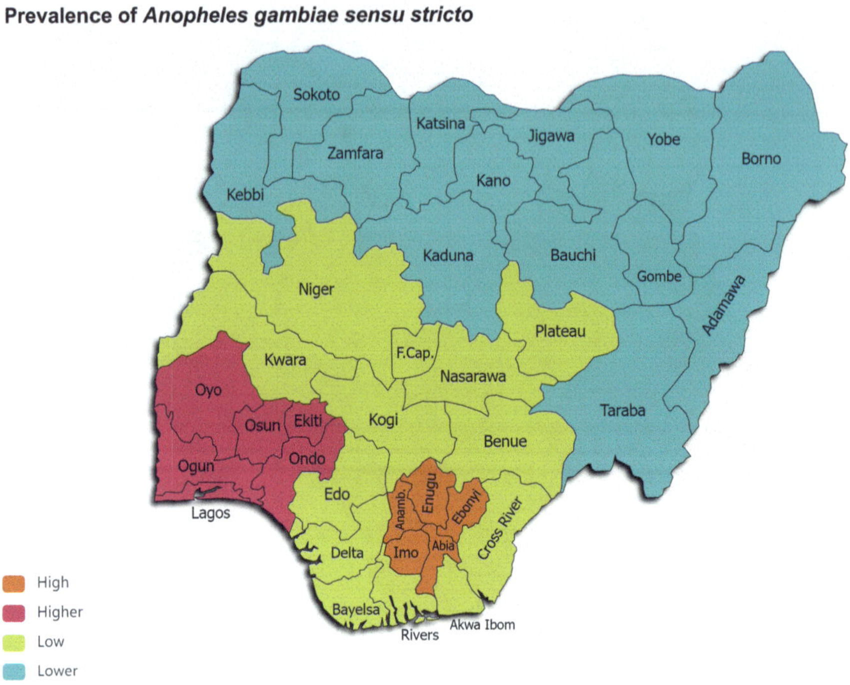

Fig. 14.3 Prevalence of *Anopheles gambiae* sensu stricto under current climate in Nigeria

inside or near residential areas (Ismail et al., 2024). This invasive mosquito species was discovered in Nigeria in 2020, just 8 years after its initial discovery in Djibouti in 2012 (WHO, 2023).

1.2.2 Aedes

A. aegypti and *A. albopictus* are the two very common species that spread viruses. *A. aegypti* is distinguished by lyre-shaped markings on its scutum and also its medium size and dark color. There are patches of white scale on the clypeus and dark scales on the proboscis. Around the world, this species can be found in warm temperate, tropical, and subtropical zones. They lay eggs on the water surface which are resistant to desiccation. Larvae live in a wide range of tiny, naturally occurring and artificially created containers, with low-to-moderate levels of organic matter in the breeding water (Agboli et al., 2021). Adult species appear about 10 days after egg laying at 27–30 °C. Females typically rest indoors and feed on humans in urban environments, biting frequently and aggressively during the day; however, they do not migrate long distances. It originated in sub-Saharan Africa and was transported to the New World via slave trade ships in the sixteenth century (Agboli et al., 2021). The lighter-colored domestic subspecies *Aedes aegypti aegypti* (Aaa) breeds in man-made containers (e.g., tires and tins) and mainly bites people (Agboli et al., 2021). It breeds in close proximity to human settlements (Powell, 2016). Chikungunya, dengue fever, yellow fever, and Zika virus have all been linked to *A. aegypti* in Nigeria and other countries (Becker et al., 2020).

Aedes albopictus is another mosquito species with significant medical importance in Nigeria. The distinctive features of this species include a bare, completely dark clypeus and stripes of narrow white scales on the scutum. *A. albopictus* has expanded worldwide and invaded many different nations by passively laying eggs in old tires and bamboo. Their eggs can travel great distances because they are resistant to desiccation. The immature stages take place in a broad range of man-made and natural environments and they breed all year round in tropical countries, including Nigeria. Although other mammals such as dogs, cows, squirrels, and rabbits can also serve as hosts, female *Aedes* primarily feed on humans. They cause a significant nuisance by feeding outside during the day and inside at dusk and night (Becker et al., 2020).

1.2.3 Culex

The *Culex* genus contains a number of species that can transmit serious diseases to humans, animals, and birds. *Culex* species are typically small or medium-sized and on the surface of the water, females deposit up to 300 egg rafts. Egg laying can occur in an assortment of habitats, from large collections of contaminated wastewaters to tiny bodies of standing water. The most common species in tropical Africa is *Culex quinquefasciatus*, belonging to the *Culex pipiens* complex. There have been reports of *C. quinquefasciatus* larvae in a variety of habitat types, such as brackish, clear, contaminated, sewage, lavatories, and synthetic containers. Being endophagic at night and exophagic from dusk to sunrise, it is one of the most bothersome mosquitoes. They often bite birds or other household animals (Becker et al., 2020). Chikungunya virus, western and eastern equine encephalitis virus, and West Nile virus (WNV) among others are all known to be transmitted by *C. quinquefasciatus*. *C. antennatus* is a zoophagic vector that primarily feeds on humans and household animals. It is a significant epizootic vector of West Nile Virus (Agboli et al., 2021). Amao et al. (2018) conducted a study in Lagos, South West Nigeria, to determine Culex species' abundance and relative distribution. Four Local Government Areas were considered (Ojo, Epe, Kosofe, and Eti-Osa) for the study. The mosquitoes were identified morphologically and with Polymerase Chain Reaction (PCR). A cumulative of 12,367 *Culex* larvae were collected ranging across six subspecies. The highest abundance of these species was observed between the end of the rainy season and the beginning of the dry season. This large number in only one location suggests the abundance of these species across the country.

2 Climate Change in Nigeria

Climate is the local atmospheric conditions of a given location over an extended period (Ani et al., 2021). It is also the prolonged accumulation of atmospheric variables, such as solar radiation, temperature, relative humidity, and precipitation. Climate change, as defined by the United Nations (UN), is any alteration in the statistical characteristics of the climate system over extended periods, irrespective of the underlying cause (Intergovernmental Panel on Climate Change, 2001; World Meteorological Organization, 2022). Also, Ani et al. (2021) define climate change as alterations in the average variability characteristics of the climate that endure for a considerable amount of time, usually several decades or more. According to the Intergovernmental Panel on Climate Change (IPCC) (2014), human activity has expanded the greenhouse effect, which is the prime origin of current climate change. Most societal and environmental health factors (such as clean air, safe housing, safe food, and potable drinking water) are negatively affected by climate change, which impacts all regions and their populations' health (United Nations Organization, 2019). Although being the least contributor, Africa is bearing the brunt of the impact of climate change having the least resilience to its effects. Based on current projections, the mean annual temperature of Africa will rise by more than 2 °C by 2050 (Intergovernmental Panel on Climate Change, 2014; Giesen et al., 2020).

Climate change is affecting Nigeria just like it is affecting the rest of the world. Throughout the vegetative regions of Nigeria, the effects of climate change are being felt. According to research, agricultural productivity is becoming increasingly threatened by climate change. Alongside the growing aridity of the Sahel and Sudan savannah of Nigeria, some formerly dry agricultural grasslands have recently been flooded (Ani et al., 2021). According to Ughaelu (2017), Nigeria is among the sub-Saharan African countries most prone to changing climate patterns.

2.1 Climatic Impact on Mosquito Populations

The distribution of mosquitoes is controlled by climatic factors such as humidity, rainfall, and temperature (Asgarian et al., 2021). As a result of the low temperature and rainfall, and especially the prolonged winter season experienced by other continents such as Asia, Europe, and America, the population of mosquitoes are very low, almost entirely absent. However, the opposite is true for Africa. The climatic conditions of most African countries (temperature >15 °C and rainfall for half of the year) favorably encourage the reproduction of a variety of mosquitoes (Ngouanet et al., 2022). Climate change is anticipated to alter mosquito population in Africa in a number of ways. This alteration occurs because survival, reproduction, transmission, and behavior of these vectors are easily influenced by environmental factors (Agyekum et al., 2021; Ferguson & Govella, 2024). Even though economic growth and control measures have successfully lowered the transmission rate, seasonality and mean intensity of malaria transmission are affected by alterations in the climate (Chemison et al., 2021).

2.2 *Impacts of Climatic Factors on Mosquito Abundance*

2.2.1 Temperature

In the most severely affected areas of Africa, *Anopheles gambiae*, *Anopheles arabiensis*, and *Anopheles funestus* act as the principal vectors of malaria. The mosquito can spread its infection to more people after the sporogonic cycle, which is the time during which it must incubate before becoming infectious. The length of this incubation period decreases with rising temperatures. *A. gambiae* and *P. falciparum* were found to require a minimum temperature of 17 °C for sporogonic development (Waite et al., 2019). As a result, temperature is imperative to complete the parasites' sporogonic cycle. The ideal range for the gonotrophic cycle of *Aedes aegypti* is between 26 °C and 30 °C, and it shortens with higher mean temperatures although excessively high temperature decreases the risk of malaria transmission owing to a rise in vector mortality. It is imperative to note that research and discussion on temperature thresholds and hydrological relationships are still ongoing (Chemison et al., 2021). The impact of rising temperatures on physiological characteristics in ectotherms, for instance, is typically non-linear and can have unfavorable effects like decreased larval development and smaller adult mosquitoes due to reduced survival. Also, reduced fecundity, blood meal size, and immunocompetence have been linked to small adult mosquito sizes. Thus, it is ironic that increasing temperatures linked to climate change can lower the threat of transmission in some areas because rising temperatures have a negative impact on vector competence, or capacity to spread infection (Franklinos et al., 2019). Agyekum et al. (2021) describe the effects of temperature as curvilinear; which means that high temperature favors mosquito populations up to a threshold beyond which a rapid decline is noticed.

Variations in temperature can also impact the duration of pathogen development in mosquitoes, also known as the extrinsic incubation period. For instance, it has been demonstrated that low temperatures in *Aedes* vectors shorten extrinsic incubation and increase the number of viral infections by inhibiting mosquito antiviral immunity. Other research has demonstrated that, in comparison with more consistent conditions, considerable temperature fluctuations at low mean temperatures result in quicker extrinsic incubation periods and higher infection in mosquitoes. To accurately forecast the risk of mosquito-borne disease, more empirical research is necessary to fill in the critical knowledge gap regarding the association between temperature and the extrinsic incubation period (Franklinos et al., 2019). The laboratory data-based model of Stopard et al. (2021) was able to demonstrate that the extrinsic incubation period of the parasite *Plasmodium falciparum* in *Anopheles stephensi* drastically reduced from 16 to about 9 days when temperature rose to 34 °C from 21 °C. Consequently, studies on the connection between temperature and transmission of mosquito-borne diseases consistently point to global warming as a key factor (Shragai et al., 2017; Mordecai, 2019; Franklinos et al., 2019; Giesen et al., 2020; Khezzani et al., 2023). Temperature affects the dynamics of mosquito-borne diseases as it directly influences mosquitoes' physiology, life cycles, behavior, and competence to transmit diseases (Cator et al., 2020).

Mosquito species react to rising temperatures in three main ways: the first is by expanding their optimal temperature ranges; avoiding or temporarily adjusting to high temperatures using phenotypic plasticity (by moving biting activities to cooler times); and lastly enduring rising temperatures via genetic evolutionary adaptation (Couper et al., 2021).

2.2.2 Precipitation

Whether there is a drought or an inundation (flood), rainfall patterns will influence mosquito-borne diseases. The formation and maintenance of new mosquito breeding habitats are facilitated by precipitation, which increases the population of vectors, particularly in their immature stages. Elevated rainfall improves the water quality in small water bodies, attracting more mosquitoes because mosquito eggs' survival depends on breeding water availability (Khezzani et al., 2023). Higher rainfall in African coastal zones can expand the range of vectors, including *A. gambiae* (Chandra & Mukherjee, 2022). According to a Zambian study, there is a 0.2% and 2.0% increase in malaria infection for every 1 mm increase in rainfall and 10.0% increase in humidity, respectively (Mzyece, 2016). Drought-related stress may promote water storage methods that put mosquito habitats closer to people, particularly *Aedes albopictus* and *Aedes aegypti*, which interact with human habitats and breeds in household water storage containers (Khezzani et al., 2023). Reduced precipitation causes rivers to dry up and pools to form in riverbeds. This helps mosquitoes find new, suitable places to breed and makes their larval populations less attractive to predators (Khezzani et al., 2023). Conversely, excess rainfall volume can reduce mosquito population by washing off mosquito larvae in existing breeding sites or hindering the development of mosquito eggs as a result of expanding breeding pools into big streams (Asgarian et al., 2021; Irikannu et al., 2023). Irikannu et al. (2023) conducted a study to determine the distribution of mosquito species under different climatic conditions in Anambra state, South East Nigeria. They noticed a positive relationship between rainfall and the abundance of mosquitoes due to the fact that increased rainfall creates more breeding sites.

2.2.3 Humidity

Humidity is essentially a measurement of the amount of water vapor present in the atmosphere, and it is a significant factor that influences the susceptibility of an organism to desiccation. Mosquito lifespan, blood-feeding habits, mating, dispersal, and oviposition are all affected by humidity (Drakou et al., 2020; Asgarian et al.,

2021). The desiccation resistance, egg-laying rate, mortality, survival, fecundity, sex, and longevity of female and male mosquitoes are all influenced by relative humidity and temperature. For instance, mosquitoes live longer in environments with relative humidity >60%, but they die sooner when relative humidity is significantly less than 60% (Tahir et al., 2023). Adult mosquitoes die quickly over 38 °C, but they are present when humidity is higher than at least 40% (Kreppel et al., 2019). A high humidity is better conducive for adult mosquito survival (Thomson et al., 2017; Ssempiira et al., 2018). According to Tahir et al. (2023), the risk of developing clinical malaria doubled at 80% relative humidity and increased exponentially above 60%. Figure 14.4 summarizes the impact of temperature, humidity, and precipitation on mosquito populations.

2.3 *Other Factors That Influence the Abundance of Mosquitoes*

Human population growth and migration, as well as subsequent environmental and climatic changes brought on by human activity, are factors linked to the spread of mosquito-borne diseases The most significant factor in the tropics appears to be widespread urbanization, which creates favorable environments for the anthropophilic and competent arbovirus vector, *Aedes aegypti*, including other *Aedes* mosquitoes (Rose et al., 2020; Agboli et al., 2021). According to a study carried out by Fagbohun et al. (2020), the occurrence and distribution of mosquito species in four selected Local Government Areas (Eti-Osa, Badagry, Kosofe, and Alimosho) in Lagos state, Nigeria, is determined by anthropogenic activities like urbanization, migration, and poor housing conditions resulting from population growth.

2.3.1 Urbanization

About 472 million people have migrated into cities in sub-Saharan Africa, which is said to have the fastest rate of urbanization. However, in the subsequent 25 years, it is foreseen that this number will be twice as high (Saghir & Santoro, 2018). According to Rose et al. (2020), the swift urbanization occurring in Africa could prompt mosquitoes to shift their host species to humans. This will result in a shift toward human biting. Zahouli et al. (2017) showed that urbanization is the major factor propelling the *Aedes* mosquito by producing various suitable breeding sites than semi-urban and rural zones. For instance, in Ivory Coast, poor waste management practices by the Ouagadougou metropolis are linked to a rise in the risk of the population contracting dengue fever (Fournet et al., 2016; Zahouli et al., 2017). The deficiency of waste management structures and other sanitation systems in West African cities that are becoming more urbanized encourages the growth of *Culex quinquefasciatus*, a common mosquito that is acclimated to contaminated environments (Agboli et al., 2021). Urban populations are predominantly susceptible to the

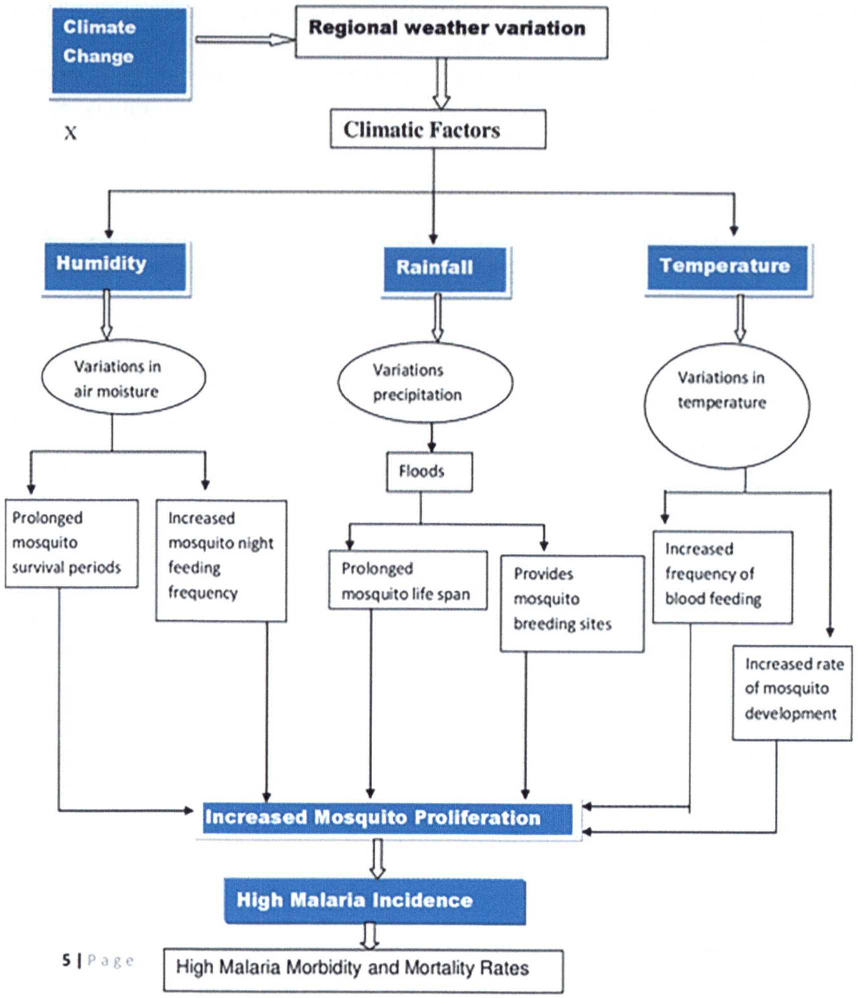

Fig. 14.4 Impact of temperature, humidity, and precipitation on mosquito population (Mzyece, 2016)

impact of climate change, especially those residing in informal settlements where flood damage is common, residents are often economically disadvantaged, and access to electricity and water is often insufficient (Agboli et al., 2021). These conditions can increase exposure to mosquitoes and mosquito-borne illnesses (Mordecai et al., 2020). For instance, the fast urbanization of Soweto resulted in an increase in the population residing in floodplain areas, which are perfect breeding areas for mosquito vectors (Agboli et al., 2021). In Lagos, Nigeria, slum areas devoid of amenities like effective drainage system, clean water, electricity, and proper waste disposal can encourage the development of several artificial breeding sites, particularly in the rainy season (Fagbohun et al., 2020).

2.3.2 Poor Housing Conditions

In addition to raising the risk of malaria and other mosquito-transmitted diseases in informal urban settlements, substandard housing may make it more challenging to apply common vector control strategies, such as installing Insecticide Treated Nets and Indoor Residual Spraying (Agboli et al., 2021). Urban settings may be better and cheaper for implementing interventions like larviciding. Moreover, urban settings are less likely to use interventions like Indoor Residual Spraying and Insecticide Treated Nets than rural ones (Agboli et al., 2021). A further risk to controlling malaria in urban African settings is the current invasion of *Anopheles stephensi* in East Africa (Sinka et al., 2020). *Anopheles stephensi* is remarkably acclimated to urban settings, in contrast to native African mosquito species (such as *A. gambiae* and *A. arabiensis*) (Mordecai et al., 2020). More adaptive vector control measures may be necessary in urban environments due to this vector's distinct ecology, behavior, and increased transmission efficiency. In Lagos state, Nigeria, overcrowding and lack of proper housing facilities, especially in rural areas can increase mosquito breeding sites, aiding occurrence of mosquitoes all year round (Fagbohun et al., 2020).

2.3.3 Human Population Growth and Migration

According to Braack et al. (2018), the danger posed by the spread of new and existing mosquito-transmitted diseases will probably grow due to the high level of international travel and human population growth (Braack et al., 2018). By 2050, the population of sub-Saharan Africa will have doubled by 2.7% annually, twice as rapidly as South Asia and Latin America combined (1.2% and 0.9% respectively). Air travel has eased global trade and human migration, aiding in disseminating the virus to non-immune regions and the movement of mosquitoes to new areas (Buchwald et al., 2020). It is acknowledged that *Aedes aegypti* was primarily brought to the Americas, and subsequently to Europe and Asia, by the slave trade in the sixteenth century (Powell et al., 2018). Particularly vulnerable to instability is the Sahel region, where population migration to cities may raise disease risk (Agboli et al., 2021).

2.3.4 Land Use

Anthropogenic changes, such as urbanization and agricultural land use, negatively impact the variety of mosquito species. Accra and Tema, two of Ghana's rapidly urbanizing cities, have seen a drop in mosquito species, from 75 to 28. Changes in vector ecology and behavior brought about by deforestation for agricultural, population settlement, or other reasons may have an effect on the incidence, distribution, and prevalence of vector-borne illnesses. Mosquito abundance and transmission in

Senegal are correlated with environmental factors like rainfall, vegetation, distance to forest, and size of landscape paths (Agboli et al., 2021).

3 Potential Hotspots for Increased Mosquito Invasiveness Due to Climate Change

The topography, climate, and patterns of human settlement in the southern parts of the north central regions all contribute to mosquito suitability. Consistent with earlier findings, the range, population, and adaptability of the members of the *A. gambiae* complex were significantly influenced by yearly rainfall, temperature fluctuations, climatic factors, and land use. Its prevalence is closely linked to the temperature, which makes the species more susceptible to climatic changes and contributes to its overall abundance. *A. gambiae* sensu stricto mostly spread in the rainy season whereas *A. arabiensis* is more widely distributed in the dry season. The limited presence of *A. arabiensis* in the cold swamps, particularly the Lake Chad basin region of Borno state may be related to its preference for warmer climates (Akpan et al., 2018). The comparatively cold highland weather, with intermediate temperatures of 21–25 °C, may cause mosquitoes' low suitability to the mid-Altitude zone. These temperatures are considerably below the 23–35 °C model optimum temperatures for the rapid proliferation of the *A. gambiae* species. High temperatures and precipitation in that region of the humid forest are also responsible for the high suitability of Lagos State for mosquitoes, according to the model. The complicated social structure, inadequate infrastructure and planning in the growing slum areas, fast population growth, and industrial activity all contribute to a warmer climate that makes it easier for malaria vectors to reproduce, survive, and bite more often, thus making Lagos state in South West a potential hotspot for the escalating spread of mosquitoes. Vector prevalence is highest in regions with constant elevated temperatures, a small daily temperature range and constant precipitation, as indicated by the gradient in mosquito distribution density observed from the coastal south to the arid north (Akpan et al., 2018).

4 Vector Control Strategies in Nigeria

4.1 Insecticide-Treated Nets (ITNs)

Insecticide-treated nets are the most popular instrument employed in most countries to control malaria transmission. Within 2004 and 2022, about 3 billion ITNs nets have been manufactured and distributed worldwide. In 2022, the national malaria programs in malaria-endemic countries distributed an aggregate of 254 million ITNs, with 93% being distributed in Africa alone. Since 2015, the use of ITNs has become widespread. In 2022, it was estimated that 56 % of children and expectant women slept under a treated net, according to the WHO 2023 report (WHO, 2023). Also in 2023, 25 million insecticide-treated nets were distributed in Nigeria (WHO, 2023).

4.2 Indoor Residual Spraying (IRS)

This involves spraying residential areas with insecticides with the aim of killing the mosquito vectors. About 47 nations adopted the IRS in 2022 in an effort to stop malaria. However, the percentage of the vulnerable people safeguarded by IRS in countries where malaria is endemic, such as Nigeria, fell from 6 % to about 1.8% (WHO, 2023).

4.3 Seasonal Malaria Chemoprevention (SMC)

This is a prophylactic measure for children at high risk of contracting malaria. Preventive doses of malaria medication are administered at regular intervals to prevent the occurrence of the disease. Seasonal malaria chemoprevention has been carried out in 17 sub-Saharan African nations to date. The number of seasonal malaria chemoprevention recipients (young children) has increased steadily from 0.2 million in 2012 to about 50 million in 2022. Nigeria alone catered to an average of 26 million children in each cycle in 2022. In Mozambique, the population of children treated with seasonal malaria chemoprevention rose by 12 times between 2021 and 2022. Also in 2022, seasonal malaria chemoprevention was implemented for the first time in Mauritania and South Sudan (WHO, 2023).

4.4 Intermittent Preventive Treatment of Malaria in Pregnancy (IPTp)

Expectant women who live in African regions with moderate-to-high malaria transmission risk are prevented from contracting the disease by using Intermittent Preventive Treatment of malaria in pregnancy. At the earliest time possible in the second trimester, the World Health Organization endorses taking a minimum of three doses of IPTp. These doses are usually given at antenatal care (ANC) visits and are spaced at least one month apart. About 35 African nations, including Nigeria, have implemented IPTp currently. Compared to 34% in 2021 and 1% in 2010, an estimated 42% of pregnant women received three doses of preventative medication in 2022. Additionally, data indicate that a greater number of expectant mothers are seeking prenatal care: the percentage of expectant mothers who visited an ANC facility at least once rose from 73% in 2021 to 78% in 2022 (WHO, 2023). Between 2018 and 2022, the Transforming Intermittent Preventive Treatment for Optimal Pregnancy Project (TIPTOP) set up by the World Health Organization, administered IPTp treatment to pregnant women in Ebonyi, Niger, and Ogun states.

4.5 Climate Change Effect on Mosquito Control Strategies

Through modifications in the effectiveness and quality of therapies, climate change may have a direct impact on mosquito and malaria control by rendering control methods ineffective.

4.5.1 Insecticide-Treated Nets and Indoor Residual Spraying

Indoor residual spraying and insecticide-treated nets require insecticides to target adult mosquitoes, especially adult females. Pyrethroids until recently were the sole insecticides authorized for ITN use. Carbamates, neonicotinoids, organochlorines, pyrethroids, and organophosphates have been authorized for use in indoor residual spraying (WHO, 2022a, b, c, d). Although ITNs and IRS are considered fundamental therapies, their effectiveness varies significantly throughout Africa. In Kenya, *A. funestus*, for instance, was shown to have decreased by 88% (Abong'o et al., 2020). In Mozambique, a smaller decrease by 48% was noticed (Wagman et al., 2021). This was due to the IRS drug Actellic (pirimiphos-methyl) being used in the trials. Variations in local vector ecology and pesticide resistance may account for some of these discrepancies, but regional environmental factors that affect product performance, durability, and efficacy may also play a role. Research in agricultural entomology shows that temperature (Arthur et al., 2019) and humidity levels (Ranabhat & Wang, 2020; Gerken et al., 2020) affect the immediate and long-term effectiveness of specific insecticide classes. The stability of the chemicals employed in IRS fluctuates with temperature, ultraviolet radiation, and other climatic conditions, as discussed by Sibanda et al. (2011). Therefore, it is safe to assume that environmental factors influence the effectiveness of vector control products (WHO, 2022a, b, c, d).

Temperature variations affect pesticide efficacy. According to the review of Glunt et al. (2013), ambient temperature frequently affects how effective insecticides are when used to control vectors (Glunt et al., 2013). This can result from how temperature affects mosquitoes' nervous system and metabolism (Glunt et al., 2013). It appears that different pesticide classes are affected differently by temperature (Glunt et al., 2013, 2018), a few possibly becoming more harmful in warmer climates and others decreasing or staying constant. Agyekum et al. (2021)

conducted a systematic review indicating a downward toxicity trend at higher temperatures. This trend may result in a decline in efficacy as a result of climate change.

5 Sustainable Mosquito Control Strategies

A variety of eco-friendly methods are used to curb the proliferation of mosquitoes and thus lessen the occurrence of mosquito-borne diseases (Okumu & Moore, 2011; Onen et al., 2023), which include the following:

5.1 Biological Control

Biological control comprises using naturally occurring species to control mosquito populations. A major advantage of this method is the fact that it is safe for the environment and is a good replacement for the harmful methods that are currently deployed. Several methods are employed for biological control including use of predatory species, use of bacterial agents, nematodes, and so on (Onen et al., 2023).

5.1.1 Use of Predatory Species

Some fish species such as *Gambusia affinis* and *Poecilia reticulata* (rainbow fish) are known to be larvivorous (larvae-loving). These fishes are introduced into water bodies suspected to be mosquito breeding sites to feed on the larvae which in turn controls the emergence of adult mosquitoes (Louca et al., 2009; Onen et al., 2023). The larvae (especially the fourth instar larva) of non-biting elephant mosquito (*Toxorhynchites* species) have demonstrated larvicidal activity against mosquito larvae of other species by feeding on them (Digma et al., 2019). The use of *Toxorhynchites* species as biological agents is making significant progress as they have been shown to be highly effective against various mosquito species, especially *Aedes aegypti*, *Aedes albopictus*, and *Culex quinquefasciatus* (Schiller et al., 2019).

5.1.2 Use of Bacterial Agents

Since the discovery of the mosquitocidal properties of *Bacillus thuringiensis israelensis* (Bti) and *Bacillus sphaericus* (Bs), it has brought widespread attention to the usage of bacteria as mosquito control agents (Valtierra-De-Luis et al., 2020). The use of Bti or Bs alone or together has shown to be very efficient against mosquito larvae; it is also non-toxic and safe for other non-target organisms that share the same habitat with mosquito larvae (Derua et al., 2018).

5.2 Genetic Control

Sterile insect technique (SIT) is a mosquito control strategy proven to be eco-friendly. It involves mass rearing of male mosquitoes, sterilizing them via radiation and releasing into the environment in a selected area to compete for mates alongside wide species. Mosquito populations start to decline as more female mosquitoes mate with the sterile males (Wilke & Marrelli, 2012; Okoro et al., 2018). Before SIT, a strategy known as RIDL (release of insects carrying a dominant lethal gene) was used. In RIDL, the expression of the lethal gene is controlled using a promoter specific to female mosquitoes while tetracycline (an antibiotic) is used to inactivate its expression, maintaining a stable population. Removal of the antibiotic in the environment leads to the death of all the female mosquitoes, leaving the males alive (Wilke & Marrelli, 2012).

5.3 Chemical Control

Chemical control includes the use of insect growth regulators (IGR) which mimic and compete with hormones resulting in underdevelopment of the insects (Pener & Dhadialla, 2012). IGRs have proven to be useful in lessening mosquito populations due to their highly selective nature and reduced environmental toxicity (Dahmana & Mediannikov, 2020). Methoprene and pyriproxyfen, two very effective hormone antagonists, are increasingly being used as IGRs. Other commonly used IGRs include diflubenzuron and novaluron (Pener & Dhadialla, 2012). Insect growth regulators are particularly actual against mosquito larvae, preventing the development and emergence of adult mosquitoes (Dennehy et al., 2010; Onen et al., 2023).

5.4 Public Sensitization

Public awareness is important for the success of mosquito control strategies. Public awareness programs should focus on educating communities on the importance of controlling mosquitoes and the role of various mosquito control strategies in achieving this aim (Achee et al., 2019).

5.5 Habitat Management

Environmental sanitation practices such as clearing drainages, gutters, clearing bushes, and getting rid of stagnant water bodies should be encouraged in communities. Other probable breeding habitats like old tires, empty plastic containers, and polythene bags should also be eradicated to reduce mosquito populations (Okwa, 2019).

6 Conclusion

Climate change profoundly impacts mosquito populations in Nigeria, as fluctuations in temperature, rainfall, and humidity directly affect their invasiveness. Increased rainfall and rising temperatures create the perfect breeding environment for mosquitoes, which in turn allows them to proliferate and spread diseases like the Zika virus, dengue fever, and malaria. Conversely, drought and lower humidity levels can impede the growth and development of mosquito populations. It is necessary to implement effective vector control strategies such as use of insect growth regulators and predatory species while considering the changing climate. Collaborative effort at international and local levels is crucial to reduce greenhouse gas emissions that aid climate change and protect the susceptible population from the mosquito-borne diseases, especially malaria. Therefore, it is imperative to put into practice efficient control strategies and stop the spread of mosquito-borne diseases.

Conflicting Interest The authors affirm no conflict of interest.

References

Abdulaziz, M. M., Ibrahim, A., Ado, M., Ameh, C., Umeokonkwo, C., Sufyan, M. B., Balogun, M. S., & Ahmed, S. A. (2020). Prevalence and factors associated with dengue fever among febrile patients attending secondary health facilities in Kano metropolis, Nigeria. *African Journal of Clinical and Experimental Microbiology, 21*(4), 340–348. https://doi.org/10.4314/ajcem.v21i4.11

Abong'o, B., Gimnig, J. E., Torr, S. J., Longman, B., Omoke, D., Muchoki, M., TerKuile, F., Ochomo, E., Munga, S., Samuels, A. M., Njagi, K., Maas, J., Perry, R. T., Fornadel, C., Donnelly, M. J., & Oxborough, R. M. (2020). Impact of indoor residual spraying with pirimiphos-methyl (Actellic 300CS) on entomological indicators of transmission and malaria case burden in Migori County, western Kenya. *Scientific Reports, 10*(1), 4518–4532. https://doi.org/10.1038/s41598-020-61350-2

Abudurexiti, A., Adkins, S., Alioto, D., Alkhovsky, S. V., Avšič-Županc, T., Ballinger, M. J., Bente, D. A., Beer, M., Bergeron, É., Blair, C. D., Briese, T., Buchmeier, M. J., Burt, F. J., Calisher, C. H., Cháng, C., Charrel, R. N., Choi, I. R., Clegg, J. C. S., de la Torre, J. C., de Lamballerie, X., & Kuhn, J. H. (2019). Taxonomy of the order Bunyavirales: Update 2019. *Archives of Virology, 164*(7), 1949–1965. https://doi.org/10.1007/s00705-019-04253-6

Abushoufa, F., Arikan, A., Sanlidag, T., Guvenir, M., Guler, E., & Suer, K. (2021). Absence of Zika virus seroprevalence among blood donors in Northern Cyprus. *The Journal of Infection in Developing Countries, 15*(7), 1032–1034. https://doi.org/10.3855/jidc.12766

Achee, N. L., Grieco, J. P., Vatandoost, H., Seixas, G., Pinto, J., Ching-NG, L., Martins, A. J., Juntarajumnong, W., Corbel, V., Gouagna, C., David, J.-P., Logan, J. G., Orsborne, J., Marois, E., Devine, G. J., & Vontas, J. (2019). Correction: Alternative strategies for mosquito-borne arbovirus control. *PLoS Neglected Tropical Diseases, 13*(3), e0007275. https://doi.org/10.1371/journal.pntd.0007275

Agarwal, A., Parida, M., & Dash, P. K. (2017). Impact of transmission cycles and vector competence on global expansion and emergence of arboviruses. *Reviews in Medical Virology, 27*(5), 1–12. https://doi.org/10.1002/rmv.1941

Agboli, E., Zahouli, J. B. Z., Badolo, A., & Jöst, H. (2021). Mosquito-associated viruses and their related mosquitoes in West Africa. *Viruses, 13*(5), 891–919. https://doi.org/10.3390/v13050891

Agyekum, T. P., Botwe, P. K., Arko-Mensah, J., Issah, I., Acquah, A. A., Hogarh, J. N., Dwomoh, D., Robins, T. G., & Fobil, J. N. (2021). A systematic review of the effects of temperature on *Anopheles* mosquito development and survival: Implications for malaria control in a future warmer climate. *International Journal of Environmental Research and Public Health, 18*(14), 7255–7277. https://doi.org/10.3390/ijerph18147255

Ahlers, L. R. H., & Goodman, A. G. (2018). The immune responses of the animal hosts of west nile virus: A comparison of insects, birds, and mammals. *Frontiers in Cellular and Infection Microbiology, 3*(8), 96–105. https://doi.org/10.3389/fcimb.2018.00096

Ahmed, A., Ali, Y., Elduma, A., Eldigail, M. H., Mhmoud, R. A., Mohamed, N. S., Ksiazek, T. G., Dietrich, I., & Weaver, S. C. (2020). Unique outbreak of Rift Valley fever in Sudan, 2019. *Emerging Infectious Diseases, 26*(12), 3030–3033. https://doi.org/10.3201/eid2612.201599

Ahn, J., Sinka, M., Irish, S., & Zohdy, S. (2023). Modeling marine cargo traffic to identify countries in Africa with greatest risk of invasion by *Anopheles stephensi*. *Scientific Reports, 13*(1), 876–888. https://doi.org/10.1038/s41598-023-27439-0

Akpan, G. E., Adepoju, K. A., Oladosu, O. R., & Adelabu, S. A. (2018). Dominant malaria vector species in Nigeria: Modelling potential distribution of *Anopheles gambiae* sensu lato and its siblings with MaxEnt A. Taylan Ozkan (ed.). *PLoS ONE, 13*(10), 1–15. https://doi.org/10.1371/journal.pone.0204233

Amanyi-Enegela, J. A., Kumbur, J., Burn, N., Sankar, G., Davies, E., Ishaya, R., Ogoshi, C., Ekweremadu, B., Omoi, S., & Qureshi, B. (2022). Assessment of the progress toward elimination of lymphatic filariasis in the Federal Capital Territory-Abuja, Nigeria. *Frontiers in Tropical Diseases, 3*(10), 1–7. https://doi.org/10.3389/fitd.2022.1033802

Amao, H., Idowu, E. T., Oyeniyim, T., Otubanjo, O. A., & Awolola, T. S. (2018). Relative abundance, distribution and diversity of Culex (Diptera Culicidae) mosquito species in Lagos, Southwest Nigeria. *Nigeria Journal of Entomology, 34*(1), 39–49. https://doi.org/10.36108/NJE/8102/43(0150)

Ani, K. J., Anyika, V. O., & Mutambara, E. (2021). The impact of climate change on food and human security in Nigeria. *International Journal of Climate Change Strategies and Management, 14*(2), 148–167. https://doi.org/10.1108/ijccsm-11-2020-0119

Arthur, F. H., Starkus, L. A., & McKay, T. (2019). Degradation and residual efficacy of beta-cyfluthrin as a surface treatment for control of *Tribolium castaneum* Herbst: Effects of temperature and environment. *Journal of Stored Products Research, 84*(5), 101–514. https://doi.org/10.1016/j.jspr.2019.101514

Asgarian, T. S., Moosa-Kazemi, S. H., & Sedaghat, M. M. (2021). Impact of meteorological parameters on mosquito population abundance and distribution in a former malaria endemic area, central Iran. *Heliyon, 7*(12), 1–7. https://doi.org/10.1016/j.heliyon.2021.e08477

Becker, N., Petric, D., Zgomba, M., Boase, C., Madon, M. B., Dahl, C., & Kaiser, A. (2020). *Mosquitoes: Identification, ecology and control*. Springer International Publishing.

Braack, L., Gouveia de Almeida, A. P., Cornel, A. J., Swanepoel, R., & De-Jager, C. (2018). Mosquito-borne arboviruses of African origin: Review of key viruses and vectors. *Parasites & Vectors, 11*(29), 2–26. https://doi.org/10.1186/s13071-017-2559-9

Brent, C., Dunn, A., Savage, H., Faraji, A., Rubin, M., Risk, I., Garcia, W., Cortese, M., Novosad, S., Krow-Lucal, E. R., Crain, J., Hill, M., Atkinson, A., Peterson, D., Christensen, K., Dimond, M., Staples, J. E., & Nakashima, A. (2016). Preliminary findings from an investigation of Zika virus infection in a patient with no known risk factors—Utah, 2016. *Morbidity and Mortality Weekly Report, 65*(36), 981–982. https://doi.org/10.15585/mmwr.mm6536e4

Buchwald, A. G., Hayden, M. H., Dadzie, S. K., Paull, S. H., & Carlton, E. J. (2020). *Aedes*-borne disease outbreaks in West Africa: A call for enhanced surveillance. *Acta Tropica, 209*(2), 105–468. https://doi.org/10.1016/j.actatropica.2020.105468

Carter, T. E., Yared, S., Gebresilassie, A., Bonnell, V., Damodaran, L., Lopez, K., Ibrahim, M., Mohammed, S., & Janies, D. (2018). First detection of *Anopheles stephensi* Liston, 1901 (Diptera: Culicidae) in Ethiopia using molecular and morphological approaches. *Acta Tropica, 188*, 180–186. https://doi.org/10.1016/j.actatropica.2018.09.001

Carter, T., Yared, S., Getachew, D., Spear, J., Sh, C., Samake, J., Mumba, P., Dengela, D., Yohannes, G., Chibsa, S., Murphy, M., Dissanayake, G., Flately, C., Lopez, K., Janies, D., Zohdy, S. I., & Balkew, M. (2021). Tracking of *Anopheles stephensi* in Ethiopia using mitochondrial DNA reveals pattern of spread. *BioRxiv (Cold Spring Harbor Laboratory), 14*(602), 2021–2104. https://doi.org/10.1101/2021.04.07.437873

Cator, L. J., Johnson, L. R., Mordecai, E. A., Moustaid, F. E., Smallwood, T. R. C., LaDeau, S. L., Johansson, M. A., Hudson, P. J., Boots, M., Thomas, M. B., Power, A. G., & Pawar, S. (2020). The role of vector trait variation in vector-borne disease dynamics. *Frontiers in Ecology and Evolution, 8*(189), 1–16. https://doi.org/10.3389/fevo.2020.00189

Center for Disease Control and Prevention. (2004). *Malaria report*. Retrieved from: http://www.cdc.gov/malaria

Center for Disease Control and Prevention. (2021). *Yellow fever in Nigeria*. Retrieved from: https://wwwnc.cdc.gov/travel/notices/level2/yellow-fever-nigeria

Chandra, G., & Mukherjee, D. (2022). Effect of climate change on mosquito population and changing pattern of some diseases transmitted by them. *Advances in Animal Experimentation and Modeling, 42*(4), 1031–1049. https://doi.org/10.1016/B978-0-323-90583-1.00030-1

Chemison, A., Ramstein, G., Tompkins, A. M., Defrance, D., Camus, G., Charra, M., & Caminade, C. (2021). Impact of an accelerated melting of Greenland on malaria distribution over Africa. *Nature Communications, 12*(1), 1–12. https://doi.org/10.1038/s41467-021-24134-4

Couper, L. I., Farner, J. E., Caldwell, J. M., Childs, M. L., Harris, M. J., Kirk, D. G., Nova, N., Shocket, M., Skinner, E. B., Uricchio, L. H., Exposito-Alonso, M., & Mordecai, E. A. (2021). How will mosquitoes adapt to climate warming? *eLife, 10*(10), 1–36. https://doi.org/10.7554/eLife.69630

Dahmana, H., & Mediannikov, O. (2020). Mosquito-borne diseases emergence/resurgence and how to effectively control it biologically. *Pathogens, 9*(4), 310. https://doi.org/10.3390/pathogens9040310

Dennehy, T. J., Degain, B. A., Harpold, V. S., Zaborac, M., Morin, S., Fabrick, J. A., Nichols, R. L., Brown, J. K., Byrne, F. J., & Li, X. (2010). Extraordinary resistance to insecticides reveals exotic Q biotype of Bemisia tabaci in the new world. *Journal of Economic Entomology, 103*(6), 2174–2186. https://doi.org/10.1603/ec10239

Derua, Y. A., Kahindi, S. C., Mosha, F. W., Kweka, E. J., Atieli, H. E., Wang, X., Zhou, G., Lee, M. C., Githeko, A. K., & Yan, G. (2018). Microbial larvicides for mosquito control: Impact of long-lasting formulations of *Bacillus thuringiensis* var. *israelensis* and *Bacillus sphaericus* on non-target organisms in western Kenya highlands. *Ecology and Evolution, 8*(15), 7563–7573. https://doi.org/10.1002/ece3.4250

Digma, J. R., Sumalde, A. C., & Salibay, C. C. (2019). Laboratory evaluation of predation of *Toxorhynchites amboinensis* (Diptera: Culicidae) on three mosquito vectors of arboviruses in the Philippines. *Biological Control, 137*(1049–9644), 104009. https://doi.org/10.1016/j.biocontrol.2019.104009

Drakou, K., Nikolaou, T., Vasquez, M., Petric, D., Michaelakis, A., Kapranas, A., Papatheodoulou, A., & Koliou, M. (2020). The effect of weather variables on mosquito activity: A snapshot of the main point of entry of Cyprus. *International Journal of Environmental Research and Public Health, 17*(4), 1–10. https://doi.org/10.3390/ijerph17041403

El-Sayed, A., & Kamel, M. (2020). Climatic changes and their role in emergence and re-emergence of diseases. *Environmental Science and Pollution Research, 27*(18), 22336–22352. https://doi.org/10.1007/s11356-020-08896-w

Fagbami, A. H., & Onoja, A. B. (2018). Dengue haemorrhagic fever: An emerging disease in Nigeria, West Africa. *Journal of Infection and Public Health, 11*(6), 757–762. https://doi.org/10.1016/j.jiph.2018.04.014

Fagbohun, K. I., Idowu, T. E., Awolola, S. T., & Otubanjo, A. O. (2020). Seasonal abundance and larval habitats characterization of mosquito species in Lagos State, Nigeria. *Scientific African, 10*(2468), 1–9. https://doi.org/10.1016/j.sciaf.2020.e00656

Ferguson, H. M., & Govella, J. N. (2024). *Impacts of climate change on malaria vector control in Africa* (pp. 387–421). Planetary health approaches to understand and control vector-borne diseases. Brill. https://doi.org/10.3920/9789004688650_015

Fouque, F., & Reeder, J. C. (2019). Impact of past and on-going changes on climate and weather on vector-borne diseases transmission: A look at the evidence. *Infectious Diseases of Poverty, 8*(51), 1–9. https://doi.org/10.1186/s40249-019-0565-1

Fournet, F., Rican, S., Vaillant, Z., Roudot, A., Meunier-Nikiema, A., Kassié, D., Dabiré, R. K., & Salem, G. (2016). The influence of urbanization modes on the spatial circulation of flaviviruses within Ouagadougou (Burkina Faso). *International Journal of Environmental Research and Public Health, 13*(12), 1226–1226. https://doi.org/10.3390/ijerph13121226

Franklinos, L. H. V., Jones, K. E., Redding, D. W., & Abubakar, I. (2019). The effect of global change on mosquito-borne disease. *The Lancet Infectious Diseases, 19*(9), 302–312. https://doi.org/10.1016/s1473-3099(19)30161-6

Gatherer, D., & Kohl, A. (2016). Zika virus: A previously slow pandemic spreads rapidly through the Americas. *Journal of General Virology, 97*(2), 269–273. https://doi.org/10.1099/jgv.0.000381

Gerken, A. R., Scully, E. D., Campbell, J. F., & Morrison, W. R. (2020). Effectiveness of long-lasting insecticide netting on *Tribolium castaneum* is modulated by multiple exposures, biotic, and abiotic factors. *Pest Management Science, 77*(3), 1235–1244. https://doi.org/10.1002/ps.6134

Giesen, C., Roche, J., Redondo-Bravo, L., Ruiz-Huerta, C., Gomez-Barroso, D., Benito, A., & Herrador, Z. (2020). The impact of climate change on mosquito borne diseases in Africa. *Pathogens and Global Health, 114*(6), 287–301. https://doi.org/10.1080/20477724.2020.1783865

Glunt, K. D., Blanford, J. I., & Paaijmans, K. P. (2013). Chemicals, climate, and control: Increasing the effectiveness of malaria vector control tools by considering relevant temperatures. *PLoS Pathogens, 9*(10), 1–5. https://doi.org/10.1371/journal.ppat.1003602

Glunt, K. D., Oliver, S. V., Hunt, R. H., & Paaijmans, K. P. (2018). The impact of temperature on insecticide toxicity against the malaria vectors *Anopheles arabiensis* and *Anopheles funestus. Malaria Journal, 17*(131), 1–8. https://doi.org/10.1186/s12936-018-2250-4

Intergovernmental Panel on Climate Change. (2001). *TAR climate change 2001: The scientific basis—IPCC*. 2019. Retrieved from https://www.ipcc.ch/report/ar3/wg1/

Intergovernmental Panel on Climate Change. (2014). *AR5 climate change 2014: Impacts, adaptation, and vulnerability—IPCC*. Retrieved from https://www.ipcc.ch/report/ar5/wg2/

Irikannu, K. C., Onyido, A. E., Ogaraku, J. C., Umeanaeto, P. U., Nzeukwu, C. I., Obiefule, I. E., Aniefuna, C. O., Elosiuba, N. V., & Onwuachusi, G. L. (2023). Seasonal distribution and diversity of mosquito species in a Rainforest Community of Southeast Nigeria. *The Bioscientist Journal, 11*(1), 71–80. Retrieved from https://bioscientistjournal.com/index.php/The_Bioscientist/article/view/140

Ismail, Y., Faramarz, B., Harold, J., Sellase, P., & Kwadwo, K. F. (2024). Predicting the environmental suitability for *Anopheles stephensi* under the current conditions in Ghana. *Scientific Reports, 14*(1), 11–16. https://doi.org/10.1038/s41598-024-51780-7

Kamgang, B., Vazeille, M., Tedjou, A. N., Wilson-Bahun, T. A., Yougang, A. P., Mousson, L., Wondji, C. S., & Failloux, A. B. (2019). Risk of dengue in Central Africa: Vector competence studies with *Aedes aegypti* and *Aedes albopictus* (Diptera: *Culicidae*) populations and dengue 2 virus. *PLoS Neglected Tropical Diseases, 13*(12), 1–12. https://doi.org/10.1371/journal.pntd.0007985

Khezzani, B., Baymakova, M., Khechekhouche, E. A., & Tsachev, I. (2023). Global warming and mosquito-borne diseases in Africa: A narrative review. *Pan African Medical Journal, 44*(70), 1–15. https://doi.org/10.11604/pamj.2023.44.70.37318

Kreppel, K., Caminade, C., Govella, N., Morse, A. P., Ferguson, H. M., & Baylis, M. (2019). Impact of ENSO 2016–17 on regional climate and malaria vector dynamics in Tanzania. *Environmental Research Letters, 14*(7), 1–12. https://doi.org/10.1088/1748-9326/ab26c7

Lee, H., Halverson, S., & Ezinwa, N. (2018). Mosquito-borne diseases. *Primary Care, 45*(3), 393–407. https://doi.org/10.1016/j.pop.2018.05.001

Louca, V., Lucas, M. C., Green, C., Majambere, S., Fillinger, U., & Lindsay, S. W. (2009). Role of fish as predators of mosquito larvae on the floodplain of the Gambia River. *Journal of Medical Entomology, 46*(3), 546–556. https://doi.org/10.1603/033.046.0320

Lourens, G. B., & Ferrell, D. K. (2019). Lymphatic filariasis. *Nursing Clinics of North America, 54*(2), 181–192. https://doi.org/10.1016/j.cnur.2019.02.007

Ma'aji, J. A., Olonitola, O. S., & Ella, E. E. (2020). Seroprevalence of West Nile virus infection among febrile patients attending selected hospitals in Kaduna state, Nigeria. *Scientific African, 10*, 2468–2276. https://doi.org/10.1016/j.sciaf.2020.e00588

Mac, A. P., Airiohuodion, P. E., Yako, A. B., Makpo, J. K., & Kreoger, A. (2022). The seroprevalence and hidden burden of Chikungunya endemicity and malaria mono and co-infection in Nigeria. *International Journal of Environmental Research and Public Health, 19*(15), 1–10. https://doi.org/10.3390/ijerph19158896

Mac, A. P., Airiohuodion, P. E., Velayudhan, R., Zubair, S., Tadele, M., Aighobahi, O. J., Anyaike, C., Kroger, A., & Planning, M. (2023). Antibody seropositivity and endemicity of Chikungunya and Zika viruses in Nigeria. *Animal Diseases, 3*(1), 1–11. https://doi.org/10.1186/s44149-023-00070-2

Masson-Delmotte, V., Zhai, P., Pirani, A., Connors, A. L., Pean, C., & Berger, S. (2021). Climate change 2021: The physical science basis. In *Contribution of working group I to the sixth assessment report of the IPCC*. Cambridge University Press.

Mboussou, F., Ndumbi, P., Ngom, R., Kamassali, Z., Ogundiran, O., Van Beek, J., Williams, G., Okot, C., Hamblion, E. L., & Impouma, B. (2019). Infectious disease outbreaks in the African region: Overview of events reported to the World Health Organization in 2018. *Epidemiology and Infection, 14*(299), 1–8. https://doi.org/10.1017/S0950268819001912

McDonald, E., Mathis, S., Martin, S. W., Erin, S. J., Fischer, M., & Lindsey, N. P. (2021). Surveillance for West Nile virus disease—United States, 2009–2018. *American Journal of Transplantation, 21*(5), 1959–1974. https://doi.org/10.1111/ajt.16595

Mencattelli, G., Ndione, M. H. D., Rosà, R., Marini, G., Diagne, C. T., Diagne, M. M., Fall, G., Faye, O., Diallo, M., Faye, O., Savini, G., & Rizzoli, A. (2022). Epidemiology of West Nile virus in Africa: An underestimated threat H.P. Fuehrer (ed.). *PLoS Neglected Tropical Diseases, 16*(1), 1–31. https://doi.org/10.1371/journal.pntd.0010075

Messina, J. P., Brady, O. J., Golding, N., Kraemer, M. U. G., Wint, G. R. W., Ray, S. E., Pigott, D. M., Shearer, F. M., Johnson, K., Earl, L., Marczak, L. B., Shirude, S., Davis Weaver, N., Gilbert, M., Velayudhan, R., Jones, P., Jaenisch, T., Scott, T. W., Reiner, R. C., & Hay, S. I. (2019). The current and future global distribution and population at risk of dengue. *Nature Microbiology, 4*(9), 508–1515. https://doi.org/10.1038/s41564-019-0476-8

Mordecai, E. A. (2019). Thermal biology of mosquito borne disease. *Ecology Letters, 22*(10), 1690–1708. https://doi.org/10.1111/ele.13335

Mordecai, E. A., Ryan, S. J., Caldwell, J. M., Shah, M. M., & LaBeaud, A. D. (2020). Climate change could shift disease burden from malaria to arboviruses in Africa. *The Lancet Planetary Health, 4*(9), 416–423. https://doi.org/10.1016/S2542-5196(20)30178-9

Mzyece, H. (2016). *Trends of climatic factors and malaria incidence in the low, moderate and high transmission zones of Zambia*. Retrieved from https://library.adhl.africa/handle/123456789/13048.

Ndiaye, E. H., Fall, G., Gaye, A., Bob, N. S., Talla, C., Diagne, C. T., Diallo, D., BA, Y., Dia, I., Kohl, A., Sall, A. A., & Diallo, M. (2016). Vector competence of *Aedes vexans* (Meigen), *Culex poicilipes* (Theobald) and Cx. *quinquefasciatus* Say from Senegal for West and East African lineages of Rift Valley fever virus. *Parasites & Vectors, 9*(94), 1–9. https://doi.org/10.1186/s13071-016-1383-y

Ngouanet, S. A., Wanji, S., Yadouleton, A., Demanou, M., Djouaka, R., & Nanfack-Minkeu, F. (2022). Factors enhancing the transmission of mosquito-borne Arboviruses in Africa. *Virus Disease, 33*(4), 477–488. https://doi.org/10.1007/s13337-022-00795-7

Nigeria Centre for Disease Control and Prevention. (2023a). *Dengue fever outbreak alert.* Retrieved from https://ncdc.gov.ng/news/506/ncdc-on-alert-following-the-confirmation-of-the-dengue-fever-outbreak-in-sokoto-state.#:~:text=The%20Nigeria%20Centre%20for%20Disease,been%20reported%20in%20the%20state.

Nigeria Centre for Disease Control and Prevention. (2023b). *Yellow fever situation report.* Retrieved from https://ncdc.gov.ng/themes/common/files/sitreps/2f8cf3ce751c92af774d62ef021c77f6.pdf.

Nsereko, G., Kadobera, D., Okethwangu, D., Nguna, J., Rutazaana, D., & Kyabayinze, D. J. (2020). Malaria outbreak facilitated by appearance of vector breeding sites after heavy rainfall and inadequate preventive measures: Nwoya District, Northern Uganda, February–May 2018. *Journal of Environmental and Public Health, 2020*(5802401), 1–10. https://doi.org/10.1155/2020/5802401

Okon, E. M., Falana, B. M., Solaja, S. O., Yakubu, S. O., Alabi, O. O., Okikiola, B. T., Awe, T. E., Adesina, B. T., Tokula, A. K., Kipchumba, A. K., & Edeme, A. B. (2021). Systematic review of climate change impact research in Nigeria: Implication for sustainable development. *Heliyon, 7*(9), 1–21. https://doi.org/10.1016/j.heliyon.2021.e07941

Okoro, O. J., Nnamonu, E. I., Ezewudo, B. I., & Okoye, I. C. (2018). Application of genetically modified mosquitoes (Anopheles Species) in the control of malaria transmission. *Asian Journal of Biotechnology and Genetic Engineering, 1*(1), 28–43. https://doi.org/10.9734/AJBGE/2018/40646

Okumu, F. O., & Moore, S. J. (2011). Combining indoor residual spraying and insecticide-treated nets for malaria control in Africa: A review of possible outcomes and an outline of suggestions for the future. *Malaria Journal, 10*(1), 1–13. https://doi.org/10.1186/1475-2875-10-208

Okwa, O. O. (2019). *Nipping the malaria vectors in the bud: Focus on Nigeria* (pp. 1–11). Malaria. IntechOpen. https://doi.org/10.5772/intechopen.83554

Omatola, C. A., Onoja, A. B., Moses, E., Mahmud, M., & Mofolorunsho, C. K. (2021). Dengue in parts of the Guinea Savannah region of Nigeria and the risk of increased transmission. *International Health, 13*(3), 248–252. https://doi.org/10.1093/inthealth/ihaa033

Onen, H., Luzala, M. M., Kigozi, S., Sikumbili, R. M., Muanga, C. J. K., Zola, E. N., Wendji, S. N., Buya, A. B., Balciunaitiene, A., Viškelis, J., Kaddumukasa, M. A., & Memvanga, P. B. (2023). Mosquito-Borne diseases and their control strategies: An overview focused on Green synthesized plant-based metallic nanoparticles. *Insects, 14*(3), 221–257. https://doi.org/10.3390/insects14030221

Oragwa, A. O., Oragwa, F. C., & Oluwayelu, D. O. (2022). Serologic evidence of silent Rift Valley virus infection among occupationally exposed persons in Northern Nigeria. *Journal of Infection in Developing Countries, 16*(5), 881–887. https://doi.org/10.3855/jidc.15367

Pener, M. P., & Dhadialla, T. S. (2012). An overview of insect growth disruptors; applied aspects. *Advances in Insect Physiology, 43*(0065-2806), 1–162. https://doi.org/10.1016/b978-0-12-391500-9.00001-2

Piovezan-Borges, A. C., Valente-Neto, F., Urbieta, G. L., Laurence, S. G. W., & de-Oliveira, R.F. (2022). Global trends in research on the effects of climate change on *Aedes aegypti*: International collaboration has increased, but some critical countries lag behind. *Parasites & Vectors, 15*(346), 1–12. https://doi.org/10.1186/s13071-022-05473-7

Plourde, A. R., & Bloch, E. M. (2016). A literature review of Zika virus. *Emerging Infectious Diseases, 22*(7), 1185–1192. https://doi.org/10.3201/eid2207.151990

Powell, J. R. (2016). Mosquitoes on the move. *Science, 354*(6315), 971–972. https://doi.org/10.1126/science.aal1717

Powell, J. R., Gloria-Soria, A., & Kotsakiozi, P. (2018). Recent history of *Aedes aegypti*: Vector genomics and epidemiology records. *Bioscience, 68*(11), 854–860. https://doi.org/10.1093/biosci/biy119

Ranabhat, S., & Wang, C. (2020). Effect of moisture on efficacy of selected insecticide dusts against the common bed bug, *Cimex lectularius* (Hemiptera: *Cimicidae*). *Journal of Economic Entomology, 113*(4), 1933–1939. https://doi.org/10.1093/jee/toaa122

Rose, N. H., Sylla, M., Badolo, A., Lutomiah, J., & Ayala, D. (2020). Climate and urbanization drive mosquito preference for humans. *Current Biology, 30*(18), 3570–3579. https://doi.org/10.1016/j.cub.2020.06.092

Roy, S. K., & Bhattacharjee, S. (2021). Dengue virus: Epidemiology, biology, and disease aetiology. *Canadian Journal of Microbiology, 67*(10), 687–702. https://doi.org/10.1139/cjm-2020-0572

Ryan, S. J., Lippi, C. A., & Zermoglio, F. (2020). Shifting transmission risk for malaria in Africa with climate change: A framework for planning and intervention. *Malaria Journal, 19*(170), 1–10. https://doi.org/10.1186/s12936-020-03224-6

Saghir, J., & Santoro, J. (2018). Urbanization in Sub-Saharan Africa. In *Meeting challenges by bridging stakeholders*. Center for Strategic & International Studies.

Sanfelice, V. (2022). Mosquito borne disease and newborn health. *Health Economics, 31*(1), 73–93. https://doi.org/10.1002/hec.4436

Sato, S. (2021). Plasmodium—A brief introduction to the parasites causing human malaria and their basic biology. *Journal of Physiological Anthropology, 40*(1), 1–3. https://doi.org/10.1186/s40101-020-00251-9

Schiller, A., Allen, M., Coffey, J., Fike, A., & Carballo, F. (2019). Updated methods for the production of *Toxorhynchites rutilus septentrionalis* (Diptera, Culicidae) for use as biocontrol agent against container breeding pest mosquitoes in Harris County, Texas. *Journal of Insect Science, 19*(2), 8. https://doi.org/10.1093/jisesa/iez011

Schuffenecker, I., Iteman, I., Michault, A., Murri, S., Frangeul, L., Vaney, M. C., Lavenir, R., Pardigon, N., Reynes, J. M., Pettinelli, F., Biscornet, L., Diancourt, L., Michel, S., Duquerroy, S., Guigon, G., Frenkiel, M. P., Bréhin, A. C., Cubito, N., Després, P., Kunst, F., & Brisse, S. (2006). Genome microevolution of Chikungunya viruses causing the Indian Ocean outbreak. *PLoS Medicine, 3*(7), 1–13. https://doi.org/10.1371/journal.pmed.0030263

Shragai, T., Tesla, B., Murdock, C., & Harrington, L. C. (2017). Zika and chikungunya: Mosquito-borne viruses in a changing world. *Annals of the New York Academy of Sciences, 1399*(1), 61–77. https://doi.org/10.1111/nyas.13306

Sibanda, M. M., Focke, W. W., Labuschagne, F. J., Moyo, L., Nhlapo, N. S., Maity, A., Muiambo, H., Massinga, P., Crowther, N. A., Coetzee, M., & Brindley, G. W. (2011). Degradation of insecticides used for indoor spraying in malaria control and possible solutions. *Malaria Journal, 10*(307), 1–12. https://doi.org/10.1186/1475-2875-10-307

Sinka, M. E., Pironon, S., Massey, N. C., Longbottom, J., Hemingway, J., Moyes, C. L., & Willis, K. J. (2020). A new malaria vector in Africa: Predicting the expansion range of Anopheles stephensi and identifying the urban populations at risk. *Proceedings of the National Academy of Sciences of the United States of America, 117*(40), 24900–24908. https://doi.org/10.1073/pnas.2003976117

Sow, A., Loucoubar, C., Diallo, D., Faye, O., Ndiaye, Y., Senghor, C. S., Dia, A. T., Faye, O., Weaver, S. C., Diallo, M., Malvy, D., & Sall, A. A. (2016). Concurrent malaria and arbovirus infections in Kedougou, Southeastern Senegal. *Malaria Journal, 15*(47), 1–7. https://doi.org/10.1186/s12936-016-1100-5

Ssempiira, J., Kissa, J., Nambuusi, B., Mukooyo, E., Opigo, J., Makumbi, F., Kasasa, S., & Vounatsou, P. (2018). Interactions between climatic changes and intervention effects on malaria spatio-temporal dynamics in Uganda. *Parasite Epidemiology and Control, 3*(3), 1–16. https://doi.org/10.1016/j.parepi.2018.e00070

Stopard, I. J., Churcher, T. S., & Lambert, B. (2021). Estimating the extrinsic incubation period of malaria using a mechanistic model of sporogony. *PLoS Computational Biology, 17*(2), 1–21. https://doi.org/10.1371/journal.pcbi.1008658

Tahir, M. A., Khan, M. A., Ikram, A., Chaudhry, T. H., Jabeen, A., Quddous, A., & Haq, I. U. (2023). Assessment of Infection Prevention and Control (IPC) implementation and strategies used for IPC preparedness at facility level in underdeveloped areas of Pakistan. *Infection and Drug Resistance, 16*(16), 1997–2006. https://doi.org/10.2147/idr.s399830

Thomson, M. C., Ukawuba, I., Hershey, C. L., Bennett, A., Ceccato, P., Lyon, B., & Dinku, T. (2017). Using rainfall and temperature data in the evaluation of national malaria control programs in Africa. *The American Journal of Tropical Medicine and Hygiene, 97*(3), 32–45. https://doi.org/10.4269/ajtmh.16-0696

Turner, H. C., Bettis, A. A., Chu, B. K., McFarland, D. A., Hooper, P. J., Mante, S. D., Fitzpatrick, C., & Bradley, M. H. (2017). Investment success in public health: An analysis of the cost-effectiveness and cost-benefit of the Global Programme to Eliminate Lymphatic Filariasis. *Clinical Infectious Diseases, 64*(6), 728–735. https://doi.org/10.1093/cid/ciw835

Ughaelu, C. M. (2017). Contemporary environmental issues respect to food production in Nigeria. *Journal of Environmental Management, 41*(2), 108–117. https://doi.org/10.1108/ijccsm-11-2020-0119

UNO (United Nations Organization). (2019). *Climate change. United nations sustainable development.* Retrieved from https://www.un.org/sustainabledevelopment/climate-change/

Valtierra-de-Luis, D., Villanueva, M., Berry, C., & Caballero, P. (2020). Potential for *Bacillus thuringiensis* and other bacterial toxins as biological control agents to combat dipteran pests of medical and agronomic importance. *Toxins, 12*(12), 773. https://doi.org/10.3390/toxins12120773

Wagman, J. M., Varela, K., Zulliger, R., Saifodine, A., Muthoni, R., Magesa, S., Chaccour, C., Gogue, C., Tynuv, K., Seyoum, A., Dengela, D., Saúte, F., Richardson, J. H., Fornadel, C., Linton, Y. M., Slutsker, L., Candrinho, B., & Robertson, M. (2021). Reduced exposure to malaria vectors following indoor residual spraying of pirimiphos-methyl in a high-burden district of rural Mozambique with high ownership of long-lasting insecticidal nets: Entomological surveillance results from a cluster-randomized trial. *Malaria Journal, 20*(1), 54–71. https://doi.org/10.1186/s12936-021-03583-8

Waite, J. L., Suh, E., Lynch, P. A., & Thomas, M. B. (2019). Exploring the lower thermal limits for development of the human malaria parasite, *Plasmodium falciparum. Biology Letters, 15*(6), 1–5. https://doi.org/10.1098/rsbl.2019.0275

Weather and Climate. (n.d.). *Nigeria climate zone, weather by month and historical data.* weatherandclimate.com. Retrieved from: https://weatherandclimate.com/nigeria#google_vignette

Whittaker, C., Hamlet, A., Sherrard-Smith, E., Winskill, P., Cuomo-Dannenburg, G., Walker, P. G. T., Sinka, M., Pironon, S., Kumar, A., Ghani, A., Bhatt, S., & Churcher, T. S. (2023). Seasonal dynamics of Anopheles stephensi and its implications for mosquito detection and emergent malaria control in the Horn of Africa. *National Academy of Sciences of the United States of America, 120*(8), 1–9. https://doi.org/10.1073/pnas.2216142120

Wilke, A. B., & Marrelli, M. T. (2012). Genetic control of mosquitoes: Population suppression strategies. *Revista do Instituto de Medicina Tropical de São Paulo, 54*(5), 287–292. https://doi.org/10.1590/s0036-46652012000500009

World Health Organization. (2020). *World malaria report 2020: 20 years of global progress and challenges.* Retrieved from https://www.who.int/publications-detail-redirect/9789240015791.

World Health Organization. (2021a). *World malaria report 2021.* Retrieved from https://www.who.int/teams/global-malaria-programme/reports/world-malaria-report-2021.

World Health Organization. (2021b). *Yellow fever—Nigeria 2021.* Retrieved from https://www.who.int/emergencies/disease-outbreak-news/item/2021-DON336.

World Health Organization. (2022a). *Dengue and severe dengue.* 2022. Retrieved from https://www.who.int/news-room/fact-sheets/detail/dengue-and-severe-dengue.

World Health Organization. (2022b). *Eliminate yellow fever epidemics (EYE) strategy 2017–2026.* Retrieved from https://www.who.int/initiatives/eye-strategy#:~:text=The%20EYE%20strategy%20is%20a,contain%20outbreaks%20rapidly.

World Health Organization. (2022c). *Report on malaria in Nigeria 2022*. Retrieved from https://www.afro.who.int/countries/nigeria/publication/report-malaria-nigeria-2022.

World Health Organization. (2022d). *World malaria report 2022*. Retrieved from https://www.who.int/teams/global-malaria-programme/reports/world-malaria-report-2022.

World Health Organization. (2023). *World malaria report 2023*. Retrieved from https://www.who.int/teams/global-malaria-programme/reports/world-malaria-report-2023.

World Meteorological Organization. (2022). Retrieved from https://wmo.int/topics/climate.

Zahouli, J. B. Z., Koudou, B. G., Müller, P., Malone, D., Tano, Y., & Utzinger, J. (2017). Urbanization is a main driver for the larval ecology of *Aedes* mosquitoes in arbovirus-endemic settings in south-eastern Côte d'Ivoire. *PLoS Neglected Tropical Diseases, 11*(7), 1–23. https://doi.org/10.1371/journal.pntd.0005751

Chapter 15
Inhibitory Effects of Selected Phytoconstituents Against Steroid 5-Alpha Reductase Type-2 Target in Prostate Cancer: An In Silico Study

Timothy Oluwatimileyin Ayeni and Shalom Nwodo Chinedu

1 Introduction

Cancer has been identified by most countries around the globe as a significant health concern due to its widespread growth, high death rate, and expensive treatment, affecting people of all ages and genders. Cancer is an abnormal growth of cells that can occur in various tissues and is a leading cause of illness globally, ranking second in morbidity (Cleanclay et al., 2023). A cancer type called prostate cancer (PC) develops in the prostate gland and affects millions of men worldwide (Sung et al., 2021). Despite advancements in diagnosis and treatment, PC still poses a significant public health concern. As a result, continuous research efforts are centered on enhancing the accuracy of prostate cancer diagnosis, identifying new and effective treatments, and creating strategies to prevent the disease.

The high cost of managing prostate cancer poses a significant financial burden on patients and the healthcare system. This cost has increased over time possibly due to over-diagnosis, over-treatment, and improved survival rates (Heijnsdijk et al., 2009). To mitigate the financial burden of this illness and minimize the side effects of chemical drugs, natural products derived from plants are considered an

T. O. Ayeni
Department of Biochemistry, College of Science and Technology, Convenant University, Ota, Ogun State, Nigeria

Covenant Applied Informatics and Communication - African Centre of Excellence (CApIC-ACE), Convenant University, Ota, Ogun State, Nigeria

S. N. Chinedu (✉)
Department of Biochemistry, College of Science and Technology, Covenant University, Ota, Ogun State, Nigeria

Covenant University Public Health and Wellbeing Research Cluster CUPHWERC), Covenant University, Ota, Ogun State, Nigeria
e-mail: shalom.chinedu@covenantuniversity.edu.ng

P. O. Isibor, P. A. Akinduti (eds.), *Harnessing Biotechnology Tools for Product Development*, https://doi.org/10.1007/978-3-031-86002-7_15

alternative option for managing PC patients (Mirzaei et al., 2019; Salehi et al., 2019). The utilization of medicinal plants in treating and curing various diseases has a long history, resulting in an increased interest in using natural products for drug discovery. Many drugs have been discovered by investigating the medicinal properties of the bioactive components of plant extracts. Plants contain a wide range of natural bioactive compounds, particularly secondary metabolites, with potential therapeutic applications (Salehi et al., 2019).

Recent investigations have centered on the examination of various plant families for their prospective roles in oncological therapies. The efficacy of these plants as anti-cancer agents is intrinsically associated with the bioactive constituents present in their extracts. Despite the overarching similarities in the bioactive profiles across numerous plant species, emergent research has uncovered distinct phytochemical constituents that may possess significant potential as novel chemotherapeutic agents (Thomas-Charles & Fennell, 2019).

The androgen biosynthesis pathway plays a critical role in PC growth and progression; therefore, targeting the genes and enzymes along this pathway is one of the therapeutic strategies against PC. The steroid 5-α-reductase type 2 (SRD5A2) gene, among other androgen biosynthesis pathway genes, encodes the SRD5A2 enzyme that transforms testosterone into dihydrotestosterone (DHT), which is an androgen receptor (AR) activator and a predominant factor in PC pathology. Studies have indicated that gene polymorphisms increase PC risk and development (Álvarez-González et al., 2023; Haffner et al., 2021; Khan & Cheng, 2022), and SRD5A2 gene polymorphisms have been studied to alter the catalytic activity of the SRD5A2 enzyme (Elzenaty et al., 2022; Ntais et al., 2003). Among these polymorphisms, A49T (a change of alanine to threonine at codon 49) and V89L (a change of valine to leucine at codon 89) have been associated with PC risk (Zeigler-Johnson et al., 2002; Cicek et al., 2004; Li et al., 2010) and also identified as probable targets for PC treatment (Montgomery et al., 2008). Targeting the SRD5A2 gene can reduce PC risk, thereby improving androgen therapy's effectiveness. Dutasteride and finasteride are clinically approved drugs used as inhibitors of SRD5A2 enzyme (Hou et al., 2021); however, the side effects associated with these drugs call for a need to identify promising lead compounds with similar or higher effectiveness but minimal side effects. Therefore, this study aimed to review and computationally screen previously identified medicinal plants and their bioactive compounds to investigate their therapeutic potentials against SRD5A2 activity.

2 Methodology

2.1 Systematic Review

2.1.1 Database Search Strategy

Using the search terms "medicinal plants," "SRD5A2," "inhibitors," "prostate cancer," and "treatment," we systematically searched five online databases, including Scopus, ScienceDirect, Google Scholar, PubMed, and Web of Science, for studies that investigated the effect of some medicinal plants' bioactive compounds on SRD5A2 activity. Thirty-six results were generated across all the searched databases, with 26 from Google Scholar, eight from Scopus, four from ScienceDirect, and none from PubMed and Web of Science.

2.1.2 Study Selection Criteria

Inclusion: (i) studies that investigated medicinal plants as SRD5A2 inhibitors, (ii) original articles, (iii) papers in English, and (iv) studies focusing on PC or BPH.

Exclusion: (i) studies on medicinal plants focusing on other genes, (ii) abstracts, case reports, editorials, or review articles, (iii) studies on other cancer types, and (iv) papers in non-English languages.

2.1.3 Data Extraction

Information regarding the author's name, year of study, medicinal plants, and their bioactive compounds, including their effects on SRD5A2, were extracted from each included study that met our criteria.

2.2 *In Silico Study*

2.2.1 Compound Retrieval and Preparation

Thirty-four selected compounds were docked, alongside two standard drugs, finasteride and dutasteride, to the active site of the proteins used as targets. The 3D conformation (in .sdf format) of the compounds and standards were downloaded from the PubChem database (https://pubchem.ncbi.nlm.nih.gov/) and imported into the Open Babel environment of the PyRx software. Energy minimization was performed on the compounds, and their conversion into pdbqt formats was followed.

2.2.2 Protein Retrieval, Preparation, and Binding Site Identification

The proteins were downloaded, in their pdb formats, from the protein data bank (PDB) (https://www.rcsb.org/) using their protein IDs and prepared for docking using the BIOVIA Discovery Studio software (1). After preparation, the proteins were loaded as macromolecules into the PyRx software and converted into their pdbqt formats (2). To identify the binding sites of the proteins, their pdb files (downloaded from the PDB) were uploaded to the ProteinsPlus web server (https://proteins.plus/), and their binding sites were detected using the DogSite scorer (Fig. 15.2a).

2.2.3 Molecular Docking and Post-docking Analyses

Before docking, a grid box was set around the binding sites using the pocket coordinates ($x = -31.88$, $y = 15.6$, and $z = 36.88$) generated from the DogSite scorer for each protein. The best conformation of the ligand–protein complex with an rmsd value of zero among each of the nine conformations generated after docking was selected for each docked ligand. The docking scores of the compounds were compared with those of the standards. The top 5 ligands with the best docking scores, alongside the standards, were selected for post-docking analysis. The interactions of the ligand–protein conformations, including hydrogen bonds and pi (π) bonds, were investigated using their 2D diagram generated from the BIOVIA Discovery Studio Visualizer software (Iheagwam et al., 2021).

2.2.4 Binding Site and Docking Validation

To ascertain the binding pockets identified by the DogSite Scorer, the co-crystallized ligand (NDX) was downloaded from the PubChem database and docked with the SRD5A2 protein using the grid box coordinates for x, y, and z as stated above. The docked ligand was then superimposed with the natural ligand (NDX) to compare the position and distance of the two ligands, as seen in Fig. 15.2b. The 2D diagram for the ligand–protein complex was also examined to compare the amino residues interacting with the ligands and those surrounding the pocket sites, as illustrated in Fig. 15.2c, d. Similar interactions of the ligands with amino acid residues and a small rmsd value of <2Å validate the binding site and make our docking more reliable.

2.2.5 Pharmacokinetics and Toxicity Prediction

The SwissADME and Protox 3.0 web servers were used to predict the pharmacokinetics and toxicity of the top 5 compounds alongside finasteride. Among the pharmacokinetics parameters examined include Lipinski's rule of five violations, gastrointestinal (GI) absorption, blood–brain barrier (BBB) and skin permeability, cytochrome P450 (CYP) enzyme inhibition, lead-likeness, synthetic accessibility (SA), and *p*-glycoprotein (*p*-gp) substrate. Ligands that satisfied these screenings were selected for further analysis (Lipinski et al., 2012). The selected compounds were also subjected to toxicity screening, including oral toxicity, carcinogenicity, cardiotoxicity, cytotoxicity, hepatotoxicity, neurotoxicity, and respiratory toxicity (Yang et al., 2019).

3 Results and Discussion

3.1 Reviewed Studies

As illustrated in Fig. 15.1, nine (9) studies out of the thirty-six (36) results generated across the databases were selected for review.

Ming et al. (2014) investigated the effect of pomegranate extracts on the androgen biosynthesis pathways in PC, using both in vitro and in vivo models. The bioactive compounds of pomegranate water-soluble extract included in their study are gallic acid, ellagic acid, caffeic acid, luteolin, hexahydroxydiphenic acid, cyanidin, and gallagyldilactone. These compounds were shown to inhibit SRD5A2, as

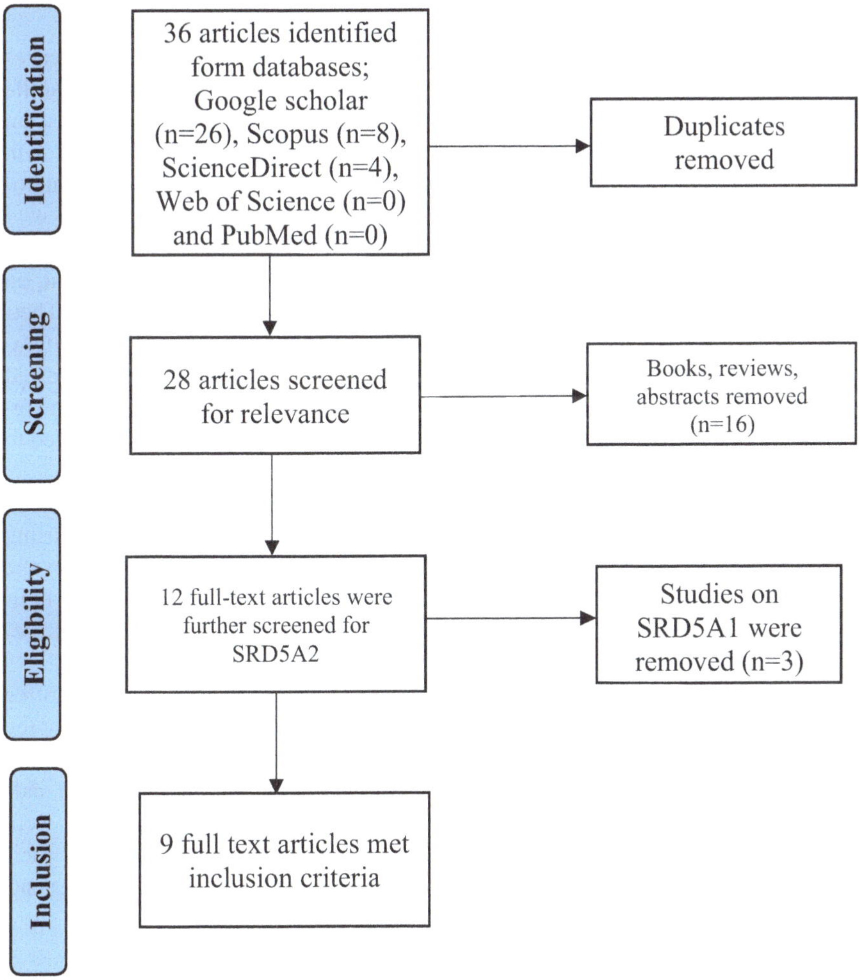

Fig. 15.1 Flow chart illustration of the study selection process

indicated by the cell lines' reduction in testosterone and DHT levels (Ming et al., 2014).

Dela Torre et al. (2023) also evaluated the effect of *Tamarindus indica* seed extracts on the microsomal SRD5A2 enzyme using an in vitro experimental design. They obtained three *Tamarindus indica* seed extracts (ethanol, ethyl acetate, and hexane extract). The bioactive compounds yielded by these extracts include catechin, catechin-(4α→8)-catechin, chrysin, 3-hydroxy baicalein, darutigenol, nigakilactoneI,24-methylcholesta-7,22-dien-3β-ol,kushenolI,4,8,12-trimethyl-tridecanoic acid, (20S)-3α,29-dihydroxylupan-27-oic acid, abrisapogenol B, ar-abietatriene, hydroxyhydromethyl abietate, Vitamin Eβ, and 3-O-[β-D-glucopyranosyl-(1→2)]-β-D-glucopyranosyl-kaempferol. Among the three extracts of *Tamarindus indica*, the ethyl acetate extract showed the highest inhibitory effect on SRD5A2 due to its bioactive components, including tocopherols, fatty acids, phytosterols, terpenoids, and phenolic compounds (Dela Torre et al., 2023).

Furthermore, Jung et al. (2022) investigated the effect of Mangifera indica leaves water extract on the SRD5A2 expression levels in human follicle dermal papilla cells. Their findings indicated a significant reduction in the SRD5A2 expression levels, following applying the extract on the papilla cells. A significant hair growth was also observed in mice treated with the extract. However, the bioactive components of the extract were not highlighted in the study (Jung et al., 2022).

Fan et al. (2023) also examined the regulatory effect of sinomenine hydrochloride, a bioactive component of the *Sinomenium acutum* Rehderett Wilson plant, on SRD5A2 expression levels, using in vitro and in vivo experimental models. Their study showed an effective inhibition of sinomenine hydrochloride on the proliferation of BPH-1 cells by downregulating the expression of SRD5A2. They also treated mice with prostate enlargement using sinomenine hydrochloride and found a significant reduction in the prostate gland weight following the treatment, compared to the non-treatment group (Fan et al., 2023).

Gopalakrishman et al. (2023) also evaluated the chemo-preventive effect of Cinnamon bark extract (CBE) on androgen receptor and SRD5A2 expression using a rat model. The bioactive compounds identified from the CBE and used in the experiment include cinnamaldehyde and procyanidin B2 (PB2). Their study indicated that these bioactive compounds decreased the expression of SRD5A2, preventing PC premalignancy in the examined rat (Gopalakrishman et al., 2023).

In addition, Muangsanguan et al. (2023) investigated the effect of Arabica coffee pulp extract on the expression of SRD5A in human fibroblast cells, to illustrate the modulating effect of this extract on hair growth. The bioactive compounds identified from the Arabica coffee pulp extract include caffeic acid, quercetin, catechin, caffeine, chlorogenic acid, gallic acid, and catechin gallate. These compounds were shown to suppress the in vitro SRD5A2 expression level significantly, and their employment could improve the treatment of androgenetic alopecia thereby preventing hair loss and promoting hair growth (Muangsanguan et al., 2023).

Also, Ruksiriwanich et al. (2022) examined the regulatory effect of *Asparagus racemosus* wild root extract on SRD5A2 mRNA expression using in vitro and in vivo models. The bioactive compounds identified from the root extract are quercetin, naringenin, *p*-coumaric acid, caffeic acid, and rosmarinic acid. Comparable to the SRD5A2 enzyme standard inhibitors, significant suppression of SRD5A2 mRNA expression was observed for the Asparagus racemosus root extract due to its rich content of polyphenols (Ruksiriwanich et al., 2022).

Ruksiriwanich et al. (2023) also assessed the inhibitory effect of the extracts of Oryza sativa by-products on SRD5A2 using an in vitro model involving mouse skin melanoma and human fibroblast cells. The compounds identified from the extract include α-tocopherols, β-tocopherol, and γ-tocopherol. The extracts, particularly the Bue Bang 4 CMU, effectively reduced SRD5A2 gene expression (Ruksiriwanich et al., 2023).

Lastly, Khantham et al. (2021) investigated the regulatory effects of four Thai rice bran extracts on SRD5A2 gene expression using in vitro and in silico models. They also identified some bioactive compounds, including α-, β-, and γ-tocopherols, oleic acid, linolenic acid, γ-linolenic acid, ferulic acid, and γ-oryzanol. These compounds were evaluated alongside the standard SRD5A2 inhibitors, finasteride and dutasteride. Among the four rice bran extracts investigated, the Tubtin Chumphae extract significantly suppressed SRD5A2 gene expression, comparable to the finasteride, and the in silico model identified tocopherol as the most potent compound for SRD5A2 inhibition (Khantham et al., 2021).

3.2 Molecular Docking Results

Among the thirty-four (34) compounds docked into the binding site of SRD5A2 protein, gamma-oryzanol had the lowest binding energy (−11.6 Kcal/mol) comparable to finasteride (−11.6 Kcal/mol). Meanwhile, cinnamaldehyde exhibited the highest binding energy at −5.8 Kcal/mol. In molecular docking, the lower the binding energy of a compound, the higher its affinity to the binding pocket of the target (protein). This means that besides the standard inhibitors, gamma-oryzanol possesses the best binding affinity with the SRD5A2 compared to other compounds. Our docking study revealed the top five (5) compounds with binding energy close to or similar to the standard inhibitors as highlighted in Table 15.1. They include gamma-oryzanol (−11.6 Kcal/mol), Kaempferol-3-O-beta-D-glucopyranosyl-7-O-alpha-L-rhamnopyranoside (−11.0 Kcal/mol), catechin gallate (−11.0 Kcal/mol), Nigakilactone1 (−10.4 Kcal/mol), and 24-Methylcholesta-7,22-dien-3beta-ol (−11.5 Kcal/mol). These compounds are, therefore, predicted as promising inhibitors of the SRD5A2 enzyme, and their incorporation into drug development may improve PC treatment by preventing

Table 15.1 The binding energy of selected ligands in binding pockets of SRD5A2

Compounds	Binding energy (kcal/mol)
Hexahydroxydiphenic acid	−8.9
Procyanidin-B2	−8.1
Nigakilactone-1	−10.4
Cyanidin	−9.3
Alpha-tocopherol	−9.6
4,8,12-Trimethyltridecanoic	−6.8
Chlorogenic acid	−9.2
Kushenol I	−8.8
Kaempferol 3-O-beta-D-glucopyranosyl-7-O-alpha-L-rhamnopyranoside	−11.0
Caffeine	−6.6
Darutigenol	−10.2
Gallic acid	−6.5
Naringenin	−9.2
Oleic acid	−6.5
Ferulic acid	−7.2
Catechin gallate	−11.0
Quercetin	−9.6
Luteolin	−9.6
Linoleic acid	−6.6
Gamma-linolenic acid	−6.8
Baicalein	−9.5
Chrysin	−9.3
Rosmarinic acid	−9.2
Ellagic acid	−10.1
Gamma-oryzanol	−11.6
24-Methylcholesta-7,22-dien-3beta-ol	−11.5
Sinomenine hydrochloride	−9.1
Finasteride	−11.6
Cinnamaldehyde	−5.8
p-coumaric acid	−6.6
ar-Abietatriene	−9.9
Beta-tocopherol	−9.1
Caffeic acid	−7.0
Catechin	−9.2
Delta-tocopherol	−8.9
Dutasteride	−12.2

Ligands within the top 7 of the docking score are bold

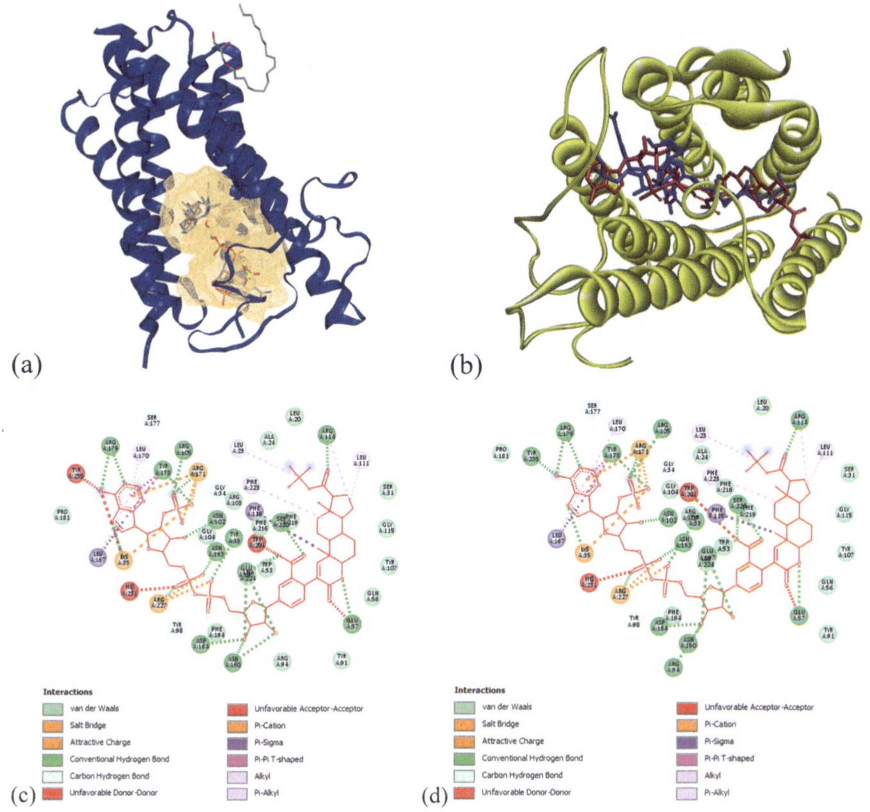

Fig. 15.2 (**a**) SRD5A2 binding site, (**b**) superimposition of ligands; co-crystallized NDX (red) and docked NDX (blue), (**c**) 2D interaction of co-crystallized NDX, (**d**) 2D interaction of docked NDX

dihydrotestosterone (DHT) production, which is significant for AR signaling and PC progression. The molecular docking study is essential in drug discovery because it helps to analyze the binding properties and predict the interactions between the ligand and targets of interest. The post-docking analysis of the top 5 ligands alongside finasteride and dutasteride revealed the molecular interactions of the ligands with the amino acid residues within and around the binding pockets of SRD5A2 protein. The 2D diagrams (Fig. 15.2a–g) of the ligand–protein complex showed that conventional hydrogen bonds, Van der Waals forces, carbon–hydrogen bonds, and pi–pi (π–π) interactions stabilize the complex. Conventional hydrogen bonds greatly influence the binding affinity and specificity during ligand–protein binding, as they involve interactions between a covalently linked hydrogen atom with an electronegative atom and another electronegative atom. Meanwhile, the Van der Waals force allows unspecific interactions between nearby atoms, allowing for the overall stability of the ligand–protein complex. Carbon–hydrogen bonds also contribute to the stability of the complex, owing to its hydrophobic interactions between the non-polar region and hydrophobic pockets of the ligand and protein, respectively, while the π–π interactions form between the aromatic rings of the ligands and proteins (Isaev, 2021; Bitencourt-Ferreira et al., 2019). Among the top 5 ligands with high binding affinity, only the compound 24-Methylcholesta-7,22-dien-3beta-ol does not exhibit conventional hydrogen bonds in its interaction with the protein binding pocket. Its high affinity with the pocket is only attributed to other bonds as stated above. Three compounds out of the top 5 selected compounds (Nigakilactone1, Kaempferol 3-O-beta-D-glucopyranosyl-7-O-alpha-L-rhamnopyranoside, and catechin gallate) showed a similar number of conventional hydrogen bond interactions compared to dutasteride, and this may contribute to their stability in the binding pocket of the SRD5A2, whereas gamma-oryzanol have similar number of interactions in total, compare to dutasteride. Its high interaction with the amino acid residues within the binding pockets of SRD5A2 suggests its high binding affinity with the pocket.

The amino acid residues interacting with the ligands are also essential in the stability of the ligand–protein complex. The amino acids that interacted with dutasteride include ARG227, TYR33, ASN160,193, PHE194,223, and LEU224, while only ARG114, LEU224, PHE223, and PHE118 interacted with finasteride. For Nigakilactone-1, the amino acids that are involved in its stability are ARG227, LYS35, ARG105, and ARG171, whereas ARG103, ASN193, PHE194, TYR33, and ARG227 interacted with Kaempferol-3-O-beta-D-glucopyranosyl-7-O-alpha-L-rhamnopyranoside to enable its stability. Catechin gallate and 24-Methylcholesta-7,22-dien-3beta-ol have ARG105, ARG227, TYR98, ARG171, ASP164, and PHE223, LEU224, TRP53, SER31, respectively, as their interacting amino acids residues. On the other hand, the amino acid residues interacting with gamma-oryzanol include PHE223, TYR33, PHE118, LEU111, ARG171,179, LEU170, and TYR178 (Fig. 15.3a–g). These amino acid residues represent potential targets for therapeutic inhibition of SRD5A2 owing to their ability to stabilize the binding pocket of the protein (Bröer, 2018).

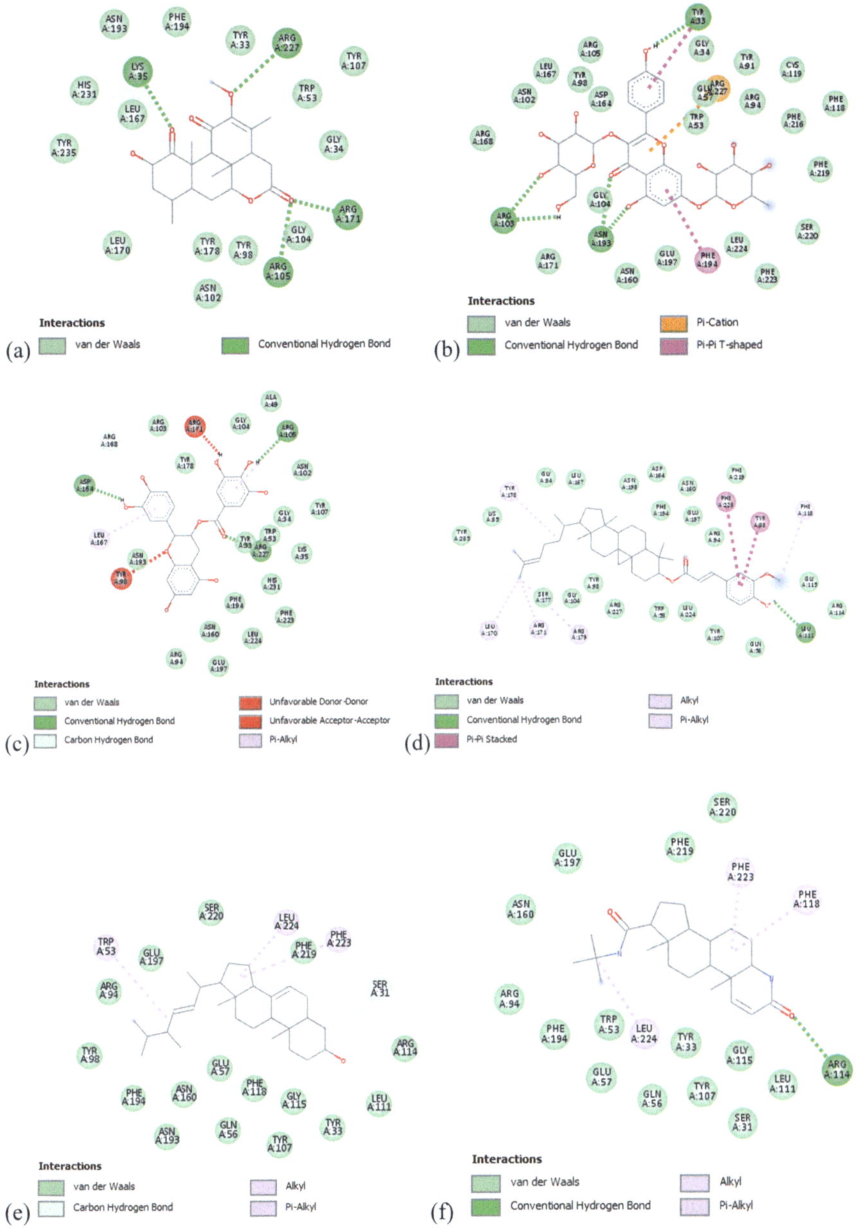

Fig. 15.3 2D diagrams showing the binding pose interaction of ligands within the top 5 alongside the standards: (**a**) Nigakilactone-1, (**b**) Kaempferol-3-O-beta-D-glucopyranosyl-7-O-alpha-L-rhamnopyranoside (**c**) catechin gallate, (**d**) gamma-oryzanol, (**e**) 24-Methylcholesta-7,22-dien-3beta-ol (**f**) finasteride, and (**g**) dutasteride

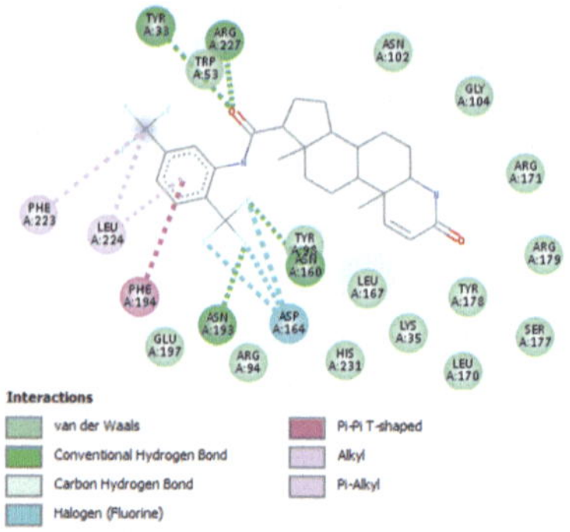

(g)

Fig. 15.3 (continued)

3.3 Pharmacokinetics and Toxicity Profiles

ADME (absorption, distribution, metabolism, and excretion) study is a vital phase in drug development that assesses the pharmacokinetics and druglike characteristics of compounds (Guan et al., 2019). In silico determination of ADME properties is a quick and cost-effective alternative to traditional experimental methods. Its early use in drug development is critical for lowering the drug failure rate during clinical pharmacokinetics studies. In addition to being effective against the therapeutic target (preferably at a low concentration), a high-quality drug candidate should have the appropriate ADME characteristics at a therapeutic dose. Compounds with projected favorable pharmacokinetic and toxicity characteristics can pass standard clinical trial requirements, making them drug candidates throughout the drug discovery process (Iheagwam & Rotimi, 2020). Table 15.2 depicts the pharmacokinetic parameters of the top 5 selected compounds compared to the standard drugs. Out of the seven compounds screened for ADME profiles, only finasteride was predicted to have the ability to pass through the blood-brain barrier (BBB) and be passively absorbed by the human intestinal tract. Other compounds, except Nigakilactone-1, lack both BBB permeability and gastrointestinal absorption. All the compounds, except for catechin gallate and 24-Methylcholesta-7,22-dien-3beta-ol, were predicted to be *p*-glycoprotein substrates. The ATP-dependent protein glycoprotein (P-gp) serves as a defense mechanism against hazardous chemicals by pumping poisons and foreign compounds out of cells (Silva et al., 2015). This indicates that these compounds use the *p*-gp transporter for drug absorption and excretion of poisonous and foreign substances from the cells. As shown in Table 15.2, no compounds inhibited more than one isoform of cytochrome P450 (CYP450) enzymes, and only CYP3A4 and CYP2C9 were predicted to be inhibited by dutasteride and 24-Methylcholesta-7,22-dien-3beta-ol, respectively. Other compounds do not inhibit any isoforms of CYP450 enzymes. Inhibition of CYP450 enzymes by the compounds might result in either drug build-up or reduced drug metabolism, potentially leading to clinical toxicity (Manikandan & Nagini, 2018;

Table 15.2 Pharmacokinetics (ADME) profiles of top 5 ligands and standards

Cmpds	GIA	BBBP	Pgp-Subs	CYP2C9 inhibitor	CYP2D6 inhibitor	CYP3A4 inhibitor	CYP1A2 inhibitor	CYP2C19 inhibitor	Log Kp (cm/s)	Lip	BS	L	SA
Nigakilactone-1	High	No	Yes	No	No	No	No	No	−7.45	0	0.55	1	5.22
K-3-O-BDG	Low	No	Yes	No	No	No	No	No	−10.35	3	0.17	1	6.52
Catechin gallate	Low	No	No	No	No	No	No	No	−7.91	1	0.55	1	4.16
Gamma-oryzanol	Low	No	Yes	No	No	No	No	No	−1.42	2	0.17	3	7.49
24-M	Low	No	No	Yes	No	No	No	No	−3.22	1	0.55	2	5.97
Finasteride	High	Yes	Yes	No	No	No	No	No	−6.42	0	0.55	1	5.37
Dutasteride	Low	No	Yes	No	No	Yes	No	No	−5.71	2	0.17	2	5.36

K-3-O-BDG Kaempferol-3-O-beta-D-glucopyranosyl-7-O-alpha-L-rhamnopyranoside, *24-M* 24 Methylcholesta-7,22-dien-3β-ol, *Cmpds* compounds, *GIA* gastrointestinal absorption, *BBBP* blood brain barrier permeability, *Pgp Sub* *p*-glycoprotein substrate, *Vb* Veber, *Mgg* Muegge, *Lip* Lipinski, *BS* bioavailability score, *L* lead-likeness, *SA* synthetic accessibility

Iheagwam et al., 2019). Among all the compounds screened, dutasteride, gamma-oryzanol, and Kaempferol-3-O-beta-D-glucopyranosyl-7-O-alpha-L-rhamnopyranoside did not obey Lipinski's rule of five (having more than one violation), and also possessed very low bioavailability scores. Our result also showed high skin permeability scores (log Kp) of all the compounds except gamma-oryzanol, predicting their ability to penetrate the skin and induce toxicity. Our study shows that the compounds' synthetic accessibility scores range from 4.16 to 7.49, indicating they can be easily synthesized and integrated into therapeutic products. The score, which runs from 1 to 10, is mainly based on the notion that the frequency of molecular fragments in readily available compounds correlates with the ease of synthesis (Daina et al., 2019).

The toxicity assessment of compounds is essential and has drawn the attention of researchers for drug discovery and development. Compounds identified to be very toxic usually do not get to clinical trials because of the possible damaging effects they might cause (Basile et al., 2019). Therefore, in our study, we predicted the toxicity properties of the compounds identified to be promising inhibitors of SRD5A2. The toxicity parameters used in our study include acute oral toxicity, neurotoxicity, mutagenicity, carcinogenicity, cardiotoxicity, immunotoxicity, hepatotoxicity, and respiratory toxicity (Banerjee et al., 2018). These toxicities can be quantified using lethal dosage (LD50) values and qualitatively measured as either active or inactive (Raies & Bajic, 2016). As shown in Table 15.3, the compounds' oral toxicity falls between class IV and VI, indicating their non-toxic properties. The acute oral toxicity classification (I–VI) of compounds is based on their likenesses with already identified toxic chemicals and the existence of toxic components in their structures (Ghosh et al., 2019) Class I compounds are known as the most toxic, and class IV compounds are the least toxic. All the screened compounds and standard drugs are non-carcinogenic, non-mutagenic, non-hepatotoxic, and non-cytotoxic. However, they are all predicted to be respiratory and immuno-toxic

Table 15.3 Toxicity profiles

Toxicity	AOT	Hep	Neur	Resp	Card	Carc	Immuno	Muta	Cyto
Nigakilactone1	IV	–	–	+	–	–	+	–	–
Kaempferol-3-O-beta-D--glucopyranosyl-7-O-alpha-L-rhamnopyranoside	V	–	–	+	+	–	+	–	–
Catechin gallate	IV	–	–	+	–	–	–	–	–
Gamma-oryzanol	VI	–	–	+	+	–	+	–	–
24-Methylcholesta-7,22-dien-3β-ol	IV	–	+	+	–	–	+	–	–
Finasteride	IV	–	+	+	–	–	+	–	–
Dutasteride	VI	–	+	+	–	–	+	–	–

AOT acute oral toxicity, *Hep* hepatotoxicity, *Hep* neurotoxicity, *Resp* respiratory, *Card* cardiotoxicity, *Carc* carcinogenicity, *Immuno* immunotoxicity, *Muta* mutagenicity, *Cyto* cytotoxicity (except catechin gallate). Kaempferol-3-O-beta-D-glucopyranosyl-7-O-alpha-L-rhamnopyranoside and gamma-oryzanol are also predicted to be cardiotoxic, while 24-Methylcholesta-7,22-dien-3β-ol, finasteride, and dutasteride were predicted to be neurotoxic. As stated above, the toxic effects of these compounds need to be considered before their incorporation into therapeutic drug development.

4 Conclusion

This study identified promising compounds suitable to bind the SRD5A2 target implicated in prostate cancer progression. These compounds have binding affinities comparable with the standard SRD5A2 inhibitors (finasteride and dutasteride), with gamma-oryzanol showing the highest binding affinity. However, as predicted, the toxic effects of these compounds may affect their therapeutic functions and cause further damage to human health. Hence, further study is recommended to ascertain their toxicities.

Acknowledgments We acknowledge the support of the Covenant University Centre for Research Innovation and Discovery (CUCRID) in covering the APC charge for this article.

Funding Source This research received no specific grant or funding from funding agencies in the public, commercial, or non-profit sectors.

Declaration of Interests The authors stated that they have no competing financial or personal interests that could have influenced the conduct of this research.

References

Álvarez-González, B., Porras-Quesada, P., Arenas-Rodríguez, V., Tamayo-Gómez, A., Vázquez-Alonso, F., Martínez-González, L. J., Hernández, A. F., & Álvarez-Cubero, M. J. (2023). Genetic variants of antioxidant and xenobiotic metabolizing enzymes and their association with prostate cancer: A meta-analysis and functional in silico analysis. *Science of The Total Environment, 898*, 165530.

Banerjee, P., Eckert, A. O., Schrey, A. K., & Preissner, R. (2018). ProTox-II: A webserver for the prediction of toxicity of chemicals. *Nucleic Acids Research, 46*(W1), W257–W263.

Basile, A. O., Yahi, A., & Tatonetti, N. P. (2019). Artificial intelligence for drug toxicity and safety. *Trends in Pharmacological Sciences., 40*(9), 624–635.

Bitencourt-Ferreira, G., Veit-Acosta, M., & De Azevedo Jr, W. (2019). Van der Waals potential in protein complexes. In *Methods in molecular biology (Clifton, N.J.)* (pp. 79–91). https://doi.org/10.1007/978-1-4939-9752-7_6

Bröer, S. (2018). Amino acid transporters as disease modifiers and drug targets. *SLAS Discovery: Advancing Life Sciences R&D, 23*(4), 303–320.

Cicek, M. S., Conti, D. V., Curran, A., Neville, P. J., Paris, P. L., Casey, G., & Witte, J. S. (2004). Association of prostate cancer risk and aggressiveness to androgen pathway genes: SRD5A2, CYP17, and the AR. *The Prostate, 59*(1), 69–76.

Cleanclay, W. D., Zakari, S., Adigun, T. O., Ayeni, T. O., Nnaji, P. C., Nnenna, A. D., Blessing, A., Adebosoye, A., Gbadebo, M., & Agbetuyi-Tayo, P. (2023). Cancer biology and therapeutics: Navigating recent advances and charting future directions. *Tropical Journal of Natural Product Research, 7*(12).

Daina, A., Michielin, O., & Zoete, V. (2019). SwissTargetPrediction: Updated data and new features for efficient prediction of protein targets of small molecules. *Nucleic Acids Research, 47*(W1), W357–W364.

Dela Torre, G. L. T., Manalo, R. A. M., & Arollado, E. C. (2023). Tamarindus indica seed extracts inhibit mouse microsomal 5α-reductase in vitro. *Journal of Research in Pharmacy, 27*(5), 1989–1997. https://doi.org/10.29228/jrp.478

Elzenaty, R. N., Du Toit, T., & Flück, C. E. (2022). Basics of androgen synthesis and action. *Best Practice & Research Clinical Endocrinology & Metabolism, 36*(4), 101665.

Fan, M.-S., Xia, Y.-F., Ye, R.-H., Sun, Z.-R., Wang, M.-Y., An, M.-F., Zhang, S.-S., Zhang, L.-J., Zhao, Y.-L., Xiang, Z.-M., & Sheng, J. (2023). Sinomenine hydrochloride can ameliorate benign prostatic hyperplasia by lowering the 5α-reductase 2 level and regulating the balance between the proliferation and apoptosis of cells. *Molecules, 28*(2). https://doi.org/10.3390/molecules28020803

Ghosh, P., Talukdar, P., & Talapatra, S. N. (2019). The Leaf alkaloid of *Catharanthus roseus* Linn. as antidiabetic potential: In silico approach through quantitative structure activity relationship modelling and molecular docking. *Journal of Advanced Scientific Research, 10*(04), 186–191.

Gopalakrishman, S., Dhaware, M., Sudharma, A. A., Mullapudi, S. V., Siginam, S. R., Gogulothu, R., Mir, I. A., & Ismail, A. (2023). Chemopreventive effect of cinnamon and its bioactive compounds in a rat model of premalignant prostate carcinogenesis. *Cancer Prevention Research, 16*(3), 139–151. https://aacrjournals.org/cancerpreventionresearch/article-abstract/16/3/139/716700

Guan, L., Yang, H., Cai, Y., Sun, L., Di, P., Li, W., Liu, G., & Tang, Y. (2019). ADMET-score—A comprehensive scoring function for evaluation of chemical drug-likeness. *Medchemcomm, 10*(1), 148–157.

Haffner, M. C., Zwart, W., Roudier, M. P., True, L. D., Nelson, W. G., Epstein, J. I., De Marzo, A. M., Nelson, P. S., & Yegnasubramanian, S. (2021). Genomic and phenotypic heterogeneity in prostate cancer. *Nature Reviews Urology, 18*(2), 79–92.

Heijnsdijk, E. A. M., der Kinderen, A., Wever, E. M., Draisma, G., Roobol, M. J., & De Koning, H. J. (2009). Overdetection, overtreatment and costs in prostate-specific antigen screening for prostate cancer. *British Journal of Cancer, 101*(11), 1833–1838. https://www.nature.com/articles/6605422.

Hou, Z., Huang, S., & Li, Z. (2021). Androgens in prostate cancer: A tale that never ends. *Cancer Letters, 516*, 1–12.

Iheagwam, F. N., & Rotimi, S. O. (2020). Computer-aided analysis of multiple SARS-CoV-2 therapeutic targets: Identification of potent molecules from African medicinal plants. *Scientifica, 2020*(1), 1878410.

Iheagwam, F. N., Ogunlana, O. O., & Chinedu, S. N. (2019). Model optimization and in silico analysis of potential dipeptidyl peptidase IV antagonists from GC-MS identified compounds in Nauclea latifolia leaf extracts. *International Journal of Molecular Sciences, 20*(23), 5913.

Iheagwam, F. N., Odiba, J. K., Iheagwam, O. T., Ogunlana, O. O., & Chinedu, S. N. (2021). Type 2 diabetes mellitus mediation by the disruptive activity of environmental toxicants on sex hormone receptors: In silico evaluation. *Toxics, 9*(10), 255.

Isaev, A. (2021). *Conventional and nonconventional hydrogen bonds. What is the difference?* https://doi.org/10.13140/RG.2.2.10134.27206

Jung, H., Jung, D.-M., Lee, S.-S., Kim, E.-M., Yoon, K., & Kim, K. K. (2022). Mangifera Indica leaf extracts promote hair growth via activation of Wnt signaling pathway in human dermal papilla cells. *Animal Cells and Systems, 26*(3), 129–136. https://doi.org/10.1080/19768354.2022.2085790

Khan, H. M., & Cheng, H. H. (2022). Germline genetics of prostate cancer. *The Prostate, 82*, S3–S12.

Khantham, C., Yooin, W., Sringarm, K., Sommano, S. R., Jiranusornkul, S., Carmona, F. D., Nimlamool, W., Jantrawut, P., Rachtanapun, P., & Ruksiriwanich, W. (2021). Effects on steroid 5-alpha reductase gene expression of Thai rice bran extracts and molecular dynamics study on SRD5A2. *Biology, 10*(4), 319. https://www.mdpi.com/2079-7737/10/4/319

Li, J., Coates, R. J., Gwinn, M., & Khoury, M. J. (2010). Steroid 5-α-reductase Type 2 (SRD5a2) gene polymorphisms and risk of prostate cancer: A HuGE review. *American Journal of Epidemiology, 171*(1), 1–13.

Lipinski, C. A., Lombardo, F., Dominy, B. W., & Feeney, P. J. (2012). Experimental and computational approaches to estimate solubility and permeability in drug discovery and development settings. *Advanced Drug Delivery Reviews, 64*, 4–17.

Manikandan, P., & Nagini, S. (2018). Cytochrome P450 structure, function and clinical significance: A review. *Current Drug Targets, 19*(1), 38–54.

Ming, D.-S., Pham, S., Deb, S., Chin, M. Y., Kharmate, G., Adomat, H., Beheshti, E. H., Locke, J., & Guns, E. T. (2014). Pomegranate extracts impact the androgen biosynthesis pathways in prostate cancer models in vitro and in vivo. *The Journal of Steroid Biochemistry and Molecular Biology, 143*, 19–28. https://www.sciencedirect.com/science/article/pii/S0960076014000351

Mirzaei, A., Khatami, F., Ebrahimi, M., Mousavibahar, S. H., & Narouie, B. (2019). The Effect of herbal extracts on the treatment and prevention of prostate cancer: A literature review. *Translational Research in Urology, 1*(2), 67–73. https://doi.org/10.22034/TRU.2020.231129.1021

Montgomery, R. B., Mostaghel, E. A., Vessella, R., Hess, D. L., Kalhorn, T. F., Higano, C. S., True, L. D., & Nelson, P. S. (2008). Maintenance of intratumoral androgens in metastatic prostate cancer: A mechanism for castration-resistant tumor growth. *Cancer Research, 68*(11), 4447–4454.

Muangsanguan, A., Linsaenkart, P., Chaitep, T., Sangta, J., Sommano, S. R., Sringarm, K., Arjin, C., Rachtanapun, P., Jantanasakulwong, K., Phimolsiripol, Y., Castagnini, J. M., & Ruksiriwanich, W. (2023). Hair growth promotion and anti-hair loss effects of by-products arabica coffee pulp extracts using supercritical fluid extraction. *Foods, 12*(22). https://doi.org/10.3390/foods12224116

Ntais, C., Polycarpou, A., & Ioannidis, J. P. A. (2003). Association of the CYP17 gene polymorphism with the risk of prostate cancer: A meta-analysis. *Cancer Epidemiology Biomarkers and Prevention, 12*(2), 120–126. https://www.scopus.com/inward/record.uri?eid=2-s2.0-0037320659&partnerID=40&md5=6408d68b33366247417f4d21e82ece14

Raies, A. B., & Bajic, V. B. (2016). In silico toxicology: Computational methods for the prediction of chemical toxicity. *Wiley Interdisciplinary Reviews: Computational Molecular Science, 6*(2), 147–172.

Ruksiriwanich, W., Khantham, C., Linsaenkart, P., Chaitep, T., Jantrawut, P., Chittasupho, C., Rachtanapun, P., Jantanasakulwong, K., Phimolsiripol, Y., & Sommano, S. R. (2022). In vitro and in vivo regulation of SRD5A mRNA expression of supercritical carbon dioxide extract from Asparagus racemosus Willd. root as anti-sebum and pore-minimizing active ingredients. *Molecules, 27*(5), 1535. https://www.mdpi.com/1420-3049/27/5/1535

Ruksiriwanich, W., Linsaenkart, P., Khantham, C., Muangsanguan, A., Sringarm, K., Jantrawut, P., Prom-U-Thai, C., Jamjod, S., Yamuangmorn, S., & Arjin, C. (2023). Regulatory effects of Thai rice by-product extracts from Oryza sativa L. cv. Bue Bang 3 CMU and Bue Bang 4 CMU on melanin production, nitric oxide secretion, and steroid 5α-reductase inhibition. *Plants, 12*(3), 653. https://www.mdpi.com/2223-7747/12/3/653

Salehi, B., Fokou, P. V. T., Yamthe, L. R. T., Tali, B. T., Adetunji, C. O., Rahavian, A., Mudau, F. N., Martorell, M., Setzer, W. N., & Rodrigues, C. F. (2019). Phytochemicals in prostate cancer: From bioactive molecules to upcoming therapeutic agents. *Nutrients, 11*(7), 1483. https://doi.org/10.3390/nu11071483p

Silva, R., Vilas-Boas, V., Carmo, H., Dinis-Oliveira, R. J., Carvalho, F., de Lourdes Bastos, M., & Remiao, F. (2015). Modulation of P-glycoprotein efflux pump: Induction and activation as a therapeutic strategy. *Pharmacology & Therapeutics, 149*, 1–123.

Sung, H., Ferlay, J., Siegel, R. L., Laversanne, M., Soerjomataram, I., Jemal, A., & Bray, F. (2021). Global cancer statistics 2020: GLOBOCAN estimates of incidence and mortality worldwide for 36 cancers in 185 countries. *CA: A Cancer Journal for Clinicians, 71*(3), 209–249.

Thomas-Charles, C., & Fennell, H. (2019). Anti-prostate cancer activity of plant-derived bioactive compounds: A review. *Current Molecular Biology Reports, 5*, 140–151. https://doi.org/10.1007/s40610-019-00123-x

Yang, H., Lou, C., Sun, L., Li, J., Cai, Y., Wang, Z., Li, W., Liu, G., & Tang, Y. (2019). admetSAR 2.0: Web-service for prediction and optimization of chemical ADMET properties. *Bioinformatics, 35*(6), 1067–1069. https://academic.oup.com/bioinformatics/article-abstract/35/6/1067/5085368

Zeigler-Johnson, C., Walker, A., Mancke, B., Spangler, E., Jalloh, M., McBride, S., Deitz, A., Malkowicz, S., Ofori-Adjei, D., & Gueye, S. (2002). Ethnic differences in the frequency of prostate cancer susceptibility alleles at SRD5A2 and CYP3A4. *Human Heredity, 54*(1), 13–21.

Chapter 16
Paratransgenesis: Overview, Current Perspectives, and Future Research Needs for Malaria Control

Olubukola Oziegbe, Chiristiana Chiamaka Okeke, and Deborah Oluwadamilola Esho

1 Introduction

Paratransgenesis is an innovative vector-borne disease control strategy that genetically modifies symbiotic microorganisms within vectors to produce effector molecules. These effector peptides target and either inhibit or completely eliminate the development of parasites within the vector, thereby reducing or preventing disease transmission. By altering the vector's internal microbiome to interfere with the parasite's lifecycle, this approach holds significant promise for controlling diseases like malaria, dengue, and Chagas, offering a novel method to interrupt transmission at the source (Aminata et al., 2022). The rapid decline in the efficacy of current control strategies against insect-borne diseases (IBD) has necessitated adopting the paratransgenesis approach in malaria control. Insect-borne diseases cause a high level of human death and illness and are transmitted by infected vectors through biting or during blood meal feeding (WHO, 2020). Malaria, an insect-borne disease, is responsible for global human mortality and morbidity (Aminata et al., 2022). The global burden of malaria increased in over 85 malaria-endemic countries from 244 million in 2021 to 249 million in 2022, with 95% of malaria-related deaths (WHO, 2023a, b). Pakistan had the highest increase in the number of malaria cases from 500,000 in 2021 to about 2.6 million cases in 2022. This increase was attributed to the devastating flooding that occurred in the country from June to October 2022, while in Iran, an increase of more than 1000 cases was recorded. A total of 608,000 deaths due to malaria occurred globally in 2022 with 50% of these deaths occurring in four African countries which are Nigeria, the Democratic Republic of Congo, Niger, and Tanzania (WHO, 2024). In Nigeria, a total of 68 million cases were recorded in 2021 with 194,000 deaths, of which 80% were children below the age

O. Oziegbe · C. C. Okeke (✉) · D. O. Esho
Department of Biological Science, Covenant University, Ota, Nigeria
e-mail: christiana.okekepgs@stu.cu.edu.ng

P. O. Isibor, P. A. Akinduti (eds.), *Harnessing Biotechnology Tools for Product Development*, https://doi.org/10.1007/978-3-031-86002-7_16

of 5 years. The burden of malaria remains the highest in Nigeria with figures reaching almost 27% of the total global burden, and transmission risk increasing all year round in the northern and eastern parts of the country (WHO, 2023a, b). The World Health Organization has considered several strategies to curb malaria in the last decade. These strategies though effective have raised concerns, especially since threats from vector resistance to insecticides and parasite resistance to Artemisinin may hinder their impact on malaria control (Lindsay et al., 2021). However, new robust technologies targeting both the parasite and the mosquito vector have been developed as additional control measures, to mitigate the impacts of these currently envisaged challenges. These technologies were developed on the premise that mosquitoes were essential to malaria transmission, and several limiting factors influenced parasite development and transmission in the mosquito (Huang et al., 2022). Approaches such as transgenesis and paratransgenesis involve the mosquito's genetic alteration, making it a poor vector for the malaria parasite. Transgenesis genetically alters the mosquito, enabling the release of antimalarial compounds, which reduces the competence of the vector and makes it refractory to disease transmission (Shengzhang et al., 2022). On the other hand, paratransgenesis utilizes genetically altered symbiotic microorganisms in the mosquito to produce and secrete antimalarial compounds that interrupt parasite development while inhibiting disease transmission (Aminata et al., 2022). However, paratransgenesis is a more considered approach as it mitigates the current challenge of introducing the genes that confer refractoriness into mosquito populations (Carballar-Lejarazú et al., 2020). Also, ethical concerns about releasing genetically transformed organisms into the environment make paratransgenesis a more preferred approach. This review, therefore, aims to discuss paratransgenesis as a malaria control strategy while highlighting the approach's current perspectives and future research needs. It encompasses a holistic examination of the requirements for a successful paratransgenesis process, while analysing the strategy for possible implementation in malaria-endemic countries.

1.1 Mosquito-Borne Diseases

These are diseases that are transmitted when an infected female mosquito bites an individual during blood meal feeding (Onen et al., 2023). Mosquitoes are carriers of several disease-causing pathogens and they act as vectors, transmitting diseases such as malaria, Zika, dengue fever, chikungunya, yellow fever, West Nile virus, and Rift Valley fever, among others to humans (Sanfelice, 2022). The most prominent vectors of these diseases belong to the *Aedes, Culex,* and *Anopheles* genera of mosquitoes, mostly found in Africa, Asia, South America, and Europe. Among these diseases, malaria, which is transmitted during blood meal feeding by an infected female Anopheles mosquito and caused by the *Plasmodium* parasite, is the most life-threatening (Venkatesh et al., 2019). West Nile virus is transmitted to birds, humans, and animals through the bite of the *Culex* mosquito, causing a range of infections. Dengue fever, yellow fever, Zika virus, and Chikungunya are transmitted by the bites of infected *Aedes* mosquitoes such as the *Aedes aegypti*, *Aedes albopictus*, and *Aedes furcifer-taylori* species, to humans and non-human primates while mosquitoes from the *Aedes, Anopheles,* and *Culex* genre transmit Rift Valley fever (Onen et al., 2023). These diseases pose significant public health concerns, especially in tropical and subtropical regions, where mosquitoes thrive.

1.2 The Malaria Parasite

In humans, malaria is caused by the malaria parasite, transmitted during feeding by an infected female *Anopheles* mosquito. *Plasmodium, a unicellular protozoan, is the cause of malaria with over 200 species* of the parasite infecting not only humans but also other animals. Notable *Plasmodium* species that infect humans include *Plasmodium falciparum*, *Plasmodium vivax*, *Plasmodium malariae*, and *Plasmodium ovale* while *Plasmodium knowlesi* infects macaque monkeys (Sato, 2021). All plasmodium species have similar life cycles, consisting of two stages (Fig. 16.1). In the first stage, the parasite infects a host, while in the second stage, it is transmitted from the infected host to another host by a vector. During the first stage of the *Plasmodium* life cycle, sporozoites from an infected mosquito are transmitted into the host after a mosquito bite. The sporozoites migrate to the liver and transform into merozoites in a schizogony process. Merozoites in the blood infect erythrocytes and multiply through schizogony, while some differentiate into gametocytes (Bancells et al., 2019). In the second stage of the parasite life cycle, gametocytes are picked up from the host's bloodstream following a blood meal, into the midgut of the mosquito. In the midgut, male and female gametocytes transform into microgametes and macrogametes respectively. The fusion of the microgametes and

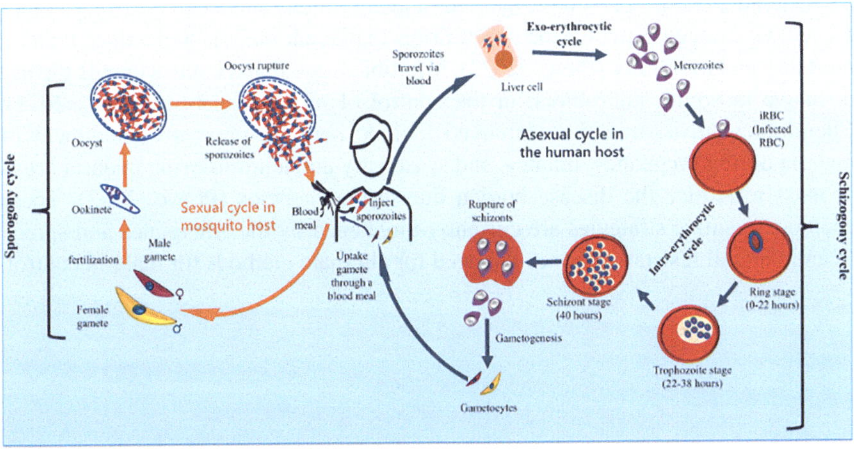

Fig. 16.1 The malaria life cycle. Source: Tripathi et al. (2023)

macrogametes produces a zygote that has a diploid genome. The zygote is differentiated by meiosis to form an ookinete, which moves into the mosquito's midgut, forming an oocyst (Siciliano et al., 2020). Following mitosis, formed sporozoites are released into the hemolymph from where they migrate to the salivary gland of the mosquito, gain the ability to infect humans, and are ready to infect another host when released during a blood meal (Araki et al., 2020).

1.3 Malaria Control Strategies

Malaria control strategies are approaches employed in the control of malaria, to reduce or prevent disease transmission by targeting either the vector or the malaria parasite. Vector control includes chemical, biological, and mechanical methods that kill or inhibit the activities of mosquitoes, preventing malaria transmission (Onen et al., 2023). Chemical methods such as treated mosquito nets, indoor residual spraying, spraying of peridomestic spaces, and mosquito repellents target the adult stages of mosquitoes. Treated mosquito nets contain chemicals such as pyrethroids that serve as physical barriers against mosquitoes (Tungu et al., 2021). Indoor residual spraying involves insecticide spraying in houses and areas that serve as mosquito resting places (Manrique-Saide et al., 2020). The spraying of pulverized insecticides such as dichlorodiphenyltrichloroethane (DDT) around houses and the use of mosquito repellents prevents mosquitoes from biting people (Dahmana & Mediannikov, 2020). Biological malaria control includes the use of agents such as *Bacillus thuringiensis*, predatory fish, and microorganisms to mitigate the hazardous effects of chemical control. Mechanical control, on the other hand, involves the use of traps such as eave tubes or attractive sugar baits laced with chemical attractants to attract adult female mosquitoes (Onen et al., 2023). This method can also be used to trap mosquitoes for laboratory and field studies. Parasite control on the other hand includes chemo-preventive methods and microbiota-mediated control. Chemo-preventive strategies involve the use of drugs to prevent malaria and reduce malaria mortality and morbidity (Plowe, 2022). Microbiota-mediated control entails the use of native mosquito microbiota, in the control of malaria (Gabrieli et al., 2021). Chemo-preventive strategies employed include periodic preventive treatment of malaria during pregnancy, infancy, and seasonally and administering malaria drugs in mass to reduce the disease burden during emergencies (Plowe, 2022). These chemo-preventive strategies are currently threatened by the emergence and spread of antimalarial resistance, hence the need for alternate methods for malaria control.

1.4 Microbiota-Mediated Control of Mosquito-Borne Diseases

Microbiota-mediated control utilizes the natural microbial communities found in the mosquito, to suppress or inhibit the transmission of disease-causing pathogens (Gabrieli et al., 2021). This approach employs the complex interactions between the mosquito and its commensal or symbiont microorganisms to control disease transmission. Colonization of mosquitoes with microorganisms and paratransgenesis are methods of microbiota-mediated control of mosquito-borne diseases. In the colonization of mosquitoes with microorganisms, members that play key roles include *Wolbachia* and *Asaia* species. *Wolbachia* are Gram-negative bacteria found in a majority of insects including mosquitoes that invade reproductive organs, altering the reproduction of their host and reducing vector competence (Manoj et al., 2021). *Asaia* are Gram-negative bacteria that colonize the midgut and male gonads of mosquitoes. *Asaia* stimulates mosquito immunity, preventing the growth and development of the malaria parasite and interrupting the parasite's transmission cycle (Cappelli et al., 2019). Genetically altered symbionts of mosquitoes in paratransgenesis produce effector molecules that eliminate or inhibit *Plasmodium* parasite development (Ratcliffe et al., 2022). Paratransgenesis also reduces the competence of the mosquito vector, making it refractory to disease transmission (Shengzhang et al., 2022).

1.4.1 Colonization of Mosquitoes with Microorganisms

The colonization of mosquitoes with microorganisms directly affects the competence of the mosquito, the success of disease transmission, and other biological traits such as mosquito developmental stages, lifespan, choice of mate, fecundity, and fertility (Romoli et al., 2021; Wu et al., 2019). The mosquito microbiota includes bacteria, viruses, protozoans, and fungi, affecting the host's phenotypic expressions. Several biotic and abiotic factors such as the host, genetics, and environmental factors determine the microbial colonization and composition of the mosquito (Cansado-Utrilla et al., 2021). A study by Hegde et al. (2019) explored the role of microbial genetics in the determination of the composition of microorganisms in the mosquito gut. The modification of the *ompA* gene influences the expression of phenotypes such as a reduced prevalence to infection by the malaria parasite in *Aedes aegypti* (Hegde et al., 2019). Microorganisms are found in various mosquito tissues such as the reproductive organs, salivary glands, midgut, and other tissues, enabling direct interaction with pathogens or indirectly affecting vector competence (Hegde et al., 2020). Vector competence determines the susceptibility of the mosquitoes to pathogen infection and mosquito gut microbiota, which can either directly interfere with pathogens through lysis or indirectly by affecting the salivary gland and midgut barriers (Gao et al., 2020). The microorganisms colonizing the adult and larval stages of the mosquito vary due to changes in environment and feeding habits. According to a study by Souza et al. (2019), *Asaia* and *Escherichia coli* fed to the adult and larval stages of *Aedes aegypti* mosquitoes resulted in better nutritional and developmental parameters for the mosquitoes and were therefore regarded as the best candidates for symbiont-based malaria control. Mosquito larvae mainly consume bacteria and plankton while the adult terrestrial mosquitoes feed mostly on nectar and blood, leading to an increase in a particular type of microbial community and a decline in others (Souza et al., 2019). Generally, the bacterial floras in mosquitoes include the Gram-negative phylum *Proteobacteria* and Bacteroidetes, Gram-positive phylum Firmicutes, and *Clostridia* and *Spirochetes*. The Gram-negative bacteria in the mosquito can reduce the development of *Plasmodium* due to the presence of lipopolysaccharides in the cell wall acting as a physical barrier (Fazal et al., 2023). The most common bacteria identified in *Anopheles* mosquitoes include *Pseudomonas, Asaia, Elizabethkingia, Wolbachia, Enterobacter, Serratia, and Pantoea with the gut being mostly dominated by Pantoea* and *Asaia. The ability of Pantoea to cross-colonize several mosquito species has been useful for paratransgenesis, while Asaia* functions by producing molecules that prevent *Plasmodium* invasion of epithelial tissues and salivary glands (Fazal et al., 2023). Fungi also form part of the mosquito microbiota, performing either mutualistic, commensal, or pathogenic roles that preserve the ecological balance of mosquitoes. Common fungi in mosquitoes include pathogenic *Aspergillus* and *Penicillium* and entomopathogenic *Beauveria* and *Metarhizium* species. Also, yeasts, such as *Candida, Pichia*, and *Wickerhamomyces* found in *Aedes* and *Anopheles* mosquitoes, have been found to mask signals in these vectors (Fazal et al., 2023). In addition, different viral families have been identified in mosquitoes such as *Bunyaviridae, Flaviviridae, Reoviridae, Chuvirus*, and *Negevirus* groups. These viruses exist as commensals because of their inability to infect cells, prolong host infection, and promote vertical transmission (Fazal et al., 2023). For instance, a study by Nasar and colleagues on the Eilat virus (EILV) found in *Anopheles coustani* indicated its inability replicate in vertebrate cells and be horizontally or vertically transmitted in *Aedes albopictus* mosquitoes (Nanfack & Vernick, 2018).

2 Paratransgenesis

The paratransgenesis approach to control mosquito-borne diseases requires the genetic alteration of symbiont microorganisms such as bacteria, fungi, and viruses in the mosquito to release transgene products that impair pathogen development and transmission (Tharaka et al., 2020). This process negatively impacts the vector's competence and reduces the transmission of mosquito-borne diseases. In paratransgenesis, transformed symbiont microorganisms are able to survive in the host by adapting to the host conditions while remaining stable, and producing effector molecules. They are also easily introduced into vectors through feeding and transmitted in vector populations through horizontal, vertical, and trans-stadial routes (Tharaka et al., 2020). Antimicrobial peptides produced in paratransgenesis can affect the mosquito microbiota, the pathogen, and even the symbiont transgene leading to indirect effects such as reduction in microbiota concentration, transgene loss, and general decline in the efficiency of the approach (Aminata et al., 2022). Hence, the choice of a stable mosquito symbiont that can persist in the vector after modification without posing any fitness cost is crucial to the success of the paratransgenesis approach (Ratcliffe et al., 2022).

The mosquito gut is the major site of microbial colonization but microorganisms are also associated with other tissues such as the salivary glands and reproductive organs (Hixson et al., 2024). Intrinsic and extrinsic factors influence the variation of microorganisms in the mosquito gut. Intrinsic factors such as age and sex have been studied in *Aedes aegypti* and *Aedes albopictus* species (Wang et al., 2018). Studies on adult *Aedes aegypti* and *Aedes albopictus* indicated more *Asaia* and *Wolbachia* in older mosquitoes of each species, respectively, compared to younger mosquitoes. Also, studies on the influence of sex on microbiota variation indicated that male *Aedes albopictus* contained more *Enterobacter, Pantoea,* and *Pseudomonas* while the female mosquitoes contained more *Acinetobacter, Enterobacter,* and *Micrococcus species.* Microbiota variation also exists between different mosquito species, such as in the larva of *Aedes albopictus* and *Anopheles species* when reared together in similar environmental and diet conditions (Saab et al., 2020). Major determinants of mosquito gut microbiota variation are environment associated extrinsic factors such as geographical locations and study sites. Studies on *Aedes aegypti* collected from different geographical locations in India showed variations in microbiota composition which included *Enterobacter* and *Bacillus species* in mosquitoes from Predash and *Acinetobacter* and Pseudomonas species in mosquitoes from Tezpur. Also, similar mosquito species reared in different laboratories showed different microbiota compositions while different species of mosquitoes reared in similar conditions showed similar microbiota compositions (Lin et al., 2021). Microbiota composition is key to the success of the paratransgenesis approach. Symbiont microorganisms in mosquitoes are more adaptable and amenable to genetic alterations as they are familiar with the host conditions, resist attack by host immune responses, and generally survive competition with natural mosquito microbiota (Shane et al., 2018).

In paratransgenesis, genetically transformed mosquito symbionts are designed to produce effector molecules that eliminate or inhibit parasite development. Studies on *Anopheles gambiae* and *Anopheles stephensi* describe the production and delivery of anti-plasmodium effector molecules that decrease *Plasmodium falciparum* sporozoites and oocyst counts (Wang et al., 2017). The production of several effector peptide from a transgene microorganism is better appreciated than when a single effector molecule is produced, as the combined effects of these peptides are more lethal to the pathogen compared to the effects of a single effector molecule (Ratcliffe et al., 2022). The mechanism of action of these effector peptides includes parasite lysis, interaction with the mosquito epithelial surfaces and parasite surface proteins to inhibit parasite binding and invasion as well as the stimulation of anti-parasite immune responses by the vector (Ratcliffe et al., 2022). Peptides involved in parasite lysis include those produced by the mosquito's immune system such as the defensins, gambicins, and cecropins. Others include peptides from non-mosquito sources which include scorpine, a scorpion antimalaria lytic peptide, angiotensin II, gemosin, and synthetic anti-parasitic peptides Shiva-1 and Shiva-2 (Wang & Jacobs-Lorena, 2013). Plasmodium enolase-plasminogen interaction peptide (EPIP) and single-chain monoclonal peptides (scFvs) such as scFv4B7 bind to the *Plasmodium falciparum* ookinete surface protein Pfs25 and interacts with the parasite to inhibit midgut invasion. Salivary gland and midgut peptides (SM1) bind to receptors on the salivary gland and midgut epithelia of the mosquito to block sporozoite and ookinete invasion while manipulated mosquito immune system genes such as the Akt and Rel 2 genes make mosquitoes resistant to infection by *Plasmodium falciparum* (Wang & Jacobs-Lorena, 2013). In a study by Wang et al. (2017), recombinant Sarratia strains were engineered to produce a fusion protein consisting of Midgut peptide 2 (MP2), Scorpine, four EPIP molecules, Shiva1, and two SM1 peptide for the inhibition of *Plasmodium falciparum* infection in Anopheles gambiae. A sugar meal containing the altered symbionts was fed to the mosquitoes for 48 hours, after which the mosquitoes were fed a *Plasmodium falciparum* infected blood meal. 93% of the *Plasmodium falciparum* oocyst load was reduced by the *Serratia* strain expressing effector molecules (Wang et al., 2017). Genetically altered symbionts are delivered into wild vector populations through feeding and they spread in the population through horizontal, vertical, or trans-stadial routes (Wang et al., 2017). The transfection of the symbiont is designed to inflict minimal fitness cost on the vector and to target only specific insect populations (Ratcliffe et al., 2022).

2.1 Requirements for Effective Paratransgenesis

A successful paratransgenesis requires the presence of a symbiont that is easily culturable and genetically manipulated (Ratcliffe et al., 2022). The symbiont association with the insect vector should be stable, its vertical or horizontal transmission should be efficiently executed while it produces effector molecules for a long period in the vector (Gabrieli et al., 2021). The symbiont should be as fit as the wild strain, not adversely affecting the host, while secreting antagonistic molecules that guarantee interaction with the target pathogen (Gabrieli et al., 2021). It is also important for the symbiont to be found in all the developmental stages of the host life cycle while being able to spread through a vast number of host species. The transmission of the modified microorganism within wild vectors in field conditions must be facilitated by specialized techniques and procedures such as electroporation, microinjection, use of viral vector mechanisms, and double- stranded RNA (dsRNA) delivery methods (Chong et al., 2021). Also, authorization to apply paratransgenesis must be obtained from relevant authorities before genetically modified organisms can be introduced into the environment. This is essential to ensure the safety of humans and non-target populations (Ratcliffe et al., 2022). The first trial to investigate the effectiveness of paratransgenesis for malaria control in semi-field conditions was carried out by Mancini et al. (2016). This study highlighted the need for risk assessment and adherence to strict regulatory protocols and guidelines for field release of genetically altered organisms.

2.1.1 Choice of Symbiont

The choice of symbionts for paratransgenesis extends from bacteria to include fungi and viruses, even though bacteria are utilized more often in studies because they are easily culturable and amenable to genetic manipulation (Ratcliffe et al., 2022). Bacteria, viruses, and fungi used in paratransgenesis are non-pathogenic strains that have been identified using high-throughput sequencing techniques (HTS) and culture-dependent methods. Bacteria symbionts used in paratransgenesis for the production of effector molecules include *Asaia, Pantoea, Serratia, Enterobacter, Escherichia, and Pseudomonas* (Ratcliffe et al., 2022). Mosquito densoviruses have also been explored for paratransgenesis studies due to their stability and ability to colonize mosquito populations through vertical, horizontal, and trans-stadial transmission (Ratcliffe et al., 2022). The presence of small genomes also makes it easy to genetically transform these viruses for effector gene expression (Johnson & Rasgon, 2018). Fungi are explored in paratransgenesis due to their long survival times as spores and their ability to directly infect insects through the exoskeleton. Yeasts such as *Wickerhamomyces anomalus and Metarhizium anisopliae* have been explored in paratransgenesis studies in the *Anopheles* mosquito, for the production of anti-plasmodium molecules such as scorpine, salivary, and midgut molecules (SMI) and PfNPNA sporozoite-binding antibody (Cappelli et al., 2019).

2.1.2 Choice of Gene Modification

The gene modification techniques employed in paratransgenesis are plasmid vector systems utilized for microbial gene transformation (Ratcliffe et al., 2022). Recent developments in gene modification include the use of the RNA- interference systems and the clustered regularly interspaced short palindromic repeats with CRISPR-associated endonuclease (CRISPR-CAS) technique for insect symbiont genome editing and transformation (Yang et al., 2020). In a study by Hegde et al. (2019), CRISPR/Cas9 was used to change the outer membrane protein A (ompA) gene of the symbiont *Cedecea neteri* symbiont obtained from *Aedes* mosquitoes. This method was employed to determine the role of bacterial genetic factors in mosquitoes' biology and gut colonization. The study indicated a reduced infection prevalence in the adult mutant mosquitoes compared to the wild type. The genome modification techniques employed therefore should be efficient and suitable for the transformation of the microorganism, producing consistent and robust knockdowns with fewer off target events (Ratcliffe et al., 2022). The process of gene modification involves using specialized tools to cut, splice, and integrate specific genes into the microbial cell, thereby endowing the modified microorganism with the desired traits for effector molecule production (Liu et al., 2022).

2.1.3 Choice of Effector Molecules

Symbiotic microorganisms should produce effector molecules to eliminate or inhibit pathogen development without adversely affecting the vector, the symbiont, and the surrounding microbiota (Aminata et al., 2022). Table 16.1 shows some effector molecules against *Plasmodium falciparum,* produced by symbiont microorganisms in the *Anopheles* mosquito (Wang & Jacobs-Lorena, 2013). These effector molecules are categorized based on their anti-parasite activities, including parasite lysis, inhibition of midgut invasion, blockage of salivary gland and midgut epithelial invasion, and immune system modulation (Ratcliffe et al., 2022).

2.1.4 Transgene Delivery Methods

These methods are required to transfer the transformed symbionts into the mosquito and facilitate their spread through wild mosquito populations, without fitness cost to the vector or transfer to non-target populations (Ratcliffe et al., 2022). Studies by Mancini et al. (2016) using wide cages to explore horizontal and vertical transfer of genetically transformed *Asaia* species showed the feasibility of transferring transformed bacteria into *Anopheles* mosquito populations through feeding, using cotton

Table 16.1 Microorganisms producing anti-plasmodium effector molecules

	Effector	Characteristics	Microorganism	Anti-parasitic activity
1	**Parasite elimination**			
	Scorpine	Scorpion venom	*Serratia, Asaia, Metarhizium anisopliae*	Lysis
	DEF1 A	*An. gambiae* defensin A	*Wolbachia*	Lysis
2.	**Parasite surface protein interaction**			
	EPIP	Enolase-Plasminogen Interaction Peptide	*Serratia*	Inhibition of midgut invasion
	Pbs21scFv-Shiva1	Single-chain immunotoxin	*Pantoea agglomerans*	Targets major ookinete surface protein
3.	**Midgut or salivary gland epithelia interaction**			
	SM1	Salivary gland and midgut peptide 1	*Pantoea agglomerans, Metarhizium anisopliae*	Blocks invasion of the midgut and salivary gland epithelium
	mPLA2	Bee venom phospholipase	*Pantoea agglomerans*	Inhibits ookinete midgut invasion
4.	**Immune system manipulation**			
	Akt	A protein kinase	*Wolbachia*	Akt boosts innate insulin immunity
	Rel2	*Anopheles* IMD pathway transcription factor	*Wolbachia*	Enhances mosquito IMD pathway

Source: Wang and Jacobs-Lorena (2013), Ratcliffe et al. (2022)

pads soaked in sucrose (Ratcliffe et al., 2022). Another study by Arora et al. (2019) showed that engineered *Pantoea agglomerans* were transferred through feeding using microencapsulation in an alginate hydrogel into *H. vitripennis* insects. The bacteria were also horizontally transmitted between *H. vitripennis* populations (Arora et al., 2019). Furthermore, Wang et al. (2017) revealed that transformed *Serratia* can be transferred horizontally, vertically, and trans-stadial through feeding in male and female *Anopheles stephensi* populations (Wang et al., 2017). These methods of transfection of engineered microorganisms have been successfully employed in the colonization and spread of genetically engineered microorganisms in mosquito populations. These methods were also employed to ensure the persistence of these transformed microorganisms in target vector populations without release or contamination of the environment (Ratcliffe et al., 2022).

3 Paratransgenesis in Various Vectors

3.1 Anopheles Mosquito

Over 400 known Anopheles species have about 30–40 competent malaria vectors. *Anopheles gamb*iae and *Anopheles funestus* are the most active malaria vectors in sub-Saharan Africa (Mwima et al., 2023). Treated mosquito nets prevent mosquito species that bite at night such as *Anopheles gambiae* and *Anopheles coluzzii* from accessing human hosts while indoor residual spraying kills mosquitoes on contact or repels them from humans (Sougoufara et al., 2020). However, there have been changes in the resting and biting behaviors of the Anopheles species with increased outdoor biting, leading to a new method to control outdoor acquired transmission (Carnevale & Manguin, 2021). In paratransgenesis, bacteria isolated from the female *Anopheles* mosquitoes antagonize the developmental stages of the pathogen in the mosquito, reducing malaria transmission. Bacteria species commonly identified in the *Anopheles* microbiome include *Serratia, Enterobacter, Asaia, Elizabethkingia*, and *Pseudomonas*. These microorganisms are exploited in malaria control and are found predominantly in the midgut and gonads of the mosquito (Egyirifa & Akorli, 2024). Identifying symbionts in female *Anopheles* mosquitoes has been of major interest because they seek blood meal and transmit diseases, hence the assumption that paratransgenesis will be more effective when microbes are obtained from the female *Anopheles* mosquito. The male *Anopheles* mosquitoes have however been considered more important for paratransgenesis because of their ability to sexually transfer microbes during mating and effectively introduce selected bacteria into the mosquito population (Ross et al., 2022).

The mechanism of paratransgenesis in the *Anopheles* mosquito includes the delivery of genetically altered symbiotic microorganisms into the mosquito gut through feeding which then produces anti-plasmodial effector molecules for inhibition of plasmodium development (Gabrieli et al., 2021). *Anopheles* symbionts, such as *Pantoea, Serratia*, and *Asaia*, are bacteria that maintain promising features for paratransgenesis. *Pantoea agglomerans* can persist in the insect gut efficiently and have been engineered to express anti-*Plasmodium* factors such as SM1, anti-Pbs21, PLA2, which prevent parasite development (Gabrieli et al., 2021). *Serratia* species found in male and female *Anopheles stephensi* mosquitoes release effector molecules such as scorpine and Plasmodium enolase–plasminogen molecules with minimal fitness cost for the insect and efficiently inhibit *Plasmodium falciparum* infection (Chiamaka et al., 2020). *Asaia* has been successfully engineered to secrete different effector proteins including scorpine that have significantly inhibited the development of *Plasmodium species* (Grogan et al., 2021). Other symbiotic organisms of the *Anopheles* mosquito such as fungi, rickettsia, and viruses have also been investigated alongside bacteria for their potential use in paratransgenesis. *Metarhizium anisopliae, a fungus,* has been engineered to produce scorpine and anti-plasmodial peptide SM1 (Qadri et al., 2020). Densonucleosis viruses in *Anopheles gambiae* have been used in viral paratransgenesis, to control malaria by stably introducing anti-plasmodium peptides or insect-specific toxins to inhibit pathogen development (Gabrieli et al., 2021). Also, Rickettsiales, such as *Wolbachia* which are maternally inherited, are potential targets for delivering effector molecules such as Defensin 1 (DEF1 A), in paratransgenic systems. Although the population of *Wolbachia* in Anopheles species is varied, its presence prevents malaria transmission by inhibiting sporozoites in *Anopheles gambiae* (Ross et al., 2020).

3.2 Triatomine Bug

Triatomine bugs transmit *Trypanosoma cruzi* which causes Chagas disease and the interactions between the parasite, its host, and the triatomine microbiome have been exploited in paratransgenesis (Schaub, 2021). The triatomine microbiota consists of Gram-negative organisms such as *Serratia*, *Pantoea*, *Pseudomonas*, and *Wolbachia as well as* Gram-positive organisms such as *Rhodococcus and Mycobacterium* (Schaub, 2021). The initial step in Paratransgenesis was carried out in the Triatomine bug *Rhodnius prolixus,* using genetically engineered *Rhodococcus rhodnii* bacterium to produce cecropin A against *Trypanosoma cruzi.* Tsetse flies transmit African trypanosomes which cause Trypanosomiasis or sleeping sickness (Holanda-Freitas et al., 2020).

3.3 Tsetse Flies

Tsetse flies are hosts to a wide range of bacterial species such as *Wigglesworthia glossinidia, Sodalis glossinidius, Wolbachia, Spiroplasma, Bacillus, Serratia, Enterobacter, Pantoea, Enterococcus, Providencia, Lactococcus, Staphylococcus,* and *Pseudomonas* (Kame-Ngasse et al., 2018). Important requirements for paratransgenesis in tsetse flies include effective symbiont transfer to offspring through vertical transmission and the stable expression of heterologous genes in the vector. Also, the identification and expression of anti-parasitic molecules by the modified symbiont that is capable of eliminating trypanosomes without any adverse effect on the symbiont and the vector is important for the success of paratransgenesis in tsetse flies.

3.4 *Sand Flies*

Sand flies transmit Leishmania which causes human leishmaniasis when the female flies seek blood meal to develop their eggs. A major requirement for paratransgenesis in sand flies is the identification of suitable symbiotic organisms for genetic modification that are non-pathogenic to humans (Papadopoulos et al., 2020). Dominant bacteria in the Sandflies are Gram-negative organisms, and some Gram-positive strains include *Bacillus subtilis* and *Enterobacter species* (Papadopoulos et al., 2020). Rodents serving as food baits have also been used as vectors of genetically modified bacteria such as *Serratia* and *Enterobacter cloacae* into sand flies to inhibit Leishmaniasis transmission (Ghassemi et al., 2023).

4 Paratransgenesis Strategy for Malaria Control

Paratransgenesis strategy for control of malaria involves a series of protocols geared towards developing a genetically modified mosquito symbiont with the potential of producing anti-plasmodial compounds that can kill or hinder the development of *Plasmodium* species. The protocol for this strategy includes the identification of a non-pathogenic culturable symbiont in the mosquito vector, the genetic modification of the symbiont, the colonization of the symbiont and its spread in the vector population, field testing, monitoring and evaluation, and the steps to obtain regulatory approval for implementation (Ratcliffe et al., 2022).

4.1 *Symbiont Stage Identification*

This stage in the paratransgenesis process involves identifying suitable symbiont microorganisms in the vector. This can be achieved using polymerase chain reaction (PCR) techniques or high-throughput sequencing (HTS), which is an advancement of the *16S* RNA technique. The HTS identifies microorganisms in tissue samples for paratransgenesis and the selected microorganisms are multiplied using culture techniques (Allaband et al., 2019). Culturomics can also be used to identify bacterial species that were previously not known in vectors using either the matrix-assisted laser desorption/ionization-time of flight (MALDI-TOF) mass spectrometry or 16S ribosomal DNA (rDNA) amplification and sequencing. Novel viruses and fungi can be identified using shotgun metagenomic sequencing or culture-dependent methods (Ratcliffe et al., 2022). The identification of non-engineered mosquito symbiont strains that have a natural tendency to block malaria transmission would be an added advantage during paratransgenesis as this increases the ease of blocking transmission, creating a lasting impact when transgenes are lost over time. *Enterobacter species* interferes with parasite development, making mosquitoes refractory to *Plasmodium falciparum* by producing reactive oxygen species (Aminata et al., 2022). *Serratia marcescens* inhibits *plasmodium species* in mosquitoes by interfering with the immunity that prevents parasite development (Bai et al., 2019), Also, *Anopheles gambiae* and *Anopheles coluzzii* impact negatively the latter stages of *Plasmodium falciparum* development (Aminata et al., 2022), while *Microsporidia* impairs the development of the parasite in *Anopheles gambiae* (Herren et al., 2020).

4.2 Genetic Modification of Symbiont

This is a process of introducing genetically transformed genes that produce anti-plasmodial compounds that kill or disrupt the developmental stages of the parasite, into selected symbionts. Parasites are reduced in mosquitoes with early Plasmodium infection, and paratransgenesis can be employed to enhance this process (Aminata et al., 2022). This can be achieved by producing efficient effector molecules that kill or block parasite development, without impairing the vector's or its microbiota's fitness. This process of genetic modification is known as transgene development (Aminata et al., 2022). Transgene expression can be achieved using type 1 secretion system HasA, as used in *Serratia*, to allow for the secretion of transgenic molecules through membranes. The expression of the transgene must be at a high-efficiency level, with no fitness cost for the microorganism. The stability of the introduced anti-parasitic gene is also of concern as the development of inducible transgenes reduces fitness costs for the microorganism (Huang et al., 2020). Also of importance is the consideration of antibiotic resistance in transgene development (Aminata et al., 2022).

4.3 Vector Colonization and Spread

Paratransgenesis does not aim to eliminate the mosquito vector but to alter it into an incompetent vector, such that the same mosquito species occupy the same environment with a reduced capacity to transmit disease (Utarini et al., 2021). The rapid spread of foreign genes must be achieved without fitness cost to the mosquito or the transfection of non-target insects. The modified microorganisms can be introduced in the vector either by oral delivery using sugar meals in soaked cotton pads or through mechanical delivery by microinjection or topical application. Transmitted microorganisms can spread in the mosquito by horizontal, vertical, or trans-stadial transmission. Horizontal spread has been recorded in *Anopheles stephensi* and *Anopheles gambiae* populations. Vertical, horizontal, and trans-stadial spread of transformed *Serratia* have also been recorded in *Anopheles stephensi* (Ratcliffe et al., 2022). These records are important in the evaluation of paratransgenesis, and data from semi-field trials are needed to develop methods to improve the transfer of transformed bacteria to target sites in the vector (Arora et al., 2019).

4.4 Field Testing, Monitoring, and Evaluation

Paratransgenesis involves an in-depth study of the environment starting from laboratory studies to semi-field trials and ultimately field trials (Aminata et al., 2022) In semi-field conditions, the ability of paratransgenic microorganisms to colonize mosquito populations is observed as a means of evaluating the paratransgenic process. Transgenic *Asaia* bacteria have been observed to efficiently spread in *Anopheles gambiae* and *Anopheles stephensi* populations. The semi-field trials provide an avenue for monitoring and evaluation, to assess the effectiveness and safety of transformed genes used in the paratransgenesis process before field trials (Aminata et al., 2022). Also, the observed dispersal and mating behavior in semi-field trial mosquitoes determine the spread of refractoriness to *Plasmodium* infection as colonization by microorganisms affects mating behavior (Romoli et al., 2021, Wu et al., 2019). Semi-field trials therefore provide an opportunity to take into account various factors the mosquitoes are likely to encounter in the field such as fluctuations in temperature and humidity, seasonality, diet, blood source, and larval breeding sites, which may affect the survival of the mosquito (Romoli & Gendrin, 2018). Field trials involving paratransgenic mosquitoes have so far not yet been performed, as this approach would require acceptance by the general population (Bilgo et al., 2018).

4.5 Regulatory Approval and Implementation

Acceptance of the paratransgenesis approach requires adequate investigations to ensure safety and applicability and frequent interactions with relevant stakeholders to mitigate possible envisaged environmental changes that may affect human health (Cisnetto & Barlow, 2020). Necessary evidence that the modified microorganisms are stable, and their release will pose no threat to humans and animals is required to gain support for implementing this approach. There is also a need to prove that the approach does not infect harmless insects through horizontal transmission (Romoli & Gendrin, 2018). Other areas such as the ease of spread of the transgenic bacteria, trans- stadial transmission, and their persistence in several vector generations without causing resistance must be properly investigated before implementation can be considered (Ratcliffe et al., 2022). Specific funding and resources must be allocated for these investigations and public engagements, beginning from the early stages of the project and throughout the project. This is required to improve the paratransgenesis strategy and foster long-lasting interactions between stakeholders while mitigating any adverse effects that may arise during the project (Thizy et al., 2019).

5 Current Perspectives

The current perspectives in paratransgenesis for malaria control involve scientific advancements that are aimed at improving the paratransgenic process and disrupting parasite infection within the mosquito. These trends, which involve a shift to more targeted and sustainable approaches towards vector control, using cutting-edge genetic engineering techniques that leverage microorganisms' anti-plasmodial properties, are effective strategies for malaria control (Ratcliffe et al., 2022). The current perspectives involving the multistage targeting of the malaria parasite, identification of symbiotic microbes with anti-plasmodial properties, and the application of genetic tools to enhance the persistence and spread of paratransgenic microorganisms provide a new and holistic approach to malaria control.

5.1 Multistage Targeting of Malaria Parasite

Multistage targeting of the malaria parasite can be achieved using *Wolbachia,* a bacterium of the Rickettsiales family, which functions as a biological control strategy against malaria transmission (Dahmana & Mediannikov, 2020). In the developmental stages of the mosquito, *Wolbachia* uses its cytoplasmic incompatibility property to kill mosquitoes or inhibit *Plasmodium* development through transmission-blocking, resulting in decreased malaria transmission (Walker et al., 2021). Cytoplasmic incompatibility causes the death of mosquito embryos when a male mosquito infected with *Wolbachia*, mates with an uninfected female mosquito (Shropshire et al., 2021). This property has led to the exploration of the possibility of carrying out paratransgenesis using *Wolbachia*, even though this area has not been explored because *Wolbachia* is not easily cultured or genetically manipulated. However, the report on the presence of plasmids in *Wolbachia* by Reveillaud et al. (2019) indicates the possibility of the use of *Wolbachia* in paratransgenesis. Presently, releasing male mosquitoes infected with *Wolbachia* to prevent offspring development and cause decline in the number of adult mosquitoes in specific areas is being considered (Zheng et al., 2019). However, pitfalls such as the cost of rearing male mosquitoes in large numbers and the subsequent recovery of the mosquito population over time if release is stopped need to be addressed (Caragata et al., 2020). Also, the issue of accidental release of female mosquitoes which may lead to the recovery of the mosquito population and render subsequent suppression strategies ineffective needs also to be considered, although the sterile insect-rearing technique has been proposed to mitigate this challenge (Zheng et al., 2019). *Wolbachia* is deleterious on mosquito sporozoites, reducing their number in the insect stage of Plasmodium development (Gomes et al., 2017). This transmission-blocking effect on *Plasmodium* sporozoites can be attributed to *Wolbachia* and its immune system influences, the competition for nutrients, how long the mosquito lives, and the interaction of *Wolbachia* with microorganisms in the mosquito (Audsley et al., 2018).

5.2 Novel Symbiont Exploration and Development

Clustered regularly interspaced short palindromic repeats with CRISPR-associated endonuclease (CRISPR-Cas) is a recent development that has been applied in paratransgenesis for malaria control. Genome editing of the gut microbiome is possible using CRISPR-Cas which makes it possible to transform bacteria by gene silencing (Yang et al., 2020). CRISPRs obtained from prokaryotes, cut chromosomes at specific sites through an endonuclease activity (Cas9) that is regulated by gRNA, which is a guide RNA (Shaw & Catterrucia, 2019). The use of the CRISPR system enables transgene introduction at a specific location and ensures high stability of the genome. The CRISPR method has successfully been used in transforming microorganisms and silencing of genes in paratransgenesis, resulting in accurate knockdowns and less off- targets compared to RNA interference. Both RNA interference (RNAi) and the CRISPR technique make it possible to inhibit multiple genes simultaneously, preventing parasite resistance (Ratcliffe et al., 2022).

5.3 RNAi-Based Paratransgenesis

RNAi-based paratransgenesis is a modified form of paratransgenesis that transforms microbes by delivering dsRNA instead of effector proteins for vector control (Asgari et al., 2020). This technique is relatively new, and the primary focus is the delivery of dsRNA instead of effector molecules to remove or replace a specific gene from the vector or the pathogen itself (Giesbrecht et al., 2020). This technique has been used in the development of transformed *Asaia* strains for RNAi-based paratransgenesis in *Anopheles* (Asgari et al., 2020). The use of this technique is fast growing though some technical challenges still exist in terms of enhancing the survival of the dsRNA in insects.

5.4 Persistence and Spread Enhancement

The persistence and spread of genetically transformed symbionts in target mosquito populations have been elucidated through various techniques. These techniques also provide relevant information on improving the spread of transformed microorganisms in paratransgenesis without fitness cost to the target vector (Ratcliffe et al., 2022). *Anopheles stephensi* and *Anopheles gambiae* have been infected with *Asaia* bacteria through vertical, horizontal, and trans-stadial transmission, indicating the feasibility of the spread and persistence of transformed symbiotic microorganisms in mosquito populations (Ratcliffe et al., 2022). Delivery of paratransgenic microorganisms, however, requires specific strategies such as microencapsulation using chitosan and the targeted delivery of symbionts using nanoparticles. These novel delivery methods ensure that the spread of genetically engineered microorganisms occurs in specific target mosquito populations, reaches specific target sites, and does not cause harm to the environment (Aminata et al., 2022).

5.5 Anti-plasmodium Compound Production

Symbiotic microorganisms are currently genetically engineered to produce anti-plasmodium compounds for malaria control using specific novel techniques such as phage display technology, which involves the presentation of peptides or proteins on the surface of bacteriophages. These peptides are screened for the determination of specific proteins to Plasmodium targets and for the development of novel anti-plasmodium compounds. Also, the combination of these effector molecules has been observed to be more effective than a single molecule and this is being leveraged, in the production of anti-plasmodium compound (Shane et al., 2018).

6 Future Research Needs in Paratransgenesis

The paratransgenesis approach holds great promise in the control of insect-borne diseases. Its application in the control of malaria has yielded tremendous results under controlled laboratory and semi-field conditions (Ratcliffe et al., 2022). However, implementing this approach is still a long way off as several concerns about releasing genetically modified organisms into the environment have been raised (Koller et al., 2023). To successfully resolve some of these issues, more attention needs to be given to some key issues such as the maintenance of accuracy of microbial genome sequencing and identification techniques, understanding host-symbiont interactions, improvement of current genetic engineering tools, assessment of safety through ecological impact studies, and the need for more research in developing countries. Other areas include assessing the cost-effectiveness of the approach and developing a strategy to garner acceptance through public enlightenment and stakeholder engagement (Aminata et al., 2022). This would provide a clearer picture of the paratransgenesis approach and eliminate possible concerns on the implementation of the strategy for malaria control.

6.1 Accuracy of Microbial Genome Sequencing and Identification

The application of high-throughput sequencing techniques has enabled better identification of the microbial genome of insect vectors (Reese & Dunn, 2018). Studies on microbial genome have been boosted with advancements in next-generation sequencing which helped to identify microbial species and their metabolic pathways (Richa & Dominik, 2021). However, frequent occurrence of error has been observed due to incomplete sequences and lack of sufficient data on the insect microbial genome. There have also been concerns about the reproducibility of microbial sequencing data already published which show areas of unknown sequences known as dark matter. Also, error in handling microbial genetic material, extraction methods, sample size, and computational analyses may result in inconsistent results (Richa & Dominik, 2021). The lack of accurate and complete information on the insect microbial genome may negatively impact the success of the paratransgenesis approach as more information is required to enhance the identification of potential symbionts with desired genetic traits. Hence, more research should be targeted towards updating and improving the insect genome sequence database.

6.2 Understanding Host-Symbiont Interactions

There is a need for a better understanding of the host-symbiont interactions in paratransgenesis and more effort should be targeted towards providing a clearer picture of the effects of these interactions on the vector, the symbiont, and the environment (Egyirifa & Akorli, 2024). Understanding the mechanism of action of anti-pathogen effector molecules produced against the malaria parasite would promote the development of newer and more specific anti-pathogen peptides for malaria control (Zach & Bianca, 2021). The knowledge of host-symbiont interactions would also provide more effective strategies to enhance symbiont colonization and persistence without any adverse effects on the vector and the genetically modified symbiont (Sandra et al., 2020). Also, the long-term effects of host immune responses on both the vector and the symbiont should be adequately investigated to ensure the stability of the transgene and the survival of the genetically altered symbiont in the vector.

6.3 Improvement of Genetic Engineering Tools

Current genetic engineering tools should be improved upon by refining and developing new methods that enhance precision and accuracy in paratransgenesis. The CRISPR-Cas system currently in use can be enhanced using alternative systems such as the Cas1 and Cas2 for more specific targeting (Wimmer & Beisel, 2020). Introducing new genetic elements such as promoters and regulatory sequences will allow for greater control of gene expression in symbiotic microorganisms specific to the host and its environment. Multiplex gene editing tools can be developed to simultaneously transform multiple genes in the symbiont, while novel delivery methods using engineered viruses, lipid nanoparticles and microencapsulation, can be explored for increased specificity and effective vector colonization.

6.4 Ecological Impact Studies

Although paratransgenesis shows potential of reducing malaria, there are concerns about the ecological impact of the approach. These concerns stem from questions about the safety of the transformed microorganisms and mosquitoes, which may disrupt natural ecological balances and alter aquatic habitats where mosquitoes breed (Ricardo et al., 2024). Also, concerns about the reduction in mosquito genetic diversity have been raised, which may increase their susceptibility to other pathogens and their ability to transfer these disease agents to humans (Hayes et al., 2020). There is also concern about the emergence of resistance to the peptides produced by genetically modified microorganisms, leading to increased malaria transmission and difficulty in controlling the disease (Aminata et al., 2022). These potential challenges posed by the possible ecological impact of paratransgenesis must be carefully addressed through research to ensure effective implementation of the approach.

6.5 Paratransgenesis Research in Developing Countries

Evidence of the applicability of the paratransgenesis approach in malaria control has been observed in recent studies carried out in Africa and other countries. Studies carried out in Africa by Egyirifa and Akorli (2024) and Claudine et al. (2024), however, have not yet advanced beyond the symbiont identification stage of paratransgenesis while this has not been the case for studies carried out in other non-African countries. Also, aside from mosquitoes, paratransgenesis studies have been carried out using other vectors to prevent disease transmission. In Iran, rodents were used as food baits to deliver *Serratia* and *Enterobacter cloacae* in sandflies to control leishmaniasis transmission (Ghassemi et al., 2023) while in Pennsylvania USA, paratransgenesis was carried out in Tsetse flies using conjugation as a means of

DNA delivery in symbiont bacteria (Kendra et al., 2020). Possible reasons for the paucity of research and data on paratransgenesis in developing countries, particularly Africa, may be because of the lack of specialized and advanced equipment, insufficient knowledge of current techniques and the lack of human and financial resources (Arbuthnot et al., 2017). These issues have negatively impacted advancements in the approach, as very little information has been recorded to validate the applicability and usage of paratransgenesis as a strategy for malaria control in these countries.

7 Strategy for Possible Implementation

The strategy for possible implementation of paratransgenesis requires a systematic approach from initial laboratory research to field studies and continuous monitoring (Aminata et al., 2022). Intensive efforts and resources should be targeted towards developing specialized techniques for enhanced identification, selection, and modification of mosquito symbionts to produce desired traits. Laboratory and semi-field studies should be designed to incorporate more accurate and reproducible techniques that can guarantee the successful implementation of paratransgenesis for malaria control. Laboratory and semi-field studies should evaluate the transgene's stability and fitness in the vectors and their reproducibility in field conditions (Williams et al., 2023). The safety of the genetically modified symbiont to the vector, host, non-target organisms and the environment should be assessed while developing contingency plans and performing containment testing to mitigate the event of accidental release to non-target populations. A major requirement for the implementation of paratransgenesis is the need for approval from regulatory bodies (Ratcliffe et al., 2022). Approval can be obtained when comprehensive data on the safety, efficacy, and environmental impact of the approach has been submitted to relevant regulatory bodies, for assessment. Collaborations with local, national, and international organizations are necessary to maintain biosafety standards and obtain necessary resources for implementation of the approach. Field trials should be carried out in designated locations after all assessments and approval are obtained. Community engagement is key to the success of this approach as it fosters compliance and eliminates resistance to the paratransgenesis approach. Education and engagement of local communities on the benefits and safety of the trials would address all ethical and cultural concerns on the use of genetically altered organisms (Roberts & Thizy, 2022). Continuous monitoring and evaluation involve the implementation of protocols to track the impact of the approach on vector populations and the environment (Ratcliffe et al., 2022). It involves observations on the stability and sustainability of the transgene in vector populations, its persistence in subsequent vector generations, the impact on disease transmission and overall impact on humans, non-target vectors, and the environment. Continuous monitoring and evaluation also study the impact of the approach when integrated with other existing malaria control strategies and ensures that the paratransgenesis approach complements these existing strategies (Ratcliffe et al., 2022). The long-term effect of this approach can be tracked through the establishment of continuous surveillance systems to monitor persistence and overall effectiveness in malaria control, while adapting to the possible changes that may occur over time (Aminata et al., 2022). Implementing paratransgenesis requires careful planning, elaborate scientific research, and effective collaboration with regulatory bodies and local communities. Applying this strategy would pave the way for possible field trials and subsequent implementation of the paratransgenesis approach in malaria control, in malaria-endemic countries.

8 Conclusion

Paratransgenesis has been explored in various studies for its applicability and usefulness in the control of various mosquito-borne diseases, especially malaria. So far, studies to improve this approach through the identification and engineering of symbiotic microorganisms with unique effector properties for the production of desired effects on the mosquito vector and malaria parasite have continued to support the relevance of this approach in the control of malaria. Therefore, advancements in the paratransgenesis approach present a promising future for malaria control globally. However, there is need for more robust and elaborate research to provide a clearer understanding of the approach and proffer possible solutions to specific unresolved issues that may affect the implementation of the approach, particularly in resource-limited countries. Therefore, the benefits of paratransgenesis cannot be overlooked; an effective and sustainable strategy for possible implementation of the approach must be developed and adopted, for malaria control in malaria-endemic countries.

References

Allaband, C., McDonald, D., Vázquez-Baeza, Y., Minich, J. J., Tripathi, A., & Brenner, D. A. (2019). Microbiome 101: Studying, analyzing, and interpreting gut microbiome data for clinicians. *Clinical Gastroenterology and Hepatology, 17*(2), 218–230.

Aminata, F., Rakiswendé, S. Y., Etienne, B., Georges, A. O., & Mathilde, G. (2022). The strategy of paratransgenesis for the control of malaria transmission. *Frontiers in Tropical Diseases, 3*, 1–11.

Araki, T., Kawai, S., Kakuta, S., Kobayashi, H., Umeki, Y., Saito-Nakano, Y., Sasaki, T., Nagamune, K., Yasutomi, Y., Nozaki, T., Franke-Fayard, B., Khan, S. M., Hisaeda, H., & Annoura, T. (2020). Three-dimensional electron microscopy analysis reveals endopolygeny-like nuclear architecture segregation in Plasmodium oocyst development. *Parasitology International, 76*, 34–45.

Arbuthnot, P., Maepa, M. B., Ely, A., & Pepper, M. S. (2017). The state of gene therapy research in Africa, its significance and implications for the future. *Gene Therapy, 24*(9), 581–589.

Arora, A. K., Miller, T. A., & Durvasula, R. (2019). Transmission of Pantoea agglomerans—A paratransgenic control agent—within a Homalodisca vitripennis population. *Journal of Applied Entomology, 144*, 232–237.

Asgari, M., Ilbeigikhamsehnejad, M., Rismani, E., Djadid, N. D., & Raz, A. (2020). Molecular characterization of Rnase III protein of Asaia Sp. for developing a robust RNAi-based paratransgenesis tool to affect the sexual life-cycle of plasmodium or anopheles fitness. *Parasites and Vectors, 13*, 42–61.

Audsley, M. D., Seleznev, A., Joubert, D. A., Woolfit, M., O'Neill, S. L., & McGraw, E. A. (2018). Wolbachia infection alters the relative abundance of resident bacteria in adult Aedes aegypti mosquitoes, but not larvae. *Molecular Ecology, 27*, 297–310.

Bai, L., Wang, L., Vega-Rodríguez, J., Wang, G., & Wang, S. (2019). A gut symbiotic bacterium Serratia marcescens renders mosquito resistance to Plasmodium infection through activation of mosquito immune responses. *Frontiers in Microbiology, 10*, 1580–1591.

Bancells, C., Llorà-Batlle, O., Poran, A., Nötzel, C., Rovira-Graells, N., & Elemento, O. (2019). Revisiting the initial steps of sexual development in the malaria parasite *Plasmodium falciparum. Nature Microbiology, 4*, 144–171.

Bilgo, E., Lovett, B., Bayili, K., Millogo, A. S., Saré, I., & Dabiré, R. K. (2018). Transgenic *Metarhizium pingshaense* synergistically ameliorates pyrethroid-resistance in wild-caught, malaria-vector mosquitoes. *PLoS One, 13*(9), 1–15.

Cansado-Utrilla, C., Zhao, S. Y., & McCall, P. J. (2021). The microbiome and mosquito vectorial capacity: rich potential for discovery and translation. *Microbiome, 9*, 111–136.

Cappelli, A., Damiani, C., Mancini, M. V., Valzano, M., Ramirez, J. L., & Favia, G. (2019). Asaia activates immune genes in mosquito eliciting an anti-plasmodium response: Implications in malaria control. *Frontiers in Genetics, 10*, 8361–8372.

Caragata, E. P., Dong, S., Dong, Y., Simões, M. L., Tikhe, C. V., & Dimopoulos, G. (2020). Prospects and pitfalls: next-generation tools to control mosquito-transmitted disease. *Annual Review in Microbiology, 74*, 455–476.

Carballar-Lejarazú, R., Ogaugwu, C., Tushar, T., Kelsey, A., Pham, T. B., Murphy, J., Schmidt, H., Lee, Y., Lanzaro, G. C., & James, A. A. (2020). Next-generation gene drives for population modification of the malaria vector mosquito, Anopheles Gambiae. *PNAS, 117*, 805–815.

Carnevale, P., & Manguin, S. (2021). Review of issues on residual malaria transmission. *Journal of Infectious Diseases, 223*(2), 61–81.

Chiamaka, L., Id, E., Abla, E., Id, A., Aboagye-Antwi, F., & Akorli, J. (2020). Mosquito midgut Enterobacter cloacae and Serratia marcescens affect the fitness of adult female Anopheles Gambiae. *PLoS One, 15*, 931–946.

Chong, Z. X., Yeap, S. K., & Ho, W. Y. (2021). Transfection types, methods and strategies: A technical review. *Peer-reviewed Journal, 21*(9), 1–37.

Cisnetto, V., & Barlow, J. (2020). The development of complex and controversial innovations. Genetically modified mosquitoes for malaria eradication. *Research Policy, 49*(3), 917–935.

Claudine, G.T.M., Maurice Marcel, S., Micareme, T., Fleuriane, M. D. K., Leon, M.J., Mugenzi, F.N., Grant, L. H. & Charles, S. W. (2024). Contrasting patterns of Asaia Sp association with Plasmodium falciparum between field-collected Anopheles gambiae and Anopheles coluzzii from Cameroon. *bioRxiv*, 377–405.

Dahmana, H., & Mediannikov, O. (2020). Mosquito-Borne diseases emergence/resurgence and how to effectively control it biologically. *Pathogens, 9*, 310–346.

Egyirifa, R. K., & Akorli, J. (2024). Two promising candidates for paratransgenesis, *Elizabethkingia* and *Asaia*, increase in both sexes of Anopheles gambiae mosquitoes after feeding. *Malaria Journal, 23*, 45–53.

Fazal, S., Manzoor Malik, R., Zafar Baig, A., Khatoon, N., Aslam, H., Zafar, A., & Ishtiaq, M. (2023). *Role of mosquito microbiome in insecticide resistance*. IntechOpen. https://doi.org/10.5772/intechopen.104265

Gabrieli, P., Caccia, S., Varotto-Boccazzi, I., Arnoldi, I., Barbieri, G., Comandatore, F., & Epis, S. (2021). Mosquito trilogy: Microbiota, immunity and pathogens, and their implications for the control of disease transmission. *Frontiers in Microbiology, 12*, 438–448.

Gao, H., Cui, C., Wang, L., Jacobs-Lorena, M., & Wang, S. (2020). Mosquito microbiota and implications for disease control. *Trends in Parasitology, 36*, 98–111.

Ghassemi, M., Akhavan, A. A., & Zahraei-Ramazani, A. (2023). Rodents as A vehicle for delivery of transgenic bacteria to make paratransgenic sand fly vectors of cutaneous leishmaniasis in field condition. *Science Report, 13*, 912–923.

Giesbrecht, D., Heschuk, D., Wiens, I., Boguski, D., LaChance, P., & Whyard, S. (2020). RNA interference is enhanced by knockdown of double-stranded Rnases in the yellow fever mosquito Aedes aegypti. *Insects, 11*, 327–333.

Gomes, F. M., Hixson, B. L., Tyner, M. D. W., Ramirez, J. L., Canepa, G. E., Alves, E., & Silva, T. L. (2017). Effect of naturally occurring Wolbachia in Anopheles gambiae mosquitoes from Mali On Plasmodium falciparum malaria transmission. *Proceedings in the National Academy of Sciences, 114*, 566–576.

Grogan, C., Bennett, M., Moore, S., & Lampe, D. (2021). Novel Asaia bogorensis signal sequences for plasmodium inhibition in Anopheles stephensi. *Frontiers in Microbiology, 12*, 667–678.

Hayes, K. R., Peel, D., Hosack, G., & Eagles, D. (2020). Structured prioritization of human and animal pathogens to scope risk assessments of genetic control strategies for malaria vectors in Sub-Saharan Africa. *CSIRO Data, 61*, 96–105.

Hegde, S., Nilyanimit, P., Kozlova, E., Anderson, E. R., & Narra, H. P. (2019). CRISPR/Cas9-mediated gene deletion of the *Ompa* gene in symbiotic Cedecea neteri impairs biofilm formation and reduces gut colonization of Aedes aegypti mosquitoes. *PLoS Neglected Tropical Diseases, 13*(12), 1–21.

Hegde, S., Voronin, D., Casas-Sanchez, A., Saldaña, M. A., Heinz, E., & Acosta-Serrano, A. (2020). Gut-associated bacteria invade the midgut epithelium of Aedes aegypti and stimulate innate immunity and suppress Zika virus infection in cells. *bioRxiv*, 866897. https://doi.org/10.1101/866897. (Accessed: 19 March 2024).

Herren, J. K., Mbaisi, L., Mararo, E., Makhulu, E. E., Mobegi, V. A., & Butungi, H. (2020). A microsporidian impairs Plasmodium falciparum transmission in Anopheles arabiensis mosquitoes. *Nature Communications, 11*(1), 2187–2197.

Hixson, B., Chen, R., & Buchon, N. (2024). Innate immunity in Aedes mosquitoes: From pathogen resistance to shaping the microbiota. *Philosophical Transactions of the Royal Society of London Series B Biological Science, 6*(1901), 379–394.

Holanda-Freitas, I. T., Cupertino, M. C., dos Santos, E. C., Oliveira, L., Geller, M., & Siqueira-Batista, R. (2020). Human African trypanosomiasis: Current standing and challenges. *Journal of Tropical Pathology, 49*(3), 133–148.

Huang, W., Wang, S., & Jacobs-Lorena, M. (2020). Self-limiting paratransgenesis. *PLoS Neglected Tropical Diseases, 14*(8), 542–552.

Huang, W., Joel, V., Chritopher, K., Sung-Jae, C., & Marcelo, J. (2022). Combining transgenesis with paratransgenesis to fight malaria. *eLife, 11*, 584–596.

Johnson, R. M., & Rasgon, J. L. (2018). Densonucleosis viruses ('Densoviruses') for mosquito and pathogen control. *Current Opinions in Insect Science, 28*, 90–97.

Kame-Ngasse, G. I., Njiokou, F., Melachio-Tanekou, T. T., Farikou, O., Simo, G., & Geiger, A. (2018). Prevalence of symbionts and trypanosome infections in tsetse flies of two villages of the "Faro And Déo" division of the Adamawa Region Of Cameroon. *BMC Microbiology, 18*, 159–163.

Kendra, C. G., Keller, C. M., Bruna, R. E., & Pontes, M. H. (2020). Conjugal DNA transfer in Sodalis glossinidius, a maternally inherited symbiont of tsetse flies. *mSphere, 5*, 64–84.

Koller, F., Schulz, M., & Juhas, M. (2023). The need for assessment of risks arising from interactions between NGT organisms from an EU perspective. *Environmental Science Europe, 35*, 27–42.

Lin, D., Zheng, X., Sanogo, B., Ding, T., Sun, X., & Wu, Z. (2021). Bacterial composition of midgut and entire body of laboratory colonies of *Aedes aegypti* and *Aedes albopictus* from Southern China. *Parasite and Vectors, 14*, 586–599.

Lindsay, S. W., Thomas, M. B., & Kleinschmidt, I. (2021). Threats to the effectiveness of insecticide-treated bednets for malaria control: Thinking beyond insecticide resistance. *Lancet Global Health, 9*(13), 25–31.

Liu, Y., Feng, J., Pan, H., Zhang, X., & Zhang, Y. (2022). Genetically engineered bacterium: Principles, practices, and prospects. *Frontiers in Microbiology, 13*, 1–15.

Mancini, M. V., Spaccapelo, R., Damiani, C., Accoti, A., Tallarita, M., Petraglia, E., Rossi, P., Cappelli, A., Capone, A., Peruzzi, G., Valzano, M., Picciolini, M., Diabaté, A., Facchinelli, L., Ricci, I., & Favia, G. (2016). Paratransgenesis to control malaria vectors: A semi-field pilot study. *Parasites and Vectors, 9*, 140–149.

Manoj, R. R. S., Latrofa, M. S., Epis, S., & Otranto, D. (2021). Wolbachia: Endosymbiont of onchocercid nematodes and their vectors. *Parasite and Vectors, 14*, 245–250.

Manrique-Saide, P., Dean, N. E., Halloran, M. E., Longini, I. M., Collins, M. H., Waller, L. A., Gomez-Dantes, H., Lenhart, A., Hladish, T. J., & Che-Mendoza, A. (2020). The TIRS trial: Protocol for a cluster randomized controlled trial assessing the efficacy of preventive targeted indoor residual spraying to reduce Aedes-borne viral illnesses in Merida, Mexico. *Trials, 21*, 839–846.

Mwima, R., Hui, T. Y. J., & Nanteza, A. (2023). Potential persistence mechanisms of the major *Anopheles gambiae* species complex malaria vectors in sub-Saharan Africa: A narrative review. *Malaria Journal, 22*, 336–348.

Nanfack, M. F., & Vernick, K. D. (2018). A systematic review of the natural virome of anopheles mosquitoes. *Viruses, 10*(5), 222–243.

Onen, H., Luzala, M. M., Kigozi, S., Sikumbili, R. M., Muanga, C. K., Zola, E. N., Wendji, S. N., Buya, A. B., Balciunaitiene, A., Viškelis, J., Kaddumukasa, M. A., & Memvanga, P. B. (2023). Mosquito-borne diseases and their control strategies: An overview focused on green synthesized plant-based metallic nanoparticles. *Insects, 14*(3), 221–227.

Papadopoulos, C., Karas, P. A., Vasileiadis, S., Ligda, P., Saratsis, A., & Sotiraki, S. (2020). Host species determines the composition of the prokaryotic microbiota in Phlebotomus sandflies. *Pathogens, 9*, 428–436.

Plowe, C. V. (2022). Malaria chemoprevention and drug resistance: A review of the literature and policy implications. *Malaria Journal, 21*, 104–108.

Qadri, M., Short, S., Gast, K., Hernandez, J., & Wong, A. C. N. (2020). Microbiome innovation in agriculture: development of microbial based tools for insect pest management. *Frontiers in Substantiable Food Systems, 4*, 751–763.

Ratcliffe, N. A., Furtado Pacheco, J. P., & Dyson, P. (2022). Overview of paratransgenesis as a strategy to control pathogen transmission by insect vectors. *Parasites and Vectors, 15*, 112–119.

Reese, A. T., & Dunn, R. R. (2018). Drivers of microbiome biodiversity: A review of general rules, feces, and ignorance. *American Society of Microbiology, 9*, 1294–1218.

Reveillaud, J., Bordenstein, S. R., Cruaud, C., Shaiber, A., Esen, O. C., & Weill, M. (2019). The Wolbachia mobilome in Culex Pipiens includes a putative plasmid. *Nature Communications, 10*(1), 1051–1062.

Ricardo, D. M., Luca, M., Cristián, B., & Juan, C. C. (2024). Gene drives, mosquitoes, and ecosystems: An interdisciplinary approach to emerging ethical concerns. *Frontiers in Environmental Science., 11*, 12–19.

Richa, B., & Dominik, G. G. (2021). Current challenges and best-practice protocols for microbiome analysis. *Briefings in Bioinformatics, 22*(1), 178–193.

Roberts, A. J., & Thizy, D. (2022). Articulating ethical principles guiding Target Malaria's engagement strategy. *Malaria Journal, 21*, 35–46.

Romoli, O., & Gendrin, M. (2018). The tripartite interactions between the mosquito, its microbiota and plasmodium. *Parasites and Vectors, 11*(1), 200–214.

Romoli, O., Schönbeck, J. C., & Hapfelmeier, S. (2021). Production of germ-free mosquitoes via transient colonisation allows stage-specific investigation of host–microbiota interactions. *Nature Communications, 12*, 942–958.

Ross, P. A., Callahan, A. G., Yang, Q., Jasper, M., Arif, M. A. K., & Afizah, A. N. (2020). An elusive endosymbiont: Does Wolbachia occur naturally in Aedes Aegypti? *Ecoland Evolution, 10*, 1581–1591.

Ross, P. A., Robinson, K. L., Yang, Q., Callahan, A. G., Schmidt, T. L., & Axford, J. K. (2022). A decade of stability for Wmel Wolbachia in natural Aedes Aegypti populations. *PLoS Pathogens, 18*, 256–262.

Saab, S. A., Dohna, H. Z., Nilsson, L. K. J., Onorati, P., Nakhleh, J., Terenius, O., & Osta, M. A. (2020). The environment and species affect gut bacteria composition in laboratory co-cultured Anopheles gambiae and Aedes albopictus mosquitoes. *Science Report, 10*(1), 3352–3365.

Sandra, Y. M., David, J. C., & Nicole, M. G. (2020). An integrative approach to symbiont-mediated vector control for agricultural pathogens. *Current Opinion in Insect Science, 39*, 57–62.

Sanfelice, V. (2022). Mosquito-Borne disease and newborn health. *Health Economics, 31*, 73–93.

Sato, S. (2021). Plasmodium-A brief introduction to the parasites causing human malaria and their basic biology. *Journal of Physiology and Anthropology, 40*(1), 1–5.

Schaub, G. A. (2021). An update on the knowledge of parasite-vector interactions of Chagas disease. *Respective Report in Tropical Medicine, 12*, 63–76.

Shane, J. L., Grogan, C. L., Cwalina, C., & Lampe, D. J. (2018). Blood meal-induced inhibition of vector-borne disease by transgenic microbiota. *Nature Communications, 9*(1), 4127–4135.

Shaw, W. R., & Catterrucia, F. (2019). Vector biology meets disease control: using basic research to fight vector-Borne diseases. *Nature Microbiology, 4*(1), 20–34.

Shengzhang, D., Yuemei, D., Maria, L. S., & George, D. (2022). Mosquito transgenesis for malaria control. *Trends in Parasitology, 38*(1), 54–66.

Shropshire, J. D., Rosenberg, R., & Bordenstein, S. R. (2021). The impacts of cytoplasmic incompatibility factor (Cifa and Cifb) genetic variation on phenotypes. *Genetics, 217*, 1–13.

Siciliano, G., Costa, G., Suárez-Cortés, P., Valleriani, A., Alano, P., & Levashina, E. A. (2020). Critical steps of plasmodium falciparum ookinete maturation. *Frontiers in Microbiology, 11*. https://doi.org/10.3389/fmicb.2020.00269/full. (Accessed: 19 March, 2024).

Sougoufara, S., Ottih, E. C., & Tripet, F. (2020). The need for new vector control approaches targeting outdoor biting Anopheline malaria vector communities. *Parasites and Vectors, 13*, 295–300.

Souza, R. S., Virginio, F., Riback, T. I. S., Suesdek, L., Barufi, J. B., & Genta, F. A. (2019). Microorganism-based larval diets affect mosquito development, size and nutritional reserves in the yellow fever mosquito *Aedes aegypti* (Diptera: Culicidae). *Frontiers in Physiology, 9*(10), 152–176.

Tharaka, W., Samadhi, G., & Nayana, G. (2020). Recent developments and future directions in the paratransgenesis based control of leishmania transmission. *Biological Control, 145*, 1–9.

Thizy, D., Emerson, C., Gibbs, J., Hartley, S., Kapiriri, L., & Lavery, J. (2019). Guidance on stakeholder engagement practices to inform the development of area-wide vector control methods. *PLoS Neglected Tropical Diseases, 13*(4), 286–293.

Tripathi, H., Bhalerao, P., Singh, S., Arya, H., Alotaibi, B. S., Rashid, S., Hasan, M. R., & Bhatt, T. K. (2023). Malaria therapeutics: Are we close enough? *Parasites & Vectors, 16*, 1–26.

Tungu, P. K., Michael, E., Sudi, W., Kisinza, W. W., & Rowland, M. (2021). Efficacy of interceptor® G2, a long-lasting insecticide mixture net treated with chlorfenapyr and alpha-cypermethrin against Anopheles funestus: Experimental hut trials in North-Eastern Tanzania. *Malaria Journal, 20*, 180–195.

Utarini, A., Indriani, C., Ahmad, R. A., Tantowijoyo, W., Arguni, E., & Ansari, M. R. (2021). Efficacy of Wolbachia-infected mosquito deployments for the control of Dengue. *New England Journal of Medicine, 384*(23), 77–86.

Venkatesh, A., Jain, A., Davies, H., Periera, L., Maki, J. N., Gomes, E., Felgner, P. L., Srivastava, S., Pantakar, S., & Rathod, P. K. (2019). Hospital-derived antibody profiles of malaria patients in Southwest India. *Malaria Journal, 18*, 138–145.

Walker, T., Quek, S., Jeffries, C. L., Bandibabone, J., Dhokiya, V., & Bamou, R. (2021). Stable high-density and maternally inherited Wolbachia infections in Anopheles moucheti and Anopheles demeilloni mosquitoes. *Current Biology, 31*, 2310–2320.

Wang, S., & Jacobs-Lorena, M. (2013). Genetic approaches to interfere with malaria transmission by vector mosquitoes. *Trends in Biotechnology, 31*(3), 185–193.

Wang, S., Santos, A. L. A., Huang, W., Liu, K. C., Oshaghi, M. A., & Wei, G. (2017). Driving mosquito refractoriness to *Plasmodium falciparum* with engineered symbiotic bacteria. *Science, 357*(6358), 1399–1402.

Wang, X., Liu, T., Wu, Y., Zhong, D., Zhou, G., Su, X., Xu, J., Sotero, C. F., Sadruddin, A. A., Wu, K., Chen, X. G., & Yan, G. (2018). Bacterial microbiota assemblage in Aedes albopictus mosquitoes and its impacts on larval development. *Molecular Ecology, 27*(14), 2972–2985.

Williams, A. E., Sanchez-Vargas, I., Martin, L. E., Martin-Martin, I., Bennett, S., Olson, K. E., & Calvo, E. (2023). Quantifying fitness costs in transgenic Aedes aegypti mosquitoes. *Journal of Visualized Experiments, 199*, 1–23.

Wimmer, F., & Beisel, C. L. (2020). CRISPR-Cas systems and the paradox of self-targeting spacers. *Frontiers in Microbiology, 10*, 3078–3095.

World Health Organization (2020). Vector-borne diseases. Available from: https://www.who.int/news-room/fact-sheets/detail/vector-borne-diseases. (Accessed 3 May 2024).

World Health Organization (2023a). Report on Malaria in Nigeria 2022. Available from: https://www.afro.who.int/sites/default/files/2023-08/WEB_7784%20WMR%20 (Accessed: 19 March 2024).

World Health Organization (2023b). World Malaria Report 2023. Available from: https://www.who.int/publications/i/item/9789240086173 (Accessed: 8 April 2024).

World Health Organization (2024). The 2023 WHO World Malaria Report 2023. Available from: https://www.thelancet.com/journals/lanmic/article/PIIS2666-5247(24)00016-8 (Accessed: 20 May 2024).

Wu, P., Sun, P., Nie, K., Zhu, Y., Shi, M., Xiao, C., Liu, H., Liu, Q., Zhao, T., Chen, X., & Zhou, H. (2019). A gut commensal bacterium promotes mosquito permissiveness to arboviruses. *Cell Host and Microbe, 25*(1), 101–112.

Yang, J., Schleicher, T. R., Dong, Y., Park, H. B., Lan, J., & Cresswell, P. (2020). Disruption of Mosgilt in Anopheles Gambiae impairs ovarian development and plasmodium infection. *Journal of Experimental Medicine, 217*(1), 682–687.

Zach, N. A., & Bianca, B. K. (2021). Malaria-Resistant Mosquitoes (Diptera: Culicidae); the principle is proven, but will the effectors be effective? *Journal of Medical Entomology, 58*(5), 1–11.

Zheng, X., Zhang, D., Li, Y., Yang, C., Wu, Y., & Liang, X. (2019). Incompatible and sterile insect techniques combined eliminate mosquitoes. *Nature, 72*, 56–61.

Chapter 17
Onychomycosis Disease Development, Treatment, and New Trends

Obinna Chukwuemeka Nwinyi, Ngozi Faith Omuekwu, and Joy Adenusi

1 Introduction

The term "Onychomycosis" encompasses fungal infections of the nail plate, nail matrix, and nail bed. The term onychomycosis is coined from two Greek words which are "Onyx" and "mykes" meaning "nail" and "fungus" (Leung et al., 2020). *Tinea unguium* is a type of onychomycosis caused by dermatophytes. This leads to nail appearance changes which can also potentially lead to discomfort and pain. It is important to recognize and treat onychomycosis to restore the health of the nails (Dinani & George, 2019). The risk of onychomycosis development exhibits a significant positive correlation with age and sex. Data suggests a potential prevalence exceeding 20% in individuals about 60 years of age, which may rise to over 50% in subjects more matured than 70 (Gupta et al., 2022). Furthermore, males within this age group demonstrate a 2.1-fold increased susceptibility to onychomycosis compared to females (Gupta et al., 2022). Susceptibility to infection can be modulated by genetic factors (Jazdarehee et al., 2022). Additionally, an infection within a household environment increases the likelihood of transmission among household members (Jazdarehee et al., 2022).

Onychomycosis represents the higher prevalent nail disease encountered in healthcare processes, affecting a stated 5.5% of the global population (Lipner & Scher, 2019a). Dermatophytes, predominantly *Trichophyton rubrum* and *Trichophyton mentagrophytes*, are the primary etiological agents chargeable for 60–70% of onychomycosis occurrences (Lipner & Scher, 2019a). Non-dermatophyte molds (20%), including *Aspergillus* spp., *Acremonium* spp., *Fusarium* spp., *Alternaria alternate, and Neoscytalidium* spp., and yeasts (10–20%), primarily

O. C. Nwinyi (✉) · N. F. Omuekwu · J. Adenusi
Department of Biological Sciences, College of Sciences and Technology, Covenant University, Ota, Ogun State, Nigeria
e-mail: obinna.nwinyi@covenantuniversity.edu.ng

P. O. Isibor, P. A. Akinduti (eds.), *Harnessing Biotechnology Tools for Product Development*, https://doi.org/10.1007/978-3-031-86002-7_17

Candida albicans and *Candida parapsilosis.*, constitute the remaining causative organisms (Albucker et al., 2023). Established risk factors associated with the development of onychomycosis encompass, mellitus, immunosuppression, diabetes, advanced age, *Tinea pedis* (athlete's foot), hyperhidrosis (excessive sweating), obesity, and exposure to fungal sources within the home environment (Albucker et al., 2023). Clinical observations suggest an increase in the incidence of fungal infections. These infections have been linked to unfavorable health results in those who are critically unwell particularly those with diabetes mellitus. In diabetic patients, onychomycosis constitutes a significant proportion of nail pathologies, affecting roughly half of all cases (Trovato et al., 2022). This predisposition to onychomycosis increases the risk of complications such as necrosis, recurrent infections caused by bacteria, and even toe or lower limb amputation (Trovato et al., 2022).

Onychomycosis exhibits a significantly lower prevalence in children (0.14%) compared to adults (6.4%) (Gupta et al., 2022). This disparity can be attributed to several factors in children: improved peripheral blood supply, a thinner nail plate, faster nail regrowth, reduced nail stress, and a lower likelihood of encountering fungal spores and hyphae (Gupta et al., 2018). These factors collectively contribute to lower susceptibility to onychomycosis in children compared to the mature form typically observed in adults (Gupta et al., 2022). Fungal acquisition in children with onychomycosis can occur through two primary routes: directly from a traumatic nail abnormality or pre-existing dystrophic or indirectly from a parent via environmental contamination.

The *Trichophyton rubrum* complex, comprising the most frequently encountered dermatophytes, includes *T. circonvolutum*, *T. fischeri*, *T. fluviomuniense*, *T. glabrum*, *T. gourvilii*, *T. kanei*, *T. kuryangei*, *T. megninii*, *T. pedis*, *T. raubitschekii*, *T. rodhainii*, *T. soudanense*, *T. violaceum*, and *T. yaoundei*.

According to historical views, *Trichophyton rubrum* may have originated in parts of Southeast Asia, West Africa, Northern Australia, and Indonesia. During the nineteenth and twentieth centuries, increased migration brought the species to Europe and the United States (Zhan et al., 2021). Recent studies, however, paint a more complex picture, suggesting that *T. rubrum* probably exhibits regionally diverse populations instead of a single origin. The use of occlusive footwear and travel during conflict may have aided in the global spread of these regional varieties (Zhan et al., 2021). The most common nail disease all over the world is onychomycosis an overall frequency of 5.5% (Gupta et al., 2020). North America's approximate prevalence of onychomycosis varies from 2 to 14% for adults to 0.44% for kids and teenagers.

2 Causative Agents of Onychomycosis

Pathogens in the fungus kingdom cause onychomycosis. They include candida (yeasts), dermatophytes, and non-dermatophytic molds which are geographically disturbed and on climate change. Dermatophytes are the primary and global fungal pathogen responsible for nail infections, especially in the temperate Western world (Moskaluk & VandeWoude, 2022). The non-dermatophytic molds and *Candida* are prevalent in tropical and subtropical parts of the world that are hot and muggy.

2.1 Dermatophytes Epidemiology

More than 50% of dermatophyte onychomycosis infections are caused by *Trichophyton rubrum*, making it the most prevalent cause of the infection. About 20% of these instances are *Trichophyton mentagrophytes* (Lipner & Scher, 2019a). *Epidermophyton floccosum* and *Trichophyton rubrum* are ubiquitous dermatophytes, with *T. rubrum* emerging as the most prevalent causative agent of onychomycosis globally (Petrucelli et al., 2020). However, regional variations exist. For instance, *Microsporum ferrugineum* is more commonly associated with onychomycosis in Asia, while *T. concentricum* and *T. megninii* are prominent in Europe and South America, respectively. Notably, Africa presents a unique case where *T. rubrum* dominance is less evident (Petrucelli et al., 2020). Here, a diverse array of dermatophytes contributes to onychomycosis, with *Microsporum audouinii* and *T. Sudanese*, the most frequently isolated species within specific microregions (Antuori et al., 2019).

2.2 Non-Dermatophytes Molds (NDM)

Clinical and epidemiological studies have demonstrated that non-dermatophyte (ND) molds constitute a significant proportion of causative agents in onychomycosis cases (Reinel, 2021). These ND molds encompass various genera, including *Acremonium, Scytalidium, Fusarium, Neoscytalidium, Aspergillus, and Scopulariopsis* (Gupta et al., 2020). A higher occurrence of onychomycosis (which affects toenails mainly) is caused by non-dermatophyte fungi, in the population older than 65 (Marcos-Tejedor et al., 2021). Compared to dermatophyte infections, onychomycosis caused by non-dermatophyte (ND) molds is frequently related with periungual inflammation (inflammation around the nail) (Reinel, 2021). For instance, *Acremonium* infections typically manifest as distolateral subungual onychomycosis, characterized by longitudinal white streaks in the central portion of the nail (Gupta et al., 2020). Additionally, superficial white onychomycosis caused by ND molds frequently involves a deeper nail plate invasion and affects a larger nail surface than dermatophyte infections. Furthermore, proximal subungual onychomycosis caused by ND molds tends to present with discoloration, periungual inflammation, and purulent discharge (Reinel, 2021).

Yeast was initially classified as a contaminant but has become a pathogenic causative agent of uncommon onychomycosis (Subramanya et al., 2019). About 70% of onychomycosis is caused by *Candida albicans*, a major example of pathogenic yeasts (Leung et al., 2020). Other species such as *Candida tropicalis* and *Candida parapsilosis* also cause about 30% of onychomycosis (Subramanya et al., 2019). Individuals with chronic mucocutaneous candidiasis (CMC) and immunodeficiency exhibit a heightened susceptibility to onychomycosis caused by yeasts, particularly affecting the fingernails (Leung et al., 2020).

2.3 Prevalence and Risk Factors

The most frequent nail condition in clinical practice is onychomycosis, with a 5.5% global prevalence (Lipner & Scher, 2019a). Some immune-based diseases such as psoriasis and psoriatic arthritis are also major risk factors for onychomycosis. Age is a predisposing risk factor, as the prevalence of the disease is higher in older people than in young individuals. However, younger people are the major group of athletes who are also exposed to onychomycosis because of some activities: the wearing of occlusive shoes; the warm, damp atmosphere brought on by sweat and socks; athletes' close quarters; and injuries to the toenails and foot (Daggett et al., 2019).

3 Classification of Clinical Manifestation

Onychomycosis is usually identified by a brown or yellow brittle nail plate with subungual hyperkeratosis which leads to nail lysis (Leung et al., 2020). It is classified into six classes according to the site they infect and the diseases they cause. The distal and lateral subungual, superficial, endonyx, proximal subungual, total dystrophic onychomycosis, and candidal onychomycosis (CO) are the main types of dermatopathy of the nail (Lim et al., 2021) (Fig. 17.1).

3.1 Distal and Lateral Subungual Onychomycosis (DLSO)

DLSO is the most common subtype of onychomycosis, which results when a fungal incursion originates from the distal or lateral surface of both toenails and fingernails (Maskan Bermudez et al., 2023). The clinical symptoms of the condition are hyperkeratosis, onycholysis, and white-yellow nail discoloration. Nevertheless, the color changes of other types, for example, black, orange, and brown, have been noted. Moreover, dermatophytosis, manifested as brown, white, yellow, orange, or longitudinal streaks or patches, can also be seen as another form of dermatomycosis complication (Maskan Bermudez et al., 2023). Although these fungi need keratin to proliferate, they can still survive on surfaces without it. In the case of a superficial infection, on the other hand, the fungus can enter the host tissue and persist because

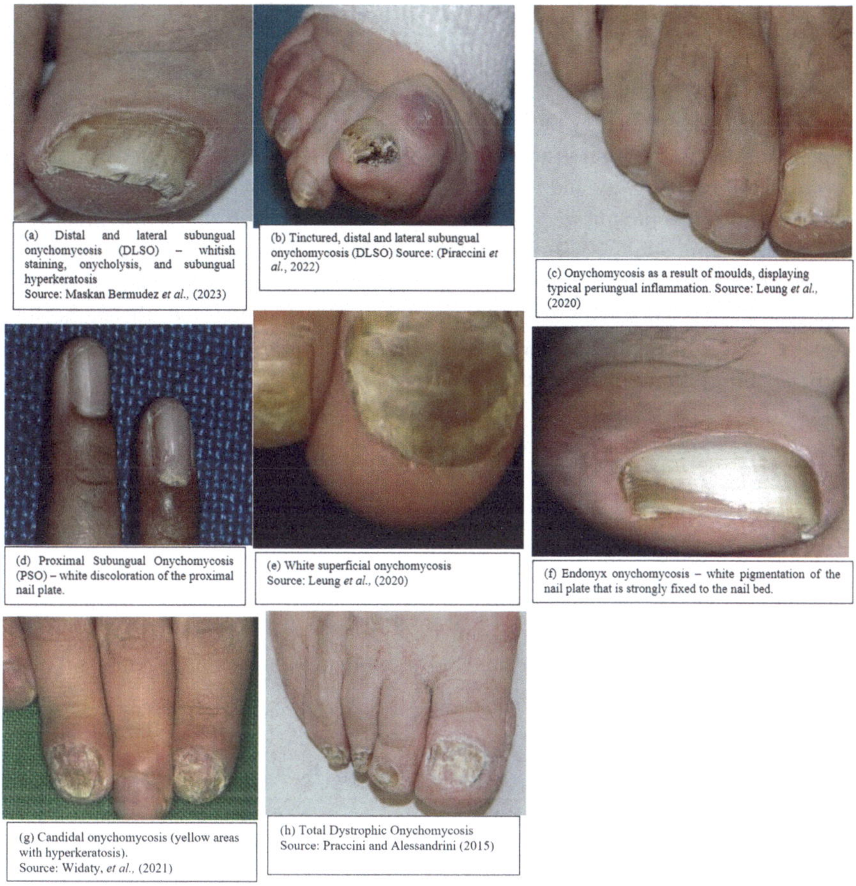

Fig. 17.1 Classification of onychomycosis clinical manifestations. (**a**) Distal and lateral subungual onychomycosis (DLSO)—whitish staining, onycholysis, and subungual hyperkeratosis. Source: Maskan Bermudez et al. (2023). (**b**) Tinctured, distal and lateral subungual onychomycosis (DLSO). Source: Piraccini et al. (2022). (**c**) Onychomycosis as a result of molds, displaying typical periungual inflammation. Source: Leung et al. (2020). (**d**) Proximal subungual onychomycosis (PSO)—white discoloration of the proximal nail plate. Source: Leung et al. (2020). (**e**) White superficial onychomycosis. Source: Leung et al. (2020). (**f**) Endonyx onychomycosis—white pigmentation of the nail plate that is strongly fixed to the nail bed. Source: Leung et al. (2020). (**h**) Total Dystrophic Onychomycosis. Source: Piraccini and Alessandrini (2015). (**g**) Candidal onychomycosis (yellow areas with hyperkeratosis). Source: Widaty et al. (2021)

of the synthesis and release of proteolytic enzymes (Klinger et al., 2021). The type of dermatophyte, the host's susceptibility, and other factors determine how severe the illness is (Klinger et al., 2021). DLSO is usually caused by *Trichophyton mentagrophytes (T. mentagrophytes)*, *Trichophyton rubrum (T. rubrum)*, and *Microsporum canis* (Sen et al., 2020) (Fig. 17.1a, b).

3.2 Proximal Subungual Onychomycosis (PSO)

Proximal subungual onychomycosis (PSO) is a highly scarce form of onychomycosis which is distinguished by the lunular area of the nail of the proximal leukonychia (Lipner & Scher, 2019a). It is an infection of the nail that affects the fingernails and toe nails in the same way and is mostly caused by *Trichophyton rubrum*, with the exceptions of *Trichophyton megnini, Trichophyton schoenleinii, and Epidermophyton floccosum* (Leung et al., 2020). Proximally, subungual onychomycosis is caused by leukonychia in the proximal nail plate and clinically grows as the nail grows distally. The closer the subungual onychomycosis is to the nail bed, the higher the probability it will happen to patients with compromised immune systems, especially acquired immunodeficiency syndrome (AIDS) (Lipner & Scher, 2019b) (Fig. 17.1d).

3.3 White Superficial Onychomycosis (WSO)

The fungus, *Trichophyton mentagrophytes*, affects the upper surface of the nail plate in cases of white superficial onychomycosis (Goldstein et al., 2019). Generally, white superficial onychomycosis develops as white dots or patches on the nail plate surface (Bodman & Krishnamurthy, 2019) (Fig. 17.1e).

3.4 Endonyx Onychomycosis

Endonyx onychomycosis is a fungal infection that exclusively targets the nail plate, leaving the nail bed uninvolved (Lipner & Scher, 2019b). Endonyx onychomycosis, typically caused by *Trichophyton Sudanese* and *Trichophyton violaceum,* is characterized by distinct clinical features (Lipner & Scher, 2019b). These include milky white patches on the nail plate, surface indentations, and a tendency to split into layers. Notably, unlike other onychomycosis subtypes, the nail plate in endonyx onychomycosis remains firmly attached to the nail bed, and subungual hyperkeratosis (thickening of the skin beneath the nail) is absent (Leung et al., 2020) (Fig. 17.1f).

3.5 Total Dystrophic Onychomycosis

Total white onychomycosis signifies the destruction of the whole nail apparatus and might mark the disease's final stage, which may go through other variants (Lipner & Scher, 2019b). It is pathologically characterized by a tremendously dystrophic and crumbled nail bed, of which the color becomes yellow, and the nail matrix which has become thickened and fragile turns into a diffuse yellow (Fig. 17.1h).

3.6 *Candidal Onychomycosis (CO)*

Candidal onychomycosis (CO) is seen in elders or patients who have any immunosuppression like diabetes mellitus or HIV or patients who are using immunosuppressive drugs or extended-spectrum antibiotics. Among CO's regional factors are poor hygiene, nail trauma, occupational exposure to chemicals (detergents, soaps), and excessive cutaneous moisture. A new classification was proposed recently, adding an extra category to the previous one. Four clinical types: chronic paronychia in addition to secondary nails: dystrophy, distal onychomycosis, chronic mucocutaneous oral candidiasis, or secondary oral candidiasis (Andrés & Alexandro, 2020).

Paronychia is the most common type of CO and is also a leading one often due to humidity or moisture (Andrés & Alexandro, 2020). It is clinically defined as erythema and edema across the nail folds which later turn into longitudinal stripes. Among the different nail patterns is pathologic (transverse depressions in the nail plate indicating the growth direction of the nails). Concerning the matrix of the nail, *Candida granuloma* is classified as a severe clinical form of onychomycosis is frequent with chronic mucocutaneous candidiasis (Widaty et al., 2021) (Fig. 17.1g).

4 Diagnosis of Onychomycosis

Diagnosis of onychomycosis is vital and accurate because the diseases affect patients' quality of life and are contagious among people. To ensure an accurate diagnosis, a combination of fungal and clinical examination is performed (Lim et al., 2021). The fungal species involved in the case can be identified using conventional diagnostic techniques such as the potassium hydroxide test, fungal culture, and histopathology of nail clippings. Recent advancements in diagnostics have led to the development of several tools that can be used to detect onychomycosis more accurately clinically, such as dermoscopy, reflectance confocal microscopy, and artificial intelligence, or mycologically, such as molecular assays (Lim et al., 2021).

4.1 *Sample Collection*

Sample collection is an essential initial step in the diagnosis of onychomycosis. The accuracy of the test results may be jeopardized by any contamination introduced during this phase. Strict pre-collection hygiene protocols are necessary to reduce the risk of contamination. Before collecting a sample, cleaning the nail plate and any surrounding soft tissue with 70% isopropyl or ethyl alcohol is necessary (Ghannoum et al., 2018). Depending on the clinical features that are present, a specific type of sample may be collected to diagnose onychomycosis (Lim et al., 2021). These could be debris gathered from beneath the nail (subungual scrapings), clippings from the nail plate, scrapings from the nail bed itself, or any combination of these (Fig. 17.2).

The field of diagnostics has advanced significantly, shifting from conventional methodologies to more sophisticated techniques, such as the application of artificial intelligence (AI). By analyzing vast datasets, AI-driven systems enhance diagnostic accuracy, speed, and predictive capabilities, enabling early and precise disease detection. However, traditional diagnostic methods remain indispensable, particularly for confirmation of results. These conventional techniques, such as microscopy, culture, and biochemical testing, continue to offer reliability and cost-effectiveness, making them integral components of diagnostic protocols, especially in resource-limited settings where newer technologies may not be easily accessible.

4.2 *Conventional Diagnostic Tool*

4.2.1 Potassium Hydroxide (KOH)

A fundamental dermatological diagnostic technique, direct potassium hydroxide (KOH) examination provides a quick and economical way to identify fungal components in nail infections (onychomycosis) (Lim et al., 2021). This method takes a sample from under the nail plate or from the nail bed and then dissolves it in a KOH solution (Lim et al., 2021). Under a microscope, septate hyphae—distinctive fungal structures—can be seen thanks to KOH's breakdown of keratin, the main component of nails (Fig. 17.3). Further staining techniques such as India ink, Calcofluor

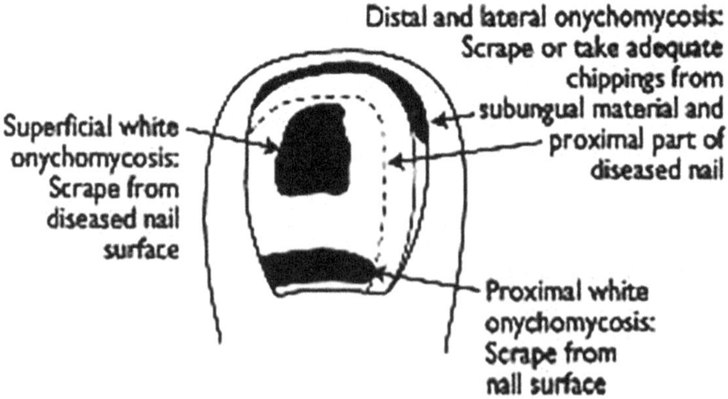

Fig. 17.2 Sites for sample collection for Onychomycosis diagnosis. Source: Gupta et al. (2018)

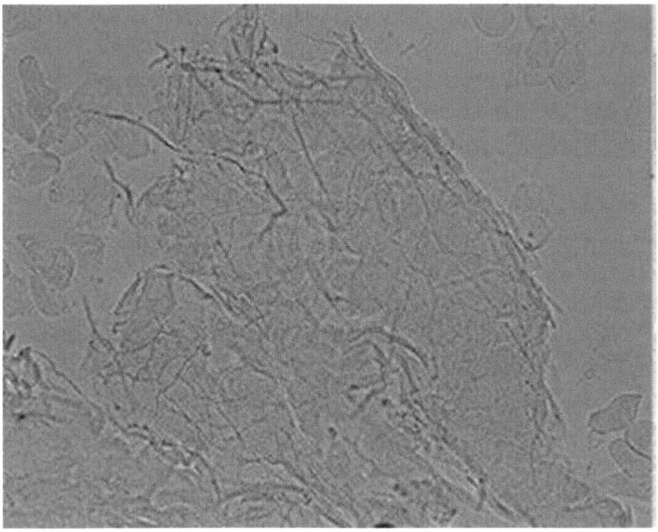

Fig. 17.3 KOH-test highlighting the presence of fungal hyphae (×200 magnification). Source: Lim et al. (2021)

White, Giemsa, Evans Blue, and Gram can be used for improved visualization (Ghannoum et al., 2018). According to Lim et al. (2021), KOH testing has a moderate sensitivity of 61% for fungal identification, but a high specificity of 95%. An hour's diagnosis is possible with this test, which has a notable advantage in terms of cost-effectiveness and quick turnaround. However, according to Leung et al. (2020), it is unable to identify the precise fungal species causing the infection. A successful KOH examination depends on obtaining a sufficient quantity of samples. It is not advisable to interpret the sample right away after KOH application to obtain the best accuracy (Leung et al., 2020).

4.2.2 Fungal Culture

The precise fungal subtype causing onychomycosis can be determined through fungal culture (Lim et al., 2021). After the surface debris is removed from the nail plate, a subungual specimen is inoculated onto Sabouraud dextrose agar (SDA) and allowed to incubate for a maximum of 30 days at a temperature between 26 °C and 30 °C (Gupta et al., 2018). The media is supplemented with antibiotics such as gentamicin and chloramphenicol to prevent bacterial growth that might impede fungal development (Lim et al., 2021). Laboratories often use SDA cultures with and without cycloheximide, an antifungal agent that inhibits the growth of NDMs, to differentiate between true fungal pathogens and non-dermatophyte molds (NDMs), which are frequently found on the skin (Gupta et al., 2018). For a conclusive diagnosis of NDM onychomycosis, however, repeat cultures may be required because of the possibility of NDM contamination (Gupta et al., 2018). According to a recent meta-analysis (Ghannoum et al., 2018), fungal culture has a high degree of specificity (99%), but a lower degree of sensitivity, averaging about 56% with a reported range of 29–82%. The main reason for this restricted sensitivity is the differences in testing centers' levels of expertise (Gupta et al., 2018).

4.2.3 Histopathology

A useful diagnostic method for onychomycosis (fungal nail infection) is the histopathological examination of a clipped nail plate section stained with periodic acid-Schiff (PAS) (Ghannoum et al., 2018). Under a microscope, this method makes it possible to see fungal components such as hyphae, pseudohyphae, spores, and yeasts (Leung et al., 2020). According to Leung et al. (2020), PAS staining has a reported sensitivity of 82–88% for the diagnosis of onychomycosis. When fungal culture is combined with PAS staining, this sensitivity can be increased even further to 96%. One major benefit of PAS staining is that it can produce results quickly. According to Gupta et al. (2018), results can usually be obtained in a day. Nevertheless, PAS staining has drawbacks, such as the inability to distinguish between the distinct fungal species responsible for the infection or assess fungal viability. Furthermore, the expense of executing a PAS stain is greater in comparison with a basic potassium hydroxide (KOH) preparation (Li et al., 2018). According to Gupta et al. (2022), PAS staining necessitates a labor-intensive processing step in comparison with the KOH method (Fig. 17.4). By enabling the visualization of fungal elements, PAS staining offers a quick and accurate diagnostic method for onychomycosis. It cannot, however, determine the precise fungal species or evaluate its viability. PAS staining can be used in conjunction with fungal culture when accurate fungal identification is required for targeted therapy.

4.3 New Diagnostic Techniques for Onychomycosis

4.3.1 Onychoscopy (Nail Dermoscopy)

Onychoscopy, or nail dermoscopy, also known as onychoscopy, emerges as a valuable non-intrusive instrument for clinicians' bedside evaluation of abnormal nails (Kayarkatte et al., 2020). This technique offers magnified visualization of microscopic features suggestive of onychomycosis (fungal nail infection). Subungual onychomycosis, both distal and lateral, is a specific type of fungal nail infection, that presents with characteristic dermoscopy findings. Among these is the severed nail area's proximal border, which is uneven and jagged. (onycholysis) with spike-like structures and extended striations (grooves) (Kayarkatte et al., 2020). Additionally, a "ruined" appearance is often associated with total dystrophic onychomycosis, a severe form of fungal nail infection (Kayarkatte et al., 2020). In

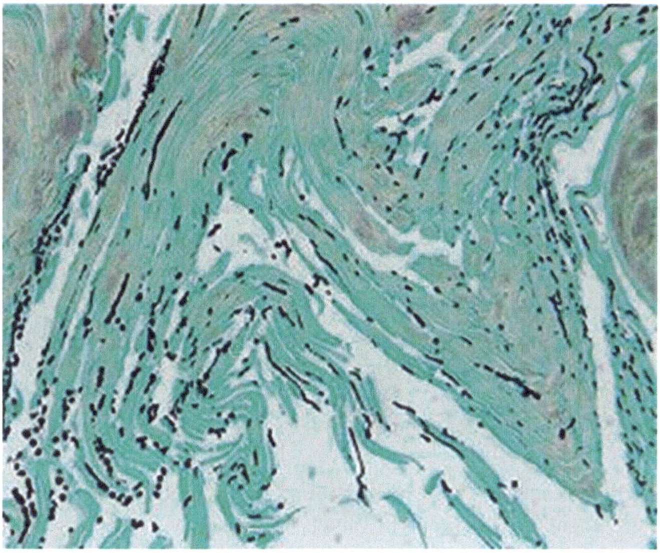

Fig. 17.4 Histopathology (nail clipping) with GMS staining showing numerous fungal hyphae in the nail plate (×400). The fungi are highlighted in black with GMS staining. Source: Lim et al. (2021)

contrast, superficial onychomycosis typically manifests uniformly opaque under dermoscopy (Kayarkatte et al., 2020). Nail dermoscopy can also help differentiate fungal melanonychia (darkening of the nail due to fungal infection) from other nail conditions. The following are considered positive indicators for fungal involvement in nail matrix nevi (moles) or subungual melanomas (skin cancers under the nail) in cases of suspected fungal melanonychia: the presence of white or yellow streaks, a non-longitudinal homogenous pattern, yellow discoloration, a reverse triangular pattern, subungual hyperkeratosis (thickening of the nail bed), a multicolor pattern, and nail scaling (Lim et al., 2021). The key advantages of nail dermoscopy lie in its speed, ease of use, and cost-effectiveness. This bedside tool has the potential to aid physicians in diagnosing onychomycosis during consultations, allowing them to decide whether further mycological assessment (diagnostic tests to identify specific fungi) is necessary (Leung et al., 2020).

4.3.2 Reflectance Confocal Microscopy (RCM)

Reflectance confocal microscopy emerges as a promising real-time bedside imaging tool for examining abnormal nails with near-histological resolution (Lim et al., 2021). This technology uses a laser with 830 nm in reflectance mode, effectively creating thin parts of the nail that are horizontal for detailed analysis (Gupta et al., 2018). In cases of onychomycosis, RCM shows characteristic networks of vibrant septate hyphae that resemble threads, indicative of fungal invasion (Leung et al.,

2020). Studies report a range of sensitivity (52.9–91.67%) and specificity (57.58–90.2%) for onychomycosis detection using RCM.

Despite its potential, RCM's limitations hinder its widespread adoption in clinical practice. The expensive nature of the technology and the dearth of subsidies in many nations restrict its accessibility (Lim et al., 2021). Further research is needed to establish its overall utility in routine clinical settings. The limited imaging depth of RCM (approximately 200 μm) also presents a challenge, as it may not be suitable for examining very thick nails. RCM offers a promising tool for diagnosing onychomycosis with high-resolution imaging capabilities. However, cost constraints, limited research on clinical integration, and the inability to effectively examine thick nails pose current hurdles to its widespread use (Lim et al., 2021).

4.3.3 Polymerase Chain Reaction

Polymerase chain reaction (PCR) testing offers a rapid and highly specific approach for detecting fungal involvement in onychomycosis by amplifying minute fragments of fungal DNA (Goldstein et al., 2019) This technique boasts high accuracy in identifying the specific causative dermatophyte fungus responsible for the infection. A significant advantage of PCR testing lies in its rapid turnaround time, delivering results within days compared to weeks required by traditional culture methods. However, limitations hinder the widespread use of PCR testing in routine clinical practice for onychomycosis diagnosis. The high cost associated with PCR assays makes them less accessible for general practitioners (Ghannoum et al., 2018). Additionally, PCR testing may not be readily available in all clinical settings, further restricting its clinical utility (Leung et al., 2020).

4.3.4 Artificial Intelligence (AI)

The field of clinical medicine has witnessed a transformative influence with the recent emergence of artificial intelligence (AI). Studies have demonstrated that AI-powered diagnostic systems can achieve performance comparable to specialist clinicians in diagnosing conditions like diabetic retinopathy and skin cancer (Zhu et al., 2022). Recognizing this potential, many dermatologists acknowledge the significant promise AI holds for advancing dermatological care (Tschandl et al., 2020).

Current explorations of AI in dermatology primarily focus on cancer of the skin detection, ulcers, psoriasis, and other inflammatory skin diseases, predicting allergy-causing substances, histopathological assessment (microscopic examination of tissue), and gene expression profiling (analyzing gene activity) (Gomolin et al., 2020). While research on AI for onychomycosis diagnosis is likely ongoing (Fig. 17.5), the provided text does not explicitly mention its application in this specific area. Due to the high specificity of this method compared to dermatoscopy, it is considered an emerging tool for diagnosis and should be included in routine diagnosis (Zhu et al., 2022).

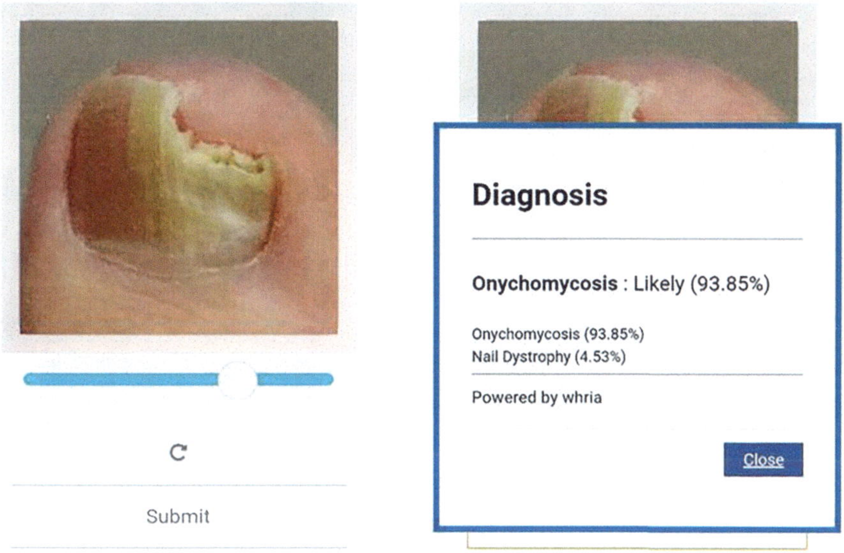

Fig. 17.5 Use of artificial intelligence for onychomycosis detection. Source: Zhu et al. (2022)

5 Treatment of Onychomycosis

Treatment aims to destroy fungal pathogens and recover the nails to typical condition (Falotico & Lipner, 2022). Treatment for onychomycosis remains a significant hurdle because the toe and fingernails grow slowly at about 1–2 mm or 2–3 mm per month (Lipner & Scher, 2019b) and the effectiveness of most antifungal therapies is demonstrably low. Moreover, the emergence of non-dermatophyte molds (NDMs) and mixed infections involving both dermatophytes and NDMs presents a growing challenge due to their frequent misdiagnosis and underestimation. A precise diagnosis is essential to successful implementing effective treatment strategies and mitigating the rise of antifungal resistance. The underestimation of NDM and mixed infections significantly contributes to this escalating resistance (Gupta et al., 2021b).

Onychomycosis treatment has several options, from traditional therapies to emerging therapies (Fig. 17.6). They include topical and oral antifungals, the use of device-based therapies for effective antifungal delivery, nail debridement, and combination therapy (which is currently very effective) (Falotico & Lipner, 2022). Current antifungal treatment strategies for onychomycosis utilize both oral and topical drugs. However, these approaches are often hindered by low efficacy, leading to relapse rates ranging from 25% to 30% (Aggarwal et al., 2020). To mitigate the limitations associated with systemic toxicity and adverse effects like gastrointestinal and liver complications observed with oral therapy, topical antifungals are frequently employed. This shift toward topical administration fosters improved patient adherence and reduces overall treatment costs but shows poor efficacy,

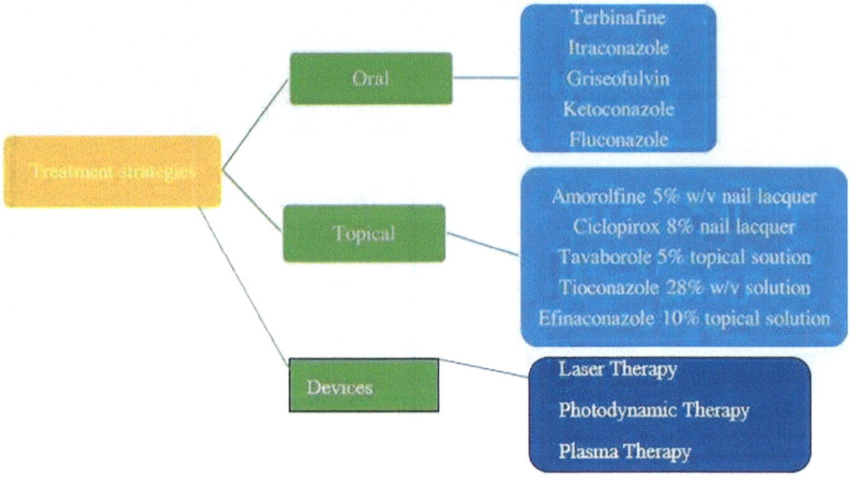

Fig. 17.6 Treatment strategies of onychomycosis. Source: Aggarwal et al. (2020)

hence the need for combination therapy (Aggarwal et al., 2020). Oral antifungals are mostly used for moderate to severe onychomycosis, while mild-to-moderate infection is treated with topical antifungals (Lipner et al., 2021). In conjunction with topical therapies, mechanical removal of infected nail tissue (debridement) can reduce fungal burden (Gupta et al., 2018). Nail avulsion, the complete removal of the affected nail, is generally reserved for severely painful or non-growing single nails due to its limited applicability (Falotico & Lipner, 2022). Laser therapies have received FDA approval for temporary improvements in clear nail appearance; however, their cure rates fall below those achieved with oral and topical antifungals. This disparity can be attributed to less stringent FDA endpoints for laser therapy approval (Lipner et al., 2021).

Furthermore, definitive guidelines for their use are still under development (Lipner & Scher, 2019b). Treatment approaches should be individualized, taking into account factors such as disease severity, the specific infecting organism, medication cost-effectiveness, underlying health conditions, past medication use, and the likelihood of patient adherence to treatment regimens (Lipner & Scher, 2019b). Combination therapy, involving the use of multiple treatment strategies simultaneously, should be reserved for individuals with poor prognostic indicators (e.g., advanced age, compromised immune system, mixed fungal infections) or those who have not responded to single-agent treatment (Falotico et al., 2022). Notably, a systematic review revealed that fewer than 5% (17.5%; 32/182) of onychomycosis. Clinical trials that compared treatment efficacy across different subgroups were the only ones that provided information on race and ethnicity (Chang et al., 2021). Thus, it is important to exercise caution when interpreting treatment recommendations derived from these trials for patients of color (Falotico & Lipner, 2022).

5.1 *Oral Therapy*

Oral therapy is also regarded as systemic medication as it enters into the system and moves from one part to another. While systemic antifungal medications are a mainstay in onychomycosis treatment due to their ease of administration, high effectiveness, and relative cost efficiency (Kayarkatte et al., 2020), their efficacy is limited by several factors. Orally administered drugs must reach the infected nail bed after systemic absorption and penetrate the ventral nail plate (Vikas et al., 2020). However, this delivery is hampered by poor blood flow to the nail bed, restricting the amount of medication that reaches the target site (Aslam et al., 2021) Additionally, biofilm formation by fungi can further hinder drug penetration (Falotico & Lipner, 2022). Furthermore, systemic antifungals carry a higher chance of significant adverse effects and possible medication interaction penetration (Falotico & Lipner, 2022).

Among the oral antifungals for onychomycosis that the US FDA has approved (Terbinafine, Itraconazole, and Griseofulvin) (Lipner & Scher, 2019b). Griseofulvin has fallen out of favor due to its less desirable treatment profile. Compared to newer options, it necessitates longer treatment durations, exhibits lower efficacy in eradicating the infection, and has a higher potential for side effects and recurrence (Lipner et al., 2021).

5.2 *Terbinafine*

Terbinafine is a type of allylamine that has some efficacy against NDMs and Candida species and broad-spectrum activity against dermatophytes by inhibiting squalene epoxidase. During 6 or 12 weeks, the daily dosage is 250 mg for nail infections in the fingers and toenails (Lipner & Scher, 2019b). Fingernails and toenails, the full cure rates are 59% and 38% respectively, whereas the rates of mycologic cure are 79% and 70%. Mild side effects that could occur include rashes, headaches, and gastrointestinal problems. Hepatotoxicity and taste abnormalities are uncommon.

5.3 *Itraconazole*

Itraconazole, a triazole antifungal medication, demonstrates a broader spectrum of activity compared to terbinafine. It exhibits efficacy against dermatophytes, candida species, and certain molds (NDMs) (Falotico & Lipner, 2022). The treatment regimen for itraconazole varies based on the infected nail: a daily dose of 200 mg for toenails is administered for 12 weeks. The treatment of fingernails needs to be pulsed, with 200 mg administered twice daily for 1 week, then for 3-week washout period. This cycle is repeated for a total of two treatment pulses (Lipner & Scher,

2019b). While itraconazole offers lower the rate of complete cure than terbinafine (47% and 14% for toenails and also fingernails, respectively), its rate of mycological cure is comparable (61% and 54% for fingernails and toenails, respectively) (Gupta et al., 2020). However, itraconazole carries a potentially wider range of side effects compared to terbinafine. These potential adverse events consist of migraines, gastrointestinal problems, elevated blood triglycerides, and upper respiratory tract infections increased liver enzyme levels, and, in rare instances, peripheral neuropathy and hepatitis (Gupta et al., 2020).

5.4 *Fluconazole*

Similar to itraconazole, fluconazole belongs to the triazole class of antifungal medications. It exerts its antifungal effect by inhibiting the enzyme lanosterol 14α-demethylase. However, its limited residual concentration within the nail necessitates extended treatment durations. Weekly dosing regimens are employed, ranging from 150 mg for 6–9 months, which is recommended for fingernails, and 12–18 months for toenails. The shift from daily to weekly dosing can present challenges in medication adherence. To address this, patients may benefit from implementing reminder systems on their mobile devices, such as text notifications or alarms, to enhance compliance (Lin & Lipner, 2019).

5.5 *Topical Therapies*

Topical antifungal therapies are gaining traction as a treatment option for onychomycosis due to several potential advantages. These advantages include their ability to potentially penetrate fungal biofilms, a lower risk of drug-drug interactions due to minimal systemic absorption, and the elimination of the need for laboratory monitoring (Costa-Orlandi et al., 2021). However, for topical medications to be effective, they have to be able to pass through the nail plate and get to the fungal infection site in the translingual space. This necessitates maintaining adequate drug concentrations at the application site to achieve a minimal inhibitory concentration (MIC) against the target fungus (Vikas et al., 2020). Unfortunately, several things can prevent this penetration, such as the dense nails' slow growth rate and natural thickness and impermeable network of keratin fibers within the nail plate, and the altered properties of diseased nails (e.g., thickened or detached from the nail bed) (Davies-Strickleton et al., 2020). Additionally, nail keratin itself can have a detrimental effect on antifungal activity (Davies-Strickleton et al., 2020). While newer topical agents offer promising results, their high cost and lengthy treatment durations (up to 48 weeks for toenail infections) can pose challenges to patient compliance, particularly for individuals with limited mobility or dexterity.

The approved topical therapies or antifungal drugs ciclopirox (8% nail lacquer), efinaconazole (10%), tavaborole (5%), and amorolfine (5%) are the medications used to treat onychomycosis in the fingers and toenails.

Hydroxypyridone in ciclopirox 8% nail lacquer chelates trivalent cations, inhibiting enzymes that are dependent on metals. Burning, periungual erythema, and application site reactions are among the side effects of ciclopirox (Gupta et al., 2020).

Efinaconazole, a topical triazole antifungal medication formulated as a 10% solution, exerts its antifungal effect by inhibiting the enzyme lanosterol 14α-demethylase. This inhibition disrupts the production of ergosterol, which is essential to the fungal cell membrane (Gupta et al., 2020). Efinaconazole demonstrates activity against a wide spectrum of fungi, including dermatophytes, Candida species, and certain molds (NDMs). Due to its efficacy, it has emerged as the preferred treatment option for dermatophytoma, a fungal nail bed infection (Lipner et al., 2021).

Tavaborole, formulated as a 5% topical solution, belongs to the benzoxazole class of antifungal medications. Its mechanism of action involves preventing the fungus from producing proteins by blocking the activity of fungal aminoacyl transfer RNA synthetase. As a consequence, there is broad-spectrum antifungal activity against specific molds (NDMs), yeasts, and dermatophytes (Lipner & Scher, 2019b). Erythema, dermatitis, and exfoliation are examples of local side effects.

5.6 *Device Therapy*

This is physical therapy that involves the use of some devices such as the laser system, and photodynamic therapy aids in the administration of specific antifungal medications to the body.

5.6.1 Laser System/Therapy

Laser therapy represents an emerging physical approach to enhance drug delivery in onychomycosis treatment. This method leverages the principles of photothermal lysis, where pulsed laser radiation is absorbed by the target tissue, generating heat (Aggarwal et al., 2020). A range of laser types are utilized, such as titanium (Ti) sapphire lasers, neodymium-doped yttrium aluminum garnet (Nd: YAG) lasers, and diode lasers. By using pulsed laser systems, the energy output can be increased significantly in a shorter amount of time.

5.6.2 Photodynamic Therapy (PDT)

Photodynamic therapy (PDT) emerges as a potential treatment option for onychomycosis, although it currently lacks US FDA approval for this specific use. This non-invasive, device-based therapy is already approved for treating actinic keratoses. PDT employs a combination of light-based techniques and photosensitizing agents, which may include methylene blue, methyl amino levulinate, or 5-aminolevulinic acid (Gupta et al., 2020). Reactive oxygen species are produced as a result of this combination (ROS) and free radicals, which exert their antifungal effects through multiple mechanisms. These mechanisms include direct antimicrobial activity, cytotoxicity (damaging fungal cells), and inducing apoptosis (programmed cell death) in the fungal cells (Falotico & Lipner, 2022). Despite its potential as an antifungal therapy, the widespread use of PDT for onychomycosis treatment is currently hindered by several limitations. These limitations include the need for multiple treatment sessions, often requiring pretreatment with urea or even nail avulsion, which can be a painful procedure (Lipner & Scher, 2019b). Furthermore, the current evidence base for PDT in onychomycosis lacks sufficient robustness. Larger patient populations require more comprehensive randomized controlled trials to establish PDT's efficacy and safety profile before its suggestion as a regular off-label therapy for this fungal infection (Lipner et al., 2021).

6 Emerging Trends

There have been emerging trends in the study of onychomycosis, from the diagnosis to the treatments. Currently, the use of artificial intelligence for fungal detection has improved the disease diagnosis. Also, the combination of antifungal drugs has helped reduce the negative impact of antifungal resistance. Most of the trends are focused on the treatment and management of onychomycosis.

6.1 New Oral Therapies

Novel oral treatments for onychomycosis have recently received attention. They include posaconazole, fosravuconazole L-lysine ethanolate (F-RVC), and oteseconazole (Falotico & Lipner, 2022). Posaconazole is a broad spectrum of triazoles for invasive fungal infections and has a higher complete cure than terbinafine of about 45–54%. F-RVC belongs to the azole group, a prodrug of ravuconazole that has increased onychomycosis treatment in Japan and Europe (Noguchi et al., 2019).

Oteseconazole, as a tetrazole, inhibits the lanosterol demethylase enzymes which helps the fungal survival. It is still undergoing clinical trials and has not been approved according to Gupta et al. (2022).

6.2 *New Topical Therapies*

The development of novel topical formulations for onychomycosis has garnered significant interest in recent years, presenting them as potential alternatives to established topical therapies (Gupta et al., 2021a). Ciclopirox hydrolacquer (P-3051) represents another emerging topical treatment for onychomycosis. Notably, it has not yet received approval for this indication within the United States but enjoys widespread use and marketing authorization in over 40 countries, particularly across Europe (Gupta et al., 2021a). This formulation incorporates hydroxypropyl chitosan, a substance that coats the surface of the nails like an invisible film. This film offers a two-pronged benefit: it hinders fungal invasion and facilitates drug penetration through the nail plate (Gupta et al., 2021a).

6.3 *Plasma Therapy*

Non-thermal plasma (NTP), also referred to as reduced-temperature plasma, represents a novel and promising tool for onychomycosis treatment. Research actively investigates its potential in this field (Bulson et al., 2019). NTP generation involves the application of short electrical pulses (around 10 ns) with high-peak electric fields (approximately 20 kV/mm). Numerous reactive chemical species, such as ions, electrons, nitric oxide, ozone, and hydroxyl radicals, are formed as a result of this process. Importantly, minimal tissue heating occurs during NTP treatment due to the low current and short pulse duration (Bulson et al., 2019). A toe/nail plate model is used in an in vitro study to explore the effectiveness of non-thermal plasma (NTP) against *Candida albicans* and *Trichophyton mentagrophytes*, two fungal species commonly associated with onychomycosis (Bulson et al., 2019). The study revealed that NTP treatment for 12 min achieved complete eradication of the fungi at doses of 39 k Pulses and 15 k Pulses for C. albicans and *T. mentagrophytes*, respectively.

6.4 *Combination Therapy*

While combination therapy for onychomycosis offers a theoretical advantage through potential drug synergy and reduced risk of antifungal resistance its use remains limited due to a lack of well-established treatment protocols and robust clinical evidence (Falotico & Lipner, 2022). A systematic review analyzing 30 onychomycosis trials (including studies with medication alone, and medication combined with procedures like laser therapy, debridement, or PDT) found that over half (60%) showed some benefit for combination therapy compared to monotherapy (Lipner et al., 2021). However, these studies often lacked robust design and adequate follow-up compared to pivotal trials investigating single medications. Notably, almost all studies (93.3%) combining medication with procedural interventions demonstrated advantages over monotherapy. Combination therapy should only be used as a last resort for patients with poor prognoses or fungal infections that are resistant to treatment, due to the higher cost of combination therapy, the possibility of drug interactions, and the dearth of well-designed trials with long-term follow-up.

6.5 Non-treatment Therapy

Beyond the use of drugs, chemical, and electrical methods of mitigating this disease, some common strategies can be encouraged to reduce the recurrence. The strategies include sanitation techniques and environmental and lifestyle changes.

6.5.1 Sanitization Techniques

Persistent exposure to fungal reservoirs, exemplified by contaminated footwear, hosiery, or fabrics, may contribute to the recurrence of fungal infections (Gupta & Versteeg, 2019). Fungi may use skin cells trapped in shoes and sweat as a source of nutrition to form and maintain fungal reservoirs (Gupta et al., 2021b). Additionally, possible fungal reservoirs are textiles like blankets, sheets, and towels. After receiving antifungal therapy to cure them, patients may continue to spread infection by wearing contaminated socks and shoes.

As part of basic infection control strategies, patients are recommended to (1) replace heavily contaminated footwear with new pairs, (2) switch to absorbent socks for enhanced moisture management, and replace these socks once or twice daily, and (3) maintain physical separation between contaminated and sterile clothing to minimize the spread of fungal elements (Gupta & Versteeg, 2019).

6.5.2 Environmental and Lifestyle Changes

Several preventive actions can be taken to minimize the transmission of fungal pathogens, such as (1) utilizing footwear (shoes or sandals) in endemic environments including swimming pools and communal baths, (2) proper wearing of sized footwear to prevent excessive moisture accumulation, (3) meticulously drying feet after bathing or showering, and (4) maintaining good personal hygiene practices, including keeping nails trimmed and clean. Furthermore, prompt diagnosis and treatment of onychomycosis (fungal nail infection) and *Tinea pedis* (athlete's foot) in cohabiting individuals (family members or housemates) is crucial, as they can serve as potential reservoirs for fungal transmission (Gupta et al., 2021b).

7 Conclusion

As the most prevalent nail disease seen in clinical settings, onychomycosis can seriously lower the quality of life for those who suffer from pain, ambulation difficulties, and social stigma. After conducting a history and physical examination, dermoscopy or dermatoscopy can distinguish onychomycosis from other nail conditions, with a confirmed diagnosis achieved through KOH and microscopy, fungal culture, histopathology, PCR, or a combination of these techniques. Artificial intelligence is a relatively new diagnostic tool that needs more testing before being applied in clinical settings. With a range of therapies available, including topical and oral medications as well as device-based treatments, onychomycosis is a highly treatable condition. Topical treatments have better safety profiles and higher cure rates than systemic treatments. Device-based therapies need to be studied using endpoints comparable to those used in oral and topical medication clinical trials to allow for accurate comparisons.

References

Aggarwal, R., Targhotra, M., Kumar, B., Sahoo, P., & Chauhan, M. (2020). Treatment and management strategies of onychomycosis. *Journal de Mycologie Médicale, 30*(2), 100949. https://doi.org/10.1016/j.mycmed.2020.100949

Albucker, S. J., Falotico, J. M., Choo, Z. N., Matushansky, J. T., & Lipner, S. R. (2023). Risk factors and treatment trends for onychomycosis: A case-control study of onychomycosis patients in the All of Us research program. *Journal of Fungi, 9*(7), 712. https://doi.org/10.3390/jof9070712

Andrés, T. S., & Alexandro, B. (2020). Candida onychomycosis: An old problem in modern times. *Current Fungal Infection Reports, 14*(3), 209–216. https://doi.org/10.1007/s12281-020-00394-3

Antuori, A., Fernández, G., Fernández, A., Alcaide, M., Boada, A., Bielsa, M. I., Romaní, N., & Matas, L. (2019). Epidemiology of dermatophytic infections between 2008 and 2017 in Barcelona, Spain. *Enfermedades Infecciosas y Microbiología Clínica, 37*(10), 642–647. https://doi.org/10.1016/j.eimc.2019.02.010

Aslam, R., Hussain, T., Yousaf, A. M., Ghori, M. U., Khan, I. U., Rizvi, S. A., & Shahzad, Y. (2021). Onychomycosis: Current understanding and strategies for enhancing drug delivery into human nail tissue. *Current Drug Research Reviews, 13*(1), 25–35. https://doi.org/10.2174/2589977512666200731171505

Bermudez, N. M., Rodríguez-Tamez, G., Perez, S., & Tosti, A. (2023). Onychomycosis: Old and new. *Journal of Fungi, 9*(5), 559. https://doi.org/10.3390/jof9050559

Bulson, J. M., Liveris, D., Derkatch, I., Friedman, G., Geliebter, J., Park, S., Singh, S., Zemel, M., & Tiwari, R. K. (2019). Non-thermal atmospheric plasma treatment of onychomycosis in an in vitro human nail model. *Mycoses, 63*(2), 225–232. https://doi.org/10.1111/myc.13030

Bodman MA, Krishnamurthy K. Onychomycosis. StatPearls [Internet]. aTreasure Island (FL): StatPearls Publishing; 2019 Jan- 2019 Jan 5.

Chang, M. J., Qiu, Y., & Lipner, S. R. (2021). Race reporting and representation in onychomycosis clinical trials: A systematic review. *Mycoses, 64*(8), 954–966. https://doi.org/10.1111/myc.13262

Costa-Orlandi, C. B., Martinez, L. R., Bila, N. M., Friedman, J. M., Friedman, A. J., Mendes-Giannini, M. J. S., & Nosanchuk, J. D. (2021). Nitric oxide-releasing nanoparticles are similar to efinaconazole in their capacity to eradicate Trichophyton rubrum biofilms. *Frontiers in Cellular and Infection Microbiology, 11*. https://doi.org/10.3389/fcimb.2021.684150

Daggett, C., Brodell, R. T., Daniel, C. R., & Jackson, J. (2019). Onychomycosis in athletes. *American Journal of Clinical Dermatology, 20*(5), 691–698. https://doi.org/10.1007/s40257-019-00448-4

Davies-Strickleton, H., Cook, J., Hannam, S., Bennett, R., Gibbs, A., Edwards, D., Ridden, C., Ridden, J., & Cook, D. (2020). Assessment of the nail penetration of antifungal agents, with different physico-chemical properties. *PLoS One, 15*(2), e0229414. https://doi.org/10.1371/journal.pone.0229414

Dinani, N., & George, S. (2019). Nail cosmetics: A dermatological perspective. *Clinical and Experimental Dermatology, 44*(6), 599–605. https://doi.org/10.1111/ced.13929

Falotico, J. M., Lapides, R., & Lipner, S. R. (2022). Combination therapy should be reserved as second-line treatment of onychomycosis: A systematic review of onychomycosis clinical trials. *Journal of Fungi, 8*(3), 279. https://doi.org/10.3390/jof8030279

Falotico, J. M., & Lipner, S. R. (2022). Updated perspectives on the diagnosis and management of onychomycosis. *Clinical, Cosmetic and Investigational Dermatology, 15*, 1933–1957. https://doi.org/10.2147/ccid.s362635

Ghannoum M, Mukherjee P, Isham N, Markinson B, Rosso JD, Leal L (2018) Examining the importance of laboratory and diagnostic testing when treating and diagnosing onychomycosis. Int J Dermatol 57:131–138.

Goldstein, A.O., Bhatia, N., & Post, T. W. ed. (2019, Up To Date). Onychomycosis: Epidemiology, clinical features, and diagnosis. Waltham, MA. Accessed on June 30, 2019

Gomolin, A., Netchiporouk, E., Gniadecki, R., & Litvinov, I. V. (2020). Artificial intelligence applications in dermatology: Where do we stand? *Frontiers in Medicine, 7*. https://doi.org/10.3389/fmed.2020.00100

Gupta, A., Stec, N., Summerbell, R., Shear, N., Piguet, V., Tosti, A., & Piraccini, B. (2020). Onychomycosis: a review. *Journal of the European Academy of Dermatology and Venereology : JEADV, 34*(9), 1972–1990. https://doi.org/10.1111/jdv.16394

Gupta, A. K., Mays, R. R., Versteeg, S. G., Shear, N. H., & Piguet, V. (2018). Update on current approaches to diagnosis and treatment of onychomycosis. *Expert Review of Anti-Infective Therapy, 16*(12), 929–938. https://doi.org/10.1080/14787210.2018.1544891

Gupta, A. K., Surprenant, M. S., Kempers, S. E., Pariser, D. M., Rensfeldt, K., & Tavakkol, A. (2021a). Efficacy and safety of topical terbinafine 10% solution (MOB-015) in the treatment of mild to moderate distal subungual onychomycosis: A randomized, multicenter, double-blind, vehicle-controlled phase 3 study. *Journal of the American Academy of Dermatology, 85*(1), 95–104. https://doi.org/10.1016/j.jaad.2020.06.055

Gupta, A. K., Talukder, M., & Venkataraman, M. (2021b). Review of the alternative therapies for onychomycosis and superficial fungal infections: Posaconazole, fosravuconazole, voriconazole, oteseconazole. *International Journal of Dermatology, 61*(12), 1431–1441. https://doi.org/10.1111/ijd.15999

Gupta, A. K., Venkataraman, M., & Talukder, M. (2022). Onychomycosis in older adults: Prevalence, diagnosis, and management. *Drugs & Aging, 39*(3), 191–198. https://doi.org/10.1007/s40266-021-00917-8

Gupta, A. K., & Versteeg, S. G. (2019). The role of shoe and sock sanitization in the management of superficial fungal infections of the feet. *Journal of the American Podiatric Medical Association, 109*(2), 141–149. https://doi.org/10.7547/17-043

Jazdarehee, A., Malekafzali, L., Lee, J., Lewis, R., & Mukovozov, I. (2022). Transmission of onychomycosis and dermatophytosis between household members: A scoping review. *Journal of Fungi, 8*(1), 60. https://doi.org/10.3390/jof8010060

Kayarkatte, M. N., Singal, A., Pandhi, D., Das, S., & Sharma, S. (2020). Nail dermoscopy (onychoscopy) findings in the diagnosis of primary onychomycosis: A cross-sectional study. *Indian Journal of Dermatology, Venereology, and Leprology, 86*(4), 341. https://doi.org/10.4103/ijdvl.ijdvl_100_19

Klinger, M., Theiler, M., & Bosshard, P. (2021). Epidemiological and clinical aspects of Trichophyton mentagrophytes/Trichophyton interdigitale infections in the Zurich area: A retrospective study using genotyping. *Journal of the European Academy of Dermatology and Venereology : JEADV, 35*(4), 1017–1025. https://doi.org/10.1111/jdv.17106

Leung, A. K., Lam, J. M., Leong, K. F., Hon, K. L., Barankin, B., Leung, A. A., & Wong, A. H. (2020). Onychomycosis: An updated review. *Recent Patents on Inflammation & Allergy Drug Discovery, 14*(1), 32–45. https://doi.org/10.2174/1872213x13666191026090713

Li, D. G., Cohen, J. M., Mikhailov, A., Williams, R. F., Laga, A. C., & Mostaghimi, A. (2018). Clinical diagnostic accuracy of onychomycosis: A multispecialty comparison study. *Dermatology Research and Practice, 2018*, 1–3. https://doi.org/10.1155/2018/2630176

Lim, S. S., Ohn, J., & Mun, J. H. (2021). Diagnosis of onychomycosis: From conventional techniques and dermoscopy to artificial intelligence. *Frontiers in Medicine, 8*. https://doi.org/10.3389/fmed.2021.637216

Lin, K., & Lipner, S. R. (2019). Mobile phone reminders for onychomycosis medication adherence. *Journal of the American Academy of Dermatology, 80*(5), e105–e107. https://doi.org/10.1016/j.jaad.2018.11.010

Lipner, S. R., Joseph, W. S., Vlahovic, T. C., Scher, R. K., Rich, P., Ghannoum, M., Daniel, C. R., & Elewski, B. (2021). Therapeutic recommendations for the treatment of toenail onychomycosis in the US. *Journal of Drugs in Dermatology, 20*(10), 1076–1084. https://doi.org/10.36849/jdd.6291

Lipner, S. R., & Scher, R. K. (2019a). Onychomycosis. *Journal of the American Academy of Dermatology, 80*(4), 853–867. https://doi.org/10.1016/j.jaad.2018.05.1260

Lipner, S. R., & Scher, R. K. (2019b). Onychomycosis. *Journal of the American Academy of Dermatology, 80*(4), 835–851. https://doi.org/10.1016/j.jaad.2018.03.062

Marcos-Tejedor, F., Mota, M., Iglesias-Sánchez, M. J., Mayordomo, R., & Gonçalves, T. (2021). Identification of fungi involved in onychomycosis in patients of a Spanish rural area. *Journal of Fungi, 7*(8), 623. https://doi.org/10.3390/jof7080623

Moskaluk, A. E., & VandeWoude, S. (2022). Current topics in dermatophyte classification and clinical diagnosis. *Pathogens, 11*(9), 957. https://doi.org/10.3390/pathogens11090957

Noguchi, H., Matsumoto, T., Kimura, U., Hiruma, M., Kano, R., Yaguchi, T., Fukushima, S., & Ihn, H. (2019). Fungal melanonychia caused by Candida parapsilosis was successfully treated with oral fosravuconazole. *Journal of Dermatology, 46*(10), 911–913. https://doi.org/10.1111/1346-8138.15024

Petrucelli, M. F., De Abreu, M. H., Cantelli, B. A. M., Segura, G. G., Nishimura, F. G., Bitencourt, T. A., Marins, M., & Fachin, A. L. (2020). Epidemiology and diagnostic perspectives of Dermatophytoses. *Journal of Fungi, 6*(4), 310. https://doi.org/10.3390/jof6040310

Piraccini, B. M., & Alessandrini, A. (2015). Onychomycosis: A review. *J Fungi (Basel), 1*(1):30–43. https://doi.org/10.3390/jof1010030. PMID: 29376897; PMCID: PMC5770011.

Piraccini, B. M., Starace, M., Rubin, A. I., Di Chiacchio, N. G., Iorizzo, M., & Rigopoulos, D. (2022). Onychomycosis: Recommendations for diagnosis, assessment of treatment efficacy, and specialist referral. The CONSONANCE Consensus Project. *Dermatology and Therapy, 12*(4), 885–898. https://doi.org/10.1007/s13555-022-00698-x

Reinel, D. (2021). Non-dermatophyte fungi in onychomycosis—Epidemiology and consequences for clinical practice. *Mycoses, 64*(7), 694–700. https://doi.org/10.1111/myc.13251

Sen, S., Borah, S. N., Bora, A., & Deka, S. (2020). Rhamnolipid exhibits anti-biofilm activity against the dermatophytic fungi Trichophyton rubrum and Trichophyton mentagrophytes. *Biotechnology Reports, 27*, e00516. https://doi.org/10.1016/j.btre.2020.e00516

Subramanya, S. H., Subedi, S., Metok, Y., Kumar, A., Prakash, P. Y., & Nayak, N. (2019). Distal and lateral subungual onychomycosis of the fingernail in a neonate: A rare case. *BMC Pediatrics, 19*(1). https://doi.org/10.1186/s12887-019-1549-9

Trovato, L., Calvo, M., De De Pasquale, R., Scalia, G., & Oliveri, S. (2022). Prevalence of onychomycosis in diabetic patients: A case-control study performed at University Hospital Policlinico in Catania. *Journal of Fungi, 8*(9), 922. https://doi.org/10.3390/jof8090922

Tschandl, P., Rinner, C., Apalla, Z., Argenziano, G., Codella, N., Halpern, A., Janda, M., Lallas, A., Longo, C., Malvehy, J., Paoli, J., Puig, S., Rosendahl, C., Soyer, H. P., Zalaudek, I., & Kittler, H. (2020). Human-computer collaboration for skin cancer recognition. *Nature Medicine, 26*(8), 1229–1234. https://doi.org/10.1038/s41591-020-0942-0

Vikas, A., Rashmin, P., Mrunali, P., Chavan, R. B., & Kaushik, T. (2020). Mechanistic insights of formulation approaches for the treatment of nail infection: Conventional and novel drug delivery approaches. *AAPS PharmSciTech, 21*(2). https://doi.org/10.1208/s12249-019-1591-9

Widaty, S., Miranda, E., & Oktarina, C. (2021). Candida onychomycosis: Mini review. In *Infectious diseases*. https://doi.org/10.5772/intechopen.96650

Zhan, P., Liang, G., & Liu, W. (2021). Dermatophytes and dermatophytic infections worldwide. In *Springer eBooks* (pp. 15–40). https://doi.org/10.1007/978-3-030-67421-2_2

Zhu, X., Zheng, B., Cai, W., Zhang, J., Lu, S., Li, X., Xi, L., & Kong, Y. (2022). Deep learning-based diagnosis models for onychomycosis in dermoscopy. *Mycoses, 65*(4), 466–472. https://doi.org/10.1111/myc.13427

Chapter 18
Potentials of Phages Against Antibiotic-Resistant Bacteria

Oranusi Uche Solomon and Adeleye Damilola

1 Introduction

The emergence and spread of antibiotic-resistant bacteria pose a significant threat to public health globally (Teng et al., 2023). Exploration of alternative treatment options, such as phage therapy, has gained renewed interest in recent years (Wang et al., 2021). Bacteriophages, also called phages, are viruses that specifically target and kill bacteria (Chan et al., 2013). Bacteriophages can potentially be potent in combating antibiotic-resistant bacteria as they can effectively target and eliminate specific bacterial strains (Yang et al., 2021). Furthermore, phages have the advantage of being naturally occurring and non-toxic, making them a favourable alternative to antibiotics. Phage therapy involves using phages to infect and kill specific bacteria, offering a targeted approach to treatment (Fabijan et al., 2023). Using phages to treat antibiotic-resistant bacteria can revolutionise healthcare by providing an effective and specific alternative to antibiotics (Pires et al., 2020). Phage therapy has shown promising results in experimental and clinical settings, with studies demonstrating successful eradication of antibiotic-resistant bacteria (Malik et al., 2017). In addition to their ability to target specific bacteria, another advantage of phages over antibiotics is their potential to prevent allergies in the host (Manyi-Loh et al., 2018). This is because the immune system tolerates phages more favourably, reducing the risk of allergic reactions (Heine et al., 2021). Nature is an almost

O. U. Solomon
Department of Biological Sciences, College of Sciences and Technology, Covenant University, Ota, Ogun state, Nigeria

A. Damilola (✉)
Department of Biological Sciences, College of Sciences and Technology, Covenant University, Ota, Ogun state, Nigeria

Covenant University, Ota, Ogun State, Nigeria
e-mail: damilola.adeleyepgs@stu.cu.edu.ng

P. O. Isibor, P. A. Akinduti (eds.), *Harnessing Biotechnology Tools for Product Development*, https://doi.org/10.1007/978-3-031-86002-7_18

infinite phage resource: new ones can be isolated for most problem bacteria as needed; bacteria and their phages constantly co-evolve. This constant co-evolution between bacteria and phages is crucial in the fight against evolving pathogenic bacterial variants and challenging situations (Uddin et al., 2021). Cocktails of multiple phages can reduce the probability of bacteria developing resistance to a specific phage and, in some cases, antibiotics (Uddin et al., 2021).

The emergence of antimicrobial resistance (AMR), characterised by the ability of microorganisms to evade the therapeutic effects of previously effective medications, poses a significant and growing public health threat (Manyi-Loh & Lues, 2023). The rise and spread of AMR bacteria, viruses, fungi, and parasites significantly undermine the effectiveness of modern medicine, jeopardising our ability to treat common infections, perform life-saving surgeries, and manage chronic diseases (Helene et al., 2022). Moreover, by 2050, antibiotic resistance could reach pandemic levels, resulting in 10 million global deaths each year (Kirtane et al., 2021).

AMR is a natural phenomenon. Microorganisms are constantly evolving and have developed mechanisms to combat the impacts of antimicrobial substances. However, the inappropriate use and overuse of antimicrobials in human medicine, agriculture, and animal husbandry have significantly accelerated the rate of AMR development.

Furthermore, unnecessary antibiotic prescriptions for viral infections or self-medication practices in humans contribute to the development of resistance (Prestinaci et al., 2015). Regularly administering antibiotics to healthy livestock to promote growth or prevent illness can inadvertently cultivate antibiotic-resistant bacteria within their gut flora. Concerns exist that these resistant strains could be transmitted to humans via the food chain (Collignon 2015). In addition, poor sanitation and hygiene practices in healthcare settings and communities facilitate the spread of resistant pathogens (Verbeek et al., 2020). The pharmaceutical industry faces economic disincentives for developing new antibiotics because of the expectation of limited return on investment and the risk of rapid emergence of resistance (Tacconelli et al., 2018).

AMR severely affects public health, the economy, and global development. AMR infections are difficult and expensive to treat, leading to more extended hospital stays, increased healthcare costs, and higher mortality rates (Karabay et al., 2019). The effectiveness of essential medical procedures like surgery, chemotherapy, and organ transplantation depends on effective antibiotics to prevent infections. Rising AMR threatens the safety and success of these procedures.

Bacteriophages, often shortened to phages, are a specific type of virus that targets and infects bacteria (Jault et al., 2019). These highly specialised structures lack a complete cellular organisation and consist solely of a single nucleic acid molecule. Notably, phages exhibit high host specificity, which can infect only particular bacterial species or even specific strains within a single species (Saral-Acarca et al., 2023).

The discovery of bacteriophages is credited to two independent researchers: Frederick W. Twort in Great Britain (1915) and Félix d'Hérelle in France (1917).

D'Hérelle, recognising the phage's ability to kill bacteria, coined the term 'bacteriophage', which translates to 'bacteria eater'. Interestingly, these viral predators can infect the archaea bacteria and the eubacteria (single-celled prokaryotic organisms).

The early twentieth century witnessed the independent discoveries of bacteriophages by Frederick William Twort and Felix d'Hérelle. In 1915, Twort observed a lytic (cell-destroying) agent in cultures of Staphylococcus while investigating the vaccinia virus (Durbas & Machnik, 2022). It is now understood that these were likely bacteriophages targeting Staphylococcus contaminants within the vaccinia culture (Ezzatpanah et al., 2022). d'Hérelle's discovery came a few years later, 1917 (Kochhar, 2020). He isolated phages effective against Shigella bacteria from the stool samples of individuals recovering from dysentery (Lauman & Dennis, 2021). Interestingly, earlier observations hinting at the existence of bacteriophages have been identified.

The therapeutic application of bacteriophages, or phage therapy, has a documented history dating back to 1919 (Fabijan et al., 2023). Following its initial use by d'Hérelle, phage therapy was introduced to the United States in 1922, and the subsequent decades, from the early 1920s to the late 1930s, witnessed a global upsurge in using phage therapy for treating infectious diseases. This trend is well-supported by the work of d'Hérelle in 1931. The landscape of phage therapy shifted significantly with the emergence of commercially available antibiotics in the 1940s. These antibiotics offered several advantages, including broader spectrum activity and greater ease of production and standardisation (Handy et al., 2012). Consequently, bacteriophages'use as therapeutic agents has declined, particularly in Western nations like the United States (Iwu et al., 2020). Research focuses on phages transitioning towards a deeper understanding of their fundamental biological properties (Wegst et al., 2014). However, the use of phages for therapeutic purposes continued unabated in Eastern Europe and Russia (Aswani & Shukla, 2021). This regional divergence in phage therapy practices provides a valuable opportunity to analyse the factors influencing its success or decline in different contexts (Jangid et al., 2024).

Following a period of successful early application, phage therapy was largely eclipsed by the widespread adoption of antibiotics (Bull, 2010). However, due to the emergence of antibiotic-resistant bacteria, research into phage therapy is experiencing a resurgence (Uddin et al., 2021). The rise of multidrug-resistant bacteria posing a global health threat, phage therapy has become a promising alternative or adjunct to antibiotics (Torfs et al., 2019). Recent success in phage genomics and bioinformatics has empowered researchers to characterise novel bacteriophages and elucidate phage–host interactions. This newfound knowledge paves the way for the development of precision phage therapy (Loureiro & Da Silva, 2019). Additionally, innovative strategies are gaining traction to enhance treatment efficacy. These include using phage cocktails targeting a broader spectrum of bacterial strains (Fabijan et al., 2020) and developing personalised phage therapy regimens tailored to individual patients.

2 Advantages of Phages over Antibiotics

Environmental reservoirs harbour diverse bacteriophages, each characterised by high host specificity, facilitating the targeted control of bacterial populations (Batinovic et al., 2019). The rise of antibiotic resistance necessitates exploring new weapons (Dhingra et al., 2020). Unlike broad-spectrum antibiotics that can disrupt healthy gut bacteria (Cryan et al., 2019), phages exhibit high target specificity (Carvalho et al., 2010), eliminating only the targeted pathogenic bacteria. This targeted approach minimises the development of antibiotic resistance in the long run (Patra et al., 2018).

Furthermore, phages possess a unique self-replication ability within the infected bacteria, amplifying their concentration at the infection site for optimal impact (Galié et al., 2018). Additionally, some phages can effectively penetrate and eradicate biofilms, offering a potential solution for chronic infections that are difficult to treat with conventional antibiotics (Hrynyshyn et al., 2022). Phage therapy offers economic advantages. Production costs are comparable to current antibiotics, with the potential for further reduction as research progresses (Wang et al., 2022). With continued research and development, phages hold significant promise for the future of bacterial infection treatment (Hsu et al., 2014).

The rise of antibiotic resistance has spurred a renewed focus on bacteriophages (phages) as a potential alternative therapeutic approach for combating bacterial infections (Seukep et al., 2022). One of the main advantages of phages over antibiotics is their specificity towards target bacteria, which can minimise disruption to the gut microbiome. Phages are highly specific in their ability to infect and kill specific strains of bacteria (Weynberg & Jaschke, 2020). This means that when phages treat an infection, they selectively target the pathogenic bacteria while leaving the beneficial bacteria in the gut undisturbed (Chan et al., 2013). Another advantage of phages is their ability to co-evolve with resistant bacteria. This co-evolutionary dynamic enables phages to continually adjust and remain potent against bacterial strains that have evolved to resist antibiotics (Murugaiyan et al., 2022). Phages can adapt and evolve alongside bacteria, making it more difficult for the bacteria to develop resistance (Brives & Pourraz, 2020). Furthermore, phages have the potential for combination therapies with traditional antibiotics. Combining phages with antibiotics can enhance the effectiveness of treatment, especially in cases where bacteria have become resistant to antibiotics alone (Kortright et al., 2019). For example, researchers have demonstrated that combining phages with antibiotics can lead to synergistic effects, where the combination is more effective at killing bacteria than the phages or antibiotics alone. Phages are generally considered safe for humans as they target bacteria and do not harm human cells (Uddin et al., 2021). Phages are easily adaptable and can be altered to target emerging dangers, making them versatile in combating bacterial infections (Muteeb et al., 2023). Phage therapy is also seen as a more environmentally friendly alternative to antibiotics. Overall, the advantages of phages over antibiotics include their specificity towards target bacteria, ability to co-evolve with resistant bacteria, and potential for combination therapies with traditional antibiotics (Uddin et al., 2021). Phages can be tailored to target specific bacterial strains causing an infection in an individual patient. By analysing the patient's bacterial culture and identifying the specific strain responsible for the infection, phages can be selected or genetically modified to target the specific bacteria causing infection effectively (Masters et al., 2019). Phages have the potential for broader applications in biotechnology, research, and therapeutics. The use of bacteriophage is beyond therapy (Kushner & Pekosz, 2021). They act as targeted biocontrol agents, potential vaccine carriers (DNA or protein), and diagnostics for pathogen detection and protein (Sohail et al., 2020). Overall, the advantages of phages over antibiotics include specificity towards target bacteria and minimising disruption to the gut microbiome—the ability to co-evolve with resistant bacteria, providing a long-term solution.

Phage therapy holds immense potential as a revolutionary approach in medicine. It can potentially address the growing challenge of antibiotic resistance and reshape how we treat bacterial infections. As ongoing research continues to uncover the diverse applications and advantages of phages, their integration into clinical practice holds great promise for the future of infectious disease management (Muteeb et al., 2023). In a time of increased environmental volatility, it is crucial to have precise weather forecasting in various socio-economic sectors. Phage therapy provides a living base for treating infections, which is both practical and environmentally friendly (Bassetti et al., 2017). The precision of phage therapy, alongside its broad potential across multiple sectors, underscores the need for further research and investment. Its ability to selectively target pathogenic bacteria, while sparing beneficial microorganisms, positions it as a promising and sustainable alternative for combating infections, particularly in an era of rising antimicrobial resistance. By focusing on the use of bacteriophages—viruses that infect and destroy specific bacterial cells—phage therapy offers a highly targeted approach that could revolutionise infection management across medical, agricultural, and environmental industries. The development of this therapeutic modality holds immense potential for addressing critical public health challenges and establishing a resilient solution to microbial infections. Compared to antibiotics, phages offer targeted action, microbiome preservation, and potential for co-evolution (Ramisetty & Sudhakari, 2019).

3 Characteristics of Phages

Phages are ubiquitous viral entities critical in regulating bacterial populations within ecosystems. Despite their potent bactericidal effects, phages exhibit remarkable structural and functional characteristics (Kim et al., 2019). Bacteriophages possess a central core of genetic material, either DNA or RNA, encapsulated by a protective protein capsid (Gurunathan et al., 2020). This capsid adopts various geometries, with icosahedral (20-sided) and filamentous morphologies most frequent (Nooraei et al., 2021). Some phages possess additional lipid envelopes. Phage genomes exhibit remarkable plasticity, existing as single-stranded or double-stranded DNA or RNA (Buzón et al., 2020). The quantity of genetic information encoded within these genomes varies considerably, ranging from a mere handful of genes to several

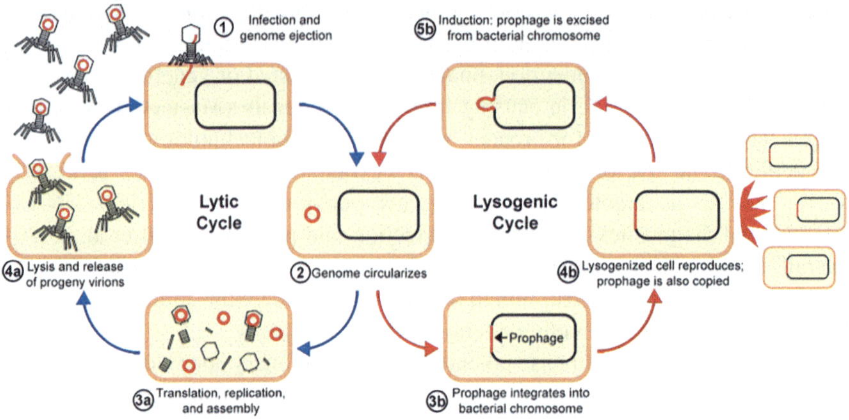

Fig. 18.1 Lytic and lysogenic cycle of phages. Source: (Mdarhri et al., 2022)

thousand, reflecting phages' diverse ecological roles. Phages demonstrate an exceptional degree of host specificity. Specific phages target only a single bacterial strain meticulously, while others exhibit a broader host range (Muteeb et al., 2023). This specificity hinges on specific receptors on the bacterial surface, which the phage recognises and binds to for successful infection.

In the lytic cycle, the phage injects its genetic material directly into the bacterium. This effectively subverts the host cell's machinery. The phage uses the cell's resources to produce new phage particles, replicating its genetic material and protein coat components (Dronina et al., 2021). This ultimately leads to the destruction, or lysis, of the host cell, releasing a new generation of phages to infect other bacteria (Murugaiyan et al., 2022). On the other hand, lysogenic phage integrates its genetic material into the host cell's chromosome. This integrated phage DNA is replicated alongside the host's DNA, creating a lysogen (Marshall & Marsella, 2023). The lysogen behaves like a typical bacterium, and the phage DNA can remain dormant for many generations (Marshall et al., 2021) (Fig. 18.1). However, under certain environmental stress conditions, the lysogenic phage can be 'induced' to switch to the lytic cycle, bursting forth and producing new phage particles (Martin et al., 2022).

Pseudolysogeny involves the phage entering the cell but not integrating or replicating actively. Also, it could persist within the host until favourable conditions return, allowing the phage to resume its attack. Chronic infection is another strategy, where the phage coexists with the host cell, producing new phage particles continuously without causing immediate cell death.

4 Quantification of Phages for Effective Therapy

Phage therapy presents unique challenges compared to traditional antibiotics. Unlike antibiotics, lytic phages are living entities that replicate inside susceptible bacteria. This biological nature translates to complex pharmacokinetic (PK) profiles and how the body absorbs, distributes, metabolises, and eliminates them.

A critical aspect of effective phage therapy is determining the optimal dosage. This hinges on the number of administered phage particles. However, a crucial, often overlooked, aspect is selecting the appropriate assays for accurately measuring phage levels. These techniques are crucial for measuring phages in both given solutions and biological samples, such as blood and urine (Nang et al., 2023).

4.1 Challenges and Advancements in Phage Therapy

The present methods for measuring phages have limitations. While widely used, traditional plaque tests underestimate the overall phage population since they only identify those that produce complete lysis and generate visible plaques (Klein et al., 2019). The concept of 'efficiency of plating' (EOP) can be a confusing one, as it involves calculating the ratio of plaque-forming units (pfu) on a specific bacterial strain in comparison to a reference strain. However, it is essential to note that using a host strain with poorer plating efficiency could potentially lead to significant underestimation of infective phages against the target strain (Mäntynen et al., 2021). This, in turn, may require some adjustments to ensure accurate counts.

Quantitative polymerase chain reaction (qPCR) is a promising technique that can be employed to monitor phage pharmacokinetics (PK) in patients receiving phage therapy, just as it is commonly used to track viral loads in humans infected with viruses such as HIV, CMV, hepatitis B & C, and SARS-Cov-2. This method can play a crucial role in assessing the disease condition and evaluating the efficacy of phage therapy (Karn et al., 2023). However, a significant challenge with qPCR-based assays lies in their inability to distinguish between infective and non-infective phages or even differentiate between intact phage particles and free phage DNA/RNA fragments. Therefore, further research is imperative to establish a clear correlation between qPCR results and therapeutic outcomes in phage therapy (Karn et al., 2023).

Standardising the measurement of phage antimicrobial activity remains a significant challenge in the area (Jagadeesan et al., 2019). Unlike antibiotics, where the minimum inhibitory concentration (MIC) is a generally acknowledged measure for measuring pharmacodynamics (PD), phage treatment does not have a standardised technique for assessing activity. Traditional methods include agar-based spot tests and plating efficiency assays, whereas planktonic killing assays are used in fluid conditions such as broth to measure phage efficacy. These various testing procedures might produce varied findings regarding the host range for the same phage. For example, direct spot testing frequently provides the most prominent coverage, followed by plating and broth-killing assays. This heterogeneity stems from varied death kinetics found under various testing settings and the possibility of phage resistance emerging under selected pressure exerted by these assays. A comprehensive evaluation of the strengths and limitations of each method is crucial for enhancing phage efficacy assessment and, ultimately, successful clinical translation of phage therapy (Nang et al., 2023).

4.2 *Administration of Phage and Interaction with the Immune System*

Bacteriophages (phages) are ubiquitous environmental regulators, exerting significant control over bacterial populations. Recent research has unveiled an additional layer of complexity in phage–host interactions, highlighting their ability to modulate the human adaptive immune system (Dąbrowska, 2019).

Antibody Response Induction: Phages demonstrably influence the adaptive immune response by stimulating the production of specific antibodies (Dąbrowska, 2019; Schultz & Frohlich, 1965). This phenomenon is well-documented in scientific literature, with studies detailing the kinetics of antibody production, the prevalence of phage-specific antibodies within the human population, and the variable immunogenicity of distinct phage proteins (Schultz & Frohlich, 1965).

The impact of phages on the adaptive immune system extends beyond antibody production. They can also influence antigen presentation processes (Schultz & Frohlich, 1965). The multifaceted functions of bacteriophages position them as essential regulators within the human microbiome. These viruses serve as potent therapeutic agents by selectively targeting pathogenic bacteria and act as natural modulators of immune function. Phages influence microbial diversity and stability within host-associated ecosystems, contributing to both immune homeostasis and the suppression of harmful bacterial populations. Their role extends beyond simple bacterial lysis, as they can interact with host immune cells, potentially enhancing immune responses and promoting tolerance to beneficial microbiota, thereby supporting both health and disease management (Kaźmierczak et al., 2021).

5 Challenges in Phage Therapy Development

Several significant limitations also marked this early period of phage therapy. First, phage preparations lacked standardisation (Brives & Pourraz, 2020). Techniques for isolating and purifying phages were not well-established, leading to inconsistencies in the potency and efficacy of phage lysates used in therapy (Berger et al., 2015). This meant that the concentration of active phages in a given preparation could vary considerably, making it difficult to determine the appropriate dosage for a patient. Additionally, contaminating bacteria or bacterial products in the lysate could have adverse effects (Hitchcock et al., 2023).

Second, treatment protocols were not standardised. There was limited understanding of phage biology, and factors like phage dosage, route of administration, and the specific bacterial strain involved were not always considered with precision (Gu & Minko, 2024). For instance, phages can be highly specific in infecting bacteria. Using a phage that targets the wrong bacterial strain would be ineffective in treating the infection. Similarly, the route of administration can be crucial. Phages delivered orally may not survive the stomach's acidic environment to reach their target in the intestines.

Third, the scientific community's understanding of phage–bacteria interactions was limited. There was a significant gap in the researchers' understanding of phage–host specificity and the possibility of the emergence of phage-resistant bacteria. This lack of comprehensive knowledge could impede the progress of related research and development activities. In some cases, bacteria can develop resistance to phages. If a bacterial population is exposed to a particular phage for a prolonged period, some bacteria may develop mutations that allow them to escape phage infection. Understanding these dynamics is essential for developing effective phage therapy strategies (Hibstu et al., 2022).

Identifying effective phages against specific bacterial strains or species can be time-consuming and challenging. The host range of bacteriophages is limited, meaning each phage has a narrow target range of bacteria it can infect (Ratnakar, 2022). Developing methods to produce and administer phage therapy products is complex and costly. The regulatory approval process for medical treatments and therapeutic compounds is complex and time-intensive, particularly for live viruses like phages. Phage resistance, similar to antibiotics, is a concern. Bacteria can evolve evasion strategies, but research is ongoing to address this and ensure phage therapy's viability (Pires et al., 2020). Removing endotoxins from phage preparations is necessary to prevent adverse patient reactions (Ratnakar, 2022). The immune system can reject phages, limiting their effectiveness (Ratnakar, 2022). Phages can be quickly cleared from the body, which can reduce their therapeutic effectiveness (Ratnakar, 2022). The cost of developing and approving phage therapy products is high, and the regulatory framework for phage therapy is still evolving. There is a concern that phages could transfer antibiotic-resistance genes or virulence factors to the microbiome (Reindel & Fiore, 2017). The development of bacterial resistance to phages is a potential challenge, as bacteria can evolve mechanisms to avoid phage predation (Caflisch et al., 2019). The development of personalised phage therapy, which involves selecting phages based on the susceptibility of bacterial isolates from infected patients, presents additional challenges regarding safety and efficacy (Pires et al., 2020).

6 The Use of Nanotechnology in Phage Therapy

The number of phages and bacteria is critical in whether phage therapy can effectively clear a bacterial infection (Kaur et al., 2021). Achieving successful bacterial control with phage therapy necessitates a two-pronged approach. Initially, a high density of phages is crucial to suppress the growth of susceptible bacteria. However, sustained phage amplification is equally essential to ensure continued bacterial multiplication and dissemination containment. This amplification process is contingent upon the bacterial concentration at the target site, necessitating a minimum threshold density for optimal therapeutic effect (Tsonos et al., 2014).

Nanotechnology enhances phage therapy by enabling precise delivery and improving the effectiveness of bacteriophages in targeting drug-resistant bacteria (Pardo-Freire & Domingo-Calap, 2023). This innovative approach utilises nanoparticles to enhance the efficacy of phage therapy, a method that harnesses bacteriophages (Uskoković, 2020) for treating bacterial infections (Villiers et al., 2009). Nanotechnology enables the targeted delivery of bacteriophages, effectively penetrating biofilms and reaching the site of infection (Górski et al., 2019). Additionally, nanotechnology can improve the stability and longevity of bacteriophages, ensuring their effectiveness in combating bacterial infections (Wojewodzic, 2020). Furthermore, nanotechnology can be used to enhance the mechanisms of phage therapy (Villiers et al., 2009). For example, nanoparticles can be engineered to carry multiple strains of bacteriophages, increasing the chances of successfully targeting diverse bacterial strains (Pardo-Freire & Domingo-Calap, 2023). By incorporating nanoparticles into phage therapy, researchers can also manipulate the properties of bacteriophages, such as their ability to lyse bacteria (Pardo-Freire & Domingo-Calap, 2023). They can evade immune detection and pharmacokinetics (Górski et al., 2019). This integration of nanotechnology with phage therapy has the potential to revolutionise the field of antibacterial treatment and combat the growing problem of antibiotic resistance (Villiers et al., 2009). The use of nanotechnology in phage therapy offers several advantages in the treatment of bacterial infections. These advantages include improved targeting and penetration of biofilms, enhanced stability and longevity of bacteriophages, and the ability to manipulate various properties of bacteria.

7 Conclusion

Phage therapy is a treatment method using bacteriophages to treat bacterial infections that can potentially combat antibiotic resistance. However, it faces challenges in addressing intellectual property frameworks and overcoming technological obstacles. Technological advancements like CRISPR-Cas could help engineer highly specific phages, allowing for patenting and commercialisation. These 'designer phages' would target antibiotic-resistant bacteria, minimising disrupting the patient's microbiome. Collaborative research to establish global phage libraries could improve the availability of these therapeutic agents. Quality control measures are crucial, to ensure sterility, stability, and the absence of harmful toxins. Phages could be integrated into existing antibiotic stewardship programs, serving as a last line of defence against multi-drug resistant (MDR) pathogens in challenging infections. The alarming rise of MDR infections calls for a re-evaluation of phage therapy and an exploration of alternative treatment methods.

References

Aswani, V. H., & Shukla, S. K. (2021). An early history of phage therapy in the United States: Is it time to reconsider? *Clinical Medicine & Research, 19*(2), 82–89. https://doi.org/10.3121/cmr.2021.1605

Bassetti, M., Poulakou, G., Ruppe, E., Bouza, E., Van Hal, S. J., & Brink, A. (2017). Antimicrobial resistance in the next 30 years, humankind, bugs and drugs: A visionary approach. *Intensive Care Medicine, 43*(10), 1464–1475. https://doi.org/10.1007/s00134-017-4878-x

Berger, B., Yang, Z., Spangler, L., Berard, V., He, Q., Lu, L., & Mart, L. (2015). Abstracts of the 29th Annual Symposium of the Protein Society. *Protein Science, 24*(S1), 1–313. https://doi.org/10.1002/pro.2823

Brives, C., & Pourraz, J. (2020). Phage therapy as a potential solution in the fight against AMR: Obstacles and possible futures. *Palgrave Communications, 6*(17). https://doi.org/10.1057/s41599-020-0478-4

Batinovic, S., Wassef, F., Knowler, S. A., Rice, D. T. F., Stanton, C. R., Rose, J., Tucci, J., Nittami, T., Vinh, A., Drummond, G. R., Sobey, C. G., Chan, H. T., Seviour, R. J., Petrovski, S., & Franks, A. E. (2019). Bacteriophages in natural and artificial environments. *Pathogens, 8*(3), 100. https://doi.org/10.3390/pathogens8030100

Bull, A. T. (2010). The renaissance of continuous culture in the post-genomics age. *Journal of Industrial Microbiology and Biotechnology, 37*(10), 993–1021. https://doi.org/10.1007/s10295-010-0816-4

Buzón, P., Maity, S., & Roos, W. H. (2020). Physical virology: From virus self-assembly to particle mechanics. *Wiley Interdisciplinary Reviews. Nanomedicine and Nanobiotechnology, 12*(4). https://doi.org/10.1002/wnan.1613

Caflisch, K. M., Suh, G. A., & Patel, R. (2019). Biological challenges of phage therapy and proposed solutions: a literature review. Expert Review of Anti-infective Therapy, 17(12), 1011–1041. https://doi.org/10.1080/14787210.2019.1694905

Carvalho, C. M., Gannon, B. W., Halfhide, D. E., Santos, S. B., Hayes, C. M., Roe, J. M., & Azeredo, J. (2010). The in vivo efficacy of two administration routes of a phage cocktail to reduce numbers of campylobacter coli and campylobacter jejuni in chickens. *BMC Microbiology, 10*(1). https://doi.org/10.1186/1471-2180-10-232

Chan, B. K., Abedon, S. T., & Loc-Carrillo, C. (2013). Phage cocktails and the future of phage therapy. *Future Microbiology, 8*(6), 769–783. https://doi.org/10.2217/fmb.13.47

Collignon, P. (2015). Antibiotic resistance: are we all doomed? *Internal Medicine Journal, 45*(11), 1109–1115. https://doi.org/10.1111/imj.12902

Cryan, J. F., O'Riordan, K. J., Cowan, C. S. M., Sandhu, K. V., Bastiaanssen, T. F. S., Boehme, M., & Dinan, T. G. (2019). The microbiota-gut-brain axis. *Physiological Reviews, 99*(4), 1877–2013. https://doi.org/10.1152/physrev.00018.2018

Dąbrowska, K. (2019). Phage therapy: What factors shape phage pharmacokinetics and bioavailability? Systematic and critical review. *Medicinal Research Reviews, 39*(5), 2000–2025. https://doi.org/10.1002/med.21572

Dhingra, S., Rahman, N. A. A., Peile, E., Rahman, M., Sartelli, M., Hassali, M. A., & Haque, M. (2020). Microbial resistance movements: An overview of global public health threats posed by antimicrobial resistance, and how best to counter. *Frontiers in Public Health, 8*. https://doi.org/10.3389/fpubh.2020.535668

Dronina, J., Samukaite-Bubniene, U., & Ramanavicius, A. (2021). Advances and insights in the diagnosis of viral infections. *Journal of Nanobiotechnology, 19*(1). https://doi.org/10.1186/s12951-021-01081-2

Durbas, I., & Machnik, G. (2022). Phage therapy: An old concept with new perspectives. *Journal of Applied Pharmaceutical Science*. https://doi.org/10.7324/japs.2022.120502

Ezzatpanah, H., Gómez-López, V. M., Koutchma, T., Lavafpour, F., Moerman, F., Mohammadi, M., & Raheem, D. (2022). Risks and new challenges in the food chain: Viral contamination and decontamination from a global perspective, guidelines, and cleaning. *Comprehensive Reviews in Food Science and Food Safety, 21*(2), 868–903. https://doi.org/10.1111/1541-4337.12899

Fabijan, A. P., Iredell, J., Danis-Wlodarczyk, K., Kebriaei, R., & Abedon, S. T. (2023). Translating phage therapy into the clinic: Recent accomplishments but continuing challenges. *PLoS Biology, 21*(5), e3002119. https://doi.org/10.1371/journal.pbio.3002119

Fabijan, A. P., Khalid, A., Maddocks, S., Ho, J., Gilbey, T., Sandaradura, I., & Iredell, J. R. (2020). Phage therapy for severe bacterial infections: A narrative review. *Medical Journal of Australia, 212*(6), 279–285. https://doi.org/10.5694/mja2.50355

Galié, S., García-Gutiérrez, C., Miguélez, E. M., Villar, C. J., & Lombó, F. (2018). Biofilms in the food industry: Health aspects and control methods. *Frontiers in Microbiology, 9*. https://doi.org/10.3389/fmicb.2018.00898
Górski, A., Bollyky, P. L., Przybylski, M., Borysowski, J., Międzybrodzki, R., Jończyk-Matysiak, E., & Weber-Dąbrowska, B. (2019). Perspectives of phage therapy in non-bacterial infections. *Frontiers in Microbiology, 9*. https://doi.org/10.3389/fmicb.2018.03306
Gu, X., & Minko, T. (2024). Targeted nanoparticle-based diagnostic and treatment options for pancreatic cancer. *Cancers, 16*(8), 1589. https://doi.org/10.3390/cancers16081589
Gurunathan, S., Qasim, M., Choi, Y., Do, J. T., Park, C., Hong, K., & Song, H. (2020). Antiviral potential of nanoparticles—Can nanoparticles fight against coronaviruses? *Nanomaterials, 10*(9), 1645. https://doi.org/10.3390/nano10091645
Handy, R. D., Van Den Brink, N., Chappell, M., Mühling, M., Behra, R., Dušinská, M., & Riediker, M. (2012). Practical considerations for conducting ecotoxicity test methods with manufactured nanomaterials: What have we learnt so far? *Ecotoxicology, 21*(4), 933–972. https://doi.org/10.1007/s10646-012-0862-y
Heine, A., Juranek, S., & Brossart, P. (2021). Clinical and immunological effects of mRNA vaccines in malignant diseases. *Molecular Cancer, 20*(1). https://doi.org/10.1186/s12943-021-01339-1
Helene, L. C. F., Klepa, M. S., & Hungria, M. (2022). New insights into the taxonomy of bacteria in the genomic era and a case study with rhizobia. *International Journal of Microbiology, 2022*, 1–19. https://doi.org/10.1155/2022/4623713
Hibstu, Z., Belew, H., Akelew, Y., & Mengist, H. M. (2022). Phage therapy: A different approach to fight bacterial infections. *Biologics :targets &therapy, 16*, 173–186. https://doi.org/10.2147/btt.s381237
Hitchcock, N. M., Nunes, D. D. G., Shiach, J., Hodel, K. V. S., Barbosa, J. D. V., Rodrigues, L. A. P., & Badaró, R. (2023). Current clinical landscape and global potential of bacteriophage therapy. *Viruses, 15*(4), 1020. https://doi.org/10.3390/v15041020
Hrynyshyn, A., Simões, M., & Borges, A. (2022). Biofilms in surgical site infections: Recent advances and novel prevention and eradication strategies. *Antibiotics, 11*(1), 69. https://doi.org/10.3390/antibiotics11010069
Hsu, P. D., Lander, E. S., & Zhang, F. (2014). Development and applications of CRISPR-CAS9 for genome engineering. *Cell, 157*(6), 1262–1278. https://doi.org/10.1016/j.cell.2014.05.010
Iwu, C. D., Korsten, L., & Okoh, A. I. (2020). The incidence of antibiotic resistance within and beyond the agricultural ecosystem: A concern for public health. *MicrobiologyOpen, 9*(9). https://doi.org/10.1002/mbo3.1035
Jagadeesan, B., Gerner-Smidt, P., Allard, M. W., Leuillet, S., Winkler, A., Xiao, Y., & Grant, K. (2019). The use of next generation sequencing for improving food safety: Translation into practice. *Food Microbiology, 79*, 96–115. https://doi.org/10.1016/j.fm.2018.11.005
Jangid, H., Kumar, D., Kumar, G., Kumar, R., & Mamidi, N. (2024). An emerging foodborne pathogen spotlight: A bibliometric analysis and scholarly review of Escherichia coli O157 research. *Antibiotics, 13*(1), 60. https://doi.org/10.3390/antibiotics13010060
Jault, P., Leclerc, T., Jennes, S., Pirnay, J. P., Que, Y., Resch, G., & Gabard, J. (2019). Efficacy and tolerability of a cocktail of bacteriophages to treat burn wounds infected by Pseudomonas aeruginosa (PhagoBurn): A randomised, controlled, double-blind phase 1/2 trial. *The Lancet. Infectious diseases, 19*(1), 35–45. https://doi.org/10.1016/s1473-3099(18)30482-1
Kaur, S., Kumari, A., Negi, A. K., Galav, V., Thakur, S., Agrawal, M., & Sharma, V. (2021). Nanotechnology based approaches in phage therapy: Overcoming the pharmacological barriers. *Frontiers in Pharmacology, 12*. https://doi.org/10.3389/fphar.2021.699054
Karabay, O., Baştuğ, A., Öztürk, R., Şencan, İ., Aksoy, M., ŞİMşek, H., Gözel, M. G., Erdoğan, H., Karlidağ, G. E., Aypak, A., Gönen, İ., Gürpinar, E. U., İşlİ, F., Yildiz, S. S., Yarsan, E., & Bodur, H. (2019). Antibiotic consumption, resistance data, and prevention strategies. *Mediterranean Journal of Infection Microbes and Antimicrobials, 2018*(7). https://doi.org/10.4274/mjima.2018.35

Karn, S. L., Gangwar, M., Kumar, R., Bhartiya, S. K., & Nath, G. (2023). Phage therapy: a revolutionary shift in the management of bacterial infections, pioneering new horizons in clinical practice, and reimagining the arsenal against microbial pathogens. *Frontiers in Medicine, 10*. https://doi.org/10.3389/fmed.2023.1209782

Kaźmierczak, Z., Majewska, J., Milczarek, M., Owczarek, B., & Dąbrowska, K. (2021). Circulation of fluorescently labelled phage in a murine model. *Viruses, 13*(2), 297. https://doi.org/10.3390/v13020297

Kim, B., Kim, E. S., Yoo, Y., Bae, H., Chung, I., & Cho, Y. (2019). Phage-derived antibacterials: Harnessing the simplicity, plasticity, and diversity of phages. *Viruses, 11*(3), 268. https://doi.org/10.3390/v11030268

Kirtane, A. R., Verma, M., Karandikar, P., Furin, J., Langer, R., & Traverso, G. (2021). Nanotechnology approaches for global infectious diseases. *Nature Nanotechnology, 16*(4), 369–384. https://doi.org/10.1038/s41565-021-00866-8

Klein, H. L., Bačinskaja, G., Che, J., Cheblal, A., Elango, R., Epshtein, A., & Malkova, A. (2019). Guidelines for DNA recombination and repair studies: Cellular assays of DNA repair pathways. *Microbial Cell (Graz, Austria), 6*(1), 1–64. https://doi.org/10.15698/mic2019.01.664

Kochhar, R. (2020). The virus in the rivers: Histories and antibiotic afterlives of the bacteriophage at the Sangam in Allahabad. *Notes and Records of the Royal Society of London, 74*(4), 625–651. https://doi.org/10.1098/rsnr.2020.0019

Kortright, K. E., Chan, B. K., Koff, J. L., & Turner, P. E. (2019). Phage therapy: A renewed approach to combat antibiotic-resistant bacteria. *Cell Host & Microbe, 25*(2), 219–232. https://doi.org/10.1016/j.chom.2019.01.014

Kushner, D. B., & Pekosz, A. (2021). Virology in the classroom: Current approaches and challenges to undergraduate- and graduate-level virology education. *Annual Review of Virology, 8*(1), 537–558. https://doi.org/10.1146/annurev-virology-091919-080047

Lauman, P., & Dennis, J. J. (2021). Advances in phage therapy: Targeting the Burkholderia cepacia complex. *Viruses, 13*(7), 1331. https://doi.org/10.3390/v13071331

Loureiro, A., & Da Silva, G. J. (2019). CRISPR-Cas: Converting a bacterial defence mechanism into a state-of-the-art genetic manipulation tool. *Antibiotics, 8*(1), 18. https://doi.org/10.3390/antibiotics8010018

Mdarhri, H. A., Benmessaoud, R., Yacoubi, H., Seffar, L., Assimi, H. G., Hamam, M., Boussettine, R., Filali-Ansari, N., Lahlou, F. A., Diawara, I., Ennaji, M. M., & Kettani-Halabi, M. (2022). Alternatives Therapeutic approaches to conventional antibiotics: Advantages, limitations and potential application in medicine. *Antibiotics, 11*(12), 1826. https://doi.org/10.3390/antibiotics11121826

Malik, D. J., Sokolov, I. J., Vinner, G. K., Mancuso, F., Cinquerrui, S., Vladisavljevic, G. T., & Kirpichnikova, A. (2017). Formulation, stabilisation and encapsulation of bacteriophage for phage therapy. *Advances in Colloid and Interface Science, 249*, 100–133. https://doi.org/10.1016/j.cis.2017.05.014

Mäntynen, S., Laanto, E., Oksanen, H. M., Poranen, M. M., & Díaz-Muñoz, S. L. (2021). Black box of phage–bacterium interactions: Exploring alternative phage infection strategies. *Open Biology, 11*(9). https://doi.org/10.1098/rsob.210188

Manyi-Loh, C. E., & Lues, R. (2023). A South African perspective on the microbiological and chemical quality of meat: Plausible public health implications. *Microorganisms, 11*(10), 2484. https://doi.org/10.3390/microorganisms11102484

Manyi-Loh, C., Mamphweli, S., Meyer, E., & Okoh, A. (2018). Antibiotic use in agriculture and its consequential resistance in environmental sources: Potential public health implications. *Molecules (Basel, Switzerland), 23*(4), 795. https://doi.org/10.3390/molecules23040795

Marshall, C. W., Gloag, E. S., Lim, C., Wozniak, D. J., & Cooper, V. S. (2021). Rampant prophage movement among transient competitors drives rapid adaptation during infection. *Science Advances, 7*(29). https://doi.org/10.1126/sciadv.abh1489

Marshall, K., & Marsella, R. (2023). Topical bacteriophage therapy for staphylococcal superficial pyoderma in horses: A double-blind, placebo-controlled pilot study. *Pathogens (Basel, Switzerland), 12*(6), 828. https://doi.org/10.3390/pathogens12060828

Martin, C. S., Jubelin, G., Darsonval, M., Leroy, S., Leneveu-Jenvrin, C., Hmidene, G., & Dubois-Brissonnet, F. (2022). Genetic, physiological, and cellular heterogeneities of bacterial pathogens in food matrices: Consequences for food safety. *Comprehensive Reviews in Food Science and Food Safety, 21*(5), 4294–4326. https://doi.org/10.1111/1541-4337.13020

Masters, E. A., Trombetta, R. P., De Mesy Bentley, K. L., Boyce, B. F., Gill, A. L., Gill, S. R., & Muthukrishnan, G. (2019). Evolving concepts in bone infection: Redefining "biofilm", "acute vs. chronic osteomyelitis", "the immune proteome" and "local antibiotic therapy.". *Bone Research, 7*(1). https://doi.org/10.1038/s41413-019-0061-z

Murugaiyan, J., Kumar, P. A., Rao, G. S., Iskandar, K., Hawser, S., Hays, J. P., & Van Dongen, M. B. M. (2022). Progress in alternative strategies to combat antimicrobial resistance: Focus on antibiotics. *Antibiotics, 11*(2), 200. https://doi.org/10.3390/antibiotics11020200

Muteeb, G., Rehman, M. T., Shahwan, M., & Aatif, M. (2023). Origin of antibiotics and antibiotic resistance, and their impacts on drug development: A narrative review. *Pharmaceuticals (Basel, Switzerland), 16*(11), 1615. https://doi.org/10.3390/ph16111615

Nang, S. C., Lin, Y., Fabijan, A. P., Chang, R. Y., Rao, G. G., Iredell, J., & Li, J. (2023). Pharmacokinetics/pharmacodynamics of phage therapy: A major hurdle to clinical translation. *Clinical Microbiology and Infection, 29*(6), 702–709. https://doi.org/10.1016/j.cmi.2023.01.021

Nooraei, S., Bahrulolum, H., Hoseini, Z. S., Katalani, C., Hajizade, A., Easton, A. J., & Ahmadian, G. (2021). Virus-like particles: Preparation, immunogenicity and their roles as nanovaccines and drug nanocarriers. *Journal of Nanobiotechnology, 19*(1). https://doi.org/10.1186/s12951-021-00806-7

Pardo-Freire, M., & Domingo-Calap, P. (2023). Phages and nanotechnology: New insights against multidrug-resistant bacteria. *Biodesign Research, 5*. https://doi.org/10.34133/bdr.0004

Patra, J. K., Das, G., Fraceto, L. F., Campos, E. V. R., Del Pilar Rodriguez-Torres, M., Acosta-Torres, L. S., & Shin, H. S. (2018). Nano based drug delivery systems: Recent developments and future prospects. *Journal of Nanobiotechnology, 16*(1). https://doi.org/10.1186/s12951-018-0392-8

Prestinaci, F., Pezzotti, P., & Pantosti, A. (2015). Antimicrobial resistance: a global multifaceted phenomenon. *Pathogens and Global Health, 109*(7), 309–318. https://doi.org/10.1179/2047773215y.0000000030

Pires, D. P., Costa, A. R., Pinto, G., Meneses, L., & Azeredo, J. (2020). Current challenges and future opportunities of phage therapy. *FEMS Microbiology Reviews, 44*(6), 684–700. https://doi.org/10.1093/femsre/fuaa017

Ratnakar, N. I. S. (2022). Phage therapy: challenges and opportunities. *Fine Focus, 8*(1), 12–35. https://doi.org/10.33043/ff.8.1.12-35

Reindel, R., & Fiore, C. R. (2017). Phage Therapy: Considerations and Challenges for development. *Clinical Infectious Diseases, 64*(11), 1589–1590. https://doi.org/10.1093/cid/cix188

Ramisetty, B. C. M., & Sudhakari, P. A. (2019). Bacterial 'grounded' prophages: Hotspots for genetic renovation and innovation. *Frontiers in Genetics, 10*. https://doi.org/10.3389/fgene.2019.00065

Saral-Acarca, E. S., Evran, E., Tayyarcan, E. K., Ilbasmis-Tamer, S., Tort, S., Boyaci, I. H., & Acartürk, F. (2023). Recent advances in phage therapy for pharmaceutical applications. *Journal of Drug Delivery Science and Technology, 90*, 105155. https://doi.org/10.1016/j.jddst.2023.105155

Schultz, I., & Frohlich, E. (1965). Viruria and viraliquoria in the dog after intravenous injection of T5 bacteriophage. *Experimental Biology and Medicine, 118*(1), 136–138. https://doi.org/10.3181/00379727-118-29778

Seukep, A. J., Mbuntcha, H. G., Kuete, V., Chu, Y., Fan, E., & Guo, M. (2022). What approaches to thwart bacterial efflux pumps-mediated resistance? *Antibiotics, 11*(10), 1287. https://doi.org/10.3390/antibiotics11101287

Sohail, H. A., Coffey, A., Debrowska, K., Meyer, I. M., Middelboe, M., Sohail, M., & Clokie, M. R. (2020). Bacteriophages: Emerging applications in medicine, food, and biotechnology. *PHAGE (New Rochelle, N.Y.), 1*(2), 75–82. https://doi.org/10.1089/phage.2020.29004.has

Tacconelli, E., Carrara, E., Savoldi, A., Harbarth, S., Mendelson, M., Monnet, D. L., Pulcini, C., Kahlmeter, G., Kluytmans, J., Carmeli, Y., Ouellette, M., Outterson, K., Patel, J., Cavaleri, M., Cox, E. M., Houchens, C. R., Grayson, M. L., Hansen, P., Singh, N., . . . Zorzet, A. (2018). Discovery, research, and development of new antibiotics: the WHO priority list of antibiotic-resistant bacteria and tuberculosis. *The Lancet Infectious Diseases, 18*(3), 318–327. https://doi.org/10.1016/s1473-3099(17)30753-3

Teng, J., Imani, S., Zhou, A., Zhao, Y., Du, L., Deng, S., & Wang, Q. (2023). Combatting resistance: Understanding multi-drug resistant pathogens in intensive care units. *Biomedicine & Pharmacotherapy, 167*, 115564. https://doi.org/10.1016/j.biopha.2023.115564

Torfs, E., Piller, T., Cos, P., & Cappoen, D. (2019). Opportunities for overcoming mycobacterium tuberculosis drug resistance: Emerging mycobacterial targets and host-directed therapy. *International Journal of Molecular Sciences, 20*(12), 2868. https://doi.org/10.3390/ijms20122868

Tsonos, J., Vandenheuvel, D., Briers, Y., De Greve, H., Hernalsteens, J., & Lavigne, R. (2014). Hurdles in bacteriophage therapy: Deconstructing the parameters. *Veterinary Microbiology, 171*(3–4), 460–469. https://doi.org/10.1016/j.vetmic.2013.11.001

Uddin, T. M., Chakraborty, A. J., Khusro, A., Zidan, B. R. M., Mitra, S., Emran, T. B., & Koirala, N. (2021). Antibiotic resistance in microbes: History, mechanisms, therapeutic strategies and future prospects. *Journal of Infection and Public Health, 14*(12), 1750–1766. https://doi.org/10.1016/j.jiph.2021.10.020

Usković, V. (2020). Why have nanotechnologies been underutilized in the global uprising against the coronavirus pandemic? *Nanomedicine, 15*(17), 1719–1734. https://doi.org/10.2217/nnm-2020-0163

Verbeek, J. H., Rajamaki, B., Ijaz, S., Sauni, R., Toomey, E., Blackwood, B., & Balci, F. S. K. (2020). Personal protective equipment for preventing highly infectious diseases due to exposure to contaminated body fluids in healthcare staff. *The Cochrane Cochrane database of systematic reviews*. https://doi.org/10.1002/14651858.cd011621.pub4

Wang, L., Wang, N., Zhang, W., Cheng, X., Yan, Z., Shao, G., & Fu, C. (2022). Therapeutic peptides: Current applications and future directions. *Signal Transduction and Targeted Therapy, 7*(1). https://doi.org/10.1038/s41392-022-00904-4

Wang, X., Xie, Z., Zhao, J., Zhu, Z., Yang, C., & Liu, Y. (2021). Prospects of inhaled phage therapy for combatting pulmonary infections. *Frontiers in Cellular and Infection Microbiology, 11*. https://doi.org/10.3389/fcimb.2021.758392

Wegst, U. G. K., Bai, H., Saiz, E., Tomsia, A. P., & Ritchie, R. O. (2014). Bioinspired structural materials. *Nature Materials, 14*(1), 23–36. https://doi.org/10.1038/nmat4089

Weynberg, K. D., & Jaschke, P. R. (2020). Building better bacteriophage with biofoundries to combat antibiotic-resistant bacteria. *PHAGE (New Rochelle, N.Y.), 1*(1), 23–26. https://doi.org/10.1089/phage.2019.0005

Wojewodzic, M. W. (2020). Bacteriophages could be a potential game changer in the trajectory of coronavirus disease (COVID-19). *PHAGE, 1*(2), 60–65. https://doi.org/10.1089/phage.2020.0014

Yang, B., Fang, D., Lv, Q., Wang, Z., & Liu, Y. (2021). Targeted therapeutic strategies in the battle against pathogenic bacteria. *Frontiers in Pharmacology, 12*. https://doi.org/10.3389/fphar.2021.673239

Chapter 19
Prevalence of Malaria Infection Among Symptomatic Patients of Selected Healthcare Centers in Ota, Ogun State Nigeria: Proactive Prevention Strategies

Solomon Uche Oranusi, Emmanuel Ojochegbe Mameh, Samuel Adeniyi Oyegbade, Daniel Oluwatobiloba Balogun, Victoria-Grace Onyekachi Aririguzoh, and Austine Atokolo

1 Introduction

Malaria is, a potentially fatal illness triggered by *Plasmodium* parasites, particularly *Plasmodium falciparum* that, persists as a major global health concern (de Sousa et al., 2020; Adegbite et al., 2023), especially in tropical and subtropical areas, including Ota, Ogun State, Nigeria. Transmitted from bites of infected Anopheles mosquitoes, malaria affects millions of people annually, causing fever, chills, and flu-like symptoms (Adedeji et al., 2020). Despite considerable progress in prevention and treatment efforts, malaria continues to take a heavy toll, especially in sub-Saharan Africa, where most malaria-related deaths occur (Umumararungu et al., 2023; Adebayo et al., 2024). *Plasmodium falciparum, Plasmodium malariae, Plasmodium ovale, Plasmodium vivax, and Plasmodium knowlesi* are the five *Plasmodium* human malaria parasites. The deadliest of all the known human malaria parasites is *Plasmodium falciparum,* and *Plasmodium vivax* are mainly implicated for prevalent malaria incidences that compromise the efficacy of contemporary antimalarials like artemisinin, sulfadoxine, pyrimethamine and artemisinin-based combination therapies (ACTs), resulting from delayed parasite clearance and multiple drug resistance leading to re-emergence of parasitic disease in Africa (Sinha et al., 2014; Olasehinde et al., 2019). This prompts newer scientific methodologies to

S. U. Oranusi(✉) · E. O. Mameh (✉) · S. A. Oyegbade · D. O. Balogun · V.-G. O. Aririguzoh · A. Atokolo
Department of Biological Sciences, College of Science and Technology, Covenant University, Ota, Ogun State, Nigeria
e-mail: solomon.oranusi@covenantuniversity.edu.ng

Covenant Applied Informatics and Communication Africa Centre of Excellence (CApIC-ACE), Covenant University, Ota, Ogun State, Nigeria
e-mail: emmanuel.mamehpgs@stu.cu.edu.ng

P. O. Isibor, P. A. Akinduti (eds.), *Harnessing Biotechnology Tools for Product Development*, https://doi.org/10.1007/978-3-031-86002-7_19

enhance therapy while exploring the various complex mechanisms that aid the parasites survival most especially the various stages in its life cycle (Bhowmik et al., 2010; Walker, et al., 2010; Duffy & Avery, 2013; de Sousa et al., 2020; Nureye & Assefa, 2020; Arendse et al., 2021).

The burden of malaria extends beyond its direct health impacts, adversely affecting socioeconomic development in endemic regions. High rates of absenteeism due to illness, decreased productivity, and healthcare expenditure strain already fragile economies. Children under five and pregnant women face the greatest risk of malaria, contributing significantly to child mortality and adverse pregnancy outcomes being the most recorded within these vulnerable population (CDC, 2016; WHO, 2023).

Malaria has plagued human societies for millennia, with historical records dating thousands of years documenting its devastating effects (Cox, 2002; Ashley et al., 2018). The disease's prevalence and impact were particularly significant in tropical regions including parts of Africa, Asia, and the Americas. Early attempts to understand malaria often attributed its cause to environmental factors like stagnant water or foul air, leading to practices such as draining swamps or burning aromatic substances to purify the air (Cowman et al., 2016).

In the late 1800s and early 1900s, significant advancements were made in understanding the function of mosquitoes in the spread of malaria, notably through the work of Sir Ronald Ross, who demonstrated the Anopheles mosquito's role as a vector for *Plasmodium* parasites (Brady et al., 2012). This discovery revolutionized malaria control efforts, paving the way for vector control strategies including the use of mosquito nets and insecticide consistently (Yang et al., 2021; World Health Organization-WHO, 2023) and well as discovering various druggable targets (Wang et al., 2015; Mathews & Odom John, 2018; Munro & McMorran, 2022).

Chemical interventions against malaria began in the early twentieth century with the discovery of quinine, derived from the bark of the cinchona tree, which had long been used by indigenous peoples in South America to treat fever (Phuong et al., 2002; Ashley et al., 2018). Quinine and its derivatives became the cornerstone of antimalarial treatment for decades until the emergence of resistance (Ecker et al., 2012; Biamonte et al., 2013; Sigala & Goldberg, 2014; Favuzza et al., 2020).

In the mid-twentieth century, chloroquine was introduced as a highly effective and affordable antimalarial drug, contributing significantly to malaria control efforts worldwide (Muraoka and Miura, 2004; Ashley et al., 2018). However, the widespread emergence of chloroquine-resistant strains of *Plasmodium falciparum* in the late twentieth century led to its gradual replacement by other antimalarial, including sulfadoxine-pyrimethamine (SP) and artemisinin-based combination therapies (ACTs) (Ecker et al., 2012; Ashley et al., 2014; Kumar et al., 2018; World Health Organization-WHO, 2023).

Artemisinin, derived from *Artemisia annua*, revolutionized malaria treatment when it was discovered in China in the 1970s (Tu, 2011). Artemisinin and its derivatives used in combination with other antimalarials to prevent resistance are currently the most effective treatments for malaria as strongly instructed by the World Health Organization.

Despite considerable progress made in malaria treatment, the emergence of drug-resistant strains underscores the ongoing need for research and innovation in developing new antimalarial drugs and preventive strategies to combat this persistent global health threat (Held et al., 2015; Abd-Rahman et al., 2022). Hence, this study aimed to conduct screening and identify *Plasmodium* species through microscopic examination of suspected malaria blood samples of febrile patients obtained from designated hospitals in the study area and present epidemiological data categorizing gender and age as variables.

2 Materials and Methods

2.1 Study Area

The study was conducted in Ota, Ado-Odo/Ota Local Government Area, Ogun State, Nigeria, with a projected population of 163,783 inhabitants. Ota is located on latitude 6.683° North of the equator and longitude 3.233° East of the Greenwich meridian, this area lies within the tropical zone and experiences two distinct climatic periods; the wet season, which usually extends from April to October and the dry season, which typically lasts from November to March. The primary occupations of the residents are trading, farming, civil services, and private employment. This study was conducted between September 2023 and February 2024.

2.2 Study Population

A total of 300 febrile patients formed the study population comprising 136 males and 164 females of age range 0 day to 60 years and above. The blood samples of consenting febrile participants were collected using random sampling technique from some selected healthcare facilities in Ota and data obtained were aggregated to achieve the research objectives. The study was conducted from September 2023 to February 2024.

2.3 Inclusion and Exclusion Criteria

Individuals presenting with prior episodes of fever as well as exhibiting a body temperature exceeding 37.5 °C, across all age groups who consented to engage in the study at the health care center were enrolled. However, individuals currently undergoing antimalaria treatment or who initiated such treatment within 3 weeks before the commencement of the study were excluded from participation.

2.4 *Ethical Clearance*

This study was approved by the Covenant Health Research Ethics Committee (CHREC), Ota Ogun State Nigeria [Reference Number: CHREC/432/2024]. Informed consent was obtained from participants across all ages and sexes; where a participant is a minor, consent was obtained from the participant's guardian.

2.5 *Sample Collection*

Blood samples was collected from each participant via venipuncture technique into an anticoagulated container of Ethylenediamine tetra-acetic acid (EDTA), adhering to protocols established by Kalu et al. (2012).

Microscopic examination recognized as the gold standard for malaria diagnosis, was utilized to ascertain the prevalence of malaria within the study area. On a clean microscope slide, blood films of varying thicknesses (thin and thick films) were made using standard laboratory procedures, stained with 10% Giemsa reagent, and investigated under a microscope using ×100 (oil immersion) objective lens. Thin films were used for the species identification of *Plasmodium* parasite using the species-specific characteristics of human *Plasmodium* species as listed by Cheesebrough (2006).

2.6 *Statistical Analysis*

The collected data was analyzed using Microsoft Excel. Differences were assessed using the chi-square (χ^2) statistical test, with significance set at $P \leq 0.05$.

3 Results

A cohort of 300 subjects was examined, comprising 136 males and 164 females. Among them, 132 individuals tested positive for malaria infection, while 168 tested negative (Fig. 19.1). Out of 132 positive cases, 71 were males and 61 were females (Fig. 19.2). The malaria prevalence rates were found to be significantly higher among males, 52.2% (71/136), compared to females, 37.2% (61/164). Statistical analysis revealed a significant difference between malaria prevalence and the gender status of participants in this study ($P \leq 0.05$) (Table 19.1).

Table 19.2 indicates the prevalence of malaria parasites across all age groups within the sampled population. Malaria prevalence rates according to age categories are: 0–10: 28.1% (16/57), 11–20: 68.2% (58/85), 21–30: 59.5% (22/37), 31–40: 31.8% (14/44), 41–50: 35.3% (12/34), 51–60: 26.3% (5/19), >60: 20.8% (5/24).

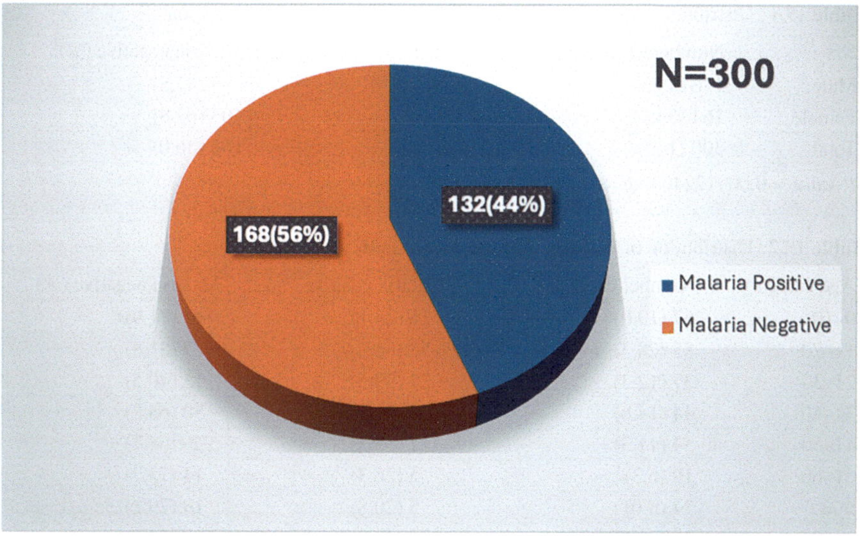

Fig. 19.1 Prevalence of malaria infection among the study population

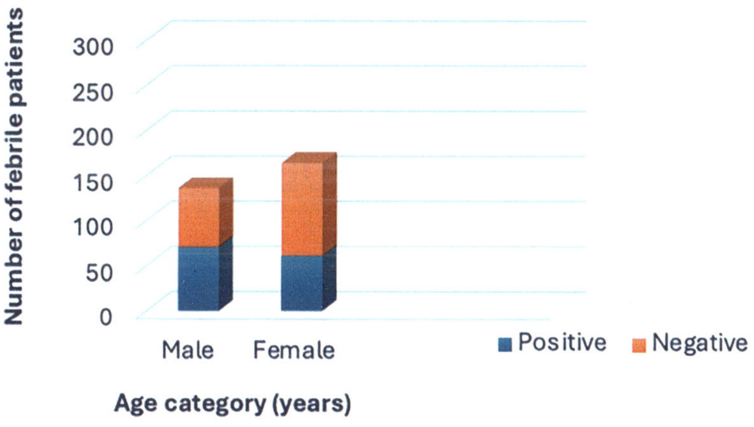

Fig. 19.2 Distribution of malaria parasite according to gender

Among the age groups investigated, 11–20 years had the highest prevalence of 58(68.2%) followed by those in the age group 21–30 years 22(59.3%). The least prevalence was observed among those aged >60 years 5(20.8%). However, there were significant differences statistically ($P \leq 0.05$) with age categories of patients.

The distribution of the *Plasmodium* species in the sampled population is shown in Table 19.3. *Plasmodium falciparum* was identified in 132 (100%) out of 132 positive individuals, making it the only cause of malaria parasitemia in the study population. No positive cases of *Plasmodium ovale, Plasmodium malariae*, and *Plasmodium vivax* were recorded in the sampled population.

Table 19.1 Distribution of malaria infection by gender in the study population

Sex	Number examined	Malaria positive (%)	Malaria negative (%)
Male	136 (45.3)	71 (52.2)	65 (47.8)
Female	164 (54.7)	61 (37.2)	103 (62.8)
Total	300 (100)	132 (44.0)	168 (56.0)

*P-value = 0.009122403, $df = 1$, $\chi^2 = 6.798704792$

Table 19.2 Distribution of malaria infection by age in the study population

Age category	Number examined (%)	Malaria positive (%)	Malaria negative (%)
0–10	57 (19.0)	16 (28.1)	41 (71.9)
11–20	85 (28.3)	58 (68.2)	27 (31.8)
21–30	37 (12.3)	22 (59.5)	15 (40.5)
31–40	44 (14.6)	14 (31.8)	30 (68.2)
41–50	34 (11.3)	12 (35.3)	22 ((64.7)
51–60	19 (6.3)	5 (26.3)	14 (73.7)
>60	24 (8.0)	5 (20.8)	19 (79.2)
Total	300 (100)	132 (44.0)	168 (56.0)

*P-value = 0.000000282367, $df = 6$, $\chi^2 = 41.05550102$

Table 19.3 Distribution of *Plasmodium* species among the study population

Plasmodium species	Number examined	No. infected with *Plasmodium* species (%)	No. infected by species	Prevalence of infection due to
Plasmodium falciparum	300	132	132	100.00
Plasmodium ovale	0	0	0	0
Plasmodium malariae	0	0	0	0
Plasmodium vivax	0	0	0	0
Plasmodium knowlesi	0	0	0	0

4 Discussion

In the current investigation, the total prevalence of malaria infection (44.0%) is less compared to the values reported in the studies previously published by Olasehinde et al. (2019) in the same region having a prevalence of 45.86%. Additionally, Kalu et al. (2012), Sam Wobo et al. (2014), Idris et al. (2017), Ajayi & Abike, (2020) and Okoroiwu, (2021) variously reported the highest malaria prevalence of 73.39%, 71.1%, 72.2%, 85.0%, 54.0%, respectively, compared to the result obtained from this current study (Fig. 19.1). The relatively low prevalence observed in this study might be attributable to the development of significant levels of acquired antimalarial immunity among residents, stemming from frequent exposure. Increased malaria awareness, intensified efforts of various health authorities in the region as well as increased vector control and preventive measures applied among residents in Ota area are also significant justifications for reduced malaria prevalence.

However, the findings from this study were high when compared with published reports within the country and globally such as 43.3% in Ilorin, Nigeria by Awoyemi (2003), 24.5% in Northeastern Nigeria by Houben et al. (2013), 9.78% in Kogi State Nigeria by studies conducted by Iyaji et al. (2015), 38.8% in North Central Nigeria by Ogbu et al. (2015), 24.1% in Ebonyi State Nigeria by Onwe, et al. (2018), 29.0% in Northwest Ethiopia by Negatu et al. (2022), respectively. Malaria prevalence in Nigeria ranged from less than 20.0% in some areas to over 70% in others, according to a recent malaria risk map estimate. There could be several reasons for this variation, including different climatic changes, less rainfall, surface water that serves as a mosquito breeding ground, socioeconomic status, and educational background of the study population, among other factors (Okoroiwu, 2021).

Within the cohort of individuals who tested positive for malaria, findings indicated a higher susceptibility to malaria infection among males, with 71 out of 136 male patients (52.2%) compared to 61 out of 164 female participants (37.2%) (Fig. 19.2; Table 19.1). Gender-based prevalence analysis revealed that despite a larger number of female participants being screened, males exhibited a notably higher prevalence of malaria illness, a finding deemed statistically significant ($P = 0.009, \chi^2 = 6.798704792$); this implies that males might exhibit a higher susceptibility to the disease compared to females. This agrees with the findings of Sam Wobo et al. (2014), Umaru and Uyaiabasi (2015), Onwe et al. (2018), Olasehinde et al. (2019), Oladele et al. (2019), and Yohanna et al. (2019). Research findings by Sam Wobo et al. (2014) suggest that females exhibit a stronger immune response against malaria and a range of other parasitic diseases, a phenomenon attributed to hormonal and genetic factors compared to males. The higher malaria prevalence in males could be attributed to factors such as increased outdoor activities, including sleeping outside during warm weather, self-medication practices, and inadequate use of protective clothing such as trousers and long-sleeve shirts, while sleeping indoors at night. Hence it was established in this current study that there is an association in malaria prevalence rate with gender, but this contradicts Sam Wobo et al. (2014) and Yohanna et al. (2019), where there were no notable disparities in the prevalence of malaria between males and females in the studied population. Also, according to studies by Okeke et al. (2016) and Ifebunandu et al. (2021), there are no marked differences in malaria parasitemia between sexes despite recording the female population as the most vulnerable, reason for the increased incidence in the female populace may stem from their regular engagement in agricultural tasks and household chores such as cooking outdoors in the evening or fetching water before sunrise, therefore, making them more vulnerable to mosquito bites thereby increasing the risk of malaria infection.

The occurrence of malaria infection detected through microscopy in this study (44%) varied across different age categories with the most notable rate of 68.2% observed among age range 11–20 years (Table 19.2) which was found to be statistically significant ($P = 0.000000282367$, $\chi^2 = 41.05550102$). This means malaria infection in the Ota area was selective for age; this agrees with previous studies recently published by Kalu et al. (2012), Umaru and Uyaiabasi (2015), Yohanna et al. (2019), and Olasehinde et al. (2019), where older children below 20 years were most vulnerable, revealing a meaningful and noticeable difference in the proportion of positive cases of malaria among different age groups but contradicts the findings of Okeke et al. (2016). Children within the age bracket 0–10 years had the lowest prevalence rate (28.1%) than older children 11–20 years (Table 19.2). This phenomenon can be attributed to the prioritization of children below the age of 10, who receive enhanced protection against mosquito bites by utilizing insecticide-treated bed nets (Okeke et al., 2016; Umaru & Uyaiabasi, 2015).

According to the World Health Organization (WHO), children under the age of 5 are particularly vulnerable to malaria infection due to their limited acquisition of protective immunity against its most severe manifestations.

The findings also revealed that *Plasmodium falciparum* was the predominant species responsible for malaria prevalence in the study area, as it was the only species detected among 132 positive cases (Table 19.3). This aligns with similar investigations conducted in other regions of the country by Ifebunandu et al. (2021). However, it contrasts the findings of related studies conducted by Sam Wobo et al. (2014) in Ogun state, where infections caused by *Plasmodium falciparum, Plasmodium Ovale, Plasmodium vivax,* and *Plasmodium malariae* were observed.

5 Conclusion

The findings of the present study suggest a relatively low prevalence of malaria in Ota area, with *Plasmodium falciparum* identified as the predominant species responsible for malaria infections. Age and gender emerged as significant risk factors for malaria incidence with males, particularly those aged 11–20 years, exhibiting the highest malaria prevalence. This trend may be attributed to various factors such as behavioral habits, lifestyle choices, inconsistent use of bed nets, greater exposure to mosquito bites in the study area, and misuse of antimalarial medications. This outcome informs the need to take preventive measures in mitigating the prevalence of falciparum malaria in the community and prioritizing the protection of children within the age bracket 11–20 years.

Acknowledgments The authors thank the management of Covenant University Centre for Research Innovation and Discovery (CUCRID), Covenant Applied Informatics and Communication Africa Center of Excellence (CApIC-ACE), Department of Biological Sciences Covenant University, and all the management and staff of healthcare facilities recruited for this study for their various support and contributions.

Conflict of Interest The authors affirm that there are no conflicts of interest regarding the publication of this chapter.

Author Contributions This study was a collaborative effort among all authors. **S.O.** supervised the work, while **E.O.M.** designed the study, managed the analysis of data and literature searches, and drafted the initial manuscript. **S.A.O.** contributed to the investigation, manuscript review, and editing. **D.O.B.** also reviewed and edited the manuscript. **V-G.O.A.** contributed to the investigation, conceptualization, and manuscript editing. **A.A.** managed the arrangement of references. All authors have reviewed and approved the final manuscript.

Funding Statement The research received funding from the management of Covenant University Center for Research Innovation and Discovery (CUCRID) and Covenant Applied Informatics and Communication Africa Center of Excellence (CApIC-ACE).

References

Abd-Rahman, A. N., Zaloumis, S., McCarthy, J. S., Simpson, J. A., & Commons, R. J. (2022). Scoping review of antimalarial drug candidates in phase I and II drug development. *Antimicrobial Agents and Chemotherapy, 66*(2). https://doi.org/10.1128/aac.01659-21

Adebayo, G. P., Oduselu, G. O., Aderohunmu, D. V., Klika, K. D., Olasehinde, G. I., Ajani, O. O., & Adebiyi, E. (2024). Structure-based design, and development of amidinyl, amidoximyl and hydroxamic acid based organic molecules as novel antimalarial drug candidates. *Arabian Journal of Chemistry, 17*(2), 105573. https://doi.org/10.1016/j.arabjc.2023.105573

Adedeji, E. O., Ogunlana, O. O., Fatumo, S., Beder, T., Ajamma, Y., Koenig, R., & Adebiyi, E. (2020). Anopheles metabolic proteins in malaria transmission, prevention and control: A review. *Parasites & Vectors, 13*(1). https://doi.org/10.1186/s13071-020-04342-5

Adegbite, G., Edeki, S., Isewon, I., Emmanuel, J., Dokunmu, T., Rotimi, S., Oyelade, J., & Adebiyi, E. (2023). Mathematical modeling of malaria transmission dynamics in humans with mobility and control states. *Infectious Disease Modelling, 8*(4), 1015–1031. https://doi.org/10.1016/j.idm.2023.08.005

Ajayi, J. B., & Abike, O. (2020). Prevalence of malaria parasites among patients attending hospitals in Igbesa, Ado/Odo-Ota Local Government. *IOSR Journal Of Pharmacy, 12*, 43–46.

Arendse, L. B., Wyllie, S., Chibale, K., & Gilbert, I. H. (2021). *Plasmodium* kinases as possible therapeutic targets for malaria: Challenges and prospects. *ACS Infectious Diseases, 7*(3), 518–534. https://doi.org/10.1021/acsinfecdis.0c00724

Ashley, E. A., Dhorda, M., Fairhurst, R. M., Amaratunga, C., Lim, P., Suon, S., Sreng, S., Anderson, J. M., Mao, S., Sam, B., Sopha, C., Chuor, C. M., Nguon, C., Sovannaroth, S., Pukrittayakamee, S., Jittamala, P., Chotivanich, K., Chutasmit, K., Suchatsoonthorn, C., et al. (2014). Spread of artemisinin resistance in *Plasmodium falciparum* malaria. *The New England Journal of Medicine, 371*(5), 411–423. https://doi.org/10.1056/nejmoa1314981

Ashley, E. A., Pyae Phyo, A., & Woodrow, C. J. (2018). Malaria. *Lancet, 391*(10130), 1608–1621. https://doi.org/10.1016/s0140-6736(18)30324-6

Awoyemi, A. O. (2003). Malaria morbidity amongst hospital workers in Ilorin. *African Journal of Clinical and Experimental Microbiology, 4*(1), 18. https://doi.org/10.4314/ajcem.v4i1.7319

Bhowmik, D. B., Chiranjib, N., Singh, J., Jaiswal, K., & Kumar, K. P. (2010). Recent advancements in prevention, treatment and medicine of malaria. *Journal of Chemical and Pharmaceutical Research, 1*(2), 83–90.

Biamonte, M. A., Wanner, J., & Le Roch, K. G. (2013). Recent advances in malaria drug discovery. *Bioorganic & Medicinal Chemistry Letters, 23*(10), 2829–2843. https://doi.org/10.1016/j.bmcl.2013.03.067

Brady, O. J., Gething, P. W., Bhatt, S., Messina, J. P., Brownstein, J. S., Hoen, A. G., Moyes, C. L., Farlow, A. W., Scott, T. W., & Hay, S. I. (2012). Refining the global spatial limits of dengue virus transmission by evidence-based consensus. *PLoS Neglected Tropical Diseases, 6*(8), 1760. https://doi.org/10.1371/journal.pntd.0001760

CDC (2016). The History of Malaria, an Ancient Disease, Division of Parasitic Diseases and Malaria, USA, 2016. Retrieved December 17, 2022.

Cheesebrough, M. (2006). *District laboratory practice in tropical countries. Part 2. Tropical health technology* (Second ed., p. 454). Cambridge University Press. Retrieved March 30, 2024.

Cowman, A. F., Healer, J., Marapana, D., & Marsh, K. (2016). Malaria: Biology and disease. *Cell, 167*(3), 610–624. https://doi.org/10.1016/j.cell.2016.07.055

Cox, F. E. G. (2002). History of human parasitology. *Clinical Microbiology Reviews, 15*(4), 595–612. https://doi.org/10.1128/cmr.15.4.595-612.2002
De Sousa, A. C. C., Combrinck, J. M., Maepa, K., & Egan, T. J. (2020). Virtual screening as a technique to uncover novel β-haematin inhibitors with effectiveness against malaria parasites. *Scientific Reports, 10*(1), 3374. https://doi.org/10.1038/s41598-020-60221-0
Duffy, S., & Avery, V. M. (2013). Identification of inhibitors of *Plasmodium falciparum* gametocyte development. *Malaria Journal, 12*(1), 408. https://doi.org/10.1186/1475-2875-12-408
Ecker, A., Lehane, A. M., Clain, J., & Fidock, D. A. (2012). PfCRT and its role in antimalarial drug resistance. *Trends in Parasitology, 28*(11), 504–514. https://doi.org/10.1016/j.pt.2012.08.002
Favuzza, P., de Lera Ruiz, M., Thompson, J. K., Triglia, T., Ngo, A., Steel, R. W. J., Vavrek, M., Christensen, J., Healer, J., Boyce, C., Guo, Z., Hu, M., Khan, T., Murgolo, N., & Cowman, A. F. (2020). Dual plasmepsin-targeting antimalarial drugs disrupt various phases of the malaria parasite life cycle. *Cell Host & Microbe, 27*(4), 642–658. https://doi.org/10.1016/j.chom.2020.02.005
Held, J., Supan, C., Salazar, C. L. O., Tinto, H., Bonkian, L. N., Nahum, A., Moulero, B., Sié, A., Coulibaly, B., Sirima, S. B., Siribie, M., Otsyula, N., Otieno, L., Abdallah, A. M., Kimutai, R., Bouyou-Akotet, M., Kombila, M., Koiwai, K., Cantalloube, C., Din-Bell, C., Djeriou, E., Waitumbi, J., Mordmüller, B., Ter-Minassian, D., Lell, B., & Kremsner, P. G. (2015). Ferroquine and artesunate in African adults and children with *Plasmodium falciparum* malaria: A phase 2, multicentre, randomised, double-blind, dose-ranging, non-inferiority study. *The Lancet Infectious Diseases, 15*(12), 1409–1419. https://doi.org/10.1016/s1473-3099(15)00079-1
Houben, C. H., Fleischmann, H., & Gückel, M. (2013). Malaria prevalence in North–Eastern Nigeria: A cross-sectional study. *Asian Pacific Journal of Tropical Medicine, 6*(11), 865–868. https://doi.org/10.1016/s1995-7645(13)60154-6
Idris, A. U., Wammanda, D. R., Frank, A. J., & Garba, D. G. (2017). Prevalence of malaria parasitaemia among febrile Nigerian children with severe malnutrition in Northwestern Nigeria. *Nigerian Journal of Basic and Clinical Sciences, 14*(2), 113–116.
Ifebunandu, N., Philip, I., Chidiogo, N., & Chinaza, U. (2021). Epidemiology of malaria in Nise community, Awka South Local Government area, Anambra state, Nigeria. *South Asian Journal of Parasitology, 5*(2), 32–44.
Iyaji, F. O., John, R., Amana, G. U., & Yaro, C. A. (2015). Epidemiological study of malaria in selected primary schools in Egume, Dekina Local Government Area. Kogi state, Nigeria. *Journal of Biological and Chemical Research, 32*(2), 914–919.
Kalu, K. M., Obasi, N. A., Nduka, F. O., & Oko, M. O. (2012). Prevalence of malaria parasitaemia in Umuchieze and Uturu communities of Abia state, Nigeria. *Asian journal of epidemiology, 5*(3), 95–102. https://doi.org/10.3923/aje.2012.95.102
Kumar, S., Bhardwaj, T. R., Prasad, D. N., & Singh, R. K. (2018). Drug targets for resistant malaria: Historic to future perspectives. *Biomedecine & Pharmacotherapie [Biomedicine & pharmacotherapy], 104*, 8–27. https://doi.org/10.1016/j.biopha.2018.05.009
Mathews, E. S., & Odom John, A. R. (2018). Tackling resistance: Emerging antimalarials and new parasite targets in the era of elimination. *F1000Research, 7*, 1170. https://doi.org/10.12688/f1000research.14874.1
Munro, B. A., & McMorran, B. J. (2022). Antimalarial drug strategies to target *Plasmodium* gametocytes. *Parasitologia (Basel, Switzerland), 2*(2), 101–124. https://doi.org/10.3390/parasitologia2020011
Muraoka, S., & Miura, T. (2004). Inhibition of xanthine oxidase by phytic acid and its antioxidative action. *Life Sciences, 74*(13), 1691–1700. https://doi.org/10.1016/j.lfs.2003.09.040
Negatu, G. A., Abebe, G. A., & Yalew, W. G. (2022). Prevalence of malaria and associated factors among malaria-suspected patients attending Hamusit health center, Northwest Ethiopia: A cross-sectional study. *Journal of Parasitology Research, 2022*, 1–7. https://doi.org/10.1155/2022/1306049
Nureye, D., & Assefa, S. (2020). Old and current achievements in life cycle, pathophysiology, diagnosis, prevention, and treatment of malaria including views in Ethiopia. *The Scientific World Journal, 2020*, 1–17. https://doi.org/10.1155/2020/1295381

Ogbu, G. I., Aimakhu, C. O., Ajen Anzaku, S., Ngwan, S., & Ogbu, D. A. (2015). Prevalence of malaria parasitaemia among asymptomatic women at booking visit in a tertiary hospital, northcentral Nigeria. *Journal of Reproductive Biology and Health, 3*(1), 1. https://doi.org/10.7243/2054-0841-3-1

Okeke, O. P., Imakwu, C. A., Eyo, J. E., & Okafor, F. C. (2016). Prevalence of malaria infection in children in Anambra state, Nigeria after change of policy from presumptive/clinical to confirmed diagnosis. *Animal Research International, 13*(1), 2385–2391.

Okoroiwu, G. I. A. (2021). Assessment of malaria parasitaemia among the residents of Abuja municipal area council (AMAC), FCT, Abuja, Nigeria. *Journal of Applied Sciences & Environmental Management, 25*(5), 787–792. https://doi.org/10.4314/jasem.v25i5.15

Oladele, O. V., Onuoha, S. C., Hamafyelto, H. S., Omisope, O., Fauziyya, A., Akindigh, M., Abdullahi, T., Ilu, M. L., & Ikeh, E. (2019). Prevalence of malaria infection among patients attending Murtala Muhammed specialist hospital Kano, Nigeria. *African Journal of Clinical and Experimental Microbiology, 19*(3). https://doi.org/10.4314/ajcem.v19i3.9

Olasehinde, G., Diji-geske, R., Fadina, I., Arogundade, D., Darby, P., Adeleke, A., Dokunmu, T., Adebayo, A. H., & Oyelade, J. (2019). Epidemiology of *Plasmodium falciparum* infection and drug resistance markers in Ota area, Southwestern Nigeria. *Infection and Drug Resistance, 12*, 1941–1949. https://doi.org/10.2147/idr.s190386

Onwe, S. O., Mbamalu, C., Ani, O. C., Anthony, I., Nwovu, A. I., & Igwe, C. C. (2018). Prevalence of malaria parasitaemia among pediatrics in Federal Teaching Hospital Abakaliki, Ebonyi State. *Middle-East Journal of Scientific Research, 26*(4), 418–423. https://doi.org/10.5829/idosi.mejsr.2018.418.423

Phuong, C. X. T., Nhan, N. T., Wills, B., Kneen, R., Ha, N. T. T., Mai, T. T. T., Huynh, T. T. T., Lien, D. T. K., Solomon, T., Simpson, J. A., White, N. J., Farrar, J. J., & the Dong Nai Paediatric Hospital Study Group. (2002). Evaluation of the World Health Organization standard tourniquet test and a modified tourniquet test in the diagnosis of dengue infection in Viet Nam. *Tropical Medicine & International Health: TM & IH, 7*(2), 125–132. https://doi.org/10.1046/j.1365-3156.2002.00841.x

Sam Wobo, S. O., Adekunle, N. O., Adeleke, M. A., Dedeke, G. A., Oke, O. A., Abimbola, W. A., & Surakat, O. A. (2014). Epidemiological factors in prevalence of malaria parasites in primary health facilities attendees, Ogun State, Nigeria. *Malaria Chemotherapy Control & Elimination, 3*, 1. https://doi.org/10.4172/2090-2778.1000111

Sigala, P. A., & Goldberg, D. E. (2014). The peculiarities and paradoxes of plasmodium heme metabolism. *Annual Review of Microbiology., 68*(1), 259–278. Available at:. https://doi.org/10.1146/annurev-micro-091313-103537

Sinha, S., Medhi, B., & Sehgal, R. (2014). Challenges of drug-resistant malaria. *Parasite (Paris, France), 21*, 61. https://doi.org/10.1051/parasite/2014059

Tu, Y. (2011). The discovery of artemisinin (qinghaosu) and gifts from Chinese medicine. *Nature Medicine, 17*(10), 1217–1220. https://doi.org/10.1038/nm.2471

Umaru, M. L., & Uyaiabasi, G. N. (2015). Prevalence of malaria in patients attending the general hospital Makarfi, Makarfi Kaduna—State, North-Western Nigeria. *American Journal of Infectious Diseases and Microbiology, 3*(1), 1–5.

Umumararungu, T., Nkuranga, J. B., Habarurema, G., Nyandwi, J. B., Mukazayire, M. J., Mukiza, J., Muganga, R., Hahirwa, I., Mpenda, M., Katembezi, A. N., Olawode, E. O., Kayitare, E., & Kayumba, P. C. (2023). Recent developments in antimalarial drug discovery. *Bioorganic & Medicinal Chemistry, 88–89*(117339), 117339. https://doi.org/10.1016/j.bmc.2023.117339

Walker, N. F., Nadjm, B., & Whitty, C. J. M. (2010). Malaria. *Medicine, 38*(1), 41–46. https://doi.org/10.1016/j.mpmed.2009.09.017

Wang, J., Zhang, C.-J., Chia, W. N., Loh, C. C. Y., Li, Z., Lee, Y. M., He, Y., Yuan, L.-X., Lim, T. K., Liu, M., Liew, C. X., Lee, Y. Q., Zhang, J., Lu, N., Lim, C. T., Hua, Z.-C., Liu, B., Shen, H.-M., Tan, K. S. W., & Lin, Q. (2015). Haem-activated promiscuous targeting of artemisinin in Plasmodium falciparum. *Nature Communications, 6*(1). https://doi.org/10.1038/ncomms10111

WHO (2023), *World malaria report 2023. Who.int*. Retrieved from February 16, 2024, From https://www.who.int/teams/global-malaria-programme/reports/world-malaria-report-2023

Yang, T., Ottilie, S., Istvan, E. S., Godinez-Macias, K. P., Lukens, A. K., Baragaña, B., Campo, B., Walpole, C., Niles, J. C., Chibale, K., Dechering, K. J., Llinás, M., Lee, M. C. S., Kato, N., Wyllie, S., McNamara, C. W., Gamo, F. J., Burrows, J., Fidock, D. A., Goldberg, D. E., Gilbert, I. H., Wirth, D. F., & Winzeler, E. A. (2021). MalDA, speeding malaria drug discovery. *Trends in Parasitology, 37*(6), 493–507. https://doi.org/10.1016/j.pt.2021.01.009

Yohanna, J., Oti, V., Amuta, E., Philip, A., & Anizoba, L. (2019). *Plasmodium falciparum* infection among febrile patients attending a tertiary healthcare facility in Central Nigeria: Prevalence, hematologic and sociodemographic factors. *International Journal of Tropical Diseases, 2*(2). https://doi.org/10.23937/ijtd-2017/1710019

Chapter 20
The Impact of Antibiotic Resistance Genes in Shigellosis

Obinna Chukwuemeka Nwinyi, Oluwayemi Esther Farotimi, and Somachi Queen Chukwukadibia

1 Introduction

Shigellosis is the second most prevalent cause of fatal diarrhoea worldwide, following rotavirus (Moreno-Mingorance et al., 2021). *Enterobacteriaceae*, specifically *Shigella* species, are Gram-negative rods known to cause disease and possess the ability to survive in both aerobic and anaerobic environments (Sati et al., 2019), facultative intracellular, non-motile, medically significant pathogens (Sabour et al., 2022). Shigellosis has become a major threat to public health, especially in areas with limited resources (Wang et al., 2019). *Shigella* species cause approximately 165 million episodes of diarrheal illness worldwide each year, resulting in 600,000 deaths, majorly occurring in children (CDC, 2024). Shigellosis has caused deaths in Nigeria, Ethiopia, and Eritrea and is rife in numerous African nations (Kakurira et al., 2019). *S. dysenteriae*, *S. flexneri*, *S. boydii*, and *S. sonnei* belonging to serogroups A, B, C, and D, respectively, are the species of *Shigella* (Pakbin et al., 2021). All four major species can cause shigellosis despite differences in virulence and antibiotic resistance (Sati et al., 2019). Diverse serotypes within each serogroup are differentiated by variations in the structure of the O-antigen (Anandan et al., 2017). Although some *Shigella* isolates are untypable (Salimiyan rizi et al., 2020), the four serogroups of *Shigella* comprise about 55 serotypes: serogroup A (15), B (19), C (20), and D (1 serotype), which have been linked to shigellosis and fatalities around the globe (Tang et al., 2019). Serotype conversion, the development of new and unfamiliar and untypable *Shigella*, is induced as an adaptive mechanism to host immune response (Wang et al., 2019). Antibiotic Resistance (AR) to common, cheap antibiotics has emerged due to unrestricted consumption of drugs. Several

O. C. Nwinyi (✉) · O. E. Farotimi · S. Q. Chukwukadibia
Department of Biological Sciences, College of Sciences and Technology, Covenant University, Ota, Ogun state, Nigeria
e-mail: obinna.nwinyi@covenantuniversity.edu.ng

P. O. Isibor, P. A. Akinduti (eds.), *Harnessing Biotechnology Tools for Product Development*, https://doi.org/10.1007/978-3-031-86002-7_20

cases of multidrug-resistant (MDR) *Shigella* have been documented globally (Nikfar et al., 2017). The effectiveness of shigellosis treatment is at risk because of the wide range of *Shigella* serotypes and the increasing resistance to common antibiotics such as tetracycline, chloramphenicol, ampicillin, sulphamethoxazole-trimethoprim, and nalidixic acid (Vubil et al., 2018). As a result, more expensive treatment options including fluoroquinolones are being utilised, yet resistance to this antibiotic class has been established (Vubil et al., 2018). The application of whole genome sequencing in regular public health evaluation will facilitate swift and extensive monitoring of antimicrobial resistance in Shigella (Sadouki et al., 2017). As a result, continual surveillance would be essential to establish the prevalence and evolution of *Shigella* resistance, as well as to comprehend the real impact of the disease and provide standards for shigellosis treatment (Wang et al., 2019).

2 Transmission of *Shigella*

The human gut is the principal host of *Shigella* species (Pakbin et al., 2021), and unsanitary activities such as faecal contamination are a means of direct or indirect transmission. These include human-human contact, faecal-oral route, exposure through fomites (Tang et al., 2019), and sexuality among MSM (Gaufin et al., 2022). It can also occur as a secondary infection in human immunodeficiency virus-infected persons (van den Beld et al., 2020; Worley et al., 2021). Furthermore, house flies are competent vectors of *Shigella* (Vubil et al., 2018). Consumption of contaminated food is linked to approximately 119 million *Shigella* infections and 30,000 deaths (CDC, 2024). Shigellosis occurs in both developed and impoverished locations. It threatens public health worldwide and is a prevalent diarrheal disease in children under 5 years of age (Kakurira et al., 2019) and children enrolled in kindergarten centres (Darton et al., 2018), and also, the elderly and immune-compromised persons, particularly in impoverished nations (Kakurira et al., 2019). In environments with high income, shigellosis is linked to visits to endemic areas and sexual activities such as MSM (Moreno-Mingorance et al., 2021). Other environmental factors such as rainfall and flooding, also contribute to the dissemination of shigellosis, as the prevalence of waterborne diseases has been shown to increase following floods (Qiu et al., 2022). Invasion of epithelial cells of the colon is vital for the replication and pathogenesis of *Shigella* (Darton et al., 2018). *Shigella* can withstand the acidic pH of the stomach and express copious medically significant virulence factors to augment their survival in the gut (Karimi-Yazdi et al., 2020). *Shigella* could migrate to the small intestine, where replication occurs, and subsequently move to invade the epithelial cells of the colon to trigger diarrhoea. The infectious dose of *Shigella* species is 10–100 cells (Gaufin et al., 2022).

3 Virulence Factors of *Shigella* and Symptoms of Shigellosis

The virulence factors may be chromosomal or plasmid-borne. Chromosomal virulence genes are *SHI*-3 of *S. boydii* and others common to the entire *Shigella* genus such as *SHI*-1 and *SHI*-2 (van den Beld et al., 2020). Type III-secretion system is encoded by *inv* plasmid (Pourakbari et al., 2022); it aids in invading host cells by delivering the four subunits of the *Ipa* complex (A, B, C, and D) (Karimi-Yazdi et al., 2020). The *sigA* gene is a prominent virulent factor in the disease process and is primarily prevalent in *S. sonnei* strains (Pourakbari et al., 2022). During infection, invasion plasmid gene D (*ipgD*), a plasmid-borne virulence gene inhibits the passage of T cells, prevents the production of adenosine triphosphate and reduces inflammation. This particular gene is more frequently found in *S. flexneri* (Karimi-Yazdi et al., 2020), and could account for its higher virulence compared to other species of *Shigella*. Other plasmid-borne virulent genes in *Shigella* are *sen*, *virF*, *ial* and *ipaBCD*, and these are expressed at different levels (Seribelli et al., 2016). *SepA* of *S. flexneri* is a main factor in the virulence gene complex linked to severe dysentery and is related to excruciating abdominal pain in *Shigella* infection. Other genes in the complex are *set* and *pic* (Medeiros et al., 2018). *Shigella* infections can lead to self-limited diarrhoea, acute dysentery, and potentially fatal systemic infections, along with other clinical symptoms. *Shigella* has the most severe clinical impact among *Enterobacteriaceae* (Vubil et al., 2018); it is characterised by cramps, abdominal pain, mucoid faeces, diarrhoea, vomiting (Karmoker et al., 2023), fever, nausea (Moreno-Mingorance et al., 2021), and neurological disorders (Nikfar et al., 2017). Karimi-Yazdi et al. (2020) reported that *S. boydii* commonly leads to mild, temporary diarrhoea. *S. sonnei* infections manifest less dysentery and abdominal pain, although these symptoms are more lethal in children and nutrient-deficient persons (Houpt et al., 2021). Oral rehydration therapy is recommended for mild cases of shigellosis while antibiotics are necessary for younger, elderly, immunocompromised, and patients with severe clinical symptoms (Sadouki et al., 2017). Multidrug-resistant *Shigella* infection is an emerging disease that threatens public health worldwide (Toro et al., 2022). This presents a significant challenge when determining the proper course of treatment and management.

4 Geographical Distribution of *Shigella*

The distribution of *Shigella* serotypes is influenced by socio-economic development, leading to variations among different geographical areas (Kakurira et al., 2019). *Shigella flexneri* and *S. sonnei* are continuously the primary causes of shigellosis, with the prevalence of different *Shigella* species tending to alternate mainly between these two (Kakurira et al., 2019). *S. flexneri* is the prevalent agent of shigellosis in underserved areas such as sub-Saharan Africa and Asia, with about two-thirds of all *Shigella* infections in these regions. In developed areas such as North

America and Europe, *S. sonnei* is the predominant pathogen for as much as 80% of all *Shigella* infections in this area (Tang et al., 2019). *S. sonnei* serotypes have replaced *S. flexneri* serotypes in areas undergoing improvements in economy and sanitation (Vubil et al., 2018). The increasing predominance of *S. sonnei* in high-income nations may be linked to improved water hygiene and sanitation standards, which hinder the natural immunity conferred by *Plesiomonas shigelloides* commonly ingested in contaminated water (Anandan et al., 2017), as well as the increased presence of *Acanthamoeba castellanii*, which serves as a reservoir for the species. Additionally, there is greater potential to develop resistance, especially in settings with minimal antimicrobial usage (Anandan et al., 2017). *Shigella boydii* is the prevalent species in Southeast Asia, while *S. dysenteriae* predominates during epidemics and outbreaks caused by crises (Sati et al., 2019). The occurrence of *S. sonnei* is increasing in regions undergoing economic growth, such as Asia. Studies revealed that *S. sonnei* is widely distributed in China, and has displaced *S. flexneri*. Gradually becoming the main cause of shigellosis worldwide (Qiu et al., 2022). Increasing prevalence of *S. sonnei* indicates progress in sanitation and health and favourable change in socio-economic conditions (Medeiros et al., 2018). A study conducted in Manhiça district in southern Mozambique, an area with limited development, disclosed the occurrence of shigellosis among children below 5 years old. It also found that *S. flexneri* infection was prevalent (Vubil et al., 2018). This highlights the impact of socio-economic development on the distribution of *Shigella*.

5 Treatment of Shigellosis

Mild cases of shigellosis can be managed effectively with oral rehydration therapy and supportive measures (Shahin et al., 2019). However, antimicrobial therapy is essential for children, the elderly, and immunocompromised patients. Antibiotics are necessary to stop the progression of the disease, alleviate severe symptoms, and prevent the spread of pathogens from the infected person into the environment. This helps to stop the transmission and inhibit the detrimental consequences of the infection (Kakurira et al., 2019; Toro et al., 2022). The common antibiotic classes recommended for treatment of shigellosis are beta-lactams (penicillin, ampicillin, oxacillin, third-generation cephalosporins-ceftriaxone); fluoroquinolones (ciprofloxacin, nalidixic acid, norfloxacin, ofloxacin); macrolides (azithromycin, erythromycin); tetracycline; colistin; sulphonamides (Sati et al., 2019; Qiu et al., 2022). Various classes of antibiotics work in different ways. Although they can generally be grouped into five main types of activity: interfering with the enzymes necessary for peptidoglycan production, synthesising nucleic acids, producing proteins, affecting metabolism, and disrupting the cytoplasmic membrane (Pulingam et al., 2022) (Fig. 20.1).

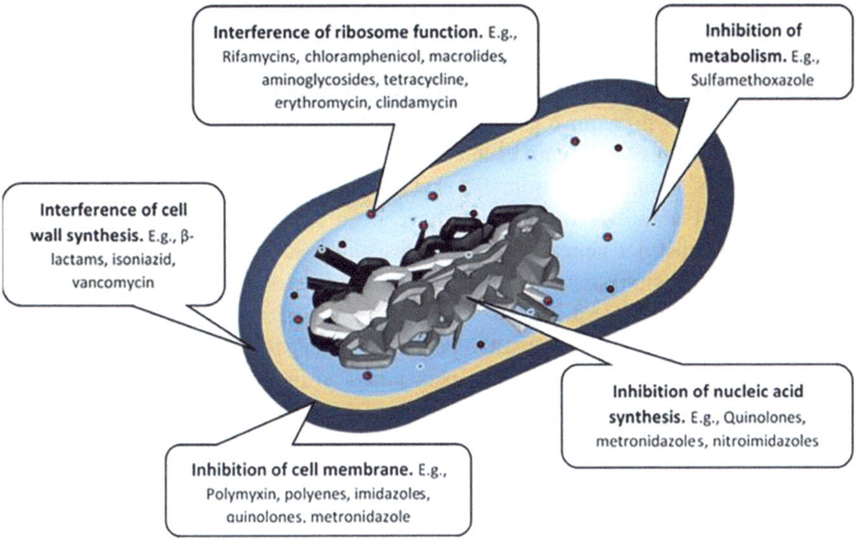

Fig. 20.1 The mechanisms of action of antibiotics in bacteria. Source: Pulingam et al. (2022)

6 Antibiotic Resistance in *Shigella*

Antibiotic Resistance (AR) occurs when microorganisms acquire the ability to survive exposure to antimicrobial drugs that previously kill or inhibit their growth. This happens as an antibiotic loses its effectiveness in inhibiting bacteria (Pulingam et al., 2022). This resistance in *Shigella* species is the consequence of extended antibiotic exposure in developing countries (Darton et al., 2018). The emergence of multidrug (MDR) and extensive drug resistance (XDR) has further complicated the management of shigellosis, causing additional challenges (Pulingam et al., 2022). XDR refers to the resistance against at least one from all except for two or fewer antibiotic classes (Pulingam et al., 2022), while MDR is characterised by the development of resistance to one or more antibiotics from a minimum of three antibiotic classes with varying modes of action (Kakurira et al., 2019). Antibiotics are widely used in agriculture; livestock animals harbour a broad range of antimicrobial-resistant microorganisms in their intestines, which can pose a contamination risk for meat during the slaughtering process. This may lead to consumers ingesting contaminated meat (Bantawa et al., 2019). MDR strains have demonstrated resistance to conventional antibiotics (ampicillin, ticarcillin, trimethoprim/sulfamethoxazole, and tetracycline) for shigellosis control. *S. flexneri* is highly resistant to fluoroquinolones, while S. *sonnei* isolates showed increased resistance to cephalosporins (Wang et al. 2019).

7 Prevalence of Antibiotic Resistance in *Shigella*

Resistant *Shigella* infection is common in younger, elderly and immunocompromised individuals who require frequent medical care; thus, it is especially prevalent in healthcare settings (Pulingam et al., 2022). MSM individuals in Belgium are at risk of MDR *S. sonnei* infection. Several separate, multi-resistant gene clusters linked to MSM activities in Belgium have been identified, each independently spreading with distinct genetic traits and resistance to antimicrobial agents. The diversity of Inc. plasmids detected in the isolates indicates that the bacteria are competent and capable of acquiring a wide array of horizontally transferred genes (Fischer et al., 2021). Shigellosis has become a threat to public health due to the emergence of drug-resistant *Shigella* species and an increase in *Shigella* infection among the MSM population (Moreno-Mingorance et al., 2021). Individuals with HIV and shigellosis showed a higher proportion of *Shigella* isolates resistant to multiple antibiotics compared to those without the virus (Gaufin et al., 2022). The expression of antibiotic-resistant genes varies from one geographic region to another in *Shigella* species (Toro et al., 2022). The problem of antibiotic-resistant *Shigella* strains is on the rise in several nations and has been documented in regions including South America, Southeast Asia, Africa (Nikfar et al., 2017), South Asia, Latin America, Europe and North America (Toro et al., 2022). *Shigella* species found in food samples in Iran were resistant to tetracycline, chloramphenicol, azithromycin, sulfamethoxazole/trimethoprim, amoxicillin, streptomycin and ampicillin (Pakbin et al., 2021) (Fig. 20.2).

8 Factors Influencing Antibiotic Resistance in *Shigella*

The occurrence of antibiotic-resistant pathogens is expanding. The trend of antimicrobial susceptibility among *Shigella* strains in a particular region varies with external factors such as climate, type and administration of the antimicrobial agent (Wang et al., 2019). The worldwide rise of antibiotic-resistant bacteria is linked to the increasing use in various healthcare and agricultural settings (Pulingam et al., 2022). Shigellosis management is challenging in settings with limited resources due to high infection rates and unrestricted access to antibiotics, resulting in a decreased availability of effective treatment and chemotherapeutic choices (Sati et al., 2019). Antimicrobial-resistant strains of *Shigella* are more common among MSM and individuals with travel-related shigellosis (Gaufin et al., 2022). Molecular surveillance conducted with whole genome sequencing (WGS) revealed that *Shigella* species commonly found in the MSM group are multidrug-resistant strains (van den Beld et al., 2020). Demographic shifts, patterns of international travel and socio-economic alteration, and genetic factors may have contributed to the exchange of resistant genes and accelerated the emergence and dissemination of resistant strains (Sati et al., 2019). Another major risk factor for the spread of resistant *Shigella* species

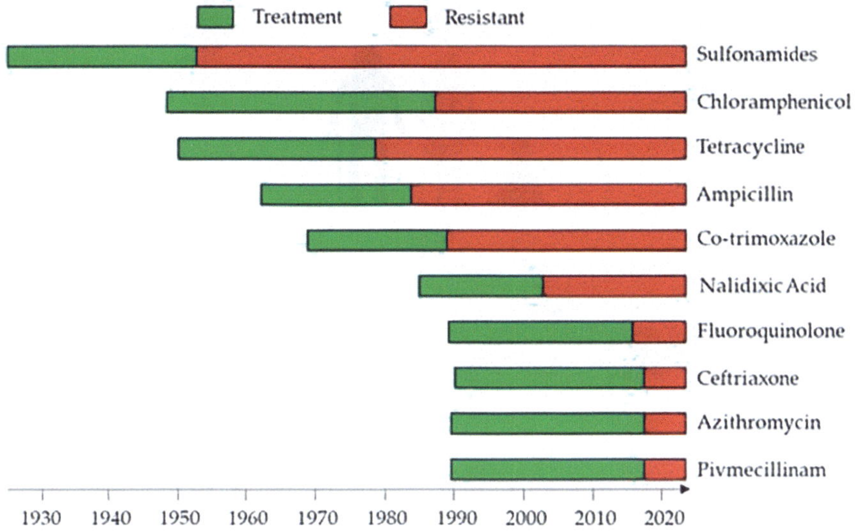

Fig. 20.2 Global treatment and antibiotic resistance in Shigellosis. Source: Raso et al. (2023)

infections and their resistance genes might be water polluted by faeces from birds. Birds and waterfowl in Bangladesh harbour *Shigella* species carrying antibiotic genes such as *tetA* and *SHV* (Karmoker et al., 2023) (Fig. 20.3).

9 Mechanisms of Antibiotic Resistance

Antibiotic Resistance (AR) in bacteria is usually mediated by biochemical and molecular mechanisms (Pulingam et al., 2022). The biochemical mechanisms include the impermeability of the cell wall to hinder drug entry, expulsion through efflux pumps, alteration of targets, and enzymatic breakdown of antibiotics (Puzari et al., 2018). The molecular mechanisms may be inherent in chromosomes or acquired through mobile genetic elements (MGEs) such as transposons, plasmids, and integrons (Rajpara et al., 2018). According to Rajpara et al. (2018), the resistance characteristic of all four *Shigella* species is due to innate genetic components, such as mutations in topoisomerases, and mobile genetic elements like plasmids, integrons, and *qnr* genes responsible for quinolone resistance. Horizontal gene transfer can occur among prokaryotes and between prokaryotes and eukaryotes, facilitated by conjugation, transformation, or transduction mechanisms (Tang et al., 2019). Horizontal gene transfer of antibiotic resistance genes (ARGs) is the primary cause of the sudden development of multidrug resistance in bacteria in the environment (Shahin et al., 2019). *Shigella* species can develop antibiotic resistance through genes acquired from MGEs in their environment (Sabour et al., 2022). This has led to the failure of many antibiotics such as cotrimoxazole, tetracycline, ampicillin,

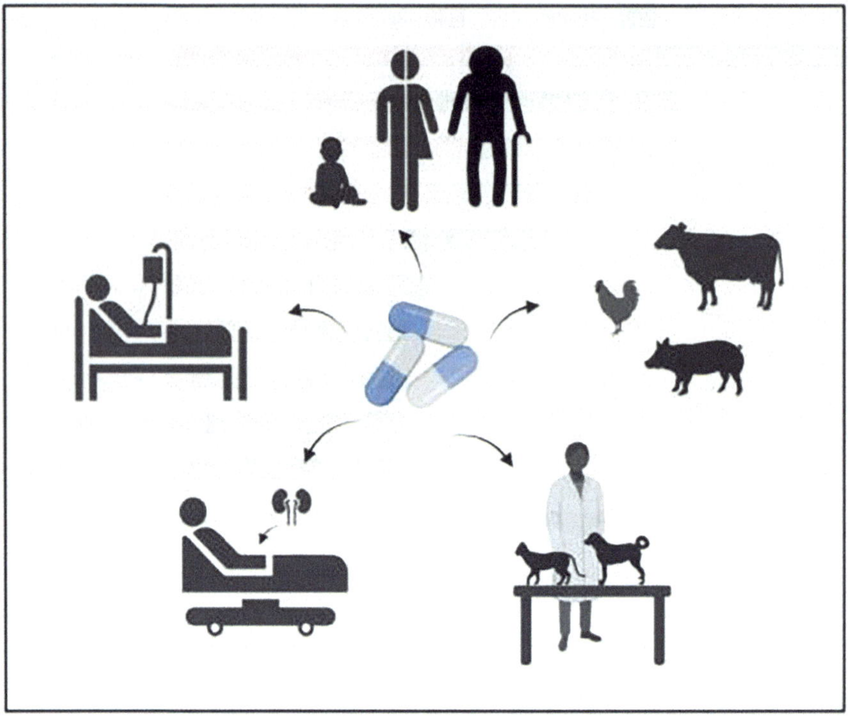

Fig. 20.3 Uses of antibiotics in several sectors such as treatment of infection by individuals, healthcare settings, veterinary centres, and livestock. Source: Pulingam et al. (2022)

and sulphonamides (Kakurira et al., 2019). Antibiotic-resistant genes have a high probability of flexibly migrating from commensals to pathogens. The size of the donor and recipient cells, the presence of an exchange process, nutrient availability, and selection pressures all influence the transfer.

Additionally, natural conjugation within and between different species is another method of transferring ARGs (Devanga Ragupathi et al., 2019). The discovery of the resistance plasmid, pABC-3, in locally acquired sub-lineages of *S. flexneri* 2a and 3a, which originated from travel-acquired IIIa *S. sonnei* in MSM-associated settings, revealed the potential for horizontal transfer of resistance genes and their contribution to the emergence of shigellosis outbreaks (Baker et al., 2018b). The repeated discovery of novel ARG in one strain and region is followed by reports of the gene in various strains and locations across the globe. This highlights the importance of horizontal gene transfer in the ongoing worldwide AMR crisis. This pattern is currently observed with *mcr*-1, which confers resistance to colistin, an antibiotic used as a last resort. Similar occurrences have been noted with genes responsible for extended-spectrum β-lactamase and carbapenemase, namely *blaNDM*-1 and *blaCTX-M* (Baker et al., 2018b) (Fig. 20.4).

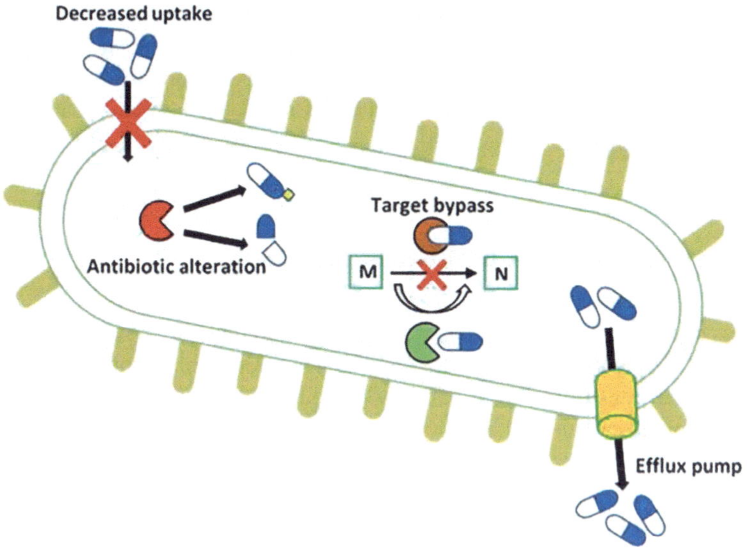

Fig. 20.4 Common mechanisms of antibiotic resistance in bacteria. Source: Pulingam et al. (2022)

10 Molecular Surveillance of Antibiotic-Resistant *Shigella*

Shigella sonnei has been categorised into five lineages using whole genome sequence analysis (WGSA). Additionally, *S. flexneri* is divided into seven phylogroups through WGSA and sub-lineages of these species have been reported in Asia, the Middle East, and Latin America. This has demonstrated the international migration of *Shigella* sub-lineages by identifying prevalent WGSA subtypes in one location and later recorded in patients in different geographical area with an epidemiological connection (Baker et al., 2018a). It has been shown that utilising genetic data to create AR profiles is a dependable and precise approach for *S. sonnei*. However, comprehending the results in terms of how they manifest phenotypically may present challenges, thus necessitating a thorough investigation of WGS data for AR identification. Forecasting susceptibility in a resistant isolate could be hindered if a new resistance mechanism not currently included in the reference database goes unnoticed. Regular updates to databases incorporating novel ARG and mutations and continuous phenotypic testing of isolates deemed "genotypically susceptible" are necessary to maintain the high sensitivity of AR profiles derived from WGS data (Sadouki et al., 2017). Over the past 30 years, a significant genetic diversity has emerged in various *Shigella* species and serotypes across Southeast Asia. Studies suggest that *S. sonnei* comprises diverse genetic populations that have spread over different periods. Furthermore, the coexistence of closely related but distinct clones forming a single serotype within *S. flexneri* highlights the challenges in identifying the genetic composition of these species through serotyping (Chung The et al., 2021). ARGs are often transmitted through plasmids, leading to variations in molecular monitoring and the phenotypic detection of resistant strains. These differences may also arise from plasmid loss during storage and sub-culturing (Sadouki et al., 2017). Plasmids or a resistant phenotype are not always linked to the presence of antibiotic-resistant genes because these genes may remain unexpressed (Muthuirulandi Sethuvel et al., 2019).

11 Antibiotic Classes and Resistance Genes

AR traits often result from either new change in the genetic material of the microorganisms or the incorporation of new AMR genes through horizontal gene transfer. These may be transmitted from parent to offspring (Baker et al., 2018b). Mobile genetic elements such as plasmids (*spA* or *pCERC1*), chromosomal integrons like the SRL-MDRE island and *In2*, and the transposon *tn7*, all play a significant role in contributing to antimicrobial resistance in *Shigella* species (van den Beld et al., 2020). Antimicrobial resistance plasmids are readily transmissible among *Shigella* and other *Enterobacteriaceae* species and have been associated with the appearance of extensively drug-resistant strains with ceftriaxone, azithromycin and colistin phenotypes (Qiu et al., 2022). Thus, it is important to identify the role of genetic factors in the emergence of multidrug-resistant and extensively drug-resistant traits (Bose et al., 2024).

Macrolides: Azithromycin resistance is mainly caused by plasmid-borne resistance genes such as *ermB* and *mphA* which act as efflux pumps for antibiotics (Fischer et al., 2021). Since 1987, documented cases of *S. sonnei* isolates carrying the *mph*(A) gene showing resistance to azithromycin have been reported in the United States (Worley et al., 2021). The plasmid pKSR100 in *Shigella* contains the IS26–*mphA*–*mrx*–*mphR*(A)–IS6100 genes, which confer resistance to macrolides (Asad et al., 2024), *mph*(E)-*msr*(E)-IS482-IS6 is an additional plasmid-borne cluster of resistant genes to azithromycin in *Shigella* strains found in Bangladesh (Nusrin et al., 2022). The plasmid pKSR100 has become a primary factor in conferring macrolide resistance in *Shigella*, linked explicitly to infections in MSM (Worley et al., 2021). Simulation analysis showed that *ermB* is found in two plasmid configurations: *IncFI* plasmid in *S. sonnei* 20094 and *IncFII* plasmid in *S. sonnei* EG430 (Darton et al., 2018). The MSM group has more azithromycin-resistant *S. flexneri* and *S. sonnei* strains compared to travellers and groups. The use of azithromycin for treating sexually transmitted infections in MSM may have played a key role in the selection and development of resistant strains, as well as the global spread of resistant strains (Moreno-Mingorance et al., 2021). The causative agent of shigellosis in children in Mashhad has changed from *S. flexneri* to *S. sonnei*. Although *S. sonnei* is less virulent, it is more resistant to azithromycin. This shift is noteworthy because of the high prevalence of azithromycin-resistant *Shigella* strains in this area (Salimiyan rizi et al., 2020). The persistence of azithromycin-resistant *Shigella* strain in a population may fail due to diversity within the population, the nature of the resistance plasmids and the absence of selection pressure from antimicrobial drugs (Darton et al., 2018).

Beta-Lactams: The *blaOXA*, *blaTEM*, *blaSHV*, and *blaCTX-M* are the β-lactamase genes found in antibiotic-resistant *Shigella* species (Bose et al., 2024). These genes are responsible for producing extended-spectrum beta-lactamase enzyme (Kakurira et al., 2019), with potent hydrolytic activity against β-lactam antibiotics such as penicillin, cephalosporins, oxacillin and ampicillin (Wang et al., 2019). These genes are acquired through plasmid-mediated gene transfer (Wang et al., 2019). The cephalosporin-resistant strains carry resistance genes such as *blaCTX-M* (-14, -55, -15, -28 and -64), *blaOXA*-1, *blaSHV* and *blaTEM*-1 (Wang et al., 2019). Polymerase chain reaction analysis showed the *blaTEM*-1 as the most frequently encountered ESBL gene in *Shigella* species (Kakurira et al., 2019). Additionally, plasmid-mediated *AmpC* β-lactamases confer cephalosporin resistance (Anandan et al., 2017). These genes have been reported in *S. flexneri*, *S. boydii* and *S. sonnei* (Kakurira et al., 2019). Patients in healthcare facilities with severe illnesses and frequent antibiotic use often carry ESBL-producing bacteria. ESBL-producing strains of *S. flexneri* and *S. sonnei* have been documented in Zimbabwe, South Africa and Nigeria (Kakurira et al., 2019).

Tetracycline: The majority of tetracycline resistance is caused by *tet*-like genes (Vubil et al., 2018). According to Bose et al. (2024), *tet*(B) is more prevalent than *tet*(A). These plasmid-borne genes encode tetracycline efflux (Hartman et al., 2003). A study conducted by Shahin et al. (2019), found *tet* (A), *tet* (B) and *tet* (C) in *S. flexneri* in Iran.

Polymyxin: The *mcr*-1 gene confers resistance to colistin (polymyxin) used as a last treatment option for shigellosis (Baker et al., 2018b). The gene is associated with the *IncI2* plasmid (Qiu et al., 2022). Horizontal gene transfer aids the spread of this gene within and across geographical areas (Baker et al., 2018b). Qiu et al. (2022), revealed that *S. sonnei* strains with *mcr*-1 gene caused an outbreak of shigellosis in China.

Quinolones: *Shigella* species have developed resistance to quinolones such as ciprofloxacin, ofloxacin, nalidixic acid, and ofloxacin. The resistance is associated with chromosomal mutations of topoisomerases at the quinolone resistance-determining regions (QRDRs, such as par and *gyrA*) (Rajpara et al., 2018), efflux pump activity, and plasmid-mediated quinolone resistance genes such as *qnr*, *aac*(6′)1b (Bose et al., 2024). Quinolone resistance genes have been found in *S. sonnei* and *S. flexneri* (Sadouki et al., 2017; Rajpara et al., 2018), particularly in travel-acquired and MSM shigellosis (Moreno-Mingorance et al., 2021). Quinolone-resistant Shigella species are globally distributed and have been reported in many countries such as India (Bose et al., 2024), Wales, England and South Africa, the later was noted as the reservoir for the worldwide dissemination of ciprofloxacin-resistant *S. sonnei* (Sadouki et al., 2017). The level of quinolone resistance is higher in *S. flexneri* than in *S. sonnei* (Rajpara et al., 2018). Conversely, Sati et al. (2019), reported a higher quinolone resistance in *S. sonnei*. This variation may be due to differences in sampling and geographical attributes. (Table 20.1).

Table 20.1 Common antibiotic resistance genes in *Shigella* species

Antibiotics	Resistance gene	References
Fluoroquinolones	*parC* *qnr*, *gryA*, *aac* (6'1B)	Sadouki et al. (2017) Bose et al. (2024)
Cotrimoxazole	*Sxt*	Sharif et al. (2023)
Colistin	*mcr*-1	Baker et al. (2018b)
Sulphonamides	*sul* 1, *sul* 2	Qiu et al. (2022)
Trimethoprim	*dfrA1* *dfrV*	Qiu et al. (2022) Rajpara et al. (2018)
Aminoglycosides	*aac*(3)-*lla*, *aadA*, *aadA*, *sat*-2, *strA*, *strB*, *Xaac*(3)-*IIa*	Qiu et al. (2022)
Tetracycline	*tetA*, *tetB*	Shahin et al. (2019)
Macrolides	*mphA*, *erm* *mphE*, *msrE*	Fischer et al. (2021) Nusrin et al. (2022)
Beta-lactams	*AmpC* *blaSHV*, *blaCTX-M*, *bla-TEM*, *blaOXA*	Anandan et al. (2017) Wang et al. (2019)
Chloramphenicol	*Cat*	Toro et al. (2022)

12 Multidrug Resistance in *Shigella*

Diverse mobile genetic elements, such as integrons, plasmids, and genomic islands, have been identified as molecular factors that influence the expression of multidrug resistance traits in *Shigella* species (Toro et al., 2022). The aggregate of AMR genes differs within and across *Shigella* species, and the lineage III of S. sonnei contains more resistant genes than other lineages in this species (Qiu et al., 2022). The marker of this lineage is the biotype g, consisting of four pulsotype (A, B, C and D). In addition, MDR may be propagated by clonal dissemination as some MDR strains often occur in clusters (Vubil et al., 2018). MSM-associated *S. sonnei* belongs to the globally distributed MDR lineage III characterized by resistance to cotrimoxazole, ciprofloxacin and azithromycin (Moreno-Mingorance et al., 2021; Qiu et al., 2022). The resistant genes, *srtA*, *strB*, *sat*-2, *aadA*, *Xaac*(3)-*IIa* encoding resistance to aminoglycosides, *dfrA1* encoding resistance to trimethoprim, *tet*(A) encoding resistance to tetracycline, *sul* encoding resistance to sulfonamides and *mcr*-1 encoding resistance to colistin have been identified in XDR shigella (Qiu et al., 2022). The SRL PAI encodes the *aadA1*, *blaOXA*-1, *cat* and *tetB* genes conferring resistance to streptomycin, ampicillin, chloramphenicol, and tetracycline respectively, in MDR *Shigella* (Toro et al., 2022). The *qnrB*, *sxt*, *mcr*-1 resistant genes of ciprofloxacin, cotrimoxazole and colistin in *Shigella* confer cross-resistance to ceftriaxone, cotrimoxazole, ciprofloxacin, tetracycline and colistin (Sharif et al., 2023). *Shigella* bacteria that are resistant to nalidixic acid and also show cross-resistance to other quinolones, such as ciprofloxacin, have been documented in regions where nalidixic acid was initially the preferred therapy (Sati et al., 2019). The *dfrV* gene on class 1 integron encodes trimethoprim resistance in only *S. sonnei*. The *dfrA1–aadA* or *blaOXA-aadA* genes on atypical class 1 integron confer resistance to trimethoprim, aminoglycosides, and β-lactams in *S. flexneri*, *S. dysenteriae* and *S. boydii* (Rajpara et al., 2018). The *dfrA1-sat-aadA* or *dfrA1-sat* genes on class 2 integrons are common in *Shigella* species. The *InsE-InsO-dfrA1-sat* on class 2 integron confer resistance to trimethoprim and streptothricin (Rajpara et al., 2018). Antibiotic resistance genes such as *mphA*, *blaTEM1*-B, *blaDHA*-1, *tetR*, *sulI*, *sulII*, *strA*, *strB* and *qnrB10* are encoded by *IncF*-type plasmid in MDR *S. flexneri* (Muthuirulandi Sethuvel et al., 2019). The lineage III *S. sonnei* has also been identified in Vietnam with resistance to macrolides, fluoroquinolones and third-generation cephalosporin (Darton et al., 2018). Ko et al., identified a group of clonal MDR *S. flexneri* of serotype 1c (7a) encoding ESBL genes, *blaCTX−M−*15 and *blaTEM−*1. Six-drug XDR *S. sonnei* from lineage III have been documented in Bangladesh (Rahman et al., 2017). MSM-associated *S. sonnei* clones in United Kingdom contain IncFII plasmid encoding multiple resistance genes (*aadA5*, *dfrA17*, *blaCTX*-M-27, *blaTEM*-1, *mphA* and *ermB*) and resistance traits (Qiu et al., 2022). The successful acquisition and dissemination of resistant genes by *Shigella* species has resulted into the emergence global spread of multidrug resistance *Shigella* infections which are difficult to manage.

13 Consequences of Antibiotic Resistance in *Shigella*

Shigella infections usually resolve on their own, but inadequate treatment can result in the emergence of resistant strains, thereby causing complications for persons with systemic infections (Murray et al., 2017). Treatment options for shigellosis caused by MDR *Shigella* strains are limited, particularly in situations involving children or when oral antibiotics are required (Darton et al., 2018). This is significant because shigellosis is one of the few enteric diseases for which antibiotics are recommended for therapy (Toro et al., 2022). There is currently no approved vaccine for *Shigella* (WHO, 2022). The evolving trends of resistance to existing antibiotics often lead to clonal substitution, influencing the evolutionary trajectory of *Shigella* species (Chung The et al., 2021). *Shigella* bacteria that produce the extended-spectrum beta-lactamase enzyme present a unique challenge for clinical microbiologists, doctors, epidemiologists, infection control experts, and researchers working on innovative drug development (Kakurira et al., 2019).

14 Management of Antibiotic Resistance in *Shigella*

Community-led efforts are essential for preventing and controlling multidrug-resistant *Shigella* strains within the community (Vubil et al., 2018). Laboratory surveillance and epidemiology are interrelated. Whole genome sequencing and several approaches in public health are crucial for monitoring outbreaks at both national and international levels (van den Beld et al., 2020)., medical personnel should consider the possibility of antibiotic-resistant strains when treatment fails in immunocompromised shigellosis patients. It is essential to enlighten multidrug-resistant carriers about prevention to minimise the transmission of infection (Worley et al., 2021). The knowledge of the antibiotic sensitivity patterns of predominant strains is crucial in the choice of efficient medication for shigellosis (Wang et al., 2019), the proper identification and characterisation of MDR clones will aid in the design of preventive measures for infection control (Vubil et al., 2018). These changes are expected to result in modification in the empiric treatment (Wang et al., 2019). The diversity of mobile genetic elements and resistant strains in different geographic areas emphasises the importance of conducting local surveillance on antimicrobial resistance. These variations are anticipated to necessitate modification in treatment approaches (Toro et al., 2022). Plasmid sequencing in Shigella species will aid the comprehension of the emergence and dissemination of antibiotic-resistant strains (Muthuirulandi Sethuvel et al., 2019). The potential spread of AMR *Shigella* strains by birds is an increasing concern in public health. This highlights the importance of monitoring bird populations and integrating potential *Shigella* outbreaks into one-health approaches (Karmoker et al., 2023). The occurrence of shigellosis varies with seasons and climatic conditions in shigellosis-endemic nations. There is a need to explore the impacts of climatic factors on the proliferation spread of drug-resistant *Shigella* strains and their significance in public health (Chung The et al., 2021).

Research and development of novel antibiotics and effective surveillance are essential (Sati et al., 2019). Phage therapy is a potential strategy for the control of antibiotic-resistant bacterial infections (Tang et al., 2019). The bacteriophage PSD9 infects multidrug-resistant *Shigella*. The lytic enzyme P9ly, encoded by the phage, can potentially replace or enhance conventional antibiotics. It can rapidly break down the peptidoglycan in the cell wall of bacteria. This action effectively kills drug-resistant *Shigella dysenteriae* (Wang et al., 2022). The development of a vaccine against *Shigella* is of significant importance in public health, and it could have an indirect impact on the evolution of antimicrobial resistance and lead to a reduced reliance on antibiotics. This is due to the potential for herd immunity and protection against all diarrhoea (WHO, 2022). Selecting the smallest possible number from the pool of complex serotypes is a challenge in developing *Shigella* vaccines based on serotypes. The intricacy and expense of vaccination rise as more serotypes are integrated (Livio et al., 2014).

15 Conclusion

The ability of *Shigella* to readily acquire and express plasmid-borne antimicrobial resistance genes has resulted in the global spread of multidrug and extensively drug-resistant *Shigella* strains. This is a serious threat to public health; these genes render the current therapy options ineffective, increasing the cost of treatment, morbidity and mortality. There is a need for further studies on environmental impacts on AMR genes and the potential benefits of *Plesiomonas shigelloides* and *Acanthamoeba castellanii* in controlling *Shigella*. Effective local and global surveillance of AMR *Shigella* should be adopted to guide the choice of treatment options and the design of new drugs, vaccines and other management approaches such as phage therapy.

References

Anandan, S., Muthuirulandi Sethuvel, D. P., Gajendiren, R., Verghese, V. P., Walia, K., & Veeraraghavan, B. (2017). Molecular characterization of antimicrobial resistance in clinical Shigella isolates during 2014 and 2015: Trends in South India. *Germs, 7*(3), 115–122.

Asad, A., Jahan, I., Munni, M. A., Begum, R., Mukta, M. A., Saif, K., Faruque, S. N., Hayat, S., & Islam, Z. (2024). Multidrug-resistant conjugative plasmid carrying mphA confers increased antimicrobial resistance in Shigella. *Scientific Reports, 14*(1), 6947–6957.

Baker, K. S., Dallman, T. J., Field, N., Childs, T., Mitchell, H., Day, M., Weill, F.-X., Lefèvre, S., Tourdjman, M., Hughes, G., Jenkins, C., & Thomson, N. (2018a). Genomic epidemiology of Shigella in the United Kingdom shows transmission of pathogen sublineages and determinants of antimicrobial resistance. *Scientific Reports, 8*(1), 7389–7396.

Baker, K. S., Dallman, T. J., Field, N., Childs, T., Mitchell, H., Day, M., Weill, F.-X., Lefèvre, S., Tourdjman, M., Hughes, G., Jenkins, C., & Thomson, N. (2018b). Horizontal antimicrobial resistance transfer drives epidemics of multiple Shigella species. *Nature Communications, 9*(1), 1462–1471.

Bantawa, K., Sah, S. N., Subba Limbu, D., Subba, P., & Ghimire, A. (2019). Antibiotic resistance patterns of Staphylococcus aureus, Escherichia coli, Salmonella, Shigella and Vibrio isolated from chicken, pork, buffalo and goat meat in eastern Nepal. *BMC Research Notes, 12*(1), 766–771.

Bose, P., Chowdhury, G., Halder, G., Ghosh, D., Deb, A. K., Kitahara, K., Miyoshi, S., Morita, M., Ramamurthy, T., Dutta, S., & Mukhopadhyay, A. K. (2024). Prevalence and changing antimicrobial resistance profiles of Shigella spp. isolated from diarrheal patients in Kolkata during 2011–2019. *PLoS Neglected Tropical Diseases, 18*(2), 1–22.

CDC, 2024. Shigellosis. CDC Yellow Book 2024. Centers for Disease Control and Prevention. Retrieved May 23, 2024, from https://wwwnc.cdc.gov/travel/yellowbook/2024/infections-diseases/shigellosis .

Chung The, H., Bodhidatta, L., Pham, D. T., Mason, C. J., Ha Thanh, T., Voong Vinh, P., Turner, P., Hem, S., Dance, D. A. B., Newton, P. N., Phetsouvanh, R., Davong, V., Thwaites, G. E., Thomson, N. R., Baker, S., & Rabaa, M. A. (2021). Evolutionary histories and antimicrobial resistance in Shigella flexneri and Shigella sonnei in Southeast Asia. *Communications Biology, 4*(1), 1–12.

Darton, T. C., Tuyen, H. T., The, H. C., Newton, P. N., Dance, D. A. B., Phetsouvanh, R., Davong, V., Campbell, J. I., Hoang, N. V. M., Thwaites, G. E., Parry, C. M., Thanh, D. P., & Baker, S. (2018). Azithromycin resistance in Shigella spp. in Southeast Asia. *Antimicrobial Agents and Chemotherapy, 62*(4), 1–11.

Devanga Ragupathi, N. K., Muthuirulandi Sethuvel, D. P., Gajendran, R., Anandan, S., Walia, K., & Veeraraghavan, B. (2019). Horizontal transfer of antimicrobial resistance determinants among enteric pathogens through bacterial conjugation. *Current Microbiology, 76*(6), 666–672.

Fischer, N., Maex, M., Mattheus, W., Van Den Bossche, A., Van Cauteren, D., Laisnez, V., Hammami, N., & Ceyssens, P.-J. (2021). Genomic epidemiology of persistently circulating MDR Shigella sonnei strains associated with men who have sex with men (MSM) in Belgium (2013–19). *Journal of Antimicrobial Chemotherapy, 77*(1), 89–97.

Gaufin, T., Blumenthal, J., Ramirez-Sanchez, C., Mehta, S., Pride, D. T., Fierer, J., & Jenks, J. D. (2022). Antimicrobial-resistant Shigella spp. in San Diego, California, USA, 2017–2020. *Emerging Infectious Diseases, 28*(6), 1110–1116.

Hartman, A. B., Essiet, I. I., Isenbarger, D. W., & Lindler, L. E. (2003). Epidemiology of tetracycline resistance determinants in Shigella spp. and enteroinvasive Escherichia coli: Characterization and dissemination of tet(A)-1. *Journal of Clinical Microbiology, 41*(3), 1023–1032.

Houpt, E. R., Ferdous, T., Ara, R., Ibrahim, M., Alam, M. M., Kabir, M., Platts-Mills, J., Ahmed, T., Faruque, A. S. G., Taniuchi, M., & Haque, R. (2021). Clinical outcomes of drug-resistant Shigellosis treated with azithromycin in Bangladesh. *Clinical Infectious Diseases, 72*(10), 1793–1798.

Kakurira, K., Gudza-Mugabe, M., Tarupiwa, A., Mashe, T., Stray-Pedersen, B., & Mbanga, J. (2019). Characterisation of clinical Shigella isolates, using serotyping and antimicrobial resistance profiles in Zimbabwe. *Global Scientific Journals, 7*, 779–806.

Karimi-Yazdi, M., Ghalavand, Z., Shabani, M., Houri, H., Sadredinamin, M., Taheri, M., & Eslami, G. (2020). High rates of antimicrobial resistance and virulence gene distribution among Shigella spp. isolated from pediatric patients in Tehran, Iran. *Infection and Drug Resistance, 13*, 485–492.

Karmoker, J., Saiful Islam, M., Liton Rana, M., Ashek Ullah, M., Neloy, F. H., Oishy, N. M., Pramanik, P. K., Siddique, M. P., Saha, S., & Tanvir Rahman, M. (2023). Molecular detection and multidrug resistance of Shigella spp. isolated from wild waterfowl and migratory birds in Bangladesh. *Veterinary Medicine International, 2023*, 1–9.

Livio, S., Strockbine, N. A., Panchalingam, S., Tennant, S. M., Barry, E. M., et al. (2014). Shigella isolates from the global enteric multicenter study inform vaccine development. *Clinical Infectious Diseases, 59*(7), 933–941.

Medeiros, P. H. Q. S., Lima, A. Â. M., Guedes, M. M., Havt, A., Bona, M. D., Rey, L. C., Soares, A. M., Guerrant, R. L., Weigl, B. H., & Lima, I. F. N. (2018). Molecular characterization of virulence and antimicrobial resistance profile of Shigella species isolated from children with moderate to severe diarrhea in Northeastern Brazil. *Diagnostic Microbiology and Infectious Disease, 90*(3), 198–205.

Moreno-Mingorance, A., Espinal, P., Rodriguez, V., Goterris, L., Fàbrega, A., Serra-Pladevall, J., Barberà, M. J., Alberny, M., Martín-González, H., Cornejo-Sánchez, T., Armas, M., Mir-Cros, A., Raventós, A., Viñado, B., Pumarola, T., Larrosa, M. N., & González-López, J. J. (2021). Circulation of multi-drug-resistant Shigella sonnei and Shigella flexneri among men who have sex with men in Barcelona, Spain, 2015–2019. *International Journal of Antimicrobial Agents, 58*(3), 106378–106385.

Murray, K., Reddy, V., Kornblum, J. S., Waechter, H., Chicaiza, L. F., Rubinstein, I., Balter, S., Greene, S. K., Braunstein, S. L., Rakeman, J. L., & Dentinger, C. M. (2017). Increasing antibiotic resistance in Shigella spp. from infected new York City residents, New York, USA. *Emerging Infectious Diseases, 23*(2), 332–335.

Muthuirulandi Sethuvel, D. P., Anandan, S., Devanga Ragupathi, N. K., Gajendiran, R., Kuroda, M., Shibayama, K., & Veeraraghavan, B. (2019). IncFII plasmid carrying antimicrobial resistance genes in Shigella flexneri: Vehicle for dissemination. *Journal of Global Antimicrobial Resistance, 16*, 215–219.

Nikfar, R., Shamsizadeh, A., Darbor, M., Khaghani, S., & Moghaddam, M. (2017). A study of prevalence of Shigella species and antimicrobial resistance patterns in paediatric medical center, Ahvaz, Iran. *Iranian Journal of Microbiology, 9*(5), 277–283.

Nusrin, S., Asad, A., Hayat, S., Islam, B., Begum, R., Nabila, F.H. & Islam, Z. (2022). Multiple mechanisms confer resistance to azithromycin in Shigella in Bangladesh: A comprehensive whole genome-based approach. M.A. Garcia-Solache (ed.). Microbiology Spectrum. 10 (4), 1–11.

Pakbin, B., Amani, Z., Allahyari, S., Mousavi, S., Mahmoudi, R., Brück, W. M., & Peymani, A. (2021). Genetic diversity and antibiotic resistance of Shigella spp. isolates from food products. *Food Science & Nutrition, 9*(11), 6362–6371.

Pourakbari, B., Ghaffari Charati, M., Mahmoudi, S., Abdolsalehi, M. R., Hosseinpour Sadeghi, R., & Mamishi, S. (2022). High frequency of antimicrobial resistance and virulence gene in Shigella species isolated from pediatric patients in an Iranian Referral Hospital. *Acta Bio Medica: Atenei Parmensis, 93*(2), 1–5.

Pulingam, T., Parumasivam, T., Gazzali, A. M., Sulaiman, A. M., Chee, J. Y., Lakshmanan, M., Chin, C. F., & Sudesh, K. (2022). Antimicrobial resistance: Prevalence, economic burden, mechanisms of resistance and strategies to overcome. *European Journal of Pharmaceutical Sciences, 170*, 106103–106120.

Puzari, M., Sharma, M., & Chetia, P. (2018). Emergence of antibiotic resistant Shigella species: A matter of concern. *Journal of Infection and Public Health, 11*(4), 451–454.

Qiu, S., Liu, K., Yang, C., Xiang, Y., Min, K., et al. (2022). A Shigella sonnei clone with extensive drug resistance associated with waterborne outbreaks in China. *Nature Communications, 13*(1), 7365–7377.

Rahman, M., Haque, A. F., Deeba, I., Ahmed, D., Zahidi, T., Rimu, A., Akter, M., Akter, F., & Talukder, K. (2017). Emergence of extensively drug-resistant Shigella sonnei in Bangladesh. *Immunology and Infectious Diseases, 5*, 1–9.

Rajpara, N., Nair, M., Chowdhury, G., Mukhopadhyay, A. K., Ramamurthy, T., Niyogi, S. K., & Bhardwaj, A. K. (2018). Molecular analysis of multidrug resistance in clinical isolates of Shigella spp. from 2001–2010 in Kolkata, India: Role of integrons, plasmids, and topoisomerase mutations. *Infection and Drug Resistance, 11*, 87–102.

Raso, M. M., Arato, V., Gasperini, G., & Micoli, F. (2023). Toward a Shigella vaccine: Opportunities and challenges to fight an antimicrobial-resistant pathogen. *International Journal of Molecular Sciences, 24*(5), 4649–4672.

Sabour, S., Teimourpour, A., Mohammadshahi, J., Peeridogaheh, H., Teimourpour, R., Azimi, T., & Hosseinali, Z. (2022). Molecular detection and characterization of Shigella spp. harboring extended-spectrum β-lactamase genes in children with diarrhea in Northwest Iran. *Molecular and Cellular Pediatrics, 9*, 19–27.

Sadouki, Z., Day, M. R., Doumith, M., Chattaway, M. A., Dallman, T. J., Hopkins, K. L., Elson, R., Woodford, N., Godbole, G., & Jenkins, C. (2017). Comparison of phenotypic and WGS-derived antimicrobial resistance profiles of Shigella sonnei isolated from cases of diarrhoeal disease in England and Wales, 2015. *Journal of Antimicrobial Chemotherapy, 72*(9), 2496–2502.

Salimiyan Rizi, K., Farsiani, H., & Sasan, M. S. (2020). High rate of resistance to ceftriaxone and azithromycin among Shigella spp. isolates at three children's referral hospitals in Northeast Iran. *Journal of Infection and Chemotherapy, 26*(9), 955–958.

Sati, H. F., Bruinsma, N., Galas, M., Hsieh, J., Sanhueza, A., Pardo, P. R., & Espinal, M. A. (2019). Characterizing Shigella species distribution and antimicrobial susceptibility to ciprofloxacin and nalidixic acid in Latin America between 2000–2015. *PLoS One, 14*(8), 1–15.

Seribelli, A. A., Frazão, M. R., Medeiros, M. I. C., & Falcão, J. P. (2016). Molecular and phenotypic characterization of strains of Shigella sonnei isolated over 31 years suggests the circulation of two prevalent subtypes in São Paulo state, Brazil. *Journal of Medical Microbiology, 65*(7), 666–677.

Shahin, K., Bouzari, M., Wang, R., & Khorasgani, M. R. (2019). Distribution of antimicrobial resistance genes and integrons among Shigella spp. isolated from water sources. *Journal of Global Antimicrobial Resistance, 19*, 122–128.

Sharif, N., Ahmed, S. N., Khandaker, S., Monifa, N. H., Abusharha, A., Vargas, D. L. R., Díez, I. T., Castilla, A. G. K., Talukder, A. A., Parvez, A. K., & Dey, S. K. (2023). Multidrug resistance pattern and molecular epidemiology of pathogens among children with diarrhea in Bangladesh, 2019–2021. *Scientific Reports, 13*(1), 13975–13986.

Tang, S.-S., Biswas, S. K., Tan, W. S., Saha, A. K., & Leo, B.-F. (2019). Efficacy and potential of phage therapy against multidrug resistant Shigella spp. *PeerJ, 7*, 1–31.

Toro, C. S., Salazar, J. C., Montero, D. A., Ugalde, J. A., Díaz, J., Cádiz, L. A., Henríquez, T., García, C., Díaz, P., Camponovo, R., Hermosilla, G., & Ulloa, M. T. (2022). Antimicrobial resistance dynamics in Chilean Shigella sonnei strains within two decades: Role of Shigella resistance locus pathogenicity Island and class 1 and class 2 integrons. *Frontiers in Microbiology, 12*, 1–15.

van den Beld, M. J. C., Reubsaet, F. A. G., Pijnacker, R., Harpal, A., Kuiling, S., Heerkens, E. M., Hoeve-Bakker, B. J. A. D., Noomen, R. C. E. A., Hendriks, A. C. A., Borst, D., van der Heide, H., Kooistra-Smid, A. M. D. M., & Rossen, J. W. A. (2020). A multifactorial approach for surveillance of Shigella spp. and entero-invasive Escherichia coli is important for detecting (inter) national clusters. *Frontiers in Microbiology, 11*, 1–15.

Vubil, D., Balleste-Delpierre, C., Mabunda, R., Acácio, S., Garrine, M., Nhampossa, T., Alonso, P., Mandomando, I., & Vila, J. (2018). Antibiotic resistance and molecular characterization of shigella isolates recovered from children aged less than 5 years in Manhiça, Southern Mozambique. *International Journal of Antimicrobial Agents, 51*(6), 881–887.

Wang, F., Xiao, Y., Lu, Y., Deng, Z.-Y., Deng, X.-Y., & Lin, L.-B. (2022). Bacteriophage lytic enzyme P9ly as an alternative antibacterial agent against antibiotic-resistant Shigella dysenteriae and Staphylococcus aureus. *Frontiers in Microbiology, 13*, 1–12.

Wang, Y., Ma, Q., Hao, R., Zhang, Q., Yao, S., Han, J., Ren, B., Fan, T., Chen, L., Xu, X., Qiu, S., & Yang, H. (2019). Antimicrobial resistance and genetic characterization of Shigella spp. in Shanxi Province, China, during 2006–2016. *BMC Microbiology, 19*(1), 116–126.

WHO, (2022). Immunization, vaccines and biologicals: Shigella. : World Health Organization. Retrieved May 18, 2024,from https://www.who.int/teams/immunization-vaccines-and-biologicals/product-and-delivery-research/full-vaccine-value-assessments.

Worley, J. N., Javkar, K., Hoffmann, M., Hysell, K., Garcia-Williams, A., Tagg, K., Kanjilal, S., Strain, E., Pop, M., Allard, M., Francois Watkins, L., & Bry, L. (2021). Genomic drivers of multidrug-resistant Shigella affecting vulnerable patient populations in the United States and abroad. *MBio, 12*(1), 1–12.

Chapter 21
Co-expression of Pfcrt and pfmdr1 Genes in Malaria Patients: Novel Treatment Techniques

Dorcas Yemisi Obafemi, Austine Atokolo, and Paul Akinniyi Akinduti

1 Introduction

Malaria is one of the most devastating public health cases in sub-Saharan Africa, primarily affecting children and pregnant women with a 94% global mortality and morbidity rate (Matrevi et al., 2022). Plasmodium species cause malaria and can be spread or initiated through the bite of an infected female *Anopheles* mosquito (Sato, 2021). The genus "Plasmodium" is known to have over a hundred species, but five are linked to human malaria (*P. falciparum*, *P. malariae, P knowlesi, P. ovale and P. vivax*). Among the five species, *P. falciparum* causes the most malaria-related mortalities. (Culleton et al., 2023). The deadliest malaria parasite known to infect humans is *P. falciparum*, which is also the cause of most malaria deaths, estimated 500,000 or more each year (WHO, 2020). The molecular markers*; Pfnhe1, Pfdhfr, Pfmrp, Pfcrt, Pfmdr1, Pfdhps, Pfk13,* and *PfATpase6* are linked with antimalarial drug resistance. Identifying the molecular markers linked to resistance in the laboratory is necessary to maintain the effectiveness of the current antimalarial drugs (Zhong et al., 2023). Treatment and control strategies against drug-resistant malaria parasites face significant challenges due to their spread, especially those that express both the *pfcrt* and *Pfmdr1* genes (Ocan et al., 2019). The co-expression of *pfcrt* and

D. Y. Obafemi · A. P. Akinduti (✉)
Department of Biological Sciences, College of Science and Technology, Covenant University, Ota, Ogun State, Nigeria
e-mail: yemisi.obafemi@covenantuniversity.edu.ng

A. Atokolo
Department of Biological Sciences, College of Science and Technology, Covenant University, Ota, Ogun State, Nigeria

Covenant Applied Informatics and Communication Africa Centre of Excellence (CApIC-ACE), Covenant University, Ota, Ogun State, Nigeria
e-mail: austine.atokolopgs@stu.cu.edu.ng

P. O. Isibor, P. A. Akinduti (eds.), *Harnessing Biotechnology Tools for Product Development*, https://doi.org/10.1007/978-3-031-86002-7_21

Pfmdr1 among symptomatic children below the age of 5 in sub-Saharan Africa has attracted attention in malaria research in recent years. The control of malaria is significantly hampered by this phenomenon (Silver et al., 2022). Co-expression of *pfcrt* and *Pfmdr1* are linked to resistance to antimalarial medications such as artemisinin-based combination therapies (ACTs). Frontline malaria treatments are less effective because of this dual resistance mechanism, which raises treatment failure rates as well as the morbidity and mortality rates of affected children (Silva et al., 2022). Malaria remains a major threat to public health due to its high prevalence. Malaria caused about 95% of cases and 96% of mortality within the sub-Saharan region in 2022. Almost half of all death cases from malaria occur in children below 5 years of age. Resistance to antimalarial drugs, especially with artemisinin-based combination therapy, and the need for more effective vaccines against malaria are among the problems that require attention (Oladipo et al., 2022). Sub-Saharan Africa has a high malaria prevalence due to several factors, including poverty, poor access to healthcare, a climate that is conducive to mosquito breeding, and antimalarial medication and insecticide resistance. In sub-Saharan Africa, efforts to control and eradicate malaria have made headway, but obstacles still need to be overcome. This highlights the significance of ongoing interventions and giving research priority (Habtamu et al., 2022).

Although malaria-related morbidity and mortality have recently decreased, 241 million cases and over 627 thousand fatalities were predicted for 2020; in the African region, 95% and 96% represent the evaluated global cases and deaths, respectively (WHO, 2020). In 2020, 77% of malaria deaths occurred worldwide in the most susceptible individuals (children below the age of 5) (WHO, 2021). The global endeavor to manage and abolish malaria is hindered by the rise and dissemination of drug-resistant *Plasmodium falciparum* (Su et al., 2019). For malaria to be controlled and eradicated, understanding the molecular processes by which the malaria parasite develops multidrug resistance is essential. The *Pfmdr1* and *Pfcrt* mutations, as well as the amplification of *Pfmdr1,* can change the susceptibility of the parasite to a variety of medications, including most antimalarial drugs currently in use and a few antimalarial candidates that are still in the development stage (Martin, 2020). The co-expression of the *pfcrt* and *Pfmdr1* genes during malaria treatment, particularly among symptomatic children below the age of 5, has a significant impact on understanding drug resistance patterns and treatment outcomes. These genes are essential for malaria caused by *Plasmodium falciparum* and are linked with drug resistance against malaria (Shafik et al., 2020).

2 Epidemiology of *Plasmodium falciparum*

Among the five medically important species of *Plasmodium* that are linked to human malaria, falciparum is the most frequent and deadly species, causing more than 90% mortality worldwide (Alemayehu, 2023). The most recent global report on malaria indicates there have been 249 million cases of malaria in 2022, contrary to 244 million in 2021 (Dwumfour et al., 2023). Malaria mortality cases were assumed to have decreased from 610,000 in 2021 to 608,000 in 2022. In 2022, the WHO Region recorded 94% morbidity and 95% of deaths. Out of all malaria deaths worldwide, slightly more than half occurred in four African countries, with Nigeria having the highest prevalence. The three other countries are the Democratic Republic of Congo, Mozambique, and Uganda. According to WHO 2019, reports estimated 228 million malaria cases worldwide in 2018, with a progressive decline in cases since 2010. These cases are thought to have killed 405,000 people, of whom 272,000 (or 67%) were children under the age of 5. In 2018, about 85% of malaria cases worldwide were reported from 19 sub-Saharan African countries and India. Children under the age of 5 are more susceptible to malaria in high-transmission areas, where they account for 67% of all malaria deaths globally. People of all ages are susceptible to low immunity in low transmission areas (Zekar & Sharman, 2023).

The prevalence of malaria is thought to be at least 50% among Nigerians, with children under the age of 5 typically suffering two to four episodes annually (Adamu et al., 2020). The *P. falciparum* is permanently spread across all regions of the country, although it is more prevalent during the rainy than the dry season (Tuedom et al., 2021). Compared with the southern parts of the country, malaria has so far been identified in higher proportions in North-West and North-East Nigeria, majorly because of climatic and environmental factors (Adamu et al., 2020). The *P. falciparum* transmission is much more common in the nation's North-West area than in other parts, such as North-East Nigeria (Muhammad et al., 2023).

Patients with minimum or completely compromised immunity are susceptible to developing severe malaria. Children below the age of 5 are the primary victims of severe malaria in regions of the world where *P. falciparum* is highly transmissible. This is because older children and adults who have repeated infections develop specific acquired immunity over time, which offers heightened protection, albeit insufficient, against the disease. However, people without immunity (such as those who travel to endemic areas) and those regions with little or unstable transmission rates can contract severe malaria at any age (Varo et al., 2020). The probability of contracting malaria is very high during pregnancy (Zekar & Sharman, 2023). While age and gravidity can also have an impact, the second trimester of pregnancy is the highest risk period for infection, with young primigravidae in high-transmission areas being the most vulnerable (Varo et al., 2020). In areas where malaria transmission is either minimal or unstable, women are more vulnerable to severe malaria and mortality resulting from infrequent infection and low immunity (Zekar & Sharman, 2023).

2.1 Plasmodium falciparum *Life Cycle*

The malaria parasite's complex and multi-stage life cycle occurs within two biological organisms: humans, rats, and chimpanzees as the host organisms and mosquitoes as the vectors (Rout & Mahapatra, 2019). A human is bitten by a mosquito that carries the Plasmodium parasite, injecting the sporozoite, the motile infectious stage of the parasite, into the skin. Then, these sporozoites travel into the circulation for a blood vessel to feed from (Nureye & Assefa, 2020). The sporozoites release vasodilators to enlarge and make blood vessels more accessible, increasing their chances of finding a suitable artery. To keep the blood flowing continuously and provide the sporozoites with food, the parasite also secretes saliva into the bloodstream to inhibit clotting. The sporozoites, which resemble microscopic threads, go toward the liver after they are within the blood vessels (Gazzinelli et al., 2014).

The merozoites enter the red blood cell (RBC) to expand by destroying hemoglobin, which starts the asexual cycle. When the parasite reaches the late trophozoite stage, it divides by mitosis until it reaches the schizont stage, which contains about 6–32 merozoites (Nureyev & Assefa, 2020). Up to this point, the parasite develops inside the host red blood cell (RBC). When the erythrocytic schizont bursts, the liberated merozoites complete their life cycle by attacking other red blood cells. Cyclic fevers usually occur just before or concurrently with red blood cell lysis as schizonts break to release new pathogenic merozoites. Tertian malaria infection occurs every 48 h, and quartan malaria infection occurs every 7 h. During this cycle, some merozoites go through sexual differentiation, thereby becoming erythrocytic gametocytes, which have a single nucleus and are then prepared to mate with a female *Anopheles mosquito seeking blood (Nureyev* and Assefa, 2020). The ingestion of gametocytes by mosquitoes causes gametogenesis. The macrogametes that produce zygotes are penetrated or fertilized by the exflagellated forms of microgametes. Before developing into a spherical oocyst, the zygotes become ookinetes. The nucleus of the oocyst divides multiple times, producing sporozoites in large numbers and growing the oocyst. When the oocyst explodes after reaching adulthood, the sporozoites are released into the internal cavity of the mosquito known as the hemocoel. The sporozoites enter the salivary glands to complete their life cycle. The malaria life cycle is sustained by the salivary gland sporozoites of the mosquito, which infect a fresh human host (Nureye and Assefa, 2020).

2.2 Classes of Antimalarial Drugs

Currently, the primary means of treating malaria is pharmacotherapy, which involves taking various medications that target different Plasmodium intracellular targets. Artemisinin derivatives, folic acid antagonists, quinoline derivatives, and antibiotics currently treat malaria (Zhong et al., 2023). Choosing effective antimalarial medications reduces the illness's duration and the risk of complications, mortality, and transmission. A wide range of drugs are available; however, the best option depends on the immune status of the patient as well as the intracellular targets of the drug (Morales-Luna, 2024).

2.2.1 Quinoline Derivatives

Derivatives of quinoline are chloroquine, quinine, primaquine, amodiaquine mefloquine, piperaquine, lumefantrine, and tafenoquine. They can be broadly grouped into four categories: 4-aminoquinolines (chloroquine, piperaquine, pyronaridine, and amodiaquine, as shown in Fig. 21.1), Arylamino alcohols (lumefantrine, mefloquine, and quinine), and 8-aminoquinoline (primaquine and tafenoquine) (Zhong et al., 2023). The antiparasitic properties of 4-aminoquinolines are exhibited by binding to hemoglobin or heme and obstructing detoxification of heme, which occurs within the food vacuoles of the parasite. Although this is questioned, lumefantrine and mefloquine are assumed to be able to prevent hemoglobin endocytosis. The complex mechanism of action of quinine most likely entails inhibiting polymers from being taken out of the food vacuole and moving them into the cytoplasm, which is the place for the formation of hemozoin (Zhong et al., 2023).

The term "Cinchona tree" refers to a native Andean tree from which quinine was extracted and utilized for centuries. For many years (between the 1940s and the late 1980s), chloroquine was the first-line antimalarial drug used because of its effectiveness, safety, and affordability.

Due to *P. falciparum*'s high level of resistance to chloroquine, regardless of its affordability, it is no longer a commonly used drug (Rout & Mahapatra, 2019). Malaria can be prevented in pregnant women by using mefloquine rather than chloroquine. Before the resistance of mefloquine emerged, it was first introduced in the 1970s and was used for 30 years (Zhong et al., 2023).

2.2.2 Folic Acid Antagonists

In the biosynthesis pathway of folate, the two essential enzymes involved are dihydrofolate reductase (DHFR) and dihydropteroate acid synthase (DHPS) (Rout & Mahapatra, 2019). In the host environment, the antagonistic relationship between biochemical metabolism and these enzymes results in inhibiting their activities, which in turn decreases the growth of gametophytes and the proliferation of Plasmodium sporozoites. Trimethoprim and pyrimethamine both inhibit the DHFR. One popular antimalarial medication is Sulfadoxine-Pyrimethamine (SP).

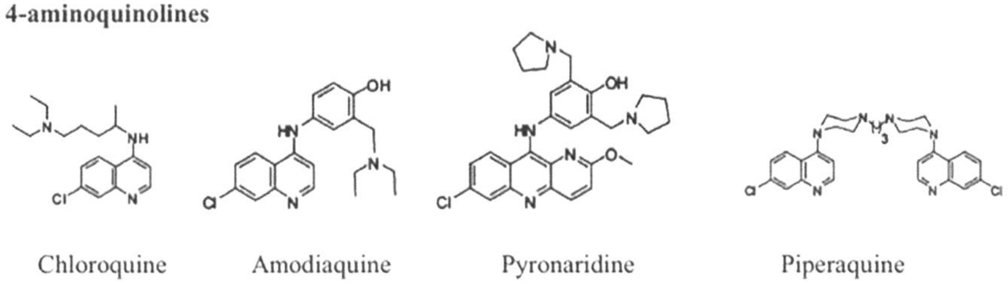

Fig. 21.1 Chemical structure of the 4-aminoquinoline. Source: Rout & Mahapatra (2019)

In regions of Africa where malaria prevalence is high, SP is currently advised for use as a preventive measure in pregnant women, newborns, and children (Desai et al., 2018). Examples of antifolates are proguanil, pyrimethamine, and sulfadoxine (Fig. 21.2).

2.2.3 Naphthoquinone

Atovaquone is a naphthoquinone antimalarial drug class member, which shares structural similarities with ubiquinol (Esser & Xia, 2024). Proguanil, also known as Malarone, is advised to be taken with Atovaquone (Blanshard & Hine, 2021). In the southeast of Thailand, this combination of medications was highly recommended. However, its mode of action and combination with proguanil are not entirely understood. A few studies revealed that Atovaquone mimics the natural substrate Ubiquinone by inhibiting the respiratory chain of malarial mitochondria at the cytochrome bc1 complex (Achan et al., 2018). Atovaquone prevents the proliferation of parasites that target a particular Plasmodium target by depolarizing mitochondria and inhibiting the transfer of mitochondrial electrons from ubiquinol to cytochrome c.

2.2.4 Artemisinin and Its Derivatives

Artemisinin (ART) molecules are unique among sesquiterpene lactone peroxides because they consist of peroxy-bridge groups (Rout & Mahapatra, 2019). A molecule known as heme-artemisinin is formed upon combining artemisinin and heme. Through homologous breaking of the hydrogen peroxide link, alkoxy radicals rearrange into alkyl radicals. In this scenario, the internal alkyl radical may alkylate the various proteins implicated in the pathways that lead to the death of the parasite by degrading hemoglobin, enhancing protein synthesis, promoting antioxidant defense, and accelerating Plasmodium glycolysis (Nsanzabana, 2019). Dihydroartemisinin (DHA), Artesunate (AS), and Artemether (ATM) are a few of the derivatives that have been created and are commonly used to treat infections that are resistant to

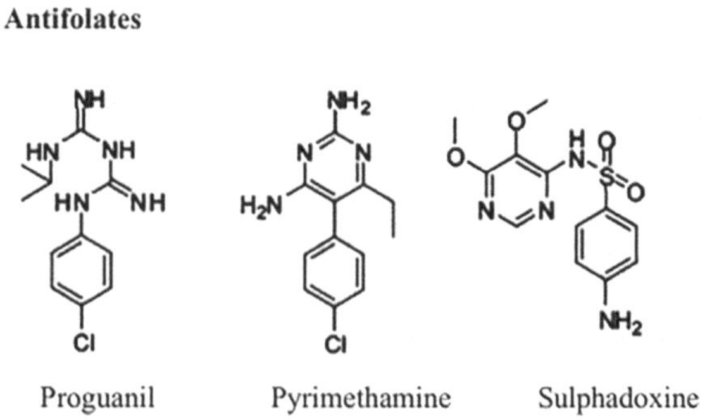

Fig. 21.2 Chemical structure of antifolates. Source: Rout and Mahapatra (2019)

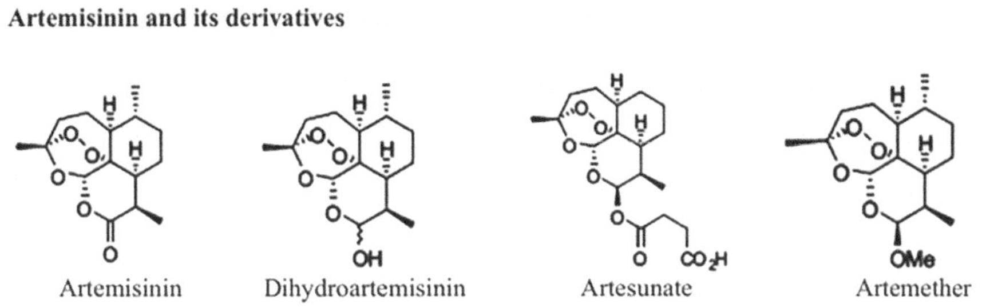

Fig. 21.3 Structure of artemisinin and its derivatives. Source: Rout and Mahapatra (2019)

other medications as well as less critical cases of malaria, such as cerebral malaria (Efferth & Oesch, 2021). Artemisinin, including its partner drugs, possesses a short half-life, and that has led to combining it with a second drug with a more prolonged half-life. Artesunate-lumefantrine, artesunate-pyronaridine, artesunate-mefloquine, artesunate-amodiaquine, and artesunate-sulfadoxine-pyrimethamine are the six first-line ACTs currently recommended by WHO artesunate-mefloquine and dihydroartemisinin-piperaquine have been the most widely used artemisinin-based medications in Southeast Asia (SEA) while artesunate-lumefantrine and artesunate-amodiaquine are the most widely used ACTs in Africa. Many countries swiftly adopted ACTs after the WHO recommendation in 2000 to be used as the first-line treatment for *P. falciparum* malaria. As a result, malaria epidemics significantly decreased (van der Pluijm et al., 2021). Figure 21.3 illustrates the chemical structure of artemisinin and its derivatives (dihydroartemisinin, artesunate, and artemether).

2.3 Artemisinin-Based Combination Therapies (ACTs)

The World Health Organization guidelines for treating malaria states that combinations of antimalarial drugs should be used to treat *Plasmodium falciparum*-caused malaria because *P. falciparum* resistance to monotherapies is becoming a greater threat (Efferth & Oesch, 2021). ACTs have taken the place of other malaria treatments as the primary treatment in nearly all endemic regions. After a 3-day ACT regimen, artemisinin derivatives have a strong and quick-acting antimalarial action; nevertheless, they require to be taken with partner medications that have a longer half-life to eliminate any remaining parasite (Nsanzabana, 2019).

2.3.1 Artemether-Lumefantrine

Artemether-lumefantrine combination therapies are commonly used to treat most uncomplicated cases of malaria caused by *Plasmodium falciparum* and *Plasmodium vivax* (Kilonzi et al., 2019). The precursor to artemether is a naturally occurring chemical called artemisinin, which is obtained from the sweet wormwood plant (*Artemisia annua*) (Verma et al., 2020). Artemether obstructs the parasite's ability to synthesize proteins and metabolize hemoglobin. Within the parasite, it produces reactive free radicals that damage the cell membranes and eventually cause the parasite to die. Lumefantrine functions by preventing the production of β-hematin, a hazardous consequence resulting from hemoglobin breakdown in the digestive vacuole (DV) of the parasite. Lumefantrine interferes with the ability of the parasite to detoxify heme by blocking the formation of this toxic compound, ultimately resulting in the parasite's death (Lopes et al., 2021). A fixed-dose combination therapy, that is, combining artemether and lumefantrine, is intended to target distinct stages of the parasite lifecycle, thereby reducing the risk of drug resistance and producing synergistic effects (Kilonzi et al., 2019). The malaria pigment known as hemozoin, which is produced during hemoglobin breakdown, cannot be detoxified by the binding of lumefantrine to hemin. In the process, harmful free radicals are released from artemether when the peroxide group binds to heme (Mishra et al., 2021).

2.3.2 Artesunate-Amodiaquine

A novel fixed-dose combination of amodiaquine (AQ) and artesunate (AS) is now available to treat mild falciparum malaria cases. It contains amodiaquine and artesunate, two potent compounds that work together to combat the parasite (Lingani et al., 2020). Artesunate primarily interferes with the process involved in heme detoxification in the DV, targeting the asexual blood stages of the parasite life cycle. This results in oxidative damage to the parasite's membranes and the buildup of harmful heme species (Ashraf, 2021). Additionally, it disrupts the mitochondrial function of the parasite (Lingani et al., 2020). In the parasite's digestive vacuole, amodiaquine, a 4-aminoquinoline molecule, prevents harmful heme species from polymerizing into hemozoin. The accumulation of dangerous heme makes it difficult for the parasite to perform its metabolic functions (de Villiers & Egan, 2021). Amodiaquine may affect the function and structure of the parasite's membranes and impede DNA replication and enzyme activity (Belete, 2020). When combined, these mechanisms help AS-AQ treat malaria effectively and lower the chance of drug resistance.

2.3.3 Dihydroartemisinin-Piperaquine (DHA-PPQ)

Plasmodium falciparum-induced malaria is treated with a combination therapy called dihydroartemisinin-piperaquine (DHA-PPQ). Dihydroartemisinin (DHA) and piperaquine are its two active ingredients (Cheviet et al., 2019). A derivative of artemisinin, dihydroartemisinin inhibits protein synthesis and causes oxidative damage to the malaria parasite during its asexual blood stages. It quickly lowers the parasite load by mainly targeting the immature, ring-shaped parasites (Rovira-Vallbona, 2020). Conversely, piperaquine prevents the parasite from producing β-hematin in its digestive vacuole, hindering its capacity to detoxify heme (Rosenthal & Ng, 2020). This ultimately results in the parasite's death because of oxidative damage to its membranes and the accumulation of toxic heme species. In combination, piperaquine and dihydroartemisinin can target different stages of the parasite life cycle and reduce the possibilities of drug resistance (de-Villiers & Egan, 2021). Due to its high efficacy, DHA-PPQ has been widely utilized in regions where malaria is resistant to drugs.

3 Resistance to Antimalarial Drugs

Following the WHO definition of drug resistance, it is a phenomenon that describes the ability of a parasite to withstand drugs and their toxic effects when administered and absorbed, regardless of the sufficiency of the drug at the site of action. The emergence of resistance to malarial medications implies the choice of parasites with gene amplification or mutations that confer reduced susceptibility (Wicht et al., 2020). Resistance to drugs in organisms seems to be caused by changes in the quantity, structure, or composition of proteins. Two genetic changes mediating the protein change are single nucleotide polymorphisms (SNPs) and gene amplification (Hodoameda, 2021). Currently, resistance to antimalarial therapy is the most significant barrier to its efficacy, so researchers are always searching for ways to mitigate it. New molecules must always be produced because there is no assurance that resistant parasite strains will not emerge. New drugs with distinct mechanisms of action are undergoing clinical trials (Kumar et al., 2018).

There are various reasons why resistance to currently available antimalarial medications can develop. Factors like the total parasite load, the rate of medication effectiveness, drug compliance and poor adherence to malaria protocol treatment are also to be considered (Wicht et al., 2020). False medications, inadequate pharmacokinetic properties, and improper dosage lead to insufficient drug exposure to parasites (Muller, 2011). Antimalarials of poor quality or those that have been fabricated and lack active pharmaceutical ingredients (APIs) increase the risk of hyperparasitemia, recrudescence, and hyper-gametocytopenia, which can enhance resistance. Halofantrine is an inappropriate API utilized when it is used in place of artemisinin since it is invisible to investigators but visible to parasites through chemical analysis (Tobing et al., 2024) (Table 21.1).

3.1 Role of Pfmdr1 and Pfcrt *in Antimalarial Drug Resistance*

Molecular markers are employed to monitor the spread of drug resistance to antimalarial drugs. It offers data on SNPs and copies of variations of genes implicated in parasite drug resistance. Utilizing molecular markers to determine drug-resistant

Table 21.1 Highlights of antimalarial compounds in their various mechanisms of resistance

Classes of antimalarials	Examples	Mode of action	Druggable targets	Resistance mechanism	Reference
Antifolates	Pyrimethamine and sulfadoxine	Inhibition of dihydrofolate reductase (DHFR) and dihydropteroate synthase (DHPS)	Cytosol	Mutation in *dhfr* and *dhps* gene	Hodoameda, (2021)
4-Aminoquinolines	Amodiaquine, chloroquine, hydroxychloroquine	Inhibit heme polymerization to hemozoin	Digestive vacuole	Elevated hemoglobinase expression in the food vacuole, including plasmepsin and falcipain	Zhong et al. (2023)
Naphthoquinones	Atovaquone	Interferes with electron transport in the mitochondria	Mitochondria	A solitary point variation in the bc1 complexes cytochrome b subunit	Achan et al. (2018)
Artemisinin	Artesunate	Alkylation of lipids and proteins	Vesicular structures	K13 mutation	Nsanzabana, (2019)
Antibiotics	Clindamycin and doxycycline	Hinder the apicoplast's ability to translate proteins	Apicoplast	Point mutation is noticeable in the apicoplast 23s Rrna	Yadav et al., (2021)

malaria would trace the profile of the disease and the emergence of varied genetic strains carrying the genetic mutation (Qidwai, 2020).

3.2 **Plasmodium falciparum** ***Chloroquine Resistance Transporter (*Pfcrt*)***

The *Plasmodium falciparum* chloroquine resistance transporter (*pfcrt*) is a drug transporter gene. It contains 424 amino acids, with 10 α-helical transmembrane domains within the digestive vacuole (DV) membrane (Roux et al., 2021). Thirty-two candidate codons within the structure are thought to be capable of point mutations that alter the *pfcrt* function (Madhav & Hoda, 2021). Most of them occur at the location facing the DV media, and Unlike P-glycoprotein molecules, ATP activation is not necessary for substrate binding to *pfcrt* (Fontecha et al., 2021). Certain ideas propose that *pfcrt* functions in the efflux of compounds that are produced when globin is broken down in the vacuoles, including divalent cations, amino acids, and peptides (Qidwai, 2020). It is also concluded to have some role in maintaining H+ homeostasis. When PfCRT is altered, the PfCRT gene accumulates in *P. falciparum*'s digestive vacuoles (DVs). In contrast to parasites expressing the wild type of PfCRT, these parasites exhibit a significantly lower chloroquine accumulation. Therefore, active transport enables drug export by chloroquine-resistant parasites, indicating that the wild and mutant types of *PFCRT* have different drug-transporting abilities.

The main mechanism responsible for chloroquine resistance is mutations in the *pfcrt*. This was shown by a genetic cross-experiment utilizing Dd2 clones that were immune to CQ and HB3 clones that were sensitive to it. A genetic study of the CQ-resistant progeny found variation in a single genetic locus on chromosome 7. A 13-exon *pfcrt* gene mutation was discovered using quantitative trait loci (QTL) research (Posayapisit et al., 2023). A genome-wide related study established that resistance to chloroquine is linked with a mutation in the *pfcrt* gene (Mu et al., 2010). One mutation that modifies an amino acid from K76T makes *P. falciparum* strains isolated from the field, and those isolated from the field and adapted in the laboratory are resistant to CQ. The chloroquine mechanism of resistance is the expulsion of deprotonated CQ from the DV, which happens when a neutral threonine (T) is substituted for a positively charged lysine (K). Facilitated diffusion or active transport are the two techniques employed for expulsion. As a result, the CQ has less access to heme, its intended target (Hodoameda, 2021).

A few *pfcrt* gene mutations add distinct amino acids to the wild-type amino acids CVMNK to compensate for the *pfcrt* K76T mutation's reduced ability to function as *Pfcrt*. This might, therefore, have an impact on the sensitivity of the parasite to drugs. These mutations are close to the K76T positions 72–76. The variations at locations 72–76 may be specific to a region. On the other hand, SVMNT mutations at positions 72–76 are found in Papua New Guinea, South America, and the

Philippines. In contrast, CVIET mutations at the same places are usually seen in African and Southeast Asian parasites (Hodoameda et al., 2022). Using *pfcrt* K76T mutations in epidemiology surveillance, additional topics covered include chloroquine resistance and specific partner drugs used in artemisinin-based combination therapy. An instance of this can be observed in the reduced vulnerability to amodiaquine and its principal metabolite, desethylamodiaquine (DEAQ), upon the insertion of mutant *pfcrt* into the CQ-sensitive GC03 strain (Hodoameda et al., 2022). Studies using field isolates provided evidence for selecting *pfcrt* K76T in the amodiaquine's regressed treatment outcome. Posayapisit et al. (2023) state that co-expression *pfcrt* and *Pfmdr1* mutation most likely cause parasite resistance to AQ or DEAQ.

The *pfcrt* K76T mutation causes resistance to both AQ and CQ in addition to elevated susceptibility to lumefantrine (Posayapisit et al., 2023), quinine, halofantrine, mefloquine, artemisinin, and its derivatives (Hodoameda, 2021).

Resistance to amodiaquine in *P. falciparum* is linked to variations in the *pfcrt* gene (Wicht et al., 2020). The *P. falciparum* chloroquine resistance transporter (PfCRT) artemisinin resistance is caused by an unclear mechanism. However, recent research suggests that point mutations in *Pfcrt* and *Pfmdr1* may be connected to resistance to artemisinin-based combination treatments (ACTs) (Wicht et al., 2022). Decreased heme iron produces free radicals that harm the cell membrane and other organelles, mediating the cleavage of the endoperoxide bridge and activating artemisinin (Noreen et al., 2021).

3.3 Plasmodium falciparum *Multidrug Resistance Transporter 1* (Pfmdr1)

Plasmodium falciparum multidrug resistance protein 1 is a member of the protein family called ATP-binding cassette (ABC) proteins, sometimes referred to as P-glycoproteins homolog 1 (*Pgh-h*) (Hodoameda, 2021). According to Bakari et al. (2024), the *Pfmdr1* transporter controls the buildup of drugs in the parasite's DV. Regarding the parasite response to various antimalarial medications, the *Pfmdr1* is crucial. Increased *Pfmdr1* copy number or gene mutations are the two ways the *Pfmdr1* gene controls the response to antimalarial drugs (Sato, 2021). Lower in vitro vulnerability to artesunate, mefloquine, quinine, halofantrine, and dihydroartemisinin has been associated with a higher copy number of *Pfmdr1* (Patel et al., 2023).

The cause of the artesunate-mefloquine combination or mefloquine monotherapy treatment failures is increased *Pfmdr1* copy number in clinical isolates (Imwong et al., 2021). The drug-resistant FCB strain was exposed to mefloquine, quinine halofantrine, lumefantrine, and artemisinin in an experiment in which one of the two copies of the strain was deleted. This resulted in the strain's reversal resistance. It was demonstrated that a higher *Pfmdr1* copy number is significant and plays a part in quinine, halofantrine, lumefantrine, mefloquine, and artemisinin resistance

(Hodoameda, 2021). The universal importer *P. falciparum* multidrug resistance transporter 1 (*Pfmdr1*) stores medications and possibly hazardous byproducts inside the parasite's digestive vacuole (Sato, 2021). *Pfmdr1* modifies pH and intracellular ions, indirectly affecting drug flux (Patel et al., 2023). Resistance to antimalarial drugs is caused by polymorphisms found in distinct haplotypes of *Pfmdr1.* These mutations change the substrate selectivity of *Pfmdr1* (Imwong et al., 2021). Treatment failure for both AQ and CQ has been linked to the *Pfmdr1* N86Y mutation; however, the correlation with CQ is weak (Hodoameda, 2021). It has been reported that the *Pfmdr1* D1246Y is involved in AQ/DEAQ resistance. In East Africa, the haplotype *Pfmdr1* 86Y-184Y-1246Y was selected based on the results of AQ recrudescence treatment. In other research, high AQ IC50 for parasites with *Pfmdr1* D1246Y 9 was observed using field isolates from Columbia (Bakari, et al., 2024). Lumefantrine resistance in Africa is linked to the *Pfmdr1* N86-F184-D1246 haplotype, while isolates from the Thai Myanmar border were linked to the *Pfmdr1* N1042D haplotype with higher in vitro lumefantrine IC50 values. Reduced quinine susceptibility is linked to the *Pfmdr1* S1034C/N1042D/D1246Y mutations (Patel et al., 2023). Higher resistance to AQ and DEAQ may be conferred by the interaction of the *Pfmdr1* and *pfcrt* alleles (Imwong et al., 2021). The mechanism by which *Pfmdr1* develops resistance to amodiaquine is due to mutations in the *Pfmdr1* gene that affect the transport of antimalarial medications. Studies have demonstrated that reduced susceptibility to amodiaquine and other antimalarial drugs is linked to mutations in *Pfmdr1,* such as N86Y and D1246Y (Wicht et al., 2020). The decreased efficacy of the medication is likely caused by these mutations, which block the transport of antimalarials from the cytosol to the digestive vacuole.

The process by which the *Plasmodium falciparum* multidrug resistance gene (*pfmdr1*) is resistant to mefloquine is due to *pfmdr1* gene amplification, which causes the resident DV membrane transporter to be overexpressed (Su et al., 2019). The *pfmdr1* overexpression causes the parasite to sequester more drugs, exacerbating multidrug resistance phenotypes (Qidwai, 2020). The mechanism of the *Plasmodium falciparum* multidrug resistance gene (*pfmdr1*) in lumefantrine resistance is influenced by several factors. Lumefantrine resistance is one of the antimalarials to which the *pfmdr1* gene contributes to P-glycoprotein resistance. The number of copies of the gene, amino acid substitutions like N86Y, and single nucleotide polymorphisms (SNPs) all affect resistance (Silva et al., 2022). Lowering the copy number of *Pfmdr1* makes parasites more susceptible to lumefantrine and other antimalarial drugs (Qidwai, 2020). Mutations in *Pfmdr1* cause a reduction in piperaquine sensitivity, which is the mechanism of resistance of *Pfmdr1* to dihydroartemisinin-piperaquine (Wicht et al., 2020). Studies have demonstrated that wild-type *Pfmdr1* transports both piperaquine and dihydroartemisinin, and that *Pfmdr1* mutations contribute to piperaquine resistance (Shafik et al., 2022).

4 Co-expression of *pfcrt* and *pfmdr1*

The emergence of resistance to antimalarial medications is one of the main obstacles in the battle against malaria in sub-Saharan Africa (Oladipo et al., 2022). Resistance to several antimalarial medications has emerged throughout the region, notably chloroquine, sulfadoxine-pyrimethamine, and artemisinin-based combination treatments (ACTs) (Blasco et al., 2017). Artemisinin resistance cases have been reported in several African regions, including Rwanda, Uganda, and the Greater Mekong subregion (Rosenthal et al., 2024). Since ACTs are presently the first-line treatment for uncomplicated *P. falciparum* malaria, there is rising concern about the development of ACT resistance (WHO, 2022). Although treatment failure due to artemisinin resistance alone is rare, high rates of treatment failure can result from resistance to both artemisinin and the partner drug in ACT drug regimens (Koehne et al., 2021).

The presence of mutations in the *pfcrt* and *pfmdr1* genes simultaneously in the same parasite is referred to as the co-expression of *pfcrt* and *pfmdr1* mutations. The co-expression of these genes can result in a higher resistance level to amodiaquine, chloroquine, and their metabolites (Silva et al., 2022). A study conducted in Southeast Nigeria assessed the frequency of *Pfcrt* and *Pfmdr1* alleles in isolates of *P. falciparum* 11 years following the cessation of chloroquine treatment. The study found that while *Pfmdr1* mutant genotypes are rare, *Pfcrt* mutant genotypes and haplotypes are highly prevalent (Ikegbunam et al., 2019). In vivo artemether-lumefantrine (AL) therapy has co-expressed *pfmdr1* and *pfcrt* in *Plasmodium falciparum*. The study discovered that the two genes had a strong positive correlation, indicating that they are functionally related to the parasite's typical physiology (Silva et al., 2022).

Co-expression of the *pfcrt* and *Pfmdr1* genes during malaria treatment in sub-Saharan Africa, mostly among symptomatic children under 5 years of age, has a significant impact on understanding drug resistance patterns and treatment outcomes. Regarding *Plasmodium falciparum* infections, these genes are important because they are associated with resistance to antimalarial medications (Silver et al., 2022). *Pfmdr1* and *pfcrt* genes significantly co-expressed before and during Artemether-Lumefantrine (ACT) treatment, according to research carried out in Tanzania on patients receiving this medication. The research discovered a positive correlation between these genes' expression levels, suggesting a physiological relationship (Silva et al., 2022). The co-expression that persisted throughout treatment indicated their important roles during drug stress in the parasite (Cheng et al., 2021).

The crucial mutation K76T that causes chloroquine resistance, associated with the *Pfcrt* gene, affects parasite resistance to a range of antimalarial medications (Shafik et al., 2020). Conversely, *Pfmdr1* plays a role in controlling drug sensitivity or tolerance to various antimalarial medications. If these genes are mutated, treatments like artemisinin-based combination therapies may be less effective (ACTs) (Shafik et al., 2022). Mutations prevalence in these genes can reveal how well antimalarial medications such as artemether-lumefantrine, amodiaquine, mefloquine, and chloroquine work in the area (Shafik et al., 2020). To guarantee successful treatment plans for symptomatic children under 5 years old, who are most susceptible to severe malaria, it is imperative to monitor these genetic markers (Cheng et al., 2021). To combat drug resistance and guide treatment protocols in malaria-endemic areas like sub-Saharan Africa, ongoing molecular surveillance of *pfcrt* and *Pfmdr1* mutations is imperative (Shafik et al., 2022).

Due to their functions as transmembrane transporters in *Plasmodium falciparum* parasites, *pfcrt*, and *Pfmdr1* expression levels influence resistance to antimalarial drugs (Shibeshi et al., 2020). The *Pfcrt* and *Pfmdr1* genes are important digestive vacuole proteins connected to sensitivity to various antimalarial drugs. When these genes are expressed more frequently, *P. falciparum* may exhibit drug-resistance traits (Yobi et al., 2021). The *Pfmdr1* resistance to several medications, including mefloquine, has been associated with elevated expression levels frequently caused by increased gene copy numbers; reduced medication efficacy may arise from this increased expression, as in the case of artesunate (Simon et al., 2022). Conversely, *Pfcrt* mutations impact parasite resistance to medications such as chloroquine and artemisinin compounds, such as the critical chloroquine resistance mutation K76T (Kim et al., 2019).

The *Pfcrt* gene, a pivotal component located in the digestive vacuole (DV) of Plasmodium falciparum, is responsible for transporting hemoglobin-derived peptides into the cytosol chloroquine builds up in sensitive strains of hemoglobin degradation products (DV), attaches itself to free heme, and prevents polymerization into non-toxic hemozoin (Heller & Roepe, 2019). Mutations such as K76T in resistant strains enable *pfcrt* to efflux chloroquine from the DV (Shafik et al., 2020). Efflux' refers to actively pumping a drug out of the cell, reducing its concentration at the heme target (Rout & Mahapatra, 2019). This efflux decreases the ability of the drug to interfere with heme detoxification, enabling the parasite to withstand the effect of antimalarials.

The *Pfmdr1*, an ATP-binding cassette (ABC) transporter in the digestive vacuole of *Plasmodium falciparum*, plays a crucial role in modifying how drugs are distributed between the DV and cytosol (Hodoameda, 2021). This function, which significantly affects drug susceptibility, is a key area of interest in the study of drug resistance. Mutations in *Pfmdr1* can influence the efficiency of drug transport into or out of the DV, further highlighting its importance and the impact of these mutations on drug resistance (Sato, 2021).

The *Pfmdr1*, a versatile player, not only exports harmful metabolites or drug molecules from their sites of action but also facilitates the entry of different antimalarials into the DV. This dual function, a remarkable adaptation, helps to maintain an equilibrium between detoxifying procedures that protect the parasite from drug-induced toxicity and drug accumulation required for efficacy (Shibeshi et al., 2020). Understanding this balance is essential in the fight against drug resistance. The co-expression of *Pfcrt* and P*fmdr1* significantly enhances the parasite's ability to withstand multiple medications simultaneously (Gibbons, 2019). Research has shown that these transporters can functionally interact, with variations in one potentially affecting the functionality of the other. This co-expression is a key factor in the complex network of drug resistance mechanisms in *Plasmodium falciparum.*

The *Pfcrt* and *Pfmdr1* co-expression is crucial in regulating drug response and resistance mechanisms. Research has demonstrated a strong relationship between these genes' expression levels, especially before and during artemether-lumefantrine (AL) administration. When exposed to drug stress, this co-expression pattern emphasizes their interdependent roles in the parasite's physiology (Shibeshi et al., 2020). Generally, the interaction between the expression levels of *pfcrt* and *Pfmdr1* affects drug transport and sensitivity mechanisms within the parasite's cellular environment, affecting the parasite's resistance to different antimalarial drugs (Silva et al., 2022).

The co-expression of these genes (*pfcrt and pfmdr1)* is not a product of genetic makeup; it is a strategic adaptation mechanism employed by *Plasmodium falciparum* to withstand drug pressure (Shafik et al., 2020). According to Silva et al. (2022), there is some correspondence in the levels of expression during artemisinin-based combination therapy treatment, which implies that they interact to optimize the response of the drug, which enables a balance between resistance and susceptibility. Co-expression of *pfcrt* and *pfmdr1* could increase resistance levels due to the compensatory action of mutation in the two genes (Shafik et al., 2020). Some mutations in *pfcrt* and *pfmdr1* may counteract each other's effects, reducing the overall resistance levels or altering drug susceptibility (Hodoameda et al., 2022). Complex drug response patterns might result from mutations in one gene that provide resistance to a particular treatment while increasing susceptibility to the same drug or vice versa in another gene (Silva et al., 2022).

5 Recent Strategies to Combat Antimalaria Resistance

Current trends in combating resistance to antimalarial drugs involve various chemoprevention techniques. These include repeated first-line treatments, triple combination therapies, and artemisinin-based combination treatments (ACT) suggested by the WHO for uncomplicated *P. falciparum malaria* (WHO, 2020). However, when the efficacy of a particular ACT drops below 90%, the WHO advises switching to an alternate ACT with over 95% efficacy (Rathmes et al., 2020). These strategies are crucial in the fight against drug resistance and highlight the importance of understanding the underlying mechanisms.

The use of multiple first-line drugs has been proposed to reduce the chances of antimalarial drug resistance and ultimately improve the duration of treatment, especially in regions with strict private-sector engagement (Dhorda et al., 2021). Moreover, two current ACT partner drugs are combined with artemisinin properties to develop triple artemisinin-containing combination treatments (TACTs). To prevent the development of drug resistance to commonly used drugs such as lumefantrine and amodiaquine research conducted in Myanmar and Tanzania investigated longer treatment regimens of ACTs and found that they were safe but did not possess any increased potency when compared to the conventional 3-day doses (Mhamilawa et al., 2020).

Chemoprevention techniques are becoming more largely identified as useful means for combating malaria, particularly considering the decline in disease progression of the recent COVID-19 pandemic. The WHO encourages several chemoprevention strategies, such as intermittent preventive therapy in pregnancy (IPTp), mass drug administration (MDA), and seasonal malaria chemoprevention (SMC) (WHO, 2021).

These strategies seek to considerably decrease malaria incidence by giving preventive chemotherapy to the most susceptible individuals during peak transmission (Plowe, 2022). However, concerns over the chances that these strategies could favor parasites resistant to antimalarial drugs linger, making cautious application alongside strict vector control protocols imperative. Recent WHO updates highlight the necessity of tailoring these treatments to local contexts to increase their efficacy while lowering resistance concerns (Plowe, 2022).

6 Future Prospect to Combat Antimalarial Drug Resistance

Repurposing drugs has become viable for developing new medications for numerous life-threatening illnesses, like cancer and malaria (Hua et al., 2022). This strategy employs drugs already licensed for humans, enhancing research and reducing expenses. Atorvastatin and itraconazole are the two notable illustrations showing effectiveness against malaria. Antibiotics such as Azithromycin and Clindamycin, alongside HIV protease inhibitors (Tipranavir and Lopinavir), are used in combination for malaria treatment (Hua et al., 2022; Pandey et al., 2023).

Drug combinations can improve treatment outcomes in clinical settings, as shown by the synergistic action between pyrimethamine and sulfadoxine (Pandey et al., 2023). This approach finds useful compounds and makes it easier to investigate novel treatment targets, especially in multidrug-resistant malaria, where the cross-resistance phenomenon renders conventional treatments less effective. Identifying novel therapeutic targets to address the issues caused by multidrug-resistant malaria is important. By focusing on important pathways such as the folate pathway and fatty acid production, scientists have identified new therapeutic targets within the malaria parasite genome (Pandey et al., 2023).

High-throughput screening methodologies are employed to maximize lead compounds and reduce the chances of resistance. Furthermore, the discovery of important proteins involved in the parasite's lifecycle is made more accessible by multi-omics technologies, which improve our understanding of the basic mechanisms behind the pathogenesis of malaria (Singh et al., 2021). Malaria and cancer are two diseases that immunotherapy is an increasingly popular and effective strategy to combat (Naran et al., 2018). The most well-known immunotherapy in recent years is checkpoint blockade, which uses monoclonal antibodies (mAbs) to prevent immune inhibitory receptor-ligand pairs from interacting (Naran et al., 2018). This blocking facilitates the immune system's regular functioning, allowing for stronger immune responses against overexpressed ligands. This comprehensive approach highlights the possibility of novel therapeutic approaches and medication repurposing in tackling intricate health issues.

7 Conclusion

Plasmodium falciparum malaria remains a serious threat to public health, accounting for over 90% of malaria deaths globally. Serious concerns have been raised by the significant increase in *Plasmodium falciparum* resistance over the past 60 years. To track parasite resistance development and comprehend resistance mechanisms, it is imperative to identify and know the function of drug-resistant genes. For instance, *pfcrt* and *Pfmdr1* discovery have made significant progress toward comprehending how parasites develop resistance to antimalarial medications. PFMDR1 and PfCRT transmembrane protein mutations are the primary causes of the resistance mechanisms to drugs that interfere with parasite DV processes like hemoglobin breakdown and hemozoin production. To prevent the rise of *Plasmodium falciparum* strains that are resistant to antimalarial drugs, the co-expression of *Pfcrt* and *Pfmdr1* highlights the complexity of resistance mechanisms in malaria parasites and emphasizes the necessity of ongoing surveillance, creative treatment approaches, and public health initiatives.

Acknowledgments The authors of this manuscript appreciate the management of CUCRID, Covenant University, Ota, Nigeria, for providing the publication funding support and Covenant Applied Informatics and Communication Africa Center of Excellence (CApIC-ACE), Covenant University, Ota, for their support.

Competing Interest The authors have no competing interests.

Authors' Contributions All authors listed have made a substantial, direct, and intellectual contribution to the work and approved it for publication.

References

Achan, J., Mwesigwa, J., Edwin, C. P., & D'alessandro, U. (2018). Malaria medicines to address drug resistance and support malaria elimination efforts. *Expert Review of Clinical Pharmacology, 11*(1), 61–70. https://doi.org/10.1080/17512433.2018.1387773

Adamu, A., Jada, M. S., Haruna, H. M. S., Yakubu, B. O., Ibrahim, M. A., Balogun, E. O., & Shuaibu, M. N. (2020). *Plasmodium falciparum* multidrug resistance gene-1 polymorphisms in Northern Nigeria: Implications for the continued use of artemether-lumefantrine in the region. *Malaria Journal, 19*(1). https://doi.org/10.1186/s12936-020-03506-z

Alemayehu, A. (2023). Biology and epidemiology of *Plasmodium falciparum* and Plasmodium vivax gametocyte carriage: Implication for malaria control and elimination. *Parasite Epidemiology and Control, 21*, e00295. https://doi.org/10.1016/j.parepi.2023.e00295

Ashraf, K. (2021. *A comprehensive study of the antiplasmodial effects of Artemisia spp. infusions against multiple parasite developmental stages including hypnozoite*. Retrieved from https://theses.hal.science/tel-03715926

Bakari, C., Mandara, C. I., Madebe, R. A., Seth, M. D., Ngasala, B., Kamugisha, E., & Ishengoma, D. S. (2024). Trends of *Plasmodium falciparum* molecular markers associated with resistance to artemisinins and reduced susceptibility to lumefantrine in Mainland Tanzania from 2016 to 2021. *Malaria Journal, 23*(1). https://doi.org/10.1186/s12936-024-04896-0

Belete, T. M. (2020). Recent progress in the development of new antimalarial drugs with novel targets. *Drug Design, Development, and Therapy, 34*(1), 3875–3889. https://doi.org/10.2147/DDDT.S265602

Blanshard, A., & Hine, P. (2021). Atovaquone-proguanil for treating uncomplicated *Plasmodium falciparum* malaria. *The Cochrane Database of Systematic Reviews, 2021*(2). https://doi.org/10.1002/14651858.cd004529.pub3

Blasco, B., Leroy, D., & Fidock, D. A. (2017). Antimalarial drug resistance: Linking *Plasmodium falciparum* parasite biology to the clinic. *Nature Medicine, 23*(8), 917–928. https://doi.org/10.1038/nm.4381

Cheng, W., Song, X., Tan, H., Wu, K., & Li, J. (2021). Molecular surveillance of anti-malarial resistance pfcrt, pfmdr1, and pfk13 polymorphisms in African *Plasmodium falciparum* imported parasites to Wuhan, China. *Malaria Journal, 20*(1). https://doi.org/10.1186/s12936-021-03737-8

Cheviet, T., Lefebvre-Tournier, I., Wein, S., & Peyrottes, S. (2019). Plasmodium purine metabolism and its inhibition by nucleoside and nucleotide analogs. *Journal of Medicinal Chemistry., 62*(18), 8365–8391. https://doi.org/10.1021/acs.jmedchem.9b00182

Culleton, R., Pain, A., & Snounou, G. (2023). *Plasmodium malariae*: The persisting mysteries of a persistent parasite. *Trends in Parasitology., 39*(1), 113–125. https://doi.org/10.1016/j.pt.2022.11.008

De Villiers, K. A., & Egan, T. J. (2021). Heme detoxification in the malaria parasite: A target for antimalarial drug development. *Accounts of Chemical Research, 54*(11), 2649–2659. https://doi.org/10.1021/acs.accounts.1c00154

Desai, M., Hill, J., Fernandes, S., Walker, P., Pell, C., Gutman, J., & Ter Kuile, F. O. (2018). Prevention of malaria in pregnancy. *The Lancet. Infectious diseases, 18*(4), e119–e132. https://doi.org/10.1016/s1473-3099(18)30064-1

Dhorda, M., Amaratunga, C., & Dondorp, A. M. (2021). Artemisinin and multidrug-resistant *Plasmodium falciparum* – a threat for malaria control and elimination. *Current Opinion in Infectious Diseases, 34*(5), 432–439.

Dwumfour, C. K., Bam, V. B., Owusu, L. B., Poku, C. A., Kpabitey, R. D., Aboagye, P., & Ibrahim, A. S. (2023). Prevalence and determinants of malaria infection among pregnant women attending antenatal clinic in Ejisu government hospital in Ghana: A cross-sectional study. *PloS One, 18*(10), e0293420. https://doi.org/10.1371/journal.pone.0293420

Efferth, T., & Oesch, F. (2021). The immunosuppressive activity of artemisinin-type drugs towards inflammatory and autoimmune diseases. *Medicinal Research Reviews, 41*(6), 3023–3061. https://doi.org/10.1002/med.21842

Esser, L., & Xia, D. (2024). Mitochondrial cytochrome bc 1 complex as validated drug target: A structural perspective. *Tropical Medicine and Infectious Disease, 9*(2), 39–48. https://doi.org/10.3390/tropicalmed9020039

Fontecha, G., Pinto, A., Archaga, O., Betancourth, S., Escober, L., Henríquez, J., & Mejía, R. E. (2021). Assessment of *Plasmodium falciparum* anti-malarial drug resistance markers in pfcrt and pfmdr1 genes in isolates from Honduras and Nicaragua, 2018–2021. *Malaria Journal, 20*(1). https://doi.org/10.1186/s12936-021-03977-8

Gazzinelli, R. T., Kalantari, P., Fitzgerald, K. A., & Golenbock, D. T. (2014). Innate sensing of malaria parasites. *Nature Reviews. Immunology, 14*(11), 744–757. https://doi.org/10.1038/nri3742

Gibbons, J. A. (2019). Genomics and transcriptomics approaches to understanding drug resistance mechanisms in the malaria parasite plasmodium falciparum (Order No. 13813216). Available from ProQuest One Academic. (2250193882). Retrieved from https://www.

proquest.com/dissertations-theses/genomics-transcriptomics-approachesunderstanding/docview/2250193882/se-2

Habtamu, K., Petros, B., & Yan, G. (2022). Plasmodium vivax: The potential obstacles it presents to malaria elimination and eradication. *Tropical Diseases, Travel Medicine and Vaccines, 8*(1). https://doi.org/10.1186/s40794-022-00185-3

Heller, L. E., & Roepe, P. D. (2019). Artemisinin-based antimalarial drug therapy: Molecular pharmacology and evolving resistance. *Tropical Medicine and Infectious Disease, 4*(2), 89. https://doi.org/10.3390/tropicalmed4020089

Hodoameda, P. (2021). *P. falciparum* and its molecular markers of resistance to antimalarial drugs. In *Plasmodium species and drug resistance*. IntechOpen. https://doi.org/10.5772/intechopen.98372

Hodoameda, P., Duah-Quashie, N. O., & Quashie, N. B. (2022). Assessing the roles of molecular markers of antimalarial drug resistance and the host pharmacogenetics in drug-resistant malaria. *Journal of Tropical Medicine, 2022*, 3492696. https://doi.org/10.1155/2022/3492696

Hua, Y., Dai, X., & Xu, Y. (2022). Drug repositioning: Progress and challenges in drug discovery for various diseases. *European Journal of Medicinal Chemistry., 20*(2), 234–239.

Ikegbunam, M. N., Nkonganyi, C. N., Thomas, B. N., Esimone, C. O., Velavan, T. P., & Ojurongbe, O. (2019). Analysis of *Plasmodium falciparum* Pfcrt and Pfmdr1 genes in parasite isolates from asymptomatic individuals in Southeast Nigeria 11 years after withdrawal of chloroquine. *Malaria Journal, 18*(1). https://doi.org/10.1186/s12936-019-2977-6

Imwong, M., Suwannasin, K., Srisutham, S., Vongpromek, R., Promnarate, C., Saejeng, A., Phyo, A. P., Proux, S., Pongvongsa, T., Chea, N., & Miotto, O. (2021). Evolution of multidrug resistance in *Plasmodium falciparum*: A longitudinal study of genetic resistance markers in the Greater Mekong Subregion. *Antimicrobial Agents and Chemotherapy, 65*(12), 10–1128. https://doi.org/10.1128/aac.01121-21

Kilonzi, M., Minzi, O., Mutagonda, R., Sasi, P., Kamuhabwa, A., & Aklillu, E. (2019). Comparison of malaria treatment outcome of generic and innovator's anti-malarial drugs containing artemether-lumefantrine combination in the management of uncomplicated malaria amongst Tanzanian children. *Malaria Journal, 18*(1). https://doi.org/10.1186/s12936-019-2769-z

Kim, J., Tan, Y. Z., Wicht, K. J., Erramilli, S. K., Dhingra, S. K., Okombo, J., Vendome, J., Hagenah, L. M., Giacometti, S. I., Warren, A. L., Nosol, K., Roepe, P. D., Potter, C.S., Carragher, B., Kossiakoff, A. A., Quick, M., Fidock, D. A. & Mancia, F. (2019). Structure and drug resistance of the Plasmodium falciparum transporter PfCRT. *Nature. 576*(4) 315–320. https://doi.org/10.1038/s41586-019-1795-x

Koehne, E., Adegnika, A. A., Held, J., & Kreidenweiss, A. (2021). Pharmacotherapy for artemisinin-resistant malaria. *Expert Opinion on Pharmacotherapy, 22*(18), 2483–2493. https://doi.org/10.1080/14656566.2021.1959913

Kumar, S., Bhardwaj, T., Prasad, D., & Singh, R. K. (2018). Drug targets for resistant malaria: Historic to future perspectives. *Biomedicine & Pharmacotherapy, 104*, 8–27. https://doi.org/10.1016/j.biopha.2018.05.009

Lingani, M., Bonkian, L. N., Yerbanga, I. I., Kazienga, A., Valéa, I., Sorgho, H., Tinto, H. (2020). In vivo/ex vivo efficacy of artemether–lumefantrine and artesunate–amodiaquine as first-line treatment for uncomplicated falciparum malaria in children: An open label randomized controlled trial in Burkina Faso. Malaria Journal, 19(1). doi: https://doi.org/10.1186/s12936-019-3089-z.

Lopes, E. A., Santos, M. M., & Mori, M. (2021). Pharmacological treatment of malaria. In E. M. Silva & J. M. Moreira (Eds.), *Antiprotozoal drug development and delivery* (pp. 219–240). Springer International Publishing. https://doi.org/10.1007/7355_2021_125

Madhav, H., & Hoda, N. (2021). An insight into the recent development of the clinical candidates for the treatment of malaria and their target proteins. *European Journal of Medicinal Chemistry, 210*, 112955. https://doi.org/10.1016/j.ejmech.2020.112955

Martin, R. E. (2020). The transportome of the malaria parasite. *Biological Reviews of the Cambridge Philosophical Society, 95*(2), 305–332. https://doi.org/10.1111/brv.12565

Matrevi, S. A., Tandoh, K. Z., Bruku, S., Opoku-Agyeman, P., Adams, T., Ennuson, N. A., & Duah-Quashie, N. O. (2022). Novel pfk13 polymorphisms in *Plasmodium falciparum* population in Ghana. *Scientific Reports, 12*(1). https://doi.org/10.1038/s41598-022-11790-9

Mhamilawa, L. E., Ngasala, B., Morris, U., Kitabi, E. N., Barnes, R., Soe, A. P., Mmbando, B. P., Björkman, A., & Mårtensson, A. (2020). Parasite clearance, cure rate, post-treatment prophylaxis and safety of standard 3-day versus an extended 6-day treatment of artemether–lumefantrine and a single low-dose primaquine for uncomplicated Plasmodium falciparum malaria in Bagamoyo district, Tanzania: a randomized controlled trial. *Malaria Journal, 19*(1). https://doi.org/10.1186/s12936-020-03287-5

Mishra, M., Mishra, V. K., Kashaw, V., & Kashaw, S. K. (2021). Molecular approaches for malaria therapy. In *Plasmodium species and drug resistance*. IntechOpen. https://doi.org/10.5772/intechopen.98396

Morales-Luna, L., Vázquez-Bautista, M., Martínez-Rosas, V., Rojas-Alarcón, M. A., Ortega-Cuellar, D., González-Valdez, A., & Gómez-Manzo, S. (2024). Fused enzyme Glucose-6-phosphate dehydrogenase::6-Phosphogluconolactonase (G6PD::6PGL) as a potential drug target in Giardia lamblia, Trichomonas vaginalis, and *Plasmodium falciparum*. *Microorganisms, 12*(1), 112. https://doi.org/10.3390/microorganisms12010112

Muhammad, I., Imam, A., & Khadijah, S. (2023). *Plasmodium falciparum* multidrug Resistance-1 (pfmdr1) gene mutation in adults malaria patients attending Murtala Muhammad Specialist Hospital Kano, Northwest Nigeria. *Journal of Biochemistry, Microbiology and Biotechnology, 11*(1), 31–36. https://doi.org/10.54987/jobimb.v11i1.803

Mu, J., Myers, R. A., Jiang, H., Liu, S., Ricklefs, S., Waisberg, M., Chotivanich, K., Wilairatana, P., Krudsood, S., White, N. J., Udomsangpetch, R., Cui, L., Ho, M., Ou, F., Li, H., Song, J., Li, G., Wang, X., Seila, S., . . . Su, X. (2010). Plasmodium falciparum genome-wide scans for positive selection, recombination hot spots, and resistance to antimalarial drugs. *Nature Genetics, 42*(3), 268–271. https://doi.org/10.1038/ng.528

Müller, O. (2011). *Malaria in Africa: challenges for control and elimination in the 21st century*. Retrieved from http://ci.nii.ac.jp/ncid/BB10295659

Naran, K., Nundalall, T., Chetty, S., & Barth, S. (2018). Principles of immunotherapy: Implications for treatment strategies in cancer and infectious diseases. *Frontiers in Microbiology, 9*(1), 31–58.

Noreen, N., Ullah, A., Salman, S. M., Mabkhot, Y., Alsayari, A., & Badshah, S. L. (2021). New insights into the spread of resistance to artemisinin and its analogues. *Journal of Global Antimicrobial Resistance, 27*, 142–149. https://doi.org/10.1016/j.jgar.2021.09.001

Nsanzabana, C. (2019). Resistance to artemisinin combination therapies (ACTs): Do not forget the partner drug! *Tropical Medicine and Infectious Disease, 4*(1), 26. https://doi.org/10.3390/tropicalmed4010026

Nureye, D., & Assefa, S. (2020, 2020). Old and recent advances in life cycle, pathogenesis, diagnosis, prevention, and treatment of malaria including perspectives in Ethiopia. *The Scientific World Journal*, 1–17. https://doi.org/10.1155/2020/1295381

Ocan, M., Akena, D., Nsobya, S., Kamya, M. R., Senono, R., Kinengyere, A. A., & Obuku, E. A. (2019). Persistence of chloroquine resistance alleles in malaria endemic countries: A systematic review of burden and risk factors. *Malaria Journal, 18*(1). https://doi.org/10.1186/s12936-019-2716-z

Oladipo, H. J., Tajudeen, Y. A., Oladunjoye, I. O., Yusuff, S. I., Yusuf, R. O., Oluwaseyi, E. M., & El-Sherbini, M. S. (2022). Increasing challenges of malaria control in sub-Saharan Africa: Priorities for public health research and policymakers. *Annals of Medicine and Surgery, 81*. https://doi.org/10.1016/j.amsu.2022.104366

Pandey, S. K., Anand, U., Siddiqui, W. A., & Tripathi, R. (2023). Drug development strategies for malaria: With the hope for new antimalarial drug discovery-an update. *Advances in Medicine, 20*(1), 23–28.

Patel, C. A., Pande, S., Shukla, P., Ranch, K., Al-Tabakha, M. M. & Boddu, S. H. (2023). Antimalarial drug resistance: Trends, mechanisms, and strategies to combat antimalarial resistance In Malarial drug delivery systems: Advances in treatment of infectious diseases. Springer International Publishing. 21 (1), 43–69. doi:https://doi.org/10.1007/978-3-031-15848-3_3.

Plowe, C. V. (2022). Malaria chemoprevention and drug resistance: A review of the literature and policy implications. *Malaria Journal, 21*(1), 104–122.

Posayapisit, N., Pengon, J., Shaw, P. J., Uthaipibull, C., Kongkasuriyachai, D., Phyo, A. P., & Kamchonwongpaisan, S. (2023). Susceptibility of Southeast Asian *Plasmodium falciparum* isolates to P218. *International Journal of Antimicrobial Agents, 62*(1), 106838. https://doi.org/10.1016/j.ijantimicag.2023.106838

Qidwai, T. (2020). Exploration of copy number variation in genes related to anti-malarial drug resistance in Plasmodium falciparum. *Gene, 736*, 144414. https://doi.org/10.1016/j.gene.2020.144414

Rathmes, G., Rumisha, S. F., Lucas, T. C., Twohig, K. A., Python, A., Nguyen, M., Nandi, A. K., Keddie, S. H., Collins, E. L., Rozier, J. A., & Gibson, H. S. (2020). Global estimation of anti-malarial drug effectiveness for the treatment of uncomplicated *Plasmodium falciparum* malaria 1991–2019. *Malaria Journal, 19*(1), 1–15.

Rosenthal, M. R., & Ng, C. L. (2020). *Plasmodium falciparum* artemisinin resistance: The effect of heme, protein damage, and parasite cell stress response. *ACS Infectious Diseases, 6*(7), 1599–1614. https://doi.org/10.1021/acsinfecdis.9b00527

Rosenthal, P. J., Asua, V., & Conrad, M. D. (2024). Emergence, transmission dynamics and mechanisms of artemisinin partial resistance in malaria parasites in Africa. *Nature Reviews. Microbiology, 22*(6), 373–384. https://doi.org/10.1038/s41579-024-01008-2

Rout, S., & Mahapatra, R. K. (2019). *Plasmodium falciparum*: Multidrug resistance. *Chemical Biology & Drug Design, 93*(5), 737–759. https://doi.org/10.1111/cbdd.13484

Roux, A. T., Maharaj, L., Oyegoke, O., Akoniyon, O. P., Adeleke, M. A., Maharaj, R., & Okpeku, M. (2021). Chloroquine and sulfadoxine-pyrimethamine resistance in Sub-Saharan Africa—A review. *Frontiers in Genetics, 12*(1), 668574–668581. https://doi.org/10.3389/fgene.2021.668574

Rovira-Vallbona, E., Van Hong, N., Kattenberg, J. H., Huan, R. M., Hien, N. T. T., Ngoc, N. T. H., & Rosanas-Urgell, A. (2020). Efficacy of dihydroartemisinin/piperaquine and artesunate monotherapy for the treatment of uncomplicated *Plasmodium falciparum* malaria in Central Vietnam. *Journal of Antimicrobial Chemotherapy*. https://doi.org/10.1093/jac/dkaa172

Sato, S. (2021). Plasmodium—A brief introduction to the parasites causing human malaria and their basic biology. *Journal of Physiological Anthropology, 40*(1). https://doi.org/10.1186/s40101-020-00251-9

Shafik, S. H., Cobbold, S. A., Barkat, K., Richards, S. N., Lancaster, N. S., Llinás, M., & Martin, R. E. (2020). The natural function of the malaria parasite's chloroquine resistance transporter. *Nature Communications, 11*(1). https://doi.org/10.1038/s41467-020-17781-6

Shafik, S. H., Richards, S. N., Corry, B., & Martin, R. E. (2022). Mechanistic basis for multidrug resistance and collateral drug sensitivity conferred to the malaria parasite by polymorphisms in PfMDR1 and PfCRT. *PLoS Biology, 20*(5), e3001616. https://doi.org/10.1371/journal.pbio.3001616

Shibeshi, M. A., Kifle, Z. D., & Atnafie, S. A. (2020). Antimalarial drug resistance and novel targets for antimalarial drug discovery. *Infection and Drug Resistance, 13*, 4047–4060. https://doi.org/10.2147/idr.s279433

Silva, M., Malmberg, M., Otienoburu, S. D., Björkman, A., Ngasala, B., Mårtensson, A., & Veiga, M. I. (2022). *Plasmodium falciparum* drug resistance genes pfmdr1 and pfcrt in vivo co-expression during artemether-lumefantrine therapy. *Frontiers in Pharmacology, 13*. https://doi.org/10.3389/fphar.2022.868723

Simon, N., Voigtländer, C., Kappes, B., Rohrbach, P., & Friedrich, O. (2022). PfMDR1 transport rates assessed in intact isolated *Plasmodium falciparum* digestive vacuoles reflect functional drug resistance relationship with pfmdr1 mutations. *Pharmaceuticals, 15*(2), 202. https://doi.org/10.3390/ph15020202

Singh, R., Singh, P. K., & Kumar, R. (2021). Multi-omics approach in the identification of potential therapeutic biomolecule for COVID-19. *Frontiers in Pharmacology, 12*(1), 234–239.

Su, X. Z., Lane, K. D., Xia, L., Sá, J. M., & Wellems, T. E. (2019). Plasmodium genomics and genetics: New insights into malaria pathogenesis, drug resistance, epidemiology, and evolution. *Clinical Microbiology Reviews, 32*(4), 345–349. https://doi.org/10.1128/cmr.00019-19

Tobing, T. C. L., Saputri, E., Wafa, N. I., Zulfianti, P. D., Sihaloho, L. I., Husna, A. R., Salsabila, D., Silalahi, F. H. H., Sitohang, A. I., Sabrina, A., & Harianja, A. D. H. (2024). Malaria in Indonesia: Current treatment approaches, future strategies, and potential herbal interventions. *Pharmacia, 37*(1), 1–15. https://doi.org/10.3897/pharmacia.71.e116095

Tuedom, A. G. B., Sarah-Matio, E. M., Moukoko, C. E. E., Feufack-Donfack, B. L., Maffo, C. N., Bayibeki, A. N., ... Nsango, S. E. (2021). Antimalarial drug resistance in the Central and Adamawa regions of Cameroon: Prevalence of mutations in P. falciparum crt, Pfmdr1, Pfdhfr and Pfdhps genes. *PloS One, 16*(8), e0256343. https://doi.org/10.1371/journal.pone.0256343

van der Pluijm, R. W., Amaratunga, C., Dhorda, M., & Dondorp, A. M. (2021). Triple artemisinin-based combination therapies for malaria - A new paradigm? *Trends in Parasitology, 37*(1), 15–24. https://doi.org/10.1016/j.pt.2020.09.011

Varo, R., Chaccour, C., & Bassat, Q. (2020). Update on malaria. *Medicina Clínica (English Edition), 155*(9), 395–402. https://doi.org/10.1016/j.medcle.2020.05.024

Verma, M. K., Sharma, S., & Kumar, S. (2020). A review on pharmacological properties of Artemisia annua. *Journal of Pharmacognosy and Phytochemistry., 9*(6), 2179–2185. https://doi.org/10.33545/2664844X

WHO (2020). Report on antimalarial drug efficacy, resistance, and response: 10 years of surveillance (2010–2019). World Health Organization. Retrieved April 30, 2024, From https://www.who.int/publications/i/item/9789240012813

WHO (2021) Word Malaria Report 2021, . World Health Organization. Retrieved April 30, 2024, From https://www.who.int/teams/global-malaria-programme/reports/world-malaria-report-2021

Wicht, K. J., Mok, S., & Fidock, D. A. (2020). Molecular mechanisms of drug resistance in *Plasmodium falciparum* malaria. *Annual Review of Microbiology, 74*(1), 431–454. https://doi.org/10.1146/annurev-micro-020518-115546

Wicht, K. J., Small-Saunders, J. L., Hagenah, L. M., Mok, S., & Fidock, D. A. (2022). Mutant PfCRT can mediate piperaquine resistance in African Plasmodium falciparum with reduced fitness and increased susceptibility to other antimalarials. *The Journal of Infectious Diseases, 226*(11), 2021–2029. https://doi.org/10.1093/infdis/jiac365

World malaria report 2019. from Who.int website: https://www.who.int/publications/i/item/9789241565721.Accessed July 30, 2023.

World Health Organization, 2022. Malaria: Artemisinin Partial Resistance. Available at: https://www.who.int/newsroom/questions-and-answers/item/artemisinin-resistance. Accessed August 21, 2023.

World Health Organization. World Malaria Report 2022. Geneva: World Health Organization; 2022. Licence: CC BYNC-SA 3.0 IGO. (accessed 3-1-2023). https://www.who.int/teams/global-malaria-programme/reports/world-malariareport-2022.

Yadav, K., Shivahare, R., Shaham, S. H., Joshi, P., Sharma, A., & Tripathi, R. (2021). Repurposing of existing therapeutics to combat drug-resistant malaria. *Biomedecine & Pharmacotherapie [Biomedicine & Pharmacotherapy], 136*(111275), 111275. https://doi.org/10.1016/j.biopha.2021.111275

Yobi, D. M., Kayiba, N. K., Mvumbi, D. M., et al. (2021). Assessment of *Plasmodium falciparum* anti-malarial drug resistance markers in *pfk13*-propeller, *pfcrt* and *pfmdr1* genes in isolates from treatment failure patients in Democratic Republic of Congo, 2018–2019. *Malar Journal, 20*, 144. https://doi.org/10.1186/s12936-021-03636-y

Zekar, L., & Sharman, T. (2023). *Plasmodium falciparum malaria*. StatPearls Publishing. Retrieved April 20, 2024, From https://www.ncbi.nlm.nih.gov/books/NBK555962/

Zhong, A., Zhang, H., & Li, J. (2023). Insight into molecular diagnosis for antimalarial drug resistance of Plasmodium falciparum parasites: A review. *Acta Tropica, 241*(1), 106–117. https://doi.org/10.1016/j.actatropica.2023.106870

Chapter 22
Differential Mosquito Attraction and Host Microbiota-Vector-Environment Complex: A Novel Malaria Mitigation Technique

Temitope Ayinde Oluwaseyi, Grace Iyabo Olasehinde, Zeendi Kwarpo, Aririguzoh Victoria, and Opeyemi Isaac Ayanda

1 Introduction

Africa is a continent laden with malaria diseases (Ajide et al., 2022). Worldwide, despite the impact of COVID-19, it is reported that malaria incidence and deaths in 2021 are still stable. In 2021, the World Health Organization reported that 619,000 people worldwide died of malaria compared to 625,000 in the first year of the epidemic. Before the outbreak, there were 568,000 cases in 2019. Compared to 245 million malaria cases in 2020 and 229 million in 2019, malaria worldwide reached 247 million in 2021 (Monroe et al., 2022). Malaria is most frequently caused by *Plasmodium falciparum*, one of the species of Plasmodium that infect Africans, with a prevalence of 99.7% (Mbama Ntabi et al., 2022). Unlike other blood protozoa, the parasite *Plasmodium* spreads by the arthropods mosquitoes (Cholewinski et al., 2015).

During host-seeking behavior, mosquitoes integrate multifaceted sensory information, including chemical, physical, and visual cues from the host. Nocturnal mosquito species primarily rely on physical and olfactory signals, such as body heat, carbon dioxide, and skin odors, for host detection. In contrast, diurnal mosquitoes place greater emphasis on visual stimuli to locate their hosts. This distinction underscores the adaptive sensory strategies employed by mosquitoes, with nocturnal

T. A. Oluwaseyi (✉) · G. I. Olasehinde · Z. Kwarpo · A. Victoria
Department of Biological Sciences, College of Science and Technology, Covenant University, Ota, Ogun State, Nigeria

Covenant Applied Informatics and Communication Africa Centre of Excellence (CApICACE), Covenant University, Ota, Ogun State, Nigeria
e-mail: temitope.oluwaseyipgs@stu.cu.edu.ng

O. I. Ayanda
Department of Biological Sciences, College of Science and Technology, Covenant University, Ota, Ogun State, Nigeria

P. O. Isibor, P. A. Akinduti (eds.), *Harnessing Biotechnology Tools for Product Development*, https://doi.org/10.1007/978-3-031-86002-7_22

species depending more on non-visual cues, while daylight-active mosquitoes leverage enhanced visual capabilities to optimize host identification (Mukabana et al., 2004). *Anopheles gambiae,* due to its preference for feeding on humans, is one of the most prevalent malaria vectors. This type of mosquito is nocturnal and locates its victim primarily by smelling it (Martinez et al., 2021). Female mosquitoes use the heat and volatile organic chemical kairomone, which humans produce in the air, as signals to find human hosts and then obtain a blood meal. However, people need to give these warnings regularly. The amount of mosquito-attracting kairomones released varies greatly depending on the individual. This diversity explains why some humans are mosquitoes' preferred predators (Ellwanger et al., 2021). The skin microorganism affects the production of odors that attract mosquitoes by metabolizing volatile substances released by the sebaceous glands, eccrine glands, and apocrine glands (Takken & Verhulst, 2017). Therefore, changes in the odor of a person may be due to changes in their skin microorganisms, which change the type and quantity of kairomones emitted into the atmosphere (Ellwanger et al., 2021).

2 African Malaria Vectors

Although there are 500 species of anopheline mosquitoes around the globe, only three of them—*Anopheles gambiae*, *Anopheles arabiensis*, and *Anopheles funestus*—are capable of transmitting human malaria (Wilkerson et al., 2021). Only in Africa can one find these three species responsible for 90% of all malaria cases worldwide (Fillinger et al., 2008). The high rate of malaria transmission in Africa results from specific human behaviors shown by Anopheles mosquitoes, which strongly prefer humans as hosts. Studies have demonstrated that this selective preference is purely odor-mediated. For example, the hypothesis that volatiles emitted by human skin microbiota drive mosquito host-seeking behavior was first proposed by Mukabana et al. (2004). This concept has since been validated by subsequent studies, which have identified specific bacterial species associated with producing these volatiles. These studies have elucidated the role of microbiota-derived compounds in enhancing the attractiveness of humans to mosquitoes, highlighting a complex interaction between microbial activity and vector behavior that influences host selection (Takken & Verhulst, 2017).

3 Host-Seeking Behavior: Genes Versus Environment

Human-seeking mosquitoes, including anautogenous species, use several cues, including host smells, in finding blood meal (Konopka et al., 2021). Since carbon dioxide is found in vertebrate exhaled breath, it is the most well-known mosquito kairomone and a good signal of the presence of a possible host. There is an upward relationship between the number of inhabitants and the number of mosquitoes captured, with occupied homes attracting considerably more *Anopheles gambiae* and *Anopheles funestus* than unoccupied ones (Kaindoa et al., 2016; Mzilahowa et al., 2016). Previous studies have shown the importance of human scents in searching for hosts by Afrotropical malaria vectors. Recent studies have revealed that malaria mosquitoes are also drawn to distant smells of individuals and animals (Takken & Knols, 1999; Robinson et al., 2018; Wolff & Riffell, 2018). Mosquitoes responded to scent signals even when the host was not present, so scientists evaluated them by pumping airborne human volatiles into a sampling device.

He et al. (2019) proposed that the development of synthetic human scents capable of attracting host-seeking mosquitoes may become feasible in the near future. This concept is grounded in the identification of key volatiles that mimic human odor, which could be engineered to effectively lure mosquitoes. By replicating these chemical signals, this innovation has the potential to revolutionize vector control strategies by intercepting mosquitoes before they reach human hosts, thus reducing the transmission of mosquito-borne diseases. According to Zhao et al. (2020), genetic variations in the chemosensory genes of different vector species account for some of the variance in host choice. For instance, *Aedes aegypti* has a domestic form that attracts people and a forest variant that favors animals (McBride et al., 2014; Rose et al., 2020). Martinez et al. (2021) found some possible genes that might have helped this behavior evolve. Additionally, *Anopheles arabiensis* preference of a particular host species has been connected to chromosomal inversions, suggesting that genetic variation influences the vector-host choice. Raji et al. (2019) compared the zoophilic *Anopheles quadriannulatus* with the highly anthropophilic *Anopheles coluzzii* using transcriptome analysis, and they proposed that variations in chemosensory genes account for the differences in host preference. The genetic sensitivity to variations in human kairomones that some mosquito species or populations may possess may affect human-mosquito interactions. A recent study showed how the distinctive human odor profile might trigger a particular neural response in the brain of *Aedes aegypti* mosquitoes, enabling them to differentiate people from other species, provides evidence in support of this theory (Zhao et al., 2020).

4 Skin Microbiota and Its Volatile Organic Compounds

Volatile organic compounds (VOCs) are thought to be produced in substantial quantities by microorganisms on human skin (Thorn & Greenman, 2012). According to research on skin bacteria, aliphatic amino acids contained in sebum, perspiration, and skin lipids are broken down into short-chain carboxylic acids, which affect the strength and kind of odor that is created on the skin (Takken & Verhulst, 2017). The type of odor generated is primarily related to the presence of particular bacteria on the skin. For instance, Bacillus species are associated with foot odor, while Corynebacteria are responsible for the apocrine odor (Grice & Segre, 2011). The characteristic malodor is caused by short-chain volatile fatty acids (VFAs), some of which are also enticing to mosquitoes. Some skin bacteria can break down

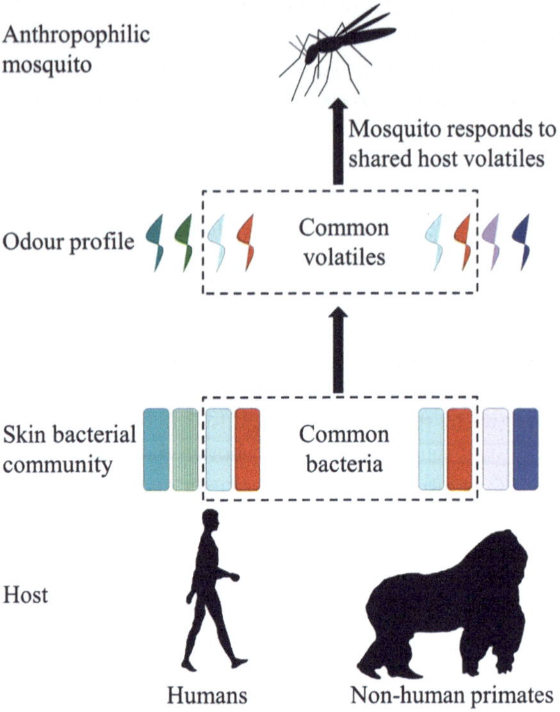

Fig. 22.1 Mosquito-host interactions and the possible response of anthropophilic mosquitoes to volatiles shared by humans and nonhuman primates (Verhulst et al., 2012)

long-chain lipids into less flammable fatty acids, while others can further break down into short-chain fatty acids, CO_2, and water (Natsch & Emter, 2020). Research has revealed that particular bacteria, namely gram-negative bacteria of the Roseomonas genus, are more transmissible than others when it comes to the human skin microbiome. Genetic variations in host genes that control the skin's environment may control how skin bacteria affect a person's attractiveness (Romano-Bertrand et al., 2016; Flowers & Grice, 2020). The skin microbiomes of humans, other primates, and cattle vary, which may influence mosquito choices for diverse host species. Although less diversified, the human skin microbiome has more Staphylococcus species, known to draw Anopheles mosquitoes (Verhulst et al., 2018; Showering et al., 2022) (Fig. 22.1).

5 Key Sensory Cues Exploited by Mosquitoes

5.1 Carbon Dioxide

Mosquitoes use the Gr1 and Gr3 subunits of the Gr1, Gr2, and Gr3 gustatory receptors in their maxillary palps to sense CO_2 (Kumar et al., 2020). CO_2 is necessary for mosquitoes to recognize a blood meal from afar and is what starts and sustains their flight. Particularly in female *Aedes aegypti*, the Gr3 subunit, is implicated in sensing CO_2 and other human kairomones (Kumar et al., 2020). Although different mosquito sensory receptors have distinctive characteristics, data suggests that the sensing mechanism that produces volatile kairomones is reliable and redundant (Ellwanger et al., 2021).

5.2 Lactic Acid

There are two types of lactic acid: L-lactic acid and D-lactic acid. Humans and other vertebrates typically create the former, whereas the other type is generally linked to bacterial illnesses (Pohanka, 2020). L-lactic acid, also known as 2-hydroxypropanoic acid, was found to be appealing to mosquitoes in tests employing the acidified component of forearm washing extracts and females of *Aedes aegypti* (Acree et al., 1968; Raji et al., 2019). Since eccrine sweat glands, which are abundant on human skin surfaces, manufacture the majority of this molecule, anthropophilic mosquitoes could use the high levels of this compound in humans as a host recognition cue. Although it is uncertain how much lactic acid (LA) is produced via perspiration versus breathing, it is believed that its varied production substantially impacts a person's attractiveness to mosquitoes (Hawkes & Hopkins, 2022).

5.3 Fatty Acids

Cutaneous microbes that metabolize glycerol, lactic acid, amino acids, and other skin lipids create the volatile fatty acids that cause skin odor. A noteworthy finding is that while other than humans, mammals or birds do not create these volatile FAs, anthropophilic mosquitoes can use their presence as a chemical signature to find a human host (Wooding et al., 2020). Numerous FAs have been found to attract various mosquito species, and they can be used alone or in combination to make efficient field-trapping baits. For instance, the BG-lure mix, which replicates human odor with hexanoic acid as the principal ingredient, and fatty acids of C3 to C5, are all artificial mixtures that are frequently utilized (Serag et al., 2021).

5.4 Human Host Odor

Anthropophilic mosquitoes can discriminate between human smells and those of nonhuman animals by recognizing particular volatile organic compounds that humans release into the environment (Wolff & Riffell, 2018; Konopka et al., 2021). Some mosquito species, however, cannot distinguish between the odors of nonhuman primates and humans (McBride, 2016). The hands and feet are frequently used in bioassays to test the attractiveness of different body parts] to mosquitoes. Studies on human body smell's effects on mosquito attractiveness also employed breath volatiles and various sweat extracts (Cummins et al., 2012).

To increase trapping success, synthetic mixtures that resemble human odors are frequently utilized in the field (Verhulst et al., 2016). These mixtures often contain compounds including lactic acid, ammonia, aldehydes, and other fatty acids that have been linked to the scent of human skin and have been proven to have an impact on mosquito behavior and electrophysiology (Dormont et al., 2021; Owino, 2021). It has been suggested that the ratio of 6-methyl-5-hepten-2-one and geranylacetone, two terpenes present in skin ejections, may help to partially explain why some people are more mosquito-attractive than others (Martinez et al., 2021). It has also been proposed that aldehydes like nonanal, decanal, or octanal play a significant role in luring anthropophilic mosquitoes (Wanjiru, 2020). The different body odor components between regions of the body may be related to variations in the distribution of skin bacteria, which are linked to the prevalence of various skin gland types. The distinctive microbial communities associated with feet may cause their pungent smell, which attracts mosquitoes in particular. Individual differences in the skin microbiota can also affect how different people smell and how attractive they are to mosquitoes.

6 Environmental Factors in Differential Attraction

Habitat alteration increasingly impacts the interaction between humans and mosquitoes (Franklinos et al., 2019). Deforestation, land use, and climate change strongly impact mosquito populations' dispersion, fostering their growth and raising the possibility of mosquito-borne diseases (Meyer Steiger et al., 2016a; Ortiz et al., 2022). Following the research by Rose et al. (2020), establishing a close relationship between humankind and mosquitoes, using human-stored water sites, and urbanization were among the factors that influenced the choosing of the anthropophilic trait seen in *Aedes aegypti*.

Abiotic factors like the wind, which affects the distribution of volatile substances, can alter the vulnerability of populations to diseases spread by mosquitoes. The wind's speed and direction influence this. In contrast to molecule diffusion from the host organism to the insect, Cummins et al. (2012) found that wind direction can cause the spread of host-derived volatile compounds such as "puffs" or "packets." A study by Endo and Eltahir (2018) also showed the effect of wind on the possibility that Anopheles mosquitoes come into touch with kairomones originating from humans, significantly affecting the risk of transmitting malaria. Climate factors like temperature, humidity, precipitation, wind, and the duration of daylight have played essential roles in mosquito ecology, development, and their rate of disease transmission. A high increase in temperature kills mosquitoes, but within a specific range of temperature, they can survive, especially at warmer temperatures, because it increases their blood-sucking activity and reproduction (Yi et al., 2014) (Fig. 22.2).

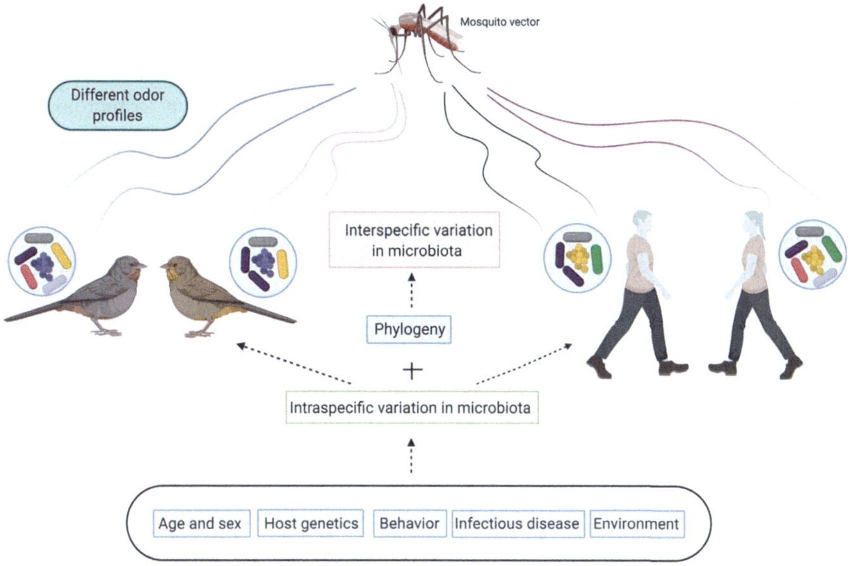

Fig. 22.2 The interplay of the environment between the vector and the microbiota as mediators Of differential attraction (Ruiz-López, 2020)

The increase in human population and activities has led to a rise in land use through deforestation, agricultural practices, road construction, hydrological transformations, and urbanization, favoring breeding sites for mosquitoes and increased exposure (Meyer Steiger et al., 2016a). Deforested habitats have features such as being open, well-lit, and warmer, increasing the survival and growth rates of mosquito larvae present in such habitats. Human-made water channeling systems, such as irrigation systems, dams, and other water-holding bodies, have also enabled mosquitoes to spread (Meyer Steiger et al., 2016b).

7 Factors Contributing to Mosquitoes-Host Attraction

Pregnant women are more attractive to *Anopheles gambiae* mosquitoes, which increases their risk of developing malaria and other mosquito-borne illnesses (Tolle, 2009; Tandina et al., 2018). They engage in behaviors that subject them to more mosquito bites, like getting up more often at night to use the restroom, as a result of the incidence of exposure. However, it is also due to increased metabolic rate and heat production; pregnant women exhale more mosquito attractants like CO_2 (Ellwanger et al., 2021).

The diet is the most frequently reported anecdotal explanation for mosquito attraction, suggesting that eating garlic, vitamin B, or beer can either repel or attract insects. In a double-blind trial, garlic was ineffective as a repellent against *Aedes aegypti*, while vitamin B had a similar result when tested with *Anopheles stephensi*. Contrarily, beer consumption has been shown to increase the likelihood of contracting *Anopheles gambiae*, compared to alcohol-free control groups (Fernández-Grandon et al., 2015), and alcohol consumption has been shown to increase the attractiveness of humans to *Anopheles gambiae* (ss) due to changes in body odor brought on by alcohol metabolism (Lefèvre et al., 2010) (Fig. 22.3).

According to a study by Fernández-Grandon et al. (2015), the genetically defined human fragrance profile may be a critical factor in mosquito attraction. It is thought that the major histocompatibility complex (MHC) genes, which are involved in mate selection and the prevention of inbreeding, control odor cues connected to genetic similarity. The skin microbiota, MHC, or peptide metabolism by the microflora are thought to be the sources of these fragrances. Given the substantial overlap between the fragrances linked to MHC and the origin of the odors that trigger behavioral responses in mosquitoes, this area of genetic research appears attractive. Human genetic variables affect the person's vulnerability to mosquitoes, clinical characteristics of illnesses, and susceptibility to diseases spread by mosquitoes. Studies on gene variants lend credence to this notion (Ellwanger et al., 2021). For instance, those who are homozygous carriers of the CCR5 gene variant CCR532 (rs333), which is a dangerous illness transmitted by mosquitoes that are on the rise in many nations, are more likely to experience the symptoms of West Nile virus infection (Glass et al., 2006). Genetic variables can also change kairomones' release patterns by affecting the production and diffusion of volatile odor molecules or, more subtly, by changing the makeup of the particular microbiota that metabolizes cutaneous gland products.

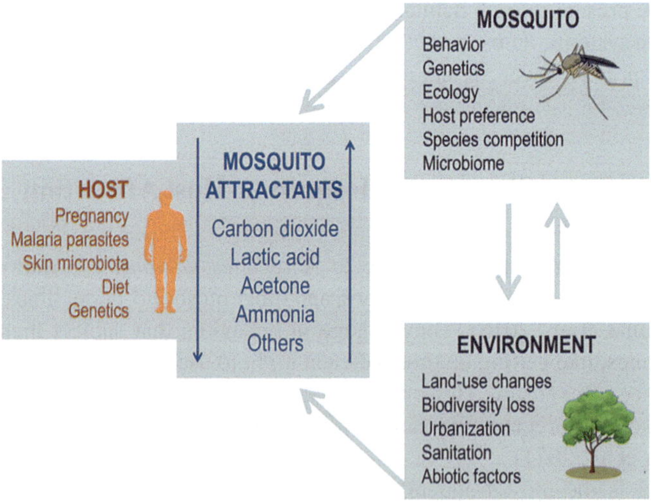

Fig. 22.3 Different factors involved in the production of Mosquito attractants in human (Ellwanger et al., 2021)

Lacroix et al. (2005) study sought to understand how malaria gametocytes lure mosquitoes to their host. In Mbita region, Kenya, where this age range (3–15) represents the largest reservoir of gametocytes, the study recruited 12 groups of three children aged 3–15. According to the findings, the higher mosquito attraction was not caused by the attractiveness of gametocyte carriers themselves but rather by the infection state associated with the presence of gametocytes. Although the specific process underpinning this alteration is unknown, the parasites likely change the host's breathing or body odor because both elements are essential for attracting mosquitoes. Interestingly, malaria infection modifies the emission of volatile molecules, creating an "odor signature" of the host or "malaria smell," which could be used for high-sensitivity diagnostic tests.

8 Conclusion

By identifying malaria super-spreaders, measures like long-lasting insecticide-treated nets (LLINs) and indoor residual spraying (IRS) coverage can combat the disease. The disproportionately high number of mosquito bites experienced by a small subset of the population highlights the existence of transmission heterogeneity, wherein mosquitoes preferentially target more attractive hosts. This selective behavior suggests that these individuals serve as focal points for disease transmission. Focusing vector control efforts on these high-risk groups, rather than implementing blanket population-wide strategies, may offer a more effective approach to reducing the transmission intensity and overall propensity of infection. Targeted interventions could disrupt the mosquito-host interactions that drive transmission cycles, thereby improving public health outcomes. Noninvasive techniques could be employed to identify individuals who exhibit elevated mosquito attraction, such as detecting specific volatile organic compounds (VOCs) associated with heightened mosquito preference or asymptomatic infections (Liao et al., 2015). While the role of host-produced volatile compounds in modulating differential mosquito attraction is widely acknowledged, effective control strategies must target specific stages of the mosquito's life cycle to disrupt transmission dynamics. Such targeted interventions could mitigate mosquito survival and reproduction, preventing them from perpetuating disease cycles and enhancing the efficacy of vector management programs. The VOCs can be harnessed to moderate mosquito affinity for human hosts and enhance noninvasive diagnostic techniques.

Acknowledgments The authors thank the management of CUCRID, Covenant University, Ota Nigeria, for providing the publication funding support and Covenant Applied Informatics and Communication Africa Center of Excellence (CApIC-ACE) for their support in data collection.

Authors Contribution OT—Conceptualization, Supervision, Methodology, Visualization, Validation.

OG and OI—Review and editing, Writing-original draft, Investigation, Visualization, project management.

OT, KZ, and AV—Data curation, Analysis, Methodology, Project administration.

Conflict of Interest The authors declare that they have no conflict of interest.

References

Acree, F., Turner, R. B., Gouck, H. K., Beroza, M., & Smith, N. (1968). L-Lactic acid: A mosquito attractant isolated from humans. *Science (New York, N.Y.), 161*(3848), 1346–1347. https://doi.org/10.1126/science.161.3848.1346

Ajide, K. B., Zhang, Q., Ibrahim, R. L., & Shah, S. A. R. (2022). The spread of and death from infectious diseases in sub-Saharan Africa: Implications for FDI attraction. *International Journal of Environmental Research and Public Health, 19*(22), Article 22. https://doi.org/10.3390/ijerph192214659

Cholewinski, M., Derda, M., & Hadas, E. (2015). Parasitic diseases in humans transmitted by vectors. *Annals of Parasitology, 61*(3), 137–157.

Cummins, B., Cortez, R., Foppa, I. M., Walbeck, J., & Hyman, J. M. (2012). A spatial model of mosquito host-seeking behavior. *PLoS Computational Biology, 8*(5), e1002500.

Dormont, L., Mulatier, M., Carrasco, D., & Cohuet, A. (2021). Mosquito attractants. *Journal of Chemical Ecology, 47*, 351–393.

Ellwanger, J. H., da Cruz Cardoso, J., & Chies, J. A. B. (2021). Variability in human attractiveness to mosquitoes. *Current Research in Parasitology & Vector-Borne Diseases, 1*, 100058. https://doi.org/10.1016/j.crpvbd.2021.100058

Endo, N., & Eltahir, E. A. (2018). Prevention of malaria transmission around reservoirs: An observational and modelling study on the effect of wind direction and village location. *The Lancet Planetary Health, 2*(9), e406–e413.

Fernández-Grandon, G. M., Gezan, S. A., Armour, J. A., Pickett, J. A., & Logan, J. G. (2015). Heritability of attractiveness to mosquitoes. *PLoS One, 10*(4), e0122716.

Fillinger, U., Kannady, K., William, G., Vanek, M. J., Dongus, S., Nyika, D., Geissbühler, Y., Chaki, P. P., Govella, N. J., Mathenge, E. M., Singer, B. H., Mshinda, H., Lindsay, S. W., Tanner, M., Mtasiwa, D., de Castro, M. C., & Killeen, G. F. (2008). A tool box for operational mosquito larval control: Preliminary results and early lessons from the Urban Malaria Control Programme in Dar es Salaam, Tanzania. *Malaria Journal, 7*(1), 20. https://doi.org/10.1186/1475-2875-7-20

Flowers, L., & Grice, E. (2020). The skin microbiota: Balancing risk and reward. *Cell Host & Microbe, 28*, 190–200. https://doi.org/10.1016/j.chom.2020.06.017

Franklinos, L. H. V., Jones, K. E., Redding, D. W., & Abubakar, I. (2019). The effect of global change on mosquito-borne disease. *The Lancet Infectious Diseases, 19*(9), e302–e312. https://doi.org/10.1016/S1473-3099(19)30161-6

Glass, W. G., McDermott, D. H., Lim, J. K., Lekhong, S., Yu, S. F., Frank, W. A., Pape, J., Cheshier, R. C., & Murphy, P. M. (2006). CCR5 deficiency increases risk of symptomatic West Nile virus infection. *The Journal of Experimental Medicine, 203*(1), 35–40. https://doi.org/10.1084/jem.20051970

Grice, E. A., & Segre, J. A. (2011). The skin microbiome. *Nature Reviews Microbiology, 9*(4), 244–253. https://doi.org/10.1038/nrmicro2537

Hawkes, F. M., & Hopkins, R. J. (2022). The mosquito: An introduction. In M. Hall & D. Tamïr (Eds.), *Mosquitopia: The place of pests in a healthy world*. Routledge. http://www.ncbi.nlm.nih.gov/books/NBK585164/

He, J., Zou, Z., & Yang, X. (2019). Measuring whole-body volatile organic compound emission by humans: A pilot study using an air-tight environmental chamber. *Building and Environment, 153*, 101–110.

Kaindoa, E. W., Mkandawile, G., Ligamba, G., Kelly-Hope, L. A., & Okumu, F. O. (2016). Correlations between household occupancy and malaria vector biting risk in rural Tanzanian villages: Implications for high-resolution spatial targeting of control interventions. *Malaria Journal, 15*(1), 199. https://doi.org/10.1186/s12936-016-1268-8

Konopka, J. K., Task, D., Afify, A., Raji, J., Deibel, K., Maguire, S., Lawrence, R., & Potter, C. J. (2021). Olfaction in Anopheles mosquitoes. *Chemical Senses, 46*, bjab021. https://doi.org/10.1093/chemse/bjab021

Kumar, A., Tauxe, G. M., Perry, S., Scott, C. A., Dahanukar, A., & Ray, A. (2020). Contributions of the conserved insect carbon dioxide receptor subunits to odor detection. *Cell Reports, 31*(2), 107510. https://doi.org/10.1016/j.celrep.2020.03.074

Lacroix, R., Mukabana, W. R., Gouagna, L. C., & Koella, J. C. (2005). Malaria infection increases attractiveness of humans to mosquitoes. *PLoS Biology, 3*(9), e298.

Lefèvre, T., Gouagna, L.-C., Dabiré, K. R., Elguero, E., Fontenille, D., Renaud, F., Costantini, C., & Thomas, F. (2010). Beer consumption increases human attractiveness to malaria mosquitoes. *PLoS One, 5*(3), e9546.

Liao, C.-M., Huang, T.-L., Cheng, Y.-H., Chen, W.-Y., Hsieh, N.-H., Chen, S.-C., & Chio, C.-P. (2015). Assessing dengue infection risk in the southern region of Taiwan: Implications for control. *Epidemiology & Infection, 143*(5), 1059–1072.

Martinez, J., Showering, A., Oke, C., Jones, R. T., & Logan, J. G. (2021). Differential attraction in mosquito–human interactions and implications for disease control. *Philosophical Transactions of the Royal Society B, 376*(1818), 20190811.

Mbama Ntabi, J. D., Lissom, A., Djontu, J. C., Diafouka-Kietela, S., Vouvoungui, C., Boumpoutou, R. K., Mayela, J., Nguiffo-Nguete, D., Nkemngo, F. N., & Ndo, C. (2022). Prevalence of non-plasmodium falciparum species in southern districts of Brazzaville in the Republic of the Congo. *Parasites & Vectors, 15*(1), 209.

McBride, C. S. (2016). Genes and odors underlying the recent evolution of mosquito preference for humans. *Current Biology, 26*(1), R41–R46. https://doi.org/10.1016/j.cub.2015.11.032

McBride, C. S., Baier, F., Omondi, A. B., Spitzer, S. A., Lutomiah, J., Sang, R., Ignell, R., & Vosshall, L. B. (2014). Evolution of mosquito preference for humans linked to an odorant receptor. *Nature, 515*(7526), 222–227. https://doi.org/10.1038/nature13964

Meyer Steiger, D. B., Ritchie, S. A., & Laurance, S. G. W. (2016a). Land use influences mosquito communities and disease risk on remote tropical islands: A case study using a novel sampling technique. *The American Journal of Tropical Medicine and Hygiene, 94*(2), rajirruiz-314–321. https://doi.org/10.4269/ajtmh.15-0161

Meyer Steiger, D. B., Ritchie, S. A., & Laurance, S. G. W. (2016b). Mosquito communities and disease risk influenced by land use change and seasonality in the Australian tropics. *Parasites & Vectors, 9*(1), 387. https://doi.org/10.1186/s13071-016-1675-2

Monroe, A., Williams, N. A., Ogoma, S., Karema, C., & Okumu, F. (2022). Reflections on the 2021 world malaria report and the future of malaria control. *Malaria Journal, 21*, 154.

Mukabana, W. R., Takken, W., Killeen, G. F., & Knols, B. G. (2004). Allomonal effect of breath contributes to differential attractiveness of humans to the African malaria vector Anopheles gambiae. *Malaria Journal, 3*(1), 1–8.

Mzilahowa, T., Luka-Banda, M., Uzalili, V., Mathanga, D. P., Campbell, C. H., Mukaka, M., & Gimnig, J. E. (2016). Risk factors for Anopheles mosquitoes in rural and urban areas of Blantyre District, southern Malawi. *Malawi Medical Journal, 28*(4), 154–158.

Natsch, A., & Emter, R. (2020). The specific biochemistry of human axilla odour formation viewed in an evolutionary context. *Philosophical Transactions of the Royal Society B, 375*(1800), 20190269.

Ortiz, D. I., Piche-Ovares, M., Romero-Vega, L. M., Wagman, J., & Troyo, A. (2022). The impact of deforestation, urbanization, and changing land use patterns on the ecology of mosquito and tick-borne diseases in Central America. *Insects, 13*(1), Article 1. https://doi.org/10.3390/insects13010020

Owino, E. (2021). Human and plant volatiles; lures for mosquito, vectors of dengue virus and malaria. *Journal of Vector Borne Diseases, 58*(1), 1–1.

Pohanka, M. (2020). D-lactic acid as a metabolite: Toxicology, diagnosis, and detection. *BioMed Research International, 2020*, 3419034. https://doi.org/10.1155/2020/3419034

Raji, J. I., Melo, N., Castillo, J. S., Gonzalez, S., Saldana, V., Stensmyr, M. C., & DeGennaro, M. (2019). Aedes aegypti mosquitoes detect acidic volatiles found in human odor using the IR8a pathway. *Current Biology, 29*(8), 1253–1262.e7. https://doi.org/10.1016/j.cub.2019.02.045

Robinson, A., Busula, A. O., Voets, M. A., Beshir, K. B., Caulfield, J. C., Powers, S. J., Verhulst, N. O., Winskill, P., Muwanguzi, J., Birkett, M. A., Smallegange, R. C., Masiga, D. K., Mukabana, W. R., Sauerwein, R. W., Sutherland, C. J., Bousema, T., Pickett, J. A., Takken, W., Logan, J. G., & de Boer, J. G. (2018). Plasmodium-associated changes in human odor attract mosquitoes. *Proceedings of the National Academy of Sciences of the United States of America, 115*(18), E4209–E4218. https://doi.org/10.1073/pnas.1721610115

Romano-Bertrand, S., Bourdier, A., Aujoulat, F., Michon, A.-L., Masnou, A., Parer, S., Marchandin, H., & Jumas-Bilak, E. (2016). Skin microbiota is the main reservoir of Roseomonas mucosa, an emerging opportunistic pathogen so far assumed to be environmental. *Clinical Microbiology and Infection: The Official Publication of the European Society of Clinical Microbiology and Infectious Diseases, 22*(8), 737.e1–737.e7. https://doi.org/10.1016/j.cmi.2016.05.024

Rose, N. H., Sylla, M., Badolo, A., Lutomiah, J., Ayala, D., Aribodor, O. B., Ibe, N., Akorli, J., Otoo, S., Mutebi, J.-P., Kriete, A. L., Ewing, E. G., Sang, R., Gloria-Soria, A., Powell, J. R., Baker, R. E., White, B. J., Crawford, J. E., & McBride, C. S. (2020). Climate and urbanization drive mosquito preference for humans. *Current Biology, 30*(18), 3570–3579.e6. https://doi.org/10.1016/j.cub.2020.06.092

Ruiz-López, M. J. (2020). Mosquito behavior and vertebrate microbiota interaction: Implications for pathogen transmission. *Frontiers in Microbiology, 11*, 573371.

Serag, A., Shakkour, Z., Halboup, A. M., Kobeissy, F., & Farag, M. A. (2021). Sweat metabolome and proteome: Recent trends in analytical advances and potential biological functions. *Journal of Proteomics, 246*, 104310.

Showering, A., Martinez, J., Benavente, E. D., Gezan, S. A., Jones, R. T., Oke, C., Tytheridge, S., Pretorius, E., Scott, D., Allen, R. L., D'Alessandro, U., Lindsay, S. W., Armour, J. A. L., Pickett, J., & Logan, J. G. (2022). Skin microbiome alters attractiveness to Anopheles mosquitoes. *BMC Microbiology, 22*(1), 98. https://doi.org/10.1186/s12866-022-02502-4

Takken, W., & Knols, B. (1999). Odour mediated behaviour of Afrotropical malaria mosquitoes. *Annual Review of Entomology, 44*, 131–157. https://doi.org/10.1146/annurev.ento.44.1.131

Takken, W., & Verhulst, N. O. (2017). Chemical signaling in mosquito–host interactions: The role of human skin microbiota. *Current Opinion in Insect Science, 20*, 68–74. https://doi.org/10.1016/j.cois.2017.03.011

Tandina, F., Doumbo, O., Yaro, A. S., Traoré, S. F., Parola, P., & Robert, V. (2018). Mosquitoes (Diptera: Culicidae) and mosquito-borne diseases in Mali, West Africa. *Parasites & Vectors, 11*(1), 467. https://doi.org/10.1186/s13071-018-3045-8

Thorn, R. M. S., & Greenman, J. (2012). Microbial volatile compounds in health and disease conditions. *Journal of Breath Research, 6*(2), 024001. https://doi.org/10.1088/1752-7155/6/2/024001

Tolle, M. A. (2009). Mosquito-borne diseases. *Current Problems in Pediatric and Adolescent Health Care, 39*(4), 97–140. https://doi.org/10.1016/j.cppeds.2009.01.001

Verhulst, N., Umanets, A., Weldegergis, B., Maas, J., Visser, T., Dicke, M., Smidt, H., & Takken, W. (2018). Do apes smell like humans? The role of skin bacteria and volatiles of primates in mosquito host selection. *Journal of Experimental Biology, 221*(Pt 22), jeb.185959. https://doi.org/10.1242/jeb.185959

Verhulst, N. O., Smallegange, R. C., & Takken, W. (2012). Mosquitoes as potential bridge vectors of malaria parasites from non-human primates to humans. *Frontiers in Physiology, 3*, 197.

Verhulst, N. O., Weldegergis, B. T., Menger, D., & Takken, W. (2016). Attractiveness of volatiles from different body parts to the malaria mosquito Anopheles coluzzii is affected by deodorant compounds. *Scientific Reports, 6*, 27141. https://doi.org/10.1038/srep27141

Wanjiru, J. J. (2020). *Responses of the malaria vector, Anopheles gambiae, to plant- and mammalian-derived odors*. Thesis, JKUAT-COHES. http://localhost/xmlui/handle/123456789/5368

Wilkerson, R. C., Linton, Y.-M., & Strickman, D. (2021). *Mosquitoes of the world*. JHU Press.

Wolff, G. H., & Riffell, J. A. (2018). Olfaction, experience and neural mechanisms underlying mosquito host preference. *The Journal of Experimental Biology, 221*(4), jeb157131. https://doi.org/10.1242/jeb.157131

Wooding, M., Naudé, Y., Rohwer, E., & Bouwer, M. (2020). Controlling mosquitoes with semiochemicals: A review. *Parasites & Vectors, 13*, 1–20.

Yi, H., Devkota, B. R., Yu, J., Oh, K., Kim, J., & Kim, H.-J. (2014). Effects of global warming on mosquitoes & mosquito-borne diseases and the new strategies for mosquito control. *Entomological Research, 44*(6), 215–235. https://doi.org/10.1111/1748-5967.12084

Zhao, Z., Zung, J. L., Kriete, A. L., Iqbal, A., Younger, M. A., Matthews, B. J., Merhof, D., Thiberge, S., Strauch, M., & McBride, C. S. (2020). Chemical signatures of human odour generate a unique neural code in the brain of Aedes aegypti mosquitoes. *bioRxiv*. https://doi.org/10.1101/2020.11.01.363861

Chapter 23
PD-1/PD-L1 Inhibitors and Triple-Negative Breast Cancer (TNBC): A Review

Olayemi Oluseun Akinnola and Oluwanifemi Omodara Ogunleye

1 Introduction

Cancer, also known as tumor or neoplasm, is the abnormal development of cells and is responsible for most deaths globally (Sitki-Copur, 2019; Iweala et al., 2015). These cancerous cells are usually uncontrolled, spreading from affected to unaffected cells until the tissues and other organs are destroyed. They are also responsible for the alterations of genetic makeup and cause epigenesis (Bray et al., 2018). This unregulated growth of cells leads to the formation of a lump or mass which, if cancerous, is referred to as a neoplasm or tumor (Saini et al., 2020).

In 2020, cancer accounted for nearly ten million deaths, whereas in 2022, there was an incidence of 20 million cases and a mortality rate of approximately 9.7 million, making it a primary public health concern. Beyond its detrimental impact on life expectancy, cancer poses a significant socioeconomic burden, incurring substantial healthcare costs that vary across cancer types, geographic regions, and genders (Bray et al., 2024). Globally, cancer of the lungs is the most frequently diagnosed tumor, followed closely by breast, colorectal, prostate, and stomach cancers. However, mortality rates differ in terms of gender. The primary cause of cancer mortality in men remains cancer of the lungs, followed by prostate and liver cancers. For women, the case is different, with breast cancer topping the list for both incidence and mortality, closely followed by cervical cancer (Bray et al., 2024).

O. O. Akinnola
Department of Biological Sciences, Covenant University, Ota, Ogun State, Nigeria

O. O. Ogunleye (✉)
Department of Biological Sciences, Covenant University, Ota, Ogun State, Nigeria

Covenant Applied Informatics and Communication Africa Centre of Excellence (CApIC-ACE), Covenant University, Ota, Ogun State, Nigeria
e-mail: oluwanifemi.ogunleyepgs@stu.cu.edu.ng

P. O. Isibor, P. A. Akinduti (eds.), *Harnessing Biotechnology Tools for Product Development*, https://doi.org/10.1007/978-3-031-86002-7_23

Globally, breast cancer has been confirmed to be the main cause of female deaths (Sung et al., 2021; Allo et al., 2019; Katz & Alsharedi, 2018). Approximately two million women suffer from breast cancer yearly, including a mortality rate of over 600,000 as a result of tumor relapse or rapid spread to distant body sites (Malabadi et al., 2024). There are widely recognized breast cancer subtypes like the HER2-positive (human epidermal growth factor 2 receptor), luminal A, luminal B, and triple-negative breast cancer subtype (TNBC). The existence of the diverse subtypes is attributed to the occurrence of unique molecular targets, including HER2, estrogen, and progesterone receptors. The several molecular diversities influence the classification of the breast cancer type, complicate the management of breast cancer in women, and determine the rate of cancer recurrence and response to treatments (Arenas et al., 2024).

2 Triple-Negative Breast Cancer (TNBC)

Although breast cancer manifests differently in each woman, it is classified based on the existence or lack of the receptors (HER2, estrogen, and progesterone). The unique triple-negative subtype, identified in 2005, remains a scientific mystery and lacks all three markers, resulting in its name. Despite ongoing research, its cause remains unclear (Tavares et al., 2021). The triple-negative breast cancer, TNBC, is considered the most violent of all breast cancer subtypes not only because it lacks the unique molecular targets present in the other subtypes but also for its rapid progression, lower chance of recovery, and occurrence in individuals below 40 years of age with BRCA1 mutation (Wesolowski et al., 2022).

Several research has been carried out on triple-negative breast cancer. A significant number, roughly 200,000 women worldwide, are affected by TNBC annually, representing nearly one-fifth of all breast cancer cases (Katz & Alsharedi, 2018). Recently, it was observed that TNBC is most prevalent among Indian women (Malabadi et al., 2024) compared to women from America or Europe (Nag et al., 2023). TNBC disproportionately affects Indian women, with prevalence ranging from 20% to 43% of all breast cancer cases (Malabadi et al., 2024) while women in the United States suffer from TNBC as a second leading cause of cancer-related deaths (Katz & Alsharedi, 2018). Several factors contribute to an increased risk of TNBC in women, including inherited mutations in BRCA genes, being African-American in premenopausal state, having had more than three children, getting pregnant at a young age, and choosing not to breastfeed, obesity (Katz & Alsharedi, 2018).

A very dangerous feature associated with TNBC is the inability to predict it. Its lack of molecular receptors makes it aggressive, with the potential to metastasize and resist treatment options, leading to worse outcomes and an average survival of only 8–15 months for an individual with metastatic TNBC (Wesolowski et al., 2022). While chemotherapy remains the primary treatment due to a lack of specific targets, it often comes with side effects, and patients often develop resistance quickly. The frequent relapses in TNBC highlight the urgent need for new treatment strategies (Wesolowski et al., 2022). This review gives an overview of the application of immunotherapy, specifically immune checkpoint inhibitors, in eradicating triple-negative breast cancer, encompassing several combination strategies and associated advantages.

3 Applicable Treatment Options of Triple-Negative Breast Cancer

The subtype, triple-negative breast carcinoma (TNBC), is distinguished by the absence of these targetable HER2 or hormone receptors, making it the most violent of all subtypes with a poor outlook. In comparison with other subtypes, it has a high risk of recurrence and often metastasizes to the brain in individuals below 40 years, leading to a reduced rate of survival (Arenas et al., 2024). Treatment of TNBC experiences setbacks as a result of its lack of estrogen, progesterone, and HER2 receptors. This renders receptor-targeting treatments ineffective, leaving surgery and chemotherapy as traditional treatment options (Malabadi et al., 2024). The current most applied treatment method to overcome TNBC is neo-adjuvant chemotherapy, which is used to make the tumor smaller before surgery. Despite neoadjuvant chemotherapy (NAC) being the standard treatment for triple-negative breast cancer (TNBC), with some success using anthracyclines and taxanes, significant variations exist in patient response (Arenas et al., 2024).

The sensitivity of TNBC to chemotherapy makes it an initial go-to option for treating this tumor type; however, there are certainly associated setbacks including recurrence and resistance to treatment that make chemotherapy a zero option for patients who experience such setbacks after an initial therapeutic administration, to prevent further complexities and metastases. Even though it accounts for about 20% of breast cancers globally (Wesolowski et al., 2022), TNBC is the most dangerous breast cancer subtype that can substantially shorten the lifespan of an individual, due to its radical nature, in terms of its ability to spread to distant organs, tumor relapse, and the absence of molecular targets which serves as a limitation to several treatment options and promotes resistance to tumor therapy (Arenas et al., 2024). However, a beacon of hope emerges with the development of immunotherapies, offering promising avenues for improved TNBC management alongside established surgical and chemotherapeutic approaches.

4 Cancer Immunotherapy: Immune Checkpoint Inhibitors as a Promising Method

Cancer immunotherapy is a remarkable therapeutic method that stimulates the immune system to eliminate tumor cells and achieve a long-lasting clinical response (Liu et al., 2024). It involves the use of substances made from living organisms, either directly by the body or in a laboratory, to train and boost the ability of the immune system to fight cancer (Abbott & Ustoyev, 2019), hence called biological therapy (Saini et al., 2020). Immunotherapy is though an emerging but promising technique undergoing rigorous testing to ensure safety and effectiveness (Abbott & Ustoyev, 2019). Current advancements in immunotherapy encompass a diverse range of strategies, including cancer vaccines, adoptive cell therapy, and immune checkpoint inhibitors (ICIs) (He & Wang, 2022).

Each immunotherapy works uniquely, but they all empower the immune system (Abbott & Ustoyev, 2019). A form of specific immunotherapy and a current promising area of research is immune checkpoint inhibitors (ICIs). Intentionally, tumor cells can avoid the immune system by activating immune checkpoints through direct binding and signaling to indicate themselves as "self" (He & Xu, 2020). Immune checkpoints, including CTLA-4 (cytotoxic T-lymphocyte antigen number 4 and programmed cell death molecule 1) and PD-1 (programmed cell death molecule 1), are mechanisms used by the immune system to control immune responses and prevent the unplanned attack of healthy cells (Esfahani et al., 2020). In other words, they control immune responses from being too strong. This natural process can become a full-blown disaster when cancer cells use it to their advantage, securing safe growth and proliferation (Wesolowski et al., 2022).

Immune checkpoint inhibitor drugs interrupt this natural process that tumors exploit to evade immune attack by interrupting the interaction between immune checkpoints and their ligands on cancer cells, disrupting tumor checkpoint binding (He & Xu, 2020). The discovery and development of checkpoint inhibitors have led to increased antitumor responses due to a drastic activation of the immune system (Esfahani et al., 2020). Over the past decade, this discovery, particularly those inhibiting the PD-1/PD-L1 pathway, has substantially improved cancer immunotherapy with their successful clinical application in treating various cancers (Liu et al., 2024; Upadhaya et al., 2022).

Ipilimumab was FDA-approved in 2011 as the first CTLA-4 checkpoint inhibitor for the treatment of advanced melanoma and is currently used to treat several other cancer types while nivolumab was FDA-approved in 2014 as the first PD-1 inhibitor to treat cancer, particularly melanoma in Japan (Dobosz & Dzieciątkowski, 2019). Currently, pembrolizumab (PD1 inhibitor) is administered to treat melanomas, metastatic breast cancer, advanced urothelial carcinomas, and cervical tumors. It can be administered alone or in combination with other treatments. (Jahangiri & Yu, 2024). Additionally, atezolizumab serves as a checkpoint inhibitor for the PD-L1 protein. It was approved in 2016 and has been effective in treating bladder cancer, melanoma, lung cancer, and breast cancer. Avelumab, dostarlimab, and durvalumab are several other immune checkpoint inhibitors that can be employed in the treatment of tumors (Dobosz & Dzieciątkowski, 2019). Notably, ICIs function by reinvigorating cytotoxic T lymphocytes (CTLs), thereby reversing the immunosuppression observed in cancer patients (Darvin et al., 2018). This mechanism employed by the PD-1/PD-L1 inhibitors is illustrated in Fig. 23.1.

5 The Role of PD-1/PD-L1 Inhibitors

Programmed cell death protein 1 (PD-1) is located on the surface of immune cells, like T lymphocytes, and acts as an "immune checkpoint" regulating their activity. This checkpoint works through interaction with its ligand, PD-L1, present in cancer cells. When activated by an antigen, T cells normally upregulate PD-1, which helps to prevent excessive immune response. Interestingly, cancer cells are known to

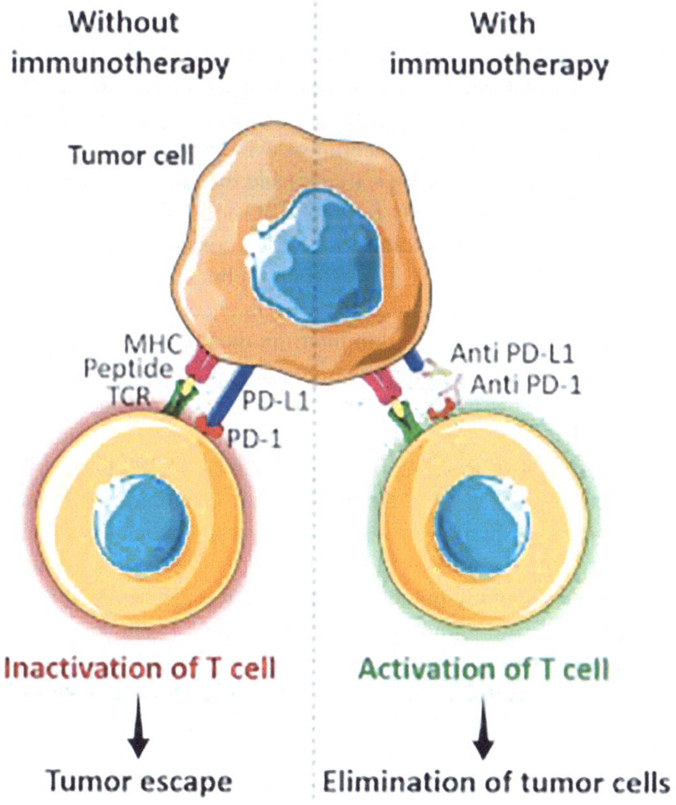

Fig. 23.1 Mechanism of checkpoint inhibitors. Source: Genoway (2019)

exploit this mechanism by expressing PD-L1 to bind with PD-1, essentially putting the brakes on T-cell function (Wesolowski et al., 2022).

PD-1/PD-L1 interaction can lead to several negative effects on the immune system including reduced production of T cells, interferons, interleukins, and overall, a lowered survival rate. These effects can lead to a state of T cell exhaustion, hindering their ability to fight cancer cells. Notably, tumor-infiltrating T cells often express higher levels of PD-1, suggesting this pathway plays a significant role in tumor immune evasion. Therefore, blocking the PD-1 pathway with specific checkpoint inhibitors holds promise for enhancing T-cell function and improving antitumor responses in cancer patients. This strategy aims to remove the "brake" applied by PD-L1 and unleash the full potential of the immune system against cancer cells (Wesolowski et al., 2022).

6 TNBC Treatment with PD-1/PD-L1 Inhibitors

Cancer progression is often associated with the development of immunosuppressive mechanisms, a strategy employed by cancer cells to evade immune system recognition and destruction. Tumor cells manipulate immune checkpoint pathways to hinder the body's natural antitumor response (Darvin et al., 2018). Triple-negative breast cancer (TNBC) cells express PD-L1 on their surface in roughly 20% of cases. This expression is linked to two concerning developments: increased resistance to a specific type of immune cell (CD8+ T cells) and the promotion of tumor growth. In simpler terms, the presence of PD-L1 on TNBC cells allows them to "turn off" the immune system's attack by binding to the PD-1 receptor found on immune cells such as natural killer (NK) cells, T and B lymphocytes. This effectively silences immune response and allows the tumor to grow unchecked (Wesolowski et al., 2022).

PD-1/PD-L1 inhibitors are used to interrupt the communication between cancer cells and T cells, thereby enabling the immune system to elicit a stronger response against neoplasms. This therapeutic method has been considered effective in the treatment of TNBC, however, it is still undergoing rigorous studies and research (Bayraktar et al., 2019). Research has revealed that TNBC tumors tend to have a greater infiltration by CD8+ T cells, proving that this tumor type has the ability to provoke an immune response. This explains why the application of immunotherapy as a therapeutic intervention in TNBC management is being studied (Katz & Alsharedi, 2018). Nivolumab and pembrolizumab are checkpoint inhibitor drugs that target the PD-1 (programmed cell death receptor 1) and have been studied for use in the treatment of TNBC. These antibodies bind to the PD-1 receptor on T cells, prevent interaction with the ligand on the tumor cell, and allow the T cells to unleash an adequate immune response against neoplasms. Both nivolumab and pembrolizumab have been approved by the FDA as effective treatment options for solid tumors; however, research is still ongoing to determine its effectiveness in treating triple-negative breast cancer (Katz & Alsharedi, 2018).

Several clinical trials on the effectiveness of ICIs on TNBC have been carried out. A certain trial called KEYNOTE-012, involving pembrolizumab, was conducted on certain patients suffering from metastatic TNBC (Katz & Alsharedi, 2018). Tumor samples of precisely 111 metastatic TNBC patients were assessed for the expression of PD-L1 which suppresses immune response. Among these patients, over half (58.6%) had tumors that expressed PD-L1 at levels of at least 1%. To evaluate the safety and effectiveness of pembrolizumab in treating TNBC, 32 women who had previously been administered chemotherapy were separated for this purpose (Tavares et al., 2021). The treatment showed promising antitumor activity, with an overall response rate of 18.5%, including complete response in 1 patient (3.7%) and partial response in 4 patients (14.8%). This supported the theory that increased expression of PD-L1 improves the effectiveness of pembrolizumab (He & Wang, 2022). The San Antonio Cancer Symposium released a report on this trial in 2016, stating that long-lasting response was established in three patients out of five who had responded to treatment. This study provided encouraging evidence for pembrolizumab's potential in treating metastatic triple-negative breast cancer (TNBC). The drug, administered every 2 weeks to patients who had previously undergone chemotherapy, demonstrated both clinical activity and an acceptable safety profile (Katz & Alsharedi, 2018). Pembrolizumab is not without side effects. Roughly 60% of patients experience treatment-related issues, with fatigue and nausea being the most common. More concerning are the potential immune-mediated side effects like hypothyroidism, hyperthyroidism, and pneumonitis. These highlight the importance of careful monitoring by healthcare professionals when considering pembrolizumab as a treatment option (Tavares et al., 2021).

Immunotherapy, specifically the use of PD-L1 (programmed death ligand 1) inhibitors such as avelumab and atezolizumab, has shown promising results in treating triple-negative breast cancer (Debien et al., 2023; Tavares et al., 2021). A study explored the effectiveness of avelumab, a PD-L1 targeting antibody, in treating metastatic breast cancer. Among the 58 patients with triple-negative breast cancer (TNBC), the overall response rate (ORR) was 5.2%, with stable disease observed in 15 patients. Interestingly, the ORR increased to 6.1% for those with PD-L1 expression (>1%) on their tumors (Katz & Alsharedi, 2018). This suggests a potential benefit for patients with TNBC whose tumors express PD-L1. The study also found avelumab to be tolerable, with fatigue, nausea and infusion reactions as the most common side effects. While these results are promising, further research is needed to confirm avelumab's efficacy in treating TNBC (Wesolowski et al., 2022).

Atezolizumab, another PD-L1 targeting antibody, shows promise in treating metastatic triple-negative breast cancer (TNBC). A study evaluated its safety and efficacy in patients with advanced solid tumors, including TNBC. Patients with at least 5% PD-L1 expression on their tumors saw a higher overall response rate (ORR) of 19% compared to those with lower expression. Interestingly, those with high levels of immune cells infiltrating the tumors (over 10%) also experienced a better response and longer survival (Tavares et al., 2021). These findings suggest that atezolizumab could be a valuable treatment option for patients with TNBC, particularly those whose tumors express PD-L1. However, side effects like nausea, cough, and neutropenia were more frequent in the atezolizumab group compared to placebo. More research is ongoing to confirm these results and explore potential ways to further improve the treatment's effectiveness (Wesolowski et al., 2022).

7 Conclusion

Chemotherapy remains the standard treatment, however, PD-1/PD-L1 inhibitor drugs offer a promising new avenue for treating triple-negative breast cancer (TNBC). These drugs boost the body's immunity against neoplasms, and early research suggests they may be particularly effective when used in the early stage of the disease. Further studies are needed to refine treatment approaches and identify optimal responders. In addition, the initial safety profile and positive results hold significant promise for the future of TNBC therapy. This paves the way for immunotherapy to potentially become a cornerstone of future TNBC treatment strategies.

8 Future Perspective

Researchers are actively investigating the use of several PD-1/PD-L1 inhibitors in the treatment of triple-negative breast cancer. These include nivolumab and pembrolizumab that target the PD-1, as well as atezolizumab, durvalumab, and avelumab that target the PD-L1 protein. Additionally, ongoing clinical trials are investigating the potential of other checkpoint inhibitors.

References

Abbott, M., & Ustoyev, Y. (2019). Cancer and the immune system: The history and background of immunotherapy. *Seminars in Oncology Nursing, 35*(5), 1–5.

Allo, T. A., Edewor, P. A., & Imhonopi, D. (2019). Assessment of perceived risks of breast cancer and breast cancer screening among women in five selected local governments in Ogun State, Nigeria. *Sage Open, 9*(2), 1–11.

Arenas, M., Fargas-Saladie, M., Moreno-Sole, M., Moyano-Femenia, L., Jimenez-Franco, A., Canela-Capdevila, M., Castane, H., Martínez-Navidad, C., Camps, J., & Joven, J. (2024). Metabolomics and triple-negative breast cancer: A systematic review. *Heliyon, 10*(1), e23628.

Bayraktar, S., Batoo, S., Okuno, S., & Glück, S. (2019). Immunotherapy in breast cancer. *Journal of Carcinogenesis, 18*, 2.

Bray, F., Ferlay, J., Soerjomataram, I., Siegel, R. L., Torre, L. A., & Jemal, A. (2018). Global cancer statistics 2018: GLOBOCAN estimates of incidence and mortality worldwide for 36 cancers in 185 countries. *CA: A Cancer Journal for Clinicians, 68*(6), 394–424.

Bray, F., Laversanne, M., Sung, H., Ferlay, J., Siegel, R. L., Soerjomataram, I., & Jemal, A. (2024). Global cancer statistics 2022: GLOBOCAN estimates of incidence and mortality worldwide for 36 cancers in 185 countries. *CA: A Cancer Journal for Clinicians, 74*, 229–263.

Darvin, P., Toor, S. M., Nair, S. V., & Elkord, E. (2018). Immune checkpoint inhibitors: Recent progress and potential biomarkers. *Experimental & Molecular Medicine, 50*(12), 1–11.

Debien, V., De Caluwé, A., Wang, X., Piccart-Gebhart, M., Tuohy, V. K., Romano, E., & Buisseret, L. (2023). Immunotherapy in breast cancer: An overview of current strategies and perspectives. *npj Breast Cancer, 9*, 7.

Dobosz, P., & Dzieciątkowski, T. (2019). The intriguing history of cancer immunotherapy. *Frontiers in Immunology, 10*, 2965–2976.

Esfahani, K., Roudaia, L., Buhlaiga, N., Del Rincon, S. V., Papneja, N., & Miller, W. H. (2020). A review of cancer immunotherapy: From the past, to the present, to the future. *Current Oncology, 27*(2), 87–97.

Genoway. (2019). *Immune checkpoint inhibitors: A strategy to tackle cancer?* Retrieved April 11, 2024, from https://www.genoway.com/science/publications/commentary/immune-checkpoint-inhibitors

He, X., & Xu, C. (2020). Immune checkpoint signaling and cancer immunotherapy. *Cell Research, 30*(8), 660–669.

He, Y., & Wang, X. (2022) Immunotherapy for triple-negative breast cancer. Trends in Immunotherapy. 6(2) 1–8.

Iweala, E. E. J., Liu, F.-F., Cheng, R.-R., Li, Y., Omonhinmin, C. A., & Zhang, Y.-J. (2015). Anticancer and free radical scavenging activity of some Nigerian food plants in vitro. *International Journal of Cancer Research, 11*(1), 41–51.

Jahangiri, S., & Yu, F. (2024). Fundamentals and applications of focused ultrasound-assisted cancer immune checkpoint inhibition for solid tumors. *Pharmaceutics, 16*(3), 411.

Katz, H., & Alsharedi, M. (2018). Immunotherapy in triple-negative breast cancer. *Medical Oncology, 35*, 13.

Liu, M. Jin, D., Yu, W., Yu, J., Cao, K., Cheng, J., Zheng, X., Wang, A., & Liu, Y. (2024) Enhancing tumor immunotherapy by multivalent anti-pd-l1 nanobody assembled via ferritin nanocage. Advanced Science. 1–12.

Malabadi, R. B., Sadiya, M. R., Kolkar, K. P., Mammadova, S. S., Chalannavar, R. K., Baijnath, H., Lavanya, L., & Munhoz, A. N. R. (2024). Triple negative breast cancer (TNBC): Signalling pathways – Role of plant-based inhibitors. *Open Access Research Journal of Biology and Pharmacy, 10*(2), 28–71.

Nag, S., Dikshit, R., Desai, S., Mane, A., Mhatre, S., Neve, R., Gurav, M., Bhosale, N., Perumal, P., Kembhavi, Y., Jethwa, D., Badwe, R., & Gupta, S. (2023). Risk factors for the development of triple-negative breast cancer versus non-triple-negative breast cancer: A case–control study. *Scientific Reports, 13*, 1–8.

Saini, A., Kumar, M., Bhatt, S., Saini, V., & Malik, A. (2020). Cancer causes and treatments. *International Journal of Pharmaceutical Sciences and Research, 11*(7), 3121–3134.

Sitki-Copur, M. (2019). State of cancer research around the globe. *Oncology Journal, 33*(5), 181–185.

Sung, H., Ferlay, J., Siegel, R. L., Laversanne, M., Soerjomataram, I., Jemal, A., & Bray, F. (2021). Global cancer statistics 2020: GLOBOCAN estimates of incidence and mortality worldwide for 36 cancers in 185 countries. *CA: A Cancer Journal for Clinicians, 71*(3), 209–249.

Tavares, D. F., Ribeiro, V. C., Andrade, M. A. V., Cardoso-Junior, L. M., Teixeira, T. R. G., Varrone, G. R., & Britto, R. L. (2021). Immunotherapy using PD-1/PDL-1 inhibitors in metastatic triple-negative breast cancer: A systematic review. *Oncology Reviews, 15*, 497–506.

Upadhaya, S., Neftelinov, S. T., Hodge, J., & Campbell, J. (2022). Challenges and opportunities in the PD1/PDL1 inhibitor clinical trial landscape. *Nature Review Drug Discovery, 21*(7), 482–483.

Wesolowski, J., Tankiewicz-Kwedlo, A., & Pawlak, D. (2022). Modern immunotherapy in the treatment of triple-negative breast cancer. *Cancers, 14*, 1–37.

Chapter 24
Local Knowledge Variation and Use Pattern of Eggplant (*Solanum* Species) in Southern Nigeria

Babafemi Ibukun Ajiboye, Jacob Olagbenro Popoola, Angela Obiageli Eni, and Conrad Asotie Omonhinmin

1 Introduction

Eggplant (*Solanum*) includes three closely related cultivated vegetable species, namely *S. melongena* L., *S. aethiopicum* L., and *Solanum macrocarpon* L. It is the third most widely grown Solanaceous vegetable. Although the value of eggplant lies in its health and nutritional properties when eaten raw (Chikaleke et al., 2020), all the parts of the plant are also used for diverse purposes (Han et al., 2021).

According to the Food and Agriculture Organization (FAO), the global total annual estimate of eggplant production is about 56 million tonnes cultivated on 1.86 million hectares (ha) of land, with an estimate of 55.1 million tonnes in 2019, 56.1 million tonnes in 2020, 56.4 million tonnes in 2021, and 56.9 million tonnes in 2022 (FAOSTAT, 2022). These numbers reflect the steady increase in global eggplant production, with notable contributions from key producers like China, India and Egypt. Globally, Asia is the world's leading eggplant producer, with China and India being significant producers of eggplant, accounting for 61% and 23% of total

B. I. Ajiboye · C. A. Omonhinmin (✉)
Department of Biological Sciences/Biotechnology Cluster, Covenant University, Ota, Nigeria
e-mail: conrad.omonhinmin@covenantuniversity.edu.ng

J. O. Popoola
Department of Biological Sciences/Biotechnology Cluster, Covenant University, Ota, Nigeria

Pure and Applied Biology Programme, College of Agriculture, Engineering and Science, Bowen University, Iwo, Nigeria

A. O. Eni
Department of Biological Sciences/Biotechnology Cluster, Covenant University, Ota, Nigeria

Regional Center of Excellence for Transboundary Plant Pathogens, Central and West African Virus Epidemiology (WAVE), Pôle Scientifique et d'Innovation, Université Félix Houphouët-Boigny, Abidjan, Ivory Coast

P. O. Isibor, P. A. Akinduti (eds.), *Harnessing Biotechnology Tools for Product Development*, https://doi.org/10.1007/978-3-031-86002-7_24

production, respectively (Bana et al., 2022). Other major producers of this crop are Egypt, Iran, Turkey, Indonesia, the Philippines, and Sri Lanka (Oladosu et al., 2021; Villanueva et al., 2021). Although there are no official figures or reports on the volume of eggplant produced in Nigeria, the crop is grown widely for its nutritional and medicinal benefits, among other uses (Naeem & Ugur, 2019), particularly in the Southern part of Nigeria.

According to Omonhinmin and Idu (2012), the overall choice of plants and plant parts consumed by locals is need-based and driven by cultures and specific characteristics. In most studies on ethnobotanical knowledge of plants, there is an unequal indigenous knowledge and perception of plant use among Indigenous populations, as it relates to differences in demographic variables such as gender, age, religion, cultural beliefs, abundance, and usefulness of the species studied (Koura et al., 2011). Therefore, understanding ethno-knowledge is relevant to ensuring sustainable and efficient implementation of future management schemes for plant species' continuous use and conservation (Irfan et al., 2023). Therefore, this study aims to gather indigenous knowledge in Southern Nigeria for the continued use and conservation of eggplants.

2 Materials and Method

2.1 Study Area

The study was conducted in 39 communities, extending from the South-West (7°15′N; 5°55′E) to the South-East (6°8′N; 6°55′E) within the humid Southern Nigeria tropical rainforest (Fig. 24.1). The selected sites fall within 14 states, cutting across three sampling areas: South-West (SW): Lagos, Oyo, Osun, Ondo, Ogun, and Ekiti; South-East (SE): Imo, Abia, Anambra and Ebonyi and South-south (SS): Edo, Akwa-Ibom, Delta, and Rivers in Southern Nigeria.

2.2 Ethics and Informed Consent

The ethical approval for the ethnobotanical survey was obtained from the Covenant Health Research Ethics Committee (CHREC/158/2022), and participation in the survey was sought through verbal consent. All participants were informed that the research was strictly for academic purposes and that they had the right to withdraw their information at any point during the study. According to the Nigerian National Code for Health Research Ethics, all information was sought after receiving oral informed consent from respondents prior to the interview, discussions, and questionnaire administration.

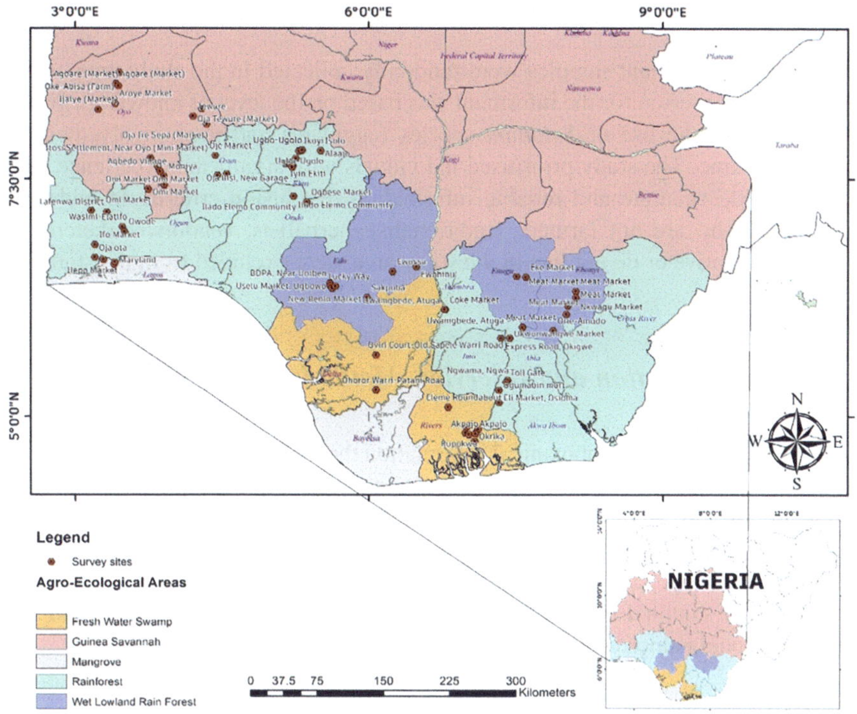

Fig. 24.1 Geographical location of communities during ethnobotanical survey

2.3 *Research Design and Sampling Method*

Three hundred and sixty-nine were randomly selected for the ethnobotanical survey. At least 50 respondents were recruited in each sampling area to obtain a representative sample. This approach aligns with the guideline suggesting that robust parametric statistical analyses require more than 35 independent random samples (Hoffman & Gallaher, 2007; Snell et al., 2021). Data was collected between January 2022 and August 2023 from 73 communities in Southern Nigeria.

The sampling method follows Omonhinmin et al. (2021), using systematic sampling along a regular transect distance of 50 km. Field visits were conducted in various agroecological zones, and consultation was based on informed consent. Ethnobotanical information and plant samples were collected concurrently at markets, open spaces, farmlands and gardens, and homesteads from urban and rural settlements within selected areas considered centres of eggplant diversity and endemism.

2.3.1 Exclusion and Inclusion Criteria

Inclusion criteria: Plant samples were randomly collected in the study area, while the inclusion criterion for the informant was based on the level of knowledge of the respondents in the use of *Solanum* species (eggplant) and their involvement in herbal medicine. The study prioritised individuals with significant experience who could provide accurate and reliable information. Exclusion criteria: In addition, informants who are not farmers, traditionalists/herbalists, vendors/marketers, or who are not native or do not live in the study area were excluded from the study.

2.4 Data Collection and Interview Methods

The survey extracted ethnobotanical information through in-depth interviews, focus group discussions (FGDs), and field observation using a participatory rural appraisal approach. Data were then recorded using administered semi-structured questionnaires (Adeoye-Olatunde & Olenik, 2021). Ethnobotanical information on local use patterns, knowledge, and accessions was collected from the different local communities in each sampled state. The information obtained includes the demography, plant data, local names of plants, parts and their uses.

2.5 Data Analysis

Data from the questionnaires were analysed using the IBM SPSS software version 25.0 (SPSS Inc., Chicago, Illinois, USA) and Excel 2016 for statistical processing.

3 Result

A total of 369 respondents participated in the study, comprising slightly more males (54.2%) than females (45.8%) (Table 24.1). The majority of respondents belonged to the Edo, Deltans, and Igbo ethnic groups, each making up approximately 29% of the total sample, while the Ibibio and Efik ethnic groups were least represented (4.34%).

Among male respondents, the Edo/Deltans and Igbo ethnic groups were the most represented, while Yoruba women constituted the largest proportion of female respondents. The Ibibio/Efik ethnic groups consistently had the least representation across both genders. Respondents aged 60 and above comprised the largest age group (45.25%), with the Igbo and Yoruba ethnic groups being the most prominent in this category. In contrast, the 24–32 age group had the least representation; the Igbo and Edo were predominant within this group.

Table 24.1 Sociodemographic characteristics of respondents across each study ethnic groups

	Edo/Deltans = 108	Igbo = 108	Ikwere/Ijaw/Ogoni = 35	Efik/Ibibio = 16	Yoruba = 102
Gender					
Female	45	51	16	4	53
Male	63	57	19	12	49
Age (years)					
24–32	17	19	02	02	16
33–59	56	32	19	06	35
≥60	35	57	14	08	51
Marital status					
Single	33	39	16	04	40
Married	73	66	19	11	58
Divorced	00	01	00	00	01
Widowed	02	02	00	01	03
Level of education					
Tertiary	23	24	01	03	15
Secondary	34	39	17	06	42
Primary	34	28	15	05	30
None	17	17	02	02	15
Occupational background					
Farmers	30	36	10	00	32
Vegetable traders	39	21	16	02	29
Researchers	14	32	09	08	25
Herbalists/traditional healers	13	02	00	00	03
Others	12	17	00	06	13

The majority of respondents were married (61.52%), with the Edo ethnic group having the most proportion of married individuals. Single respondents were mainly from the Yoruba and Igbo groups, while the divorced and widowed categories were minimal, each representing a small fraction of the overall population. In terms of education, most respondents had a secondary school education (37.4%). The Yoruba and Igbo ethnic groups were prominent at both the secondary and tertiary education levels, while the Ikwere, Ijaw, and Ogoni groups had the least representation in higher education.

Additionally, a small but significant portion of the population had no formal education. Farming was the most common occupation, with a large proportion of respondents identifying as farmers, particularly among the Igbo and Yoruba groups. Edo respondents dominated the vegetable trading sector, while the Igbo were the most represented among researchers. Other occupations varied but with limited participation from smaller ethnic groups like the Oron, Anang, and Ibibio.

3.1 Use Pattern of Eggplant Species

The three eggplant species—*Solanum aethiopicum, Solanum macrocarpon*, and *Solanum melongena*—are widely used across various ethnic groups as snacks, soup ingredients, and salad condiments. The fruit can be eaten raw, often paired with banana, groundnut, or traditional groundnut paste (Table 24.2). It can also be boiled, ground, and fried with other soup ingredients. The Igbo-speaking communities consume the leaves raw or use the tender stalks and leaves of *S. aethiopicum* and *S. macrocarpon* to prepare "Abacha" (cassava salad), while the Edo/Deltans grind seeds from all three species to make soups. Furthermore, the leaves and stems of these species are used as forage, and damaged fruits are repurposed as livestock feed.

All studied ethnic groups also use raw or dried fruits for medicinal purposes. These include treatments for digestion issues, abdominal pain, diarrhoea, diabetes,

Table 24.2 Use pattern of eggplant species in Southern Nigeria

Use category	Form of use	Mode of use
Food	Fruit	Eat raw solely or with banana, groundnut paste or groundnut
		Boiled, mashed and cooked with other soup condiments
	Leaves and stalk	Eat raw solely or with cassava slices "Abacha" mixed with steamed tender stalks and leaves of eggplant (salad)
	Seeds	Eat raw solely or with cassava slices "Abacha" mixed with steamed tender stalks and leaves of eggplant (salad)
Medicine	Leaf and seeds	Drink decoction or leaf extract: Boil the leaf in water or crush and squeeze the contents of the leaf
Economics	Fruit, leaves and seeds	Sales and farming
Traditional customs	Fruit	Served to entertain visitors or substituted for "kola" during social functions—"Orji"

oedema, hernia, hypertension, piles, constipation, blood pressure regulation, detoxification, and weight management. They are also believed to relieve excessive spitting in pregnant women. Specifically, Igbo-speaking communities use the dried, ground fruits of *S. aethiopicum* and *S. macrocarpon* to enhance male sexual health, improve vision, and treat eye infections. All ethnic groups use *S. aethiopicum* and *S. macrocarpon* to treat toothache by applying crushed dried fruits directly to the affected area.

All the studied ethnic groups use *Solanum aethiopicum* and *Solanum macrocarpon* species to treat toothache by applying crushed dried fruits directly on the affected area. The three species also heal burns, skin infections, eczema, and rashes by incorporating dried fruit pulp or juice extracts into homemade skincare products. Leaf extracts or decoctions from these species are consumed to regulate blood pressure, alleviate stomach pain, aid digestion, and act as an analgesic for chronic illnesses such as cancer. The Edo/Deltans and Yoruba communities grind dried seeds of *S. aethiopicum* and *S. macrocarpon* into a decoction to induce bowel movement, ease stomach pain, and promote digestion.

All the ethnic groups, particularly the Igbo-speaking people, cultivate the fruits and leaves of the three eggplant species for commercial and subsistence purposes as a source of income and livelihood. Additionally, the Yoruba-speaking people extract natural dye from grinding the raw fruit of *Solanum melongena* (purple) to enhance clothing colour. The Igbo and Edo/Deltans serve the fruit of *Solanum aethiopicum* and *Solanum macrocarpon* species to entertain visitors or as a substitute for "kola" during social functions for hospitality purposes.

3.2 Ecotype Preference and Trait Criteria of Eggplant

A considerable proportion of respondents (57.18%) preferred *Solanum aethiopicum*, compared to 16% favouring *S. melongena*, while 27% preferred *S. macrocarpon* (Fig. 24.2). The study identified three preference criteria: agronomic (fruit traits and yield), economic, and culinary importance, as presented in Fig. 24.3. These criteria influenced the eggplant species cultivated and consumed in the study area. Across all criteria—fruit traits, yield, economic value, and taste—a higher proportion of respondents preferred *S. aethiopicum*, with *S. melongena* being the least preferred. For fruit traits, 45.78% (169 respondents) favoured *S. aethiopicum*, followed by 30.1% (111 respondents) for *S. macrocarpon* and 24.12% (89 respondents) for *S. melongena*. Regarding fruit yield, *S. aethiopicum* was preferred by 45.53% (168 respondents), *S. macrocarpon* by 33.06% (122 respondents), and *S. melongena* by 21.41% (79 respondents). In terms of economic value, *S. aethiopicum* was favoured by 43.36% (160 respondents), followed by *S. melongena* at 29.27% (108 respondents), and *S. macrocarpon* at 27.37% (101 respondents). For fruit taste, 55.82% (206 respondents) preferred *S. aethiopicum*, with 23.58% (87 respondents) favouring *S. macrocarpon* and 20.6% (76 respondents) preferring *S. melongena*.

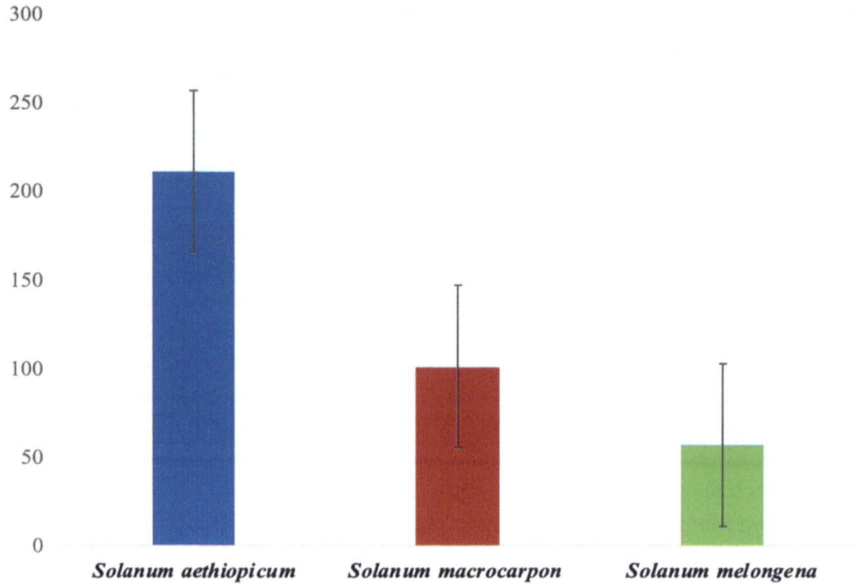

Fig. 24.2 Eggplant species preferences amongst the ethnic groups

4 Discussion

This study showed that the ethnic groups and localities studied in different regions attributed various local names to eggplant despite "eggplant" being a household name. These names, which indicate how the plant is perceived and integrated into local traditions, were consistently applied within each area. Variations in local names highlight the integration of eggplant into different cultural contexts and dialects (Aguessy et al., 2021). According to Lawal et al., 2022, local names are not chosen randomly but reflect cultural significance. For example, the Igbo name "Ofe" and the Yoruba name "Ikan yinrin" or "Igba Ijesu" reflect cultural perceptions and uses of the plant.

The age grouping is pivotal in the sociodemographic characteristics of respondents (Barkaoui et al., 2017). Older respondents, 45.26% of the study population, displayed more knowledge of eggplant use. This buttresses the fact that older individuals are better indigenous knowledge custodians (Bag, 2017). Younger respondents showed less interest in traditional knowledge. This is likely due to exposure to Western education and modern lifestyles (Tembo et al., 2021). Male respondents, who constituted a majority in this study, often formed the core of the producers (farmers) of the species, although women were reported to play an active role in making household food preparations, often with plant-based recipes (Miara et al. (2018). Furthermore, crop knowledge and use may be gender specific depending on varied conditions, particularly economic benefits (Ogwu et al., 2018). The

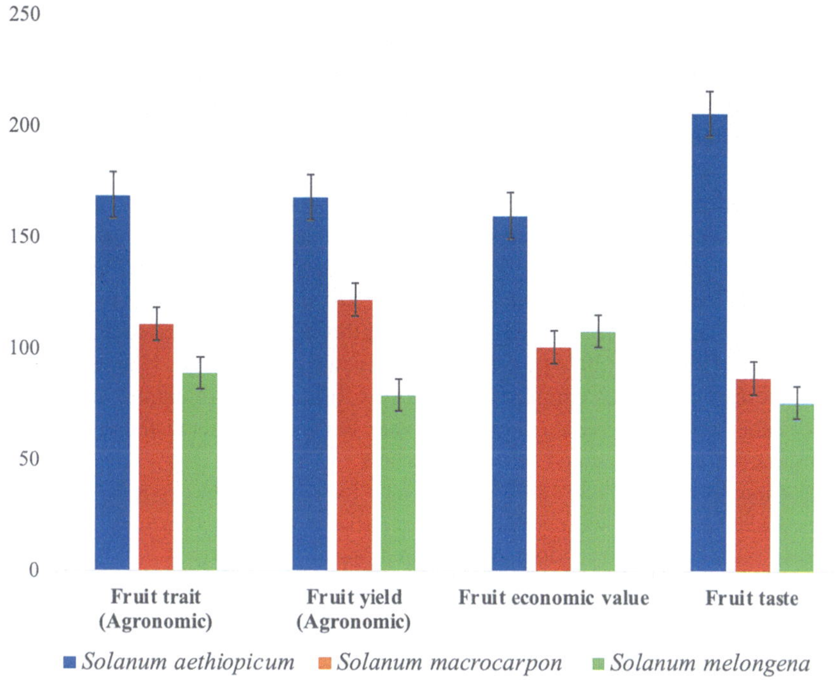

Fig. 24.3 Eggplant trait criteria employed for the ethnobotanical evaluation amongst the various ethnic groups

dominance of males of males in ethnobotanical knowledge is consistent with previous studies (Imarhiagbe, 2020; Adeniji & Aloyce, 2012).

The married individuals comprised 61.52% of the respondents studied. This supports the fact that, in African society, marriage is a significant social factor that contributes to responsibility and knowledge sharing within households (Onu & Mbah, 2018). The data on educational background indicated that more respondents in the study had a high level of formal education (85.64%). This suggests that educational attainment influences the retention and transmission of ethnobotanical knowledge (Salihu et al., 2018).

4.1 Use Pattern of Eggplant Species

There has been growing awareness of the use of eggplant as a nutritive and medicinal crop in Nigeria (Oguntolu et al., 2022), particularly in Southern Nigeria, which may account for it being widely known among all ethnic groups, as revealed in this study. The richness of species corresponds to the knowledge and use preferences of the species among populations in an area (Kunwar et al., 2018).

The results of the plant parts validate the food property of the fruit parts as the most valuable aspect of the plant. The fruit pulp is rich in nutrients, including starch, protein, dietary fibre, and various bioactive compounds and minerals, making it an excellent source of both traditional nutrition and medicinal components (Chinedu et al., 2011; Ebiringa, 2020; Henry et al., 2022). This high nutritional value of the plant pulp of the fruit likely explains the importance of the consumption of the fruit by all studied ethnic groups. The substantial size of eggplant fruits makes them satisfying and versatile, whether eaten raw or prepared. Raw fruits are eaten either solely or in combination with groundnut or groundnut paste or mashed fruit, as well as cooked with another condiment, such as soups, to eat with yams, plantain, bread, or cassava pudding as a revitalising refreshment throughout the occasions.

Opara and Udourioh (2023) noted that a significant portion of the Nigerian population frequently includes eggplants in their diet, often with groundnut paste or as part of other meals. Preparation methods vary but include dicing the fruits with onions, garden egg leaves, and other condiments.

Additionally, dishes such as "Abacha" or "local salad" incorporate eggplant leaves, stalks, and fruits with boiled or sliced cassava, creating a textured, consumable dish traditionally served to guests, often alongside palm wine.

Alternative foods from indigenous vegetables have recently attracted interest in improving food and nutrition security (Bokelmann et al., 2022). Therefore, it is unsurprising that Namo Foods International Ltd., Nigeria, has already developed and packaged eggplant into powder as an "eggplant/garden egg swallow" to be eaten with soups and stews. Although food processing initiatives for fruits are rare in Nigeria, in other countries like the United States of America, the transformation of these fruits for frozen entrees, specialty dips, and appetisers is a growing initiative (AgMRC, 2021). From this perspective, promoting the development and packaging of eggplant products can benefit the locals socioeconomically.

In the study areas, eggplants were used as a cost-effective and natural option in health care, particularly the leaves and fruits are common vegetables eaten raw or cooked and used for traditional medicine preparations to treat varied illnesses such as ulcers, abdominal pain, food indigestion, diabetes, constipation, anaemic, glaucoma, pain blood pressure, diarrhoea, oedema, hernia, hypertension, piles, constipation, blood supply, chronic sickness such as cancer and weight management. It serves as medication for the general wellness of the body.

The bitter characteristics of some eggplant fruits and leaves were associated with many beneficial medicinal properties of the plant in the study area. Hence, some respondents preferred eggplant species with bitter characteristics. Nakyewa et al. (2021) highlighted that taste is an important trait for eggplants, as the market generally favours a mildly bitter flavour perceived as more palatable. Consequently, farmers also prioritise this trait since they grow eggplants primarily for market demand. Previous studies have shown that the slight bitterness in eggplant is attributed to the nicotinoid alkaloids found in the numerous edible soft seeds (Nergiz et al., 2018).

While the most commonly used parts of the plant are the leaves and fruits, the use of seeds for both culinary and medicinal purposes was particularly prevalent among the Edo/Delta ethnic groups. In this study, eggplant had implications for the enhancement of vision or sight; according to Aguessy et al. (2021), Igbo-speaking people can hardly do without eating eggplant because it is good for their sight. Eggplants were used in this study to treat diabetes, hypertension, and obesity. According to Yarmohammadi et al. (2021), eggplant can be useful in treating patients with metabolic syndrome through its antioxidative, inhibition and protective properties. Metabolic syndrome patients often develop diabetes, abdominal obesity and hypertension.

Additionally, eggplants are rich in dietary fibre, promoting digestive health (Gürbüz et al., 2018). Biochemical analysis has recorded active compounds in different organs, explaining and confirming several indigenous medicinal uses, such as laxatives, aiding digestion and reducing the risk of hypertension (Chikaleke et al., 2020; Edeke et al., 2021). These properties support the use of eggplants in traditional medicine and the increasing interest of the pharmaceutical industry in plants (Han et al., 2021; Oladosu et al., 2021). The species' therapeutic potential constitutes credible research and development on-the road to developing standardised phytomedicines.

The many ethnomedicinal uses of the plant by local communities in Nigeria are not surprising. 80% of Africans use or depend on a form of medicinal plant or other for their primary healthcare (Borokini & Omotayo, 2012; Miara et al., 2018). Ethnomedicinal research among various Nigerian ethnic groups shows a deluge of vast medicinal plant knowledge exists among locals; also, the widespread usage is indicative of efficiency, affordability, acceptability and accessibility by the people (Enebeli-ekwutoziam et al., 2021). Despite the abundance of eggplant species, this study targets only three notable species (*Solanum melongena, Solanum macrocarpon* and *Solanum aethiopicum*). In the towns and villages visited, eggplant was grown, sold, and eaten. However, *S. aethiopicum* and *S. aethiopicum* were more abundant than other species, and this may be due to the two species have been reported to be domesticated in Africa and are referred to as African eggplant (Rakha et al., 2021). Although the fruit of eggplants has high commercial potential, the crop has yet to enjoy a good level of commercial exploitation. The fewer potential uses cited by respondents show the limited awareness of the species' potential amongst local people. The use of eggplant remains traditional, and the species is considered an underutilised crop since its nutritional and economic potential is mostly underexploited (Padulosi et al., 2019).

Despite the use pattern based on the different cultural attributes of the investigated ethnic group and localities, for example, the Edo/Deltans ethnic group reported eating cooked ground seeds with solid staple foods and giving eggplant leaf extracts or decoction to hasten the postpartum healing process and relieve mothers after child delivery. The Igbo ethnic group cooks fruit slices with egg and eats them with yam, cassava slices, rice, and plantains and also uses raw fruit for ornamental purposes by displaying the raw fruit creatively as decorative pieces during important social gatherings or functions. Similarities in the use categories of eggplant parts are also observed; for example, the use of vegetables as food and medicine was observed in all the ecogeographical localities. Medicine and food homology are commonly observed in the traditional applications of plants (Hou & Jiang, 2013).

In Nigeria, eggplant farming has taken a great turn towards commercial production, given that almost all ethnic groups in Nigeria consume the species, thus making its farming very profitable (Punch, 2020). A considerable respondent consensus relies on their farms as a primary source of income, with a significant portion serving as a source of income for the locals, particularly Igbo-speaking and other easterners, who cultivate substantial quantities of eggplant in many of their local communities (Onunka et al., 2011). Similarly, in Ghana, Wahaga and Arimeyaw (2021) stated that eggplant is a reliable and important source of income for several farmers.

Eggplant is most prevalent in Southern Nigeria, particularly in Igbo land, because of its traditional and cultural importance (Omotesho et al., 2017). As a gesture of goodwill and honour, the Igbo and Edo/Deltans ethnic groups serve raw fruit as "kola" to entertain visitors as "Orji" during social gatherings, visits, and ceremonial or festive periods. Eggplant arranged on a tray or plate is commonly served alongside alligator pepper, fried groundnuts, or paste to each participant throughout the occasion (Ogunka-Nnoka & Ekrika, 2018). Other forms of use, such as local skin care and veterinary, are indigenous practices in the studied areas. Different forms of use of the fruit were natural dye and male sexual prowess, which were indigenous practices in the studied area.

According to Sutjaritjai et al. (2019), ethnobotanical studies focused on a particular plant group can help discover species with previously unreported uses that may be underutilised. This statement is validated by the documentation of unreported uses (improved male sexual enhancement, ornament, skincare and use as fodder) of some species in this study, showing that focusing on the traditional uses of a taxonomic group (e.g. *Solanum*) may help respondents remember more uses of species in a group.

4.2 Ecotype Preference and Trait Criteria of Eggplant

The increased consumption and use of *S. aethiopicum* in this study aligns with a previous report by Aguessy et al. (2021), who reported high production and consumption of *S. aethiopicum*. Although no official figures are recorded for its production in Nigeria, the crop is widely distributed as a garden crop (Omotesho et al., 2017). The species is one of the most consumed fruit vegetables and has nutritive value in the leaf and fruit. It is the third most consumed, after tomato and onion and before okra, in tropical Africa and the study area, the species are largely used as food and consumed widely by the locals (Oguntolu et al., 2022). In this study, the fruit traits (size, shape and colour) interestingly showed increased preference and importance compared to fruit yield; Adeniji and Aloyce (2012) also recorded similar results. This trend may be associated with the fact that the fruit trait is important for assessing quality. It is usually the first trait farmers wish to promote and potential vendors and consumers seek (Omonhinmin & Idu, 2012). The ecotype preference criteria recorded in this study are demand-driven traits. Hence, selection and breeding programmes can improve eggplant species based on these preferred traits that are important for the local populations.

5 Conclusion

Ethnobotanical studies often reveal the diversity of plant species used by local communities for different purposes such as food, medicine, traditional rituals, ornamentation, etc. Wahaga and Arimeyaw (2021).

Understanding the cultural significance of eggplant provides insights into preferred crop varieties and landraces cultivated by local communities (Sesay et al., 2023). For instance, knowledge of eggplant preference and trait criteria from this study can be used to identify and prioritise eggplant species with desirable traits for breeding programmes. Incorporating this ethno-knowledge can help identify useful eggplant species, enhance the resilience of eggplant, and result in the development of improved eggplant species that are well adapted to specific agroecological conditions across Nigeria and the West and Central regions of Africa.

A common Igbo saying "Ndi mba ozo, na-azu na-anwu n' aguu", which is interpreted as "people who depend on foreign food eventually die of hunger". In this context, it is necessary to implement strategies to support local communities in actively participating in the conservation and use of eggplant species to promote food security and preserve plant biodiversity.

References

Adeniji, O. T., & Aloyce, A. (2012). Farmer's knowledge of horticultural traits and participatory selection of African eggplant varieties (Solanum aethiopicum) in Tanzania. *Tropicultura, 30*(3), 185–191.

Adeoye-Olatunde, O. A., & Olenik, N. L. (2021). Research and scholarly methods: Semi-structured interviews. *Journal of the American College of Clinical Pharmacy, 4*(10), 1358–1367.

Agriculture Marketing Resource Center (AgMRC). (2021). Retrieved from https://www.agmrc.org/commodities-products/vegetables/eggplants

Aguessy, S., Idossou, R., Dassou, A. G., Yêyinou Laura Estelle, L., Yelome, O. I., Gbaguidi, A. A., Agre, P. A., Dansi, A., & Agbangla, C. (2021). Ethnobotanical characterization of scarlet eggplant (Solanum aethiopicum L.) varieties cultivated in Benin (West Africa). *Journal of Agriculture and Food Research, 5*, 100173.

Bag, M. (2017). Industrialization: A threat to the indigenous knowledge of ethno-medicine. *ACME Intellects International Journal of Research in Management, Social Science & Technology, 20*(20), 1–13.

Bana, R. S., Jat, G. S., Grover, M., Bamboriya, S. D., Singh, D., Bansal, R., Choudhary, A. K., Kumar, V., Laing, A. M., Godara, S., Bana, C. R., Kumar, H., Kuri, R. B., Yadav, A., & Singh, T. (2022). Foliar nutrient supplementation with micronutrient-embedded fertilizer increases biofortification in eggplant fruit and soil biological activity while enhancing plant productivity. *Scientific Reports, 12*(1), 5146.

Barkaoui, M., Katiri, A., Boubaker, H., & Msanda, F. (2017). Ethnobotanical survey of medicinal plants used in the traditional treatment of diabetes in Chtouka Ait Baha and Tiznit (Western Anti-Atlas), Morocco. *Journal of Ethnopharmacology, 198*, 338–350.

Bokelmann, W., Huyskens-Keil, S., Ferenczi, Z., & Stöber, S. (2022). The role of indigenous vegetables to improve food and nutrition security: Experiences from the project HORTINLEA in Kenya (2014–2018). *Frontiers in Sustainable Food Systems, 6*, 806420.

Borokini, T. I., & Omotayo, F. O. (2012). Phytochemical and ethnobotanical study of some selected medicinal plants from Nigeria. *Journal of Medicinal Plants Research, 6*(7), 1106–1118.

Chikaleke, V., Ojo, D., Adebisi, M., & Bodunde, J. (2020). Morphological diversity assessment in seventy-eight accessions of eggplants (*Solanum* sp.). *Nigerian Journal of Horticultural Science, 25*(1), 108.

Chinedu, S. N., Olasumbo, A. C., Eboji, O. K., Emiloju, O. C., Arinola, O. K., & Dania, D. I. (2011). Proximate and phytochemical analysis of *Solanum aethiopicum* L and *Solanum macrocarpon* L. fruits. *Research Journal of Chemical Sciences, 1*, 63–71.

Ebiringa, D. C. (2020). Proximate and micronutrient compositions of four different cultivars of aubergine (*Solanum melongena*). *Journal of Agriculture and Food Sciences, 18*(2), 135–142.

Edeke, A., Uchendu, N., Omeje, K., & Odiba, A. S. (2021). Nutritional and pharmacological potentials of *Solanum melongena* and *Solanum aethiopicum* fruits. *The Journal of Phytopharmacology, 10*(1), 61–67.

Enebeli-Ekwutoziam, K. C., Aruah, C. B., Ogbonna, B. O., Eze, U. J., Egedeye-Fubura, F. S., Nwankwo, C. F., Oliseyenum, I. N., Udoha, N. W., Afuye, T. N., Asogwa, G. N., & Chinweokwu, K. A. (2021). A review of ethnomedicinal plant resources in southern Nigeria. *Ethnobotany Research and Applications, 22*, 1–40.

Food and Agriculture Organization Corporate Statistical Database (FAOSTAT). (2022). *Production database*. Retrieved March 3, 2022, from https://www.fao.org/faostat/en/#data/QI

Gürbüz, N., Uluişik, S., Frary, A., Frary, A., & Doğanlar, S. (2018). Health benefits and bioactive compounds of eggplant. *Food Chemistry, 268*, 602–610.

Han, M., Opoku, K. N., Bissah, N. A. B., & Su, T. (2021). Solanum aethiopicum: The nutrient-rich vegetable crop with great economic, genetic biodiversity and pharmaceutical potential. *Horticulturae, 7*(6), 126.

Henry, M. U., Dogun, O., Ogenyi, R. A., Obidola, S. M., & Henry, U. I. (2022). Phytochemical, nutritional and trace element of some Solanum (Garden Egg). *Nigerian Journal of Biotechnology, 39*(2), 44–52.

Hoffman, B., & Gallaher, T. (2007). Ethnobotany research and applications. *Research Methods, 5*, 201–218.

Hou, Y., & Jiang, J. G. (2013). Origin and concept of medicine food homology and its application in modern functional foods. *Food & Function, 4*(12), 1727–1741.

Imarhiagbe, O. (2020). Prospects of ethnobotanical uses of Thonningia sanguinea Vahl. (Balanophoraceae) among selected tribes in southern Nigeria. *Journal of Medicinal Plants Studies, 8*(2), 126–132.

Irfan, M., Ullah, F., & Haq, I. U. (2023). Ethnomedicinal and traditional uses of the flora of District Lower Dir, Khyber Pakhtunkhwa, Pakistan. *Ethnobotany Research and Applications, 26*, 1–22.

Koura, K., Ganglo, J. C., Assogbadjo, A. E., & Agbangla, C. (2011). Ethnic differences in use values and use patterns of Parkia biglobosa in Northern Benin. *Journal of Ethnobiology and Ethnomedicine, 7*, 1–12.

Kunwar, S., Iriarte, F., Fan, Q., Evaristo da Silva, E., Ritchie, L., Nguyen, N. S., Freeman, J. H., Stall, R. E., Jones, J. B., Minsavage, G. V., & Colee, J. (2018). Transgenic expression of EFR and Bs2 genes for field management of bacterial wilt and bacterial spot of tomato. *Phytopathology, 108*(12), 1402–1411.

Lawal, I. O., Rafiu, B. O., Ale, J. E., Majebi, O. E., & Aremu, A. O. (2022). Ethnobotanical survey of local flora used for medicinal purposes among indigenous people in five areas in Lagos State, Nigeria. *Plants, 11*(5), 633.

Miara, M. D., Bendif, H., Hammou, M. A., & Teixidor-Toneu, I. (2018). Ethnobotanical survey of medicinal plants used by nomadic peoples in the Algerian steppe. *Journal of Ethnopharmacology, 219*, 248–256.

Nakyewa, B., Sseremba, G., Kabod, N. P., Rwothtimutung, M., Kyebalyenda, T., Waholi, K., Buteme, R., Nakanwangi, M. J., Bishop, G., & Kizito, E. B. (2021). Farmer preferred traits and genotype choices in *Solanum aethiopicum* L., Shum group. *Journal of Ethnobiology and Ethnomedicine, 17*, 1–9.

Naeem, M. Y. & Ugur, S. (2019). Nutritional Content and Health Benefits of Eggplant. *Turkish Journal of Agriculture - Food Science and Technology, 7*, 31–36.

Nergiz, G., Selman, U., Anne, F., Amy, F., & Sami, D. (2018). Health benefits and bioactive compounds of eggplant. *Food Chemistry, 268*, 602–610.

Ogunka-Nnoka, C. U., & Ekrika, D. (2018). Nutritional profiling of garden egg *stalks Solanum macrocarpon* L. (Solanales: Solanaceae). *Brazilian Journal of Biological Sciences, 5*(10), 265–275.

Oguntolu, O. O., Anyaoha, C. O., Chikaleke, V. A., & Olofintoye, J. A. T. (2022). Nutritional and phenotypic variations among newly selected African eggplant (Solanum aethiopicum L.). *Genetic Resources, 3*(6), 38–48.

Ogwu, M. C., Chime, A. O., & Oseh, M. O. (2018). Ethnobotanical survey of tomato in some cultivated regions in southern Nigeria. *The Maldives National Journal of Research, 6*(1), 19–29.

Oladosu, Y., Rafii, M. Y., Arolu, F., Chukwu, S. C., Salisu, M. A., Olaniyan, B. A., Fagbohun, I. K., & Muftaudeen, T. K. (2021). Genetic diversity and utilisation of cultivated eggplant germplasm in varietal improvement. *Plants, 10*, 1714.

Omonhinmin, C. A., Popoola, J. O., Ejoh, A., Mordi, R. (2021). Sustainability and Genetic Diversity of Under-exploited African Plants: A Collection Expedition in Nigeria-Facts, Challenges, and Prospects. *IOP Conference Series Earth and Environmental Science*.

Omonhinmin, C. A., & Idu, M. (2012). Genetic diversity, domestication, and conservation implications of fruit morphometric data analyses for *Dacryodes edulis* in southern Nigeria. *Journal of Plant Development Sciences, 4*(2), 167–173.

Omotesho, O., Muhammad-Lawal, A., Amolegbe, K., & Abubakar, T. (2017). Assessment of dry season garden egg production among small-scale farmers in Edu Local Government Area of Kwara State, Nigeria. *Journal of Agricultural Sciences, Belgrade, 62*, 89–103.

Onu, S. E., & Mbah, G. O. (2018). Extent of utilization of medicinal plants among rural households in Imo State, Nigeria. *Journal of Agricultural Extension, 22*(1), 151–160.

Onunka, B. N., Chinaka, C. C., & Ezeh, C. I. (2011). Socio-economic determinants of African eggplant (Solanum gilo) marketers in Abia State, Nigeria. *Journal of Agriculture and Social Research (JASR), 11*(2), 147–154.

Opara, E., & Udourioh, G. A. (2023). Garden egg (Solanum aethiopicum) as a mystical plant in Akabo, South Eastern Nigeria: Health and economic implications. *Journal of Applied Sciences and Environmental Management, 27*(11), 2651–2660.

Padulosi, S., Phrang, R., & Rosado-May, F. J. (2019). *Soutenir une agriculture axée sur la nutrition grâce aux espèces négligées et sous-utilisées: Cadre opérationnel* (pp. 978–992). Bioversity International and IFAD.

Punch Newspaper. (2020, May 15). Cultivating garden egg for commercial purpose. *Punch Newspaper*. Retrieved from https://punchng.com/cultivating-garden-egg-for-commercial-purpose/#google_vignette

Rakha, M., Prohens, J., Taher, D., Wu, T. H., & Solberg, S. Ø. (2021). Eggplant (*Solanum melongena, S. aethiopicum* and *S. macrocarpon*) breeding. In *Advances in plant breeding strategies: Vegetable crops. Fruits and young shoots* (Vol. 9, pp. 163–203).

Salihu, T., Olukunle, J. O., Adenubi, O. T., Mbaoji, C., & Zarma, M. H. (2018). Ethnomedicinal plant species commonly used to manage arthritis in North-West Nigeria. *South African Journal of Botany, 118*, 33–43.

Sesay, J. V., Lebbie, A., Wadsworth, R., & Chakanda, R. T. M. (2023). Ethnobotanical knowledge of local cassava varieties and characteristics desired by farmers in Sierra Leone. *American Journal of Plant Sciences, 14*, 74.

Snell, K. I., Archer, L., Ensor, J., Bonnett, L. J., Debray, T. P., Phillips, B., Collins, G. S., & Riley, R. D. (2021). External validation of clinical prediction models: Simulation-based sample size calculations were more reliable than rules-of-thumb. *Journal of Clinical Epidemiology, 135*, 79–89.

Sutjaritjai, N., Wangpakapattanawong, P., Balslev, H. & Inta, A. (2019) Traditional Uses of Leguminosae among the Karen in Thailand. *Plants, 8*(12), 600. https://doi.org/10.3390/plants8120600

Tembo, N., Lampiao, F., Mwakikunga, A., & Chikowe, I. (2021). Ethnobotanical survey of medicinal plants used for cervical cancer management in Zomba District, Malawi. *Scientific African, 13*, e00941.

Wahaga, E., & Arimeyaw, S. I. S. (2021). Post-harvest losses of eggplant in Ghana. *Nigeria Agricultural Journal, 52*(3), 316–324.

Villanueva, G., Rosa-Martínez, E., Şahin, A., García-Fortea, E., Plazas, M., Prohens, J. & Vilanova, S. (2021). Evaluation of advanced backcrosses of eggplant with Solanum elaeagnifolium introgressions under low N conditions. *Agronomy, 11*(9), 1770.

Yarmohammadi, F., Rahbardar, M. G., & Hosseinzadeh, H. (2021). Effect of eggplant (Solanum melongena) on the metabolic syndrome: A review. *Iranian Journal of Basic Medical Sciences, 24*(4), 420.

Chapter 25
Beyond Foes and Friends: Deciphering the Gut Microbiome's Bidirectional Role in Breast Cancer

Olayemi Oluseun Akinnola and Moyosoreoluwa Mary Williams

1 Introduction

The human body harbors a diverse community of trillions of symbiotic microbes, including archaea, bacteria, fungi, viruses, and protists (Malczewski et al., 2020). These microbes reside primarily on the surface of human epithelia and other internal surfaces, such as the gastrointestinal tract, urogenital tract, lungs, and breast (Ogunrinola et al., 2020). Collectively, the microorganisms within the gut and their collective genomes are referred to as the "gut microbiome"(Vemuri et al., 2018). Studies have recently highlighted the importance of enteric bacteria as a significant component of the human microbial community. It comprises the highest abundance and diversity of bacteria compared to other body sites and is also considered a crucial organ due to its extensive interconnections and communication networks with other organs, facilitated through multiple pathways, including neural, hormonal, chemical, immune, and metabolic links (Liu et al., 2022; Afzaal et al., 2022). While the gut microbiome composition differs significantly from person to person, certain dominant bacterial groups are consistently present. These core phyla include *Firmicutes, Bacteroides, Actinobacteria,* and *Proteobacteria* (Nandi et al., 2023). The distinctive balance of microorganisms in the gut is influenced by a combination of genetic and lifestyle influences, including nutrient intake, level of kinesis,

O. O. Akinnola
Department of Biological Sciences, College of Science and Technology, Covenant University, Ota, Ogun State, Nigeria

M. M. Williams (✉)
Department of Biological Sciences, College of Science and Technology, Covenant University, Ota, Ogun State, Nigeria

Covenant Applied Informatics and Communication Africa Centre of Excellence (CApIC-ACE), Covenant University, Ota, Ogun State, Nigeria
e-mail: moyosore.williamspgs@stu.cu.edu.ng

P. O. Isibor, P. A. Akinduti (eds.), *Harnessing Biotechnology Tools for Product Development*, https://doi.org/10.1007/978-3-031-86002-7_25

exposure to antibiotics, and even the microbial profile of inherited genes (Jaye et al., 2022). The gut microbiome is significantly influenced by dietary preferences, as evidenced by recent research (Simões et al., 2022). To optimize gut health and reduce disease risk, it is essential to consider how socio-economic factors, cultural diversity, and population-specific nutrition shape dietary patterns. A novel approach to prevention involves leveraging biotherapeutic substances from prebiotic- and probiotic-rich indigenous foods to mitigate gut colitis, cancer, and diabetes in adults with minimal or no adverse effects (Isibor et al., 2021).

The gut microbiome is involved in numerous functions, including nutrient digestion and absorption, the production of vital compounds such as short-chain fatty acids (SCFAs), the synthesis of essential vitamins and hormones, regulation of the immune system, and the maintenance of the gut epithelial barrier (Liu et al., 2022). Dysbiosis, or an imbalance of the gut microbiome, has been revealed to produce molecules with deleterious effects, which, in turn, may implicate a range of health conditions, including autoimmune diseases, inflammatory bowel disease (IBD), obesity, diabetes, and cancer (Álvarez-Mercado et al., 2023). The metabolic products of the gut microbiome, known as gut metabolites, have shown a significant connection with the host in suppressing various types of cancer, including breast cancer. Although the exact communication mechanism between the gut and organs remains unclear, research revealed a remarkable interplay between the two (Jaye et al., 2022; Afzaal et al., 2022).

Breast cancer (BC) is a primary global health concern, ranking as the most commonly diagnosed cancer among women, affecting one in eight women during their lifetime and a leading cause of female cancer deaths (Ferlay et al., 2020). Breast cancer cases have increased significantly over the past few decades. Further analysis has identified five distinct subtypes based on gene expression and receptor profiles, including estrogen receptor-positive (ER+), progesterone receptor-positive (PR+), and human epidermal growth factor receptor 2-positive (HER2+); these subtypes are further categorized into five main groups: Luminal A, Luminal B, normal-like, HER2-enriched (HER2+ without ER or PR), and triple-negative or basal-like (negative for all three receptors: ER, PR, and HER2) that does not express any of the three receptors, with varying prognoses. Breast cancer develops through a series of genetic and epigenetic changes that lead to uncontrolled cell growth. Key risk factors include genetic predispositions (e.g., BRCA genetic variant), hormonal fluctuations, lifestyle choices (such as diet and alcohol use), and environmental toxins (American Cancer Society, 2023). While mortality rates have shown improvement due to advancements in early detection and treatment, the root cause of most BC cases remains unclear (Chen et al., 2019). However, metabolomics and metagenomics studies have revealed a potential link between gut bacteria (microbiota) and cancer development, including breast cancer. This relationship appears complex and two-way, with some gut microbes potentially suppressing tumors while others may promote cancer initiation and development. Understanding how these factors interplay with the gut microbiome can provide insights into breast cancer pathogenesis. Studies suggest that the gut microbiome's influence on breast cancer risk might be through mechanisms of various types, with both harmful and beneficial properties, and modulating the immune system's response.

This review aims to comprehensively explore the potential link between gut flora and breast cancer development. This investigation will aim to decipher the complex interactions between the microbial community of the gastrointestinal and mammary tumor development, identifying specific microbial signatures associated with either increased or decreased risk.

1.1 Rationale for the Study

The gut microbiome's influence on systemic health suggests it could significantly impact breast cancer risk. However, the exact nature of this relationship still needs to be better understood. Some studies suggest beneficial gut bacteria may protect against cancer, while others indicate that dysbiosis, or microbial imbalance, can promote cancer development (Zhu et al., 2018). This bidirectional role necessitates a comprehensive investigation to elucidate the mechanisms by which gut microbiota affect breast cancer.

2 Gut Microbiome and Breast Cancer: Mechanistic Insights

Preclinical studies investigating gut microbial metabolites revealed promising antitumorigenic properties. While established risk factors contribute to breast cancer initiation, specific lifestyle modifications demonstrate protective effects. These include regular physical activity, lactation, and parity (Mikó et al., 2019). Emerging evidence suggests a direct modulation of breast cancer risk by gut microbiota through alterations in host metabolism, estrogen enterohepatic circulation, and immune response regulation (Alizadehmohajer et al., 2020). Additionally, recent research proposes the potential for gut bacteria translocation to breast tissue via the integumentary route, potentially impacting mammary health (Zhang et al., 2020). This translocation is hypothesized to occur through various routes of transmission, encompassing intimate physical contact and the nipple-areolar interaction during breastfeeding and openings within the nipple-areolar complex (Zhang et al., 2020). The multifaceted association between gut bacteria and breast cancer highlights the potential for this microbial ecosystem to act as both a foe and a friend.

2.1 Estrogen Metabolism

Studies have proven that disruptions in sex hormone regulation, particularly estrogen and progesterone, are well-established contributors to breast cancer (BC) development. These hormonal imbalances manifest differently across different BC subtypes, with clinical and molecular evidence supporting this link (Laborda-Illanes et al., 2020). Intriguingly, research suggests a link between the gut microbiome and estrogen levels. Emerging evidence points to a specific group of gut microbiome called the astrobleme, which can influence estrogen metabolism (Laborda-Illanes et al., 2020). These microbes produce enzymes like beta-glucuronidase, which convert conjugated estrogen (a less active form) into its more potent form, facilitating its binding to estrogen receptors and potentially increasing their availability for reabsorption into the bloodstream through the gut (enterohepatic circulation) (Laborda-Illanes et al., 2020). Several bacteria with β-glucuronidase activity fall into two main groups: *Clostridium leptum* and the *Clostridium coccoides* clusters, both part of the *Firmicutes* phylum. Additionally, bacteria within the *Escherichia/Shigella* group, found within the *Proteobacteria* phylum, also possess these enzymes (Álvarez-Mercado et al., 2023).

Consequently, free estrogens are distributed to reach various organs, including the breast. This process can disrupt the breast microenvironment, trigger chronic inflammation, and contribute to DNA damage, uncontrolled cell proliferation, blood vessel formation, and cancer spread (Alizadehmohajer et al., 2020; Viswanathan et al., 2023). Furthermore, gut bacteria can produce other estrogen-like metabolites through oxidative and reductive reactions and potentially induce the synthesis of estrogen-promoting growth factors, all with potential carcinogenic effects (Fernández et al., 2018). Additionally, bacterial enzymes (beta-glucuronidase) might reactivate certain environmental toxins or xenoestrogens, increasing their effects and lifespan in the body (Fernández et al., 2018).

Furthermore, Bernardo et al. (2023) described that increased concentrations of naturally occurring or circulating estrogen directly correlate with a higher susceptibility to mammary tumors, especially in postmenopausal females and estrogen receptor-positive (HR+) tumors. These HR+ tumors undergo metabolic processing in the liver, and their growth is fueled by other established factors like aging, genetic predisposition, and abnormal mammary cell growth (Song et al., 2022). Adding to the complexity, specific gut bacteria like *Enterolactone*, *Sesamin*, and *Enterodiol* have also been identified to have the ability to influence estrogen levels further. These bacteria can synthesize estrogen-like compounds by breaking down plant substances in the diet called lignans (Parida & Sharma, 2019). These bacterial estrogens then interact with the nuclear receptor alpha (ER-α) in breast tissue, stimulating the generation of molecules that encourage the proliferation of breast tumor cells. However, estrogens can also have a paradoxical effect. While they can stimulate growth, estrogens can also induce cytotoxic (cell-killing) effects via cytochrome P450 complexes, independent of ER-α signaling (Bernardo et al., 2023). This produces unstable molecules and metabolic byproducts, leading to oxidative stress, genetic alterations, and hindered DNA repair mechanisms within tumor cells (Bernardo et al., 2023).

Interestingly, some dietary estrogens, such as phytoestrogens, have been described to compete with the body's natural estrogen receptor 1 (ESR1). However, research suggests that enterolactone, a phytoestrogen produced by gut bacteria, might inhibit breast cancer cell growth (Song et al., 2022). The complex interaction involving estrogens and gut bacteria emphasizes the critical function of intestinal microbes in estrogen metabolism and, perhaps, mammary tumor development. The ability of gut bacteria to release active estrogens underscores the potential for manipulating the microbiome as a strategy for influencing the risk of breast cancer.

2.2 *Bacterial Metabolites*

Preliminary studies suggest a link between gut microbiome metabolite and the development of breast carcinoma. This finding has potential applications in both risk assessment and prevention strategies. For example, cadaverine, a biogenic amine, is formed when gut bacteria break down the amino acid L-lysine. This breakdown process might be one way that gut bacteria influence breast cancer risk (Eslami-S et al., 2020). A reduction in cadaverine biosynthesis in the gastrointestinal tract and early-stage breast cancer has been reported, resulting in decreased production of an anticancer bacterial metabolite and reduced breast cancer invasion (Eslami-S et al., 2020).

In addition to estrogens, gut flora plays a complex role in breast cancer development by influencing various metabolic pathways. The mode of mechanism is through the breakdown of fats and bile acids by bacteria. These bacterial processes generate molecules that can directly or indirectly affect the growth of tumor cells (Mikó et al., 2018). For example, certain gut bacteria lacking oxygen (intestinal anaerobic bacteria), such as *Clostridia* species, produce a molecule called lithocholic acid (LCA). This secondary bile acid acts as a brake on breast cancer cell proliferation by activating a specific receptor (TGR5) on the cell surface (Sampsell et al., 2020). Findings imply that LCA can disrupt the transition of cells to switch between different shapes (mesenchymal-to-epithelial transition), the infiltration of immune cells into tumors, energy production pathways (tricarboxylic acid cycle and oxidative phosphorylation), and the balance between molecules that promote and prevent cellular damage (oxidants and antioxidants) (Mikó et al., 2018; Kovács et al., 2019). Interestingly, research indicates that breast cancer patients tend to have lower levels of LCA and a reduced abundance of specific bacterial groups, such as *Clostridiales* and *Bacteroidales*, that have the ability to secrete it (Mikó et al., 2018). Beyond LCA, other bacterial products can also influence breast cancer.

Furthermore, the gut microbiome's influence on breast cancer remains multifaceted. Nisin, the most abundant gut bacteriocin produced by *Lactococcus lactis*, exemplifies this complexity. Studies show promising anticancer potential for nisin in various contexts, but its effectiveness against breast cancer remains largely unexplored. Under normal conditions, nisin interacts with the cell membrane, modifying its integrity and forming short-lived pores, which alters the cell's electrical potential (Mohammadi et al., 2024; Wang et al., 2024). The positively charged regions of the nisin molecule protrude through the cell membrane, binding directly to the negatively charged phospholipid heads on one side of the membrane.

It is observed that the hydrophobic regions of nisin interact with the membrane core, modulating cell membrane activity and enabling ion influx, including calcium, which halts cell division (Jaye et al., 2022). It is also noted that the combination of nisin with doxorubicin, a standard chemotherapy drug, exhibits synergistic cytotoxic activity at sub-inhibitory concentrations, as demonstrated by Akbari et al. 2018. This synergy is further supported by an in vitro study (Baindara & Mandal, 2020), which showed improved outcomes for skin cancer cells when treated with the combination compared to individual drug administration. The potential synergy between nisin and standard chemotherapy drugs is being explored for improved treatment outcomes across various carcinomas. Moreover, it is found that nisin interacts directly with the innate immune system, increasing chemokine secretion and inhibiting lipopolysaccharide-stimulated cytokines. This is supported by preclinical studies in vitro and in vivo.

Moving forward, the focus on nisin research can be prioritized on in vitro studies using a more comprehensive range of breast cancer cell lines. Additionally, studies could employ animal models that encompass breast cancer types with limited treatment options, such as triple-negative breast cancer. Investigating the potential for combining nisin with immunotherapies to see if this compound would yield more effect together is also crucial.

Short-chain fatty acids (SCFAs) constitute another set of molecules secreted by the digestive system bacteria through the breakdown of nutritional fiber. Examples of SCFA-producing bacteria include *Eubacterium hallii*, *Eubacterium rectale*, *Faecalibacterium prausnitzii*, *Anaerostipes*, and *Clostridium leptum*. These SCFAs also appear to have an impact on breast cancer development, although the exact mechanisms are still being investigated (Jaye et al., 2022). By exploring the vast array of bacterial metabolites produced in the gut, valuable insights can be gained into their potential roles as allies or foes in the fight against breast cancer.

2.3 Inflammation and the Immune Response

The immune system acts as a critical line of defense against cancer, eliminating abnormal cells with the potential to become cancerous (Scott et al., 2019). The gut microbiome plays a complex role in this fight. While some gut bacteria collaborate with the immune system to combat cancer through mechanisms like enhancing T-cell activity and boosting the overall immune response, others may have immunosuppressive effects, allowing tumors to evade detection by the body's defenses (Gori et al., 2019).

The gut microbiome's dysbiotic expansion and microbial product extravasation are implicated in malignancy promotion through chronic inflammation induction, host cell growth, and cell mortality balance disruption, inducing excessive and unregulated immune reactions, which hinder depletion of unwanted cellular population, thereby fostering tumor growth (Biragyn & Ferrucci, 2018). Chronic, persistent, and dysregulated inflammation is associated with increased breast cancer (BC) risk (Fernández et al., 2018). One proposed mechanism involves the increased activity of two molecules, COX2 and PGE2, which in turn leads to higher levels of aromatase in fat tissue. This results in more androgen precursors being converted into estrogens, potentially contributing to the development of hormone-sensitive cancers (Fernández et al., 2018). Intestinal microflora alterations are linked to local and systemic cancer development through changes in the microenvironment, persistent inflammatory response initiation, and metabolic changes (Song et al., 2022).

The gut microbiome plays a crucial role in shaping the immune system. Dysbiosis can lead to a pro-inflammatory state, promoting cancer initiation and progression. Conversely, beneficial microbes can enhance anti-tumor immunity. For example, specific bacterial strains have been shown to activate immune cells that target cancer cells (Liu et al., 2024).

Breast milk has been found to contain an abundance of specific bacterial components, including species of *Firmicutes* (such as *Clostridium* and *Lactobacillus*), *Actinobacteria* (including *Propionibacterium*), and *Bacteroidetes* (such as *Prevotella*), which are involved in the metabolic production gut metabolites and hormones (Parida & Sharma, 2019). Reduced microbial heterogeneity and low concentrations of *Bifidobacterium* and *Bacteroides* species have been associated with altered immune system stimulation, potentially leading to a predisposition to Th2-mediated allergies in infancy (Tanaka & Nakayama, 2017). It was revealed in another study that the composition of gut bacteria may influence the development of BC, possibly through interactions with immune cells such as the T-lymphocyte cells and neutrophils, along with associated inflammatory cytokines (Mir et al., 2024). Research by Rutkowski et al. (2015) provided evidence for such interactions in breast cancer patients, highlighting a connection between gut bacteria, a molecule called IL-6, and neutrophils. Animal models further support this link. A mouse model of breast cancer was used to investigate the effects of *Helicobacter hepaticus* infection, a type of gut bacteria, on mammary tumor progression. This effect was associated with an enhanced migration and penetration of neutrophils into the tumors.

Conversely, depleting neutrophils in these mice led to a reduction in mammary tumor formation (Lin et al., 2020). It has also been established through studies that high concentrations of beneficial or protective bacteria in gut microbiota correlate with a robust immune system leading to a more effective defense against pathogenic microbes from the body's microenvironment (Wang et al., 2020). It has been found that metabolites generated by gut microbes can enter the host's bloodstream and influence organs distant from the gut, including the immune system, highlighting the systemic impact of gut microbiota on host physiology (Wang et al., 2020). Collectively, the intricate link between the gut microbiota and the host's immune system necessitates exploring the potential role of gut metabolites in breast cancer, both as therapeutic agents and causative factors.

3 Future Perspective

The nuanced interactions between intestinal bacteria and tumor of breast risk represent a rapidly evolving frontier in cancer research. Emerging evidence suggests that the gut microbiome can influence breast cancer development and progression through various mechanisms, including modulation of systemic inflammation, estrogen metabolism, and immune function. Conversely, breast cancer and its treatments can also alter the gut microbiome composition. The research and results so far entail a multifaceted approach with the potential to revolutionize the understanding of this tumor with the scope of precision therapeutic strategies, novel preventive, and diagnoses based on microbiome profiles, advanced biomarkers for early detection, microbiome modulation interventions, exploring microbiome-immunotherapy interactions, longitudinal studies, and integrating multi-omics data for improved breast cancer management.

3.1 Advanced Biomarkers for Early Detection

Current research is actively exploring the connection between gut microbial communities and breast cancer, seeking to pinpoint specific microbial patterns that correlate with the development and progression of the disease. This knowledge will potentially lead to the creation of new biomarkers for breast cancer diagnosis and monitoring. Notably, a study conducted by Banerjee et al. in 2018 discovered unique microbial profiles associated with triple-negative breast cancer, a subtype of breast cancer known for its aggressive behavior and limited treatment options (Banerjee et al., 2018). These findings offer valuable insights into the complex relationship between the gut microbiome and breast cancer and may ultimately contribute to developing more effective diagnostic tools and treatments.

3.2 Precision Therapeutics Based on Microbiome Profiles

As we delve deeper into the interaction between these gut microbial communities and carcinoma of the breast, personalized treatment strategies can be devised based on individual microbiome compositions. This could involve targeted interventions aimed at modulating the microbiome to enhance treatment efficacy and patient outcomes (Fu et al., 2022).

3.3 Gut Microbiome Intervention Trial

Understanding how alterations in the gut microbiome contribute to breast cancer risk allows for the exploration of preventive measures. Intervention trials assessing the efficacy of probiotics, prebiotics, and dietary modifications in modulating the gut microbiome may emerge as a viable influence for reducing breast cancer risk (Rebersek et al., 2021). Recent research by Parida and Sharma (2020) shows that probiotics can cause essential changes in the gut microbiome. Their review compiled evidence from various studies suggesting that probiotics may be beneficial in combating breast cancer. One study investigated the effects of milk fermented with *Lactobacillus helveticus* R389 on mice with breast tumor cells. The results were promising: mice fed this fermented milk exhibited increased levels of interleukin-10 (IL-10), an anti-inflammatory molecule, alongside decreased levels of interleukin-6 (IL-6), a pro-inflammatory molecule, in both their blood serum and mammary cells. This shift in the immune response ultimately led to inhibiting breast tumor cell growth. Another study highlighted the potential of *Lactobacillus casei* strain CRL431. When administered to mice, this probiotic strain significantly reduced breast cancer development and metastasis. The mechanism appears to involve a decrease in the infiltration of macrophages, a type of immune cell, into the tumor site while simultaneously enhancing the anti-tumor activity of helper T cells (CD4+) and cytotoxic T cells (CD8+) subtypes of T lymphocytes critical players in the adaptive immune system (Nandi et al., 2023). These trials can provide evidence for the clinical application of microbiome-based therapies.

3.4 Integration of Omics Approaches

The integration of omics technologies such as genomics, transcriptomics, proteomics, and metabolomics holds immense promise in unraveling the intricate mechanisms underlying microbiome-cancer interactions. By elucidating novel therapeutic targets and pathways, these approaches could lead to the development of innovative treatment modalities (Nejman et al., 2020).

3.5 Ethical Considerations and Public Awareness

With the rapid advancements in microbiome research, addressing ethical considerations surrounding microbiome-based interventions in breast cancer is crucial. Additionally, public awareness and education campaigns are essential to inform individuals about the potential influence of the microbial community on breast cancer risk and the importance of preventive measures (American Cancer Society, 2023).

3.6 Advanced Analytical Techniques

Advanced analytical techniques, such as metagenomics and metabolomics, can provide deeper insights into the gut microbiome's composition and functions. These techniques can help identify novel microbial markers and therapeutic targets (Song et al., 2020).

3.7 Multi-Omics Approaches

Integrating multi-omics methods such as metagenomics, transcriptomics, proteomics, genome sequencing, and metabolomics can offer a comprehensive understanding of the gut microbiome's composition and functions and its role in breast cancer. These approaches can elucidate complex interactions between the microbiome and host physiology and novel microbial markers (Liu et al., 2024).

4 Conclusion

Over time, research has revealed incontrovertible data confirming the importance of the gut microbial community in maintaining health and avoiding pathological alterations such as cancer. Breast cancer remains a significant global health challenge, and existing treatment options for breast malignancies often have significant drawbacks and are limited. While the gut microbiome's link to breast cancer requires further exploration, imbalances in the gut flora have increasingly been identified as a critical factor in the genesis of breast cancer, from the initiation stage down to the metastasis stage. These imbalances in the gut microbiota can favor certain microorganisms that inhabit our intestinal tract to increase the risk of breast cancer through various mechanisms, such as estrogen metabolism, modulate the immune system and inflammation, production of bacterial metabolites with anti-tumor properties, interaction with the cancer cells, and the tumor microenvironment, and influence response to breast cancer therapies. The query of whether reversing this gut commensal dysbiosis reduces breast cancer is a captivating route for future study.

Harnessing gut microbe involvement in future breast cancer strategies, studies look toward precision therapies, multi-omics approaches, and randomized controlled trials as essentials in uncovering the complex relationships between the microbial community, immunity, and breast cancer. Ultimately, the gut flora also holds excellent capability as an analytical biomarker, target for treatment, and prognostic indicator in breast cancer management. Harnessing the power of the gut microbiome might lead to innovative, personalized, and practical strategies for breast cancer prevention, treatment, and survival.

Acknowledgments The authors thank and acknowledge the management of the Covenant University Centre for Research, Innovation and Discovery (CUCRID) and Covenant Applied Informatics and Communication Africa Centre of Excellence (CApIC-ACE), Ota, Nigeria, for their support in data collection and also financial publication of this review study.

Competing Interests The authors declare that they have no conflict of interest.

References

Afzaal, M., Saeed, F., Shah, Y. A., Hussain, M., Rabail, R., Socol, C. T., Hassoun, A., Pateiro, M., Lorenzo, J. M., Rusu, A. V., & Aadil, R. M. (2022). Human gut microbiota in health and disease: Unveiling the relationship. *Frontiers in Microbiology, 13*, 999001.

Alizadehmohajer, N., Shojaeifar, S., Nedaeinia, R., Esparvarinha, M., Mohammadi, F., Ferns, G. A., Ghayour-Mobarhan, M., Manian, M., & Balouchi, A. (2020). Association between the microbiota and women's cancers - Cause or consequences? *Biomedicine & Pharmacotherapy, 127*, 110203.

Álvarez-Mercado, A. I., Del Valle Cano, A., Fernández, M. F., & Fontana, L. (2023). Gut microbiota and breast cancer: The dual role of microbes. *Cancers, 15*(2), 443.

American Cancer Society. (2023). *Breast cancer causes and risk factors*. Retrieved from https://www.cancer.org/cancer/types/breast-cancer/risk-and-prevention.html

Akbari, V., Avand, A., & Shafizadegan, S. (2018). In vitro cytotoxic activity of a Lactococcus lactis antimicrobial peptide against breast cancer cells. *Iranian Journal of Biotechnology, 16*(3), e1867.

Baindara, P., & Mandal, S. M. (2020). Bacteria and bacterial anticancer agents as a promising alternative for cancer therapeutics. *Biochimie, 177*, 164–189.

Banerjee, S., Tian, T., Wei, Z., Shih, N., Feldman, M. D., Peck, K. N., DeMichele, A. M., Alwine, J. C., & Robertson, E. S. (2018). Distinct microbial signatures associated with different breast cancer types. *Frontiers in Microbiology, 9*, 951.

Bernardo, G., Le Noci, V., Di Modica, M., Montanari, E., Triulzi, T., Pupa, S. M., & Sfondrini, L. (2023). The emerging role of the microbiota in breast cancer progression. *Cells, 12*(15), 1945.

Biragyn, A., & Ferrucci, L. (2018). Gut dysbiosis: A potential link between increased cancer risk in ageing and inflammaging. *Lancet Oncology, 19*(6), e295–e304.

Chen, J., Douglass, J., Prasath, V., Neace, M., Atrchian, S., Manjili, M. H., Shokouhi, S., & Habibi, M. (2019). The microbiome and breast cancer: A review. *Breast Cancer Research and Treatment, 178*(3), 493–496.

Eslami-S, Z., Majidzadeh-A, K., Halvaei, S., Babapirali, F., & Esmaeili, R. (2020). Microbiome and breast cancer: New role for an ancient population. *Frontiers in Oncology, 10*, 120.

Ferlay J., Ervik M., Lam F., Colombet M., Mery L., Piñeros M., Znaor A., Soerjomataram I., & Bray F. (2020). *Global cancer obstetrics*. Cancer Today. International Agency for Research on Cancer. Retrieved from https://gco.iarc.fr/today

Fernández, M. F., Reina-Pérez, I., Astorga, J. M., Rodríguez-Carrillo, A., Plaza-Díaz, J., & Fontana, L. (2018). Breast cancer and its relationship with the microbiota. *International Journal of Environmental Research and Public Health, 15*(8), 1747.

Fu, A., Yao, B., Dong, T., Chen, Y., Yao, J., Liu, Y., Li, H., Bai, H., Liu, X., Zhang, Y., Wang, C., Guo, Y., Li, N., & Cai, S. (2022). Tumor-resident intracellular microbiota promotes metastatic colonization in breast cancer. *Cell, 185*(8), 1356–1372.e26.

Gori, S., Inno, A., Belluomini, L., Bocus, P., Bisoffi, Z., Russo, A., & Arcaro, G. (2019). Gut microbiota and cancer: How gut microbiota modulates activity, efficacy and toxicity of antitumoral therapy. *Critical Reviews in Oncology/Hematology, 143*, 139–147.

Isibor, P. O., Akinduti, P. A., Aworunse, O. S., Oyewale, J. O., Oshamika, O., Ugboko, H. U., Taiwo, O. S., Ahuekwe, E. F., Obafemi, Y. D., Onibokun, E. A., Oziegbe, O., Oniha, M. I., Olopade, B. K., Atolagbe, O. M., Adekeye, B. T., Ajiboye, I. B., Bello, O. A., Popoola, J. O., Ayanda, O. I., Akinnola, O. O., & Obembe, O. O. (2021). Significance of African diets in biotherapeutic modulation of the gut microbiome. *Bioinformatics and Biology Insights, 15*, 11779322211012697.

Jaye, K., Chang, D., Li, C. G., & Bhuyan, D. J. (2022). Gut metabolites and breast cancer: The continuum of dysbiosis, breast cancer risk, and potential breast cancer therapy. *International Journal of Molecular Sciences, 23*(16), 9490.

Kovács, T., Mikó, E., Vida, A., Sebő, É., Toth, J., Csonka, T., Boratkó, A., Ujlaki, G., Lente, G., Kovács, P., Tóth, D., Árkosy, P., Kiss, B., Méhes, G., Goedert, J. J., & Bai, P. (2019). Cadaverine, a metabolite of the microbiome, reduces breast cancer aggressiveness through trace amino acid receptors. *Scientific Reports, 9*(1), 1–14.

Laborda-Illanes, A., Sanchez-Alcoholado, L., Dominguez-Recio, M. E., Jimenez-Rodriguez, B., Lavado, R., Comino-Méndez, I., Alba, E., & Queipo-Ortuño, M. I. (2020). Breast and gut microbiota action mechanisms in breast cancer pathogenesis and treatment. *Cancers, 12*(9), 2465.

Lin, Y.-H., Yamazaki, S., Goel, S., Xu, X., Liu, X., Gao, Y., Luo, M., Wang, Y., & Fu, Y.-X. (2020). Listeria monocytogenes infection accelerates breast cancer growth in mice by promoting immunosuppression. *PLoS One, 15*(7), e0236224.

Liu, Y., Baba, Y., Ishimoto, T., Gu, X., Zhang, J., Nomoto, D., Okadome, K., Baba, H., & Qiu, P. (2022). Gut microbiome in gastrointestinal cancer: A friend or foe? *International Journal of Biological Sciences, 18*(10), 4101–4117.

Liu, C., Fu, L., Wang, Y., & Yang, W. (2024). Influence of the gut microbiota on immune cell interactions and cancer treatment. *Journal of translational medicine, 22*(1), 939.

Malczewski, A. B., Navarro, S., Coward, J. I., & Ketheesan, N. (2020). Microbiome-derived metabolome as a potential predictor of response to cancer immunotherapy. *Journal for Immunotherapy of Cancer, 8*(2), e001383.

Mikó, E., Kovács, T., Sebő, É., Tóth, J., Csonka, T., Ujlaki, G., Sipos, A., Szabó, J., Méhes, G., & Bai, P. (2019). Microbiome-microbial metabolome-cancer cell interactions in breast cancer-familiar, but unexplored. *Cells, 8*(4), 293.

Mikó, E., Vida, A., Kovács, T., Ujlaki, G., Trencsényi, G., Márton, J., Sári, Z., Kovács, P., Boratkó, A., Hujber, Z., Csonka, T., Antal-Szalmás, P., Watanabe, M., Gombos, I., Csoka, B., Kiss, B., Vígh, L., Szabó, J., Méhes, G., Sebestyén, A., & Bai, P. (2018). Lithocholic acid, a bacterial metabolite reduces breast cancer cell proliferation and aggressiveness. *Biochimica et Biophysica Acta (BBA)—Bioenergetics, 1859*, 958–974.

Mir, R., Albarqi, S. A., Albalawi, W., Alatwi, H. E., Alatawy, M., Bedaiwi, R. I., Almotairi, R., Husain, E., Zubair, M., Alanazi, G., Alsubaie, S. S., Alghabban, R. I., Alfifi, K. A., & Bashir, S. (2024). Emerging Role of Gut Microbiota in Breast Cancer Development and Its Implications in Treatment. *Metabolites, 14*(12), 683.

Mohammadi, A., Kazemipour, N., Ghorbankhani, G. A., Morovati, S., & Hashempour Sadeghian, M. (2024). Glycated nisin enhances nisin's cytotoxic effects on breast cancer cells. *Scientific reports, 14*(1), 17808.

Nandi, D., Parida, S., & Sharma, D. (2023). The gut microbiota in breast cancer development and treatment: The good, the bad, and the useful! *Gut Microbes, 15*(1), 2221452.

Nejman, D., Livyatan, I., Fuks, G., Gavert, N., Zwang, Y., Geller, L. T., Rotter-Maskowitz, A., Weiser, R., Mallel, G., Gigi, E., Meltser, A., Douglas, G. M., Kamer, I., Gopalakrishnan, V., Dadosh, T., Levin-Zaidman, S., Avnet, S., Atlan, T., Cooper, Z. A., Arora, R., & Straussman, R. (2020). The human tumor microbiome is composed of tumor type-specific intracellular bacteria. *Science, 368*, 973–980.

Ogunrinola, G. A., Oyewale, J. O., Oshamika, O. O., & Olasehinde, G. I. (2020). The human microbiome and its impacts on health. *International Journal of Microbiology, 2020*, 8045646.

Parida, S., & Sharma, D. (2019). The power of small changes: Comprehensive analyses of microbial dysbiosis in breast cancer. *Biochimica et Biophysica Acta Reviews on Cancer, 1871*, 392–405.

Parida, S., & Sharma, D. (2020). Microbial alterations and risk factors of breast cancer: Connections and mechanistic insights. *Cells, 9*(5), 1091.

Rebersek, M., Petriz, J., Nikolić, M., Knežević, A., Carić, H., & Mikolić, B. (2021). Gut and breast microbiome: A potential link in breast cancer development and therapy. *International Journal of Molecular Sciences, 22*(1), 439.

Rutkowski, M. R., Stephen, T. L., Svoronos, N., Allegrezza, M. J., Tesone, A. J., Perales-Puchalt, A., Brencicova, E., Escovar-Fadul, X., Nguyen, J. M., Cadungog, M. G., Zhang, R., Salatino, M., Tchou, J., Rabinovich, G. A., & Conejo-Garcia, J. R. (2015). Microbially driven TLR5-dependent signaling governs IL-10 production in human breast cancer. *Cancer Research, 75*(10), 2155–2164.

Sampsell, K., Hao, D., & Reimer, R. A. (2020). The gut microbiota: A potential gateway to improved health outcomes in breast cancer treatment and survivorship. *International Journal of Molecular Sciences, 21*, 9239.

Scott, A. J., Alexander, J. L., Merrifield, C. A., Cunningham, D., Jobin, C., Brown, R., Alverdy, J., O'Keefe, S. J., Gaskins, H. R., Teare, J., Yu, J., Hughes, D. J., Verstraelen, H., Burton, J., O'Toole, P. W., Rosenberg, D. W., Marchesi, J. R., & Kinross, J. M. (2019). International Cancer Microbiome Consortium consensus statement on the role of the human microbiome in carcinogenesis. *Gut, 68*(9), 1624–1632.

Simões, C. D., Maganinho, M., & Sousa, A. S. (2022). FODMAPs, inflammatory bowel disease and gut microbiota: Updated overview on the current evidence. *Europe Journal Nutrition, 61*, 1187–1198.

Song, M., Chan, A. T., & Sun, J. (2020). Influence of the gut microbiome, diet, and environment on risk of colorectal cancer. *Gastroenterology, 158*(2), 322–340.

Song, X., Wei, C., & Li, X. (2022). The relationship between microbial community and breast cancer. *Frontiers in Cellular and Infection Microbiology, 12*, 849022.

Tanaka, M. & Nakayama, J. (2017). Development of the gut microbiota in infancy and its impact on health in later life, *Allergology International, 4*(66), 515–522.

Vemuri, R., Gundamaraju, R., Shastri, M. D., Shukla, S. D., Kalpurath, K., Ball, M., Tristram, S., Shankar, E. M., Ahuja, K., & Eri, R. (2018). Gut microbial changes, interactions, and their implications on human life cycle: An ageing perspective. *BioMed Research International, 2018*, 4178607.

Viswanathan, S., Parida, S., Lingipilli, B. T., Krishnan, R., Podipireddy, D. R., & Muniraj, N. (2023). Role of gut microbiota in breast cancer and drug resistance. *Pathogens (Basel, Switzerland), 12*(3), 468.

Wang, J., Yang, H. R., Wang, D. J., & Wang, X. X. (2020). Association between the gut microbiota and patient responses to cancer immune checkpoint inhibitors. *Oncology Letters, 20*(6), 342.

Wang, Y., Sun, T., & Xu, J. (2024). Bacteriocins in Cancer Treatment: Mechanisms and Clinical Potentials. *Biomolecules, 14*(7), 831.

Zhang, J., Xia, Y., & Sun, J. (2020). Breast and gut microbiome in health and cancer. *Genes & Diseases, 8*(5), 581–589.

Zhu, J., Liao, M., Yao, Z., Liang, W., Li, Q., Liu, J., Yang, H., Ji, Y., Wei, W., Tan, A., Liang, S., Chen, Y., Lin, H., Zhu, X., Huang, S., Tian, J., Tang, R., Wang, Q., & Mo, Z. (2018). Breast cancer in postmenopausal women is associated with an altered gut metagenome. *Microbiome, 9*(1), 1–15.

Chapter 26
Comparative Lipid Profile and Oxidative Stress in Synodontis clarias and Tilapia zillii Exposed to Polybrominated Diphenyl Ethers (PBDEs) in Lekki Lagoon, Nigeria

Patrick Omoregie Isibor and Bamidele Akinsanya

1 Introduction

Polybrominated diphenyl ethers (PBDEs) represent a class of brominated flame retardants that are extensively employed in a variety of consumer products, including electronics, textiles, and furniture, to mitigate the risk of fire. Chemically, PBDEs are characterized by the presence of two phenyl rings linked by one to ten bromine atoms, resulting in varying degrees of bromination among different congeners (Abdallah & Harrad, 2014). The environmental persistence of PBDEs, in conjunction with their hydrophobic properties, has facilitated their widespread distribution in ecological systems. These compounds are released into the environment through multiple pathways, including manufacturing processes, product degradation, and improper disposal, subsequently contaminating air, water, and soil matrices (O'Driscoll et al., 2016; Okoffo et al., 2021; Oloruntoba et al., 2021; Olujimi et al., 2024).

Polybrominated diphenyl ethers (PBDEs) present substantial ecological and human health risks due to their persistence in the environment, potential for long-range atmospheric transport, and bioaccumulative characteristics. In aquatic systems, PBDEs adsorb onto particulate matter and sediments, establishing a persistent source of contamination. Their lipophilic nature enhances their accumulation in biota, where they undergo biomagnification through the food web, resulting in elevated concentrations in apex predators. Chronic exposure to PBDEs has been linked to neurodevelopmental deficits, endocrine disruption, and hepatotoxicity in both wildlife and humans (Olutona et al., 2016; Olutona et al., 2017).

P. O. Isibor (✉)
Department of Biological Sciences, Covenant University, Ota, Ogun State, Nigeria

B. Akinsanya
Department of Zoology, Faculty of Life Sciences, University of Lagos, Yaba, Lagos, Nigeria

P. O. Isibor, P. A. Akinduti (eds.), *Harnessing Biotechnology Tools for Product Development*, https://doi.org/10.1007/978-3-031-86002-7_26

The environmental burden of PBDEs necessitates comprehensive monitoring and mitigation strategies to alleviate their impact. Investigating bioaccumulation in aquatic organisms is critical for understanding the ecological and health risks associated with persistent organic pollutants (POPs) such as PBDEs. Bioaccumulation refers to the process by which organisms absorb and retain contaminants from their environment and diet at rates exceeding their elimination capabilities. Aquatic organisms, including fish, mollusks, and crustaceans, are particularly susceptible due to their direct exposure to contaminated water and sediments (Adewuyi & Adeleye, 2013; Luo et al., 2014; McGrath et al., 2018; Margolis et al., 2020).

Bioaccumulation studies offer insights into the trophic transfer of contaminants and the potential for biomagnification, where contaminant concentrations intensify at higher trophic levels. This is particularly pertinent for humans who consume aquatic species, directly influencing food safety and public health. Understanding the determinants of bioaccumulation—including species-specific metabolic capacities, environmental conditions, and the chemical properties of the contaminants—is essential for risk assessment and the formulation of effective regulatory policies. Furthermore, bioaccumulation data contribute to the conservation of biodiversity and the protection of vulnerable species within aquatic ecosystems (Zhu et al., 2015; Yu et al., 2016; Babayemi et al., 2018; Zhang et al., 2020).

Lekki Lagoon, situated in the southwestern region of Nigeria, is a brackish water body encompassing an area of approximately 247 km^2. It constitutes a component of the larger Lagos Lagoon system and is interconnected with the Atlantic Ocean through a network of waterways, including the Lagos Lagoon and the Ogun River. The lagoon's hydrology is subject to seasonal fluctuations in freshwater inflow, tidal exchanges, and anthropogenic activities, rendering it a dynamic and complex ecosystem.

Ecologically, Lekki Lagoon harbors a diverse assemblage of flora and fauna, encompassing numerous fish species, crustaceans, and aquatic plants (Bergayou et al., 2009; Bodin et al., 2011; Bamidele et al., 2020). It serves as a critical habitat for the breeding, feeding, and nursery stages of many commercially significant fish species, thereby contributing to regional fisheries and local livelihoods. The lagoon's mangrove forests and wetlands provide vital ecosystem services, including water filtration, carbon sequestration, and shoreline protection. The ecological significance of Lekki Lagoon extends beyond its biodiversity (Banni et al., 2014; Wu et al., 2020). It plays an essential role in nutrient cycling, sediment transport, and the maintenance of regional hydrological balance. However, the lagoon is increasingly subjected to environmental pressures stemming from urbanization, industrialization, and pollution, resulting in habitat degradation and deteriorating water quality. Thus, understanding and preserving the ecological integrity of Lekki Lagoon is imperative for sustaining its ecological functions, supporting local communities, and enhancing regional environmental health (Bobescu et al., 2021).

The contamination of Lekki Lagoon by polybrominated diphenyl ethers (PBDEs) primarily arises from a spectrum of anthropogenic activities within the surrounding catchment area. These activities facilitate the introduction of PBDEs into the environment via multiple pathways, thereby exacerbating the lagoon's overall pollution levels. Manufacturing facilities involved in the production of electronics, textiles, plastics, and other consumer products frequently employ PBDEs as flame retardants. Effluents released from these industrial operations often exhibit elevated concentrations of PBDEs, which are discharged directly into adjacent water bodies. Moreover, inadequately treated industrial wastewater and sewage from treatment facilities represent significant sources of PBDE contamination (Bramwell et al., 2017; Cao et al., 2018). These facilities frequently lack the necessary infrastructure to effectively eliminate complex organic pollutants, resulting in their unintentional release into surrounding aquatic ecosystems. Additionally, rainwater runoff from urban areas further complicates the issue by transporting PBDEs from diverse sources, including building materials, vehicles, and household dust. This runoff typically enters the lagoon untreated, particularly during heavy precipitation events. PBDEs may accumulate in road dust derived from vehicle interiors, tires, and road surfaces, which are subsequently mobilized into the lagoon by rainwater.

Improper disposal practices concerning PBDE-containing materials—such as discarded electronics, furniture, and textiles—also exacerbate the problem. When these materials are disposed of in landfills, they can leach contaminants into the soil and groundwater, ultimately facilitating their transport to the lagoon (Winid, 2015; Chakraborty et al., 2019). Another significant contributor to PBDE contamination in Lekki Lagoon is the open burning of waste. Suboptimal waste management practices, including the incineration of both household and industrial waste, can result in the atmospheric release of PBDEs. These airborne contaminants may subsequently settle into the lagoon via atmospheric deposition, further compounding the lagoon's contamination challenges. Polybrominated diphenyl ethers (PBDEs) are incorporated into a diverse array of consumer products, including televisions, computers, mattresses, and upholstered furniture (Guo et al., 2016). Over time, these compounds can leach from such products into household dust and subsequently enter the environment via improper waste disposal. Activities such as laundering clothing treated with flame retardants and other cleaning processes can release PBDEs into wastewater, where they may not be effectively removed during sewage treatment. The utilization of lagoon water for irrigation can further facilitate the introduction of PBDEs into agricultural fields, allowing for their re-entry into aquatic systems through runoff and soil leaching (Onyeche & Akankali, 2013; Park et al., 2014). Additionally, certain agricultural chemicals may be contaminated with PBDEs, thereby exacerbating the pollution load when these substances are applied to crops.

PBDEs emitted into the atmosphere from various sources can traverse considerable distances before being deposited into water bodies, including Lekki Lagoon. This atmospheric transport is enhanced by their semi-volatile characteristics, permitting repeated cycles of evaporation and deposition (Pereira et al., 2013; Choi et al., 2014; Pellacani et al., 2014; Ukenye et al., 2016; Tang & Zhai, 2017; Wang et al., 2018).

Understanding these sources is essential for implementing strategies to mitigate PBDE contamination through enhanced waste management practices, stringent industrial regulations, and public awareness initiatives. Addressing these anthropogenic activities is imperative for reducing the environmental and health impacts of PBDEs in Lekki Lagoon and its surrounding ecosystems (Daso et al., 2013; Hahladakis et al., 2018). Effective strategies may include improving the treatment of industrial effluents, promoting the proper disposal and recycling of PBDE-containing products, and regulating agricultural practices to minimize contamination. Additionally, public education on the environmental ramifications of PBDEs can play a critical role in diminishing pollution and safeguarding the ecological integrity of Lekki Lagoon (Wepener et al., 2012; Ssebugere et al., 2014; Doherty et al., 2019).

The primary objective of this study is to elucidate the effects and extent of polybrominated diphenyl ethers (PBDEs) bioaccumulation in two pivotal fish species, *Synodontis clarias* and *Tilapia zillii*, inhabiting the Lekki Lagoon in Lagos, Nigeria (Davis et al., 2012; Ge et al., 2014; Su et al., 2018). This investigation seeks to quantify the concentrations of PBDEs within the tissues of these species, thereby providing essential data on the degree of environmental contamination and the associated risks linked to their bioaccumulation. Specifically, the study will measure and compare the bioaccumulation levels of PBDEs in the tissues of *S. clarias*, a benthic catfish, and *T. zillii*, a widely consumed cichlid, both of which are significant for their ecological roles and economic importance. By evaluating these species, the study aims to provide a comprehensive understanding of the distribution of PBDEs across different trophic levels and ecological niches within the lagoon (Dazy et al., 2009; Gonzalez et al., 2013; Guo et al., 2015; Kanaya et al., 2019).

Moreover, the study intends to assess the safety of consuming these fish species by evaluating the PBDE concentrations relative to established human health guidelines. This aspect is crucial for public health, given that fish from Lekki Lagoon serve as an essential protein source for the local population. The findings will contribute to risk assessments and inform regulatory policies to ensure food safety and safeguard consumer health (Johnson et al., 2013; Huang et al., 2020; Khan et al., 2022).

This dual-species approach enhances the rigor of the study, offering a broader perspective on the bioaccumulation dynamics of PBDEs within the aquatic ecosystem of Lekki Lagoon. The outcomes will further contribute to environmental monitoring initiatives, supporting the management and conservation of this vital water body.

2 Materials and Methods

2.1 Study Area

Lekki Lagoon, located in Lagos State, southwestern Nigeria, is geographically delineated between longitudes 4°00′ and 4°15′ E and latitudes 6°25′ and 6°37′ N (Fig. 26.1). The lagoon encompasses a surface area of approximately 247 km^2 and

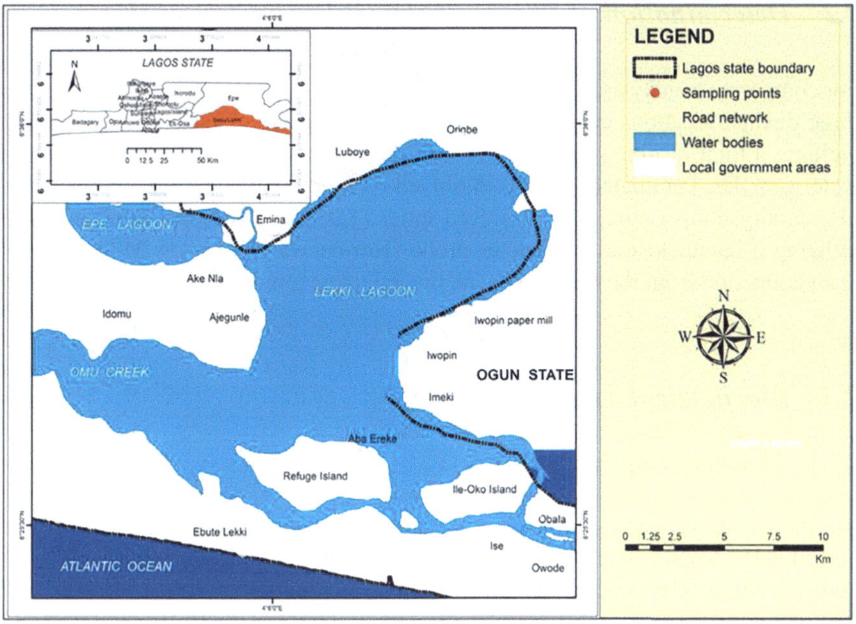

Fig. 26.1 Map of Lekki-Epe Lagoon, Lagos, Nigeria

attains a maximum depth of 6.4 m, with the majority of its expanse exhibiting relatively shallow depths of <3.0 m. As part of a vast and complex network of lagoons and creeks that extends approximately 200 km along the southwestern coast of Nigeria—from the Dahomey border to the Niger Delta—Lekki Lagoon plays a critical role in regional hydrodynamics and biodiversity.

The hydrology of Lekki Lagoon is significantly influenced by several river systems; notably, the Oni River contributes inflow from the northeast, while the Rivers Oshun and Saga discharge into the lagoon from the northwest. This fluvial input is essential for sustaining the lagoon's ecological equilibrium, fostering a diverse assemblage of aquatic and terrestrial species. The interconnectedness of this lagoon system underscores its importance for nutrient cycling and as a habitat for various wildlife.

Moreover, Lekki Lagoon is a focal point for commercial activities, with local communities engaging in multiple economic pursuits, including fishing, aquaculture, and tourism. Fishing holds particular significance, serving as a vital source of livelihood for many residents and contributing substantially to both the local and regional economy. The lagoon's proximity to Lagos, a major economic and industrial center, further enhances its role in trade and commerce within the region. The lush vegetation and unique ecosystem of Lekki Lagoon underscore its ecological significance, supporting various species of fish, aquatic plants, and avifauna, including oil palms (*Elaeis guineensis*) and coconuts (*Cocos nucifera*), among others.

2.2 Determination of Physicochemical Parameters of Water

For comparative analysis, surface water samples were systematically collected from three distinct locations corresponding to each fish species over a 3-month period, yielding a total of nine replicates per fish species and an aggregate of 18 surface water samples. Parameters such as dissolved oxygen (DO), electrical conductivity, pH, salinity, temperature, total dissolved solids (TDS), and turbidity were measured utilizing a handheld multiparameter probe (Horiba Water Checker Model U-10). Data collection from the lagoon was facilitated using a motorized canoe.

2.3 Biochemical Analysis

2.3.1 Homogenizing Sample

The dissected liver and intestine were excised and weighed. Each organ was homogenized in 0.1 M phosphate buffer (pH 7.2) using a mortar and pestle to ensure thorough blending. The resulting homogenate was then subjected to centrifugation at 2500 rpm for 15 min. Subsequently, the supernatant was carefully decanted and stored at −20 °C for further analysis.

2.3.2 Determination of Catalase Activity (CAT)

Catalase activity will be assessed utilizing the methodology established. The reaction mixture comprises 5% potassium heptaoxodichromate, 0.2 M hydrogen peroxide, a dichromate/acetic acid solution, and 0.01 M phosphate buffer (pH 7.0). To prepare the sample, 0.1 mL of the biological specimen will be diluted in 4.9 mL of distilled water, achieving a 1:5 dilution. The assay will be conducted in a 10 mL flat-bottom flask containing 4 mL of 0.2 M hydrogen peroxide and 5 mL of 0.01 M phosphate buffer. A 1-mL aliquot of the appropriately diluted enzyme preparation (test sample) will be swiftly introduced into the reaction mixture, ensuring thorough mixing through gentle swirling. The reaction will proceed at ambient temperature.

To monitor the reaction progress, 1 mL portions of the reaction mixture will be extracted at 60-s intervals and transferred into test tubes containing 2 mL of the dichromate/acetic acid reagent over a duration of 3 min. The extinction coefficient for hydrogen peroxide at 240 nm, quantified as 40.0 M^{-1} cm^{-1}, will be utilized for subsequent calculations of catalase activity.

2.3.3 Determination of Reduced Glutathione (GSH)

The quantification of total sulfhydryl groups, protein-bound sulfhydryl groups, and free sulfhydryl groups, including glutathione (GSH), in biological samples can be achieved using Ellman's reagent, 5,5′-dithio-bis-2-nitrobenzoic acid (DTNB), as detailed. This methodology relies on the formation of a stable yellow complex resulting from the reaction between Ellman's reagent and free sulfhydryl groups. A sample volume of 0.2 mL will be combined with 1.8 mL of distilled water to create a 1:10 dilution. Subsequently, approximately 3 mL of precipitating reagent (4% sulfosalicylic acid) will be introduced to the diluted sample, which will then be allowed to stand for 10 min to facilitate precipitate formation. An aliquot of 0.5 mL of the supernatant will be extracted and mixed with 4 mL of phosphate buffer, followed by the addition of 0.5 mL of Ellman's reagent. The blank will be prepared with 4 mL of 0.1 M phosphate buffer (pH 7.4), 1 mL of diluted precipitating solution, and 0.5 mL of Ellman's reagent. Absorbance will be recorded within 20 min of color development at 412 nm using a spectrophotometer.

Additionally, the chromic acetate produced in the assay will be measured spectrophotometrically at wavelengths between 570 and 610 nm. It is noteworthy that dichromate exhibits no absorbance in this wavelength range; thus, its presence in the assay mixture will not interfere with the determination of chromic acetate. Catalase preparations within the samples will be permitted to decompose H_2O_2 over varying periods, after which the reaction will be terminated, and the mixture will be heated. The residual hydrogen peroxide concentration will subsequently be assessed by measuring the chromic acetate, as outlined by Klinčić et al. (2020).

2.3.4 Determination of Glutathione Peroxidase (GPx)

To evaluate glutathione peroxidase (GPx) activity, the reaction mixture comprised 0.2 mL of 0.4 M phosphate buffer (pH 7.0), 0.1 mL of 10 mM sodium azide, 0.2 mL of plasma, 0.2 mL of glutathione salt (GSH), and 0.1 mL of 0.2 mM hydrogen peroxide (H_2O_2). The mixture was incubated at 37 °C for a duration of 10 min. Following the incubation, the reaction was terminated by the addition of 0.4 mL of 10% trichloroacetic acid (TCA), and the mixture was subsequently centrifuged to separate the precipitate. The glutathione content in the resulting supernatant was quantified using Ellman's reagent, which facilitates the spectrophotometric determination of thiol groups.

2.3.5 Malondialdehyde (MDA)

The assay for lipid peroxidation will be conducted by quantifying the thiobarbituric acid reactive substances (TBARS) in the test sample, following the protocol of, and expressed as micromoles of malondialdehyde (MDA) per gram of tissue. The assay is based on the reaction between the chromogenic reagent, 2-thiobarbituric acid (TBA), and MDA, a terminal product of lipid peroxidation. Under acidic conditions, MDA forms a stable pink chromophore with TBA, exhibiting a maximum absorbance at 532 nm.

The reaction mixture will consist of 30% trichloroacetic acid (TCA) solution, 0.75% TBA solution in 0.1 M HCl, and 0.15 M Tris-KCl buffer (pH 7.4). A 0.4-mL aliquot of the test sample will be added to 1.6 mL of Tris-KCl buffer and pre-pipetted into the test tube. This will be followed by the addition of 0.5 mL of 30% TCA and 0.5 mL of 0.75% TBA. The mixture will be incubated in a water bath for 1 h at 90–95 °C, then rapidly cooled on ice. The samples will be centrifuged at 3000 rpm for 15 min, and the clear pink supernatant will be collected.

The absorbance of the supernatant will be measured spectrophotometrically at 532 nm against a blank containing distilled water. The concentration of malondialdehyde (MDA), an index of lipid peroxidation (LPO), will be calculated using an extinction coefficient of 1.5×10^5 M^{-1} cm^{-1}, allowing for precise quantification of oxidative damage within the sample.

2.3.6 Determination of Superoxide Dismutase (SOD) Activity

Superoxide dismutase (SOD) activity will be quantified following the protocol established. The reaction mixture consists of 0.05 M carbonate buffer at pH 10.2 and 0.3 mM epinephrine as the substrate. A 0.1-mL aliquot of the sample will be diluted in 0.9 mL of distilled water, achieving a 1:10 dilution. From this diluted enzyme preparation, 0.2 mL will be added to 2.5 mL of the carbonate buffer (0.05 M, pH 10.2) in a spectrophotometer cuvette to equilibrate the system.

The enzymatic reaction will be initiated by the addition of 0.3 mL of freshly prepared 0.3 mM epinephrine to the reaction mixture, which will then be rapidly mixed by inversion. A reference cuvette will contain 2.5 mL of 0.05 M carbonate buffer, 0.3 mL of epinephrine, and 0.2 mL of distilled water, serving as a control for the nonenzymatic oxidation of epinephrine. The absorbance increase at 480 nm, attributed to the oxidation of epinephrine to adrenochrome, will be recorded at 30-s intervals over a period of 150 s. The SOD activity will be calculated based on the extinction coefficient of epinephrine at 480 nm, which is 4020 M^{-1} cm^{-1}. This approach enables precise measurement of SOD activity by monitoring the inhibition of epinephrine auto-oxidation.

2.3.7 Determination of PBDES in Tissues

The KOH refluxing/vortex extraction technique, as outlined in EPA Method 3611C, was employed for sample preparation and cleanup during the analysis. Initially, 15 g of wet tissue, comprising excised liver and intestine samples from the fish specimens, was weighed into a crucible, macerated, and homogenized thoroughly. From the homogenized tissue, 10 g was transferred into a 50-mL centrifuge tube, followed by the addition of 15 mL of 6N KOH. The tubes were securely sealed and incubated in a 35 °C water bath for 18 h, with vigorous shaking for 30 s every half-hour during the first 4 h. The samples were then allowed to cool.

Subsequently, 15 mL of methylene chloride or ethyl ether was added to each centrifuge tube, vortexed for 1 min, and centrifuged at 2000 rpm for 5 min to promote efficient phase separation. The upper layer (organic phase) was carefully extracted using a Pasteur pipette and transferred into a 250-mL round-bottom flask. This solvent extraction and centrifugation process was repeated twice, with all aliquots combined in the round-bottom flask.

To concentrate the sample extracts, the combined aliquots were reduced to a volume of approximately 5–10 mL using a rotary evaporator. The concentrated extracts were then subjected to fractionation and cleanup via an alumina gel column prior to analysis using gas chromatography-mass spectrometry (GC-MS). This ensured the removal of interfering substances, allowing for accurate quantification and characterization of the target analytes.

2.3.8 Histopathological Assessment of Infected Tissues

The intestines of the fish were initially fixed in Bouin's fluid for 6 h to initiate tissue preservation, followed by immersion in 10% phosphate-buffered formalin to enhance fixation. After fixation, the tissues were sectioned and placed into labeled tissue embedding cassettes for processing. Using an automated tissue processor, samples were dehydrated over a period of 17–19 h through a graded series of alcohol concentrations, including 70%, 95%, and two treatments in absolute ethanol, each for 30 min.

Subsequent to dehydration, the tissues were impregnated with molten paraffin in three consecutive treatments and then embedded in molten paraffin wax to allow for tissue solidification. Each paraffin block was mounted on a rotary microtome and sectioned into thin slices of 4–5 μ. These sections were carefully transferred to labeled slides and floated on a 45 °C water bath to prevent folding or creasing, utilizing curved floating forceps for precision. The sections were then picked up on labeled slides, ensuring central adhesion, and dried in a 60 °C hot air oven to promote secure attachment, following the protocol established.

For further preparation, the slide-mounted sections were dewaxed in xylene and rehydrated through a descending alcohol series (absolute ethanol, 95%, 80%, and 70%) and finally in water. Hematoxylin and eosin (H&E) staining was applied, where hematoxylin highlighted the cell nuclei, and eosin stained the cytoplasmic components. Post-staining, the sections were washed in tap water, and excess staining was rectified with 1% acid alcohol. The tissues were then mounted using Di-*n*-butyl phthalate in xylene (DPX) and left to dry.

The prepared slides were subsequently examined microscopically, with photomicrographs taken for documentation. The pathological lesions identified in the tissues were analyzed at the Pathology Laboratory, Department of Veterinary Pathology, University of Ibadan, Nigeria.

2.4 Quality Assurance

Each tested medium was subjected to analysis using ten sample replicates, with the mean values reported to enhance analytical accuracy. Prior to sample analysis, the instrument underwent a rigorous calibration procedure, involving the injection of a series of calibration standards, each with an injection volume of 1 μL. A five-point calibration curve was constructed using commercially sourced calibration standards, with the curve's validity confirmed by achieving an R^2 value of ≥ 0.995.

For each analyte/component, the response factor (RF) was determined by correlating the area response with the corresponding amount of standard material. The relative standard deviation percentage (%RSD) of the RF was calculated for each analyte across the calibration curve, adhering to a strict criterion that the %RSD should not exceed 15% for the calibration to be considered valid. The average RF within the specified weight ranges was then calculated and applied for quantification of the analytes in the samples. In instances where any alkane component exhibited an %RSD exceeding the 15% threshold, the response factors within the respective weight ranges were reassessed and utilized, contingent on their adherence to the predefined quality criteria.

To ensure robust sample analysis and quality control, a systematic approach was employed, incorporating initial calibration standards, continuous calibration standards (5 μg/mL or 10 μg/mL), standard reference materials (SRMs), and instrument blanks. These measures ensured the reliability and precision of the analytical data generated during the study.

2.5 Statistical Analysis

The data obtained from bioaccumulation analysis, oxidative stress analysis, and physicochemical parameters were evaluated using analysis of variance (ANOVA), statistical package for the social sciences (SPSS), and IBM 20.0 version.

3 Results

The results of the physicochemical parameters of the surface water indicated that the turbidity of the water samples at the location of *T. zillii* was significantly higher than that of *S. clarias* ($p < 0.001$). No significant difference occurred among other physicochemical parameters tested (Table 26.1). Notably, the biological oxygen demand (BOD) of both surface water samples was higher than the adopted FEPA regulatory limit of ≤ 5.

Table 26.1 Mean physicochemical parameters of the two sampling points along Lekki Lagoon

	T. zillii water		*S. clarias* water				
Parameters	mean	SD	mean	SD	Summary	*p*-value	FEPA limit
Temp (°C)	30.01	0.13	29.40	0.03	ns	0.159	≤40
pH	7.43	0.20	7.04	0.07	ns	0.359	6–9
Current (ms/cm)	0.42	0.01	0.45	0.01	ns	0.944	–
Turbidity (NTU)	15.87	0.81	8.94	1.46	***	<0.001	≤30
BOD (g/L)	6.58	0.62	6.27	0.04	ns	0.463	≤5
TDS (g/L)	0.27	0.01	0.29	0.01	ns	0.957	≤30

***Highly significant

Table 26.2 Comparative bioaccumulation of PBDE congeners (mg/g) in the intestine of *T. zillii* and *S. clarias* in Lekki Lagoon

	T. zillii		*S. clarias*			
PBDE congeners	mean	SD	mean	SD	Summary	*p*-value
BDE-28	0.02	0.02	0.02	0.02	ns	0.96
BDE-47	0.01	0.01	0.01	0.01	ns	0.51
BDE-100	0.01	0.01	0.01	0.01	ns	0.37
BDE-99	0.01	0.01	0.00	0.00	ns	0.74
BDE-154	0.00	0.00	0.00	0.00	ns	0.98
BDE-153	0.00	0.00	0.00	0.00	ns	0.91
BDE-183	0.02	0.05	0.00	0.00	**	0.00
BDE-209	0.00	0.00	0.00	0.00	ns	>0.999

Keys: BDE-28 (2,4,4′-tribromodiphenyl ether); BDE-47 (2,2′,4,4′-tetrabromodiphenyl ether); BDE-100 (2,2′,4,4′,6-pentabromodiphenyl ether); BDE-99 (2,2′,4,4′,5,-pentabromodiphenyl ether); BDE-154 (2,2′,4,4′,5,6′-hexabromodiphenyl ether); BDE-153 (2,2′,4,4′,5,5′-hexabromodiphenyl ether); BDE-183 (2,2′,3,4,4′,5′,6-heptabromodiphenyl ether); BDE-209 (2,2′,3,3′,4,4′,5,5′,6,6′-decabromodiphenyl ether). Sample size (*N*) = 10, ns = not significant

**Significant

The comparative analysis of PBDE congeners accumulation in the intestines of *T. zillii* and *S. clarias* indicates that most congeners were similarly accumulated, except PBDE-183 (2,2′,3,4,4′,5′,6-heptabromodiphenyl ether) that recorded 0.02 mg/g, while the PBDE congener was not detected in the intestine of *S. clarias* (Table 26.2).

The comparative assessment of the oxidative stress initiated in the intestines of *T. zillii* and *S. clarias* as a result of bioaccumulation of the PBDEs indicated that there was no significant difference in the concentrations of the PBDEs across all the congeners analyzed (Table 26.3).

A comparative analysis of the lipid contents in the intestine of both fish species indicated that all the lipid parameters assessed in *T. zillii* were significantly much higher ($p < 0.001$) than the concentrations detected in *S. clarias* (Table 26.4).

Table 26.3 Comparative oxidative stress assessment in the intestine of *T. zillii* and *S. clarias* in Lekki Lagoon

	T. zillii		*S. clarias*			
Antioxidants	Mean	SD	Mean	SD	Summary	*p*-value
PRO	10.97	2.84	9.18	1.06	ns	0.93
CAT	6.47	2.83	2.15	1.42	ns	0.83
SOD	679.88	114.46	714.12	100.66	ns	0.09
MDA	4.03	2.08	1.36	0.79	ns	0.89
GSH	13.47	10.03	1.39	0.52	ns	0.55
GPX	211.24	12.63	188.68	16.95	ns	0.26

Keys: Sample size (*N*) = 10, ns = not significant

Table 26.4 Comparative lipid profile (mg/dL) in the intestine of *T. zillii* and *S. clarias* in Lekki Lagoon

	T. zillii		*S. clarias*			
Lipids	Mean	SD	Mean	SD	Summary	*p*-value
CHOL	145.94	7.54	1.97	0.39	***	<0.001
TRIG	54.96	6.08	0.72	0.22	***	<0.001
HDL	49.11	8.70	0.54	0.13	***	<0.001
LDL	85.84	8.85	1.10	0.38	***	<0.001

***Highly significant

4 Discussion

The sole discernible variation in ambient conditions between the habitats of *Tilapia zillii* and *Sarotherodon clarias* was the markedly elevated turbidity levels observed in the former, as reported by Kofi et al. (2018). This disparity is likely attributable to localized anthropogenic disturbances in the *T. zillii* habitat, with potential drivers including dredging, shoreline construction, or the discharge of untreated wastewater, all of which can significantly elevate suspended particulate matter in the water column, thereby increasing turbidity (Kostić et al., 2017). The observed turbidity in the *T. zillii* habitat appears to be confined to the specific sampling location, indicating that the increased turbidity is not a result of upstream water influx or broader hydrological shifts but is rather due to site-specific anthropogenic activities (Akortia et al., 2016; Yang et al., 2018).

Understanding the origins and ecological ramifications of heightened turbidity is critical, as such conditions can impair photosynthetic activity in aquatic plants, reduce light penetration, and diminish foraging efficiency in fish by restricting visibility. Additionally, elevated turbidity levels can exacerbate the transport of pollutants and pathogens adsorbed onto suspended particles, thereby heightening the risk of contamination within the aquatic ecosystem (Lana et al., 2010; La Guardia et al., 2012). Given these risks, the identification and mitigation of localized anthropogenic drivers are essential for preserving the ecological integrity, habitat quality, and water clarity within the ecosystems inhabited by *T. zillii* and *S. clarias*.

Temperature influences the growth and distribution of flora and fauna as it is the most variable factor in the environment and plays a vital role in chemical and biological activities. pH shows the intensity of the acidity or basicity character of a solution and is controlled by the dissolved chemical compounds and biochemical processes in the solution. Salinity is one of the important factors that influence the abundance and distribution of the animals in the water body. It is an important ecological variable, important in some chemical processes and also has a strong effect on nutrient levels in the coastal areas. The physicochemical properties of selected points of the lagoon surface water were carried out and compared with the Federal Environmental Protection Agency standard (FEPA). The temperature, pH, conductivity, dissolved oxygen, salinity, and turbidity results were all within the FEPA limit for physicochemical parameters in lagoon surface waters. According to the study, an increase or change in the level of physicochemical parameters in surface water influences the fate of contaminants in an aquatic environment.

Furthermore, the concentration of 0.02 mg/g of PBDE-183 (2,2′,3,4,4′,5′,6-Heptabromodiphenyl ether) detected in the intestine of *Tilapia zillii* was significantly higher than that observed in *Synodontis clarias*. This marked difference in PBDE-183 bioaccumulation was the only discernible variation between the two fish species. The differential bioaccumulation of PBDE-183 in *T. zillii* may have profound implications on its lipid profile, as PBDEs are known to be lipophilic and can integrate into lipid-rich tissues, potentially altering lipid metabolism and storage. The relatively higher susceptibility of *T. zillii* to accumulate PBDE-183 could be attributed to its physiological and metabolic characteristics, which differ from the more robust and tenacious *S. clarias* (Xin et al., 2014; Xu et al., 2018).

T. zillii, being considered more fragile, might exhibit a higher rate of lipid turnover or have a different lipid composition that facilitates the greater accumulation of PBDE-183. This enhanced bioaccumulation could lead to disruptions in its lipid homeostasis, affecting vital biological functions such as energy storage, membrane fluidity, and hormone production. In contrast, the tenacity and possibly more efficient detoxification mechanisms in *S. clarias* might confer it a resilience against the bioaccumulation of such contaminants.

The observed variance in PBDE-183 levels underscores the importance of understanding species-specific differences in contaminant uptake and metabolism. These differences are crucial for assessing ecological risks and potential health impacts on fish populations and, by extension, on the human consumers of these fish. The higher bioaccumulation of PBDE-183 in *T. zillii* necessitates targeted studies to elucidate its effects on the lipid profile and overall health of this species, thereby informing conservation and public health strategies (Lenhardt et al., 2015; Li et al., 2018).

The impacted lipid profile observed in *Tilapia zillii* might not have inflicted any oxidative damage at present; however, the findings of this study underscore the necessity for proactive mitigation of anthropogenic activities around the catchment area of Lekki Lagoon to forestall the release of PBDE-183 (2,2′,3,4,4′,5′,6-Heptabromodiphenyl ether).

While no immediate oxidative damage has been detected, the elevated levels of PBDE-183 in *T. zillii* highlight a potential risk for future metabolic disturbances. PBDEs, being lipophilic, tend to bioaccumulate in lipid-rich tissues, and prolonged exposure could lead to chronic health effects, including oxidative stress, endocrine disruption, and impaired reproductive functions. Therefore, it is crucial to address the sources of PBDE-183 contamination to prevent long-term ecological and health consequences (Liagkouridis et al., 2014; Linares et al., 2015).

To mitigate the anthropogenic release of PBDE-183 into Lekki Lagoon, a comprehensive set of proactive strategies must be implemented. Industries around the lagoon should adopt advanced effluent treatment technologies to effectively eliminate PBDEs and other hazardous contaminants from their discharges, while strict disposal and recycling protocols for PBDE-containing materials, such as electronics, textiles, and furniture, must be enforced to prevent environmental leaching. Additionally, the deployment of efficient stormwater management infrastructure is essential to curb the inflow of PBDEs from urban areas, and advanced sediment control measures should be utilized to limit the transport of PBDE-bound particles into aquatic systems, minimizing water body contamination. Public engagement initiatives should be prioritized to educate local communities on the sources, pathways, and risks of PBDEs, encouraging collective action to reduce pollutant release while fostering collaboration among government agencies, industry, and nongovernmental organizations to develop, enforce, and monitor effective PBDE pollution regulations. Continuous monitoring of PBDE levels in the lagoon and its biota is vital to assess contamination trends and evaluate the effectiveness of mitigation efforts, while research into the long-term ecological and physiological impacts of PBDEs on aquatic organisms will provide the foundation for informed management and future conservation strategies. Implementing these measures will significantly reduce PBDE-183 pollution in Lekki Lagoon, protecting its aquatic ecosystem and the health of species like *Tilapia zillii* while ensuring the lagoon remains a crucial ecological and economic resource for surrounding communities and its long-term sustainability.

5 Conclusion

This study has elucidated the significant presence and differential bioaccumulation of PBDE-183 (2,2′,3,4,4′,5′,6-heptabromodiphenyl ether) in *T. zillii* and *S. clarias* within the Lekki Lagoon ecosystem. The detection of higher concentrations of PBDE-183 in *T. zillii* underscores the species-specific vulnerabilities and metabolic differences that contribute to the varying levels of contaminant uptake. Although no immediate oxidative damage was observed in the impacted lipid profiles of *T. zillii*, the potential long-term health and ecological implications of such bioaccumulation warrant urgent and proactive intervention.

The study highlights the necessity for comprehensive mitigation strategies aimed at curbing the release of PBDE-183 and other similar pollutants. Addressing anthropogenic activities, particularly industrial discharges, improper waste management, and urban runoff, is critical in preserving the ecological integrity of Lekki Lagoon. Enhanced waste treatment practices, strict regulatory frameworks, and widespread public awareness campaigns are essential to prevent further contamination and safeguard the lagoon's aquatic life.

Furthermore, continuous monitoring and research are imperative to understand the full spectrum of PBDE impacts on the lagoon's biodiversity and to develop targeted conservation efforts. The findings of this study serve as a clarion call for immediate and concerted action to mitigate environmental pollution, protect vulnerable species, and ensure the sustainability of Lekki Lagoon as a vital ecological and economic resource for future generations.

References

Abdallah, M. A. E., & Harrad, S. (2014). Polybrominated diphenyl ethers in UK human milk: Implications for infant exposure and relationship to external exposure. *Environment International, 63*, 130–136.

Adewuyi, G. O., & Adeleye, A. (2013). Evaluation of polybrominated diphenyl ethers in sediment of Lagos Lagoon, Nigeria. *African Journal of Environmental Science and Technology, 7*(7), 686–693.

Akortia, E., Okonkwo, J. O., Lupankwa, M., Osae, S. D., Daso, A. P., Olukunle, O. I., & Chaudhary, A. (2016). A review of sources, levels, and toxicity of polybrominated diphenyl ethers (PBDEs) and their transformation and transport in various environmental compartments. *Environmental Reviews, 24*(3), 253–273.

Babayemi, J. O., Osibanjo, O., Sindiku, O., & Weber, R. (2018). Inventory and substance flow analysis of polybrominated diphenyl ethers in the Nigerian transport sector—End-of-life vehicles policy and management. *Environmental Science and Pollution Research, 25*, 31805–31818.

Bamidele, A., Kuton, M. P., Iniobong, A. D., Uchenna, N. D., Saliu, J. K., & David, U. U. (2020). Bioaccumulation of polychlorinated biphenyls (PCBs) in fish host-parasite bentho-pelagic food chain in Epe lagoon, Lagos, Nigeria. *Bulletin of Environmental Contamination and Toxicology, 105*, 770–776.

Banni, M., Hajer, A., Sforzini, S., Oliveri, C., Boussetta, H., & Viarengo, A. (2014). Transcriptional expression levels and biochemical markers of oxidative stress in Mytilus galloprovincialis exposed to nickel and heat stress. *Comparative Biochemistry and Physiology Part C: Toxicology & Pharmacology, 160*, 23–29.

Bergayou, H., Mouneyrac, C., Pellerin, J., & Moukrim, A. (2009). Oxidative stress responses in bivalves (Scrobicularia plana, Cerastoderma edule) from the Oued Souss estuary (Morocco). *Ecotoxicology and Environmental Safety, 72*(3), 765–769.

Bobescu, E., Marceanu, L. G., Dima, L., Balan, A., Strempel, C. G., & Covaciu, A. (2021). Trimetazidine therapy in coronary artery disease: The impact on oxidative stress, inflammation, endothelial dysfunction, and long-term prognosis. *American Journal of Therapeutics, 28*(5), e540–e547.

Bodin, N., Ka, R. G., Le Loc'h, F., Raffray, J., Budzinski, H., Peluhet, L., & de Morais, L. T. (2011). Are exploited mangrove molluscs exposed to persistent organic pollutant contamination in Senegal, West Africa? *Chemosphere, 84*(3), 318–327.

Bramwell, L., Harrad, S., Abdallah, M. A. E., Rauert, C., Rose, M., Fernandes, A., & Pless-Mulloli, T. (2017). Predictors of human PBDE body burdens for a UK cohort. *Chemosphere, 189*, 186–197.

Cao, L. Y., Ren, X. M., Yang, Y., Wan, B., Guo, L. H., Chen, D., & Fan, Y. (2018). Hydroxylated polybrominated biphenyl ethers exert estrogenic effects via non-genomic G protein-coupled estrogen receptor mediated pathways. *Environmental Health Perspectives, 126*(5), 057005.

Chakraborty, P., Zhang, G., Li, J., Sampathkumar, P., Balasubramanian, T., Kathiresan, K., et al. (2019). Seasonal variation of atmospheric organochlorine pesticides and polybrominated diphenyl ethers in Parangipettai, Tamil Nadu, India: Implication for atmospheric transport. *Science of the Total Environment, 649*, 1653–1660.

Choi, G., Kim, S., Kim, S., Kim, S., Choi, Y., Kim, H. J., et al. (2014). Occurrences of major polybrominated diphenyl ethers (PBDEs) in maternal and fetal cord blood sera in Korea. *Science of the Total Environment, 491*, 219–226.

Daso, A. P., Fatoki, O. S., Odendaal, J. P., & Olujimi, O. O. (2013). Polybrominated diphenyl ethers (PBDEs) and 2,2′,4,4′,5,5′-hexabromobiphenyl (BB-153) in landfill leachate in Cape Town, South Africa. *Environmental Monitoring and Assessment, 185*, 431–439.

Davis, E. F., Klosterhaus, S. L., & Stapleton, H. M. (2012). Measurement of flame retardants and triclosan in municipal sewage sludge and biosolids. *Environment International, 40*, 1–7.

Dazy, M., Masfaraud, J. F., & Férard, J. F. (2009). Induction of oxidative stress biomarkers associated with heavy metal stress in *Fontinalis antipyretica* Hedw. *Chemosphere, 75*(3), 297–302.

Doherty, F. V., Aneyo, I., & Otitoloju, A. A. (2019). Histopathological and biochemical alterations in Eudrilus eugeniae (Kinberg 1867) as biomarkers of exposure to monocyclic aromatic hydrocarbons in oil impacted site. *The Journal of Basic and Applied Zoology, 80*, 1–15.

Ge, J., Yun, X., Liu, M., Yang, Y., Zhang, M., & Wang, J. (2014). Distribution, potential source and ecotoxicological risk of polychlorinated biphenyls and polybrominated diphenyl ethers in the surface water of the Three Gorges Dam region of the Yangtze River, China. *Ecotoxicology, 23*, 978–987.

Gonzalez, M., Miglioranza, K. S., Grondona, S. I., Barni, M. F. S., Martinez, D. E., & Peña, A. (2013). Organic pollutant levels in an agricultural watershed: The importance of analyzing multiple matrices for assessing streamwater pollution. *Environmental Science: Processes & Impacts, 15*(4), 739–750.

Guo, J., Zhang, R., & Xu, Z. (2015). PBDEs emission from waste printed wiring boards during thermal process. *Environmental Science and Technology, 49*(5), 2716–2723.

Guo, W., Holden, A., Smith, S. C., Gephart, R., Petreas, M., & Park, J. S. (2016). PBDE levels in breast milk are decreasing in California. *Chemosphere, 150*, 505–513.

Hahladakis, J. N., Velis, C. A., Weber, R., Iacovidou, E., & Purnell, P. (2018). An overview of chemical additives present in plastics: Migration, release, fate and environmental impact during their use, disposal and recycling. *Journal of Hazardous Materials, 344*, 179–199.

Huang, Y., Rajput, I. R., Sanganyado, E., Yajing, S., Yu, F., Liang, B., & Liu, W. (2020). Immune stimulation effect of PBDEs via prostaglandin pathway in pantropical spotted dolphin: An in vitro study. *Chemosphere, 254*, 126717.

Johnson, P. I., Stapleton, H. M., Mukherjee, B., Hauser, R., & Meeker, J. D. (2013). Associations between brominated flame retardants in house dust and hormone levels in men. *Science of the Total Environment, 445*, 177–184.

Kanaya, N., Bernal, L., Chang, G., Yamamoto, T., Nguyen, D., Wang, Y. Z., et al. (2019). Molecular mechanisms of polybrominated diphenyl ethers (BDE-47, BDE-100, and BDE-153) in human breast cancer cells and patient-derived xenografts. *Toxicological Sciences, 169*(2), 380–398.

Khan, S., Naushad, M., Govarthanan, M., Iqbal, J., & Alfadul, S. M. (2022). Emerging contaminants of high concern for the environment: Current trends and future research. *Environmental Research, 207*, 112609.

Klinčić, D., Dvoršćak, M., Jagić, K., Mendaš, G., & Herceg Romanić, S. (2020). Levels and distribution of polybrominated diphenyl ethers in humans and environmental compartments: A comprehensive review of the last five years of research. *Environmental Science and Pollution Research, 27*, 5744–5758.

Kofi, E. D., Kweku, A. J., Kweku, C. S., Essel, G., & Osei-Fosu, P. (2018). Levels of polybrominated diphenyl ethers (PBDEs) in some Ghanaian water body environments. *Research Journal of Environmental Sciences, 12*, 73–82.

Kostić, J., Kolarević, S., Kračun-Kolarević, M., Aborgiba, M., Gačić, Z., Paunović, M., & Vuković-Gačić, B. (2017). The impact of multiple stressors on the biomarkers response in gills and liver of freshwater breams during different seasons. *Science of the Total Environment, 601*, 1670–1681.

La Guardia, M. J., Hale, R. C., Harvey, E., Mainor, T. M., & Ciparis, S. (2012). In situ accumulation of HBCD, PBDEs, and several alternative flame-retardants in the bivalve (Corbicula fluminea) and gastropod (Elimia proxima). *Environmental Science & Technology, 46*(11), 5798–5805.

Lana, N. B., Fontana, A. R., Ciocco, N. F., & Altamirano, J. C. (2010). Determination of polybrominated diphenyl ethers in water samples from Mendoza River Basin by HS-SPME-GC-MS/MS. *Revista de la Facultad de Ciencias Agrarias, 42*(2), 85–98.

Lenhardt, M., Poleksić, V., Vuković-Gačić, B., Rašković, B., Sunjog, K., Kolarević, S., & Gačić, Z. (2015). Integrated use of different fish related parameters to assess the status of water bodies. *Slovenian Veterinary Research, 52*(1), 5–13.

Li, P., Liu, L., Zhou, G., Tian, Z., Luo, C., Xia, T., et al. (2018). Perigestational exposure to low doses of PBDE-47 induces excessive ER stress, defective autophagy and the resultant apoptosis contributing to maternal thyroid toxicity. *Science of the Total Environment, 645*, 363–371.

Liagkouridis, I., Cousins, I. T., & Cousins, A. P. (2014). Emissions and fate of brominated flame retardants in the indoor environment: A critical review of modelling approaches. *Science of the Total Environment, 491*, 87–99.

Linares, V., Bellés, M., & Domingo, J. L. (2015). Human exposure to PBDE and critical evaluation of health hazards. *Archives of Toxicology, 89*(3), 335–356.

Luo, P., Ni, H. G., Bao, L. J., Li, S. M., & Zeng, E. Y. (2014). Size distribution of airborne particle-bound polybrominated diphenyl ethers and its implications for dry and wet deposition. *Environmental Science & Technology, 48*(23), 13793–13799.

Margolis, A. E., Banker, S., Pagliaccio, D., De Water, E., Curtin, P., Bonilla, A., et al. (2020). Functional connectivity of the reading network is associated with prenatal polybrominated diphenyl ether concentrations in a community sample of 5 year-old children: A preliminary study. *Environment International, 134*, 105212.

McGrath, T. J., Morrison, P. D., Ball, A. S., & Clarke, B. O. (2018). Concentrations of legacy and novel brominated flame retardants in indoor dust in Melbourne, Australia: An assessment of human exposure. *Environment International, 113*, 191–201.

O'Driscoll, K., Robinson, J., Chiang, W. S., Chen, Y. Y., Kao, R. C., & Doherty, R. (2016). The environmental fate of polybrominated diphenyl ethers (PBDEs) in western Taiwan and coastal waters: Evaluation with a fugacity-based model. *Environmental Science and Pollution Research, 23*, 13222–13234.

Okoffo, E. D., O'Brien, S., Ribeiro, F., Burrows, S. D., Toapanta, T., Rauert, C., et al. (2021). Plastic particles in soil: State of the knowledge on sources, occurrence and distribution, analytical methods and ecological impacts. *Environmental Science: Processes & Impacts, 23*(2), 240–274.

Oloruntoba, K., Sindiku, O., Osibanjo, O., Herold, C., & Weber, R. (2021). Polybrominated diphenyl ethers (PBDEs) concentrations in soil and plants around municipal dumpsites in Abuja, Nigeria. *Environmental Pollution, 277*, 116794.

Olujimi, O. O., Adegbite, K. I., Sojinu, S. O., Daso, A. P., Oyebanji, F. F., Oluwadare, I., et al. (2024). Occurrence, compositional profiles and health risks assessment of polybrominated diphenyl ethers (PBDEs) in groundwater in and around dumpsites in Lagos State, Nigeria. *Environmental Forensics, 25*(1–2), 49–58.

Olutona, G. O., Oyekunle, J. A., Ogunfowokan, A. O., & Fatoki, O. S. (2016). Assessment of polybrominated diphenyl ethers in sediment of Asunle stream of the Obafemi Awolowo University, Ile-Ife, Nigeria. *Environmental Science and Pollution Research, 23*, 21195–21205.

Olutona, G. O., Oyekunle, J. A., Ogunfowokan, A. O., & Fatoki, O. S. (2017). Concentrations of polybrominated diphenyl ethers (PBDEs) in water from Asunle Stream, Ile-Ife, Nigeria. *Toxics, 5*(2), 13.

Onyeche, L. A., & Akankali, J. A. (2013). Determination of some environmental parameters of Anwai–Stream, Niger-Delta, Nigeria. *Research Journal of Agricultural and Environmental Management, 2*(6), 142–149.

Park, J. E., Kang, Y. Y., Kim, W. I., Jeon, T. W., Shin, S. K., Jeong, M. J., & Kim, J. G. (2014). Emission of polybrominated diphenyl ethers (PBDEs) in use of electric/electronic equipment and recycling of e-waste in Korea. *Science of the Total Environment, 470*, 1414–1421.

Pellacani, C., Tagliaferri, S., Caglieri, A., Goldoni, M., Giordano, G., Mutti, A., & Costa, L. G. (2014). Synergistic interactions between PBDEs and PCBs in human neuroblastoma cells. *Environmental Toxicology, 29*(4), 418–427.

Pereira, S., Pinto, A. L., Cortes, R., Fontaínhas-Fernandes, A., Coimbra, A. M., & Monteiro, S. M. (2013). Gill histopathological and oxidative stress evaluation in native fish captured in Portuguese northwestern rivers. *Ecotoxicology and Environmental Safety, 90*, 157–166.

Ssebugere, P., Sillanpää, M., Wang, P., Li, Y., Kiremire, B. T., Kasozi, G. N., et al. (2014). Polychlorinated dibenzo-p-dioxins, polychlorinated dibenzofurans and polybrominated diphenyl ethers in sediments and fish species from the Murchison Bay of Lake Victoria, Uganda. *Science of the Total Environment, 500*, 1–10.

Su, P. H., Tomy, G. T., Hou, C. Y., Yin, F., Feng, D. L., Ding, Y. S., & Li, Y. F. (2018). Gas/particle partitioning, particle-size distribution of atmospheric polybrominated diphenyl ethers in Southeast Shanghai rural area and size-resolved predicting model. *Chemosphere, 197*, 251–261.

Tang, J., & Zhai, J. X. (2017). Distribution of polybrominated diphenyl ethers in breast milk, cord blood and placentas: A systematic review. *Environmental Science and Pollution Research, 24*, 21548–21573.

Ukenye, E. A., Taiwo, I. A., Oguntade, O. R., Oketoki, T. O., & Usman, A. B. (2016). Molecular characterization and genetic diversity assessment of *Tilapia zilli* from some coastal rivers in Nigeria. *African Journal of Biotechnology, 15*(1), 20–28.

Wang, J., Wang, Y., Shi, Z., Zhou, X., & Sun, Z. (2018). Legacy and novel brominated flame retardants in indoor dust from Beijing, China: Occurrence, human exposure assessment and evidence for PBDEs replacement. *Science of the Total Environment, 618*, 48–59.

Wepener, V., Smit, N., Covaci, A., Dyke, S., & Bervoets, L. (2012). Seasonal bioaccumulation of organohalogens in tigerfish, Hydrocynus vittatus Castelnau, from Lake Pongolapoort, South Africa. *Bulletin of Environmental Contamination and Toxicology, 88*, 277–282.

Winid, B. (2015). Environmental threats of natural water contamination with polybrominated diphenyl ethers (PBDEs). *Polish Journal of Environmental Studies, 24*(1), 47.

Wu, Z., He, C., Han, W., Song, J., Li, H., Zhang, Y., et al. (2020). Exposure pathways, levels and toxicity of polybrominated diphenyl ethers in humans: A review. *Environmental Research, 187*, 109531.

Xin, J., Liu, X., Liu, W., & Zheng, X. (2014). Effects of biochar–BDE-47 interactions on BDE-47 bioaccessibility and biodegradation by Pseudomonas putida TZ-1. *Ecotoxicology and Environmental Safety, 106*, 27–32.

Xu, B., Wu, M., Wang, M., Pan, C., Qiu, W., Tang, L., & Xu, G. (2018). Polybrominated diphenyl ethers (PBDEs) and hydroxylated PBDEs in human serum from Shanghai, China: A study on their presence and correlations. *Environmental Science and Pollution Research, 25*, 3518–3526.

Yang, J., Huang, D., Xue, W., Wei, X., Qin, J., Ou, S., et al. (2018). Multiple-life-stage probabilistic risk assessment for the exposure of Chinese population to PBDEs and risk managements. *Science of the Total Environment, 643*, 1178–1190.

Yu, G., Bu, Q., Cao, Z., Du, X., Xia, J., Wu, M., & Huang, J. (2016). Brominated flame retardants (BFRs): A review on environmental contamination in China. *Chemosphere, 150*, 479–490.

Zhang, T., Zhou, X., Xu, A., Tian, Y., Wang, Y., Zhang, Y., & Wang, Z. (2020). Toxicity of polybrominated diphenyl ethers (PBDEs) on rodent male reproductive system: A systematic review and meta-analysis of randomized control studies. *Science of the Total Environment, 720*, 137419.

Zhu, C., Li, Y., Wang, P., Chen, Z., Ren, D., Ssebugere, P., et al. (2015). Polychlorinated biphenyls (PCBs) and polybrominated biphenyl ethers (PBDEs) in environmental samples from Ny-Ålesund and London Island, Svalbard, the Arctic. *Chemosphere, 126*, 40–46.

Chapter 27
Effects of Insecticide Resistance on Malaria Vector Competence

Oniha Margaret, Omonigbehin Comfort, Kwarpo Zeendi, and Emmanuel Omonigbehin

1 Introduction

Malaria remains the deadliest vector-borne disease, having claimed millions of lives throughout history (Suh et al., 2023). In 2020 alone, 241 million victims were promulgated, resulting in 627,000 fatalities, with Africa bearing most brunt of the disease, accounting for 96% of all malaria-related fatalities. The ailment is effectuated by *Plasmodium* parasites, which are transmitted to humans by the bites of infected female *Anopheles* mosquitoes. In Africa, *Plasmodium falciparum* is the preponderance common parasite causing malaria in humans, while *Anopheles gambiae*, *Anopheles coluzzii*, *Anopheles funestus*, and *Anopheles arabiensis* serve as the major vectors. Malaria control strategies focus on both treating infected individuals and preventing human exposure to mosquitoes, using methods like insecticide-treated nets (ITNs), indoor residual spraying, and enhanced case management. This interposition significantly depleted malaria incidence from 81.1 cases per 1000 people in 2000 to 58.9 in 2015. Nonetheless, the rise of insecticide-resistant mosquito populations in many endemic regions has diminished the effectiveness of these strategies. Insecticide resistance was first identified in Africa in the 1950s, with four major groups of insecticides, pyrethroids, organochlorines, organophosphates, and carbamates, used in vector control. Mosquitoes have developed four key

O. Margaret · O. Comfort · K. Zeendi
Department of Biological Sciences, College of Science and Technology, Covenant University, Ota, Ogun State, Nigeria

Covenant Applied Informatics and Communication Africa Centre of Excellence (CApIC-ACE), Covenant University, Ota, Ogun State, Nigeria

E. Omonigbehin (✉)
Department of Biological Sciences, College of Science and Technology, Covenant University, Ota, Ogun State, Nigeria
e-mail: comfort.omonigbehinpgs@stu.cu.edu.ng

P. O. Isibor, P. A. Akinduti (eds.), *Harnessing Biotechnology Tools for Product Development*, https://doi.org/10.1007/978-3-031-86002-7_27

resistance mechanisms: (1) metabolic resistance in which insecticides are broken down by detoxifying enzymes; (2) target site resistance, caused by altering the insecticide's action site; (3) cuticular resistance, involving reduced insecticide penetration; and (4) behavioral resistance, where treated areas are avoided by mosquitoes. Of the aforementioned, target site and metabolic resistance has the likelihood to lead to control failures. Target site resistance occurs when mutations in the voltage-gated sodium channel (vgsc) reduce the sensitivity of mosquitoes to pyrethroids and organochlorines. This mutation, where leucine is replaced by phenylalanine or serine at position 1014, is known as knockdown resistance (kdr). Similarly, ace-1 resistance develops when glycine is replaced by serine at position 119 of acetylcholinesterase, resulting in resistance to organophosphates and carbamates. Metabolic resistance takes place when mosquitoes synthesize extra detoxifying enzymes, e.g., cytochrome P450 monooxygenases, glutathione S-transferases (GSTs), and esterases, that neutralize insecticides before they can take effect. Mutations in these detoxifying enzymes can also alter their affinity for insecticides (Shah et al., 2020). For instance, in *A. funestus*, the cytochrome P450 genes (CYP6P9a, CYP6P9b, and CYP6M7) contribute to pyrethroid resistance, while the exchange of leucine for phenylalanine at position 119 in the epsilon class of GST (GST2-L119F) brings about cross-resistance to both DDT and pyrethroids in this species. Although insecticide resistance is widespread, its full influence on malaria epidemiology is still uncertain and contentious. Research into how insecticide resistance influences vector fitness and the capability of mosquitoes to transfer *Plasmodium* parasites has become increasingly important, as it could clarify whether resistance affects transmission from the ookinete stage to the infective sporozoite stage. This review explores existing research on how insecticide resistance impacts mosquito infectivity with *Plasmodium* and identifies knowledge gaps that need further exploration.

2 Materials and Methods

A thorough analysis of the literature was conducted utilizing the PubMed database to gather studies related to insecticide resistance in malaria vectors. Five key topics were targeted: (1) the presence of *Plasmodium* infection in insecticide-resistant mosquitoes; (2) the sialome (salivary proteins) of these vectors; (3) the effect of insecticide resistance on vector immunity; (4) the microbiota of these vectors and its role in infection; and (5) the aptness costs associated with insecticide resistance. The search was carried out between January and April 2024, without restrictions on the publication date. From an initial pool of 560 articles, studies that did not specifically address insecticide resistance in *Anopheles* mosquitoes in relation to *Plasmodium* infection, vector competence, sialome, or longevity were excluded. Ultimately, 30 articles that met the inclusion criteria were selected for the review.

3 Competency of Vectors in Malaria Transmission

Vector competence encompasses the intrinsic propensity of *Anopheles* species to support *Plasmodium* growth, enabling the parasite to progress from the ookinete stage to infective sporozoites (Phillips et al., 2017). Once a mosquito ingests infected blood, gametocytes in the blood begin their development within the mosquito's midgut. Male gametocytes undergo mitotic divisions to produce microgametes, while female gametocytes mature into macrogametes. These gametes fuse, forming zygotes that develop into ookinetes in the gut lumen. Ookinetes then invade the midgut lining and differentiate into oocysts on the basal side of the epithelium (Cox, 2010). Each oocyst undergoes sporogony, an asexual reproductive process yielding thousands of haploid sporozoites. Upon maturation, the oocysts rupture, releasing sporozoites into the hemocoel, allowing them to drift to the salivary glands. The parasite's extrinsic incubation period is approximately 14 days, with the conversion from ookinete to fully developed oocyst taking the longest at about 10 days. During development, *Plasmodium* parasites encounter numerous immune-related obstacles that hinder their progression from the early stages in the midgut and the sporozoite phase in the salivary glands. Successful infection depends on genetic compatibility between the mosquito and *Plasmodium*. Another key factor is whether the parasite's developmental timeline aligns with the mosquito's lifespan. Only mosquito species, that *Plasmodium* reaches its infective stage, are capable of transmitting malaria and are recognized as competent vectors. The influence of vector competence on malaria transmission is often measured using the fundamental reproductive number (R_0), first described by McDonald in 1957. R_0 estimates whether a disease will persist or die out, depending on whether R_0 is greater than or less than one. It stands for the number of new infections expected in a susceptible population from one infected person. McDonald's model incorporates two critical parameters: the likelihood of mosquito infection (b) and mosquito longevity (p). Variations in these parameters can significantly affect disease transmission, either enhancing or reducing it. An increase in b raises R_0, while a decrease in p lowers it.

$$R_0 = \ln(p)\, pvma^2bc$$

Mosquito to human ratio (m)
Human biting rate (a^2)
Probability of mosquito infection (b)
(Physiology, immunity, and bacterial diversity)
Probability of human infection (c)
Mosquito longevity (p)
(Physiology and bacterial diversity)
Extrinsic incubation period (v)

4 Insecticide Resistance and Malaria Vector Susceptibility to *Plasmodium* Infection

The prompt proliferation of insecticide resistance among malaria vectors in malaria-prevalent areas over previous years has prompted inquiries into how this resistance affects mosquito susceptibility to *Plasmodium* infection (Barnes et al., 2017). Although limited research has explored this relationship, some studies have compared the infection rates of *Plasmodium falciparum* in insecticide-resistant *Anopheles* mosquitoes with their susceptible counterparts, both in field-collected specimens and in mosquitoes experimentally infected with the parasite (see Table 27.1). For instance, *Anopheles gambiae* strains carrying the kdr resistance allele (Vgsc-L1014S) have been observed to have naturally higher sporozoite

Table 27.1 Overview of research analyzing the effect of insecticide resistance on malaria vectors susceptibility to *P. falciparum* infection

Mosquito species	Experiment performed	Infection type	Infection outcome		References
			Reoccurrence	Intensity	
A. gambiae[a]	Susceptible strain matched to *kdr*-resistant strain	Natural	Increased sporozoite infection incidence in resistant strain[s]		(Cens & Bertaud, 2022)
A. funestus s.s[a]	*GSTe2*-resistant genotypes contrasted to prone genotypes	Natural	No variance in oocysts infection incidence[ns] Greater sporozoite infection incidence in resistant strain[s]		
A. gambiae s.l.[a]	*ace-1*-resistant strain in contrast to prone strain	Natural	Increased oocyst infection occurrence in resistant strain[s]		
A. gambiae s.l.[a]	Prone strain matched to *kdr*-resistant strain	Natural	Not any variance in oocysts infection prevalence[ns]		
A. gambiae[a]	*kdr*-resistant strain compared to susceptible strain	Experimental	Greater oocyst infection prevalence in resistant strain[s] Heightened sporozoite infection prevalence in resistant strain[s]	Greater oocyst burden in the resistant strain[s] Increased sporozoite infection burden in the resistant strain[s]	

(continued)

Table 27.1 (continued)

Mosquito species	Experiment performed	Infection type	Infection outcome		References
			Reoccurrence	Intensity	
A. coluzzii[a]	*kdr*-resistant strain matched to disposed strain	Experimental	Greater oocyst occurrence in the resistant strain[s] Greater sporozoite infection prevalence in resistant strain[s]	Greater oocyst burden in the resistant strain[s] Advanced sporozoite burden prevalence in the resistant strain[s]	
A. funestus s.s[a]	*GSTe2*-resistant genotypes equated to prone genotypes	Experimental	Lesser oocyst infection occurrence in homozygous resistant genotypes[s]	Higher oocysts burden in homozygous and heterozygous resistant genotypes[s]	
A. gambiae[b]	*ace-1*-resistant strain compared to prone strain	Experimental	Greater oocyst-prone infection in resistant strain[s] Not any variance in sporozoite infection prevalence[ns]	Not any variance in oocyst and sporozoite infection load[ns]	
A. gambiae[b]	*kdr*-resistant strain matched to disposed strain	Experimental	Higher oocyst infection occurrence in resistant strain[s] Upper sporozoite infection incidence in resistant strain[s]	Lesser oocyst and sporozoite infection load in resistant strain[s]	
A. gambiae[b]	*ace-1*-resistant strain likened to prone strain		Greater oocyst infection incidence in resistant strain[s]		
A. gambiae[b]	Susceptible strain related to *kdr*-resistant strain	Experimental	Upper oocyst infection incidence in resistant strain[s]		
A. gambiae[b]	*kdr*-resistant strain in comparison to *kdr*-resistant strain introduced to insecticides		Reduced oocyst infection occurrence in resistant strain[s]	Insignificant difference in oocyst infection load[ns]	(Taylor-Wells et al., 2018)

(continued)

Table 27.1 (continued)

Mosquito species	Experiment performed	Infection type	Infection outcome		References
			Reoccurrence	Intensity	
A. gambiae[b]	*ace-1*-resistant strain likened to *ace-1*-resistant strain introduced to insecticides	Experimental	Lesser oocyst infection occurrence in resistant strain[s]	Low oocyst infection burden in resistant strain[s]	
A. gambiae[a]	*kdr*-resistant strain likened to *kdr*-resistant strain introduced to insecticide-treated nets	Experimental	Lesser oocyst infection occurrence in resistant strain[s]	Lesser oocyst infection burden in resistant strain[s]	

s significant; *ns* not significant at P = 0.05

infections compared to susceptible mosquitoes. These discoveries have also been established experimentally in *A. gambiae* and *Anopheles coluzzii*. In contrast, *A. gambiae* mosquitoes with the ace-1 resistance allele failed to show a significant difference in infection rates compared to wild-type mosquitoes in spite of a higher prevalence of oocysts in the resistant strain. Further investigation, particularly using field populations, is necessary to understand whether reduced longevity or other factors contribute to these patterns (Phulera et al., 2018). Experimental infection studies have shown that *A. funestus* mosquitoes with the L119F-GSTe2 resistance genotype exhibited a higher susceptibility to oocyst infections compared to their susceptible counterparts. Similarly, field studies on naturally infected *A. funestus* populations showed that mosquitoes homozygous for the L119F-GSTe2 genotype are prospective to be infected with sporozoites although no significant differences were found at the oocyst stage. In contrast, experiments by Lo and Coetzee, which involved infecting two sub-colonies of *FUMOZ* with *Plasmodium berghei*, demonstrated that insecticide-resistant colonies were less susceptible to infection than susceptible mosquitoes. However, no studies have yet examined the connection between cytochrome P450 genes associated with insecticide resistance and *Plasmodium falciparum* infection in *A. funestus*. Similar studies on *A. gambiae* s.l. are also lacking due to the deficiency of known metabolic resistance markers in these species.

5 Impact of Insecticide Resistance on Mosquito Sialome

Blood-feeding arthropods like mosquitoes have grown a refined array of salivary molecules to counter the host's hemostatic and immune responses, facilitating successful blood-feeding. These salivary components could also influence

Plasmodium development within the mosquito's salivary glands. The proteins produced by mosquito salivary glands are members of numerous families (D7, mucin, gSG1, gSG2, gSG6 peptide, gSG7, cE5, 8.2-kDa, 6.2-kDa, etc.) and have various roles, such as in cytoskeletal support, digestion, chemosensation, immunity, and metabolism (Cane et al., 2016). Physiological changes associated with insecticide resistance may influence the composition of the sialome, potentially impacting the mosquito's vector competence (Cansado-Utrilla et al., 2021). However, research exploring how insecticide resistance influences the sialome is still scarce. One notable protein, a 100-kDa secretory protein encoded by the *Saglin* gene in *A. gambiae* salivary glands, has been identified as a binding target for *Plasmodium falciparum* and *Plasmodium berghei* sporozoites prior to their penetration of the salivary glands. In mosquitoes with the ace-1 resistance allele, this protein was downregulated, suggesting a potential effect on vector infectivity. Nevertheless, more recent studies have shown that this 100-kDa protein is unequally dispersed across salivary gland lobes, with its nonappearance in the main sporozoite-occupancy site (the distal lateral lobes), indicating a more secondary role in infection. The D7 protein family, which is vastly expressed in malaria vectors, plays a key role in anti-hemostatic activity and possibly in blood meal digestion (Early et al., 2019). Two short forms of D7 family genes (D7r3 and D7r4) were overexpressed in pyrethroid-resistant *A. funestus* (L119F-GSTe2), while most D7 genes were underexpressed in pyrethroid-resistant *A. gambiae* (kdr, L1014F). Similar findings were observed in insecticide-resistant *Culex quinquefasciatus* (ace-1 resistant) and two strains of *Aedes aegypti* (homozygous resistant C1534 and G1016 kdr), suggesting that insecticide resistance mechanisms may differentially have an effect on the sialome across mosquito species. In addition to the D7 proteins, various immune proteins like cecropins and defensins have also been identified in mosquito saliva, highlighting the potential involvement of the salivary glands in the mosquito's immune response (Hegde et al., 2018). However, studies on insecticide resistance alleles and their effect on the manifestation of immune genes in the salivary glands have not reported significant changes between resistant and susceptible mosquitoes. This propounds that insecticide resistance may not substantially affect the immune aspect of the sialome. It is yet to be determined if these immune proteins' expression changes in response to *Plasmodium* infection, particularly based on mosquito genotype. Previous studies by The gene expression in mosquito salivary glands is modified by blood meal ingestion and permutate all through the period of *Plasmodium* growth in the mosquito. These factors warrant further investigation in future studies aiming to identify differentially expressed salivary gland genes and their effects on vector competence.

6 Effect of Insecticide Resistance on Mosquito Immune Responses

Once female *Anopheles* mosquito ingests an infectious blood meal, its immune system activates to defend against infection (Juma et al., 2020). *Plasmodium falciparum* initially encounters the peritrophic membrane in the mosquito midgut, which acts as a barrier, protecting the mosquito from infection and regulating the activity of digestive enzymes. These digestive enzymes, such as trypsin, chymotrypsin, and serine proteases, are upregulated to break down the blood meal's proteins and also aid in eliminating *Plasmodium* parasites. Numerous research have explored the influence of insecticide resistance on mosquito immunity. Genes involved in *Anopheles coluzzii* infectivity and found that the kdr-carrying *para* gene, associated with insecticide resistance through voltage-gated sodium channel mutations, was not directly linked to infection. However, the *ClipC9* gene, responsible for synthesizing serine protease, was identified as a key factor in determining mosquito resistance to infection. Serine protease is vital for activating the three vital mosquito immune pathways (Toll, Imd, and JAK/STAT), which make available antimicrobial peptides (AMPs) like defensins, cecropins, and gambicin that are potent against malaria parasites. Defensin and cecropin were augmented in *A. gambiae* post-exposure to permethrin, suggesting that insecticide-resistant mosquitoes may have an enhanced capacity to battle infections. However, these immune effectors alone may not be sufficient to fully block malaria infections, as multiple immune pathways are triggered during parasite invasion. Another study on *Culex pipiens* found no significant changes in defensin or cecropin expression after exposure to lipopolysaccharide (LPS) although gambicin levels increased. This suggests that *Plasmodium* infections may selectively upregulate certain immune factors while leaving others unchanged. Reactive oxygen species (ROS), which is crucial to *Plasmodium* parasites, play another crucial role in mosquito immunity. The *A. coluzzii* VK7 colony, which harbored kdr resistance along with CYP6M6 and CYP6P3 genes involved in insecticide metabolism, exhibited ROS overexpression after sub-lethal exposure to deltamethrin. This indicates a potential mechanism for increased resistance to *Plasmodium* infection. Further findings are paramount to deciding if ROS overexpression happens naturally in resistant mosquitoes.

Cellular immune responses, involving hemocytes that combat pathogens through phagocytosis and melanization, have also been affected by insecticide resistance. Although some insecticides have been shown to alter hemocyte populations in other insect species, the effects of insecticide resistance on mosquito hemocytes and their impact on *Plasmodium* infectivity are still under investigation. Finally, phenoloxidase (PO), responsible for melanization and regulated by serine protease inhibitors, is another key immune factor. In insecticide-resistant *Culex pipiens* strains, PO levels were found to be comparable to those in susceptible mosquitoes, suggesting that some immunity-related genes may not be impacted by insecticide resistance. However, more studies are tantamount to exploring the importance of PO in malaria vectors (Greenberg & Bourc'his, 2019; Guo et al., 2020; Hegde et al., 2019; Henry et al., 2020; Herren et al., 2020; Saab et al., 2020).

7 Impact of Insecticide Resistance on Mosquito Gut Microbiota

Mosquitoes host a variety of microorganisms, such as bacteria, fungi, and viruses, which populate their gut, salivary glands, and reproductive organs. These microorganisms are primarily obtained from the environment, and their structure is shaped by the aquatic breeding habitats of mosquitoes. The mosquito microbiome is dynamic and varies by location and season, which may explain differences in infection rates observed among field-collected mosquitoes. The microbiota within mosquitoes has the potential to disrupt malaria transmission by influencing vector competence. Additionally, microbiota affects various features of mosquito biology, including immunity, digestion, nutrition, vector competence, reproduction, and resistance to insecticides (Holder et al., 2018; Jones, 2018; Sgolastra et al., 2018; Shane et al., 2018; Shen et al., 2018; Siviter et al., 2018; Smith et al., 2018; Solomon et al., 2019; Shen et al., 2019; Sheng et al., 2019; Sine et al., 2019; Leroy et al., 2019; Levitan et al., 2019; Hegde et al., 2019; Saha et al., 2019, Saab et al., 2020; Henry et al., 2020; Herren et al., 2020; Huang et al., 2021; Martin et al., 2021). Given the expeditious expansion of insecticide resistance in *Anopheles* mosquitoes, multiple research have inspected the role of gut microbial communities in the advancement of resistance. Specific microbiota populations have been linked to resistance to organochlorines in *A. arabiensis* and *A. albimanus* as well as resistance to pyrethroids in *A. gambiae* and *A. stephensi*, indicating a possible role for the microbiota in resistance mechanisms. Altered feeding behavior in insecticide-resistant mosquitoes could increase microbial diversity, potentially affecting vector competence by altering protective bacteria populations or those that enhance pathogen transmission (Chen & Kubo, 2017; Cannon et al., 2018; Chen et al., 2019; Caragata et al., 2019; Case et al., 2019; Chang et al., 2019; Burton-Smith et al., 2019; Cai et al., 2020). Recent research by Bassene et al. found significant differences in the microbiota between *Plasmodium falciparum*-infected and uninfected *A. gambiae* and *A. funestus* mosquitoes although their insecticide resistance status was not examined. Further research is mandatory to fully characterize the microbiota of insecticide-resistant malaria vectors and understand their role in vector competence.

8 Impact of Insecticide Resistance on Mosquito Longevity

Mosquito continuity plays a critical role in malaria transmission, as the *Plasmodium* parasite requires 11–14 days to develop within its mosquito host. Only mosquitoes that survive long enough to allow *Plasmodium* to reach the sporozoite stage can transmit the disease. Given the increasing prevalence of insecticide resistance, researchers have examined its effects on mosquito lifetime and its potential epidemiological implications. Research on the effects of insecticide resistance on mosquito lifetime has mainly concentrated on four species: *A. gambiae*, *A. arabiensis*, *A. coluzzii*, and *A. funestus*. The discoveries suggest that insecticide resistance has pleiotropic effects on mosquito lifespan (Table 27.2). Most laboratory studies (10 out of 14) have shown that pyrethroid-resistant *A. funestus* and *A. gambiae* live inveterate than their prone counterparts. In field studies, a longer lifespan was observed in organochlorine- and pyrethroid-resistant *A. funestus* populations, while shorter lifespans were seen in organochlorine-resistant *A. gambiae* strains. Pre-exposure to insecticides, simulating field conditions, has been shown to decrease the lifespan of insecticide-resistant *A. gambiae*, with delayed mortality depending on the intensity of resistance (Kyrou et al., 2018). This highlights the significance of interpreting laboratory findings cautiously, as they may not fully represent field conditions. Nevertheless, laboratory studies offer valuable insights into possible mechanisms influencing mosquito longevity, including resource allocation trade-offs and oxidative stress (Zhu et al., 2017; Wilson et al., 2019; Zhang et al., 2019; Zheng et al., 2019).

The resource-based trade-off concept in evolving ecology suggests that environmental limitations necessitate increased allocation of resources to one biological trait leading to other traits experiencing reduced vigor availability (Frank, 2018; Fux et al., 2018; Gao et al., 2020). In mosquitoes, the growth of insecticide resistance through detoxification mechanisms requires higher production of detoxifying enzymes, demanding additional resources to maintain this function (Figan et al., 2018). Insecticide-resistant *Anopheles gambiae* had lower metabolic rates and reduced longevity compared to susceptible strains, along with higher levels of reactive oxygen species (ROS), which contribute to oxidative stress. ROS are multifunctional molecules produced during normal cellular metabolism in organisms and are key drivers of aging, including in mosquitoes (Adair & Douglas, 2017; Barratt et al., 2019; Beadle et al., 2019; Beatriz et al., 2019; Bora et al., 2019; Burman et al., 2022). Mosquitoes that develop mechanisms to manage oxidative stress may have an increased likelihood of living longer. In experiments conducted on *A. arabiensis* and *A. funestus*—species with the kdr and Cytochrome P450 resistance technique, respectively—Oliver and Brooke found that these species had longer lifespans. Their findings suggest that cytochrome P450 activity provides greater protection against oxidative stress. Different detoxifying enzymes may affect vector longevity in varying ways. For instance, glutathione S-transferase (GST) is thought to protect against oxidative stress, and studies on *A. funestus* showed that mosquitoes with elevated GST levels lived longer, both with and without insecticide exposure. In

Table 27.2 Summary of studies evaluating the effect of insecticide resistance on the longevity of malaria vectors

Study species	Origin of strains	Class of insecticides	Resistance mechanism(s)	Pre-exposure[a]/exposure[b] to insecticide	Effect on longevity	References
A. gambiae s.l	Field Field Field Laboratory Laboratory Laboratory Laboratory Laboratory Laboratory Laboratory	PY PY and OC PY PY and OC PY PY OC CA OC OC OC	Unavailable *kdr* Unavailable *kdr* and P450 Unavailable *kdr*, P450, and esterase *kdr* *ace*-1 *kdr*, *GSTe*, P450, and Es *GSTe*, P450, and Es *GSTe*, P450, Es; *kdr*	Deltamethrin and permethrin[a] Not used Permethrin (Net)[b] Permethrin (Hut-net)[b] Deltamethrin[b] Not used DDT[a] Bendiocarb[a] Not used Not used Not used DDT[a] Permethrin[a] Deltamethrin[a] Malathion	RR durability > SS durability RR durability < SS durability RR^{e} longevity = RR^{ne} RR durability < SS durability RR durability > SS durability RR durability = SS durability RR durability > SS durability RR durability = SS durability RR^{b} durability > RR^{s} RR durability < SS durability RR^{e} durability = RR^{ne} RR^{e} durability < RR^{ne} RR^{e} durability < RR^{ne} RR^{e} longevity < RR^{ne}	(Han et al., 2018) (Dia et al., 2018)

(continued)

Table 27.2 (continued)

Study species	Origin of strains	Class of insecticides	Resistance mechanism(s)	Pre-exposure[a]/exposure[b] to insecticide	Effect on longevity	References
A. funestus	Field Field Laboratory Laboratory Laboratory	OC PY and OC PY PY and CA PY and CA	*GSTe2* *GSTe2* Unavailable *P450-a* *P450-a* *P450-b* *P450-a/P450-b*	Permethrin (Net)[b] Not used Not used Not used Not used	RR durability > SS durability RR durability > SS durability RR durability > SS durability RR durability = SS durability RR durability = SS durability RR durability = SS durability RR durability = SS durability	(Isaacs et al., 2018)

[a,b,e,ne,s] significantly different at 0.05

contrast, while monooxygenase is associated with increased oxidative stress, it has not caused the expected reduction in longevity in *A. funestus*. More research is needed, particularly using field populations under natural conditions, to fully understand how insecticide resistance impacts mosquito longevity (Strand, 2018; Tchouakui et al., 2019; Tessema et al., 2019; Wang et al., 2019; Watson et al., 2021; Tasman et al., 2021).

9 Conclusion

A thorough understanding of how insecticide resistance impacts malaria vector fitness is urgently needed, as the present body of knowledge is both limited and contradictory. However, there is a growing indication that insecticide resistance may enhance the infectivity of malaria vectors to *Plasmodium*, potentially increasing their vector competence. This may be linked to alterations in the expression of genes associated with blood-feeding and immunity. Furthermore, resistant mosquitoes exhibit differences in their microbiota communities in comparison to their susceptible equals, but the exact effects of these changes on the infection process and vector infectivity to *P. falciparum* remain largely unexplored. In terms of longevity, insecticide resistance does not consistently affect the lifespan of vectors. It is important to note that studies on vectors with metabolic resistance have been underrepresented, partly because molecular markers to identify this resistance type have only recently been developed, particularly in *Anopheles funestus*. From an ecological immunology perspective, vectors with metabolic resistance may experience impairments in biological functions, including immunity, which could make them less resistant to *Plasmodium* infection. With recent advancements in genomics, transcriptomics, and the molecular characterization of insecticide resistance, more detailed studies can be conducted. These investigations are crucial for addressing knowledge gaps as regards the effects of insecticide resistance on key factors that influence vector competence. Such insights will help predict potential shifts in malaria epidemiology as insecticide resistance continues to rise.

Acknowledgements I acknowledge the opportunity the Covenant University Centre of Research Innovation and Development gave me to present the review during its 11th International Biotechnology Conference Exhibition & Workshop 2024 (IBCE&W) tagged "Harnessing Biotechnology Tools for Product Development."

Funding This research work is sponsored by the Covenant Applied Informatics and Communication African Centre of Excellence (CApIC-ACE) of Covenant University, Ota, Ogun State, Nigeria.

References

Adair, K. L., & Douglas, A. E. (2017). Making a microbiome: The many determinants of host-associated microbial community composition. *Current Opinion in Microbiology, 35*, 23–29. https://doi.org/10.1016/j.mib.2016.11.002

Barnes, K. G., Weedall, G. D., Ndula, M., Irving, H., Mzihalowa, T., Hemingway, J., & Wondji, C. S. (2017). Genomic footprints of selective sweeps from metabolic resistance to pyrethroids in African malaria vectors are driven by scale up of insecticide-based vector control. *PLoS Genet, 13*(2), 1–22. https://doi.org/10.1371/journal.pgen.1006539

Barratt, J., Lane, M., Talundzic, E., Richins, T., Robertson, G., Formenti, F., Pritt, B. S., Verocai, G. G., de Souza, J. N., Soares, N. M., Traub, R. J., Buonfrate, D., & Bradbury, R. B. (2019). A global genotyping survey of Strongyloides stercoralis and Strongyloides fuelleborni using deep amplicon sequencing. *PLoS Neglected Tropical Diseases, 13*(9), e0007609. https://doi.org/10.1371/journal.pntd.0007609

Beadle, K., Singh, K. S., Troczka, B. J., Randall, E., Zaworra, M., Zimmer, C. T., Hayward, A., Reid, R., Kor, L., Kohler, M., Buer, B., Nelson, D. R., Williamson, M. S., Davies, T. G. E., Field, L. M., Nauen, R., & Bass, C. (2019). Genomic insights into neonicotinoid sensitivity in the solitary bee Osmia bicornis. *PLoS Genetics, 15*(2), e1007903. https://doi.org/10.1371/journal.pgen.1007903

Beatriz, A., Trisnadi, N., Ramirez, J. L., & Barillas-Mury, C. (2019). Mosquito midgut prostaglandin release establishes systemic immune priming. *IScience, 19*, 54–62. https://doi.org/10.1016/j.isci.2019.07.012

Bora, A., Suzuki, T., & Funar-Timofei, S. (2019). Neonicotinoid insecticide design: Molecular docking, multiple chemometric approaches, and toxicity relationship with Cowpea aphids. *Environmental Science and Pollution Research, 26*(14), 14547–14561. https://doi.org/10.1007/s11356-019-04662-9

Burman, R. J., Rosch, R. E., Wilmshurst, J. M., Sen, A., Ramantani, G., Akerman, C. J., & Raimondo, J. V. (2022). Why won't it stop? The dynamics of benzodiazepine resistance in status epilepticus. *Nature Reviews Neurology, 18*, 428. https://doi.org/10.1038/s41582-022-00664-3

Burton-Smith, R. N., Watanabe, A., Tokutsu, R., Song, C., Murata, K., & Minagawa, J. (2019). Structural determination of the large photosystem II–light-harvesting complex II supercomplex of Chlamydomonas reinhardtii using nonionic amphipol. *Journal of Biological Chemistry, 294*(41), 15003–15013. https://doi.org/10.1074/jbc.ra119.009341

Cai, T., Zhang, Y., Liu, Y., Deng, X., He, S., Li, J., & Wan, H. (2020). *Wolbachia* enhances expression of *NlCYP4CE1* in *Nilaparvata lugens* in response to imidacloprid stress. *Insect Science, 28*(2), 355–362. https://doi.org/10.1111/1744-7917.12834

Cane, J. H., Dobson, H. E. M., & Boyer, B. (2016). Timing and size of daily pollen meals eaten by adult females of a solitary bee (Nomia melanderi) (Apiformes: Halictidae). *Apidologie, 48*(1), 17–30. https://doi.org/10.1007/s13592-016-0444-8

Cannon, M. V., Bogale, H., Rutt, L., Humphrys, M., Korpe, P., Duggal, P., Ravel, J., & Serre, D. (2018). A high-throughput sequencing assay to comprehensively detect and characterize unicellular eukaryotes and helminths from biological and environmental samples. *Microbiome, 6*(1), 195. https://doi.org/10.1186/s40168-018-0581-6

Cansado-Utrilla, C., Zhao, S. Y., McCall, P. J., Coon, K. L., & Hughes, G. L. (2021). The microbiome and mosquito vectorial capacity: Rich potential for discovery and translation. *Microbiome, 9*(1), 111. https://doi.org/10.1186/s40168-021-01073-2

Caragata, E. P., Tikhe, C. V., & Dimopoulos, G. (2019). Curious entanglements: Interactions between mosquitoes, their microbiota, and arboviruses. *Current Opinion in Virology, 37*, 26–36. https://doi.org/10.1016/j.coviro.2019.05.005

Case, L. B., Ditlev, J. A., & Rosen, M. K. (2019). Regulation of transmembrane signaling by phase separation. *Annual Review of Biophysics, 48*(1), 465–494. https://doi.org/10.1146/annurev-biophys-052118-115534

Cens, T., & Bertaud, A. (2022). Molecular targets of neurotoxic insecticides in Apis mellifera. *European Journal of Organic Chemistry, 2022*(21), 150–167. https://doi.org/10.1002/ejoc.202101531

Chang, H.-H., Wesolowski, A., Sinha, I., Jacob, C. G., Mahmud, A., Uddin, D., Zaman, S. I., Hossain, M. A., Faiz, M. A., Ghose, A., Sayeed, A. A., Rahman, M. R., Islam, A., Karim, M. J., Rezwan, M. K., Shamsuzzaman, A. K. M., Jhora, S. T., Aktaruzzaman, M. M., Drury, E., & Gonçalves, S. (2019). Mapping imported malaria in Bangladesh using parasite genetic and human mobility data. *eLife, 8*, e43481. https://doi.org/10.7554/elife.43481

Chen, I.-S., & Kubo, Y. (2017). Ivermectin and its target molecules: Shared and unique modulation mechanisms of ion channels and receptors by ivermectin. *The Journal of Physiology, 596*(10), 1833–1845. https://doi.org/10.1113/jp275236

Chen, Z., Yao, X., Dong, F., Duan, H., Shao, X., Chen, X., Yang, T., Wang, G., & Zheng, Y. (2019). Ecological toxicity reduction of dinotefuran to honeybee: New perspective from an enantiomeric level. *Environment International, 130*, 104854. https://doi.org/10.1016/j.envint.2019.05.048

Cox, F. E. (2010). History of the discovery of the malaria parasites and their vectors. *Parasites & Vectors, 3*(1), 5–14. https://doi.org/10.1186/1756-3305-3-5

Dia, A. K., Guèye, O. K., Niang, E. A., Diédhiou, S. M., Sy, M. D., Konaté, A., Samb, B., Diop, A., Konaté, L., & Faye, O. (2018). Insecticide resistance in Anopheles arabiensis populations from Dakar and its suburbs: Role of target site and metabolic resistance mechanisms. *Malaria Journal, 17*(1), 116. https://doi.org/10.1186/s12936-018-2269-6

Early, A. M., Daniels, R. F., Farrell, T. M., Grimsby, J., Volkman, S. K., Wirth, D. F., MacInnis, B. L., & Neafsey, D. E. (2019). Detection of low-density Plasmodium falciparum infections using amplicon deep sequencing. *Malaria Journal, 18*(1), 219. https://doi.org/10.1186/s12936-019-2856-1

Figan, C. E., Sá, J. M., Mu, J., Melendez-Muniz, V. A., Liu, C. H., & Wellems, T. E. (2018). A set of microsatellite markers to differentiate Plasmodium falciparum progeny of four genetic crosses. *Malaria Journal, 17*(1), 60. https://doi.org/10.1186/s12936-018-2210-z

Frank, J. (2018). Single-particle reconstruction of biological molecules-story in a sample (nobel lecture). *Angewandte Chemie, 57*(34), 10826–10841. https://doi.org/10.1002/anie.201802770

Fux, J. E., Mehta, A., Moffat, J., & Spafford, J. D. (2018). Eukaryotic voltage-gated sodium channels: On their origins, asymmetries, losses, diversification and adaptations. *Frontiers in Physiology, 9*, 1406. https://doi.org/10.3389/fphys.2018.01406

Gao, H., Cui, C., Wang, L., Jacobs-Lorena, M., & Wang, S. (2020). Mosquito microbiota and implications for disease control. *Trends in Parasitology, 36*(2), 98–111. https://doi.org/10.1016/j.pt.2019.12.001

Greenberg, M. V. C., & Bourc'his, D. (2019). The diverse roles of DNA methylation in mammalian development and disease. *Nature Reviews. Molecular Cell Biology., 20*(10), 590–607. https://doi.org/10.1038/s41580-019-0159-6

Guo, H., Franken, E., Deng, Y., Benlekbir, S., Singla Lezcano, G., Janssen, B., Yu, L., Ripstein, Z. A., Tan, Y. Z., & Rubinstein, J. L. (2020). Electron-event representation data enable efficient cryoEM file storage with full preservation of spatial and temporal resolution. *IUCrJ, 7*(5), 860–869. https://doi.org/10.1107/s205225252000929x

Han, W., Wang, Y., Gao, J., Wang, S., Zhao, S., Liu, J., Zhong, Y., & Zhao, D. (2018). Acute toxicity and sublethal effects of myclobutanil on respiration, flight and detoxification enzymes in Apis cerana cerana. *Pesticide Biochemistry and Physiology, 147*, 133–138. https://doi.org/10.1016/j.pestbp.2017.11.001

Hegde, S., Khanipov, K., Albayrak, L., Golovko, G., Pimenova, M., Saldaña, M. A., Rojas, M. M., Hornett, E. A., Motl, G. C., Fredregill, C. L., Dennett, J. A., Debboun, M., Fofanov, Y., & Hughes, G. L. (2018). Microbiome interaction networks and community structure from laboratory-reared and field-collected Aedes aegypti, Aedes albopictus, and Culex quinquefasciatus mosquito vectors. *Frontiers in Microbiology, 9*, 2160. https://doi.org/10.3389/fmicb.2018.02160

Hegde, S., Voronin, D., Casas-Sánchez, A., Saldaña, M. A., Heinz, E., Acosta-Serrano, Á., Popov, V. L., Chopra, A. K., & Hughes, G. L. (2019). Gut-associated bacteria invade the midgut epithelium of *Aedes aegypti* and stimulate innate immunity and suppress Zika virus infection in cells. *BioRxiv (Cold Spring Harbor Laboratory)*. https://doi.org/10.1101/866897

Henry, C., Cens, T., Charnet, P., Cohen-Solal, C., Collet, C., van-Dijk, J., Guiramand, J., de Jésus-Ferreira, M.-c., Menard, C., Mokrane, N., Roussel, J., Thibaud, J.-B., Vignes, M., & Rousset, M. (2020). Heterogeneous expression of GABA receptor-like subunits LCCH3 and GRD reveals functional diversity of GABA receptors in the honeybee *Apis mellifera*. *British Journal of Pharmacology, 177*(17), 3924–3940. https://doi.org/10.1111/bph.15135

Herren, J. K., Mbaisi, L., Mararo, E., Makhulu, E. E., Mobegi, V. A., Butungi, H., Mancini, M. V., Oundo, J. W., Teal, E. T., Pinaud, S., Lawniczak, M. K. N., Jabara, J., Nattoh, G., & Sinkins, S. P. (2020). A microsporidian impairs Plasmodium falciparum transmission in Anopheles arabiensis mosquitoes. *Nature Communications, 11*(1), 2187. https://doi.org/10.1038/s41467-020-16121-y

Holder, P. J., Jones, A., Tyler, C. R., & Cresswell, J. E. (2018). Fipronil pesticide as a suspect in historical mass mortalities of honey bees. *Proceedings of the National Academy of Sciences, 115*(51), 13033–13038. https://doi.org/10.1073/pnas.1804934115

Huang, Q. T., Sheng, C. W., Jones, A. K., Jiang, J., Tang, T., Han, Z. J., & Zhao, C. Q. (2021). Functional characteristics of the lepidopteran ionotropic GABA receptor 8916 subunit interacting with the LCCH3 or the RDL subunit. *Journal of Agricultural and Food Chemistry, 69*(39), 11582–11591. https://doi.org/10.1021/acs.jafc.1c00385

Isaacs, A. T., Mawejje, H. D., Tomlinson, S., Rigden, D. J., & Donnelly, M. J. (2018). Genome-wide transcriptional analyses in Anopheles mosquitoes reveal an unexpected association between salivary gland gene expression and insecticide resistance. *BMC Genomics, 19*(1), 225. https://doi.org/10.1186/s12864-018-4605-1

Jones, A. K. (2018). Genomics, cys-loop ligand-gated ion channels and new targets for the control of insect pests and vectors. *Current Opinion in Insect Science, 30*, 1–7. https://doi.org/10.1016/j.cois.2018.07.016

Juma, E. O., Kim, C.-H., Dunlap, C., Allan, B. F., & Stone, C. M. (2020). Culex pipiens and Culex restuans egg rafts harbor diverse bacterial communities compared to their midgut tissues. *Parasites & Vectors, 13*(1), 532. https://doi.org/10.1186/s13071-020-04408-4

Kyrou, K., Hammond, A. M., Galizi, R., Kranjc, N., Burt, A., Beaghton, A. K., Nolan, T., & Crisanti, A. (2018). A CRISPR–Cas9 gene drive targeting doublesex causes complete population suppression in caged Anopheles gambiae mosquitoes. *Nature Biotechnology, 36*(11), 1062–1066. https://doi.org/10.1038/nbt.4245

Leroy, D., Macintyre, F., Adoke, Y., Ouoba, S., Barry, A., Mombo-Ngoma, G., Ndong Ngomo, J. M., Varo, R., Dossou, Y., Tshefu, A. K., Duong, T. T., Phuc, B. Q., Laurijssens, B., Klopper, R., Khim, N., Legrand, E., & Ménard, D. (2019). African isolates show a high proportion of multiple copies of the plasmodium falciparum plasmepsin-2 gene, a piperaquine resistance marker. *Malaria Journal, 18*(1), 126. https://doi.org/10.1186/s12936-019-2756-4

Levitan, O., Chen, M., Kuang, X., Cheong, K. Y., Jiang, J., Banal, M., Nambiar, N., Gorbunov, M. Y., Ludtke, S. J., Falkowski, P. G., & Dai, W. (2019). Structural and functional analyses of photosystem II in the marine diatom Phaeodactylum tricornutum. *Proceedings of the National Academy of Sciences, 116*(35), 17316–17322. https://doi.org/10.1073/pnas.1906726116

Martin, R. J., Robertson, A. P., & Choudhary, S. (2021). Ivermectin: An anthelmintic, an insecticide, and much more. *Trends in Parasitology, 37*(1), 48–64. https://doi.org/10.1016/j.pt.2020.10.005

Phillips, M. A., Burrows, J. N., Manyando, C., van Huijsduijnen, R. H., Van Voorhis, W. C., & Wells, T. N. C. (2017). Malaria. *Nature Reviews. Disease Primers, 3*(17050), 17050–17074. https://doi.org/10.1038/nrdp.2017.50

Phulera, S., Zhu, H., Yu, J., Claxton, D. P., Yoder, N., Yoshioka, C., & Gouaux, E. (2018). Cryo-EM structure of the benzodiazepine-sensitive α1β1γ2S tri-heteromeric GABAA receptor in complex with GABA. *eLife, 7*, e39383. https://doi.org/10.7554/elife.39383

Saab, S. A., Dohna, H. Z., Nilsson, L. K. J., Onorati, P., Nakhleh, J., Terenius, O., & Osta, M. A. (2020). The environment and species affect gut bacteria composition in laboratory co-cultured Anopheles gambiae and Aedes albopictus mosquitoes. *Scientific Reports, 10*(1), 3352. https://doi.org/10.1038/s41598-020-60075-6

Saha, S., Ramesh, A., Kalantar, K., Malaker, R., Hasanuzzaman, M., Khan, L. M., Mayday, M. Y., Sajib, M. S. I., Li, L. M., Langelier, C., Rahman, H., Crawford, E. D., Tato, C. M., Islam, M., Juan, Y.-F., de Bourcy, C., Dimitrov, B., Wang, J., Tang, J., & Sheu, J. (2019). Unbiased metagenomic sequencing for pediatric meningitis in Bangladesh reveals neuroinvasive chikungunya virus outbreak and other unrealized pathogens. *mBio, 10*(6), e02877-19. https://doi.org/10.1128/mbio.02877-19

Sgolastra, F., Hinarejos, S., Pitts-Singer, T. L., Boyle, N. K., Joseph, T., Lūckmann, J., Raine, N. E., Singh, R., Williams, N. M., & Bosch, J. (2018). Pesticide exposure assessment paradigm for solitary bees. *Environmental Entomology, 48*(1), 22–35. https://doi.org/10.1093/ee/nvy105

Shah, Z., Adams, M., Moser, K. A., Shrestha, B., Stucke, E. M., Laufer, M. K., Serre, D., Silva, J. C., & Takala-Harrison, S. (2020). Optimization of parasite DNA enrichment approaches to generate whole genome sequencing data for Plasmodium falciparum from low parasitaemia samples. *Malaria Journal, 19*(1), 135. https://doi.org/10.1186/s12936-020-03195-8

Shane, J. L., Grogan, C. L., Cwalina, C., & Lampe, D. J. (2018). Blood meal-induced inhibition of vector-borne disease by transgenic microbiota. *Nature Communications, 9*(1), 4127. https://doi.org/10.1038/s41467-018-06580-9

Shen, H., Li, Z., Jiang, Y., Pan, X., Wu, J., Cristofori-Armstrong, B., Smith, J. J., Chin, Y. K. Y., Lei, J., Zhou, Q., King, G. F., & Yan, N. (2018). Structural basis for the modulation of voltage-gated sodium channels by animal toxins. *Science, 362*(6412), eaau2596. https://doi.org/10.1126/science.aau2596

Shen, L., Huang, Z., Chang, S., Wang, W., Wang, J., Kuang, T., Han, G., Shen, J.-R., & Zhang, X. (2019). Structure of a C2S2M2N2-type PSII–LHCII supercomplex from the green alga Chlamydomonas reinhardtii. *Proceedings of the National Academy of Sciences, 116*(42), 21246–21255. https://doi.org/10.1073/pnas.1912462116

Sheng, C., Huang, Q., Liu, G., Ren, X., Jiang, J., Jia, Z., Han, Z., & Zhao, C. (2019). Neurotoxicity and mode of action of fluralaner on honeybee *Apis mellifera* L. *Pest Management Science, 75*(11), 2901–2909. https://doi.org/10.1002/ps.5483

Sine, S. M., Strikwerda, J. R., & Mazzaferro, S. (2019). Structural basis for α-bungarotoxin insensitivity of neuronal nicotinic acetylcholine receptors. *Neuropharmacology, 160*, 107660. https://doi.org/10.1016/j.neuropharm.2019.05.037

Siviter, H., Brown, M. J. F., & Leadbeater, E. (2018). Sulfoxaflor exposure reduces bumblebee reproductive success. *Nature, 561*(7721), 109–112. https://doi.org/10.1038/s41586-018-0430-6

Smith, C., Morandin, C., Noureddine, M. A., & Pant, S. C. (2018). Conserved roles of Osiris genes in insect development, polymorphism and protection. *Journal of Evolutionary Biology, 31*(4), 516–529. https://doi.org/10.1111/jeb.13238

Solomon, V. R., Tallapragada, V. J., Chebib, M., Johnston, G. A. R., & Hanrahan, J. R. (2019). GABA allosteric modulators: An overview of recent developments in non-benzodiazepine modulators. *European Journal of Medicinal Chemistry, 171*, 434–461. https://doi.org/10.1016/j.ejmech.2019.03.043

Strand, M. R. (2018). Composition and functional roles of the gut microbiota in mosquitoes. *Current Opinion in Insect Science, 28*, 59–65. https://doi.org/10.1016/j.cois.2018.05.008

Suh, P. F., Elanga-Ndille, E., Tchouakui, M., Sandeu, M. M., Tagne, D., Wondji, C., & Ndo, C. (2023). Impact of insecticide resistance on malaria vector competence: A literature review. *Malaria Journal, 22*(1), 1–11. https://doi.org/10.1186/s12936-023-04444-2

Tasman, K., Rands, S. A., & Hodge, J. J. L. (2021). The power of Drosophila melanogaster for modeling neonicotinoid effects on pollinators and identifying novel mechanisms. *Frontiers in Physiology, 12*, 659440. https://doi.org/10.3389/fphys.2021.659440

Taylor-Wells, J., Senan, A., Bermudez, I., & Jones, A. K. (2018). Species specific RNA A-to-I editing of mosquito RDL modulates GABA potency and influences agonistic, potentiating and antagonistic actions of ivermectin. *Insect Biochemistry and Molecular Biology, 93*, 1–11. https://doi.org/10.1016/j.ibmb.2017.12.001

Tchouakui, M., Chiang, M.-C., Ndo, C., Kuicheu, C. K., Amvongo-Adjia, N., Wondji, M. J., Tchoupo, M., Kusimo, M. O., Riveron, J. M., & Wondji, C. S. (2019). A marker of glutathione S-transferase-mediated resistance to insecticides is associated with higher plasmodium infection in the African malaria vector Anopheles funestus. *Scientific Reports, 9*(1), 5772. https://doi.org/10.1038/s41598-019-42015-1

Tessema, S. K., Wesolowski, A., Chen, A., Murphy, M., Wilheim, J., Mupiri, A. R., Ruktanonchai, N. W., Alegana, V. A., Tatem, A. J., Tambo, M., Didier, B., Cohen, J. M., Bennett, A., Sturrock, H. J., Gosling, R., Hsiang, M. S., Smith, D. L., Mumbengegwi, D. R., Smith, J. L., & Greenhouse, B. (2019). Using parasite genetic and human mobility data to infer local and cross-border malaria connectivity in Southern Africa. *eLife, 8*, e43510. https://doi.org/10.7554/elife.43510

Wang, X., O'Reilly, A. O., Williamson, M. S., Puinean, A. M., Yang, Y., Wu, S., & Wu, Y. (2019). Function and pharmacology of glutamate-gated chloride channel exon 9 splice variants from the diamondback moth Plutella xylostella. *Insect Biochemistry and Molecular Biology, 104*, 58–64. https://doi.org/10.1016/j.ibmb.2018.12.005

Watson, G. B., Siebert, M. W., Wang, N. X., Loso, M. R., & Sparks, T. C. (2021). Sulfoxaflor – A sulfoximine insecticide: Review and analysis of mode of action, resistance and cross-resistance. *Pesticide Biochemistry and Physiology, 178*, 104924. https://doi.org/10.1016/j.pestbp.2021.104924

Wilson, M. R., Sample, H. A., Zorn, K. C., Arevalo, S., Yu, G., Neuhaus, J., Federman, S., Stryke, D., Briggs, B., Langelier, C., Berger, A., Douglas, V., Josephson, S. A., Chow, F. C., Fulton, B. D., DeRisi, J. L., Gelfand, J. M., Naccache, S. N., Bender, J., & Dien Bard, J. (2019). Clinical metagenomic sequencing for diagnosis of meningitis and encephalitis. *New England Journal of Medicine, 380*(24), 2327–2340. https://doi.org/10.1056/nejmoa1803396

Zhang, R., Huang, Z.-D., Yu, G., & Zhang, Z. (2019). Characterization of microbiota diversity of field-collected Haemaphysalis longicornis (Acari: Ixodidae) with regard to sex and blood meals. *Journal of Basic Microbiology, 59*(2), 215–223. https://doi.org/10.1002/jobm.201800372

Zheng, X., Zhang, D., Li, Y., Yang, C., Wu, Y., Liang, X., Liang, Y., Pan, X., Hu, L., Sun, Q., Wang, X., Wei, Y., Zhu, J., Qian, W., Yan, Z., Parker, A. G., Gilles, J. R. L., Bourtzis, K., Bouyer, J., & Tang, M. (2019). Incompatible and sterile insect techniques combined eliminate mosquitoes. *Nature, 572*(7767), 56–61. https://doi.org/10.1038/s41586-019-1407-9

Zhu, Y. C., Yao, J., Adamczyk, J., & Luttrell, R. (2017). Feeding toxicity and impact of imidacloprid formulation and mixtures with six representative pesticides at residue concentrations on honey bee physiology (Apis mellifera). *PLoS One, 12*(6), e0178421. https://doi.org/10.1371/journal.pone.0178421

Chapter 28
Infliction of Oxidative Stress by Bioaccumulated Polybrominated Diphenyl Ethers (PBDEs) in Tilapia zilli Sampled from Lekki Lagoon, Lagos

Patrick Omoregie Isibor and Bamidele Akinsanya

1 Introduction

The aquatic environment is subject to persistent exposure to a diverse array of anthropogenic pollutants. Numerous aquatic organisms, particularly invertebrates, are vulnerable to these contaminants, leading to their bioaccumulation within biological tissues (Van Dyk et al., 2012; Wepener et al., 2012; Winid, 2015; Wu et al., 2020). Anthropogenic pollutants encompass toxic chemicals, substances, and energies of human origin that adversely affect air, water, and food quality, ultimately harming living organisms (Akortia et al., 2016). The generation and emission of these environmental pollutants primarily stem from human activities, including urbanization, industrialization, agricultural practices, and resource extraction (Guo et al., 2015). Such activities result in the release of a broad spectrum of both synthetic and natural pollutants, ranging from plasticizers to pharmaceuticals and substances utilized in agriculture and industry (Khan et al., 2022).

Recent years have seen increased scrutiny of open dumpsites for municipal waste in relation to environmental contamination (Ukenye et al., 2016). This issue is particularly pronounced in developing nations, where the vast majority of products containing polybrominated diphenyl ethers (PBDEs) are disposed of alongside other plastic waste and refuse in landfills or open waste dumps. Furthermore, the uncontrolled incineration of solid waste, predominantly fueled by plastic materials, has become a common practice (Babayemi et al., 2018; Oloruntoba et al., 2021).

Polybrominated diphenyl ethers (PBDEs) are persistent organic pollutants with widespread environmental distribution, particularly in aquatic ecosystems (Wang

P. O. Isibor (✉) · B. Akinsanya
Department of Biological Sciences, Covenant University, Ota, Ogun State, Nigeria

Department of Zoology, University of Lagos, Akoka, Lagos, Nigeria
e-mail: patrickisibor007@gmal.com

P. O. Isibor, P. A. Akinduti (eds.), *Harnessing Biotechnology Tools for Product Development*, https://doi.org/10.1007/978-3-031-86002-7_28

et al., 2018). The predominant pathway for PBDE bioaccumulation in aquatic organisms, including fish and mollusks, is through dietary exposure (Luo et al., 2014; Margolis et al., 2020). These compounds are absorbed from contaminated water or sediment and undergo bioconcentration as they ascend the food web. Lower brominated congeners preferentially accumulate in lipid-rich tissues, whereas higher brominated forms exhibit a tendency to associate with proteins. This partitioning behavior is influenced by the hydrophobicity and molecular size of PBDEs, with their lipid affinity facilitating the transfer and accumulation of these lipophilic pollutants (Miranda et al., 2011; Nishida, 2011; Roberts & Rodger, 2012; O'Driscoll et al., 2016; McGrath et al., 2018; Okoffo et al., 2021; Olujimi et al., 2022).

Moreover, several environmental factors, including salinity, temperature, and metabolic processes of the organisms, can modulate the extent of PBDE bioaccumulation (Tang & Zhai, 2017). Biotransformation mechanisms, such as debromination, may also alter the PBDE congener profiles present in biological tissues. Certain PBDE congeners have been observed to biomagnify at higher trophic levels due to increased dietary intake coupled with slow elimination rates. The primary route of PBDE bioaccumulation is through the diet, especially in benthic organisms and filter feeders (Lushchak, 2011; Liagkouridis et al., 2014; Linares et al., 2015). Contaminated sediments represent a significant source of PBDE exposure, as these chemicals can be ingested alongside sediment particles or through benthic organisms residing in the sediment matrix. Additionally, fish and other aquatic organisms uptake PBDEs through their intestines and gills during feeding activities.

In aquatic biota, polybrominated diphenyl ethers (PBDEs) exhibit a strong affinity for lipid-rich compartments, such as cell membranes and adipose tissues, attributable to their lipophilic characteristics. Lower brominated PBDE congeners demonstrate a greater propensity to associate with lipids, whereas higher brominated variants tend to bind more readily to proteins (Schreck et al., 2008; Slaninova et al., 2009; Ssebugere et al., 2014; Su et al., 2018; Johnson et al., 2013; Li et al. 2018). The patterns of PBDE accumulation are contingent upon the specific tissue composition. PBDEs undergo biomagnification throughout the aquatic food web, culminating in elevated concentrations in apex predators. This phenomenon arises from dietary ingestion combined with slow elimination rates, whereby longer-lived species accumulate greater PBDE burdens (Nnorom & Osibanjo, 2008). Additionally, metabolic pathways such as debromination may modify the congener profile of PBDEs present in tissues. Species possessing enhanced biotransformation capabilities may facilitate more rapid elimination of PBDEs, contributing to the observed disparities among aquatic organisms (Bergayou et al., 2009; Bobescu et al., 2021). Bioaccumulation transpires when an organism assimilates contaminants at a rate surpassing their elimination or metabolic degradation. A variety of biotic and abiotic factors can influence the uptake, accumulation, and elimination of pollutants such as PBDEs in aquatic species, making the understanding of these dynamics essential for contamination risk assessment. The molecular structure and physicochemical properties of a contaminant significantly influence its potential for bioaccumulation. Lipophilic compounds with elevated octanol-water partition coefficients preferentially accumulate in fatty tissues, thereby enhancing overall bioaccumulation. Moreover, high molecular weight and chemical stability further promote bioaccumulation by diminishing elimination rates. Conversely, the size of a compound can impede its uptake, and certain metabolites may exhibit greater accumulation than their parent compounds.

Polybrominated diphenyl ethers (PBDEs) were first detected within the ecosystem of South San Diego Bay, a component of the San Diego National Wildlife Refuge (Adewuyi & Adeleye, 2013). Law et al. (2006) documented a remarkable approximately 60-fold increase in PBDE concentrations in human breast milk in Sweden. Additionally, Tlustos et al. (2007) demonstrated that PBDE exposure can adversely affect brain development in murine models, leading to impairments in learning capabilities. Similar to polychlorinated biphenyls (PCBs), exposure to PBDEs during critical periods of brain development—specifically during pregnancy and early childhood—may confer significant neurodevelopmental risks. While pentabrominated compounds are considered the most toxic, many of these persistent organic pollutants have not been subjected to comprehensive research (Labadie et al., 2010; Lana et al., 2010; Huang et al., 2010; Guerra et al., 2011; La Guardia et al., 2012; Guo et al., 2016; Kostić et al., 2017; Hahladakis et al., 2018; Kanaya et al., 2019; Klinčić et al., 2020; Huang et al., 2020).

Nigeria's extensive water resources, particularly Lagos Lagoon, are significantly impacted by environmental stressors stemming from human population growth, urbanization, and industrialization (Pereira et al., 2013). The lagoon's waste disposal and management practices present severe environmental challenges, as conventional waste disposal methods—including landfilling, open dumping, and incineration—result in the contamination of both underground and surface water bodies. Lagos Lagoon serves as a multipurpose resource, supporting fishing, recreational activities, solid waste disposal, non-living resource extraction (such as sand and coral), effluent discharge, and various other anthropogenic pressures. Previous investigations have revealed that the lagoon is heavily polluted due to a range of anthropogenic activities in its vicinity (Onyeche & Akankali, 2013; Park et al., 2014; Pellacani et al., 2014; Bamidele et al., 2020).

Aquatic environments adjacent to urban and industrial areas often exhibit elevated levels of polychlorinated diphenyl ether (PBDE) contamination, primarily due to proximity to point sources of pollution. This poses significant ecological challenges, as various fish, shellfish, and other organisms residing in contaminated waters are prone to bioaccumulating PBDEs, potentially leading to toxic effects from chronic exposure (Munschy et al., 2011). Documented adverse effects associated with PBDE toxicity encompass impaired reproductive and developmental processes, immunosuppression, and disruptions in thyroid function among aquatic organisms.

Despite these concerns, considerable knowledge gaps persist regarding the environmental fate and transport mechanisms of PBDEs within specific aquatic systems (Davis et al., 2012). Notably, there is a dearth of published data addressing PBDE contamination in aquatic ecosystems of developing nations. Additionally, the effects of chronic PBDE exposure on native fish and invertebrate populations remain inadequately explored (Olukunle et al., 2011; Olutona et al., 2016; Olutona et al., 2017;

Xu et al., 2018). Consequently, there is an urgent need for research focused on the bioaccumulation of PBDE congeners in aquatic organisms and the resultant adverse effects stemming from polluted waters in understudied regions (Dazy et al., 2009; Johnson et al., 2013; Zhu et al., 2015). Such investigations will yield essential data to bolster remediation and conservation initiatives. This study aims to elucidate the bioaccumulation levels of polychlorinated diphenyl ethers (PBDEs) in the tissues of *Tilapia zilli* collected from Lekki Lagoon. Given the paucity of ecotoxicity data regarding these pervasive pollutants, particularly in developing nations, this research is poised to make significant contributions to the understanding of PBDE contamination and its effects on aquatic biota. Despite their widespread detection in environmental samples, PBDEs, which have been extensively manufactured as flame retardants (Davis et al., 2012), exhibit a poorly characterized distribution and impact within Nigeria's aquatic ecosystems (Nnorom & Osibanjo, 2008). Quantifying PBDE bioaccumulation alongside histopathological alterations in indigenous fish species is essential for effective biomonitoring and health assessments of vulnerable lagoons and fisheries. The results of this investigation will inform regulatory frameworks and remediation strategies aimed at controlling PBDE pollution, which remains inadequately addressed in numerous regions. Furthermore, elucidating the human health risks associated with PBDE accumulation in seafood is critical; this study will also examine cellular-level effects pertinent to dietary exposure (Munschy et al., 2011).

On a broader scale, the research endeavors to establish standardized methodologies that utilize oxidative stress biomarkers and histopathology as rapid ecotoxicity screening tools for emerging contaminants in economically significant fish species (Daso et al., 2013).

2 Materials and Methods

2.1 *Study Area*

Lekki Lagoon is located in Lagos State, Southwestern Nigeria, positioned between longitudes 4°00′ and 4°15′ E and latitudes 6°25′ and 6°37′ N. The lagoon encompasses a surface area of approximately 247 km^2 and reaches a maximum depth of 6.4 m, with the majority of its expanse being shallow, measuring <3.0 m in depth. Lekki Lagoon is an integral component of a complex network of lagoons and creeks that extends 200 km along the southwestern coastline of Nigeria, spanning from the Dahomey border to the Niger Delta. It receives hydrological input from the Oni River, which flows northeast, along with the Oshun and Saga Rivers, which flow northwest.

Physicochemical parameters, including dissolved oxygen (DO), electrical conductivity, pH, salinity, temperature, total dissolved solids (TDS), and turbidity, were assessed utilizing a handheld multiparameter probe (Horiba Water Checker Model U-10). Data collection was facilitated through the use of a motorized canoe, enabling comprehensive spatial sampling across the lagoon.

2.2 *Biochemical Analysis*

2.2.1 Homogenizing Sample

The excised liver and intestine were collected and weighed. Each organ was subsequently homogenized in 0.1 M phosphate buffer (pH 7.2) using a mortar and pestle to ensure thorough blending. The resultant homogenate was then subjected to centrifugation at 2500 rpm for 15 min. Following centrifugation, the supernatant was carefully decanted and stored at −20 °C for further analysis.

2.2.2 Determination of Catalase Activity (CAT)

The reaction mixture comprises 5% potassium heptaoxodichromate, 0.2 M hydrogen peroxide, a dichromate/acetic acid solution, and 0.01 M phosphate buffer at pH 7.0. A 0.1-mL aliquot of the sample will be diluted with 4.9 mL of distilled water to achieve a 1:5 dilution.

The assay mixture will be prepared in a 10-mL flat-bottom flask, containing 4 mL of 0.2 M hydrogen peroxide and 5 mL of 0.01 M phosphate buffer. A 1-mL aliquot of the appropriately diluted enzyme preparation (test sample) will be rapidly introduced into the reaction mixture and gently swirled to ensure homogeneity. The reaction will be conducted at ambient temperature.

At 60-s intervals over a 3-min period, 1 mL portions of the reaction mixture will be withdrawn and transferred into test tubes containing 2 mL of the dichromate/acetic acid reagent. The extinction coefficient for hydrogen peroxide at 240 nm, quantified at 40.0 M, will be employed for subsequent calculations.

2.2.3 Determination of Reduced Glutathione (GSH)

The quantification of total sulfhydryl groups, protein-bound sulfhydryl groups, and free sulfhydryl groups, such as glutathione (GSH), in biological samples can be achieved using Ellman's reagent, 5,5′-dithio-bis-2-nitrobenzoic acid (DTNB). This methodology relies on the formation of a relatively stable yellow complex resulting from the reaction between Ellman's reagent and free sulfhydryl groups.

To prepare the sample, 0.2 mL of the biological sample will be diluted with 1.8 mL of distilled water to achieve a 1:10 dilution. Subsequently, 3 mL of precipitating reagent (4% sulfosalicylic acid) will be added to the diluted sample, which will be allowed to stand for 10 min to facilitate precipitate formation. Following this, 0.5 mL of the supernatant will be extracted and mixed with 4 mL of phosphate buffer, followed by the addition of 0.5 mL of Ellman's reagent. The blank control will consist of 4 mL of 0.1 M phosphate buffer (pH 7.4), 1 mL of the diluted precipitating solution, and 0.5 mL of Ellman's reagent. Absorbance will be measured at 412 nm within 20 min of color development using a spectrophotometer.

Additionally, chromic acetate produced in the reaction will be quantified spectrophotometrically at wavelengths between 570 and 610 nm. Notably, dichromate does not exhibit absorbance at these wavelengths, ensuring that its presence in the assay mixture does not interfere with the determination of chromic acetate. Catalase preparations in the samples will be allowed to decompose hydrogen peroxide for varying durations. The reaction will be terminated by heating the mixture, and the residual hydrogen peroxide will be quantified by measuring chromic acetate levels.

2.2.4 Determination of Glutathione Peroxidase (GPx)

To assess glutathione peroxidase (GPx) activity, a reaction mixture was prepared containing 0.2 mL of 0.4 M phosphate buffer (pH 7.0), 0.1 mL of 10 mM sodium azide, 0.2 mL of plasma, 0.2 mL of glutathione (GSH) salt, and 0.1 mL of 0.2 mM hydrogen peroxide. The mixture was incubated at 37 °C for 10 min to facilitate the enzymatic reaction. Subsequently, the reaction was terminated by the addition of 0.4 mL of 10% trichloroacetic acid (TCA), followed by centrifugation to separate the precipitate. The glutathione content in the supernatant was quantified using Ellman's reagent, enabling the determination of GPx activity.

2.2.5 Malondialdehyde (MDA)

The assay for assessing lipid peroxidation will be conducted by quantifying the thiobarbituric acid (TBA) reactive products in the test sample. The results will be expressed as micromolar concentrations of malondialdehyde (MDA) per gram of tissue. This assay is predicated on the reaction between the chromogenic reagent, 2-TBA, and MDA, a terminal product of lipid peroxidation, under acidic conditions to form a stable pink chromophore that exhibits maximum absorbance at 532 nm.

The reaction mixture will consist of a 30% solution of trichloroacetic acid (TCA), a 0.75% solution of thiobarbituric acid (TBA) in 0.1 M hydrochloric acid (HCl), and a 0.15 M Tris-KCl buffer (pH 7.4). An aliquot of 0.4 mL of the test sample will be combined with 1.6 mL of the Tris-KCl buffer, which will be introduced into the test tube prior to the addition of the test sample. Subsequently, 0.5 mL of the 30% TCA solution and 0.5 mL of the 0.75% TBA solution will be added. The resulting mixture will then be incubated in a water bath at 90–95 °C for 1 h.

Following incubation, the mixture will be cooled on ice and centrifuged at 3000 rpm for 15 min. The clear pink supernatant will be carefully collected, and absorbance will be measured at 532 nm against a reference blank of distilled water using a spectrophotometer. The concentration of malondialdehyde (MDA), serving as an index of lipid peroxidation, will be calculated using an extinction coefficient of $1.5 \times 105\ M^{-1}\ cm^{-1}$.

2.2.6 Determination of Superoxide Dismutase (SOD) Activity

The activity of superoxide dismutase (SOD) will be quantified utilizing the methodology. The reaction mixture will consist of 0.05 M carbonate buffer (pH 10.2) and 0.3 mM epinephrine. A sample volume of 0.1 mL will be diluted in 0.9 mL of distilled water to achieve a 1:10 dilution. An aliquot of 0.2 mL of the diluted enzyme preparation will then be introduced into 2.5 mL of 0.05 M carbonate buffer (pH 10.2) for equilibration in the spectrophotometer. The reaction will be initiated by the addition of 0.3 mL of freshly prepared 0.3 mM epinephrine to the mixture, which will be mixed rapidly by inversion.

The reference cuvette will contain 2.5 mL of 0.05 M carbonate buffer, 0.3 mL of epinephrine (acting as the substrate), and 0.2 mL of distilled water. The increase in absorbance will be monitored at 480 nm at 30-s intervals for a total duration of 150 s. An extinction coefficient of 4020 M for epinephrine at 480 nm will be utilized for the calculation of enzyme activity.

2.2.7 Determination of PBDES in Tissues

The analysis utilized the KOH refluxing/vortex extraction method in accordance with EPA method 3611C for sample cleanup. A total of 15 g of wet-weight excised liver and intestinal tissue from the fish specimens was weighed and subsequently macerated and homogenized in a crucible. From the homogenized tissue, a 10-g aliquot was transferred to a 50-mL centrifuge tube, to which 15 mL of 6 N KOH was added. The tubes were sealed and incubated for 18 h in a water bath maintained at 35 °C, with vigorous shaking for 30 s every half hour during the initial 4 h, fol lowed by cooling.

Subsequently, 15 mL of either methylene chloride or ethyl ether was added to the centrifuge tube, followed by vortexing for 1 min and centrifugation at 2000 rpm for 5 min to promote phase separation. The upper organic layer was carefully extracted using a Pasteur pipette and transferred into a 250-mL round-bottom flask. This solvent extraction process was repeated twice, and all aliquots were combined in the round-bottom flask. The sample extracts were concentrated to approximately 5–10 mL using a rotary evaporator before undergoing fractionation cleanup with an alumina gel column, followed by analysis using gas chromatography-mass spectrometry (GC-MS).

2.3 Statistical Analysis

The data obtained from bioaccumulation analysis, oxidative stress analysis, and physicochemical parameters were evaluated using analysis of variance (ANOVA), statistical package for the social sciences (SPSS), and IBM 20.0 version.

3 Results

3.1 Physicochemical Parameters

The mean values of selected physicochemical parameters at various sampling points along the Lekki Lagoon are presented in Table 28.1. The average temperature at Station 1 was notably higher at 30.2 °C compared to 29.7 °C at Station 2. Both stations exhibited neutral average pH levels, ranging from 7.3 to 7.5. The electrical conductivity of surface water was significantly greater at Station 2 (0.93 mS/cm) than at Station 1 (0.42 mS/cm). Conversely, turbidity levels were higher at Station 1, averaging 15 NTU compared to 13 NTU at Station 2. Dissolved oxygen concentrations were greater at Station 2, with an average of 8.2 mg/L compared to Station 1. Total dissolved solids were consistent across both sampling stations, averaging 0.27 g/L. However, salinity levels were elevated at Station 2, averaging 0.7 ppt compared to 0.2 ppt at Station 1. Statistical analysis indicated no significant differences ($p > 0.05$) between the two stations for the selected physicochemical parameters (Table 28.1).

3.2 Polybrominated Diphenyl Ether (PBDE) Concentration in Uninfected and Infected Intestine of Tilapia zilli

The average concentration of BDE-28 (2,4,4′-tribromodiphenyl ether) in the tissues of uninfected fish was significantly higher ($p < 0.001$) compared to that in infected fish, measuring 0.017 ng/g and 0.0012 ng/g, respectively. The concentrations of BDE-47 (2,2′,4,4′-tetrabromodiphenyl ether) and BDE-100 (2,2′,4,4′,6-pentabromodiphenyl ether) did not exhibit significant differences between uninfected and infected fish ($p > 0.05$). Notably, BDE-99 (2,2′,4,4′,5-pentabromodiphenyl ether), BDE-154 (2,2′,4,4′,5,6-hexabromodiphen yl ether), BDE-153 (2,2′,4,4′,5,5′-hexabromodiphenyl ether), and BDE-183 (2,2′,3,4,4′,5,6-Heptabromodiphenyl ether) were not detected in the tissues of

Table 28.1 Physicochemical parameters of sampling points along the Lekki Lagoon

Parameters	Station 1	Station 2	*p*-value	Summary	FEPA limit
Temperature °C	30.2	29.7	$p > 0.05$	ns	≤40
pH	7.5	7.3	$p > 0.05$	ns	6–9
Conductivity (ms/cm)	0.42	0.93	$p > 0.05$	ns	–
Turbidity (NTU)	15	13	$p > 0.05$	ns	≤30
Dissolved oxygen (mg/L)	6.2	8.2	$p > 0.05$	ns	≥5
Total dissolved solids (g/L)	0.27	0.27	$p > 0.05$	ns	≤30
Salinity (ppt)	0.2	0.7	$p > 0.05$	ns	<34

ns not significant

infected fish. Furthermore, BDE-209 (2,2′,3,3′,4,4′,5,5′,6,6′-decabromodiphenyl ether) was undetectable in both infected and uninfected fish (Table 28.2). Overall, the total concentrations of polybrominated diphenyl ethers (PBDEs) differed significantly ($p < 0.001$) between infected and uninfected fish.

3.3 *Biochemical Analysis*

3.3.1 Antioxidant Enzymes Level in Tissues of Uninfected and Infected *Tilapia zilli*

The results of antioxidant enzyme activity in the tissues of infected and uninfected *Tilapia zilli* are presented in Table 28.3 and Fig. 28.1. Protein (PRO) concentration was higher in the tissues of uninfected fish, with an average of 14 mg/mg compared to 11 mg/mg in infected fish. Catalase (CAT) activity was significantly elevated ($p < 0.001$) in uninfected fish, with an average of 21 mg/mg, compared to 6.5 mg/mg in infected tissues. Superoxide dismutase (SOD) levels were higher in infected fish (680 mg/mg) compared to uninfected fish (620 mg/mg). Malondialdehyde (MDA) concentrations, an indicator of lipid peroxidation, were elevated in uninfected fish (7.4 μmol/mL) relative to infected fish (4 μmol/mL). Conversely, glutathione (GSH) levels were higher in infected fish tissues (13 μmol/mL) compared to uninfected tissues (6.3 μmol/mL). Glutathione peroxidase (GPx) activity was greater in uninfected fish (260 μmol/mL) compared to infected fish (210 μmol/mL). Among the measured enzymes, catalase activity showed a statistically significant

Table 28.2 Comparative concentrations of PBDE congeners in tissues of infected and uninfected *Tilapia zilli*

PBDE congeners	Uninfected	Infected	*p*-value	Summary
BDE-28 (244′-tribromodiphenyl ether)	0.017	0.0012	$p < 0.001$	***
BDE-47 (22′44′-tetrabromodiphenyl ether)	0.0094	ND	$p > 0.05$	ns
BDE-100 (22′44′6-pentabromodiphenyl ether)	0.0078	0.003	$p > 0.05$	ns
BDE-99 (22′44′5-penabromodiphenyl ether)	0.0059	ND	$p > 0.05$	ns
BDE-154 (22′44′56′-hexabromodiphenyl ether)	0.0002	ND	$p > 0.05$	ns
BDE-153 (22′44′55′-hexabromodiphenyl ether)	0.0008	ND	$p > 0.05$	ns
BDE-183 (22′344′5′6-heptabromodiphenyl ether)	0.02	ND	$p > 0.05$	ns
BDE-209 (22′33′44′55′66′-decabromodiphenyl ether)	ND	ND	$p > 0.05$	ns
Total PBDEs	0.061	0.0043	$p < 0.001$	***

ND not detected, *ns* not significant, ***Significant

BDE-28 = 2,4,4′-tribromodiphenyl ether, BDE-47 = 2,2′,4,4′-tetrabromodiphenyl ether, BDE-100 = 2,2′,4,4′,6-pentabromodiphenyl ether, BDE-99 = 2,2′,4,4′,5,-penabromodiphenyl ether, BDE-154 = 2,2′,4,4′,5,6′-hexabromodiphenyl ether, BDE-153 = 2,2′,4,4′,5,5′ hexabromodiphenyl ether, BDE-183 = 2,2′,3,4,4′,5′,6-heptabromodiphenylether, BDE-209 = 2,2′,3,3′,4,4′,5,5′, 6,6′ decabromodiphenyl ether

Table 28.3 Comparative analysis of oxidative stress enzyme level in tissues of *Tilapia zilli*

Enzymes	Uninfected	Infected	*p*-value	Summary
PRO min/mg protein	14	11	$p < 0.001$	ns
CAT min/mg protein	21	6.5	$p < 0.001$	***
SOD min/mg protein	620	680	$p > 0.05$	ns
MDA μmol/mL	7.4	4	$p > 0.05$	ns
GSH μmol/mL	6.3	13	$p > 0.05$	ns
GPX μmol/mL	260	210	$p > 0.05$	ns

*** Highly significant

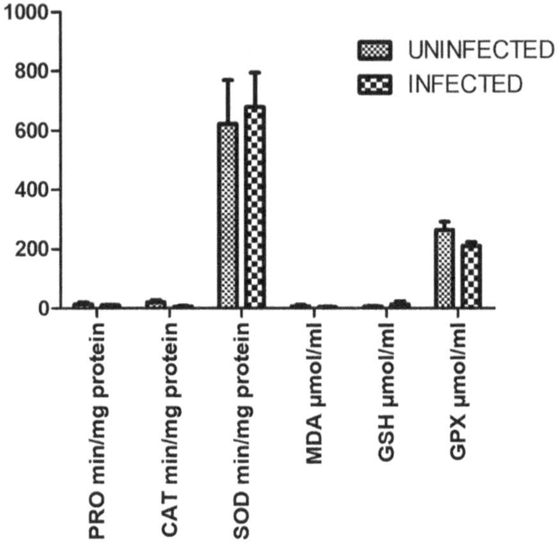

Fig. 28.1 Level of oxidative stress enzymes in tissues of *Tilapia zilli*

difference between infected and uninfected fish ($p < 0.001$), while other enzymes exhibited no significant differences ($p > 0.05$).

3.4 *Lipid Profile Levels*

3.4.1 Lipid Profile Level in Tissues of Uninfected and Infected *Tilapia zilli*

The lipid profile levels in the tissues of infected and uninfected fish are presented in Table 28.4 and Fig. 28.2. Cholesterol concentrations were significantly elevated ($p < 0.001$) in the tissues of uninfected fish compared to their infected counterparts. Similarly, high-density lipoprotein (HDL) and low-density lipoprotein (LDL) levels were markedly higher in uninfected tissues relative to those of infected fish. In contrast, triglyceride (TRIG) concentrations did not show a significant difference ($p > 0.05$) between infected and uninfected fish, with mean values of 49 mmol/L and 51 mmol/L, respectively.

Table 28.4 Comparative analysis of lipid profile level in tissues of *Tilapia zilli*

Enzymes (mmol/L)	Uninfected	Infected	*p*-value	Summary
CHOL	170	150	$p < 0.001$	***
HDL	84	55	$p < 0.001$	***
TRIG	51	49	$p > 0.05$	ns
LDL	100	86	$p < 0.01$	**

ns not significant, ** Significant, ***Highly significant

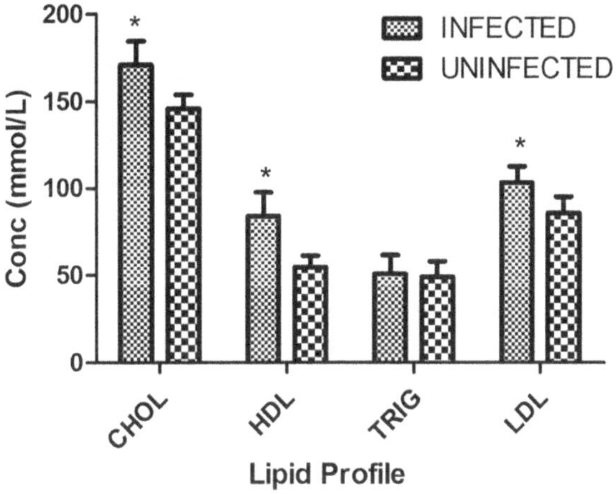

Fig. 28.2 Lipid profile level in the tissue of infected and uninfected *Tilapia zilli*

4 Discussion

This study investigated the bioaccumulation of polybrominated diphenyl ether (PBDE) congeners in the tissues of *Tilapia zilli* sourced from Epe Fish Market, located along the coastal region of Lekki Lagoon. The primary objective was to generate data on the contamination levels of PBDEs within this aquatic ecosystem. Given the mobility of fish species, their presence or absence in contaminated areas serves as a valuable bioindicator of environmental quality, making them integral to biological assessments. In addition to quantifying PBDE levels, this study also evaluated the health status of the fish by analyzing key biomarkers, including antioxidant enzyme activity and lipid profiles, to assess oxidative stress and potential disruptions to metabolic regulation (Daso et al., 2013).

Temperature exerts a profound influence on the growth and distribution of aquatic flora and fauna, as it represents one of the most variable environmental factors, driving key chemical and biological processes. pH reflects the acidity or basicity of a solution and is regulated by dissolved chemical compounds and biochemical reactions within the aquatic medium. Salinity is another critical ecological parameter that directly affects the abundance and distribution of aquatic organisms (Xin et al., 2014; Yang et al., 2018). It plays a pivotal role in various chemical processes and significantly influences nutrient dynamics, particularly in coastal zones.

This study assessed the physicochemical properties—temperature, pH, conductivity, dissolved oxygen, salinity, and turbidity—of selected points in the lagoon surface water and compared the results with the Federal Environmental Protection Agency (FEPA) standards. All parameters fell within the FEPA-recommended limits for surface water quality in lagoons.

Changes in physicochemical conditions in surface waters can significantly affect the fate and behavior of contaminants within aquatic ecosystems (Yu et al., 2012; Yu et al., 2016; Zhang et al., 2020). Bioaccumulation occurs due to an imbalance between chemical uptake and elimination, the latter involving biotransformation and excretion processes. Pollutants that are poorly water-soluble often accumulate in lipid-rich tissues, leading to bioaccumulation, particularly in aquatic organisms (Bergayou et al., 2009; Bobescu et al., 2021).

The average concentration of BDE-28 (2,4,4′-Tribromodiphenyl ether) in the tissues of uninfected fish was significantly higher ($p < 0.001$) compared to infected fish, potentially due to the parasites absorbing a portion of the contaminant load. Despite this, traceable levels of PBDE congeners were detected in both infected and uninfected *Tilapia zilli* tissues, with the exception of BDE-209, which was undetected in both groups. Overall, the levels of PBDE congeners were higher in uninfected fish tissues, contrasting with the findings of Kuton et al. (2021), who reported that infected *Parachanna obscura* and *Sarotherodon melanotheron* accumulated more heavy metals in their organs than noninfected individuals. Parasite infections in fish can impact feeding behavior, thereby reducing the uptake of lipophilic contaminants such as PCBs, which leads to decreased bioaccumulation. Parasites may modulate pollutant bioaccumulation through complex interactions involving host physiology, feeding patterns, and contaminant properties, highlighting the variable influence of parasitism on contaminant dynamics (Di Giulio & Meyer, 2008; Frederiksen et al., 2009; Ge et al., 2014; Lenhardt et al., 2015; Doherty et al., 2019).

Studies have also shown the capacity of intestinal parasites to accumulate high levels of persistent organic pollutants (PAHs) which in turn reduces the pollutant accumulation load of the fish host. Abdallah & Harrad (2014) also reported a similar observation on the bioaccumulation of PCB congeners on tissues of fish and their parasites with the results indicating that some parasitic organisms may influence positively their hosts in PCB-contaminated environment. The absence of BDE-209 in the tissues of both infected and uninfected fish may be attributed to its low water solubility and high molecular weight, which limits its bioavailability and uptake from water but binds strongly to sediments rather than dissolving into aquatic environments. BDE-28 levels in uninfected and infected fish were significantly different. Parasitic infection may alter membrane lipid composition and increase lipid content which could enhance partitioning and accumulation of some PBDE congeners such as BDE-28 and BDE-100.

Under normal conditions, the physiologically important intracellular levels of reactive oxygen species (ROS) are maintained at low levels by various enzyme systems participating in the in vivo redox homeostasis. Therefore, oxidative stress can also be viewed as an imbalance between the prooxidants and antioxidants in the body. Antioxidant enzymes are frequently used as markers of oxidative stress. Antioxidant enzymes are proteins involved in the catalytic transformation of reactive oxygen species and their by-products into stable nontoxic molecules, therefore representing the most important defense mechanism against oxidative stress-induced cell damage (Ghedira et al., 2011; Gonzalez et al., 2013; Johnson et al., 2013).

Elevated levels of reactive oxygen species (ROS) in tissues result in cellular damage when their production exceeds the capacity of antioxidant defense systems to neutralize them. The measurement of lipid peroxides, particularly malondialdehyde (MDA), in plant and animal tissues exposed to environmental pollutants, is widely recognized as a reliable early indicator of oxidative stress and is often incorporated into environmental monitoring programs (Valavanidis et al., 2006). In this study, uninfected fish exhibited higher MDA levels compared to their infected counterparts. MDA, a widely used biomarker for lipid peroxidation, reflects oxidative damage to cellular lipids and membranes. The elevated MDA levels observed in infected fish indicate more extensive oxidative damage, which is consistent with prior research showing increased ROS production and heightened antioxidant enzyme activity during parasitic infections in fish. These findings underscore the role of parasitic infections in exacerbating oxidative damage, contributing to the overall physiological stress on the host organism (Bergayou et al., 2009; Abdallah & Harrad, 2014; Banni et al., 2014; Choi et al., 2014; Bergayou et al., 2021).

The pronounced oxidative stress observed in parasitized fish likely results from several converging mechanisms. Parasitic attachment and feeding inflict direct tissue damage, compromising cellular integrity and initiating oxidative injury. The host's immune response to the parasitic invasion triggers inflammation, driving the excessive production of reactive oxygen species (ROS) as part of the host's defense strategy. The parasites also impose elevated metabolic demands, leading to enhanced electron leakage during mitochondrial respiration, which surpasses the host's antioxidant capacity to effectively neutralize ROS. Furthermore, parasitic feeding often induces hemorrhaging, potentially releasing malondialdehyde (MDA), a marker of lipid peroxidation, from compromised tissues into the bloodstream, exacerbating oxidative damage.

Superoxide dismutase (SOD) activity was significantly elevated in infected fish tissues compared to uninfected ones, reflecting an adaptive response to counteract the heightened superoxide flux. However, despite the increased SOD activity, oxidative damage persisted, as evidenced by lipid peroxidation, suggesting that antioxidant defenses were insufficient to fully mitigate the oxidative burden. The extent of oxidative stress is likely modulated by parasite load and interspecific differences in host physiology (Daso et al., 2009; Bodin et al., 2011; Daso et al., 2013; Bramwell et al., 2017; Cao et al., 2018; Chakraborty et al., 2019).

This study identified significantly elevated levels of cholesterol, low-density lipoprotein (LDL), high-density lipoprotein (HDL), and triglycerides in the tissues of fish infected with parasites compared to their parasite-free counterparts. These increases are likely attributable to the metabolic demands imposed by parasitic infection as well as disruptions in lipid regulation and transport mechanisms.

Parasites extract essential nutrients from their host to support their growth and reproduction, potentially stimulating hepatic lipogenesis to compensate for lipid loss. This process may enhance liver synthesis of lipids and lipoproteins, contributing to altered lipid homeostasis. Furthermore, parasitic infection induces a systemic inflammatory response, which can modulate lipid metabolism and result in altered lipid profiles.

4.1 Conclusion

This study identified detectable levels of polybrominated diphenyl ethers (PBDEs) in tissue samples of *Tilapia zilli* from Lekki Lagoon, with the PBDE burden showing a strong correlation with elevated oxidative stress markers, including increased activity of superoxide dismutase (SOD), catalase, and malondialdehyde (MDA) concentrations. These findings provide compelling evidence that PBDE contamination in Lekki Lagoon induces oxidative damage in native fish species. The biomarkers of oxidative stress demonstrated sensitivity and reliability as indicators of cellular impacts resulting from chronic PBDE exposure.

The study establishes *Tilapia zilli* as an effective bioindicator species, reflecting the extent of PBDE pollution within Lekki Lagoon. This data fills a crucial gap in understanding bioaccumulation patterns and the toxicity of PBDEs in West African aquatic ecosystems. Continued biomonitoring efforts are essential for guiding strategies to mitigate PBDE contamination and protect the health of both wildlife and human populations reliant on local fish for sustenance. Furthermore, this research offers a model for integrating pollutant bioaccumulation assessments with biochemical effects in environmentally exposed organisms, providing a comprehensive approach to ecological toxicology.

References

Abdallah, M. A. E., & Harrad, S. (2014). Polybrominated diphenyl ethers in UK human milk: Implications for infant exposure and relationship to external exposure. *Environment International, 63*, 130–136.

Adewuyi, G. O., & Adeleye, A. (2013). Evaluation of polybrominated diphenyl ethers in sediment of Lagos Lagoon, Nigeria. *African Journal of Environmental Science and Technology, 7*(7), 686–693.

Akortia, E., Okonkwo, J. O., Lupankwa, M., Osae, S. D., Daso, A. P., Olukunle, O. I., & Chaudhary, A. (2016). A review of sources, levels, and toxicity of polybrominated diphenyl ethers (PBDEs) and their transformation and transport in various environmental compartments. *Environmental Reviews, 24*(3), 253–273.

Babayemi, J. O., Osibanjo, O., Sindiku, O., & Weber, R. (2018). Inventory and substance flow analysis of polybrominated diphenyl ethers in the Nigerian transport sector—End-of-life vehicles policy and management. *Environmental Science and Pollution Research, 25*, 31805–31818.

Bamidele, A., Kuton, M. P., Iniobong, A. D., Uchenna, N. D., Saliu, J. K., & David, U. U. (2020). Bioaccumulation of polychlorinated biphenyls (PCBs) in fish host-parasite bentho-pelagic food chain in Epe lagoon, Lagos, Nigeria. *Bulletin of Environmental Contamination and Toxicology, 105*, 770–776.

Banni, M., Hajer, A., Sforzini, S., Oliveri, C., Boussetta, H., & Viarengo, A. (2014). Transcriptional expression levels and biochemical markers of oxidative stress in Mytilus galloprovincialis exposed to nickel and heat stress. *Comparative Biochemistry and Physiology Part C: Toxicology & Pharmacology, 160*, 23–29.

Bergayou, H., Mouneyrac, C., Pellerin, J., & Moukrim, A. (2009). Oxidative stress responses in bivalves (Scrobicularia plana, Cerastoderma edule) from the Oued Souss estuary (Morocco). *Ecotoxicology and Environmental Safety, 72*(3), 765–769.

Bobescu, E., Marceanu, L. G., Dima, L., Balan, A., Strempel, C. G., & Covaciu, A. (2021). Trimetazidine therapy in coronary artery disease: The impact on oxidative stress, inflammation, endothelial dysfunction, and long-term prognosis. *American Journal of Therapeutics, 28*(5), e540–e547.

Bodin, N., Ka, R. G., Le Loc'h, F., Raffray, J., Budzinski, H., Peluhet, L., & de Morais, L. T. (2011). Are exploited mangrove molluscs exposed to persistent organic pollutant contamination in Senegal, West Africa? *Chemosphere, 84*(3), 318–327.

Bramwell, L., Harrad, S., Abdallah, M. A. E., Rauert, C., Rose, M., Fernandes, A., & Pless-Mulloli, T. (2017). Predictors of human PBDE body burdens for a UK cohort. *Chemosphere, 189*, 186–197.

Cao, L. Y., Ren, X. M., Yang, Y., Wan, B., Guo, L. H., Chen, D., & Fan, Y. (2018). Hydroxylated polybrominated biphenyl ethers exert estrogenic effects via non-genomic G protein–coupled estrogen receptor mediated pathways. *Environmental Health Perspectives, 126*(5), 057005.

Chakraborty, P., Zhang, G., Li, J., Sampathkumar, P., Balasubramanian, T., Kathiresan, K., et al. (2019). Seasonal variation of atmospheric organochlorine pesticides and polybrominated diphenyl ethers in Parangipettai, Tamil Nadu, India: Implication for atmospheric transport. *Science of the Total Environment, 649*, 1653–1660.

Choi, G., Kim, S., Kim, S., Kim, S., Choi, Y., Kim, H. J., et al. (2014). Occurrences of major polybrominated diphenyl ethers (PBDEs) in maternal and fetal cord blood sera in Korea. *Science of the Total Environment, 491*, 219–226.

Daso, A. P., Fatoki, O. S., Odendaal, J. P., & Olujimi, O. O. (2013). Polybrominated diphenyl ethers (PBDEs) and 2,2′,4,4′,5,5′-hexabromobiphenyl (BB-153) in landfill leachate in Cape Town, South Africa. *Environmental Monitoring and Assessment, 185*, 431–439.

Davis, E. F., Klosterhaus, S. L., & Stapleton, H. M. (2012). Measurement of flame retardants and triclosan in municipal sewage sludge and biosolids. *Environment International, 40*, 1–7.

Dazy, M., Masfaraud, J. F., & Férard, J. F. (2009). Induction of oxidative stress biomarkers associated with heavy metal stress in *Fontinalis antipyretica* Hedw. *Chemosphere, 75*(3), 297–302.

Di Giulio, R. T., & Hinton, D. E. (2008). Reactive oxygen species and oxidative stress. In *The toxicology of fishes* (pp. 273–324). CRC Press.

Doherty, F. V., Aneyo, I., & Otitoloju, A. A. (2019). Histopathological and biochemical alterations in Eudrilus eugeniae (Kinberg 1867) as biomarkers of exposure to monocyclic aromatic hydrocarbons in oil impacted site. *The Journal of Basic and Applied Zoology, 80*, 1–15.

Frederiksen, M., Vorkamp, K., Thomsen, M., & Knudsen, L. E. (2009). Human internal and external exposure to PBDEs–a review of levels and sources. *International Journal of Hygiene and Environmental Health, 212*(2), 109–134.

Ge, J., Yun, X., Liu, M., Yang, Y., Zhang, M., & Wang, J. (2014). Distribution, potential source and ecotoxicological risk of polychlorinated biphenyls and polybrominated diphenyl ethers in the surface water of the three gorges dam region of the Yangtze River, China. *Ecotoxicology, 23*, 978–987.

Ghedira, J., Jebali, J., Banni, M., Chouba, L., Boussetta, H., López-Barea, J., & Alhama, J. (2011). Use of oxidative stress biomarkers in Carcinus maenas to assess littoral zone contamination in Tunisia. *Aquatic Biology, 14*(1), 87–98.

Gonzalez, M., Miglioranza, K. S., Grondona, S. I., Barni, M. F. S., Martinez, D. E., & Peña, A. (2013). Organic pollutant levels in an agricultural watershed: The importance of analyzing multiple matrices for assessing streamwater pollution. *Environmental Science: Processes & Impacts, 15*(4), 739–750.

Guerra, P., Alaee, M., Eljarrat, E., & Barcelo, D. (2011). Introduction to brominated flame retardants: Commercially products, applications, and physicochemical properties. In *Brominated flame retardants* (pp. 1–17). Springer.

Guo, J., Zhang, R., & Xu, Z. (2015). PBDEs emission from waste printed wiring boards during thermal process. *Environmental Science and Technology, 49*(5), 2716–2723.

Guo, W., Holden, A., Smith, S. C., Gephart, R., Petreas, M., & Park, J. S. (2016). PBDE levels in breast milk are decreasing in California. *Chemosphere, 150*, 505–513.

Hahladakis, J. N., Velis, C. A., Weber, R., Iacovidou, E., & Purnell, P. (2018). An overview of chemical additives present in plastics: Migration, release, fate and environmental impact during their use, disposal and recycling. *Journal of Hazardous Materials, 344*, 179–199.

Huang, H., Zhang, S., Christie, P., Wang, S., & Xie, M. (2010). Behavior of decabromodiphenyl ether (BDE-209) in the soil–plant system: Uptake, translocation, and metabolism in plants and dissipation in soil. *Environmental Science & Technology, 44*(2), 663–667.

Huang, Y., Rajput, I. R., Sanganyado, E., Yajing, S., Yu, F., Liang, B., & Liu, W. (2020). Immune stimulation effect of PBDEs via prostaglandin pathway in pantropical spotted dolphin: An in vitro study. *Chemosphere, 254*, 126717.

Johnson, P. I., Stapleton, H. M., Mukherjee, B., Hauser, R., & Meeker, J. D. (2013). Associations between brominated flame retardants in house dust and hormone levels in men. *Science of the Total Environment, 445*, 177–184.

Kanaya, N., Bernal, L., Chang, G., Yamamoto, T., Nguyen, D., Wang, Y. Z., et al. (2019). Molecular mechanisms of polybrominated diphenyl ethers (BDE-47, BDE-100, and BDE-153) in human breast cancer cells and patient-derived xenografts. *Toxicological Sciences, 169*(2), 380–398.

Khan, S., Naushad, M., Govarthanan, M., Iqbal, J., & Alfadul, S. M. (2022). Emerging contaminants of high concern for the environment: Current trends and future research. *Environmental Research, 207*, 112609.

Klinčić, D., Dvoršćak, M., Jagić, K., Mendaš, G., & Herceg Romanić, S. (2020). Levels and distribution of polybrominated diphenyl ethers in humans and environmental compartments: A comprehensive review of the last five years of research. *Environmental Science and Pollution Research, 27*, 5744–5758.

Kostić, J., Kolarević, S., Kračun-Kolarević, M., Aborgiba, M., Gačić, Z., Paunović, M., & Vuković-Gačić, B. (2017). The impact of multiple stressors on the biomarkers response in gills and liver of freshwater breams during different seasons. *Science of the Total Environment, 601*, 1670–1681.

Kuton, M. P., Ayanda, I, O., Uzoalu, I, A., Akhiromen, D, I., George, A., & Akinsanya, B. (2021). Studies on heavy metals and fish health indicators in Malapterurus electricus from Lekki Lagoon, Lagos, Nigeria. *Veterinary and Animal Science, 12*, 100169, ISSN 2451-943X, https://doi.org/10.1016/j.vas.2021.100169

La Guardia, M. J., Hale, R. C., Harvey, E., Mainor, T. M., & Ciparis, S. (2012). In situ accumulation of HBCD, PBDEs, and several alternative flame-retardants in the bivalve (Corbicula fluminea) and gastropod (Elimia proxima). *Environmental Science & Technology, 46*(11), 5798–5805.

Labadie, P., Tlili, K., Alliot, F., Bourges, C., Desportes, A., & Chevreuil, M. (2010). Development of analytical procedures for trace-level determination of polybrominated diphenyl ethers and tetrabromobisphenol a in river water and sediment. *Analytical and Bioanalytical Chemistry, 396*, 865–875.

Lana, N. B., Fontana, A. R., Ciocco, N. F., & Altamirano, J. C. (2010). Determination of polybrominated diphenyl ethers in water samples from Mendoza River basin by HS-SPME-GC-MS/MS. *Revista de la Facultad de Ciencias Agrarias, 42*(2), 85–98.

Law, R. J., Allchin, C. R., De Boer, J., Covaci, A., Herzke, D., Lepom, P., et al. (2006). Levels and trends of brominated flame retardants in the European environment. *Chemosphere, 64*(2), 187–208.

Lenhardt, M., Poleksić, V., Vuković-Gačić, B., Rašković, B., Sunjog, K., Kolarević, S., & Gačić, Z. (2015). Integrated use of different fish related parameters to assess the status of water bodies. *Slovenian Veterinary Research, 52*(1), 5–13.

Li, P., Liu, L., Zhou, G., Tian, Z., Luo, C., Xia, T., et al. (2018). Perigestational exposure to low doses of PBDE-47 induces excessive ER stress, defective autophagy and the resultant apoptosis contributing to maternal thyroid toxicity. *Science of the Total Environment, 645*, 363–371.

Liagkouridis, I., Cousins, I. T., & Cousins, A. P. (2014). Emissions and fate of brominated flame retardants in the indoor environment: A critical review of modelling approaches. *Science of the Total Environment, 491*, 87–99.

Linares, V., Bellés, M., & Domingo, J. L. (2015). Human exposure to PBDE and critical evaluation of health hazards. *Archives of Toxicology, 89*(3), 335–356.

Luo, P., Ni, H. G., Bao, L. J., Li, S. M., & Zeng, E. Y. (2014). Size distribution of airborne particle-bound polybrominated diphenyl ethers and its implications for dry and wet deposition. *Environmental Science & Technology, 48*(23), 13793–13799.

Lushchak, V. I. (2011). Environmentally induced oxidative stress in aquatic animals. *Aquatic Toxicology, 101*(1), 13–30.

Margolis, A. E., Banker, S., Pagliaccio, D., De Water, E., Curtin, P., Bonilla, A., et al. (2020). Functional connectivity of the reading network is associated with prenatal polybrominated diphenyl ether concentrations in a community sample of 5-year-old children: A preliminary study. *Environment International, 134*, 105212.

McGrath, T. J., Morrison, P. D., Ball, A. S., & Clarke, B. O. (2018). Concentrations of legacy and novel brominated flame retardants in indoor dust in Melbourne, Australia: An assessment of human exposure. *Environment International, 113*, 191–201.

Miranda, S. G., Purdie, N. G., Osborne, V. R., Coomber, B. L., & Cant, J. P. (2011). Selenomethionine increases proliferation and reduces apoptosis in bovine mammary epithelial cells under oxidative stress. *Journal of Dairy Science, 94*(1), 165–173.

Munschy, C., Héas-Moisan, K., Tixier, C., Olivier, N., Gastineau, O., Le Bayon, N., & Buchet, V. (2011). Dietary exposure of juvenile common sole (Solea solea L.) to polybrominated diphenyl ethers (PBDEs): Part 1. Bioaccumulation and elimination kinetics of individual congeners and their debrominated metabolites. *Environmental Pollution, 159*(1), 229–237.

Nishida, Y. (2011). The chemical process of oxidative stress by copper (II) and iron (III) ions in several neurodegenerative disorders. *Monatshefte für Chemie-Chemical Monthly, 142*, 375–384.

Nnorom, I. C., & Osibanjo, O. (2008). Sound management of brominated flame retarded (BFR) plastics from electronic wastes: State of the art and options in Nigeria. *Resources, Conservation and Recycling, 52*(12), 1362–1372.

O'Driscoll, K., Robinson, J., Chiang, W. S., Chen, Y. Y., Kao, R. C., & Doherty, R. (2016). The environmental fate of polybrominated diphenyl ethers (PBDEs) in western Taiwan and coastal waters: Evaluation with a fugacity-based model. *Environmental Science and Pollution Research, 23*, 13222–13234.

Okoffo, E. D., O'Brien, S., Ribeiro, F., Burrows, S. D., Toapanta, T., Rauert, C., et al. (2021). Plastic particles in soil: State of the knowledge on sources, occurrence and distribution, analytical methods and ecological impacts. *Environmental Science: Processes & Impacts, 23*(2), 240–274.

Oloruntoba, K., Sindiku, O., Osibanjo, O., Herold, C., & Weber, R. (2021). Polybrominated diphenyl ethers (PBDEs) concentrations in soil and plants around municipal dumpsites in Abuja, Nigeria. *Environmental Pollution, 277*, 116794.

Olujimi, O. O., Adegbite, K. I., Sojinu, S. O., Daso, A. P., Oyebanji, F. F., Oluwadare, I., et al. (2022). Occurrence, compositional profiles and health risks assessment of polybrominated diphenyl ethers (PBDEs) in groundwater in and around dumpsites in Lagos State, Nigeria. *Environmental Forensics, 25*(1–2), 49–58.

Olukunle, O., Okonkwo, J., & Odusanya, O. (2011). *Accelerated solvent extraction of common polybrominated diphenyl ethers from river sediment* (pp. 1–9). Tshwane University of Technology and Department of Water Affairs.

Olutona, G. O., Oyekunle, J. A., Ogunfowokan, A. O., & Fatoki, O. S. (2016). Assessment of polybrominated diphenyl ethers in sediment of Asunle stream of the Obafemi Awolowo University, Ile-Ife, Nigeria. *Environmental Science and Pollution Research, 23*, 21195–21205.

Olutona, G. O., Oyekunle, J. A., Ogunfowokan, A. O., & Fatoki, O. S. (2017). Concentrations of polybrominated diphenyl ethers (PBDEs) in water from Asunle Stream, Ile-Ife, Nigeria. *Toxics, 5*(2), 13.

Onyeche, L. A., & Akankali, J. A. (2013). Determination of some environmental parameters of Anwai–Stream, Niger-Delta, Nigeria. *Research Journal of Agricultural and Environmental Management, 2*(6), 142–149.

Park, J. E., Kang, Y. Y., Kim, W. I., Jeon, T. W., Shin, S. K., Jeong, M. J., & Kim, J. G. (2014). Emission of polybrominated diphenyl ethers (PBDEs) in use of electric/electronic equipment and recycling of e-waste in Korea. *Science of the Total Environment, 470*, 1414–1421.

Pellacani, C., Tagliaferri, S., Caglieri, A., Goldoni, M., Giordano, G., Mutti, A., & Costa, L. G. (2014). Synergistic interactions between PBDEs and PCBs in human neuroblastoma cells. *Environmental Toxicology, 29*(4), 418–427.

Pereira, S., Pinto, A. L., Cortes, R., Fontaínhas-Fernandes, A., Coimbra, A. M., & Monteiro, S. M. (2013). Gill histopathological and oxidative stress evaluation in native fish captured in Portuguese northwestern rivers. *Ecotoxicology and Environmental Safety, 90*, 157–166.

Roberts, R. J., & Rodger, H. D. (2012). The pathophysiology and systematic pathology of teleosts. *Fish Pathology, 12*, 62–143.

Schreck, E., Geret, F., Gontier, L., & Treilhou, M. (2008). Neurotoxic effect and metabolic responses induced by a mixture of six pesticides on the earthworm *Aporrectodea caliginosa nocturna*. *Chemosphere, 71*(10), 1832–1839.

Slaninova, A., Smutna, M., Modra, H., & Svobodova, Z. (2009). REVIEWS oxidative stress in fish induced by pesticides. *Neuroendocrinology Letters, 30*(1), 2.

Ssebugere, P., Sillanpää, M., Wang, P., Li, Y., Kiremire, B. T., Kasozi, G. N., et al. (2014). Polychlorinated dibenzo-p-dioxins, polychlorinated dibenzofurans and polybrominated diphenyl ethers in sediments and fish species from the Murchison Bay of Lake Victoria, Uganda. *Science of the Total Environment, 500*, 1–10.

Su, P. H., Tomy, G. T., Hou, C. Y., Yin, F., Feng, D. L., Ding, Y. S., & Li, Y. F. (2018). Gas/particle partitioning, particle-size distribution of atmospheric polybrominated diphenyl ethers in Southeast Shanghai rural area and size-resolved predicting model. *Chemosphere, 197*, 251–261.

Tang, J., & Zhai, J. X. (2017). Distribution of polybrominated diphenyl ethers in breast milk, cord blood and placentas: A systematic review. *Environmental Science and Pollution Research, 24*, 21548–21573.

Tlustos, C., McHugh, B., Pratt, I., Tyrrell, L., & McGovern, E. (2007). *Investigation into levels of dioxins, furans, polychlorinated biphenyls and brominated flame retardants in fishery produce in Ireland*. Marine Institute.

Ukenye, E. A., Taiwo, I. A., Oguntade, O. R., Oketoki, T. O., & Usman, A. B. (2016). Molecular characterization and genetic diversity assessment of *Tilapia zilli* from some coastal rivers in Nigeria. *African Journal of Biotechnology, 15*(1), 20–28.

Valavanidis, A., Vlahogianni, T., Dassenakis, M., & Scoullos, M. (2006). Molecular biomarkers of oxidative stress in aquatic organisms in relation to toxic environmental pollutants. *Ecotoxicology and Environmental Safety, 64*(2), 178–189.

Van Dyk, J. C., Cochrane, M. J., & Wagenaar, G. M. (2012). Liver histopathology of the sharptooth catfish Clarias gariepinus as a biomarker of aquatic pollution. *Chemosphere, 87*(4), 301–311.

Wang, J., Wang, Y., Shi, Z., Zhou, X., & Sun, Z. (2018). Legacy and novel brominated flame retardants in indoor dust from Beijing, China: Occurrence, human exposure assessment and evidence for PBDEs replacement. *Science of the Total Environment, 618*, 48–59.

Wepener, V., Smit, N., Covaci, A., Dyke, S., & Bervoets, L. (2012). Seasonal bioaccumulation of organohalogens in tigerfish, Hydrocynus vittatus Castelnau, from Lake Pongolapoort, South Africa. *Bulletin of Environmental Contamination and Toxicology, 88*, 277–282.

Winid, B. (2015). Environmental threats of natural water contamination with polybrominated diphenyl ethers (PBDEs). *Polish Journal of Environmental Studies, 24*(1), 47.

Wu, Z., He, C., Han, W., Song, J., Li, H., Zhang, Y., et al. (2020). Exposure pathways, levels and toxicity of polybrominated diphenyl ethers in humans: A review. *Environmental Research, 187*, 109531.

Xin, J., Liu, X., Liu, W., & Zheng, X. (2014). Effects of biochar–BDE-47 interactions on BDE-47 bioaccessibility and biodegradation by Pseudomonas putida TZ-1. *Ecotoxicology and Environmental Safety, 106*, 27–32.

Xu, B., Wu, M., Wang, M., Pan, C., Qiu, W., Tang, L., & Xu, G. (2018). Polybrominated diphenyl ethers (PBDEs) and hydroxylated PBDEs in human serum from Shanghai, China: A study on their presence and correlations. *Environmental Science and Pollution Research, 25*, 3518–3526.

Yang, J., Huang, D., Xue, W., Wei, X., Qin, J., Ou, S., et al. (2018). Multiple-life-stage probabilistic risk assessment for the exposure of Chinese population to PBDEs and risk managements. *Science of the Total Environment, 643*, 1178–1190.

Yu, G., Bu, Q., Cao, Z., Du, X., Xia, J., Wu, M., & Huang, J. (2016). Brominated flame retardants (BFRs): A review on environmental contamination in China. *Chemosphere, 150*, 479–490.

Yu, Y. X., Pang, Y. P., Li, C., Li, J. L., Zhang, X. Y., Yu, Z. Q., et al. (2012). Concentrations and seasonal variations of polybrominated diphenyl ethers (PBDEs) in in-and out-house dust and human daily intake via dust ingestion corrected with bioaccessibility of PBDEs. *Environment International, 42*, 124–131.

Zhang, T., Zhou, X., Xu, A., Tian, Y., Wang, Y., Zhang, Y., & Wang, Z. (2020). Toxicity of polybrominated diphenyl ethers (PBDEs) on rodent male reproductive system: A systematic review and meta-analysis of randomized control studies. *Science of the Total Environment, 720*, 137419.

Zhu, C., Li, Y., Wang, P., Chen, Z., Ren, D., Ssebugere, P., et al. (2015). Polychlorinated biphenyls (PCBs) and polybrominated biphenyl ethers (PBDEs) in environmental samples from Ny-Ålesund and London Island, Svalbard, the Arctic. *Chemosphere, 126*, 40–46.

Chapter 29
The Role of Infection and Inflammation in Prostate Cancer Development

Olayemi Oluseun Akinnola, Abosede Samuel, and Sharon Akinpelu

1 Introduction

Prostate cancer is the most prevalent type of cancer diagnosed in males worldwide and the second leading cause of death from cancer-related illnesses (Rotimi et al., 2019). Age, ethnicity, genetics, diet, and lifestyle have been connected to the cause of prostate cancer (Cavarretta et al., 2017).

Chronic inflammation has also been suggested to promote the initiation and advancement of prostate cancer (Crocetto et al., 2020). The inflammatory microenvironment is considered to play a role in prostate cancer development through a few mechanisms, including the production of reactive molecule species with an established capacity to induce DNA damage. Furthermore, elevated levels of specific cytokines and chemokines in chronically inflamed tissue have been linked to prostate cancer initiation, development, and progression. Androgens are crucial for the growth of advanced prostate cancer as well as the creation of both benign and malignant prostatic cells (Rotimi et al., 2019). Therefore, it has been hypothesized that androgens play a causal role in prostate tumorigenesis.

It has been reported that infections can lead to chronic inflammation. Studies have also reported the occurrence of microorganisms such as bacteria, viruses, and parasites in the prostate (Cavarretta et al., 2017; Feng et al., 2019a; Choto et al.,

O. O. Akinnola · A. Samuel (✉)
Department of Biological Sciences, College of Science and Technology, Covenant University, Ota, Ogun State, Nigeria

Covenant Applied Informatics and Communication Africa Centre of Excellent (CApIC-ACE), Covenant University, Ota, Ogun State, Nigeria
e-mail: abosede.samuelpgs@stu.cu.edu.ng

S. Akinpelu
Department of Biological Sciences, College of Science and Technology, Covenant University, Ota, Ogun State, Nigeria

P. O. Isibor, P. A. Akinduti (eds.), *Harnessing Biotechnology Tools for Product Development*, https://doi.org/10.1007/978-3-031-86002-7_29

2020). Microorganisms produce virulence factors that can induce chronic inflammation resulting in uncontrolled cell proliferation. These microorganisms include sexually transmitted pathogens like *Neisseria gonorrhoeae*, *Chlamydia trachomatis*, *Trichomonas vaginalis*, opportunistic *Enterobacteriaceae-Escherichia coli* or *Pseudomonas* spp. (Lee et al., 2020), *Propionibacterium acnes*, and urogenital schistosomiasis have been specifically identified in PCa samples and associated with prostate inflammation (Davidsson et al., 2016).

Since the beginning of the twentieth century, it has been understood that some infections might contribute to cancer. Several human cancer forms have lately been linked to infections with certain viruses, bacteria, and parasites. These infections are attributed to 15–20% of malignancies worldwide. A portion of the body may have chronic inflammation as a result of some infections. This may result in modifications in the impacted and immune cells, ultimately resulting in cancer. A person's immune system, which typically works to protect the body from several malignancies, might be suppressed by certain illnesses (Brüggemann & Al-Zeer, 2020).

Two extensive meta-analyses revealed a significant association between prostatitis exposure and prostate cancer (Jiang et al., 2013; Perletti et al., 2017). Evidence suggests a correlation between infections, the chronic inflammation induced by these infections, and the subsequent development of cancer. In prostate cancer, extensive regions of prostatic atrophy are often associated with inflammatory cell infiltrates, likely triggered by unidentified stimuli, which may or may not involve prostatic infections (Gong et al., 2008; Manasseh et al., 2009; Namiki et al., 2009; Saha et al., 2010; Porter et al., 2018; Kabiru et al., 2020; Rotimi et al., 2021).

It is hypothesized that prostatic intraepithelial neoplasia (PIN) serves as the primary precursor lesion for prostate cancer, with proliferative inflammatory atrophy (PIA), or inflammation-associated atrophy, potentially acting as an alternative precursor in certain cases (Samanta et al., 2003; Sindhwani & Wilson (2005)). Dysplastic alterations were more noticeable 26 weeks after infection and resembled high-grade dysplasia and prostatic intraepithelial neoplasia. When compared to control prostate glands, the reactive dysplasia-affected prostate glands showed higher staining for oxidative DNA damage, increased proliferation of epithelial cells, and decreased expression of the androgen receptor, GSTP1, p27Kip1, and PTEN. According to these findings, a persistent bacterial infection may have caused a PIN lesion linked to severe chronic inflammation. This prostate lesion resembles the PIN lesion that has previously been reported as occurring near chronic inflammation foci and possibly serving as a precursor to human prostate adenocarcinoma. A separate investigation indicated that a human prostate isolate led to chronic inflammation, marked by an influx of macrophages and Th17 lymphocytes, which in turn accelerated cancer progression in a murine model. Cancer represents the second leading cause of death globally, with its continuously escalating incidence and mortality rates rapidly across Africa (Rotimi et al., 2019). Prostate cancer is the most prevalent non-cutaneous malignancy, affecting approximately one in nine men in Western nations (Wu & Dai, 2017; Ugge et al., 2018; Taha et al., 2020). It stands as the leading cause of cancer-related mortality among males globally. It is a very diverse illness and one of the most challenging cancers in urologic oncology. About

120 forms of cancer are currently known to exist, and more than 30 have been reported in Nigeria (Ferlay et al., 2021). Nigeria recorded 124,815 new cases of cancer in 2020, with breast cancer and prostate cancer accounting for 22.7% and 12.2%, respectively. Prostate cancer constitutes one-third of new cases of cancer among African American men (Crocetto et al., 2021).

Advanced age, African ethnicity, familial history, and genetic mutations are well-established risk factors for PCa. Beyond its role in promoting health and well-being, diet is fundamental to human growth and development. Research indicates that the initiation of prostate cancer in sub-Saharan Africa may be influenced by socioeconomic factors, cultural perspectives, and dietary patterns (Isibor et al., 2021). A clear connection has been established between chronic inflammation and both the onset and advancement of cancers. Inflammation that promotes tumour growth as a potential new hallmark of cancer has been identified. The initial observations that associated inflammation with cancer can be traced back to the nineteenth century when Rudolf Virchow noted that tumours were often found in areas characterized by chronic inflammation and the infiltration of leukocytes within these malignant tissues. Dysplastic alterations were more noticeable 26 weeks after infection and resembled high-grade dysplasia and prostatic intraepithelial neoplasia. When compared to control prostate glands, the reactive dysplasia-affected prostate glands showed higher staining for oxidative DNA damage, increased proliferation of epithelial cells, and decreased expression of the androgen receptor, GSTP1, p27Kip1, and PTEN. According to these findings, a persistent bacterial infection may have caused a PIN lesion linked to severe chronic inflammation. This prostate lesion resembles the PIN lesion that has previously been reported as occurring near chronic inflammation foci and possibly serving as a precursor to human prostate adenocarcinoma. A separate investigation indicated that a human prostate isolate led to chronic inflammation, marked by an influx of macrophages and Th17 lymphocytes, which in turn accelerated cancer progression in a murine model. There are some of the pathways of prostate cancer emergence, mutations in tumour suppressor genes like TP53 and RB1 as well as oncogenes such as MYC that contribute to the dysregulation of cell growth. Additionally, mutations in DNA repair genes, including BRCA1 and BRCA2, increase susceptibility to PCa. Prostate cancer is highly dependent on androgen receptor (AR) signalling for growth and survival. Androgens, such as testosterone and dihydrotestosterone (DHT), activate ARs, promoting cell proliferation. Over time, cancer cells may develop mechanisms to persist despite androgen deprivation, leading to castration-resistant prostate cancer (CRPC). DNA methylation, histone modification, and chromatin remodelling influence gene expression without altering the DNA sequence. Hypermethylation of tumour suppressor genes, such as GSTP1, silences their expression, facilitating cancer progression.

Numerous environmental cancer risk factors and causes are linked to chronic inflammation in some way. Up to 20% of cancer cases are related to chronic illnesses, 30% to cigarette use and airborne contaminants (including asbestos and silica), and 35% to dietary variables (20% of cancer cases are related to obesity).

The prostate gland is located between the bladder and the penis, in front of the rectum (Tonry et al., 2020). The prostate gland is a male accessory sex gland with a walnut-like shape. It has an average length of 3 cm, width of 4 cm, thickness between 2 and 5 cm, and a weight of 20–25 g, the size increasing with age. The prostate contains 30–35 small glands. Its function is to synthesize and secrete proteins and fluid to support the semen. The prostate gland requires androgen (testosterone) to function optimally (Banerjee et al., 2018).

The clinical manifestation of prostate cancer ranges from a slowly progressing disease to a rapidly progressing aggressive disease that may metastasize causing severe pain and death. Patients may be asymptomatic, but symptoms include difficulty in urinating, straining to start, high frequency of urination, and nocturia (Rawla, 2019). PCa may metastasize to the bones and regional lymph nodes where the axial skeleton is the most common location of metastases in individuals who first complain of back discomfort (Nwadi et al., 2021).

Diagnosis of PCa is usually done by assessing the serum PSA 4 ng/mL as well as with the use of a transrectal ultrasound-guided (TRUS) biopsy of the prostate gland (Ikuerowo et al., 2013).

The Gleason system is a prostate adenocarcinoma histologic grading system based on cell architectural patterns (Short et al., 2019). This grading system ranges from Gleason grade 1 (cells that are well-differentiated and have a favourable prognosis) through grade 5 (poorly differentiated cells with the worst prognosis). The Gleason score, which can vary from 2 to 10, denoting the level of aggressiveness of prostate cancer, is calculated by adding the two most common Gleason grades (Chen & Zhou, 2016).

Treatment options include radical prostatectomy, hormone therapy, and radiotherapy which have been associated with high rates of cure (Katongole et al., 2020). PCa mortality has reduced globally due to advancements in diagnostics and treatment strategies. Early detection and interventions have prevented the clinical manifestation of disease in many ways. However, the disease is incurable in advanced stages (Puhr et al., 2016). Metastatic disease is often treated using androgen deprivation therapy. However, many patients experience relapse within 10 years. Castration-resistant PCa may also develop in many patients (Rizzo et al., 2021).

The specific causes of the development of PCa are not well-elucidated. The role of androgens and lack of apoptosis of prostate cells have been described as important processes in the development and progression of PCa. Many approaches have been explored to study the biology of cancer including the tumour genome, epigenome, and microenvironment (Hui & Chen, 2015). Inflammation has gained significant attention as a possible cause of PCa (Crocetto et al., 2020).

2 The Role of Inflammation in Prostate Cancer

Based on some theories, inflammatory trauma may trigger the progression of cancer by inducing cellular stress and repetitive damage of the gene, several studies have concentrated their emphasis on the effects of inflammation in the initiation of PCa (de Bono et al., 2020).

Prostatitis is inflammation of the prostate gland. The NIH classifies prostatitis into four types: acute bacterial prostatitis, chronic bacterial prostatitis, chronic non-bacterial prostatitis, and asymptomatic noninflammatory prostatitis. Bacterial prostatitis has been linked to mostly Gram-negative pathogens, but Gram-positive pathogens and sexually transmitted pathogens have also been associated with the disease (Cai et al., 2011). The causative agents for acute/chronic bacterial prostatitis have been identified as mostly *Enterobacteriaceae* family—*Escherichia coli*, *Klebsiella* spp., *Enterobacter* spp., *Pseudomonas* spp., and Gram-positive bacteria—*Enterococcus* spp. and *Staphylococcus* spp. (Brede & Shoskes, 2011). *Chlamydia trachomatis* has also been identified in patients with prostatitis (Cai et al., 2014). Acute prostatitis results in a transient rise in the expression of prostate-specific antigen (PSA). Prostatic epithelial cells express prostate-specific antigen, a glycoprotein that is encoded by the KLK3 gene in humans and belongs to the kallikrein-related peptidase family. Prostate-specific antigen serves a variety of purposes during copulation and fertilization (Iweala & Ogidigo, 2014). It has been suggested that the presence of bacteria may enhance the proinflammatory response or modify the prostate extracellular environment (Ma et al., 2019). It has been proposed that persistent inflammation of the prostate has a role in the onset and advancement of prostate cancer (Crocetto et al., 2020). Inflammation is a common characteristic of tumour development. Chronic inflammation may result in malignant transformation by inducing repeated cycles of cellular damage, cell replication, and tissue repair (Puhr et al., 2016). Inflammation is proposed to contribute to cancer development by inducing cellular stress and consequent repetitive genomic damage (Cai et al., 2019).

Chronic inflammation is characteristic of benign and metastatic prostatic growth although the source of the inflammation remains poorly characterized (Grivennikov et al., 2010). Factors that may contribute to inflammation in the prostate include urine reflux, diet, hormonal changes, corpora amylacea, and prostatic calculi, leading to physical trauma (Hyun, 2018). Corpora amylacea and prostatic calculi are made up of inflammatory proteins found especially in neutrophil granules, e.g., lactoferrin and calprotectin which are common features in the adult prostate. It has been suggested they represent past infections and inflammation that occurred in the prostate.

It is unclear whether prostatitis increases the risk of PCa. Perletti et al. (2017) discovered a connection between the onset of PCa and a history of prostatitis in 5821 men >65 years old (Daniels et al., 2005; Jiang et al., 2013). A substantial link between exposure to prostatitis and PCa in a comprehensive meta-analysis was discovered by Daniels et al. (2005); a possible area of future research.

An increased incidence of inflammation in African American men has been observed in comparison to European-American men (Esatham et al., 2010; Rybicki et al., 2016). This is of interest because the aggressiveness of PCa in men of African ancestry has been well established. People who use non-steroidal anti-inflammatory medications (NSAIDs) regularly have been observed to have a lower risk of developing PCa (Salinas et al., 2010; Doat et al., 2017). An increased risk of PCa has also been linked to the presence of other inflammatory conditions in men such as osteoporosis, diabetes mellitus, and arthritis (Dzutsev et al., 2015).

It is well known that there is a link between persistent inflammation and cancer, and that chronic inflammation is a prominent factor in 20% of occurrences of human cancer (Siegel et al., 2017). Numerous contributing variables, including dietary practices, hormonal impacts, urine reflux, physical injury, and viral or bacterial infection (STIs) can cause inflammation in the prostate gland (Zhang et al., 2020). However, acute inflammatory responses are crucial for eradicating pathogenic infections, but prolonged and unregulated inflammatory cells could damage cells and tissues (Sutcliffe et al., 2006). The risk of getting cancers including prostate cancer and the onset of tumours has been linked to chronic inflammation (Crocetto et al., 2021). The immune system of the body includes inflammation and through this mechanism, the immune system recognizes harmful and foreign stimuli, rejects them, and starts the healing process; both acute and enduring irritation and inflammation could occur.

Acute inflammation may result from aggressive chemicals, pathogenic invasion, or trauma-induced tissue injury. Acute pneumonia or cellulitis are two conditions where symptoms can begin unexpectedly, get worse swiftly, and persist for several days. Subacute inflammation can linger for 2–6 weeks, in contrast to acute and chronic inflammation. Long-lasting inflammation that persists for several months or years is referred to as chronic inflammation. The cause of the harm and the body's capacity to reverse and repair the damage frequently affect the intensity and outcomes of chronic inflammation.

3 Sexually Transmitted Infections and Prostate Cancer

According to certain research, sexually transmitted infections enhance the risk of PCa by causing inflammation in the prostate (Nwadi et al., 2021). The prostate-specific antigen (PSA) increase has also been noted during infections with sexually transmitted microorganisms suggesting an association of the prostate with STIs (Sutcliffe et al., 2006). It has been reported that men with a history of gonorrhoea or syphilis infections had a 60% greater risk of having PCa (Sfanos et al., 2009).

Viruses are estimated to cause about 10–15% of all human cancers. Infection with DNA viruses and tumour development has been well-established in many cancer types. These viruses are generally referred to as tumour viruses. They can insert themselves into the genome of the host causing the target cell to express the viral genes. This promotes cell growth, proliferation, and apoptosis prevention. Viruses have been shown to cause cell transformation by interfering with the interferon signalling pathway. A classic example is the increased risk of hepatocellular carcinoma due to chronic hepatitis B infection (Chung et al., 2013). Some studies have implicated viruses in the chronic inflammation of the prostate and the promotion of carcinogenesis (Klein & Silverman, 2008; Aghakhani et al., 2011).

Viruses that have been associated with prostate infections or prostate cancer include human papillomavirus (HPV), polyomaviruses and herpesviruses, human cytomegalovirus (HCMV), and Epstein-Barr virus (EBV).

Because of its physical location, the prostate can be a target for human papillomavirus (HPV) infections (Aghakhani et al., 2011). The 8000 bp genome of human papillomaviruses (HPVs) is contained in a tiny, encapsulated DNA virus. One prevalent sexually transmitted virus is HPV. High-risk strains of HPV have been linked to cancers of the cervix, vulva, penis, and anus (Chaturvedi, 2010). The oncoproteins E6 and E7 play important roles in oncogenesis. The E6 protein interacts with the tumour suppressor gene p53, leading to its degradation, while E6 inactivates the retinoblastoma protein (Münger et al., 2004). In high-risk HPVs, they cause the release of inflammatory cytokines and chemokines (Nahand et al., 2021).

Many studies have explored a possible connection between HPV infection and the development of PCa; however, the results are inconclusive. Some studies have reported an association of PCa with HPV infection (Leiros et al., 2005; Martinez-Fierro et al., 2010), while others have found no link (Bergh et al., 2007; Chen et al., 2011; Aydin et al., 2017) and investigated the presence of HPV in prostate tissues of 60 PCa and 36 BPH patients using real-time PCR. HPV DNA was found in one (1.7%) of PCa tissues, while the BPH tissues were HPV-negative. Serth et al. (1999) detected HPV16 in 21% of PCa tissues and 3% of BPH tissues. Leiros et al. (2005) reported a 41.5% prevalence of HPV in PCa and 0% in BPH tissues. HPV DNA was found to be present in 20% of PCa tissues and 5.3% of BPH tissues in a study by Martinez-Fierro et al. (2010). Banerjee et al. (2018) also found human papillomavirus (HPV16 and HPV18) in prostate cancer tissues.

Herpesviruses can establish latency in the body for extended periods. However, they may reactivate producing different symptoms and causing different malignancies. In healthy people, human cytomegalovirus (HCMV) infection is typically asymptomatic but can be fatal in immunocompromised patients such as infants, HIV-infected patients, and organ transplant recipients. The presence of human cytomegalovirus was reported in prostate cancer tissues (Banerjee et al., 2018). It has been reported that HCMV may be linked to prostate cancer.

The Epstein-Barr virus (EBV) causes infectious mononucleosis and some lymphomas. In a study by Ali and Al-Alwany (2014), EBV IgG antibodies were detected in 56.16% of PCa patients and 30.84% of patients with BPH. Similar results were obtained by

Mycoplasmas are regarded as commensal members of the human urogenital system which may be opportunistic pathogens (Barykova et al., 2011). Studies have demonstrated the ability of *Mycoplasma* spp. to induce genetic instability and malignant transformation in cell cultures. Infection with *Mycoplasma* increased the invasiveness and metastasis of human tumour cells in vitro and in vivo

(Martinez-Fierro et al., 2010). A common characteristic of tumour cells is the suppression of p53 activity and activation of NF-κB, and *Mycoplasma* can induce these responses (Logunov et al., 2008). Therefore, infection of the prostate with *Mycoplasma* may be involved in chronic prostatic inflammation and subsequent PCa development.

Barykova et al. (2011) screened prostate tissues of 123 Russian PCa patients for the presence of *M. hominis*, *M. genitalium*, and *U. urealitycum* using PCR and quantitative real-time PCR. *Mycoplasma hominis* had a higher prevalence of 15.2%, while the prevalence of *M. genitalium* and *U. urealitycum* was 5.6% and 0.8%, respectively. *Mycoplasma hominis* was found in 37.4% of patients suspected of PCa and 0% of healthy controls using qPCR. The presence of *M. hominis* was associated with the development of PCa.

Miyake et al. (2019) also performed PCR on 33 BPH and 45 PCa tissues to detect the presence of sexually transmitted infections, including *Mycoplasma hominis*, *Mycoplasma genitalium*, *Ureaplasma urealyticum*, *Neisseria gonorrhoeae*, *Chlamydia trachomatis*, and HPV16 and HPV18. *Mycoplasma genitalium* was the only species linked to PCa.

The presence of *Trichomonas vaginalis* has been significantly associated with PCa development (Kim et al., 2019; Langston et al., 2019). *Trichomonas vaginalis* may create a conducive microenvironment for cell proliferation, inflammation, and invasiveness (by initiating epithelial-mesenchymal transition) in the prostate gland (Twu et al., 2014) *Trichomonas vaginalis* was reported to secrete a protein-like human macrophage migration inhibitory factor which is found at high levels in PCa patients (Feng et al. 2019b; Göbel et al., 2021).

Propionibacterium acnes has a well-established function in human health and illness (McDowell et al., 2005). *Propionibacterium acnes* is a Gram-positive bacterium that produces propionic acid from fermentation, from which its name is derived (Achermann et al., 2014). It forms part of the normal human skin flora although it has been recognized to contribute to acne vulgaris. It protects the skin by lowering the pH to prevent colonization by pathogens. *Propionibacterium acnes* can also be found in the oral cavity, intestinal tract, and conjunctiva (Mollerup et al., 2016).

However, it is an opportunistic pathogen, causing diseases in immunocompromised patients or patients with injury or trauma. It has been implicated in implant-related infections, dental infections, shunt-related central nervous systems, and prosthetic valve endocarditis (Perry & Lambert, 2011). The virulence properties of *P. acnes* include haemolytic activity, cytotoxicity, biofilm formation, and immune evasion (Fehri et al., 2011).

Propionibacterium acnes's existence in PCa patients has been highlighted in numerous studies (Shannon et al., 2006; Bae et al., 2014; Davidsson et al., 2016). *Propionibacterium acnes* has also been found in significant numbers in prostatitis, BPH, and PCa tissue samples (Alexeyev et al., 2007; Sfanos et al., 2009). Chronic *P. acnes* infection was found in prostate samples obtained up to 6 years apart. Chen et al. (2011) detected *Propionibacterium acnes* in all PCa samples tested. Cavarretta et al. (2017) also found a significant abundance of *P. acnes* in tumour (60%), peri-tumour (60%), and non-tumour regions (49%) of prostate tissues. *Propionibacterium acnes* was found in 95% of samples examined by Yow et al. (2017). *Propionibacterium acnes* was also detected in 100% of samples analysed in two studies (Feng et al., 2019a).

The presence of *P. acnes* in prostate tissues has also been linked to evidence of inflammation proposing a relationship with the development of cancer (Cohen et al., 2005; Fehri et al., 2011; Olsson et al., 2012) and a higher level of inflammation was found in samples infected with *P. acnes* than uninfected *P. acnes* samples. This shows the involvement of *P. acnes* in prostatic inflammation.

Propionibacterium acnes induced a robust inflammatory response and cell transformation in a prostate epithelial cell line which could contribute to cancer initiation or progression (Perry & Lambert, 2011). In a murine model, Olsson et al. (2012) demonstrated that *P. acnes* can induce a chronic infection in previously healthy prostate tissue as well as induce persistent histological inflammation in the infected tissue. *Propionibacterium acnes* has been demonstrated to promote the release of cytokines and chemokines such as interleukin 6 (IL-6) and interleukin 8 (IL-8) in addition to increased cell proliferation and reduction of E-cadherin [87]. These are pro-inflammatory cytokines that have been linked to the development of cancer including PCa (Araki et al., 2007).

Human schistosomiasis is a neglected, tropical disease caused by parasitic flatworms called Schistosoma species currently affecting over 200 million people worldwide. It is only second to malaria in terms of public health relevance with a mortality rate of about 280,000 people in Africa (Akinneye et al., 2018).

Their lifecycle begins in infected water snails from which cercarial forms of the parasite are released and penetrate the human skin. These parasites reproduce sexually in the human circulatory system (Akinneye et al., 2018). The adult flatworms inhabit the human organs and produce eggs in the intestines and bladder (Adebayo et al., 2017). *Schistosoma mansoni* and *Schistosoma japonicum* inhabit the gut venules, while *Schistosoma haematobium* resides in the genitourinary system (Akinneye et al., 2018). *Schistosoma haematobium* is responsible for urogenital manifestations such as granulomatous inflammation, urinary tract infections, and bladder cancer. *Schistosoma haematobium* causes urinary schistosomiasis and is the most prevalent in Africa (Aula et al., 2021).

Clinical manifestations of urinary schistosomiasis include hematuria, dysuria, and suprapubic pain. *Schistosoma haematobium* has been implicated in bladder cancer. Risks associated with this endemic disease include anaemia, nutritional deficiencies, stunted growth, delayed puberty, and bladder cancer (Akinneye et al., 2018).

Socioeconomic factors that predispose to schistosomiasis include poor hygiene and sanitation, poor housing, rural dwellings, and poverty (Blaheta et al., 2006). People who engage in agricultural and other human activities near freshwater sources such as bathing, fishing, and swimming where the water snails are present are at risk for the infection. This results in continual parasite exposure and significant re-infection rates (Mazigo et al., 2010). Diagnosis of this parasite is usually done by microscopy analysis to check for schistosome eggs in urine or faeces. Praziquantel, an anti-helminth medication, is used to treat schistosomiasis (Akinneye et al., 2018).

Prostatic schistosomiasis has been associated with prostate cancer with the causative agent being *S. haematobium* which causes urinary schistosomiasis (Mazigo et al., 2010). The clinical manifestations of Schistosomiasis are like those of prostate adenocarcinoma-nocturia, haematuria, and pain during urination. Cohen et al. (1995) reported advanced prostate cancer in three patients aged 27–29 years old with no family history of prostate cancer. The presence of multiple eggs of *S. haematobium* (viable and calcified) was associated with prostate cancer. A 62-year-old Angolan presenting with symptoms such as urine dripping and poor streaming had elevated PSA levels and was discovered to have prostate adenocarcinoma, Gleason score 8, and scattered Schistosoma eggs were found in the prostate (Metrogos et al., 2017). The coexistence of prostatic adenocarcinoma and the presence of *Schistosoma haematobium* eggs in the prostate was also reported in a 66-year-old and 70-year-old man in Tanzania and Jos, Nigeria, respectively (Lodhia et al., 2020).

Gomes et al. (2014) reported that a 51-year-old male presented with prostate enlargement and elevated PSA levels. Biopsy revealed the presence of *S. mansoni* eggs in the prostate with moderate severity of infection. Gomes et al. (2014) and Basilio-De-Oliveira et al. (2002) detected the presence of various *S. mansoni* eggs in 68-year-old Brazilian men discovered to have prostate adenocarcinoma after biopsy. This indicates a possible relationship between these two conditions. Bacelar et al. (2007) also reported on a 47-year-old Brazilian patient who underwent radical prostatectomy and was diagnosed with prostate cancer associated with *S. mansoni. Schistosoma mansoni* is highly prevalent in Latin America.

High PSA levels have been observed in studies reporting on schistosomiasis and PCa. Therefore, the PSA levels could be due to inflammation in the prostate due to this parasitic infection or a result of PCa. Aula et al. (2021) reported that having schistosome eggs caused individuals to develop bladder cancer at a younger age earlier than those who didn't have the eggs. It can be deduced that individuals living in endemic regions for schistosomiasis should be screened for prostate tumours at a younger age. Evidence of schistosome eggs in the prostate and PCa suggests a need to investigate the association between Schistosoma infection and the development of PCa. The presence of schistosomiasis infection has also been correlated with infertility in men (Aliyu et al., 2021).

4 Conclusion

Prolonged inflammation encourages the start and progression of cancer in humans by modifying the extracellular matrix, the tumour microenvironment, and the epithelial-mesenchymal transition. Recent reviews and meta-analyses have linked a history of clinical chronic prostatitis to an elevated risk of PCa in the general population. Numerous factors, including bacteria that cause prostatitis and STDs, an imbalance in oestrogen levels, physical injury, urine reflux into the prostate gland, and elements in the environment, such as food can result in prostate inflammation. The gold standard for diagnosing prostate inflammation is a prostate biopsy, but when prostate inflammation is suspected, other factors are, in standard clinical practice, it is possible to employ variables like clinical measures (prostatic calcifications, symptom severity, and response to therapy), cytokine laboratory biomarkers, and/or clinical information (cytokines). These elements show how prostatic inflammation and inflammatory mediators affect the development, spread, and growth of tumours.

Every aspect of tumorigenesis, progression, and therapeutic response can be impacted by inflammation. Identified unique microbial signatures such as high abundance of *Pseudomonas*, *Streptococcus*, and *Escherichia–Shigella* spp. associated with prostate cancer development, and this may serve as potential indicative biomarker and influence the development of microbiome-targeted therapeutic options by which cancer and inflammation interact. Prostate cancer remains a disease of interest, particularly in men of African descent. This study reveals the prevalence of pro-inflammatory bacteria and uropathogens in the urinary tract of Nigerian men with prostate cancer. Uropathogenic bacteria such as *Pseudomonas* and *Escherichia–Shigella* have been identified as indicative biomarkers of prostate cancer. These inflammatory genotypes and their expression levels can potentially serve as predictive markers for the risk of prostate cancer.

References

Achermann, Y., Goldstein, E. J., Coenye, T., & Shirtliff, M. E. (2014). *Propionibacterium acnes*: From commensal to opportunistic biofilm-associated implant pathogen. *Clinical Microbiology Reviews, 27*(3), 419–440. https://doi.org/10.1128/CMR.00092-13

Adebayo, A. S., Survayanshi, M., Bhute, S., Agunloye, A. M., Isokpehi, R. D., Anumudu, C. I., & Shouche, Y. S. (2017). The microbiome in urogenital schistosomiasis and induced bladder pathologies. *PLoS Neglected Tropical Diseases, 11*(8), e0005826. https://doi.org/10.1371/journal.pntd.0005826

Aghakhani, A., Hamkar, R., Parvin, M., Ghavami, N., Nadri, M., Pakfetrat, A., Banifazi, M., Eslamifar, A., Izadi, N., Jam, S., & Ramezani, A. (2011). The role of human papillomavirus infection in prostate carcinoma. *Scandinavian Journal of Infectious Diseases, 43*, 64–69.

Akinneye, J. O., Fasidi, M. M., Afolabi, O. J., & Adesina, F. P. (2018). Prevalence of urinary schistosomiasis among secondary school students in Ifedore local government, Ondo State, Nigeria. *International Journal of Tropical Diseases, 1*, 004.

Alexeyev, O. A., Marklund, I., Shannon, B., Golovleva, I., Olsson, J., Andersson, C., Eriksson, I., Cohen, R., & Elgh, F. (2007). Direct visualization of *Propionibacterium acnes* in prostate tissue by multicolor fluorescent in situ hybridization assay. *Journal of Clinical Microbiology, 45*(11), 3721–3728.

Ali, S. H. M., & Al-Alwany, M. (2014). Molecular localization of Epstein Barr virus and Rb tumor suppressor gene expression in tissues from prostatic adenocarcinoma and benign prostatic hyperplasia. *Iraqi Journal of Biotechnology, 13*(2), 161–172.

Aliyu, M., Ibrahim, K. H., & Abubakar, M. A. (2021). Sexual dysfunction and infertility amongst spouses in Adamawa state, Nigeria. *American Journal of Health Research, 9*(1), 1–8. https://doi.org/10.11648/j.ajhr.20210901.11

Araki, S., Omori, Y., Lyn, D., Singh, R. K., Meinbach, D. M., Sandman, Y., Lokeshwar, V. B., & Lokeshwar, B. L. (2007). Interleukin-8 is a molecular determinant of androgen independence and progression in prostate cancer. *Cancer Research, 67*(14), 6854–6862.

Aula, O. P., McManus, D. P., Jones, M. K., & Gordon, C. A. (2021). Schistosomiasis with a focus on Africa. *Tropical Medicine and Infectious Disease, 6*, 109.
Aydin, M., Bozkurt, A., Cikman, A., Gulhan, B., Karabakan, M., Gokce, A., Alper, M., & Kara, M. (2017). Lack of evidence of HPV etiology of prostate cancer following radical surgery and higher frequency of the Arg/Pro genotype in Turkish men with prostate cancer. *International Braz J Urol, 43*(1), 36.
Bacelar, A., Castro, L. G. M., Queiroz, A. C., & Café, E. (2007). Association between prostate cancer and schistosomiasis in young patients: Case report and literature review. *The Brazilian Journal of Infectious Diseases, 11*, 520.
Bae, Y., Ito, T., Iida, T., Uchida, K., Sekine, M., Nakajima, Y., Kumagai, J., Yokoyama, T., Kawachi, H., Akashi, T., & Eishi, Y. (2014). Intracellular *Propionibacterium acnes* infection in glandular epithelium and stromal macrophages of the prostate with or without cancer. *PLoS One, 9*, e90324.
Banerjee, P. P., Banerjee, S., Brown, T. R., & Zirkin, B. R. (2018). Androgen action in prostate function and disease. *American Journal of Clinical and Experimental Urology, 6*(2), 62–77.
Barykova, Y. A., Logunov, D. Y., Shmarov, M. M., Vinarov, A. Z., Fiev, D. N., Vinarova, N. A., Rakovskaya, I. V., Baker, P. S., Shyshynova, I., Stephenson, A. J., Klein, E. A., Naroditsky, B. S., Gintsburg, A. L., & Gudkov, A. V. (2011). Association of *Mycoplasma hominis* infection with prostate cancer. *Oncotarget, 2*, 289–297.
Basilio-De-Oliveira, C. A., Aquino, A., Simon, E. F., & Eyer-Silva, W. A. (2002). Concomitant prostatic schistosomiasis and adenocarcinoma: Case report and review. *The Brazilian Journal of Infectious Diseases, 6*, 45–49.
Bergh, J., Marklund, I., Gustavsson, C., Wiklund, F., Grönberg, H., Allard, A., Alexeyev, O., & Elgh, F. (2007). No link between viral findings in the prostate and subsequent cancer development. *British Journal of Cancer, 96*, 137–139.
Blaheta, R. A., Weich, E., Marian, D., Bereiter-Hahn, J., Jones, J., Jonas, D., Michaelis, M., Doerr, H. W., & Cinatl, J., Jr. (2006). Human cytomegalovirus infection alters PC3 prostate carcinoma cell adhesion to endothelial cells and extracellular matrix. *Neoplasia, 8*(10), 807–816. https://doi.org/10.1593/neo.06379
Brede, C. M., & Shoskes, D. A. (2011). The etiology and management of acuteprostatitis. *Nature Reviews. Urology, 8*, 207–212.
Brüggemann, H., & Al-Zeer, M. A. (2020). Bacterial signatures and their inflammatory potentials associated with prostate cancer. *APMIS: Acta Pathologica, Microbiologica, et Immunologica Scandinavica, 128*, 80–91.
Cai, T., Mazzoli, S., Meacci, F., Boddi, V., Mondaini, N., Malossini, G., & Bartoletti, R. (2011). Epidemiological features and resistance pattern in uropathogens isolated from chronic bacterial prostatitis. *Journal of Microbiology, 49*, 448–454.
Cai, T., Pisano, F., Magri, V., Verze, P., Mondaini, N., D'Elia, C., Malossini, G., Mazzoli, S., Perletti, G., Gontero, P., Mirone, V., & Bartoletti, R. (2014). *Chlamydia trachomatis* infection is related to premature ejaculation in chronic prostatitis patients: Results from a cross-sectional study. *The Journal of Sexual Medicine, 11*, 3085–3092.
Cai, T., Santi, R., Tamanini, I., Galli, I. C., Perletti, G., Johansen, T. B., & Nesi, G. (2019). Current knowledge of the potential links between inflammation and prostate cancer. *International Journal of Molecular Sciences, 20*, 3833. https://doi.org/10.3390/ijms20153833
Cavarretta, I., Ferrarese, R., Cazzaniga, W., Saita, D., Lucianò, R., Ceresola, E. R., Locatelli, I., Visconti, L., Lavorgna, G., Briganti, A., Nebuloni, M., Doglioni, C., Clement, M., Montorsi, F., Canducci, F., & Salonia, A. (2017). The microbiome of the prostate tumor microenvironment. *European Urology, 72*, 625–631.
Chaturvedi, A. K. (2010). Beyond cervical cancer: Burden of other HPV-related cancers among men and women. *The Journal of Adolescent Health, 46*, S20–S26.
Chen, A. C., Waterboer, T., Keleher, A., Morrison, B., Jindal, S., McMillan, D., Nicol, D., Gardiner, R. A., McMillan, N. A., & Antonsson, A. (2011). Human papillomavirus in benign

prostatic hyperplasia and prostatic adenocarcinoma patients. *Pathology Oncology Research, 17*, 613–617.

Chen, N., & Zhou, Q. (2016). The evolving Gleason grading system. *Chinese Journal of Cancer Research, 28*(1), 58–64. https://doi.org/10.3978/j.issn.1000-9604.2016.02.04

Choto, E. T., Mduluza, T., Sibanda, E. N., Mutapi, F., & Chimbari, M. J. (2020). Possible association and co-existence of schistosome infection and prostate cancer: A systematic review. *African Journal of Reproductive Health, 24*(4), 185–197. https://doi.org/10.29063/ajrh2020/v24i4.19

Chung, S. D., Lin, Y. K., Huang, C. C., & Lin, H. C. (2013). Increased risk of prostate cancer following sexually transmitted infection in an Asian population. *Epidemiology and Infection, 141*, 2663–2670.

Cohen, R. J., Edgar, S. G., & Cooper, K. (1995). Schistosomiasis and prostate cancer. *Pathology, 27*(2), 115–116.

Cohen, R. J., Shannon, B. A., McNeal, J. E., Shannon, T., & Garett, K. L. (2005). *Propionibacterium acnes* associated with inflammation in radical prostatectomy specimens: A possible link to cancer evolution? *The Journal of Urology, 173*, 1969–1974.

Crocetto, F., Arcaniolo, D., Napolitano, L., Barone, B., La Rocca, R., Capece, M., Caputo, V. F., Imbimbo, C., De Sio, M., Calace, F. P., & Manfredi, C. (2021). Impact of sexual activity on the risk of male genital tumors: A systematic review of the literature. *International Journal of Environment Responds Public Health, 18*, 8500.

Crocetto, F., Boccellino, M., Barone, B., Di Zazzo, E., Sciarra, A., Galasso, G., Settenbre, G., Quagliuolo, L., Imbimbo, C., Boffo, S., Angelillo, I. F., & Di Domenico, M. (2020). The crosstalk between prostate cancer and microbiota inflammation: Nutraceutical products are useful to balance this interplay? *Nutrients, 12*, 2648. https://doi.org/10.3390/nu12092648

Daniels, A. N., Ewing, S. K., Zmuda, J. M., Wilt, T. J., & Bauer, D. C. (2005). Correlates and prevalence of prostatitis in a large community-based cohort of older men. *Urology, 66*, 964–970.

Davidsson, S., Mölling, P., Rider, J. R., Unemo, M., Karlsson, M. G., Carlsson, J., Andersson, S. O., Elgh, F., Söderquis, B., & Andrén, O. (2016). Frequency and typing of *Propionibacterium acnes* in prostate tissue obtained from men with and without prostate cancer. *Infectious Agents and Cancer, 11*, 26. https://doi.org/10.1186/s13027-016-0074-9

de Bono, J. S., Guo, C., Gurel, B., De Marzo, A. M., Sfanos, K. S., Mani, R. S., Gil, J., Drake, C. G., & Alimonti, A. (2020). Prostate carcinogenesis: Inflammatory storms. *Nature Review Cancer, 20*, 455–469.

Doat, S., Cénée, S., Trétarre, B., Rebillard, X., Lamy, P. J., Bringer, J. P., Iborra, F., Murez, T., Sanchez, M., & Menegaux, F. (2017). Nonsteroidal anti-inflammatory drugs (NSAIDs) and prostate cancer risk: Results from the EPICAP study. *Cancer Medicine, 6*(10), 2461–2470.

Dzutsev, A., Goldszmid, R. S., Viaud, S., Zitvogel, L., & Trinchieri, G. (2015). The role of the microbiota in inflammation, carcinogenesis, and cancer therapy. *European Journal of Immunology, 45*, 17–31.

Fehri, F. L., Mak, T. N., Laube, B., Brinkmann, V., Ogilvie, L. A., Mollenkopf, H., Lein, M., Schmid, T., Meyer, T. F., & Brüggemann, H. (2011). Prevalence of *Propionibacterium acnes* in diseased prostates and its inflammatory and transforming activity on prostate epithelial cells. *International Journal of Medical Microbiology, 301*, 69–78.

Feng, Y., Jaratlerdsiri, W., Patrick, S. M., Lyons, R. J., Haynes, A.-M., Collins, C. C., Stricker, P. D., Bornman, R., & Hayes, V. M. (2019a). Metagenomic analysis reveals a rich bacterial content in high-risk prostate tumors from African men. *The Prostate, 79*, 1731–1738. https://doi.org/10.1002/pros.23897

Feng, Y., Ramnarine, V. R., Bell, R., Volik, S., Davicioni, E., Hayes, V. M., Ren, S., & Collins, C. C. (2019b). Metagenomic and metatranscriptomic analysis of human prostate microbiota from patients with prostate cancer. *BMC Genomics, 20*, 146.

Ferlay, J., Colombet, M., Soerjomataram, I., Parkin, D. M., Piñeros, M., Znaor, A., & Bray, F. (2021). Cancer statistics for the year 2020: An overview. *International Journal of Cancer, 149*, 778. https://doi.org/10.1002/ijc.33588

Göbel, A., Dell'Endice, S., Jaschke, N., Pählig, S., Shahid, A., Hofbauer, L. C., & Rachner, T. D. (2021). The role of inflammation in breast and prostate cancer metastasis to bone. *International Journal Molecular Science, 22*, 5078.
Gomes, E. C. S., Leal Neto, O. B., Oliveira, F. J., Campos, J. V., Souza-Santos, R., & Barbosa, C. S. (2014). Risk analysis for occurrences of schistosomiasis in the coastal area of Porto de Galinhas, Pernambuco, Brazil. *BMC Infectious Diseases, 14*, 101.
Gong, M., Meng, L., Jiang, B., Zhang, J., Yang, H., Wu, J., & Shou, C. (2008). p37 from *Mycoplasma hyorhinis* promotes cancer cell invasiveness and metastasis through activation of MMP-2 and followed by phosphorylation of EGFR. *Molecular Cancer Therapeutics, 7*, 530–537.
Grivennikov, S. I., Greten, F. R., & Karin, M. (2010). Immunity, inflammation and cancer. *Cell, 140*, 883–899.
Hui, L., & Chen, Y. (2015). Tumor microenvironment: Sanctuary of the devil. *Cancer Letters, 368*, 7–13. https://doi.org/10.1016/j.canlet.2015.07.039
Hyun, J. S. (2018). Clinical significance of prostatic calculi: A review. *The World Journal of Men's Health, 36*(1), 15–21. https://doi.org/10.5534/wjmh.17018
Ikuerowo, S. O., Omisanjo, O. A., Bioku, M. J., Ajala, M. O., Mordi, V. P. N., & Esho, J. O. (2013). Prevalence and characteristics of prostate cancer among participants of a community-based screening in Nigeria using serum prostate specific antigen and digital rectal examination. *Pan African Medical Journal, 15*, 129. https://doi.org/10.11604/pamj.2013.15.129.2489
Isibor, P., Akinduti, P., Aworunse, O., Oyewale, J., Oshamika, O., Ugboko, H., et al. (2021). Significance of African diets in biotherapeutic modulation of the gut microbiome. *Bioinformatics and Biology Insights, 15*, 117793222110126. https://doi.org/10.1177/11779322211012697
Iweala, E. E. J., & Ogidigo, J. O. (2014). Prostate specific antigen, antioxidant and hematological parameters in prostatic rats fed *Solanum macrocarpon* L. leaves. *Asian Journal of Biological Sciences, 8*(1), 30–41. https://doi.org/10.3923/ajbs.2015.30.41
Jiang, J., Li, J., Zhang, Y., Zhu, H., Liu, J., & Pumill, C. (2013). The role of prostatitis in prostate cancer: Meta-analysis. *PLoS One, 8*, e85179.
Kabiru, A., Wemimo, R. M., Umar, M., Taiwo, A. M., Abiodun, A. E., & Salihu, A. (2020). Spectrum of histopathological diagnosis of schistosomiasis as a systemic parasitic infection in the North-West Nigeria. *American Journal of Laboratory Medicine, 5*(6), 180–184. https://doi.org/10.11648/j.ajlm.20200506.15
Katongole, P., Sande, O. J., Joloba, M., Reynolds, S. J., & Niyonzima, N. (2020). The human microbiome and its link in prostate cancer risk and pathogenesis. *Infectious Agents and Cancer, 15*, 53.
Kim, J., Moon, H., Kim, S., Hwang, S., Ryu, S., & Park, S. (2019). Comparison of seropositivity to *Trichomonas vaginalis* between men with prostatic tumor and normal men. *The Korean Journal of Parasitology, 57*, 21–25.
Klein, E. A., & Silverman, R. (2008). Inflammation, infection, and prostate cancer. *Current Opinion in Urology, 18*(3), 315–319. https://doi.org/10.1097/MOU.0b013e3282f9b3b7
Langston, M. E., Bhalla, A., Alderete, J. F., Nevin, R., Pakpahan, R., Hansen, J., Elliott, D., De Marzo, A. M., Gaydos, C. A., Isaacs, W. B., Nelson, W. G., Sokoli, L. J., Zenilman, J. M., Platz, E. A., & Sutcliffe, S. (2019). *Trichomonas vaginalis* infection and prostate-specific antigen concentration: Insights into prostate involvement and prostate disease risk. *Prostate, 79*, 1622–1628.
Lee, K. W., Song, H. Y., & Kim, Y. H. (2020). The microbiome in urological diseases. *Investigative and Clinical Urology, 61*, 338–348. https://doi.org/10.4111/icu.2020.61.4.338
Leiros, G. J., Galliano, S. R., Sember, M. E., Kahn, T., Schwarz, E., & Eiguchi, K. (2005). Detection of human papillomavirus DNA and p53 codon 72 polymorphisms in prostate carcinomas of patients from Argentina. *BMC Urology, 5*, 15.
Lodhia, J., Mremi, A., Pyuz, J. J., Bartholomeo, N., & Herman, A. M. (2020). Schistosomiasis and cancer: Experience from a zonal hospital in Tanzania and opportunities for prevention. *Journal of Surgical Case Reports, 5*(44), 1–5.

Logunov, D. Y., Scheblyakov, D. V., Zubkova, O. V., Shmarov, M. M., Rakovskaya, I. V., Gurova, K. V., Tararova, N. D., Burdelya, L. G., Naroditsky, B. S., Ginzburg, A. L., & Gudkov, A. V. (2008). *Mycoplasma* infection suppresses p53, activates NF-kappaB and cooperates with oncogenic Ras in rodent fibroblast transformation. *Oncogene, 27*, 4521–4531.

Ma, X., Chi, C., Fan, L., Dong, B., Shao, X., Xie, S., Li, M., & Xue, W. (2019). The microbiome of prostate fluid is associated with prostate cancer. *Frontiers in Microbiology, 10*, 1664. https://doi.org/10.3389/fmicb.2019.01664

Manasseh, A. N., Echejoh, G. O., Tanko, M. N., Silas, O. O., Dakum, N. K., & Mandong, B. M. (2009). Prostatic adenocarcinoma coexisting with schistosomiasis. A case report and review of literature. *International Journal of Medicine and Medical Sciences, 1*(3), 33–37.

Martinez-Fierro, M. L., Leach, R. J., Gomez-Guerra, L. S., Garza-Guajardo, R., Johnson-Pais, T., Beuten, J., Morales-Rodriguez, I. B., Hernandez-Ordonez, M. A., Calderon-Cardenas, G., Ortiz-Lopez, R., Rivas-Estilla, A. M., Ancer-Rodriguez, J., & Rojas-Martinez, A. (2010). Identification of viral infections in the prostate and evaluation of their association with cancer. *BMC Cancer, 10*, 326.

Mazigo, H. D., Zinga, M., Heukelbach, J., & Rambau, P. (2010). Case series of adenocarcinoma of the prostate associated with *Schistosoma haematobium* infection in Tanzania. *Journal of Global Infectious Diseases, 2*(3), 307–309.

McDowell, A., Valanne, S., Ramage, G., Tunney, M. M., Glenn, J. V., McLorinan, G. C., Bhatia, A., Maisonneuve, J. F., Lodes, M., Persing, D. H., & Patrick, S. (2005). *Propionibacterium acnes* types I and II represent phylogenetically distinct groups. *Journal of Clinical Microbiology, 43*(1), 326–334.

Metrogos, V., Ramos, N., Marialva, C., & Bastos, J. (2017). Rare association between prostate adenocarcinoma and schistosomiasis: A case report. *ACTA Urológica Portuguesa, 34*(3–4), 41–43.

Miyake, M., Ohnishi, K., Hori, S., Nakano, A., Nakano, R., Yano, H., Ohnishi, S., Owari, T., Morizawa, Y., Itami, Y., Nakai, Y., Inoue, T., Anai, S., Torimoto, K., Tanaka, N., Fujii, T., Furuya, H., Rosser, C. J., & Fujimoto, K. (2019). *Mycoplasma genitalium* infection and chronic inflammation in human prostate cancer: Detection using prostatectomy and needle biopsy specimens. *Cells, 8*(3), 212. https://doi.org/10.3390/cells8030212

Mollerup, S., Friis-Nielsen, J., Vinner, L., Hansen, T. A., Richter, S. R., Fridholm, H., Herrera, J. A. R., Lund, O., Brunak, S., Izarzugaza, J. M. G., Mourier, T., Nielsen, L. P., & Hansen, A. J. (2016). *Propionibacterium acnes*: Disease-causing agent or common contaminant? Detection in diverse patient samples by next-generation sequencing. *Journal of Clinical Microbiology, 54*, 980–987. https://doi.org/10.1128/JCM.02723-15

Münger, K., Baldwin, A., Edwards, K. M., Hayakawa, H., Nguyen, C. L., Owens, M., Grace, M., & Huh, K. (2004). Mechanisms of human papillomavirus-induced oncogenesis. *Journal of Virology, 78*, 11451–11460.

Nahand, J. S., Khanaliha, K., Mirzael, H., Moghoofe, M., Baghi, H. B., Esghaei, M., Khatami, A. R., Fatemipour, M., & Bokharaei-Salim, F. (2021). Possible role of HPV/EBV coinfection in anoikis resistance and development in prostate cancer. *BMC Cancer, 21*, 926. https://doi.org/10.1186/s12885-021-08658-y

Namiki, K., Goodison, S., Porvasnik, S., Allan, R. W., Iczkowski, K. A., Urbanek, C., Reyes, L., Sakamoto, N., & Rosser, C. J. (2009). Persistent exposure to *Mycoplasma* induces malignant transformation of human prostate cells. *PLoS One, 4*, e6872.

Nwadi, U. V., Nwofor, A., Oranusi, C. K., Orakwe, J. C., Obiesie, E. A., Mbaeri, T. U., Abiahu, J. A., & Mbonu, O. O. (2021). Correlation between body mass index and Gleason score in men with prostate cancer in South-Eastern Nigeria. *Nigerian Journal of Surgery: Official Publication of the Nigerian Surgical Research Society, 27*(1), 22–27. https://doi.org/10.4103/njs.NJS_66_20

Olsson, J., Drott, J. B., Laurantzon, L., Laurantzon, O., Bergh, A., & Elgh, F. (2012). Chronic prostatic infection and inflammation by *Propionibacterium acnes* in a rat prostate infection model. *PLoS One, 7*(12), e51434. https://doi.org/10.1371/journal.pone.0051434

Perletti, G., Monti, E., Magri, V., Cai, T., Cleves, A., Trinchieri, A., & Montanari, E. (2017). The association between prostatitis and prostate cancer. Systematic review and meta-analysis. *Archivio Italiano di Urologia, Andrologia, 89*, 259–265.

Perry, A., & Lambert, P. (2011). *Propionibacterium acnes*: Infection beyond the skin. *Expert Review of Anti-Infective Therapy, 9*(12), 1149–1156. https://doi.org/10.1586/eri.11.137

Porter, C. M., Shrestha, E., Peiffer, L. B., & Sfanos, K. S. (2018). The microbiome in prostate inflammation and prostate cancer. *Prostate Cancer and Prostatic Diseases, 21*, 345–354. https://doi.org/10.1038/s41391-018-0041-1

Puhr, M., De Marzo, A., Isaacs, W., Lucia, M. S., Sfanos, K., Yegnasubramanian, S., & Culig, Z. (2016). Inflammation, microbiota, and prostate cancer. *European Urology Focus, 2*, 374–382.

Rawla, P. (2019). Epidemiology of prostate cancer. *World Journal of Oncology, 10*(2), 63–89. https://doi.org/10.14740/wjon1191

Rizzo, A., Santoni, M., Mollica, V., Florentino, M., Brandi, G., & Massari, F. (2021). Microbiota and prostate cancer. *Seminars in Cancer Biology, 86*, 1058. https://doi.org/10.1016/j.semcancer.2021.09.007

Rotimi, S. O., Rotimi, O. A., Salako, A. A., Jibrin, P., Oyelade, J., & Iweala, E. E. J. (2019). Gene expression profiling analysis reveals putative phytochemotherapeutic target for castration-resistant prostate cancer. *Frontiers in Oncology, 9*, 714. https://doi.org/10.3389/fonc.2019.00714

Rotimi, S. O., Rotimi, O. A., & Salhia, B. (2021). A review of cancer genetics and genomics studies in Africa. *Frontiers in Oncology, 10*, 606400. https://doi.org/10.3389/fonc.2020.606400

Rybicki, B. A., Kryvenko, O. N., Wang, Y., Jankowski, M., Trudeau, S., Chitale, D. A., Gupta, N. S., Rundle, A., & Tang, D. (2016). Racial differences in the relationship between clinical prostatitis, presence of inflammation in benign prostate and subsequent risk of prostate cancer. *Prostate Cancer and Prostatic Diseases, 19*(2), 145–150. https://doi.org/10.1038/pcan.2015.54

Saha, A., Kaul, R., Murakami, M., & Robertson, E. S. (2010). Tumor viruses and cancer biology: Modulating signaling pathways for therapeutic intervention. *Cancer Biology and Therapy, 10*(10), 961–978.

Salinas, C. A., Kwon, E. M., FitzGerald, L. M., Feng, Z., Nelson, P. S., Ostrander, E. A., Peters, U., & Stanford, J. L. (2010). Use of aspirin and other nonsteroidal antiinflammatory medications concerning prostate cancer risk. *American Journal of Epidemiology, 172*(5), 578–590. https://doi.org/10.1093/aje/kwq175

Samanta, M., Harkins, L., Klemm, K., Britt, W. J., & Cobbs, C. S. (2003). High prevalence of human cytomegalovirus in prostatic intraepithelial neoplasia and prostatic carcinoma. *Journal of Urology, 170*(3), 998–1002. https://doi.org/10.1097/01.ju.0000080263.46164.97

Serth, J., Panitz, F., Paeslack, U., Kuczyk, M. A., & Jonas, U. (1999). Increased levels of human papillomavirus type 16 DNA in a subset of prostate cancers. *Cancer Research, 59*, 823–825.

Sfanos, K. S., Wilson, B. A., De Marzo, A. M., & Isaacs, W. B. (2009). Acute inflammatory proteins constitute the organic matrix of prostatic corpora amylacea and calculi in men with prostate cancer. *Proceedings of the National Academy of Sciences of the United States of America, 106*, 3443–3448.

Shannon, B. A., Garrett, K. L., & Cohen, R. J. (2006). Links between *Propionibacterium acnes* and prostate cancer. *Future Oncology, 2*(2), 225–232. https://doi.org/10.2217/14796694.2.2.225

Short, E., Warren, A. Y., & Varma, M. (2019). Gleason grading of prostate cancer: A pragmatic approach. *Diagnostic Histopathology, 25*(10), 371–378.

Siegel, R. L., Miller, K. D., & Jemal, A. (2017). Cancer statistics. *CA: a Cancer Journal for Clinicians, 67*, 7–30.

Sindhwani, P., & Wilson, C. M. (2005). Prostatitis and serum prostate-specific antigen. *Current Urology Reports, 6*, 307–312.

Sutcliffe, S., Zenilman, J. M., Ghanem, K. G., Jadack, R. A., Sokoll, L. J., Elliot, D. J., Nelson, W. G., De Marzo, A. M., Cole, S. R., Isaacs, W. B., & Platz, E. A. (2006). Sexually transmitted infections and prostatic inflammation/cell damage as measured by serum prostate specific antigen concentration. *The Journal of Urology, 175*, 1937–1942.

Taha, Z. N., Saadoon, I. H., & Hadi, A. M. (2020). Detection of Epstein-Barr virus in patients with prostate cancer and benign prostatic hyperplasia. *Biochemical and Cellular Archives, 20*(2), 1–3.

Tonry, C., Finn, S., Armstrong, J., & Pennington, S. R. (2020). Clinical proteomics for prostate cancer: Understanding prostate cancer pathology and protein biomarkers for improved disease management. *Clinical Proteomics, 17*(1), 41. https://doi.org/10.1186/s12014-020-09305-7

Twu, O., Dessí, D., Vu, A., Mercer, F., Stevens, G. C., de Miguel, N., Rappelli, P., Cocco, A. R., Clubb, R. T., Fiori, P. L., & Johnson, P. J. (2014). *Trichomonas vaginalis* homolog of macrophage migration inhibitory factor induces prostate cell growth, invasiveness, and inflammatory responses. *Proceedings of the National Academy of Sciences of the United States of America, 111*, 8179–8184.

Ugge, H., Carlsson, J., Söderquist, B., Fall, K., Anden, O., & Davidsson, S. (2018). The influence of prostatic *Cutibacterium acnes* infection on serum levels of IL6 and CXCL8 in prostate cancer patients. *Infectious Agents and Cancer, 13*, 34–40. https://doi.org/10.1186/s13027-018-0204-7

Wu, T., & Dai, Y. (2017). Tumor microenvironment and therapeutic response. *Cancer Letters, 28*, 61–68. https://doi.org/10.1016/j.canlet.2016.01.043

Yow, M. A., Tabrizi, S. N., Severi, G., Bolton, D. M., Pedersen, J., Giles, G. G., & Southey, M. C. (2017). Characterisation of microbial communities within aggressive prostate cancer tissues. *Infectious Agents and Cancer, 12*(1), 4. https://doi.org/10.1186/s13027-016-0112-7

Zhang, L., Wang, Y., Qin, Z., Gao, X., Xing, Q., Li, R., et al. (2020). Correlation between prostatitis, benign prostatic hyperplasia and prostate cancer: A systematic review and meta-analysis. *Journal of Cancer, 11*, 177–189.

Chapter 30
Nutritional Composition, GC-MS Evaluation and In Vitro Antioxidant Activity of *Prosopis africana* (Guill. and Perr.) Leaves

Okukwe Christa Obode, Abiodun Humphrey Adebayo, Omolara Faith Yakubu, and Chunyang Li

1 Introduction

Plant materials are indispensable natural resources that have played significant roles in the survival of humans as well as animals. They are a source of food, shelter as well as medicine. Medicinal plants have long been in use for the management and treatment of numerous ailments and diseases, before the advent of synthetic drugs. Medicinal plants continue to serve as essential therapeutic alternatives to conventional pharmacological agents, particularly in cases where the latter are associated with significant adverse effects. They were used mainly for nutritional purposes, but later on, medicinal properties were discovered and they became essential in managing various ailments to improve health (Azmir et al., 2013). Historically, Egyptian papyruses were indicative of the use of essential oils from coriander and castor plants in medicine and cosmetic industries as well as preservative purposes

O. C. Obode
Department of Biochemistry Department, College of Science and Technology, Covenant University, Ota, Ogun State, Nigeria

Nutrition and Toxicology Division, Food Technology Department, Federal Institute of Industrial Research, Oshodi, Lagos, Nigeria

A. H. Adebayo (✉) · O. F. Yakubu
Department of Biochemistry Department, College of Science and Technology, Covenant University, Ota, Ogun State, Nigeria
e-mail: abiodun.adebayo@covenantuniversity.edu.ng

C. Li
Department of Biochemistry Department, College of Science and Technology, Covenant University, Ota, Ogun State, Nigeria

Institute of Agro-Products Processing, Jiangsu Academy of Agricultural Sciences, Nanjing, Jiangsu, China

P. O. Isibor, P. A. Akinduti (eds.), *Harnessing Biotechnology Tools for Product Development*, https://doi.org/10.1007/978-3-031-86002-7_30

(Kamwamba, 2017). Also, many scholars, including Hippocrates and Theophrastus, have illustrated numerous therapeutic uses of medicinal plants all through the Greek and Roman periods. The existence of pharmacologically active compounds (mostly secondary metabolites) was linked to these therapeutic advantages of medicinal plants. They are compounds other than primary metabolites, which do not contribute to the energy needs of plants but are believed to be helpful to plants for survival from predators. Different species over time produced different metabolites based on the particular needs of those species during evolution. These secondary metabolites of plants have been shown to exert both pharmacological and toxicological effects in humans as well as animals and are thus harnessed for the development of pharmaceutical as well as nutraceutical therapies. The majority now view the usage of nutraceuticals as an effective alternative to medicines because of their high price, limited availability and unwanted side effects in some cases (Ademosun et al., 2021).

Numerous qualitative and quantitative techniques for phytochemical analysis are used to identify the presence of plant phytochemical substances. The scientific validation of traditional medicines and the development of novel drugs depend on the study of plant bioactive components. Additionally, it aids in locating more affordable sources of bioactive chemicals with pharmacological relevance (Sermakkani & Thangapandian, 2012). A high throughput method for identifying bioactive molecules as well as other plant ingredients, volatile materials, hydrocarbons, acid alcohols, alkaloids and other substances is coupling gas chromatography to mass spectroscopy (GC-MS).

Prosopis africana (Guill. and Perr.), a member of the Fabaceae family, is a Prosopis species indigenous to Africa (Henciya et al., 2017). Conventional names include 'Iron tree' and African mesquite with local names including 'kiriya' (in Hausa), 'ubwo' (in Igbo), 'ayan' (in Yoruba) and 'sanchi' (in Nupe), respectively (Dosumu et al., 2012). Prosopis species have found use as wood, gum and food conditions as well as medicine. While the roots are used as chewing sticks to prevent tooth decay, the hardwood is used for various furniture needs and charcoal (Obode et al., 2021). The fermented seeds are utilized in producing a food condiment (commonly called 'okpehe'). *Prosopis africana* is mentioned in ethnobotanical use for treating migraines, high blood pressure, dysentery and rheumatism (Raji et al., 2013; Bekoe et al., 2017). Reported phytochemicals in *P. africana* leaves include terpenes, saponins, tannins, anthraquinones, steroids and cardiac glycosides; while the biological activities published include analgesic, anti-inflammatory, wound healing, anti-trypanosomal and antibacterial (Obode et al., 2021). As far as we know, information about the detailed phytochemical composition of the leaves of *P. africana* is limited. Therefore, this research was conducted to evaluate the proximate and phytochemical composition of *P. africana* leaves by subjecting the various solvent fractions to GC-MS and to verify the presence of bioactive compounds that could confer health benefits in disease management.

2 Materials and Methods

2.1 *Materials*

The leaves of *Prosopis africana* were collected from the University of Ilorin premises. The tree branches were cut, and the leaves plucked out and transported in an aerated sack. Some of the leaves were sent to the Forestry Research Institute of Nigeria for identification and validation, while some portions were deposited at the herbarium of the institute. The voucher referencing number FHI.112370 was assigned to the deposited sample.

2.2 *Sample Preparation*

The leaves were gently cleaned and dried at room temperature for 4 weeks before grinding to fine particles using a laboratory mill. Powdered leaves of *P. africana* (2 kg) were immersed for 72 h in 15 L of 96% v/v ethanol and subsequently filtered with the aid of cheesecloth and cotton wool based on the method described by Obode et al. (2021). The extraction was done 3 times before the combined extract was distilled by means of a rotary evaporator (50 °C) under reduced pressure. The crude ethanolic extracts (ELPA; 200 g dissolved in 10% Tween 20 solution) were suspended in 1 L of distilled water and thereafter sequentially partitioned with 1 L *n*-hexane, 1 L ethyl acetate and 1 L *n*-butanol. The fractions were concentrated with the aid of a rotary evaporator to obtain the ethyl acetate (EAPA), *n*-butanol (BLPA), and water-soluble (ALPA) fractions, respectively.

2.3 *Proximate Composition of* P. africana *Leaves*

The proximate composition of the milled dried leaves of *P. africana* was analysed chemically using the analytical methods of the Association of Official Analytical Chemists (AOAC) (2016). All analyses were carried out in duplicates.

2.4 *Preliminary Phytochemical Evaluation*

The preliminary phytochemical evaluation was conducted on the differential solvent fractions of *P. africana* following the standard protocols for the analysis of phytochemicals described by Harborne (1984).

2.5 *GC-MS Analysis of Various Solvent Fractions of* P. africana

GC-MS was used to assess the quantitative phytochemical analysis of *P. africana's* differential solvent fractions following the protocol described by Padma et al. (2019). The GCMS-QP2010SE Shimadzu, Japan instrument was used for this evaluation. After dissolving the samples in methanol, they were injected into the column using a split-less injection system (separately), with injection and column temperatures of 60 and 250 °C, respectively. 0.5 μL of ultra-high-quality helium was injected as the carrier gas, with a flow rate of 3.22 mL/min. 230 °C was used as the source ionization temperature. The positive ion mode was used for the investigations. MS detection took 18.9 min, and the mass spectrum scan speed was 1428. The National Institute of Standards and Technology (NIST) library from 2011 was used for the detection.

2.6 *Antioxidant Activity of Solvent Fractions of* P. africana

The antioxidant property of the plant was evaluated using Trolox equivalent and 2, 2-diphenyl-1-picryl-hydrazyl (DPPH) antioxidant capacity following the methods illustrated by Singh et al. (2019). Trolox equivalent antioxidant assay evaluated the power of antioxidant compounds to douse 2, 2-azino-bis (3-ethylbenzothiazoline-6-sulphonic acid) (ABTS) radical cation. Free radical scavenging activity was assessed by adding 300 μL of plant extract to 3.0 mL of working standard solution of ABTS in a microcuvette. The solution was vortexed, and the reduction in absorbance was evaluated thereafter from 1 to 6 min. The percentage inhibition was determined using the formula:

$$\text{Scavenging activity}\left(\%\right)=\frac{\text{Control absorbance}-\text{Sample absorbance}}{\text{Control absorbance}}\times 100$$

For DPPH, a 3 mL portion of the solution (DPPH) was added to 100 μL of the sample at different concentrations (25–100 μg/mL). The solution was vortexed and incubated for 15 min at room temperature away from light. Absorbance was measured at 517 nm. The experimental control was prepared using the same method without the sample. The scavenging activity was determined using the formula:

$$\text{Scavenging activity}\left(\%\right)=\frac{\text{Control absorbance}-\text{Sample absorbance}}{\text{Control absorbance}}\times 100$$

2.7 Statistical Analysis

One-way analysis of variance (ANOVA) was used to analyse the data using the Statistical Package for Social Sciences (SPSS) for Windows, version 21.0 (SPSS Inc., Chicago, IL, USA). Data were given as the mean ± standard error of the mean (SEM), with a significance level of $p < 0.05$.

3 Results

3.1 Proximate Composition of P. africana Leaves

The proximate composition of *P. africana* leaves includes carbohydrates (2.58 ± 0.01%), protein (10.5 ± 0.09%), fat (3.42 ± 0.03%), crude fibre (76.02 ± 0.03%), ash (3.50 ± 0.04%) and moisture content (3.98 ± 0.01%).

3.2 Phytochemical Composition of the Differential Solvent Fractions of P. africana

Phytochemicals present in the differential solvent fractions of *P. africana* are shown in Table 30.1. Tannins, saponins, flavonoids, phenols and coumarins were detected in all differential solvent fractions of *P. africana*.

Table 30.1 Phytochemical composition of the differential solvent fractions of *P. africana* leaves

	Results		
Phytochemicals	EAPA	BLPA	ALPA
Tannins	+	+	+
Saponins	+	+	+
Flavonoids	+	+	+
Alkaloids	+	–	–
Anthocyanins	–	+	+
Betacyanins	+	–	–
Quinones	–	+	+
Glycosides	–	–	–
Cardiac glycosides	–	–	–
Terpenoids	–	+	–
Triterpenoids	+	–	–
Phenols	+	+	+
Coumarins	+	+	+
Steroids	+	–	–

EAPA, BLPA and ALPA represent n-hexane, ethyl-acetate, butanol and aqueous extracts, respectively. + represents a positive result, while – represents a negative result

3.3 *GC-MS Analysis of Various Solvent Fractions of* P. africana

GC-MS is a major technique employed in the identification of the bioactive as well as other constituents of plants, volatile materials, hydrocarbons, acid alcohols, etc. The total ion chromatogram (TIC) of the various fractions (ethanol, ethyl acetate, butanol and aqueous) of *P. africana* is presented in Figs. 30.1, 30.2, 30.3 and 30.4, while the compounds identified are shown in Tables 30.2, 30.3, 30.4 and 30.5. The retention time, peak area percentage and compound names are presented. The ethanol, ethyl acetate, butanol and aqueous fractions contain 32, 53, 37 and 52

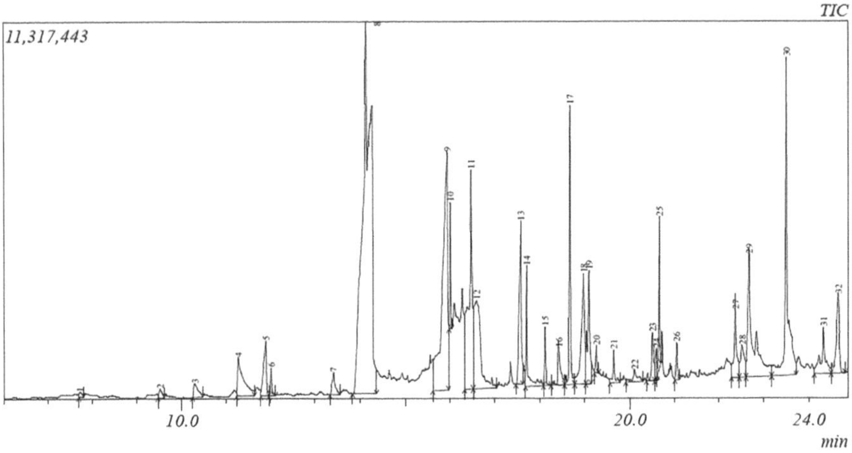

Fig. 30.1 Total ion chromatogram of the ethanol fraction of *P. africana* (ELPA)

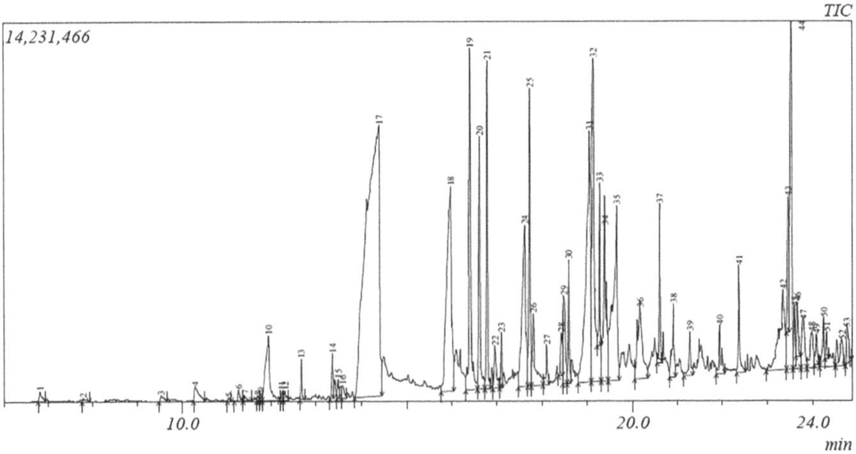

Fig. 30.2 Total ion chromatogram of the ethyl-acetate fraction of *P. africana* (EAPA)

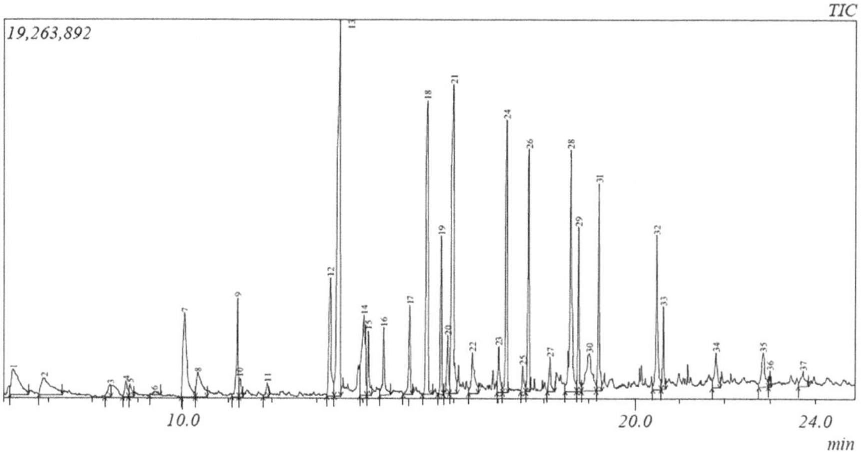

Fig. 30.3 Total ion chromatogram of butanol fraction of *P. africana* (BLPA)

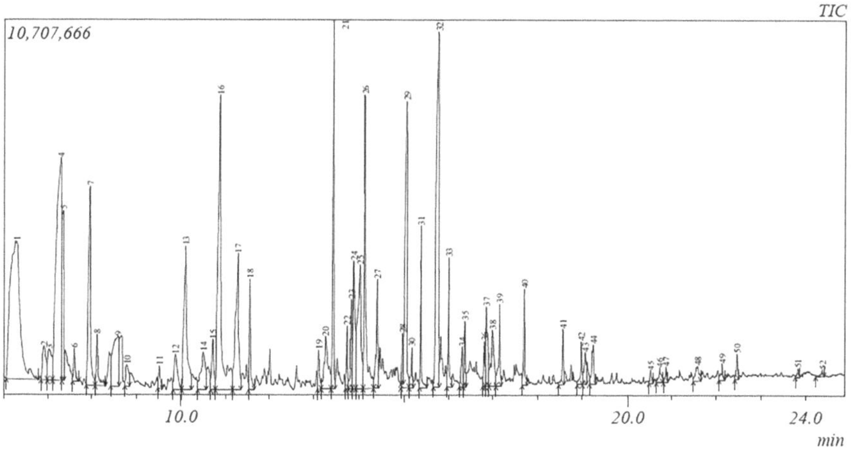

Fig. 30.4 Total ion chromatogram of aqueous fraction of *P. africana* (ALPA)

phytochemical constituents, respectively. The first compounds identified with lesser retention times for the ethanol, ethyl acetate, butanol and aqueous fractions include 1,2,3,4-butanetetrol (7.708 min), 2-oxabicyclo[2.2.2]octan-6-ol, 1,3,3-trimethyl (6.811 min), diethylene glycol monomethyl ether (6.218 min) and 2-methoxyethyl ethylphosphonofluoridate (6.289 min), respectively. (Tables 30.2, 30.3, 30.4, and 30.5, Figs. 30.1, 30.2, 30.3, and 30.4); while the last compounds identified based on

Table 30.2 Compounds identified in the ethanol fraction of *P. africana*

Peak no.	Area %	Molecular weight	Similarity index (%)	Formula	Compound name	Classification of compound
1	0.17	122	76	$C_4H_{10}O_4$	1,2,3,4-Butanetetrol	Sugar alcohol
2	0.26	144	84	$C_8H_{16}O_2$	*n*-Caprylic acid	Fatty acid
3	0.61	120	87	C_8H_8O	Coumarin	Flavonoid
4	2.15	110	86	$C_6H_6O_2$	Resorcinol	Aromatic alcohol
5	1.40	172	95	$C_{10}H_{20}O_2$	*n*-Capric acid	Fatty acid
6	0.33	200	90	$C_{12}H_{24}O_2$	Ethyl caprate	Fatty acid ester
7	0.64	176	82	$C_8H_{16}O_4$	1,4,7,10-Tetraoxacyclododecane	Crown amine
8	27.41	200	87	$C_{12}H_{24}O_2$	Lauric acid	Fatty acid
9	11.79	228	95	$C_{14}H_{28}O_2$	Myristic acid	Fatty acid
10	1.18	242	88	$C_{15}H_{30}O_2$	Ethyl tridecanoate	Fatty acid ester
11	7.26	270	89	$C_{17}H_{34}O_2$	Isoamyl laurate	Fatty alcohol ester
12	5.43	256	95	$C_{16}H_{32}O_2$	Palmitic acid	Fatty acid
13	3.39	284	91	$C_{18}H_{36}O_2$	Ethyl palmitate	Fatty acid ester
14	1.73	270	89	$C_{17}H_{34}O_2$	Isoamyl laurate	Fatty alcohol ester
15	0.65	326	89	$C_{14}H_{30}O_8$	Heptaethylene glycol	Polyether
16	1.04	296	96	$C_{20}H_{40}O$	Phytol	Terpenoid
17	3.23	280	88	$C_{18}H_{32}O_2$	cis,cis-Linoleic acid	Fatty acid
18	3.38	352	86	$C_{21}H_{36}O_4$	α-Glyceryl linoleate	Fatty acid
19	2.44	298	86	$C_{19}H_{38}O_2$	Ethyl 14-methyl-hexadecanoate	Fatty acid methyl ester
20	0.28	270	85	$C_{17}H_{34}O_2$	2-Methylbutyl laurate	Fatty acid ester

21	0.54	430	61	$C_{29}H_{50}O_2$	Cholestan-7-one, cyclic 1,2-ethanediyl acetal	–
22	0.51	264	85	$C_{12}H_{24}O_6$	Ethylene oxide cyclic hexamer	Crown ether
23	1.11	318	74	$C_{18}H_{38}O_4$	Triethylene glycol monododecyl ether	Polyether
24	0.48	155	84	$C_8H_{13}NO_2$	Carbonic acid	Organic acid
25	1.59	155	80	$C_8H_{13}NO_2$	Carbonic acid	Organic acid
26	0.52	450	86	$C_{24}H_{50}O_7$	Hexaethylene glycol monododecyl ether	Polyether
27	1.85	186	66	$C_{10}H_{18}OS$	δ-Thionodecalactone	–
28	1.16	243	68	$C_{12}H_{21}NO_4$	Formamide, *N*-{4-[2-(1,1-dimethylethyl)-5-oxo-1,3-dioxolan-4-yl]butyl}	–
29	5.82	410	96	$C_{30}H_{50}$	Squalene	Triterpene
30	7.54	362	72	$C_{20}H_{42}O_5$	Tetraethylene glycol monododecyl ether	Polyether
31	1.80	246	81	$C_{18}H_{30}$	d-Norandrostane (5-α,14-α)	–
32	2.29	194	83	$C_7H_{14}O_6$	3-O-Methyl-D-glucose	Sugar

Table 30.3 Compounds identified in the ethyl acetate fraction of *P. africana*

Peak no.	Area %	Molecular weight	Similarity index (%)	Formula	Compound name	Classification of compound
1	0.18	170	79	$C_{10}H_{18}O_2$	2-Oxabicyclo[2.2.2]octan-6-ol, 1,3,3-trimethyl	–
2	0.06	208	72	$C_{14}H_{24}O$	6,11-Dimethyl-2,6,10-dodecatrien-1-ol	–
3	0.14	144	93	$C_8H_{16}O_2$	*n*-Caprylic acid	Fatty acid
4	0.45	120	83	C_8H_8O	Coumaran	Flavonoid
5	0.03	120	76	C_9H_{12}	Spiro[4.4]nona-1,6-diene,	
6	0.14	150	88	$C_9H_{10}O_2$	2-Methoxy-4-vinylphenol	Phenolics
7	0.10	166	78	$C_{12}H_{22}$	cis,trans-4,5-Diethyl-3,5-octadiene	–
8	0.06	250	74	$C_{12}H_{17}F_3O_2$	1-Methyl-1-(4-methyl-3-cyclohexen-1-yl)ethyl trifluoroacetate	–
9	0.07	172	69	$C_{13}H_{16}$	1,1,6-Trimethyl-1,2-dihydronaphthalene	Naphthalenes
10	1.76	172	96	$C_{10}H_{20}O_2$	*n*-Capric acid	Fatty acid
11	0.09	224	68	$C_{14}H_{24}O_2$	2′-Acetonaphthone, 1′,2′.α,3′,4′,4’a,5′,6′,	–
12	0.14	200	67	$C_{12}H_{24}O_2$	Lauric acid	Fatty acid
13	0.28	194	90	$C_{13}H_{22}O$	Nerylacetone	Acyclic monoterpenoid
14	0.41	206	83	$C_{14}H_{22}O$	2,4-Di-tert-butylphenol	Phenolics
15	0.17	170	69	$C_{10}H_{18}O_2$	8-Methyl-6-nonenoic acid	–
16	0.23	162	86	$C_6H_{10}O_5$	Anhydro-D-mannosan	Anhydrosugar
17	22.19	200	97	$C_{12}H_{24}O_2$	Lauric acid	Fatty acid
18	6.24	228	96	$C_{14}H_{28}O_2$	Myristic acid	Fatty acid
19	3.89	338	92	$C_{22}H_{42}O_2$	Phytol, acetate	Diterpene alcohol
20	2.16	338	90	$C_{20}H_{40}O$	3,7,11,15-Tetramethyl-2-hexadecen-1-ol	Diterpene alcohol
21	2.82	296	93	$C_{20}H_{40}O$	3,7,11,15-Tetramethyl-2-hexadecen-1-ol	Diterpene alcohol

22	0.64	240	75	$C_{15}H_{28}O_2$	2-(3-Methyl-undec-3-enyl)-[1,3]dioxolane	–
23	0.40	290	89	$C_{20}H_{34}O$	Trans-geranylgeraniol	Diterpenoid
24	3.63	256	95	$C_{16}H_{32}O_2$	Palmitic acid	Fatty acid
25	2.72	284	94	$C_{18}H_{36}O_2$	Ethyl palmitate	Fatty acid ester
26	1.30	320	54	$C_{20}H_{32}O_3$	Butyric acid, 2-(2,4-di-tert-pentylphenoxy)-	Carboxylic acid
27	0.46	328	79	$C_{16}H_{32}O_3$	Myristic acid, 2-hydroxyethyl ester	Fatty acid ester
28	0.48	406	80	$C_{22}H_{46}O_6$	Pentaethylene glycol monododecyl ether	Hydroxypolyether
29	1.23	282	77	$C_{18}H_{34}O_2$	Ethyl 9-hexadecenoate	Fatty acid ester
30	0.92	288	77	$C_{16}H_{32}O_4$	Diethylene glycol monolaurate	–
31	6.58	280	92	$C_{18}H_{32}O_2$	cis,cis-Linoleic acid	Fatty acid
32	6.96	306	94	$C_{20}H_{34}O_2$	Ethyl linolenate	Fatty acid ester
33	1.07	312	91	$C_{20}H_{40}O_2$	Ethyl stearate	Fatty acid ethyl ester
34	2.81	298	91	$C_{21}H_{42}O_2$	1-Acetoxynonadecane	–
35	4.79	396	80	$C_{29}H_{48}$	Stigmastan-3,5-diene	Fatty acid
36	2.38	430	88	$C_{29}H_{50}O_2$	Vitamin E	Terpenoid
37	1.07	362	77	$C_{20}H_{42}O_5$	Tetraethylene glycol monododecyl ether	Polyether
38	0.93	334	87	$C_{22}H_{38}O_2$	Butyl 9,12,15-octadecatrienoate	Fatty acid ester
39	0.55	283	80	$C_{16}H_{30}O$	(Z)-7-Hexadecenal	–
40	0.61	318	78	$C_{18}H_{38}O_4$	Triethylene glycol monododecyl ether	Polyether
41	0.98	538	82	$C_{28}H_{58}O_9$	Octaethylene glycol monododecyl ether	Polyether
42	2.89	426	78	$C_{30}H_{50}O$	Oxirane, 2,2-dimethyl-3-(3,7,12,16,20-pentamethyl-3,7,11,15,19-heneicosapentaenyl)-, (all-E)-	–
43	1.50	506	91	$C_{30}H_{51}BrO$	6,10,14,18,22-Tetracosapentaen-2-ol,3-bromo-2,6,10,15,19,23-hexamethyl-(all-E)-	–
44	6.35	410	96	$C_{30}H_{50}$	Squalene	Triterpene

(continued)

Table 30.3 (continued)

Peak no.	Area %	Molecular weight	Similarity index (%)	Formula	Compound name	Classification of compound
45	1.36	428	90	$C_{30}H_{52}O$	2,2,4-Trimethyl-3-(3,8,12,16-tetramethyl-heptadeca-3,7,11,15-tetraenyl)-cyclohexanol	–
46	0.50	290	89	$C_{20}H_{34}O$	Trans-geranylgeraniol	Diterpenoid
47	0.97	452	57	$C_{26}H_{44}O_4S$	Methanesulfonic acid, 2-(3-hydroxy-4,4,10,13,14-pentamethyl-2,3,4,5,6,7,10,11,12,13,14,15,16,17-tetradecahydro-1H-cyclopenta[a]phenanthryl)-	–
48	1.23	618	65	$C_{20}H_{23}F_{17}O_2$	Heptadecafluorononanoic acid, undecyl ester	–
49	0.36	914	80	$C_{54}H_{108}Br_2$	1,54-Dibromotetrapentacontane	–
50	1.35	400	73	$C_{28}H_{48}O$	Pollinastanol	Steroid
51	0.23	428	78	$C_{30}H_{52}O$	Ambrein	Triterpenoid alcohol
52	0.67	452	82	$C_{33}H_{56}$	1,1,6-Trimethyl-3-methylene-2-(3,6,9,13-tetramethyl-6-ethenye-10,14-dimethylene-pentadec-4-enyl)cyclohexane	Menthane monoterpenoids
53	0.45	380	60	$C_{23}H_{44}O_2Si$	cis-11,14-Eicosadienoic acid, trimethylsilyl ester	–

Table 30.4 Compounds identified in the butanol fraction of *P. africana*

Peak no.	Area %	Molecular weight	Similarity index (%)	Formula	Compound name	Classification of compound
1	2.48	120	89	$C_5H_{12}O_3$	Diethylene glycol monomethyl ether	–
2	2.19	124	82	$C_4H_9ClO_2$	2-Propanol,1-chloro-3-methoxy-	–
3	0.47	125	72	$C_2H_7NO_3S$	Taurine	Organosulfonic
4	0.53	98	74	$C_5H_6O_2$	2-Furanmethanol	Furfural alcohol
5	0.44	162	81	$C_6H_{10}O_5$	Butanedioic acid, hydroxy-, dimethyl ester	–
6	0.21	154	71	$C_8H_{10}O_3$	5-Methoxy-3,3a,4,6a-tetrahydro-1H-cyclopenta[c]furan-1-one	–
7	4.14	164	97	$C_7H_{16}O_4$	Methoxytriethylene glycol	Polyethylene glycol
8	1.66	150	95	$C_6H_{14}O_4$	Triethylene glycol	Polyethylene glycol
9	1.97	186	88	$C_{11}H_{22}O_2$	Decanoic acid, methyl ester	Fatty acid methyl ester
10	0.46	176	91	$C_8H_{16}O_4$	12-Crown-4 ether	Crown ether
11	0.35	172	84	$C_7H_8O_3S$	Methyl benzenesulfonate	–
12	3.48	208	97	$C_9H_{20}O_5$	Tetraethyleneglycol monomethylether	Polyethylene glycol
13	13.26	214	97	$C_{13}H_{26}O_2$	Methyl laurate	Fatty acid methyl ester
14	3.70	200	96	$C_{12}H_{24}O_2$	Laurostearic acid	Fatty acid
15	1.25	270	78	$C_{17}H_{34}O_2$	Ethyl pentadecanoate	Fatty acid ester
16	1.44	264	97	$C_{12}H_{24}O_6$	Ethylene oxide cyclic hexamer	Crown ether
17	1.82	256	75	$C_{14}H_{28}N_2O_2$	Hexanoic acid, (2-hexanoylaminoethyl)-amide	–
18	7.78	242	97	$C_{15}H_{30}O_2$	Myristic acid, methyl ester	Fatty acid methyl ester
19	2.40	146	77	$C_6H_{10}O_4$	Dianhydromannitol	Sugar alcohol
20	1.21	228	91	$C_{14}H_{28}O_2$	Myristic acid	Fatty acid
21	9.94	256	93	$C_{16}H_{32}O_2$	*n*-Butyl laurate	Fatty acid ester
22	1.55	272	81	$C_{16}H_{32}O_3$	Ethylene glycol monomyristate	Glycol
23	1.11	264	91	$C_{12}H_{24}O_6$	Ethylene oxide cyclic hexamer	Crown ether
24	5.63	270	97	$C_{17}H_{34}O_2$	Palmitic acid, methyl ester	Fatty acid methyl ester
25	0.61	256	93	$C_{16}H_{32}O_2$	Palmitic acid	Fatty acid
26	4.80	284	95	$C_{18}H_{36}O_2$	Butyl myristate	Fatty acid ester

(continued)

Table 30.4 (continued)

Peak no.	Area %	Molecular weight	Similarity index (%)	Formula	Compound name	Classification of compound
27	0.76	104	83	$C_3H_{12}Si_2$	1,1,2-Trimethyldisilane	–
28	6.72	268	91	$C_{17}H_{32}O_2$	Methyl (7E)-7-hexadecenoate	Fatty acid methyl ester
29	2.29	298	96	$C_{19}H_{38}O_2$	Methyl stearate	Fatty acid methyl ester
30	2.84	270	73	$C_{15}H_{26}O_4$	Ethylene brassylate	Macrolide
31	3.64	312	92	$C_{20}H_{40}O_2$	*n*-Butyl palmitate	Fatty acid ester
32	3.66	338	91	$C_{22}H_{42}O_2$	Butyl 9-octadecenoate	Fatty acid ester
33	1.21	340	88	$C_{22}H_{44}O_2$	*n*-Butyl stearate	Fatty acid ester
34	1.32	412	61	$C_{27}H_{40}O_3$	7-Dehydrodiosgenin	Steroid saponin
35	1.58	552	70	$C_{36}H_{72}O_3$	Palmitic acid, 2-(octadecyloxy)ethyl ester	Fatty acid ester
36	0.28	538	82	$C_{28}H_{58}O_9$	Octaethylene glycol monododecyl ether	Polyether
37	0.84	322	71	$C_{21}H_{38}O_2$	[1,1′-Bicyclopropyl]-2-octanoic acid, 2′-hexyl-, methyl ester	–

the duration of retention time include 3-O-methyl-d-glucose (24.673 min), cis-11,14-eicosadienoic acid, trimethylsilyl ester (24.792 min), [1,1′-bicyclopropyl]-2-octanoic acid, 2′-hexyl-, methyl ester (23.742 min) and [1,1′-bicyclopropyl]-2-octanoic acid, 2′-hexyl-, methyl ester (24.383 min), respectively.

3.4 *Antioxidant Activity of Various Solvent Fractions of* P. africana

The radical scavenging activity of the extracts was executed using DPPH and ABTS assays. The results are presented in Fig. 30.5. DPPH scavenging activity of ethanol (80.17 ± 0.24%), ethyl acetate (76.05 ± 0.48%), butanol (77.74 ± 0.40%) and aqueous fractions (88.44 ± 0.40%) at 100 μg/L was comparable to that of ascorbic acid (76.94 ± 0.54) at 75 μg/mL.

Table 30.5 Compounds identified in the aqueous fraction of *P. africana*

Peak no.	Area %	Molecular weight	Similarity index (%)	Formula	Compound name	Classification of compound
1	11.29	170	76	$C_5H_{12}FO_3P$	2-Methoxyethyl ethylphosphonofluoridoate	–
2	1.00	130	84	$C_6H_{10}O_3$	Methyl levulinate	Carboxylic acid methyl ester
3	1.07	106	83	$C_4H_{10}O_3$	Glycerol 1-monomethyl ether	Glycerol ester
4	9.97	134	83	$C_6H_{14}O_3$	Carbitol	Polyether
5	2.29	154	92	$C_4H_{10}O_4S$	Diethyl sulphate	–
6	0.41	110	81	$C_7H_{10}O$	2-Methyl-2-cyclohexen-1-one	–
7	3.84	144	95	$C_7H_{12}O_3$	Ethyl levulinate	Keto acid ester
8	0.79	162	90	$C_8H_{18}O_3$	Diethyl carbitol	Polyether
9	2.63	125	75	$C_2H_7NO_3S$	Taurine	Organosulfonic
10	0.51	142	70	$C_6H_{10}N_2O_2$	Acetic acid, 2-(2-pyrrolidinylideneamino)-	–
11	0.36	136	80	$C_8H_8O_2$	Ethylene o-phenylene dioxide	–
12	0.98	120	78	$C_5H_{12}O_3$	Diethylene glycol monomethyl ether	Polyether
13	3.81	164	97	$C_7H_{16}O_4$	Methoxytriethylene glycol	Polyether
14	1.36	166	74	$C_6H_{14}O_3S$	Diisopropyl sulphite	–
15	0.89	90	76	$C_4H_{10}O_2$	Ethyl cellosolve	Hydroxyether
16	7.95	178	97	$C_8H_{18}O_4$	Ethyltriethylene glycol	Polyether
17	3.47	102	87	$C_6H_{14}O$	1-Ethylpropyl methyl ether	–
18	1.12	250	94	$C_{12}H_{26}O_5$	Tetraethylene glycol diethyl ether	Polyether
19	0.49	116	75	$C_6H_{12}O_2$	5-Methoxy-2-pentanone	–
20	1.76	252	94	$C_{11}H_{24}O_6$	Pentaethylene glycol monomethyl ether	Polyether
21	3.65	214	96	$C_{13}H_{26}O_2$	Methyl laurate	Fatty acid ester

(continued)

Table 30.3 (continued)

Peak no.	Area %	Molecular weight	Similarity index (%)	Formula	Compound name	Classification of compound
22	0.66	162	77	$C_8H_{18}O_3$	Diethyl carbitol	Polyether
23	0.98	252	87	$C_{11}H_{24}O_6$	Pentaethylene glycol monomethyl ether	Polyether
24	2.11	222	96	$C_{10}H_{22}O_5$	Tetraethylene glycol monoethyl ether	Polyether
25	3.59	200	92	$C_{12}H_{24}O_2$	Lauric acid	Fatty acid
26	4.15	228	90	$C_{14}H_{28}O_2$	Ethyl laurate	Fatty acid ester
27	1.51	250	93	$C_{12}H_{26}O_5$	Tetraethylene glycol diethyl ether	Polyether
28	0.53	218	74	$C_{10}H_{18}O_5$	Diethyl 3-hydroxy-2,2-dimethylsuccinate	–
29	5.33	256	75	$C_{14}H_{28}N_2O_2$	*N*-[2-(hexanoylamino)ethyl]hexanamide	–
30	0.51	206	73	$C_8H_{14}O_6$	4,6-O-ethylidenehexopyranose	Sugar
31	1.31	242	96	$C_{15}H_{30}O_2$	Myristic acid, methyl ester	Fatty acid ester
32	8.06	173	77	$C_7H_{11}NO_2S$	Ethyl 3-isothiocyanatobutyrate	–
33	1.22	242	91	$C_{15}H_{30}O_2$	*n*-Tridecanoic acid ethyl ester	Fatty acid ester
34	0.52	250	74	$C_{12}H_{26}O_5$	Tetraethylene glycol diethyl ether	Polyether
35	0.63	296	85	$C_{13}H_{28}O_7$	Hexaethylene glycol monomethyl ether	Polyether
36	0.43	538	74	$C_{28}H_{58}O_9$	Octaethylene glycol monododecyl ether	Polyether
37	1.24	250	92	$C_{12}H_{26}O_5$	Tetraethylene glycol diethyl ether	Polyether
38	1.36	158	78	$C_{10}H_{22}O$	3-Ethyl-4-methyl-3-heptanol	Ethyl alcohol

39	0.94	270	93	$C_{17}H_{34}O_2$	Palmitic acid, methyl ester	Fatty acid ester
40	0.92	284	89	$C_{18}H_{36}O_2$	Ethyl palmitate	Fatty acid ester
41	0.61	538	76	$C_{28}H_{58}O_9$	Octaethylene glycol monododecyl ether	Polyether
42	0.49	250	84	$C_{12}H_{26}O_5$	Tetraethylene glycol diethyl ether	Polyether
43	0.53	282	81	$C_{18}H_{34}O_2$	Ethyl 9-hexadecenoate	Fatty acid ester
44	0.71	268	77	$C_{24}H_{48}O_2$	Ethyl docosanoate	Fatty acid ester
45	0.20	538	76	$C_{28}H_{58}O_9$	Octaethylene glycol monododecyl ether	Polyether
46	0.33	356	71	$C_{22}H_{44}O_3$	2-Hydroxy-1-methylpropyl stearate	–
47	0.23	538	78	$C_{28}H_{58}O_9$	Octaethylene glycol monododecyl ether	Polyether
48	0.45	322	69	$C_{21}H_{38}O_2$	[1,1′-Bicyclopropyl]-2-octanoic acid, 2′-hexyl-, methyl ester	–
49	0.26	274	66	$C_{15}H_{34}O_2Si$	1-Methoxy-3-(2-trimethylsilyloxyethyl)nonane	–
50	0.30	356	69	$C_{22}H_{44}O_3$	2-Hydroxy-1-methylpropyl stearate	–
51	0.11	322	69	$C_{21}H_{38}O_2$	[1,1′-Bicyclopropyl]-2-octanoic acid, 2′-hexyl-, methyl ester	–
52	0.12	322	70	$C_{21}H_{38}O_2$	[1,1′-Bicyclopropyl]-2-octanoic acid, 2′-hexyl-, methyl ester	–

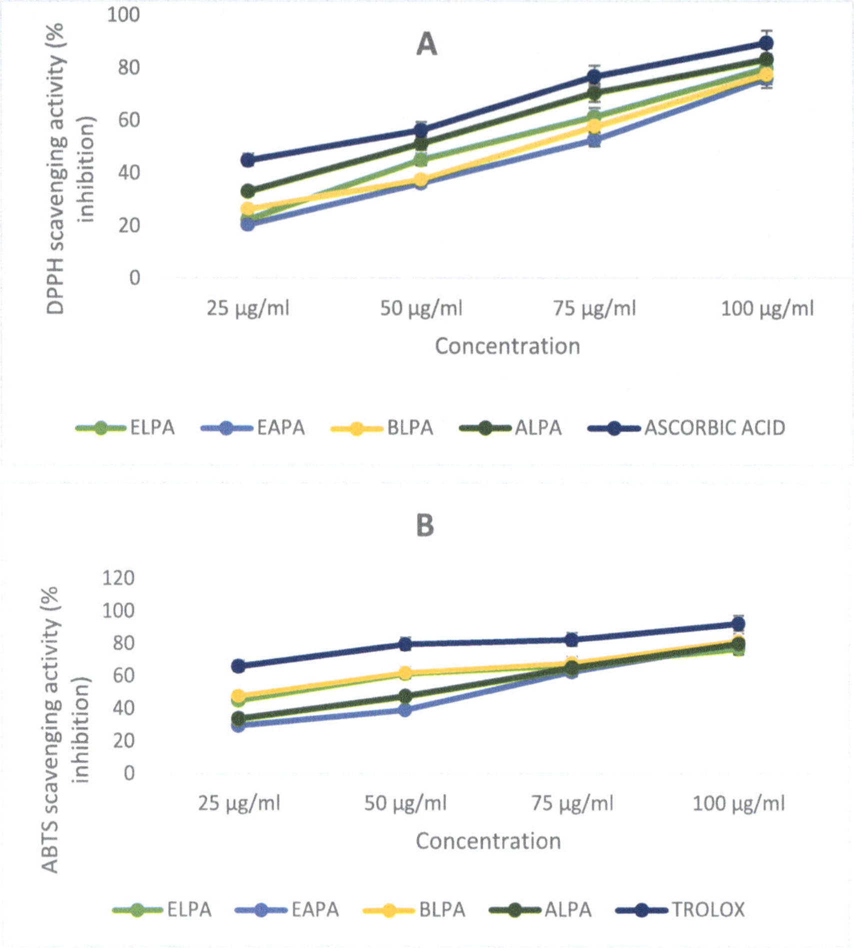

Fig. 30.5 Antioxidant activity of *P. africana* extracts. (**a**) DPPH scavenging activity of the solvent extract and differential fractions of *P. africana*; (**b**) ABTS scavenging activity of the solvent extract and differential fractions of *P. africana*. Values are presented as mean ± SEM of triplicate readings. ELPA, EAPA, BLPA and ALPA represent ethanolic, ethyl acetate, butanol and aqueous fractions of *P. africana*

4 Discussion

Plants that are high in protein content have been used to manage cases of protein-energy malnutrition. Similarly, plants rich in fibre have been employed in the management of metabolic diseases such as obesity and cardiovascular diseases (Lockyer et al., 2016). In recent times, bioactive peptides derived from food proteins have been reported to possess both nutritional and functional properties (Lee & Hur, 2017), and some peptides of plant proteins are beneficial in hypertension management. Plant peptides have been reported to possess antioxidant, hypotensive, anti-thrombotic and immunomodulatory effects (Lee & Hur, 2017). The proteins of this plant could be further studied to develop drugs and functional foods needed for disease management.

A preliminary phytochemical evaluation carried out on the ethanol extract of *P. africana* leaves showed the presence of saponins, flavonoids, tannins, alkaloids, quinones, anthocyanin, steroids and terpenoids (Obode et al., 2021). In this study, tannins, saponins, flavonoids, phenols and coumarins were present in all differential solvent fractions of *P. africana*; whereas anthocyanins and quinones were detected in the butanol and aqueous fractions. Alkaloids, betacyanins, triterpenoids and steroids were seen in the ethyl-acetate fraction of *P. africana*. Previous reports by Ajiboye et al. (2013) showed that the seed extract contains saponins, steroids, alkaloids, phlobatannins, flavonoids and tannins. The complete phytochemical profile of *P. africana* leaf using GC-MS is reported for the first time in this study. The analysis revealed the presence of 32, 53, 37 and 52 phytochemicals in the ethanol, ethyl acetate, butanol and aqueous leaf fractions, respectively. The most abundant compounds present in the ethanol, ethyl acetate, butanol and aqueous fractions based on area per cent include lauric acid (27.41), myristic acid (11.79), tetraethylene glycol monododecyl ether (7.54), isoamyl laurate (7.26), squalene (5.82) and palmitic acid (5.43); lauric acid (22.19), ethyl linolenate (6.96), cis-cis-linoleic acid (6.58), squalene (6.35) and myristic acid (6.24); methyl laurate (13.26), *n*-butyl laurate (9.94), myristic acid methyl ester (7.78), methyl (7E)-7-hexadecenoate (6.72) and palmitic acid methyl ether (5.63); 2-methoxyethyl ethylphosphonofluoridoate (11.29), carbitol (9.97), ethyl 3-isothiocyanatobutyrate (8.06), ethyl triethyleneglycol (7.95) and *N*-[2-(hexanoylamino)ethyl] hexanamide (5.33), respectively.

Lauric acid is found in numerous foods including seeds and fruits as well as human milk (Silberstein et al., 2013). It is a saturated fatty acid; however, various studies indicate that lauric acid possesses anti-inflammatory and cardio-protective properties. It has been reported that lauric acid supplementation in diet using human subjects enhances high-density lipoprotein (HDL) with unaltered plasma homocysteine concentration (an indicator of cardiovascular risk) and inflammatory markers (Voon et al., 2011; Alves et al., 2017). Also, lauric acid has been reported to reduce blood pressure and confer antioxidant and vasorelaxant benefits (Braga et al., 2011; Alves et al., 2017). The antihypertensive potentials of dietary linolenic and conjugated linoleic acid have been documented (Caligiuri et al., 2014). Linolenic acids are precursors of prostaglandins (such as $PG1_3$). $PG1_3$ is a vasodilatory molecule that could reduce blood pressure through modulation of vascular tone (Rieg et al., 2018). Also, the antihypertensive mechanism of linolenic acid could occur via the anti-inflammatory route owing to reduced expression of interleukin-6 and C-reactive protein (Alfaidi et al., 2018). Other biological activities reported for linolenic acid include hepatoprotective, hypocholesterolaemia, cancer-preventive, antiacne, anti-arthritic, etc. (Sermakkani & Thangapandian, 2012).

Squalene, predominantly found in sharks (Squalidae family) is also found in appreciable quantities in plant sources. It is a naturally occurring lipid with a symmetrical 30-carbon polyprenyl chemical structure that belongs to the triterpene hydrocarbon group. Squalene is a potent natural antioxidant, and studies have linked higher oil squalene concentrations to better antioxidant activity. Strong in vivo antioxidant effects of squalene have been seen in mice, and it has also been proposed that it may be able to control inflammatory and oxidative stress responses (Jayathilake et al., 2019; Lee et al., 2022). Phytol is diterpene alcohol and is a precursor for vitamins E and K. It has been reported as an anticancer, antimicrobial, diuretic and anti-inflammatory molecule (Gliszczyńska et al., 2021).

Taurine also is known as 2-aminoethanesulphonic acid, and it is the highest-occurring nonessential, sulphur-containing amino acid in the heart, brain, skeletal muscle, blood, liver and other organs (Rahman et al., 2011). Numerous biological and pharmacological as well as physiological roles have been reported on taurine. These include antioxidants, detoxification, membrane stabilizing, protection against DNA damage, hypertension regulation and ion flux modulation (Ibrahim et al., 2020). In an earlier study, taurine administration significantly reduced high blood pressure, enhancing nitric oxide bioavailability and mitigating renal dysfunction in rats (Ibrahim et al., 2020). Other researchers have reported similar compounds in alcoholic extracts of plants. For instance, Sermakkani and Thangapandian (2012) reported the presence of phytol and squalene in methanol extracts of *Cassia italica*, while the same compounds were found in GC-MS analysis of *Aloe vera* (Rath et al., 2020).

The capacity of *P. africana* to quench radical molecules was determined using DPPH and ABTS radical scavenging capacity. The percentage inhibition of the various extracts was compared with standard antioxidants—ascorbic acid and Trolox, respectively. Higher DPPH inhibition was observed for the aqueous extract (33.12 ± 0.90, 51.27 ± 0.57, 70.52 ± 0.66 and 88.44 ± 0.40%) at all four concentrations used (25, 50, 75 and 100 μg/mL, respectively) than that obtained for the ethanol (22.62 ± 0.42, 45.20 ± 0.40, 61.60 ± 0.51 and 80.17 ± 0.24%), ethyl acetate (20.62 ± 0.71, 36.34 ± 1.05, 52.74 ± 0.75 and 76.05 ± 0.48%) and butanol fractions (26.74 ± 0.32, 37.82 ± 0.88, 57.96 ± 0.33 and 77.74 ± 0.40%); whereas ethanol (45.2 ± 0.52, 61.61 ± 0.35, 66.95 ± 0.51 and 76.83 ± 0.17%) and butanol (48.10 ± 0.60, 62.29 ± 0.71, 68.25 ± 0.35 and 82.00 ± 0.52%) fractions had higher ABTS scavenging activities than the ethyl acetate (29.75 ± 0.60, 39.24 ± 0.43, 62.75 ± 0.52 and 80.18 ± 0.26%) and water (33.84 ± 0.43, 47.53 ± 0.51, 64.80 ± 0.43 and 79.73 ± 0.4%) fractions. Also, the ABTS scavenging capacity of all extracts was comparable at higher concentrations (75 and 100 μg/mL). Although the radical scavenging capacity of the standard antioxidants used is higher than that obtained for the extracts, some of the bioactive compounds present in the leaves of *P. africana* (lauric acid and taurine) could be responsible for the antioxidant action of the plant. These compounds have been reported to confer antioxidant benefits (Alves et al., 2017; Obasi & Ogugua, 2021). Similar studies conducted on another Prosopis species showed that extracts of *P. cineraria* possess antioxidant activity which was, however, lower than that of the standard antioxidant used (Gupta et al., 2015).

5 Conclusion

The present study reported the proximate composition, GC-MS characterization, and antioxidant capacities of the leaves of *P. africana*. There is a continuous search for new biological compounds that could be effective in the management of diseases. Our findings revealed that extracts of *P. africana* contain bioactive compounds that could be beneficial to health as well as the management of diseases. Further research is recommended in the isolation and purification of these compounds as well as the study of their mechanisms of action.

Acknowledgments The authors acknowledge the support of the technical staff at the Department of Biochemistry, Covenant University, Ota, as well as Mr. Godffery Asieba of the Analytical section of Federal Institute of Industrial Research, Oshodi, Lagos, for their technical assistance. Also, we appreciate the Covenant University Centre for Research, Innovation, and Discovery for paying the cost of publication.

Funding Not applicable.

Declarations Not applicable.

Competing Interests There are no competing interests.

References

Ademosun, O. T., Adebayo, A. H., & Ajanaku, K. O. (2021). Solanum lycopersicum and Daucus carota: Effective anticancer agents (a mini review). *Journal of Physics: Conference Series, 1943*(1), 012169.

Ajiboye, A. A., Agboola, D. A., Fadimu, O. Y., & Afolabi, A. O. (2013). Antibacterial, phytochemical and proximate analysis of *Prosopis africana* (Linn) seed and pod extract. *FUTA Journal of Research in Sciences, 1*, 101–109.

Alfaidi, M. A., Chamberlain, J., Rothman, A., Crossman, D., Villa-Uriol, M. C., Hadoke, P., et al. (2018). Dietary docosahexaenoic acid reduces oscillatory wall shear stress, atherosclerosis, and hypertension, most likely mediated via an IL-1-mediated mechanism. *Journal of the American Heart Association, 7*(13), e008757.

Alves, N. F. B., de Queiroz, T. M., de Almeida Travassos, R., Magnani, M., & de Andrade Braga, V. (2017). Acute treatment with lauric acid reduces blood pressure and oxidative stress in spontaneously hypertensive rats. *Basic and Clinical Pharmacology and Toxicology, 120*(4), 348–353.

Association of Official Analytical Chemists (AOAC). (2016). *Official methods of analysis of AOAC international* (20th ed.). Association of Official Analytical Chemists International.

Azmir, J., Zaidul, I. S. M., Rahman, M. M., Sharif, K. M., Mohamed, A., Sahena, F., Jahurul, M. H. A., Ghafoor, K., Norulaini, N. A. N., & Omar, A. K. M. (2013). Techniques for extraction of bioactive compounds from plant materials: A review. *Journal of Food Engineering, 117*(4), 426–436.

Bekoe, E., Kretchy, I., Sarkodie, J., Okraku, A., Sasu, C., Adjei, D., & Twumasi, M. (2017). Ethnomedicinal survey of plants used for the management of hypertension sold in the Makola market, Accra, Ghana. *European Journal of Medicinal Plants, 19*(3), 1–9.

Braga, V. A., Medeiros, I. A., Ribeiro, T. P., França-Silva, M. S., Botelho-Ono, M. S., & Guimarães, D. D. (2011). Angiotensin-II-induced reactive oxygen species along the SFO-PVN-RVLM pathway: Implications in neurogenic hypertension. *Brazilian Journal of Medical and Biological Research, 44*(9), 871–876.

Caligiuri, S. P., Aukema, H. M., Ravandi, A., Guzman, R., Dibrov, E., & Pierce, G. N. (2014). Flaxseed consumption reduces blood pressure in patients with hypertension by altering circulating oxylipins via an α-linolenic acid–induced inhibition of soluble epoxide hydrolase. *Hypertension, 64*(1), 53–59.

Dosumu, O. O., Oluwaniyi, O. O., Awolola, G. V., & Oyedeji, O. O. (2012). Nutritional composition and antimicrobial properties of three Nigerian condiments. *Nigerian Food Journal, 30*(1), 43–52.

Gliszczyńska, A., Dancewicz, K., Gabryś, B., Świtalska, M., Wietrzyk, J., & Maciejewska, G. (2021). Synthesis of novel phytol-derived γ-butyrolactones and evaluation of their biological activity. *Scientific Reports, 11*(1), 1–14.

Gupta, A., Verma, S., Gupta, A. K., Jangra, M., & Pratap, R. (2015). Evaluation of *Prosopis cineraria* (Linn.) Druce leaves for wound healing activity in rats. *Annals of Pharma Research, 3*, 70–74.

Harborne, J. B. (1984). Methods of plant analysis. In *Phytochemical methods* (pp. 1–36). Springer.

Henciya, S., Seturaman, P., James, A. R., Tsai, Y. H., Nikam, R., Wu, Y. C., Dahms, H. U., & Chang, F. R. (2017). Biopharmaceutical potentials of Prosopis spp. (Mimosaceae, Leguminosa). *Journal of Food and Drug Analysis, 25*(1), 187–196.

Ibrahim, M. A., Eraqi, M. M., & Alfaiz, F. A. (2020). Therapeutic role of taurine as antioxidant in reducing hypertension risks in rats. *Heliyon, 6*(1), e03209.

Jayathilake, A. G., Kadife, E., Luwor, R. B., Nurgali, K., & Su, X. Q. (2019). Krill oil extract suppresses the proliferation of colorectal cancer cells through activation of caspase 3/9. *Nutrition & Metabolism, 16*(1), 1–15.

Kamwamba, T. E. (2017). *Synergistic interactions of selected medicinal plants traditionally used to treat meningitis* (Doctoral dissertation). Nelson Mandela Metropolitan University.

Lee, S. C., Nkurunziza, D., Kim, S. Y., Surendhiran, D., Singh, A. A., & Chun, B. S. (2022). Supercritical carbon dioxide extraction of squalene rich cod liver oil: Optimization, characterization and functional properties. *The Journal of Supercritical Fluids, 188*, 105693.

Lee, S. Y., & Hur, S. J. (2017). Antihypertensive peptides from animal products, marine organisms, and plants. *Food Chemistry, 228*, 506–517.

Lockyer, S., Spiro, A., & Stanner, S. (2016). Dietary fibre and the prevention of chronic disease–should health professionals be doing more to raise awareness? *Nutrition Bulletin, 41*(3), 214–231.

Obasi, D. C., & Ogugua, V. N. (2021). GC-MS analysis, pH and antioxidant effect of Ruzu herbal bitters on alloxan-induced diabetic rats. *Biochemistry and Biophysics Reports, 27*, 101057.

Obode, O. C., Adebayo, A. H., & Li, C. (2021). Phytochemical and toxicological evaluations of Prosopis africana (GUILL. and PERR.) extract on albino wistar rats. *Toxicological Research, 37*(2), 183–195.

Padma, M., Ganesan, S., Jayaseelan, T., Azhagumadhavan, S., Sasikala, P., Senthilkumar, S., & Mani, P. (2019). Phytochemical screening and GC–MS analysis of bioactive compounds present in ethanolic leaves extract of *Silybum marianum* (L). *Journal of Drug Delivery and Therapeutics, 9*(1), 85–89.

Rahman, M. M., Park, H. M., Kim, S. J., Go, H. K., Kim, G. B., Hong, C. U., Lee, Y. U., Kim, S. Z., Kim, J. S., & Kang, H. S. (2011). Taurine prevents hypertension and increases exercise capacity in rats with fructose-induced hypertension. *American Journal of Hypertension, 24*(5), 574–581.

Raji, N. O., Adebisi, I. M., & Bello, S. O. (2013). Ethnobotanical survey of antihypertensive agents in Sokoto, Northwest Nigeria. *International Journal of Innovative Research and Development, 2*(5), 1820–1836.

Rath, M., Bhattacharya, A., Rath, K., Santra, S., Ghosh, G., & Nanda, B. B. (2020). A comprehensive study of the neuropharmacological profile of methanol leaf extract of aloe vera and identification of associated neuroprotective compounds through gas chromatography-mass spectrometry analysis. *Indian Journal of Pharmaceutical Sciences, 82*(6), 996–1005.

Rieg, A. D., Suleiman, S., Anker, C., Verjans, E., Rossaint, R., Uhlig, S., & Martin, C. (2018). PDGF-BB regulates the pulmonary vascular tone: Impact of prostaglandins, calcium, MAPK- and PI3K/AKT/mTOR signalling and actin polymerisation in pulmonary veins of Guinea pigs. *Respiratory Research, 19*(1), 1–18.

Sermakkani, M., & Thangapandian, V. (2012). GC-MS analysis of *Cassia italica* leaf methanol extract. *Asian Journal of Pharmaceutical and Clinical Research, 5*(2), 90–94.

Silberstein, T., Burg, A., Blumenfeld, J., Sheizaf, B., Tzur, T., & Saphier, O. (2013). Saturated fatty acid composition of human milk in Israel: A comparison between Jewish and Bedouin women. *The Israel Medical Association Journal, 15*(4), 156–159.

Singh, M., Dubey, R. K., Koley, T. K., Maurya, A., Singh, P. M., & Singh, B. (2019). Valorization of winged bean (*Psophocarpus tetragonolobus* (L) DC) by evaluation of its antioxidant activity through chemometric analysis. *South African Journal of Botany, 121*, 114–120.

Voon, P. T., Ng, T. K. W., Lee, V. K. M., & Nesaretnam, K. (2011). Diets high in palmitic acid (16: 0), lauric and myristic acids (12: 0+ 14: 0), or oleic acid (18: 1) do not alter postprandial or fasting plasma homocysteine and inflammatory markers in healthy Malaysian adults. *The American Journal of Clinical Nutrition, 94*(6), 1451–1457.

Chapter 31
Jobelyn® Reverses Hypertension and Improves Selected Biochemical Parameters in Rats Induced with Sodium Chloride

Abiodun Humphrey Adebayo, Okukwe Christa Obode, Kosiso Chiekweta, Oluwatobiloba Adesewa Oriowo, and Olajuwon Okubeno

1 Introduction

Hypertension, also known as systemic arterial hypertension, is characterized by a persistent elevation in arterial blood pressure. Typically expressed as a ratio of systolic to diastolic blood pressure, hypertension has become a global health concern, with its prevalence shifting from industrialized nations to developing and underdeveloped countries. For instance, recent statistics show that Tanzania and Kenya have recorded the highest rates (Okello et al., 2020). Hypertension has been associated with significant health consequences (Adal et al., 2024). This is mainly rooted in the fact that hypertension acts as a major risk factor for cardiovascular diseases (CVDs), which include coronary heart diseases, heart failure, stroke, and myocardial infarction (Oparil et al., 2018). Hypertension is caused by numerous factors, including genetic and environmental causes. While some of the risk factors are modifiable

A. H. Adebayo (✉) · K. Chiekweta
Department of Biochemistry, College of Science and Technology, Covenant University, Ota, Ogun State, Nigeria
e-mail: abiodun.adebayo@covenantuniversity.edu.ng

O. C. Obode
Department of Biochemistry, College of Science and Technology, Covenant University, Ota, Ogun State, Nigeria

Nutrition and Toxicology Division, Food Technology Department, Federal Institute of Industrial Research, Oshodi, Lagos, Nigeria

O. A. Oriowo
Covenant Applied Informatics and Communication Africa Center of Excellence (CApIC-ACE), Covenant University, Ota, Ogun State, Nigeria

O. Okubeno
Health Forever International, Ikeja, Lagos, Nigeria

P. O. Isibor, P. A. Akinduti (eds.), *Harnessing Biotechnology Tools for Product Development*, https://doi.org/10.1007/978-3-031-86002-7_31

(exercise/level of activity and dietary habits), others are non-modifiable like family history, diabetes, and chronic renal disease. Eating and drinking habits are well-known factors that control high blood pressure in an individual. Khalili et al. (2022) stated that hypertension is associated with low potassium and high sodium consumption; this has resulted in a salt substitution strategy in China which has effectively reduced their risk of cardiovascular disease (Wang et al., 2023). According to the World Health Organization (WHO), people with hypertension and those taking more than 5 g of salt daily are likely to develop cardiovascular diseases. Furthermore, fluid retention in body tissues is indicative of kidney malfunction which worsens the condition (Kehrenberg & Bachmann, 2022).

In the management of hypertension, diuretics have long been considered the first line of treatment. These medications function by eliminating excess water and electrolytes, particularly sodium, from the body. Their mechanism of action involves inhibiting the reabsorption of salt, which is abundant in the body, by the kidneys, and consequently reducing water reabsorption (Arumugham & Shahin, 2022). The principle underlying diuretic action is the inhibition of sodium reabsorption at various points along the nephron. By blocking sodium reabsorption, diuretics increase sodium excretion, leading to increased water excretion and ultimately increased urine output. This process directly affects the vascular system by decreasing fluid volume, thereby reducing blood pressure.

While sodium plays a vital role in cellular action potentials, its excessive intake has been linked to numerous health issues. The WHO's recommendation of limiting salt intake to <5 g/day is based on substantial evidence linking high sodium consumption to increased risk of hypertension and cardiovascular diseases (Grillo et al., 2019). Despite their efficacy, the use of diuretics is not without drawbacks. A notable study associated diuretic use with an increased mortality rate in patients suffering from acute kidney injury (AKI) (Hegde, 2020). Furthermore, prolonged or excessive use of diuretics has been linked to various side effects and risks, including arrhythmias, hypovolemic shock, and renal complications. These adverse effects, coupled with economic considerations, have spurred interest in alternative treatment modalities, particularly the use of medicinal plants. The appeal of plant-based remedies lies in their perceived effectiveness, affordability, and rich content of natural compounds, including antioxidants, when compared to synthetic drugs (Chaachouay & Zidane, 2024). The field of ethnopharmacology has seen significant advancements in recent years, with medicinal plants being isolated and utilized as active ingredients in the development of novel treatment options. These plants produce active chemicals with diverse physiological effects, including medicinal and synergistic properties. One such promising plant-based product is Jobelyn® (JB), a nutritional supplement derived from the polyphenol-rich leaf sheath of Sorghum bicolor (L) Moench, belonging to the Poaceae family. Sorghum bicolor has garnered attention for its applications in African traditional medicine (Okubena et al., 2022). Jobelyn® is typically available in capsule form, with each capsule containing approximately 250 mg of the active ingredient. The efficacy of Jobelyn® can be attributed to its rich phytochemical profile. The key active constituents include proanthocyanidins, anthocyanidins, apigenins, apigenidins, proapigenidins, luteolins, and naringenins (Omorogbe et al., 2018). These compounds contribute to the supplement's diverse pharmacological activities. One of the notable mechanisms of action of Jobelyn® is its ability to enhance lipid metabolism by activating the AMP-activated protein kinase-sterol regulatory enzyme-binding protein (AMPK-SREBP) pathway (Lu et al., 2019). This activation plays a crucial role in regulating cellular energy homeostasis and lipid metabolism, which are often implicated in cardiovascular health.

Sorghum bicolor, the source plant for Jobelyn®, has a long history of use in traditional medicine. It has been employed to treat various ailments, including hypertension, cancer, and inflammatory conditions (Lin et al., 2008). The transition from traditional use to modern pharmacological applications represents a significant advancement in ethnomedicinal research. Among the active compounds in Jobelyn®, luteolin has emerged as a compound of particular interest in cardiovascular health. Research has demonstrated that luteolin can effectively decrease blood pressure following oral administration (Su et al., 2015). This anti-hypertensive effect is complemented by luteolin's potent antioxidant capabilities. The antioxidant properties of luteolin are primarily attributed to its molecular structure, which enables it to scavenge both oxygen and nitrogen species effectively. Beyond its antioxidant role, luteolin exhibits anti-inflammatory properties by inhibiting cytokine expression and TLR4 signaling in immune cells (Nabavi et al., 2015). At a cellular level, luteolin's protective effects are further enhanced by its ability to attach to membranes in various organs through hydrogen bonding. This mechanism helps shield cellular structures from harmful chemicals, potentially mitigating oxidative stress-induced damage associated with hypertension and other cardiovascular conditions.

The therapeutic potential of Jobelyn® is not limited to the actions of individual compounds. Recent research has highlighted synergistic effects between its components and other pharmacological agents. For instance, luteolin and metformin have been shown to work synergistically in inhibiting pro-inflammatory mediators (such as interleukins and tumor necrosis factors) and proapoptotic factors (including caspases and Bax) in liver cells (Taheri et al., 2021). Such synergistic interactions suggest the potential for combination therapies that could enhance efficacy while potentially reducing side effects associated with conventional treatments.

While the preliminary evidence supporting the antihypertensive and cardioprotective effects of Jobelyn® is promising, there remains a critical need for comprehensive, well-designed studies to fully elucidate its mechanisms of action and clinical efficacy. The shift toward plant-based remedies in managing hypertension and other cardiovascular conditions represents a promising avenue for more holistic, potentially safer treatment options. This study aims to contribute to this growing body of knowledge by investigating the effects of Jobelyn® on biochemical, hematological, and histological assessment of sodium chloride (NaCl)-induced hypertension in a rat model. By elucidating the potential efficacy of this plant-derived supplement, new avenues for alternative or complementary treatments for hypertension are opened, ultimately contributing to improved patient outcomes and quality of life.

2 Materials and Method

NaCl was acquired from Kernel Chemicals, China. Other reagents were purchased from Randox laboratories (Crumlin, UK) unless stated otherwise. Jobelyn herbal formulation was obtained in powder form from a Lagos pharmaceutical outlet.

Twenty-four male albino rats weighing 175–200 g were obtained from the rat Animal Laboratory, College of Medicine, University of Lagos (CUML), Idi Araba, and acclimatized for 2 weeks. All experimental animals were handled following standard protocols officially stated by the Covenant Research Ethics Committee of Covenant University, Ota, Ogun State, Nigeria.

2.1 Experimental Design

Twenty-four male albino rats were weighed and grouped randomly into four groups of four animals per group for drug administration. They were administered a standard diet (feed) and water. The animal feed was formulated with 8% NaCl and administered orally with a dose of 250 mg/kg and a concentration of 2.5 mg/mL. Group 1 served as the normal control and were given standard rat feed without sodium, group 2 animals served as the negative control, induced with sodium chloride with no treatment, group 3 animals were induced with sodium chloride and treated with a standard drug- Amlodipine (10 mg/kg body weight), and group 4 animals were induced with sodium chloride and treated with the test drug, Jobelyn (250 mg/kg body weight). After 3 weeks of administration, the animals were fasted overnight, their blood pressure was taken after which they were sacrificed.

2.2 Biochemical and Hematological Assays

After the treatment time, the animals were sacrificed under euthanasia. Fresh blood was obtained and kept for blood chemistry analysis. The liver and kidney were excised and homogenized. Serum obtained from each blood sample collected from the mice were used to ascertain the activities of aspartate aminotransferase (AST) (Bergmeyer et al., 1986a), alkaline phosphatase (ALP) (Bergmeyer et al., 1986b), alanine aminotransferase (ALT) (Tietz et al., 1983), bilirubin (Doumas et al., 1971), urea, creatinine, triglycerides (Zoppi & Fellini, 1976), and albumin (Doumas et al., 1973), while the liver homogenate was used to determine the level of triglycerides (Doumas et al., 1973), total cholesterol, HDL-cholesterol, and LDL-cholesterol (Zoppi & Fellini, 1976).

The blood samples of experimental animals were collected in an EDTA tube for evaluation using the Cobas biochemical analyzer and different blood parameters such as hemoglobin, red blood cell (RBC) count, white blood cell (WBC) count, packed cell volume (PCV), neutrophils, monocytes, etc. were carried out using hematology analyzer from Best Choice™ Hub merchandise Laboratories RC 2382463.

Histological evaluation was done following the method reported by Obode et al. (2021). After blood collection, the liver was carefully dissected and fixed in normal saline for 72 h and sliced into a thickness of 2.1 mm. The tissues were dehydrated with alcohol of varying concentrations, further treated with paraffin wax and cast into blocks; sections of the tissues were then cut on a microtome to 5 μm. These were later placed on a slide and allowed to dry. The sample slides were subsequently stained in hematoxylin-eosin and examined under a light microscope (Olympus CX21); photomicrographs of the samples were then recorded.

2.3 *Statistical Analysis*

Interpretation of results and statistical analysis of the results were done with GraphPad Prism statistical software version 9.5. The graphs show the mean and standard error of the mean (SEM) of the results and Turkey's multiple comparisons using the one-way analysis of variance (ANOVA). Statistical significance between groups is represented by alphabets. No alphabets signify there is no significant difference between the groups. The results were considered to be of statistical significance where the values of p were <0.05.

3 Results

4 Discussion

Hypertension, a significant global health concern, is particularly prevalent in Sub-Saharan African regions and underdeveloped countries. Interestingly, its occurrence in low-income countries often differs from developed nations, as the typical risk factors such as reduced physical activity and obesity are not always present (Kohler et al., 2022). The gravity of this condition is underscored by its contribution to over 8.5 million deaths globally, primarily due to vascular diseases, stroke, and renal complications (Zhou et al., 2021).

Our study investigated the potential of Jobelyn®, a supplement derived from Sorghum bicolor, as a novel approach to hypertension management. The results reveal a complex interplay of effects on various physiological parameters, which will be discussed in detail.

Jobelyn® demonstrated significant antihypertensive properties, as evidenced by reductions in systolic blood pressure, diastolic blood pressure, arterial blood pressure, mean arterial blood pressure, and heart rate compared to the negative control (Fig. 31.1). These findings align with a similar study with *Clausena anisata* leaf extract containing comparable phytochemicals such as alkaloids and flavonoids (Kamyab et al., 2021). The parallel effects suggest that Jobelyn® may possess antihypertensive properties through similar mechanisms, potentially involving vasodilation or modulation of the renin-angiotensin-aldosterone system. Figure 31.2c shows a significant increase in cholesterol levels in the Jobelyn® treated group when compared to the positive control (Group I) although the increase in the low-density lipoprotein cholesterol (LDLC) is negligible as it had no significant ($p > 0.05$) difference between the groups. This increase may be suggested to be associated with the parent plant from which Jobelyn was derived. Sorghum bicolor, the parent plant of Jobelyn, contains phytosterols that compete with cholesterol for absorption in the intestine. In a review article by Poli et al. (2021), phytosterols were reported as highly lipophilic and have a characterized membrane transport protein called Niemann-Pick C1–Like 1 (NPC1L1), cholesterol is absorbed through the same pathway as phytosterol. The phytosterol, therefore, competes with the cholesterol in the transporter resulting in inhibition of cholesterol absorption and increased circulating cholesterol. Although HDLC (Fig. 31.2b) and triglyceride (Fig. 31.2d) did not show a significant difference with other treatment groups, their decrease compared to the other treatment groups, especially the control, was negligible, and generally lower triglyceride levels are associated with reduced risk of coronary heart disease.

Jobelyn effectively down-regulated liver marker enzymes AST as the AST level of the Jobelyn-treated group was significantly lower than the group induced with sodium chloride. AST is an enzyme that is involved in the cellular metabolic process, namely the amino acid metabolic process. It can also be located in various organs like the heart, liver, and kidneys. Interpreting the results in Fig. 31.3, elevated concentrations of AST in the blood may indicate some severe injuries such as cirrhosis of the liver. Accordingly, in this study, Jobelyn® was able to overcome the increased risk of heart disease by lowering AST levels (Fig. 31.3a). This correlates with a study where Jobelyn (5, 10, and 20 mg/kg) reversed monosodium glutamate-induced hepatic damage (Omogbiya et al., 2021). The levels of ALT, ALP, and total bilirubin remained constant compared to the baseline levels in the patients. This may imply that Jobelyn has no effects on the functional capacity of the bile ducts based on its effect on ALP. So, lowering AST without changes in ALT suggests that the Jobelyn could have a selective effect on non-hepatic tissue because ALT is specific to hepatic tissue, and therefore the significant reduction may be because one of the tissues that is rich in AST is the heart; therefore, Jobelyn exerted its antioxidant property in the heart.

Urea levels were significantly ($p < 0.05$) reduced after treatment with Jobelyn® (Fig. 31.4b). Urea is the metabolic waste product present in urine and released during amino acid metabolism. An increased level of urea signifies a problem related to the kidney which could result in kidney damage and failure. The decreased levels of

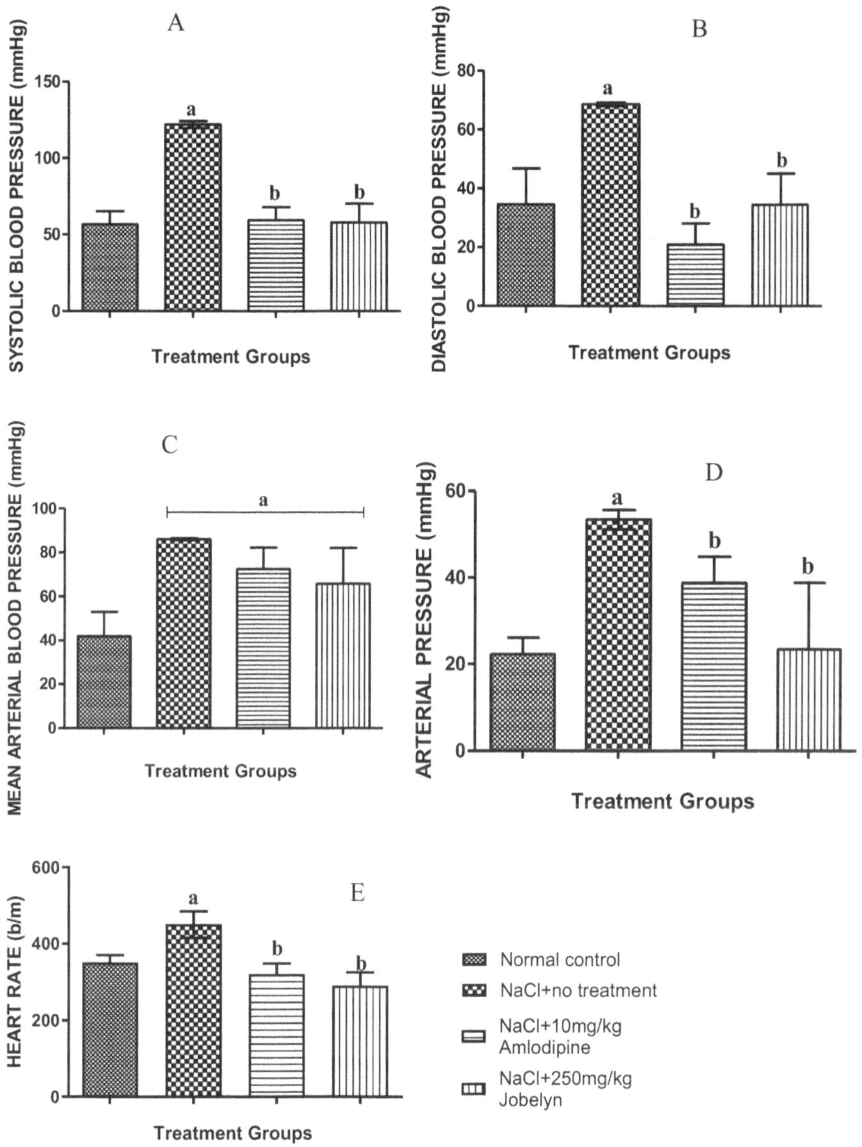

Fig. 31.1 Effect of Jobelyn® on blood pressure parameters of NaCl-induced hypertension in rats. (**a**) Systolic blood pressure. (**b**) Diastolic blood pressure. (**c**) Mean arterial blood pressure. (**d**) Arterial pressure. (**e**) Heart rate. Values are expressed as mean ± SEM of six replicates. $p < 0.05$ significantly different from the control group; $^{a}p < 0.05$

urea by Jobelyn suggest the protective effects of Jobelyn® on the kidney as decreased urea level has been associated with healthy functioning of the kidney (Imo et al., 2021). The creatinine levels also indicate renal health and function. An increased

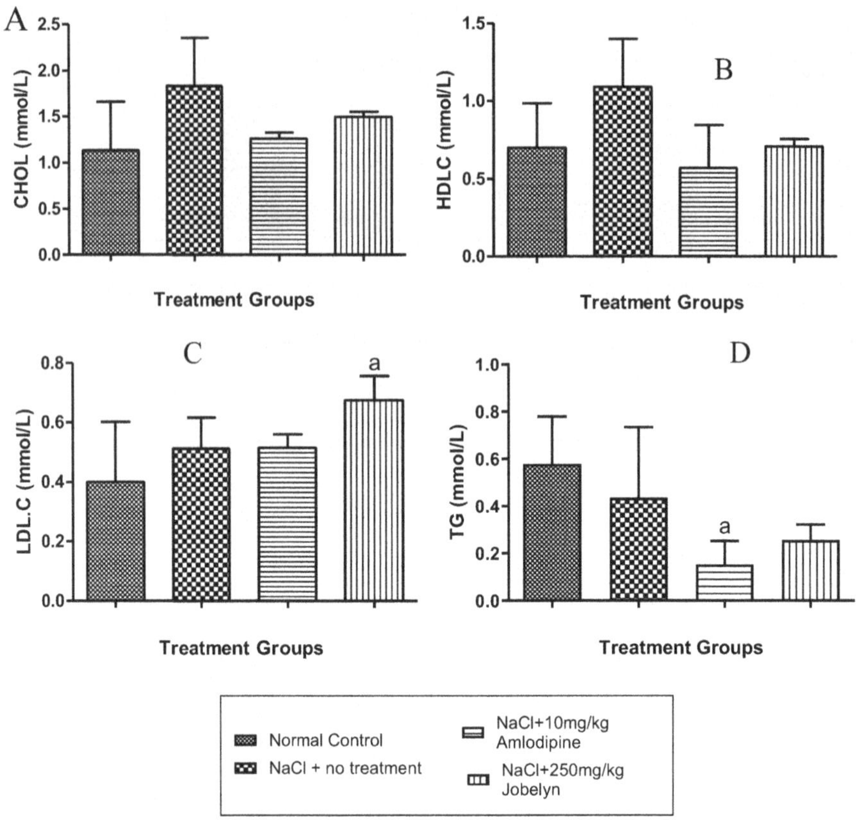

Fig. 31.2 Effect of Jobelyn® on lipid profile parameters in liver homogenate of hypertensive rats. (**a**) Total cholesterol. (**b**) High-density lipoprotein cholesterol (HDL-C). (**c**) Low-density lipoprotein cholesterol (LDL-C). (**d**) Triglyceride (TG). Results are expressed as mean ± standard error of the mean. *N* value equals 6. (**) indicates $p < 0.05$

level of creatinine in the blood may also suggest kidney impairment (Cheng et al., 2020) although the creatinine levels are not significantly ($p < 0.05$) reduced or increased from the result. This may further suggest or confirm the specificity of Jobelyn to some tissues.

The analyses involving hematological parameters give essential information on the effects of toxins in the blood (Table 31.1). The white blood cells (WBC) are major components of the immune system; they function in times of infection and invasion of pathogens, and their replication process in the body signals the effectiveness of the immune system. However, in this study, the white blood cells were significantly ($p < 0.05$) reduced after treatment with Jobelyn®. Lower WBCs are associated with reduced cardiovascular risk as chronic inflammation has been associated with increased WBC. The lymphocytes function in antibody production and immune response and increased levels of lymphocytes are associated with

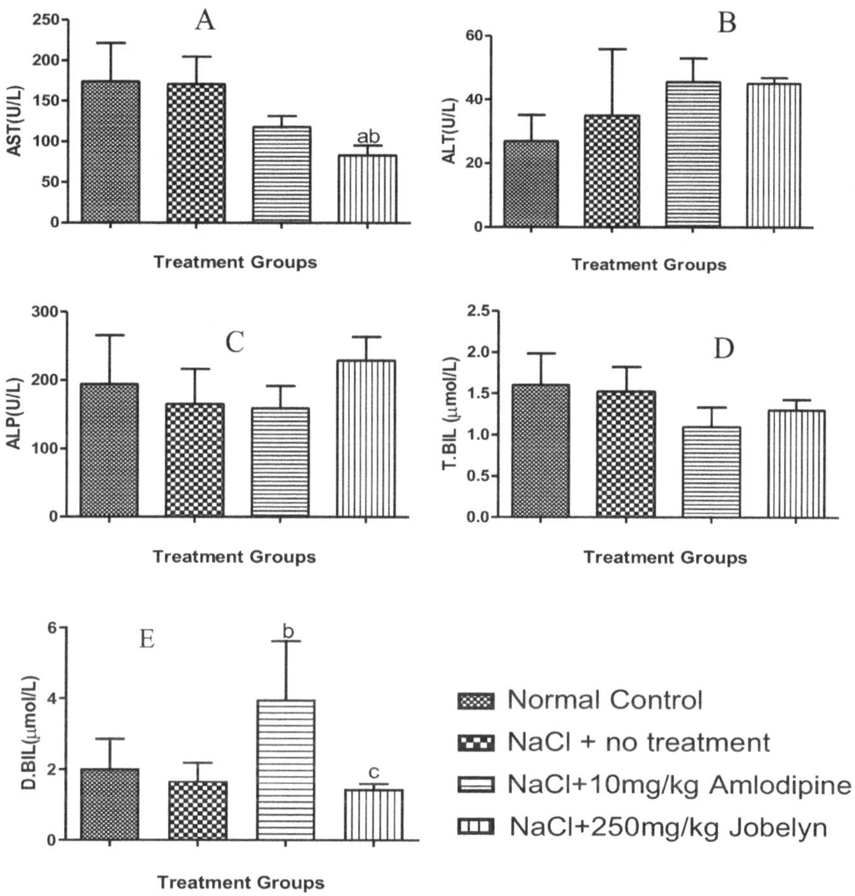

Fig. 31.3 Effect of Jobelyn® on liver function parameters of NaCl-induced hypertensive rats. (**a**) AST aspartate aminotransferase. (**b**) ALT alanine aminotransferase. (**c**) ALP alkaline phosphatase. (**d**) Total bilirubin T.BIL. (**e**) Direct bilirubin. Results are expressed as mean ± standard error of the mean. N value equals 6. (*) indicates $p < 0.05$

cardiovascular diseases. Treatment with Jobelyn® significantly ($p < 0.05$) decreased the lymphocyte levels; preventing possible risk of heart challenges as supported in a study by Menni et al. (2021). However, it has been suggested that the relationship between lymphocytes and heart diseases is complex as the effect of lymphocyte can be both pro-atherogenic and anti-atherogenic, depending on the specific lymphocyte subsets. The anti-atherogenic effect, especially the regulatory T cells (Tregs), can have protective effects against atherosclerosis; it can suppress inflammation and promote plaque stability by regulating the activity of pro-atherogenic immune cells (Churov et al., 2023), while others promote coronary heart diseases (Fig. 31.5).

The histology result in Fig. 31.6 showed a normal liver when treated with Jobelyn® as compared to the mild vascular congestion of the negative control group

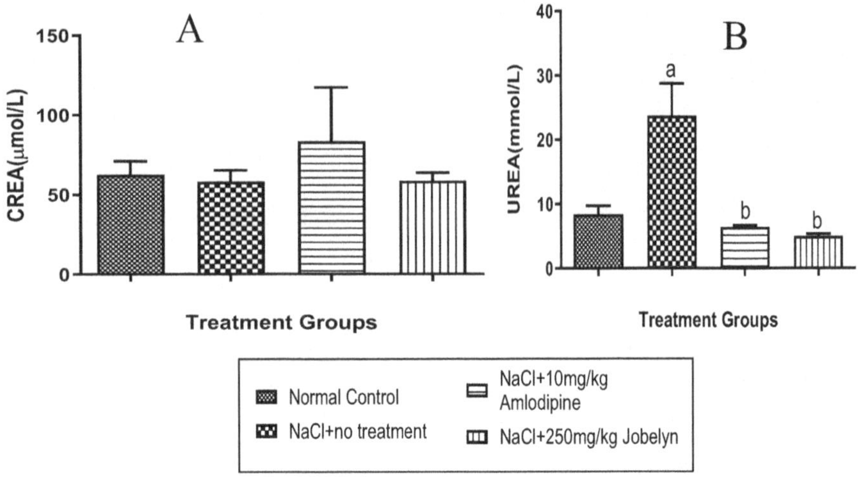

Fig. 31.4 Effect of Jobelyn® on kidney function of NaCl-induced hypertension in rats. (**a**) Creatinine. (**b**) Urea. Results are expressed as mean ± standard error of the mean. *N* value equals 6. (*) indicates $p < 0.05$

Table 31.1 Effects of Jobelyn® on hematological parameters of NaCl-induced hypertensive rats

Hematology parameters	Positive control	Negative control	NaCl + 10 mg/kg amlodipine	Nacl + 250 mg/kg Jobelyn
WBC	4.300 ± 0.4690	13.50 ± 3.643^{a}	4.400 ± 0.8888^{b}	1.733 ± 0.2082^{b}
LYMPH	0.5667 ± 0.1155	5.067 ± 0.9074^{a}	1.167 ± 0.4163^{b}	0.5000 ± 0.08165^{b}
MID	2.225 ± 0.2630	1.525 ± 0.2217	1.700 ± 0.9539	0.7333 ± 0.2517^{a}
GRANULOCYTE %	39.93 ± 4.952	50.47 ± 3.508^{a}	48.87 ± 3.522	53.33 ± 3.742^{a}
RBC	9.233 ± 0.3215	9.167 ± 0.2309	7.375 ± 0.4113^{ab}	6.9 ± 0.2646^{ab}
HB	15.70 ± 1.140	17.07 ± 0.05774	14.50 ± 0.2000^{b}	13.13 ± 0.05773^{ab}
HEMATOCRIT	48.08 ± 3.487	52.2 ± 0.6557	45.97 ± 1.387^{b}	42.6 ± 0.9899^{ab}
MCV	53.37 ± 3.317	56.03 ± 3.097	55.05 ± 2.899	57.87 ± 2.501
MCH	18.03 ± 0.9074	19.15 ± 0.07071	19.07 ± 1.422	19.00 ± 0.6928
MID %	53.37 ± 1.405	21.00 ± 1.637^{a}	33.87 ± 3.014^{ab}	43.63 ± 4.484^{ab}
MCHC	31.93 ± 0.5620	32.63 ± 0.2630	31.48 ± 0.4349^{b}	31.45 ± 0.5972^{b}
LYMPH %	18.20 ± 0.4243	23.40 ± 0.4243^{a}	18.45 ± 0.07071^{b}	14.05 ± 1.202^{ab}
GRANULOCYTE	2.125 ± 0.9979	5.400 ± 0.4583^{a}	2.167 ± 1.102^{b}	1.375 ± 0.3775^{b}
PLATELET COUNT	847.0 ± 20.69	869.0 ± 48.46	618.3 ± 47.39^{ab}	754.0 ± 53.99^{b}

Values are represented as mean ± SD of 6 replicates. Values on the same row followed by different superscript letters differ significantly ($^{a}p > 0.05$, $^{b}p < 0.05$ vs. positive and normal control, respectively). White blood cell count (WBC), Lymphocyte (LYMPH), percentage granulocyte (GRANULOCYTR%), red blood cell count (RBC), hemoglobin (HB), hematocrit (HCT), mean cell volume (MCV), mean corpuscular hemoglobin (MCH), mean corpuscular hemoglobin concentration (MCHC), percentage lymphocyte, (% LYM), granulocyte, and platelet coun

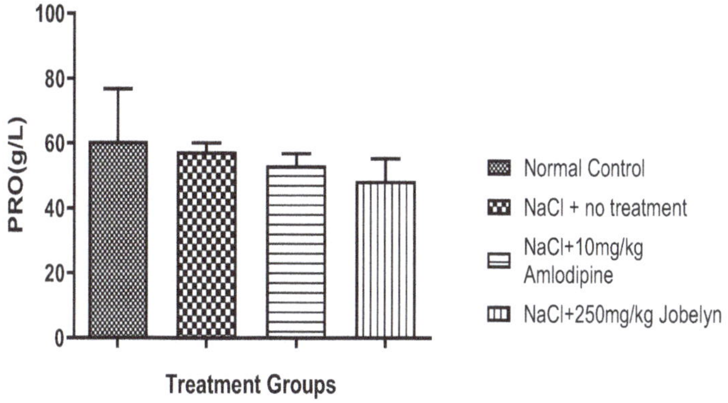

Fig. 31.5 Effects of Jobelyn® on total protein levels in sodium chloride-induced hypertensive rats. Results are expressed as mean ± standard error of the mean. *N* value = 6. (*) indicates $p < 0.05$

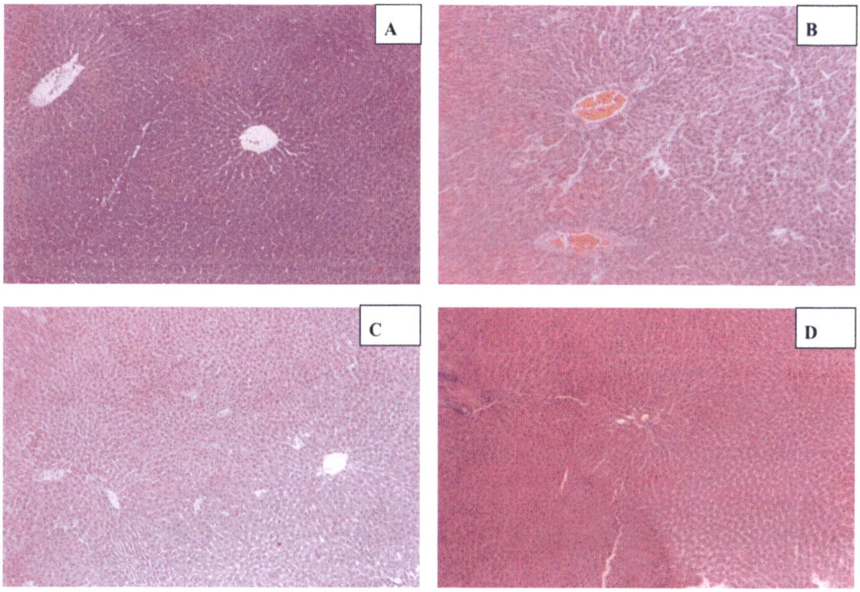

Fig. 31.6 Photomicrograph of the liver sections of Wistar rats. (**a**) Group 1 (showing no abnormalities and normal features). (**b**) Group 2 (administered 8% NaCl with no treatment showing mild congestion). (**c**) Group 3 (administered 10 mg/kg amlodipine showing no abnormality). (**d**) Group 4 (administered 250 mg/kg Jobelyn showing no congestion)

induced with NaCl following no treatment. This result aligns with a study that reported hyperchloremic metabolic acidosis in the liver section of animals given a continuous salt load (Omage & Azeke, 2019). The result shows the liver samples treated with hematoxylin and eosin (magnification x100) of each test group. The control group that was given normal feed with no induction had a normal liver with no abnormalities seen; the group induced with 8% NaCl and was given no treatment had moderate vascular congestion with edema; however, the group treated with Jobelyn and Amlodipine (the standard drug), both were normal with no abnormalities seen. This showed that Jobelyn was able to reverse the toxic effect caused by sodium chloride induction.

5 Conclusion

Results from this study revealed that Jobelyn® could reduce hypertension and impact positively various other aspects of physiology. From this study, Jobelyn is considered safe at a dose of 250 mg/kg body weight. There is therefore a need for other studies at different levels to understand the mechanism of antihypertensive action and identify the bioactive compound responsible for that effect. Furthermore, given that the product is already in use by humans, studies involving human hypertensive participants are needed to support the present outcomes and assess Jobelyn®'s efficacy as a hypertension remedy.

References

Adal, O., Kebede, N., Delie, A. M., Bogale, E. K., Anagaw, T. F., Tiruneh, M. G., et al. (2024). Hypertensive crisis and its predictors in Africa: Systematic review and meta-analysis, 2024. *International Journal of Cardiology Cardiovascular Risk and Prevention., 21*, 200285. https://doi.org/10.1016/j.ijcrp.2024.200285

Arumugham, V. B., & Shahin, M. H. (2022). Therapeutic uses of diuretic agents. In *StatPearls*. StatPearls Publishing. Retrieved from http://www.ncbi.nlm.nih.gov/books/NBK557838/

Bergmeyer, H. U., Herder, M., & Rej, R. (1986a). International Federation of Clinical Chemistry (IFCC) scientific committee, analytical section: approved recommendation (1985) on IFCC methods for the measurement of catalytic concentration of enzymes. Part 2. IFCC method for aspartate aminotransferase (L-aspartate: 2-oxoglutarate aminotransferase, EC 2.6 1:1). *Journal of Clinical Chemistry and Clinical Biochemistry, 24*(7), 497–510.

Bergmeyer, H. U., Herder, M., & Rej, R. (1986b). International Federation of Clinical Chemistry (IFCC) scientific committee, analytical section: approved recommendation (1985) on IFCC methods for the measurement of catalytic concentration of enzymes. Part 3. IFCC method for alanine aminotransferase (L-alanine: 2-oxoglutarate aminotransferase, EC 2.6 1:2). *Journal of Clinical Chemistry and Clinical Biochemistry, 24*(7), 481–495.

Chaachouay, N., & Zidane, L. (2024). Plant-derived natural products: A source for drug discovery and development. *Drugs and Drug Candidates, 3*(1), 184–207. https://doi.org/10.3390/ddc3010011

Cheng, Y., Luo, R., Wang, K., Zhang, M., Wang, Z., Dong, L., et al. (2020). Kidney disease is associated with in-hospital death of patients with COVID-19. *Kidney International, 97*(5), 829–838.

Churov, A. V., Chegodaev, Y. S., Khotina, V. A., Ofitserov, V. P., & Orekhov, A. N. (2023). Regulatory T cells in atherosclerosis: is adoptive cell therapy possible?. *Life, 13*(9), 1931.

Doumas, B. T., Perry, B. W., Sasse, E. A., & Straumfjord, J. V., Jr. (1973). Standardization in bilirubin assays: Evaluation of selected methods and stability of bilirubin solutions. *Clinical Chemistry, 19*(9), 984–993.

Doumas, B. T., Watson, W. A., & Biggs, H. G. (1971). Albumin standards and the measurement of serum albumin with bromocresol green. *Clinica Chimica Acta, 31*(1), 87–96.

Grillo, A., Salvi, L., Coruzzi, P., Salvi, P., & Parati, G. (2019). Sodium intake and hypertension. *Nutrients, 11*(9), 1970. https://doi.org/10.3390/nu11091970

Hegde, A. (2020). Diuretics in acute kidney injury. *Indian Journal of Critical Care Medicine, 24*(Suppl 3), S98–S99. https://doi.org/10.5005/jp-journals-10071-23406

Imo, C., Arowora, K. A., Ezeonu, C. S., Ikwebe, J., Yakubu, O. E., Imo, N. G., & Danlami, G. C. (2021). Biochemical and histological effects of ethanolic extracts of fruits of Xylopia aethiopica and seeds and leaves of Piper guineense on liver and kidney function in male albino rats. *Future Journal of Pharmaceutical Sciences, 7*(1), 35. https://doi.org/10.1186/s43094-021-00187-6

Kamyab, R., Namdar, H., Torbati, M., Ghojazadeh, M., Araj-Khodaei, M. & Fazljou, S.M.B. (2021) Medicinal Plants in the Treatment of Hypertension: A Review. *Advanced Pharmaceutical Bulletin. 11*(4), 601–617. https://doi.org/10.34172/apb.2021.090

Kehrenberg, M. C. A., & Bachmann, H. S. (2022). Diuretics: A contemporary pharmacological classification? *Naunyn-Schmiedeberg's Archives of Pharmacology, 395*(6), 619–627. https://doi.org/10.1007/s00210-022-02228-0

Khalili, P., Abdollahpoor, S., Ayoobi, F., Vakilian, A., Hakimi, H., Rajabi, Z., & Jamali, Z. (2022). Evaluation of relationship between serum liver enzymes and hypertension: A cross-sectional study based on data from Rafsanjan Cohort Study. *International Journal of Hypertension, 2022*, 5062622. https://doi.org/10.1155/2022/5062622

Kohler, I. V., Sudharsanan, N., Bandawe, C., & Kohler, H.-P. (2022). Aging and hypertension among the global poor—Panel data evidence from Malawi. *PLOS Global Public Health, 2*(6), e0000600. https://doi.org/10.1371/journal.pgph.0000600

Lin, Y., Shi, R., Wang, X., & Shen, H.-M. (2008). Luteolin, a flavonoid with potential for cancer prevention and therapy. *Current Cancer Drug Targets, 8*(7), 634. https://doi.org/10.2174/156800908786241050

Lu, J., Meng, Z., Cheng, B., Liu, M., Tao, S., & Guan, S. (2019). Apigenin reduces the excessive accumulation of lipids induced by palmitic acid via the AMPK signaling pathway in HepG2 cells. *Experimental and Therapeutic Medicine, 18*(4), 2965–2971. https://doi.org/10.3892/etm.2019.7905

Menni, C., Louca, P., Berry, S. E., Vijay, A., Astbury, S., Leeming, E. R., et al. (2021). High intake of vegetables is linked to lower white blood cell profile and the effect is mediated by the gut microbiome. *BMC Medicine, 19*(1), 37. https://doi.org/10.1186/s12916-021-01913-w

Nabavi, S. F., Braidy, N., Gortzi, O., Sobarzo-Sanchez, E., Daglia, M., Skalicka-Woźniak, K., & Nabavi, S. M. (2015). Luteolin as an anti-inflammatory and neuroprotective agent: A brief review. *Brain Research Bulletin, 119*, 1–11. https://doi.org/10.1016/j.brainresbull.2015.09.002

Obode, O. C., Adebayo, A. H., & Li, C. (2021). Phytochemical and toxicological evaluations of Prosopis africana (GUILL. and PERR.) extract on albino wistar rats. Toxicological research, 37, 183–195.

Okello, S., Muhihi, A., Mohamed, S. F., Ameh, S., Ochimana, C., Oluwasanu, A. O., et al. (2020). Hypertension prevalence, awareness, treatment, and control and predicted 10-year CVD risk: A cross-sectional study of seven communities in East and West Africa (SevenCEWA). *BMC Public Health, 20*(1), 1706. https://doi.org/10.1186/s12889-020-09829-5

Okubena, O., Adebesin, A., Omogbiya, A., Oluwole, O., Asomadu, R., Afolabi, M., Makanjuola, S., Ajonuma, L., Dosunmu, A., Otitoloju, O. & Umukoro, S. (2022) Polyphenol Rich Nutritional Supplement Derived from the West African Sorghum Bicolor Leaf Sheaths Has Evidence-Based Efficacy and Health Promoting Effects.

Omage, K., & Azeke, M. A. (2019). ACALYPHA WILKESIANA regulates fluid volume but affects selected tissues in salt loaded rabbits. *Clinical Phytoscience, 5*(1), 10. https://doi.org/10.1186/s40816-019-0103-5

Omogbiya, A. I., Ben-Azu, B., Eduviere, A. T., Eneni, A. O., Nwokoye, P. O., Ajayi, A. M., & Umukoro, S. (2021). Monosodium glutamate induces memory and hepatic dysfunctions in mice: Ameliorative role of Jobelyn® through the augmentation of cellular antioxidant defense machineries. *Toxicological Research, 37*, 323–335.

Omorogbe, O., Ajayi, A. M., Ben-Azu, B., Oghwere, E. E., Adebesin, A., Aderibigbe, A. O., et al. (2018). Jobelyn® attenuates inflammatory responses and neurobehavioural deficits associated with complete Freund-adjuvant-induced arthritis in mice. *Biomedicine & Pharmacotherapy, 98*, 585–593. https://doi.org/10.1016/j.biopha.2017.12.098

Oparil, S., Acelajado, M. C., Bakris, G. L., Berlowitz, D. R., Cífková, R., Dominiczak, A. F., et al. (2018). Hypertension. *Nature Reviews. Disease Primers, 4*, 18014. https://doi.org/10.1038/nrdp.2018.14

Poli A, Marangoni F, Corsini A, Manzato E, Marrocco W, Martini D, …, & Visioli F. Phytosterols, cholesterol control, and cardiovascular disease. Nutrients 2021;13(8):2810. doi: https://doi.org/10.3390/nu13082810.

Su, J., Xu, H.-T., Yu, J.-J., Gao, J.-L., Lei, J., Yin, Q.-S., et al. (2015). Luteolin ameliorates hypertensive vascular remodeling through inhibiting the proliferation and migration of vascular smooth muscle cells. *Evidence-based Complementary and Alternative Medicine: eCAM, 2015*, 364876. https://doi.org/10.1155/2015/364876

Taheri, Y., Sharifi-Rad, J., Antika, G., Yılmaz, Y. B., Tumer, T. B., Abuhamdah, S., et al. (2021). Paving luteolin therapeutic potentialities and agro-food-pharma applications: Emphasis on in vivo pharmacological effects and bioavailability traits. *Oxidative Medicine and Cellular Longevity, 2021*, 1987588. https://doi.org/10.1155/2021/1987588

Tietz, N. W., Rinker, A. D., & Shaw, L. M. (1983). International Federation of Clinical Chemistry. IFCC methods for the measurement of catalytic concentration of enzymes, Part 5. IFCC method for alkaline phosphatase (orthophosphoric-monoester phosphohydrolase, alkaline optimum, EC 3.1.3.1). *Journal of Clinical Chemistry and Clinical Biochemistry, 21*(11), 731–748.

Wang, J. G., Zhang, W., & Li, Y. (2023). Hypertension in China: Epidemiology and treatment initiatives. *Natural Reviews Cardiology, 20*, 531–545.

Zhou, B., Carrillo-Larco, R. M., Danaei, G., Riley, L. M., Paciorek, C. J., Stevens, G. A., et al. (2021). Worldwide trends in hypertension prevalence and progress in treatment and control from 1990 to 2019: A pooled analysis of 1201 population-representative studies with 104 million participants. *The Lancet, 398*(10304), 957–980. https://doi.org/10.1016/S0140-6736(21)01330-1

Zoppi, F., & Fellini, D. (1976). Enzymatic colorimetric cholesterol determination. *Clinical Chemistry, 22*, 690–691.

Index

P. O. Isibor, P. A. Akinduti (eds.), *Harnessing Biotechnology Tools for Product Development*, https://doi.org/10.1007/978-3-031-86002-7

The manufacturer's authorised representative in the EU is Springer Nature Customer Service Centre GmbH, Europaplatz 3, 69115 Heidelberg, Germany. If you have any concerns regarding our products, please contact ProductSafety@springernature.com

Printed and bound by CPI Group (UK) Ltd, Croydon, CR0 4YY
30/09/2026
02225596-0017